ATOMIC COLLISIONS

ATOMIC COLLISIONS

Electron and Photon Projectiles

Earl W. McDaniel

Georgia Institute of Technology
Atlanta, Georgia

WILEY

A WILEY-INTERSCIENCE PUBLICATION

JOHN WILEY & SONS

New York / Chichester / Brisbane / Toronto / Singapore

Library of Congress Cataloging in Publication Data:

McDaniel, Earl Wadsworth, 1926–
Atomic collisions: electron and photon projectiles/Earl W. McDaniel.
p. cm.
"A Wiley-Interscience publication."
Bibliography: p.
1. Collisions (Nuclear physics) 2. Photons. I. Title.
QC794.6.C6M43 1989
539.7′54--dc19 89-5427
ISBN 0-471-85307-0 CIP

Printed in the United States of America

10 9 8 7 6 5 4 3 2 1

This book is dedicated to my wife, Frances,
and our grandchildren, Johnny and Faith

PREFACE

In 1964 my book *Collision Phenomena in Ionized Gases* was published by John Wiley & Sons. It purported to cover, in 775 pages, the entire field of atomic collisions on the experimental side, as well as relevant topics in gaseous electronics. Some elementary scattering kinematics and collision theory were included, along with a few more advanced theoretical discussions. *Collision Phenomena* was designed to serve both as a text and as a reference, and it appears to have been widely used as both. However, in the quarter-century since its publication, the field of atomic collisions has matured very rapidly; *Collision Phenomena* is now being replaced by this volume (*Atomic Collisions: Electron and Photon Projectiles*) and another volume (*Atomic Collisions: Heavy Particle Projectiles*) that is in preparation for publication by Wiley several years hence. (By "heavy particles" we mean atoms, molecules, and ions.)

The present volume deals mainly with collisions of electrons and photons with heavy particles, with some discussion of electron-electron and photon-electron collisions. The energy range covered in *Electron and Photon Projectiles* extends from a few meV up to a few MeV (in rare cases). The upper cutoff excludes collisions in which nuclear forces are important, so the discussion here is limited to atomic, molecular, and chemical phenomena. The forthcoming book, *Heavy Particle Projectiles*, will include collisions between heavy particles, negative ions, electron and ion swarm phenomena, recombination, collisions in strong fields, and surface impact. The objectives of these two books taken together are much the same as those of *Collision Phenomena*. However, the field of atomic collisions has now progressed to the point that the phenomena are understood at a much deeper level; hence the heavy emphasis here on the *mechanisms* by which the various kinds of collisions take place, and the inclusion of much more theory than appeared in *Collision Phenomena*.

One part of the additional theoretical discussion that may be of special utility is the enumeration and brief description (in Sections 3-23 and 3-24) of almost all of the scattering approximations used in atomic collision theory. The author gratefully acknowledges the guidance and constructive comments of Professors

A. Dalgarno, M. R. Flannery, and T. Uzer in the preparation of this new material. These experts are responsible for whatever success I have had in avoiding the rocks and shoals of this vast and difficult subject.

Another difference between the present book and *Collision Phenomena* relates to the inclusion of data. In the earlier book it was possible to present a large fraction of the available experimental data considered reliable at the time, but this is no longer possible. One must now rely on the many extensive data indices and compilations to access the enormous quantity of cross sections and reaction rate coefficients that are on hand. Here, I include, for the most part, only samples of data that are representative, or that illustrate trends, or that illuminate points of special interest. Similarly, the present availability of many review articles and books on various detailed aspects of atomic collisions has led me to rely heavily on these sources and not to attempt a more exhaustive coverage of the primary literature.

An effort has been made to stay abreast of current developments during the preparation of this book. Topics included here, but not in *Collision Phenomena*, include scattering resonances, coincidence measurements, merged-beam experiments, positron collisions, collisions between spin-polarized particles, GaAs polarized electron sources, alignment and orientation parameters, "complete" experiments, position-sensitive detection, synchrotron radiation sources, free electron lasers, cyclotron resonance masers, laser cooling and trapping, electron beam cooling, and multiphoton processes.

This book contains more than 200 problems, at the ends of the chapters, that deal with scattering kinematics, collision theory, radiation theory, charged particle optics, spin dynamics, optics, and experimental techniques. In most cases, one or more sources are cited for a solution to the problem at hand. Also included at the end of each chapter is a set of Notes designed to amplify points in the text or to cover some peripheral topics.

I have attempted to make this book as nearly self-contained as possible. I believe that the reader who knows quantum mechanics and special relativity at about the level of graduate school entry can effectively use this book without having to resort to standard physics texts to an excessive degree.

Actually, I hope that *Electron and Photon Projectiles* may prove as useful to chemists and engineers as *Collision Phenomena* apparently did. Atomic collisions impinges on many other fields, such as astrophysics, planetary atmospheres, controlled fusion, laser development, isotope separation, magnetogasdynamics, high-velocity projectiles and vehicles, surface physics, chemical kinetics, laser spectroscopy, plasma dynamics, electrical discharges, gaseous electronics, and gas dynamics. Most of this book is accessible to chemists and engineers working in these fields even if they have not had the full theoretical background described above.

I am indebted to many people for their help during the planning and writing of this book. Sir David Bates and the late Sir Harrie Massey played a special role in encouraging me to undertake this task. Helpful criticism of portions of the manuscript were given by C. Bottcher, R. N. Compton, S. Datz, H. Ehrhardt, D.

C. Gregory, R. Johnsen, J. Kessler, W. C. Lineberger, M. S. Lubell, E. J. Mansky, S. T. Manson, J. T. Moseley, W. R. Newell, W. Raith, F. H. Read, A. C. H. Smith, B. Van Zyl, R. N. Zare, and J. C. Weisheit. It is also a pleasure to acknowledge productive discussions with other experts: J. W. Gallagher, T. C. Griffith, D. W. O. Heddle, J. W. Humberston, W. E. Kauppila, M. H. Kelley, J. W. McConkey, M. A. Morrison, D. T. Pierce, R. Roy, and T. S. Stein. I wish to thank Ms. Audrey Ralston for expertly typing the manuscript, and Prof. E. W. Thomas and Dean Les Karlovitz for their support. Finally, sincere thanks to my family, especially my wife, Frances, for their help, which was so important during the years occupied by the writing of this book.

EARL W. MCDANIEL

Atlanta, Georgia

CONTENTS

ATOMIC COLLISIONS

1

FUNDAMENTAL CONCEPTS

We begin by defining some terms. By *atomic collisions* we mean collisions involving electrons, photons, heavy particles, and surfaces at energies such that the phenomena are electronic, atomic, and chemical, not nuclear. The maximum energy considered in each case is below that at which nuclear forces (*n-n*, *p-p*, and *n-p*) become significant. By *heavy particles* we mean atoms, molecules, and ions. A *swarm* is an ensemble of electrons or ions moving in a gas with a significant random component of velocity. In this volume the emphasis is on collisions of electrons and photons with heavy particles. A future volume[1] will deal with collisions between heavy particles, and with swarm and surface impact phenomena.

The discovery of X-rays in 1895 marked the beginning of quantitative studies of atomic collisions. These investigations have made important contributions to the growth of physics, both in the development of experimental apparatus and techniques and in the formulation of modern concepts and theory. They have also led to a host of applications, many of which are discussed in Ref. 2. The continuing and growing interest in atomic collisions may be partially explained in terms of these applications. Recent studies of planetary atmospheres and astrophysical phenomena have revealed that atomic collisions play a much more important role in nature than was previously suspected, and present knowledge is insufficient to explain many of the observations. Another strong motivation arises from the fact that the world's conventional energy resources, including fissionable materials, are expected to be largely exhausted within a few centuries. The development of techniques for the controlled fusion of light nuclei in thermonuclear plasmas appears to offer the greatest hope for the long-range solution of our fuel problem. Finally, the development and detailed understanding of many useful devices have depended on our acquisition of new information about many kinds of atomic collisions. Here we refer to such devices as gas lasers, gas-filled radiation detectors, and ion sources, which are important in industry and in scientific research. Much of the interest in atomic collisions lies in the context of ionized gases, so it is appropriate to digress here and briefly consider this subject.

1-1. THE SCOPE OF THE STUDY OF IONIZED GASES

The phenomena of interest in ionized gases are of two basically different types: *noncollective processes*, which involve binary or ternary collisions of the particles composing the gas with one another and with surfaces, and *collective phenomena* based on the simultaneous interaction of a large number of particles and the interaction of the ionized gas with externally applied fields. In the first category lie such topics as elastic scattering, excitation, ionization, dissociation, charge transfer, electron attachment, drift velocities, energy distributions, diffusion, recombination, and secondary and photoemission. The noncollective processes are thus seen to fall mainly within the realms of atomic collisions and transport phenomena. The second category encompasses electrical discharges, space charge effects, plasma oscillations and waves, and magnetohydrodynamics. Collective phenomena usually are significant only if the ionized gas is a true *plasma*, that is, if its dimensions are large compared with its debye shielding length.* Our attention is centered here on the noncollective phenomena, which are important in plasmas and in ordinary ionized gases.

1-2. LABORATORY AND CENTER-OF-MASS COORDINATES AND THE ASYMPTOTIC ASPECTS OF NONRELATIVISTIC ELASTIC COLLISIONS

Two different coordinate systems are commonly used in discussions of collision problems, the laboratory (Lab) and the center of mass (CM). The Lab system is the frame of reference of a laboratory observer; the CM system moves with respect to the Lab system so that its origin is always coincident with the center of mass of the colliding particles. It is obviously convenient to make physical measurements in the Lab system, but as we show in Chapter 3, the mathematical analysis of a collision is considerably simpler in the CM frame. Therefore, it is desirable to establish general relationships between the two systems so that transformations from one to the other may be easily made. *All the transformations we shall obtain apply only to the special case in which one of the collision partners is initially at rest in the Lab system.* Other cases are discussed in Note 1-1 and in Chapter 7.

To derive the desired relationships, we shall consider a nonrelativistic elastic collision between two particles and focus our attention on its ultimate effects on the motion of the particles. We are not interested here in the detailed nature of the interaction or in what takes place "during" the collision. The pair of particles is first viewed long before the collision occurs, while the separation distance is

*The debye length is a measure of the distance over which deviations from charge neutrality can occur in an ionized gas. It is also a measure of the thickness of the sheaths which develop at plasma boundaries. The debye length is directly proportional to the square root of the energy and inversely proportional to the square root of the number density of the charged particles in the ionized gas. The expression for the debye length is derived in Appendix I.

still large and the interaction between the collision partners negligible. It is next viewed long after the collision has taken place, as the particles recede from one another with their final fixed velocities. The scattering angles and changes in speed are our major concerns. The particles are assumed to interact through a *central force* here and elsewhere unless otherwise indicated (see Problem 1-1).

A. Velocities and Kinetic Energy

Assume that a particle of mass m is moving with initial speed v_0 in the Lab system toward a remote particle of mass M which is at rest in the Lab system. If the particles were incapable of interaction, the closest distance of approach of m to M would be b, the impact parameter, shown in Fig. 1-2-1. Long after the collision has occurred, m is observed to be moving with constant speed v at an angle ϑ with respect to its original path, whereas M is traveling with constant speed V at an angle θ with respect to the original direction; ϑ and θ are the scattering angles of m and M, respectively, in the Lab system. In this frame of reference m may be regarded as the projectile and M as the target particle.

In the Lab system the center of mass of the two particles moves with velocity V_{CM} in a direction parallel to the initial direction of m; V_{CM} is constant in magnitude and direction before, during, and after the collision. From conservation of momentum considerations, the magnitude of V_{CM} is given by the equation*

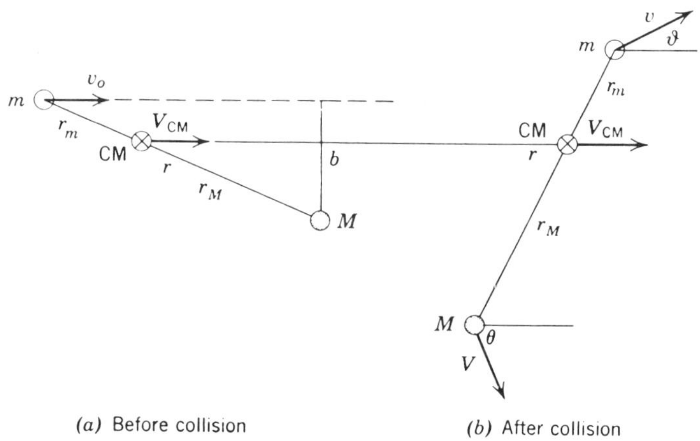

Figure 1-2-1. Asymptotic views of an elastic collision in the Lab system. In each of the figures the distance between the particles r is much greater than the range of the interaction between the particles; m and M are at distances r_m and r_M, respectively, from the center of mass.

*Equation (1-2-1) also follows from the equation defining the instantaneous position of the center of mass. If r_m and r_M denote the distances of m and M, respectively, from CM, the position of CM is specified by the equation $mr_m = Mr_M$. From Fig. 1-2-1 it is apparent that $V_{CM} = v_0(r_M)/(r_m + r_M)$. By substituting $r_M = r_m(m/M)$, we then obtain (1-2-1). The velocity of the center of mass is given by (2-2-20) when both m and M are moving before the collision in the Lab system.

$$(m + M)V_{\mathrm{CM}} = mv_0. \tag{1-2-1}$$

This result may be expressed in the form

$$V_{\mathrm{CM}} = v_0 \frac{M_r}{M}, \tag{1-2-2}$$

where M_r is the reduced mass of the pair of particles, defined by the equation

$$M_r = \frac{mM}{m + M}. \tag{1-2-3}$$

Note that the reduced mass always lies between 0.5 and 1.0 times the mass of the lighter particle.

Since kinetic energy and linear momentum are conserved in the collision,

$$\frac{mv_0^2}{2} = \frac{mv^2}{2} + \frac{MV^2}{2}, \tag{1-2-4}$$

$$mv_0 = mv \cos \vartheta + MV \cos \theta, \tag{1-2-5}$$

$$0 = mv \sin \vartheta - MV \sin \theta. \tag{1-2-6}$$

It should be noted that (1-2-1) to (1-2-6) are valid regardless of the form of the interaction potential between the particles, provided that the collision is elastic. Equations (1-2-4) to (1-2-6) enable us to obtain the final velocities, v and V, in terms of the initial velocity and the scattering angles:

$$\left(\frac{v}{v_0}\right)^2 - 2\left(\frac{v}{v_0}\right)\left(\frac{m}{m + M}\right)\cos \vartheta - \left(\frac{M - m}{m + M}\right) = 0, \tag{1-2-7}$$

$$v^2 = v_0^2\left(1 - \frac{4M_r^2}{mM}\cos^2 \theta\right), \tag{1-2-8}$$

$$V = 2v_0 \frac{M_r}{M} \cos \theta. \tag{1-2-9}$$

Now examine the collision in the center-of-mass coordinate system (Fig. 1-2-2). In this frame of reference the velocity of the center of mass is zero. The mutual velocity of approach of the two particles is seen to be the same in the CM system as it was in the Lab system, having the value v_0.

The total linear momentum of the system in the CM frame is zero at all times. Therefore, we may equate the magnitudes of the momenta of the particles and obtain the equation

$$m(v_0 - V_{\mathrm{CM}}) = MV_{\mathrm{CM}}. \tag{1-2-10}$$

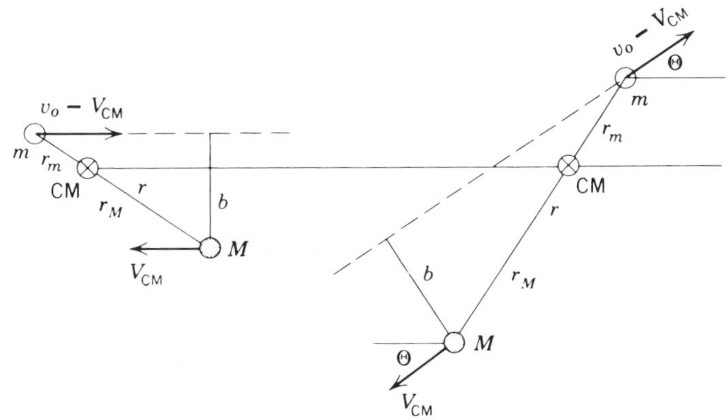

(a) Before collision (b) After collision

Figure 1-2-2. Asymptotic views of an elastic collision in the center-of-mass system.

The use of (1-2-2) and (1-2-3) shows that each particle has a CM momentum equal to $v_0 M_r$. Since the linear momenta of the particles are always equal and opposite, the particles must be traveling in opposite directions to one another before and after the collision. The distinction between projectile and target vanishes, and the particles are now properly regarded as initially approaching one another. The common angle of scattering in the CM system is denoted by Θ.

Since kinetic energy is conserved (except "during" the collision, when the forces acting between the particles and, consequently, their mutual potential energy, are appreciable), each of the collision partners has the same speed in the CM system after the collision as it had initially.* The total kinetic energy in the CM frame before and after the collision is given by

$$T_{CM} = \frac{m}{2}(v_0 - V_{CM})^2 + \frac{M}{2}V_{CM}^2 = \frac{m}{2}v_0^2 - \frac{(m + M)}{2}V_{CM}^2. \tag{1-2-11}$$

Thus the total CM kinetic energy is the initial Lab energy of m minus the energy corresponding to the motion of the center of mass in the Lab system. Again using (1-2-2), we see that

$$T_{CM} = \tfrac{1}{2}M_r v_0^2. \tag{1-2-12}$$

Here we have a relationship between the total CM kinetic energy, the reduced

*If the CM speed of m after the collision differed from its initial value by some factor α, the speed of M would also have to change by the same factor as the result of the collision. This would mean that the total kinetic energy of the system would have to change in violation of the principle of conservation of energy.

mass of the system, and the mutual velocity of approach of the collision partners.†

B. Angular Momentum and Moment of Inertia

Since no external torques act on the pair of particles, the angular momentum, J, of the system is constant before, during, and after the collision. The angular momentum about the center of mass in the CM coordinates is

$$J_{\text{CM}} = m(v_0 - V_{\text{CM}})\frac{r_m}{r_m + r_M}b + MV_{\text{CM}}\frac{r_M}{r_m + r_M}b. \tag{1-2-13}$$

This result may also be expressed as

$$J_{\text{CM}} = M_r v_0 b. \tag{1-2-14}$$

The moment of inertia, I, of the system about the center of mass is

$$I = mr_m^2 + Mr_M^2. \tag{1-2-15}$$

Making use of the fact that

$$mr_m = Mr_M = M_r r, \tag{1-2-16}$$

where r is the total separation distance between m and M, we may write

$$I = M_r r^2. \tag{1-2-17}$$

C. Scattering Angles

Relationships between the scattering angles may be obtained by vector addition of the velocities of Fig. 1-2-1*b* to those of Fig. 1-2-2*b* to obtain Fig. 1-2-3. An immediate result is

$$\theta = \tfrac{1}{2}(\pi - \Theta). \tag{1-2-18}$$

Another relationship is obtained by examination of the triangle ABC:

$$\tan\vartheta = \frac{AB}{BC} = \frac{(v_0 - V_{\text{CM}})\sin\Theta}{V_{\text{CM}} + (v_0 - V_{\text{CM}})\cos\Theta} \tag{1-2-19}$$

†This result actually applies regardless of the initial motion of the particles in the Lab system. In the preceding development we assumed the target particle to be initially at rest in the Lab frame of reference.

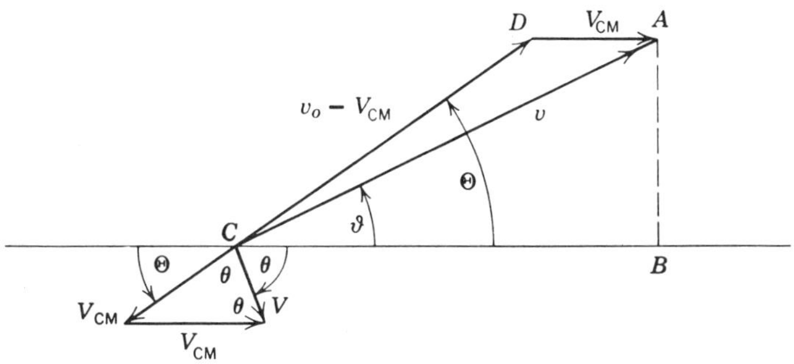

Figure 1-2-3. Vector addition of the respective velocities of the particles in the CM system to the velocity of the center of mass in the Lab system to obtain the velocities of the particles in the Lab system and the relationships between the CM and Lab scattering angles: $\mathbf{v}_0 - \mathbf{V}_{CM}$ is the final velocity of m in the CM system, and $\mathbf{V}_{CM}$ is the velocity of CM in the Lab frame. The sum of these vectors must equal $\mathbf{v}$, the final velocity of m in the Lab system. Similarly, the vector sum of the final velocity of M in the CM frame and the velocity of CM in the Lab frame must yield $\mathbf{V}$, the final velocity of M in the Lab system.

or

$$\tan \vartheta = \frac{\sin \Theta}{\gamma + \cos \Theta}, \qquad (1\text{-}2\text{-}20)$$

where

$$\gamma = \frac{m}{M} = \frac{V_{CM}}{v_0 - V_{CM}}. \qquad (1\text{-}2\text{-}21)$$

Also, consideration of the line segment AB in Fig. 1-2-3 shows that

$$v \sin \vartheta = v_0 \frac{M_r}{m} \sin \Theta. \qquad (1\text{-}2\text{-}22)$$

Finally, we note three useful relationships between ϑ and Θ for three special values of the ratio m/M.

$$\text{If} \quad m \ll M, \quad \vartheta \approx \Theta. \qquad (1\text{-}2\text{-}23)$$

The Lab and CM systems are almost identical; ϑ increases monotonically from 0 to π as Θ increases from 0 to π.

$$\text{If} \quad m = M, \quad \vartheta = \frac{\Theta}{2}. \qquad (1\text{-}2\text{-}24)$$

ϑ varies from 0 to $\pi/2$ as Θ varies from 0 to π. No particles are back-scattered in the Lab system. Head-on scattering corresponds to $\Theta = \pi$, $\vartheta = \pi/2$, $\theta = 0$.

$$\text{If} \quad m \gg M, \quad \vartheta \approx \frac{M}{m} \sin \Theta. \tag{1-2-25}$$

When $m > M$, ϑ first increases from 0 to a maximum value $\vartheta_{\text{max}} = \sin^{-1}(M/m)$, which is less than $\pi/2$, as Θ increases from 0 to

$$\Theta = \cos^{-1}(-M/m);$$

ϑ then decreases to 0 as Θ increases further to π, so that Θ is a double-valued function of ϑ. No particles are scattered beyond ϑ_{max} in the Lab system. We may distinguish between the two values of Θ that give a particular value of ϑ between 0 and $\sin^{-1}(M/m)$ by comparing the energies of m after scattering in the two cases—the energy is greater for the smaller Θ.

The geometrical relationships we have presented apply in a quantum mechanical collision as well as in the classical case. The reason for this is that the equations are essentially relationships of momentum vectors and are applied only in the asymptotic region, remote from the collision site, where the particle positions do not have to be precisely known and where consequently their momenta can be precisely determined.

D. Relationship between Elements of Solid Angle in the Lab and CM Systems

A valuable relationship in the discussion of collision problems is the transformation law between elements of solid angle in the Lab and CM systems. If the scattering distribution is independent of the azimuthal angle, it can be shown that

$$d\Omega_{\text{CM}} = \frac{[(m/M)^2 + 2(m/M)\cos\Theta + 1]^{3/2}}{1 + (m/M)\cos\Theta} \, d\Omega_{\text{Lab}}, \tag{1-2-26}$$

where $d\Omega_{\text{CM}} = \sin\Theta \, d\Theta \, d\varphi_{\text{CM}}$, $d\Omega_{\text{Lab}} = \sin\vartheta \, d\vartheta \, d\varphi_{\text{Lab}}$, and $\varphi_{\text{Lab}} = \varphi_{\text{CM}} = \varphi$. Equation (1-2-26) applies when the scattering is described by a spherically symmetric potential. More complex situations in which (1-2-26) does not hold are mentioned in Section 1-7-B.

1-3. INELASTIC AND RELATIVISTIC COLLISIONS

Section 1-2 dealt only with the nonrelativistic elastic collision, the simplest type to analyze. In an *elastic collision* the total kinetic energy of the system of colliding particles is the same before and after the collision—there are no permanent changes in the internal excitation energies of the collision partners. Linear and angular momenta are conserved at all times throughout the collision.

Another important type of collision is the *inelastic collision*, in which momentum is still conserved but in which the total kinetic energy of the system decreases (or increases) as the result of the excitation (or de-excitation) of one or both of the particles.* The analysis of this case is evidently more difficult than that of the elastic case, but the discussion in Section 1-2 immediately provides one useful and important result: The maximum amount of kinetic energy of the system that can be transferred in the collision to the form of internal excitation energy is the center-of-mass kinetic energy, which is given by (1-2-12). This fact results from the constancy of the total linear momentum of the system throughout the event.

Now let us inquire into the relation between the Lab and CM scattering angles in an inelastic collision. Suppose, as before, that a projectile of mass m with initial velocity v_0 in the Lab system strikes a target of mass M which is at rest in the Lab system. The projectile is scattered through an angle ϑ in the Lab system, and the target recoils at angle θ. If, in the collision, a quantity of energy ΔE is converted from kinetic to internal excitation energy, (1-2-20) is still valid:

$$\tan \vartheta = \frac{\sin \Theta}{\gamma + \cos \Theta}. \tag{1-2-20}$$

However, γ is no longer equal to m/M but is now given by

$$\gamma = \frac{m}{M}\left(\frac{T_{\mathrm{CM}}}{T_{\mathrm{CM}} - \Delta E}\right)^{1/2} \tag{1-3-1}$$

(γ is still equal to the ratio of V_{CM} to the CM speed of m after the collision).

Several additional useful relationships among the energies and scattering angles may also be derived. From the laws of conservation of total energy and linear momentum, we may easily show that

$$\cos \theta = \frac{\Delta E + [1 + (M/m)]T_{\mathrm{Lab},M}}{2\sqrt{(M/m)T_{\mathrm{Lab},i}T_{\mathrm{Lab},M}}} \tag{1-3-2}$$

and

$$\cos \vartheta = \frac{T_{\mathrm{Lab},i} + T_{\mathrm{Lab},f} - (M/m)T_{\mathrm{Lab},M}}{2\sqrt{T_{\mathrm{Lab},i}T_{\mathrm{Lab},f}}}, \tag{1-3-3}$$

*Many examples of inelastic collisions between particles are given in Chapters 5, 6, and 7.

whereas application of the law of sines and the law of cosines to a diagram similar to that in Fig. 1-2-3 yields

$$T_{\text{Lab},f}\frac{\sin^2\vartheta}{\sin^2\Theta} = T_{\text{Lab},f} + \left(\frac{m}{m+M}\right)^2 T_{\text{Lab},i} - \frac{2m}{m+M}\sqrt{T_{\text{Lab},i}T_{\text{Lab},f}}\cos\vartheta \tag{1-3-4}$$

and

$$T_{\text{Lab},M}\frac{\sin^2\theta}{\sin^2\Theta} = T_{\text{Lab},M} + \frac{mMT_{\text{Lab},i}}{(m+M)^2} - \frac{2\sqrt{mM}}{m+M}\sqrt{T_{\text{Lab},i}T_{\text{Lab},M}}\cos\theta. \tag{1-3-5}$$

In these equations $T_{\text{Lab},i}$ denotes the kinetic energy of the projectile in the Lab system before the collision and $T_{\text{Lab},f}$, the corresponding energy after the collision; $T_{\text{Lab},M}$, the kinetic energy of the recoiling target in the Lab system, is equal to $T_{\text{Lab},i} - T_{\text{Lab},f} - \Delta E$.

Further complications are presented if one or both of the colliding particles have relativistic velocities. The kinematics of inelastic and relativistic collisions are treated in Ref. 3.

1-4. THE COLLISION CROSS SECTION CONCEPT

Before proceeding, it would be desirable to introduce the concept of the collision cross section. This will enable us to discuss more quantitatively the various types of reactions that can occur when collisions take place. The cross section is basically a measure of the probability that a given type of reaction will occur under given conditions. Its value depends on the nature of the particles and reaction considered, the mutual velocity of approach of the collision partners, and the impact parameter.

A. The Microscopic Elastic Scattering Cross Section,* q_s

As before, we begin our discussion with the conceptually simple case of elastic scattering. The ideas to be developed are easily generalized to other situations. Consider a parallel beam of monoenergetic projectiles approaching the origin of the laboratory coordinate system, as shown in Fig. 1-4-1. The beam is directed along the Z axis and is composed of N_p particles per "cm^2-sec." Imagine these projectiles to be point particles but ascribe a finite size to the N_t target particles which are clustered at rest close to the origin O. We shall assume for the present that elastic scattering is the only reaction of which the projectiles and targets are capable. We shall further assume that there are sufficiently few targets present that none is shielded by another and that no projectile is scattered more than once.

*The German word for cross section is *querschnitt*—hence the frequent use of the symbol q to denote this quantity.

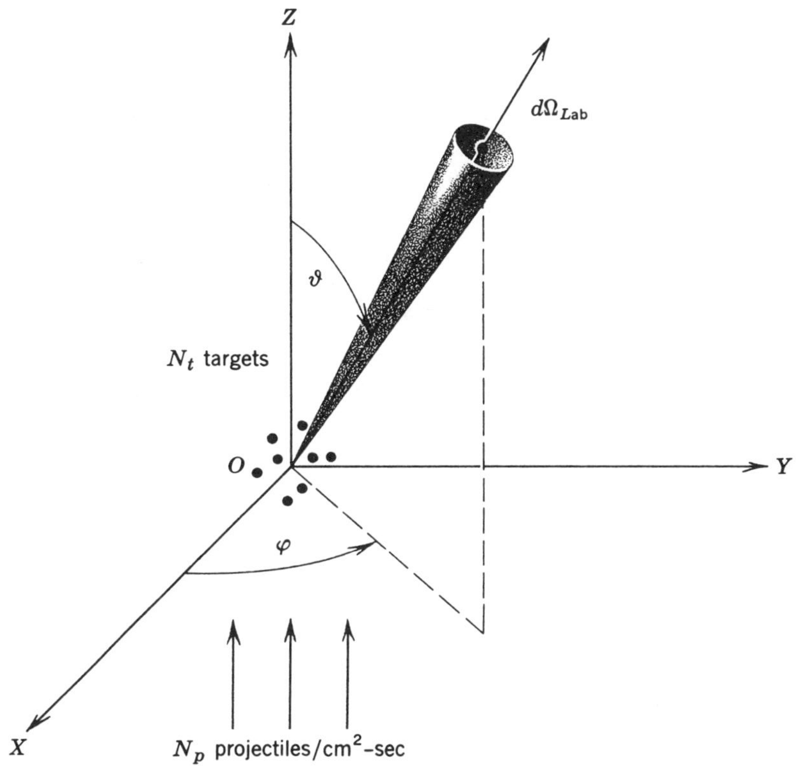

Figure 1-4-1. The laboratory coordinate system used for discussion of the scattering of a beam of projectiles by N_t target particles clustered about the origin.

Let $d\Omega_{\text{Lab}}$ represent the element of solid angle oriented in the spherical polar coordinate system by the angles ϑ and φ, and let $N_s(\vartheta, \varphi)\, d\Omega_{\text{Lab}}$ be the number of projectiles scattered into $d\Omega_{\text{Lab}}$ per second. Obviously,

$$N_s(\vartheta, \varphi)\, d\Omega_{\text{Lab}} \sim N_p N_t\, d\Omega_{\text{Lab}}.$$

Insert a constant of proportionality $I_s(\vartheta, \varphi)$ into this expression so that

$$N_s(\vartheta, \varphi)\, d\Omega_{\text{Lab}} = I_s(\vartheta, \varphi) N_p N_t\, d\Omega_{\text{Lab}} = dI_s(\vartheta, \varphi) N_p N_t. \tag{1-4-1}$$

The quantity

$$I_s(\vartheta, \varphi)\, d\Omega_{\text{Lab}} \equiv dI_s(\vartheta, \varphi) \tag{1-4-2}$$

is the *differential microscopic elastic scattering cross section in the Lab system.* Solving for $dI_s(\vartheta, \varphi)$, we obtain

$$dI_s(\vartheta, \varphi) = \frac{N_s(\vartheta, \varphi)\, d\Omega_{\text{Lab}}}{N_p N_t}, \tag{1-4-3}$$

which has the dimensions of cm^2 per target particle; $dI_s(\vartheta, \varphi)$ accordingly may be considered to be the area presented by each target particle for scattering of the projectiles into the element of solid angle $d\Omega_{\text{Lab}}$. To show that this interpretation is correct, assume that the incident beam impinges normally on a thin foil of 1 cm^2 area which contains N_t targets. The fraction of the incident beam scattered into $d\Omega_{\text{Lab}}$ at (ϑ, φ) with respect to the beam axis is equal to the ratio of the opaque area to the total area of the foil. Thus the number of projectiles scattered into $d\Omega_{\text{Lab}}$ per second must be

$$N_s(\vartheta, \varphi)\, d\Omega_{\text{Lab}} = N_p \frac{I_s(\vartheta, \varphi)\, d\Omega_{\text{Lab}} N_t}{1\ \text{cm}^2}.$$

But this is just what is predicted by (1-4-1), so our interpretation of $dI_s(\vartheta, \varphi)$ must be correct.

Now we define the *integral microscopic elastic scattering cross section*, q_s:

$$q_s = \int dI_s(\vartheta, \varphi) = \iint I_s(\vartheta, \varphi)\, d\Omega_{\text{Lab}} = \int_0^{2\pi} \int_0^{\pi} I_s(\vartheta, \varphi) \sin\vartheta\, d\vartheta\, d\varphi. \tag{1-4-4}$$

Evidently, q_s represents the area presented by each of the target particles for scattering into the total solid angle, 4π steradians. The integral in (1-4-4) does not converge for all projectile-target scattering potentials. The condition for convergence is that $I_s(\vartheta, \varphi)$ increase less rapidly than $1/\vartheta^2$, for small ϑ, as ϑ decreases. The subject of convergence of the cross-section integral is treated further in Chapter 3, in which techniques for calculating the scattering cross section for various assumed interaction potentials are also discussed.

Let $p(\vartheta, \varphi)\, d\Omega_{\text{Lab}}$ denote the probability that in an elastic collision the projectile is scattered into the element of solid angle $d\Omega_{\text{Lab}}$. Then the differential scattering cross section may be written

$$I_s(\vartheta, \varphi)\, d\Omega_{\text{Lab}} = q_s p(\vartheta, \varphi)\, d\Omega_{\text{Lab}}. \tag{1-4-5}$$

It follows that

$$p(\vartheta, \varphi) = \frac{I_s(\vartheta, \varphi)}{q_s}. \tag{1-4-6}$$

The symbol $\sigma_s(\vartheta, \varphi)$ is frequently used in place of $I_s(\vartheta, \varphi)$, and the integral elastic scattering cross section is written as

$$\sigma_s = \int_0^{2\pi} \int_0^{\pi} \sigma_s(\vartheta, \varphi) \sin\vartheta\, d\vartheta\, d\varphi. \tag{1-4-7}$$

Microscopic cross sections are usually expressed in units of cm^2 or in units of $\pi a_0^2 = 0.88 \times 10^{-16}\ \text{cm}^2$, where $a_0 = 0.53 \times 10^{-8}$ cm is the radius of the first Bohr orbit of the hydrogen atom.

B. The Macroscopic Scattering Cross Section, Q_s, and the Scattering Mean Free Path, λ_s

Consider a beam of projectiles impinging normally on a semi-infinite medium containing N stationary target particles per cm^3. Of those projectiles that penetrate to depth x in the medium without experiencing scattering, a fraction equal to $Nq_s\,dx$ will be scattered in passing from x to $x + dx$. This quantity is also equal to the probability that a given projectile that reaches depth x without scattering will then suffer scattering between x and $x + dx$. If I_0 represents the initial intensity of the incident beam in projectiles per cm^2-sec and I is the intensity of the unscattered component at depth x, then

$$dI = -INq_s\,dx \tag{1-4-8}$$

and

$$I = I_0 e^{-Nq_s x} = I_0 e^{-Q_s x}, \tag{1-4-9}$$

where $Q_s = Nq_s$ is termed the *macroscopic or bulk scattering cross section.* If the scattering medium is a gas of molecular weight M and density ρ g/cm^3,

$$N = \frac{\rho N_A}{M}, \tag{1-4-10}$$

where N_A is Avogadro's number, 6.02×10^{23} molecules per mole. Evidently, Q_s represents the effective cross-sectional area for elastic scattering of all the target particles in 1 cm^3 of the scattering medium; Q_s has dimensions of cm^{-1}.

We now see that the probability that a given projectile will survive scattering until it penetrates the medium at least to depth x is $e^{-Q_s x}$. As shown above, the probability that it will then be scattered in traveling a further distance dx is $Nq_s\,dx = Q_s\,dx$. Consequently, $\bar{x}$, the average depth of penetration before scattering, or equivalently, the average distance between scattering sites for a given projectile, is given by

$$\int_0^\infty x e^{-Q_s x} Q_s\,dx = \bar{x} \int_0^\infty e^{-Q_s x} Q_s\,dx. \tag{1-4-11}$$

Performing the integration, we find that

$$\bar{x} = \frac{1}{Q_s}. \tag{1-4-12}$$

Now denote by λ_s the *mean free path for elastic scattering in the medium.* By definition

$$\lambda_s = \bar{x}, \tag{1-4-13}$$

so that

$$\lambda_s = \frac{1}{Q_s}. \tag{1-4-14}$$

It should be emphasized that in these derivations we considered the target particles to be at rest and the projectiles to constitute a parallel, monoenergetic, and homogeneous beam.

C. Relationship between Cross Sections in the Lab and CM Systems

For a given beam and target combination, the same number of projectiles must be scattered into the Lab element of solid angle, $d\Omega_{\text{Lab}}$, as are scattered into the corresponding CM element of solid angle, $d\Omega_{\text{CM}}$. Let us assume that the scattering distribution is independent of the azimuth angle and put $\varphi_{\text{Lab}} = \varphi_{\text{CM}} = \varphi$. Then

$$I_s(\vartheta, \varphi)\, d\Omega_{\text{Lab}} = I_s(\Theta, \varphi)\, d\Omega_{\text{CM}}, \tag{1-4-15}$$

and from (1-2-26) it follows that the relationship between the cross sections in the two coordinate systems is given by

$$I_s(\vartheta, \varphi) = \frac{[(m/M)^2 + 2(m/M)\cos\Theta + 1]^{3/2}}{1 + (m/M)\cos\Theta}\, I_s(\Theta, \varphi). \tag{1-4-16}$$

Evidently, the total cross section must be the same in either frame of reference; that is,

$$(q_s)_{\text{Lab}} = (q_s)_{\text{CM}}. \tag{1-4-17}$$

D. Cross Sections for Reactions Other Than Elastic Scattering

The cross section concept obviously applies to types of reactions other than elastic scattering, and the equations in the foregoing sections may be used to describe inelastic collisions if the appropriate cross sections are used. If we now inquire about the likelihood that a given type of projectile will undergo *some* reaction, regardless of type, in traversing a medium we are led to the concept of a total microscopic cross section

$$q = q_i + q_j + q_k + \cdots \tag{1-4-18}$$

and a total macroscopic cross section

$$Q = Q_i + Q_j + Q_k + \cdots, \tag{1-4-19}$$

where the subscripts serve to label the individual constituent cross sections. Evidently,

$$\lambda = \frac{1}{Q} \tag{1-4-20}$$

and

$$\frac{1}{\lambda} = \frac{1}{\lambda_i} + \frac{1}{\lambda_j} + \frac{1}{\lambda_k} + \cdots, \tag{1-4-21}$$

where λ is the total mean free path, considering collisions of all types.

When the physical problem at hand involves the passage of a homogeneous beam through a material medium, the total mean free path is sometimes referred to as the relaxation length. It is the distance in which the intensity of the beam is reduced to $1/e$ of its initial value, projectiles undergoing any type of reaction being considered as lost from the beam.

E. Reaction Rates and Particle Flux for Monoenergetic Projectiles

Consider a monoenergetic parallel beam containing n projectiles per cm^3 which passes with velocity v through a medium composed of N stationary particles per cm^3. Each projectile makes, on the average, v/λ collisions per second. Thus $nv/\lambda = Qnv$ is the average number of collisions per cm^3 per second, or the *reaction rate.*

The projectile flux is defined as

$$\varphi = nv \tag{1-4-22}$$

and has the dimensions of particles/cm^2-sec. The reaction rate in this case is seen to be equal to $Q\varphi$/cm^3-sec.*

The flux nv is usually described geometrically as the number of projectiles crossing 1 cm^2 of area per second. This description is correct for a parallel beam of number density n and velocity v crossing a plane perpendicular to the beam axis. Under other conditions, however, this description is not correct. Suppose, for example, that the surface element is located in a medium in which the projectiles are moving isotropically in all directions. Then the number of projectiles crossing a plane surface of 1 cm^2 area in 1 sec is $nv/2$, not nv, as shown in Section 2-6.

*In the more general case in which the motion of the target particles cannot be neglected, the reaction rate is expressed by an integral containing the relative velocity of the projectiles and targets [see (2-12-2) and R. V. Meghreblian and D. K. Holmes, *Reactor Analysis*, McGraw-Hill, New York (1960)].

F. Reactions Involving Polyenergetic Projectiles

In our treatment of cross sections the assumption was made that all the projectiles have the same energy, but this is frequently not even approximately so in practice. Since the cross section for a particular reaction generally varies with the energy, we must extend our treatment to cover the general case.

Let $n(E)$ be the number of projectiles of energy E per cm^3 per unit energy interval, so that $n(E)\,dE$ is the number per cm^3 in the range E to $E + dE$. The total flux for projectiles of all energies is

$$\varphi = \int_0^\infty n(E)v\,dE = \int_0^\infty \varphi(E)\,dE. \tag{1-4-23}$$

Here $\varphi(E)\,dE$ is the flux in the energy range E to $E + dE$. The reaction rate for stationary targets is

$$\int_0^\infty Q(E)n(E)v\,dE = \int_0^\infty Q(E)\varphi(E)\,dE = \bar{Q}\varphi, \tag{1-4-24}$$

where $Q(E)$ is the macroscopic cross section for particles of energy E and $\bar{Q}$ is the average macroscopic cross section, given by

$$\bar{Q} = \frac{\int_0^\infty Q(E)n(E)v\,dE}{\int_0^\infty n(E)v\,dE} = \frac{\int_0^\infty Q(E)\varphi(E)\,dE}{\int_0^\infty \varphi(E)\,dE}. \tag{1-4-25}$$

The corresponding average mean free path is

$$\bar{\lambda} = \frac{\int_0^\infty \lambda(E)\varphi(E)\,dE}{\int_0^\infty \varphi(E)\,dE}. \tag{1-4-26}$$

In general, $\bar{\lambda}$ is not equal to the reciprocal of $\bar{Q}$, although for any E,

$$\lambda(E) = \frac{1}{Q(E)}. \tag{1-4-27}$$

G. The Probability of Collision, *P*, the Collision Frequency, ν, and the Mean Free Time, τ

Some writers phrase the discussion of collision phenomena in terms of the probability of collision of the particles rather than of collision cross sections or mean free paths. The relationships among these quantities are expressed later.

We shall consider the target material to be gaseous and denote its number density by N. N_A denotes *Avogadro's number*, 6.02×10^{23} particles per mole, which is the number of molecules of an ideal gas contained in 22.4 liters at the standard conditions of 760 mm Hg pressure and 0°C. The *Loschmidt number* N_L is the number density at standard pressure and temperature and has the value 2.69×10^{19} per cm^3. The number density at 1 mm Hg and 0°C, $N_{1\,\mathrm{mm}}$, equals 3.54×10^{16} per cm^3.

Let $Q_{1\,\mathrm{mm}}$ be the macroscopic collision cross section at 1 mm pressure and 0°C. Evidently,

$$Q_{1\,\mathrm{mm}} = qN_{1\,\mathrm{mm}} = \frac{1}{\lambda_{1\,\mathrm{mm}}}, \tag{1-4-28}$$

where $\lambda_{1\,\mathrm{mm}}$ is the collision mean free path at 1 mm pressure and 0°C.

The *probability of collision* P is defined as the average number of collisions experienced by a projectile in 1 cm of path through the gas at 1 mm Hg pressure and 0°C:

$$P = \frac{1}{\lambda_{1\,\mathrm{mm}}} = Q_{1\,\mathrm{mm}}. \tag{1-4-29}$$

It is convenient to introduce the "reduced pressure," p_0, of the gas. If we denote the actual pressure by p mm Hg and the temperature by T K and assume ideal gas behavior, p_0 is given by the equation

$$p_0 = p\frac{273}{T}. \tag{1-4-30}$$

Thus p_0 is the pressure of the gas, in mm Hg, reduced to 0°C at constant density.

The collision mean free path is now given by

$$\lambda = \frac{p_{1\,\mathrm{mm}}}{p_0 P} \tag{1-4-31}$$

and the *collision frequency* ν by

$$\nu = \frac{v}{\lambda} = vQ = vp_0Q_{1\,\mathrm{mm}} = vp_0P. \tag{1-4-32}$$

The *mean free time* τ is defined as the average time between successive collisions experienced by a single projectile; τ is the reciprocal of ν:

$$\tau = \frac{1}{\nu} \tag{1-4-33}$$

The attenuation of a monoenergetic beam passing through a gas may now be expressed by the equation

$$I = I_0 e^{-Qx} = I_0 e^{-p_0 Q_{1\,\mathrm{mm}} x} = I_0 e^{-p_0 P x} \tag{1-4-34}$$

The unit of pressure used in this discussion, 1 mm Hg, has been named the *torr*. We shall henceforth use this newer designation.

1-5. THE AVERAGE ENERGY LOSS AND ANGULAR SCATTERING DISTRIBUTION IN THE CLASSICAL COLLISION OF SMOOTH ELASTIC SPHERES

Consider two smooth elastic spheres, of masses m and M, which are about to collide (Fig. 1-5-1). The spheres are considered to interact only at the instant of impact; m has velocity v_0 before the collision, measured in the Lab system; M is at rest. After the collision, m rebounds with velocity v and M moves along the

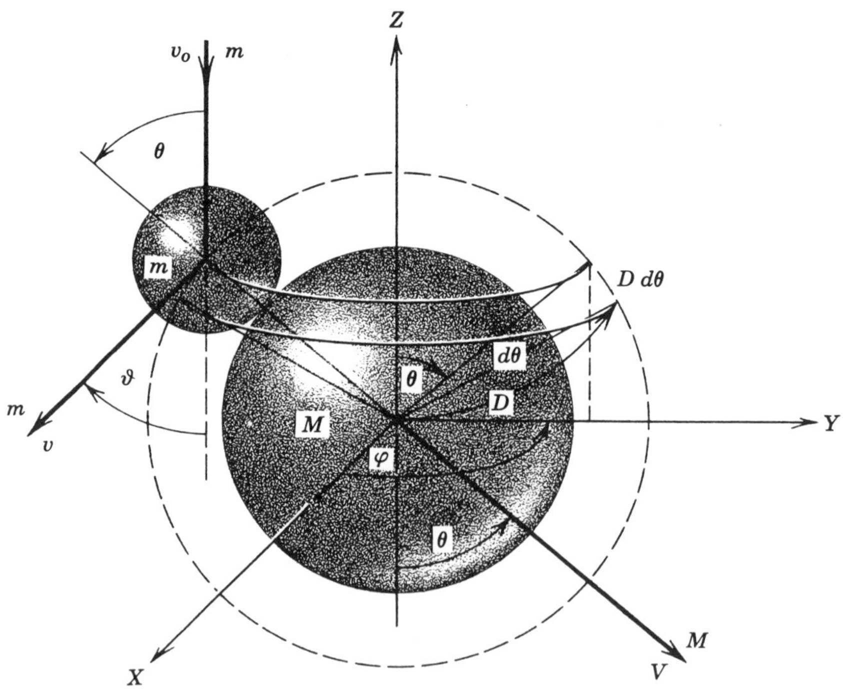

Figure 1-5-1. The collision of elastic spheres in the Lab system. The projectile m impinges on the stationary target M with velocity v_0 at incidence angle θ and is scattered with velocity $\mathbf{v}$ at angle ϑ. The target recoils at angle θ with velocity V. The sum of the radii of the two elastic spheres is denoted by D.

impact line of centers with velocity V.* Let D be the distance between centers when the spheres are in contact; θ is the incidence angle and ϑ is the scattering angle of m in the Lab system. The usual polar angle in the spherical coordinate system shown is θ, whereas φ is the azimuthal angle, measured in the X-Y plane; θ is obviously not constrained to lie in the Y-Z plane—it is shown in the Y-Z plane here for clarity. Evidently, the scattering angular distribution will be independent of the angle φ in this example.

The fractional loss of energy of m in a collision at incidence angle θ is

$$\Delta(\theta) = \frac{v_0^2 - v^2}{v_0^2}. \tag{1-5-1}$$

The momentum and energy conservation equations are

$$mv_0 = mv \cos \vartheta + MV \cos \theta, \tag{1-5-2}$$

$$0 = mv \sin \vartheta - MV \sin \theta, \tag{1-5-3}$$

$$\frac{mv_0^2}{2} = \frac{mv^2}{2} + \frac{MV^2}{2}. \tag{1-5-4}$$

Therefore,

$$\Delta(\theta) = \frac{MV^2}{mv_0^2}, \tag{1-5-5}$$

an expected result. The final velocity of M in the Lab system, which appears in this equation for $\Delta(\theta)$, may be expressed as

$$V = \frac{2mv_0 \cos \theta}{m + M}. \tag{1-5-6}$$

[The validity of (1-5-2) through (1-5-6) does not depend on the assumption of smoothness of the spheres.]

Now imagine m to be one of the projectiles in a monoenergetic beam which has uniform composition in a plane normal to the Z axis. Assuming that m undergoes a collision, let $p(\theta)\, d\theta$ be the probability that the impact occurs at an incidence angle between θ and $\theta + d\theta$. The total effective area presented for collision of m against M is πD^2. The area of the surface element defined by the cones θ and $\theta + d\theta$ is $(2\pi D \sin \theta)(D\, d\theta)$, but only a fraction, $\cos \theta$, is presented directly to projectiles approaching along $+Z$. The probability of the collision

*The collision is specular because the spheres are assumed to be smooth. The smoothness assumption guarantees that no energy goes into rotation.

taking place between θ and $\theta + d\theta$ is therefore

$$p(\theta)\,d\theta = \begin{cases} \dfrac{2\pi D^2 \sin\theta \cos\theta\,d\theta}{\pi D^2} = \sin 2\theta\,d\theta & \left(0 \leqslant \theta \leqslant \dfrac{\pi}{2}\right) \\ 0 & \left(\dfrac{\pi}{2} < \theta \leqslant \pi\right). \end{cases} \tag{1-5-7}$$

This is also the probability that the target will recoil at an angle between θ and $\theta + d\theta$ because the spheres have been assumed to be smooth.

Δ, the mean fractional loss of energy allowing for collisions at all possible angles, is then

$$\Delta = \overline{\Delta(\theta)} = \frac{\displaystyle\int_0^{\pi/2} p(\theta)\,\Delta(\theta)\,d\theta}{\displaystyle\int_0^{\pi/2} p(\theta)\,d\theta}.$$

Using (1-5-4), (1-5-6), and (1-5-7), we obtain

$$\Delta = \frac{2mM}{(m+M)^2}. \tag{1-5-8}$$

From (1-5-5), (1-5-6), and (1-2-18), it follows that

$$\Delta(\Theta) = \frac{2mM}{(m+M)^2}(1 - \cos\Theta), \tag{1-5-9}$$

which is a general result. In the present case of smooth elastic spheres,

$$\Delta(\Theta) = \Delta(1 - \cos\Theta). \tag{1-5-10}$$

We shall now consider several special cases.

A. $m = M$

From (1-5-8) it is apparent that here

$$\Delta = \tfrac{1}{2}, \tag{1-5-11}$$

whereas (1-5-9) gives

$$\Delta(\vartheta) = \tfrac{1}{2}(1 - \cos 2\vartheta). \tag{1-5-12}$$

From the conservation equations and (1-5-6) it follows that

$$\vartheta + \theta = \frac{\pi}{2}. \tag{1-5-13}$$

Therefore, the spheres leave the collision site at right angles to one another in the Lab system.

Now, assuming that m suffers a collision, let $p(\vartheta)\,d\vartheta$ be the probability that m is scattered between ϑ and $\vartheta + d\vartheta$. The use of (1-5-13) in (1-5-7) yields

$$p(\vartheta)\,d\vartheta = \begin{cases} \sin(\pi - 2\vartheta)\,d\vartheta & \left(0 \leqslant \vartheta \leqslant \dfrac{\pi}{2}\right) \\ 0 & \left(\dfrac{\pi}{2} < \vartheta \leqslant \pi\right) \end{cases}$$

or

$$p(\vartheta)\,d\vartheta = \begin{cases} \sin 2\vartheta\,d\vartheta & \left(0 \leqslant \vartheta \leqslant \dfrac{\pi}{2}\right) \\ 0 & \left(\dfrac{\pi}{2} < \vartheta \leqslant \pi\right). \end{cases}$$

Note that

$$\int_0^{\pi/2} p(\vartheta)\,d\vartheta = \int_0^{\pi/2} \sin 2\vartheta\,d\vartheta = 1.$$

All the projectiles undergoing collisions are scattered between 0 and $\pi/2$—there is no backward scattering in the Lab system.

Let $F(\vartheta)$ denote the probability per unit solid angle that if scattering occurs m will be deflected between ϑ and $\vartheta + d\vartheta$. $F(\vartheta)$ also equals the fraction of the scattered component of the beam that is deflected between ϑ and $\vartheta + d\vartheta$, per unit solid angle.

$$F(\vartheta) = \begin{cases} \dfrac{p(\vartheta)\,d\vartheta}{\displaystyle\int_0^{2\pi} d\Omega_{\text{Lab}}} = \dfrac{\sin 2\vartheta\,d\vartheta}{2\pi \sin\vartheta\,d\vartheta} = \dfrac{\cos\vartheta}{\pi} & \left(0 \leqslant \vartheta \leqslant \dfrac{\pi}{2}\right) \\ 0 & \left(\dfrac{\pi}{2} < \vartheta \leqslant \pi\right). \end{cases}$$

As must be the case,

$$\int_0^{2\pi}\int_0^{\pi/2} F(\vartheta)\,d\Omega_{\text{Lab}} = \int_0^{\pi/2} F(\vartheta) 2\pi \sin\vartheta\,d\vartheta = \int_0^{\pi/2} p(\vartheta)\,d\vartheta = 1.$$

$F(\vartheta)\,d\Omega_{\text{Lab}}$ is related to the differential microscopic scattering cross section $I_s(\vartheta)\,d\Omega_{\text{Lab}}$ by the equation

$$I_s(\vartheta)\,d\Omega_{\text{Lab}} = q_s F(\vartheta)\,d\Omega_{\text{Lab}}, \tag{1-5-14}$$

where

$$q_s = \pi D^2 \tag{1-5-15}$$

is the integral microscopic scattering cross section.

Note that the scattering is not isotropic in the Lab system, that is,

$$I_s(\vartheta) = \begin{cases} q_s \dfrac{\cos\vartheta}{\pi} & \left(0 \leqslant \vartheta \leqslant \dfrac{\pi}{2}\right) \\ 0 & \left(\dfrac{\pi}{2} < \vartheta \leqslant \pi\right) \end{cases} \tag{1-5-16}$$

is not independent of ϑ. However, the scattering is isotropic in the CM system. Equation (1-2-24) shows that when $m = M$,

$$\vartheta = \frac{\Theta}{2}, \tag{1-5-17}$$

where Θ is the scattering angle of m in the CM system. The use of this result in (1-2-26) gives

$$d\Omega_{\text{CM}} = 4\cos\vartheta\; d\Omega_{\text{Lab}}, \tag{1-5-18}$$

and so by (1-4-15)

$$I_s(\Theta) = \frac{I_s(\vartheta)}{4\cos\vartheta} = \frac{q_s}{4\pi} = \frac{D^2}{4} \qquad (0 \leqslant \Theta \leqslant \pi). \tag{1-5-19}$$

Equation (1-5-19) holds even if $m \neq M$; $I_s(\Theta)\,d\Omega_{\text{CM}}$ is the differential scattering cross section in the CM system.

We observe finally that our results are compatible with the requirement that

$$q_s = \int_0^{2\pi}\int_0^{\pi/2} I_s(\vartheta)\,d\Omega_{\text{Lab}} = \int_0^{2\pi}\int_0^{\pi} I_s(\Theta)\,d\Omega_{\text{CM}}. \tag{1-5-20}$$

B. $m \ll M$

Equation (1-5-8) gives

$$\Delta \approx \frac{2m}{M}, \tag{1-5-21}$$

another useful result, whereas ϑ and θ are now related by the equation

$$\theta \approx \frac{\pi}{2} - \frac{\vartheta}{2}. \tag{1-5-22}$$

From (1-5-6) it follows that

$$V \approx \frac{2m}{M} v_0 \cos\theta, \tag{1-5-23}$$

and therefore from (1-5-5)

$$\Delta(\vartheta) \approx \frac{2m}{M}(1 - \cos\vartheta). \tag{1-5-24}$$

We may also get (1-5-24) directly from (1-5-9). We also find that

$$p(\vartheta)\,d\vartheta = \sin(\pi - \vartheta)\frac{d\vartheta}{2} = \frac{\sin\vartheta}{2}\,d\vartheta \tag{1-5-25}$$

and

$$F(\vartheta) = \frac{1}{4\pi} \qquad (0 \leqslant \vartheta \leqslant \pi). \tag{1-5-26}$$

Therefore,

$$I_s(\vartheta) = \frac{q_s}{4\pi} \qquad (0 \leqslant \vartheta \leqslant \pi) \tag{1-5-27}$$

and the scattering is now isotropic in the Lab system.

When $m \ll M$, $\vartheta \approx \Theta$, and the Lab and CM systems are almost identical. Consequently, the scattering is also isotropic in the CM system. All directions of motion after collision are equally probable. (Actually, the scattering of smooth elastic spheres is isotropic in the CM system regardless of the ratio of m to M.)

Attention should now be directed to the fact that according to the elastic sphere model used in this section the angular scattering distributions are all independent of the velocity of the projectiles.

C. Two Species of Particles with Maxwellian Velocity Distributions Corresponding to Different Temperatures

Bates[4] has solved the problem of an ensemble of ions of mass m_1 moving in a gas composed of molecules of mass m_2 under the following assumptions: (1) The ions and molecules interact through an attractive $1/r^4$ polarization potential (Section 1-8) with a hard repulsive core; (2) an average amount of energy $\bar{\Delta}$ is released per collision; and (3) the two species have Maxwellian velocity distributions, corresponding to absolute temperatures T_1 and T_2, respectively. Bates shows that

$$f = \frac{2m_1 m_2}{(m_1 + m_2)^2} \frac{T_1 - T_2}{T_1} - \frac{2m_2 \bar{\Delta}}{3(m_1 + m_2)kT_1}, \tag{1-5-28}$$

where f is the average energy lost per collision by the ions expressed as a fraction of the average energy of these particles. (Here k is the Boltzmann constant.) Bates's proof goes as follows:

Let $\mathbf{u}_1$ and $\mathbf{u}_2$ be the respective Lab velocities of the ionic and neutral species before the collision, and let $(\mathbf{w} + \mathbf{v}_1)$ and $(\mathbf{w} - m_1\mathbf{v}_1/m_2)$ be their velocities after the collision, where

$$\mathbf{w} = \frac{m_1 \mathbf{u}_1 + m_2 \mathbf{u}_2}{m_1 + m_2} \tag{1-5-29}$$

is the velocity of their center of mass. If internal energy Δ is released in the collision, conservation of energy and momentum demands that

$$\frac{1}{2} m_1 u_1^2 + \frac{1}{2} m_2 u_2^2 + \Delta = \frac{1}{2}\left[(m_1 + m_2)w^2 + m_1 \left(\frac{m_1 + m_2}{m_2}\right) v_1^2\right]. \tag{1-5-30}$$

Then, using (1-5-29), we obtain

$$\frac{m_1(m_1 + m_2)}{m_2} v_1^2 = 2\Delta + \frac{m_1 m_2}{m_1 + m_2}(u_1^2 + u_2^2 + 2\mathbf{u}_1 \cdot \mathbf{u}_2). \tag{1-5-31}$$

For the special case of an ion-induced dipole $1/r^4$ interaction, the rate $k = \langle \sigma v \rangle$ at which collisions occur is independent of the velocity of relative motion. To be consistent with this fact, we put

$$\overline{\mathbf{u}_1 \cdot \mathbf{u}_2} = 0, \tag{1-5-32}$$

where the bar indicates an average over all colliding pairs. Now the energy lost by the ion in the collision is

$$\Delta E_1 = \tfrac{1}{2} m_1 [u_1^2 - w^2 - v_1^2 - 2\mathbf{w} \cdot \mathbf{v}_1]. \tag{1-5-33}$$

The direction of motion of the ion in the CM frame after collision is randomly oriented with respect to the velocity of the CM, so we put

$$\mathbf{w} \cdot \mathbf{v}_1 = 0. \tag{1-5-34}$$

Then (1-5-33) with (1-5-31) and (1-5-32) gives

$$\Delta E_1 = \frac{m_1 m_2}{(m_1 + m_2)^2}[u_1^2 - u_2^2] - \frac{m_2 \Delta}{m_1 + m_2}. \tag{1-5-35}$$

Integration over Maxwellian distributions in u_1 (at temperature T_1) and in u_2 (at temperature T_2) shows that the mean energy loss for a mean internal energy release of $\bar{\Delta}$ is

$$\overline{\Delta E}_1 = \frac{3m_1 m_2}{(m_1 + m_2)^2} k(T_1 - T_2) - \frac{m_2 \bar{\Delta}}{m_1 + m_2} \tag{1-5-36}$$

and (1-5-28) follows immediately.

1-6. THE DIFFUSION AND VISCOSITY CROSS SECTIONS

It is now appropriate to introduce two additional cross sections: the diffusion cross section q_D and the viscosity cross section q_η; q_D is useful in discussions of the diffusion of neutral and charged particles in gases and the mobility of gaseous ions, whereas q_η appears in the treatment of the viscosity and heat conductivity of gases.[5] As we shall show in Section 2-10, the diffusion coefficient is inversely related to the diffusion cross section. The coefficients of viscosity and heat conductivity are inversely related to the viscosity cross section. These cross sections are actually of greater significance in the kinetic theory of gases than is q_s, which never appears directly in rigorous discussions of transport phenomena.

A. The Diffusion Cross Section, q_D

Consider a large number of particles of mass m swarming about in a gas composed of molecules of mass M and assume that the energies of the two types of particles are sufficiently low that internal excitation on impact is impossible. Thus we are dealing only with elastic collisions. Now view, in the CM system, a collision between one of the particles m and a gas molecule M. The particle has linear momentum equal to $M_r v_0$ before the collision, where M_r is the reduced mass of m and M, and v_0 is the relative velocity of approach. If the CM scattering angle is Θ, m suffers a change in its forward momentum of $M_r v_0(1 - \cos\Theta)$. The *diffusion cross section* is defined by the equation

$$q_D = \int (1 - \cos\Theta) I_s(\Theta)\, d\Omega_{\text{CM}} = 2\pi \int_0^\pi I_s(\Theta)(1 - \cos\Theta)\sin\Theta\, d\Theta. \tag{1-6-1}$$

It is apparent that q_D is a measure of the average forward momentum lost by the particles m in collisions with the molecules M, and q_D is therefore often called the *momentum transfer cross section.*

Scattering at small angles produces only a small contribution to q_D. The quantity $(1-\cos\Theta)$ vanishes as Θ^2 for small deflections, and for this reason the diffusion cross section often has a finite value when the integral elastic cross section is (mathematically) infinite. The matter of theoretical predictions of infinitely large cross sections is discussed in Section 3-11.

The diffusion cross section q_D differs appreciably from q_s, the integral elastic scattering cross section, only when the scattering is distinctly anisotropic. When the backward scattering dominates the forward scattering, $q_D > q_s$, whereas $q_D < q_s$ when the scattering is concentrated in the forward direction. If the differential elastic scattering cross section is independent of Θ, $q_D = q_s$.

Using the diffusion cross section, we may define the *momentum transfer mean free path* λ_m by the equation

$$\lambda_m = \frac{1}{Nq_D}. \tag{1-6-2}$$

λ_m is frequently called the transport mean free path. We may also define the *collision frequency for momentum transfer* ν_m by the equation

$$\nu_m = \frac{\bar{v}_0}{\lambda_m}, \tag{1-6-3}$$

where $\bar{v}_0$ is the mean velocity of the particles.

Let us now suppose that the particles of mass m moving about in the gas are electrons, which, of course, are much lighter than the gas molecules. The average energy of the electrons is assumed to be considerably greater than that of the molecules, but the assumption of elastic collisions is retained. Since, now, $m \ll M$, the fractional energy loss by an electron in a collision with a molecule which results in the electron being scattered at angle ϑ in the Lab system is

$$\Delta(\vartheta) \approx \frac{2m}{M}(1 - \cos\vartheta),$$

as seen from (1-5-24). Thus the mean fractional energy loss by the electrons per collision is

$$\Delta \approx \frac{2m}{Mq_s}\int_0^{\pi}\int_0^{2\pi}(1 - \cos\vartheta)I_s(\vartheta)\sin\vartheta\, d\vartheta\, d\varphi = \frac{2m}{M}\frac{q_D}{q_s}. \tag{1-6-4}$$

We were able to replace the integral in the middle of (1-6-4) by q_D, since here $m \ll M$ and therefore the Lab and CM scattering angles are almost equal.

The mean fractional loss of energy experienced by an electron in traveling a distance dx through a gas containing N molecules/cm^3 is then

$$\left(\frac{2m}{M}\frac{q_D}{q_s}\right)(q_s N\,dx) = \frac{2m}{M} q_D N\,dx, \tag{1-6-5}$$

since $q_s N\,dx$ is the probability that m will undergo a collision in moving a distance dx. It is thus evident that electrons moving about in a gas lose energy in elastic collisions with the molecules at the rate that they would if the elastic scattering cross section were equal to q_D and the fractional energy loss per collision were equal to $2m/M$.

Actually, the analysis in the foregoing paragraph does not depend much on m being very small compared with M. If we use (1-5-9) and work in CM coordinates, we get in place of (1-6-5) the result $[2mM/(m + M)^2]q_D N\,dx$, so that the energy loss in general corresponds to a cross section of q_D and a fractional energy loss per collision of $2mM/(m + M)^2$.

B. The Viscosity Cross Section, q_η

The *viscosity cross section* is defined by the equation

$$q_\eta = \int \sin^2\Theta\; I_s(\Theta)\,d\Omega_{\text{CM}} = 2\pi \int_0^\pi I_s(\Theta)\sin^3\Theta\,d\Theta. \tag{1-6-6}$$

It may be noted that collisions involving CM scattering at nearly $\pi/2$ are weighted more heavily in evaluating q_η than are small-angle or backward-scattering events. The physical basis for this weighting is easily explained by observing that the greater the rate of collisional equalization of energy, the smaller the viscosity and heat conductivity. The energy is equalized in the collision of two identical molecules if the CM scattering angle is $\pi/2$. Collisions at angles near $\pi/2$ are therefore more effective than any others in inhibiting conductivity; hence $\sin^2\Theta$ is the appropriate weighting factor in the integral for q_η. By contrast, diffusion is retarded most effectively by back-scattering collisions, and $(1 - \cos\Theta)$ is the proper weighting factor for q_D. The viscosity cross section converges for the usual intermolecular force fields because $\sin^2\Theta$ goes to 0 as Θ^2 for small deflections, as did $(1 - \cos\Theta)$ in the expression for q_D.

1-7. POTENTIAL FUNCTIONS USED TO DESCRIBE THE INTERACTION BETWEEN PARTICLES

Many different potential functions are used to describe the interaction between various types of particles. The desirability of choosing a function that realistically represents the interaction in a given problem is obvious, but it is perhaps just as obvious that in many cases realism must be sacrificed for mathematical tractability. We shall enumerate some of the potentials most frequently used in practice. These potentials are of two types—those that depend only on the distance between centers of the interacting particles and are thus angle independent and those that contain a dependence on the relative angular orientation of the particles as well as the separation distance. Potentials involving an explicit dependence on the relative velocity are also occasionally required. Electron-neutral interaction potentials are discussed in Chapters 4 and 5. Additional material on neutral-neutral potentials appears in Chapters 9-11 of our companion volume.[1]

A. Potentials Depending Only on the Separation Distance

The potential V is plotted in Fig. 1-7-1 for several types of spherically symmetric interaction, each of which is discussed.

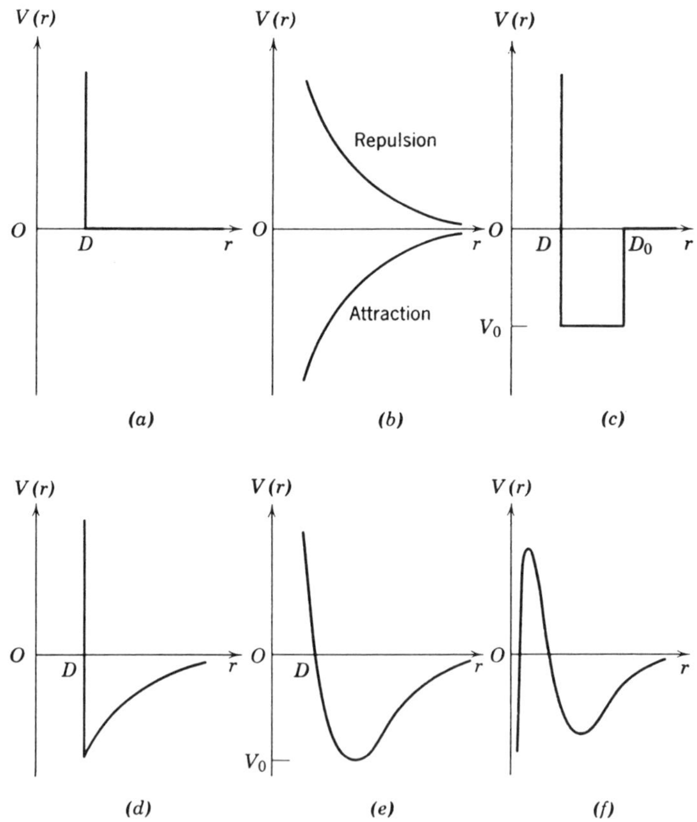

Figure 1-7-1. Commonly used spherically symmetric potential functions: (*a*) smooth, elastic spheres; (*b*) point centers; (*c*) the square well; (*d*) the Sutherland potential; (*e*) the Lennard-Jones potential; (*f*) the Buckingham potential.

1. *Smooth elastic spheres*

$$V(r) = \begin{cases} \infty & (r < D) \\ 0 & (r > D) \end{cases} \tag{1-7-1}$$

Here the model consists of two rigid impenetrable spheres, the sum of whose radii is D. This potential function is crude but easy to apply in calculations.

2. *Point centers of attraction or repulsion*

$$V(r) = \begin{cases} -\dfrac{a}{r^n} & \text{(attraction)} \\ \dfrac{a}{r^n} & \text{(repulsion)} \end{cases} \tag{1-7-2}$$

Here a is a positive constant and n is the index of attraction or repulsion. When $n = 1$, we have *Coulomb interaction.* The interaction between molecules is frequently fitted by the repulsive potential of (1-7-2) with a value of n between 9 and 15. Particles assumed to repel one another according to a potential of the form $+a/r^4$ are referred to as "Maxwellian"—the r^{-4} interaction has some important special properties which were pointed out by Maxwell[6] and which are discussed in Section 1-8. The attractive r^{-4} potential is particularly important in the study of ionized gases because, as shown in Section 1-8, the point-charge-induced dipole interaction between an ion (or electron) and a neutral molecule is of this type.

The mathematical difficulty associated with the use of an r^{-n} potential depends to a considerable degree on the particular value of n chosen.

3. *The square well*

$$V(r) = \begin{cases} \infty & (r < D) \\ -V_0 & (D < r < D_0) \\ 0 & (r > D_0) \end{cases} \tag{1-7-3}$$

This potential function represents an impenetrable core of radius D surrounded by an attractive well of depth V_0 which extends radially out to a distance D_0. Here we have a short-range repulsion and a longer range attraction. The square-well potential is relatively easy to use.

4. *The Sutherland potential*

$$V(r) = \begin{cases} \infty & (r < D) \\ -\dfrac{a}{r^n} & (r > D) \end{cases} \tag{1-7-4}$$

This model applies to hard spheres of combined radius D which attract one another according to an inverse power law. It is reasonably tractable in calculations. When $n = 6$, (1-7-4) gives the *van der Waals interaction.*

5. *The Lennard-Jones potential*

$$V(r) = \frac{a}{r^m} - \frac{b}{r^n} \tag{1-7-5}$$

Both constants, a and b, are positive, so that the first term is repulsive and the second attractive. When the potential takes the form

$$V(r) = 4V_0\left[\left(\frac{D}{r}\right)^{12} - \left(\frac{D}{r}\right)^{6}\right] \tag{1-7-6}$$

it is called the *Lennard-Jones* (6-12) *potential.* The inverse-sixth-power attraction represents an induced dipole-induced dipole interaction. The choice of $m = 12$ in the repulsive term is largely a matter of mathematical expediency, although it does prove to be fairly realistic for spherical nonpolar molecules.

6. *The Buckingham potentials* The basic Buckingham potential takes the form

$$V(r) = ae^{-br} - \frac{c}{r^6} - \frac{d}{r^8}, \tag{1-7-7}$$

where a, b, c, and d are constants. Its attractive terms represent the induced dipole-induced dipole and induced dipole-induced quadrupole interactions, whereas the repulsive contribution is approximated by the exponential term. The potential of (1-7-7), as well as the Buckingham-Corner potential and the modified Buckingham (6-exp) potential, are discussed in Hirschfelder et al.[7]

7. *The* 12-6-4 *potential* A reasonable model for the interaction of an ion and a molecule is obtained by the superposition of a Lennard-Jones (6-12) interaction and the interaction of a point charge with a polarizable molecule, which gives

$$V(r) = \frac{a}{r^{12}} - \frac{b}{r^6} - \frac{c}{r^4}, \tag{1-7-8}$$

where a, b, and c are positive constants. The polarization attraction due to the ionic charge contributes the r^{-4} term and part of the r^{-6} term, as discussed in Section 1-8.

B. Potentials Involving the Relative Angular Orientation

Hirschfelder et al.[7] also treat four angle-dependent models: hard ellipsoids of revolution, spherocylindrical molecules, hard spheres in which point dipoles are embedded, and the Stockmayer model. The Stockmayer represents a superposition of a Lennard-Jones (6-12) interaction and the interaction of two point dipoles.

1-8. THE POLARIZATION ATTRACTION OF MOLECULES BY CHARGED PARTICLES

We now wish to determine the law of force between a charged particle, which for convenience we refer to as an ion, and a neutral gas molecule. We must first calculate the dipole moment induced in a molecule by a nearby ion. Let us consider a parallel-plate capacitor and the polarization that occurs in the dielectric separating its plates. Let the surface density of free charge on the plates be σ. In the absence of a dielectric the uniform electric field intensity between the plates is $E = 4\pi\sigma$. With a gas of dielectric constant* K filling the capacitor, the field intensity is reduced to $E' = E/K$. This reduction in field intensity is due to the termination of lines of force by charges induced at the surface of the dielectric by the field. If σ' denotes the charge density induced at the surface of the dielectric, $E - E' = 4\pi\sigma'$. Then

$$\frac{E - E'}{E'} = K - 1 = \frac{4\pi\sigma'}{E'}$$

and

$$\sigma' = \frac{(K - 1)E'}{4\pi}.$$

Now the electric dipole moment of all the molecules in the gas is given by the product of σ' and the area of the capacitor plate, multiplied by the plate separation, or σ' times the volume defined by the capacitor plates. Thus σ' is the total dipole moment divided by the volume, but this is just the dipole moment per cm^3, or the polarization P. Consequently,

$$P = \frac{(K - 1)E'}{4\pi}.$$

*The dielectric constant is a tensor quantity that reduces to scalar form when the medium is isotropic. Here it is a positive scalar whose magnitude is slightly greater than unity.

If N is the number of molecules per cm^3, the dipole moment per molecule is

$$\mu = \frac{P}{N} = \frac{(K-1)E'}{4\pi N}.$$

In the present problem E' is the Coulomb field of an ion, $E' = e/r^2$, except at small distances. Therefore,

$$\mu = \frac{K-1}{4\pi N}\frac{e}{r^2}. \qquad (1\text{-}8\text{-}1)$$

Now the force with which a dipole of moment μ is attracted by a point charge e located at a distance r from the dipole is

$$f = \frac{2\mu e}{r^3}\cos\beta, \qquad (1\text{-}8\text{-}2)$$

where β is the angle between the dipole axis and the line joining the dipole and the point charge. Since the dipole here is induced by the charge, β always equals zero. Therefore, the force between the ion and molecule is

$$f = \frac{(K-1)e^2}{2\pi N r^5}. \qquad (1\text{-}8\text{-}3)$$

The mutual potential energy of the ion and molecule, accordingly, is

$$V(r) = -\int_r^\infty \frac{(K-1)e^2}{2\pi N r^5}\,dr$$

or

$$V(r) = -\frac{(K-1)e^2}{8\pi N r^4}. \qquad (1\text{-}8\text{-}4)$$

Equation (1-8-3) gives only an approximate expression for the force of attraction of a molecule by an ion. It is obtained by assuming that the field produced by the ion is uniform over the volume occupied by the molecule. It is possible to obtain the complete solution of the problem in the case in which the molecules are assumed to be spherical and to show that the approximate solution indeed furnishes the principal term.[8] The next term corresponds to the interaction between the charge and the quadrupole it induces in the molecule and varies as r^{-6}.[9]

The equations in this section may be expressed in terms of the polarizability of the gas α, rather than the dielectric constant K, by use of the relationship

$$\alpha = \frac{K - 1}{4\pi N}. \tag{1-8-5}$$

In Chapter 3 it is shown by dimensional arguments that if the interaction potential between two collision partners varies as r^{-n}, the variation of the CM differential scattering cross section is given classically by the expression

$$I_s(\Theta) \sim v_0^{-4/n}, \tag{1-8-6}$$

where v_0 is the relative velocity of approach. Thus the frequency at which the particles are scattered at any angle Θ for which the classical calculation is valid is proportional to

$$I_s(\Theta)v_0 \sim v_0^{1-4/n}. \tag{1-8-7}$$

(Limitations of the classical scattering theory are discussed in Chapter 3.) Evidently in the present case in which $n = 4$ the collision frequency is independent of velocity, and thus the mean free time is constant. Maxwell[6,10] showed that this fact greatly simplifies many calculations and that it is possible to calculate all the transport properties for the r^{-4} potential without knowing the distribution function (see Section 2-12).

The combination of a charged point particle and a polarizable sphere is often referred to as a *constant mean free time model*. By contrast, the combination of two elastic spheres that interact only at the instant of impact is a *constant mean free path model*, since the scattering cross section in this case is independent of the velocity.

1-9. NOTES

1-1. More Scattering Kinematics. In Sections 1-2, 1-3, and 1-4 we discussed the kinematics of elastic and inelastic collisions for the special case of a beam of projectiles incident on target particles stationary in the Lab frame (*single beam–stationary target collisions*). Other cases involving two beams intersecting at an angle θ in the Lab frame are also important and must be considered.[11] The first experiments of this type were *crossed-beam* experiments, with $\theta = 90°$; the basic techniques were highly developed by the early 1960s. Soon afterward came *merging-beam* experiments ($\theta = 0°$) and then *inclined-beam* experiments ($0 < \theta < 90°$). (The *colliding-beam* technique of elementary particle physics, used to obtain the highest possible CM energy, is not needed in atomic collisions because reaching high energies on the atomic scale is not a problem. In fact, it is the extremely low energies that are difficult to reach in atomic collisions experiments.)

Problems 1-4, 1-6, and 1-7 deal with the kinematics of intersecting beams. They are solved by straightforward application of the laws of conservation of energy and linear momentum. Experimental techniques based on intersecting beams are discussed in Chapters 4 to 8.

Problem 1-9 is concerned with the two-body *reactive collision*, which we encounter in the study of ion-molecule reactions and reactions involving only neutral molecules.

1-2. Spin Polarized Beams. With reference to Fig. 1-4-1, the scattering can depend on the azimuthal angle φ if the beam or target is spin polarized (see Section 4-6). Note that in (1-4-3), the right-hand side involves only observable quantities and is independent of the model used in the discussion.

1-3. Notation. Several alternative sets of notation are commonly used to represent cross sections, two of the most frequently employed being the following:

Angular differential elastic scattering cross section

$$I_s(\vartheta, \varphi)\, d\Omega \equiv \frac{d\sigma(\vartheta, \varphi)}{d\Omega}\, d\Omega \text{ (page 11)}$$

Integral elastic scattering cross section $q_s = \sigma_s$ (page 12)

Total cross section $q_t \equiv \sigma_t$ (page 14)

Momentum transfer (diffusion) cross section $q_m \equiv \sigma_m \equiv q_D \equiv \sigma_D$ (page 25)

We shall use the various notations interchangeably, and introduce minor variations as required to denote other cross sections that will appear later.

1-4. Transformations of Distribution Functions. Frequently, it is necessary to transform a cross section expressed in terms of one set of experimental variables to a corresponding cross section involving another set of variables. A general procedure for doing this (transformation of distribution functions) is described in Section 3-8 of Sard[3] and in Chapter 5 of Hagedorn.[3] The same procedure also applies to transformations of cross sections from one Lorentz frame to another.

1-5. Three-Body Collisions. Three-body collisions are very important under certain conditions in atomic collisions. The concepts of three-body reaction rates and cross sections are discussed in Ref. 12.

1-6. The Point-Charge-Induced Dipole Potential. The electron-molecule interaction potential is treated in detail in Sections 4-7, 4-8, and 5-2. The *point-charge-induced dipole potential* is usually expressed in the form

$$V(r) = -\frac{\alpha e^2}{2r^4}, \qquad (1\text{-}9\text{-}1)$$

where α is the scalar dipole polarizability of the gas molecule in which the dipole is induced, e the electronic charge, and r the separation distance between the point charge and the molecule. If the molecule is not spherically symmetric, the polarizability is a tensor, and a more complicated form for the potential emerges (see Section 5-2-B).

1-10. PROBLEMS*

Solutions for most of these problems appear in the references indicated.

1-1. A *central force* acts between two particles if the force on each particle acts along the line of centers and if the magnitude of the force depends only on the separation distance, r. The force is given by the gradient of the potential: $\mathbf{F} = -\nabla V$. Assume now that V is a function of the three spherical polar coordinates r, θ, and φ, and write down the corresponding components of the force.

1-2. Show that it is impossible to satisfy both energy and linear momentum conservation in a collision of the type $A + B \rightarrow AB$ if no third bodies, photons, or excited states are involved.

1-3. **(a)** Justify the statements immediately following (1-2-25).

(b) Show that the following equations are valid: (1-2-26), (1-3-1) through (1-3-5). "Newton diagrams" are useful in solving such problems.[13] The three diagrams in Fig. 1-10-1 are relevant to this discussion.

1-4. Calculate the relative energy and relative energy spread in a *merging-beam experiment.*

(a) A beam of atoms (each of mass m) and a beam of ions (each of mass m) are traveling in the same direction along the same path in the Lab system. In this reference frame, the kinetic energy of each of the ions is 3000 eV; the kinetic energy of each of the atoms is 3085 eV. Collisions result because atoms in one of the beams overtake ions in the other beam. In such collisions, what is the impact energy in the center-of-mass coordinate system? That is, how much energy per collision can be transferred from kinetic to potential form? *Ans.:* 0.297 eV.

(b) Actually, it is impossible to prepare perfectly monoenergetic beams. Make the realistic assumption that because of experimental factors, the energy spread in each beam is 3 eV, centered about the mean

*Most of these problems deal with the kinematics of collisions. This subject is discussed in the context of low-energy heavy particle collisions in G. Scoles (Ed.), *Atomic and Molecular Beam Methods*, Vol. I, Oxford University Press, New York (1988). There, Chapters 18, 20, and 21 (by U. Buck) are concerned mainly with elastic and inelastic processes, and Chapter 22 (by Y. T. Lee) with reactive collisions.

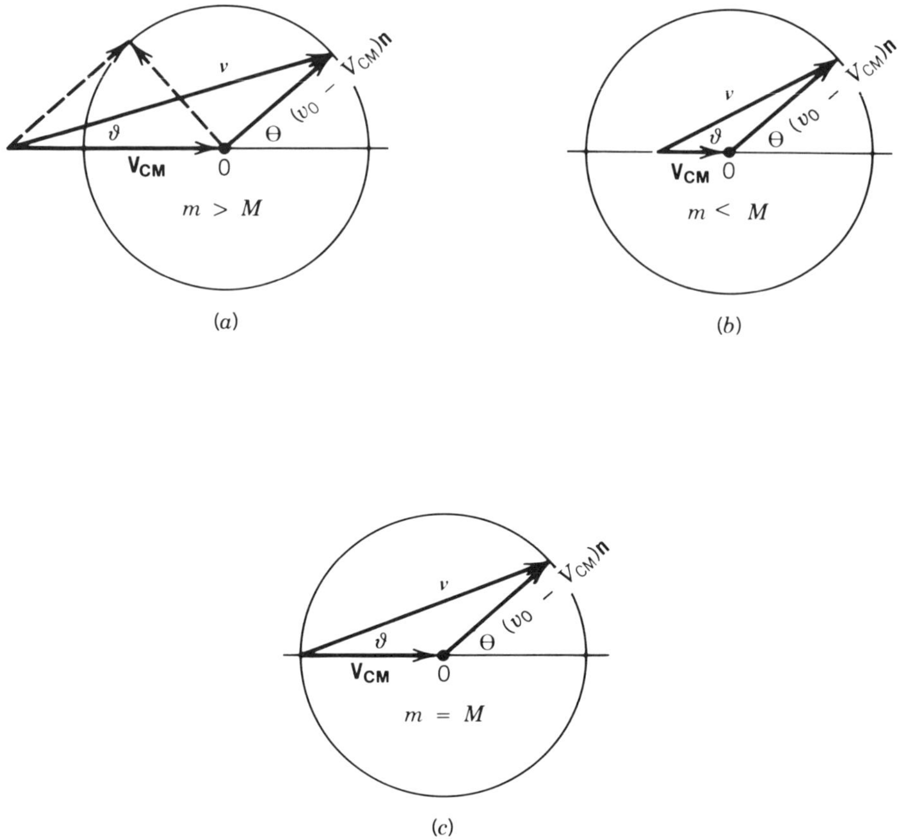

Figure 1-10-1. Newton diagrams.

energy. Thus the ions in the slower beam have energies ranging from (3000 − 1.5) eV to (3000 + 1.5) eV. Calculate the resulting spread in the impact energy in the CM system. *Ans.:* 40 meV.

(c) Show that the relative energy is approximately $(\Delta E)^2/8E$, where ΔE is the difference in the Lab energies of the ions and atoms, and E is the mean of these energies.

1-5. Show that in an *inelastic collision,* the energy loss ΔE by a projectile impinging on a stationary target is

$$\Delta E = T_0\left[1 - \frac{\sin^2\theta + (m/M)\sin^2\vartheta}{\sin^2(\theta + \vartheta)}\right], \qquad (1\text{-}10\text{-}1)$$

where $T_0 = T_{\text{Lab},i}$ is the initial kinetic energy of the projectile (mass m), θ the recoil angle of the target (mass M), and ϑ the scattering angle of the projectile.

1-6. Show that the relative energy of two particles in beams inclined at an angle θ with respect to each other, and traveling at speeds v_1 and v_2 is

$$E_R = \tfrac{1}{2} M_r v_R^2, \tag{1-10-2}$$

where M_r is the reduced mass and $\mathbf{v}_R$ is the relative velocity, with

$$v_R = (v_1^2 + v_2^2 - 2v_1 v_2 \cos\theta)^{1/2}. \tag{1-10-3}$$

1-7. Calculate the relative energy and energy spread in an *inclined-beam experiment.*

(a) A beam of electrons with a Lab energy of 10 eV collides at an acute angle of 20° with a beam of H^- ions having a Lab energy of 10 keV. Calculate the relative kinetic energy in the collisions. *Ans.:* 1.6 eV.

(b) Assume that the ions are monoenergetic but that the electrons have an energy spread of ± 0.3 eV. Estimate the resulting spread in the CM energy. *Ans.:* 0.09 eV.

1-8. Calculate the maximum possible energy transfer in a collision that is inelastic but in which momentum is still conserved for the following cases: (a) a 100-eV electron colliding with a stationary H atom, and (b) a 100-eV proton colliding with a stationary H atom.

1-9. The chemical identities, masses, and internal energies of particles may change in a reaction of the type A + B → C + D. Suppose that B is at rest in the Lab frame. Let $T_{\text{CM},i}$ be the initial kinetic energy in the CM frame, and Q be the decrease in the total internal energy of the system occurring in the collision. Show that the CM and Lab differential cross sections for observation of the product C in a given direction are related by

$$\frac{d\sigma_{\text{Lab}}}{d\Omega_{\text{Lab}}}(\theta_{\text{Lab}}, \varphi_{\text{Lab}}) = \frac{(1 + x^2 + 2x\cos\theta_{\text{CM}})^{3/2}}{|1 + x\cos\theta_{\text{CM}}|} \frac{d\sigma_{\text{CM}}}{d\Omega_{\text{CM}}}(\theta_{\text{CM}}, \varphi_{\text{CM}}), \tag{1-10-4}$$

where

$$x = \left(\frac{m_A m_C}{m_B m_D} \frac{T_{\text{CM},i}}{T_{\text{CM},i} + Q}\right)^{1/2}. \tag{1-10-5}$$

This problem is solved on pages 20 and 21 of C. J. Joachain, *Quantum Collision Theory*, North-Holland, Amsterdam (1975) (see also the problems at the end of Joachain's Chapter 2).

1-10. The model described in Section 1-5 is useful in calculations of the slowing down of neutrons in matter.[14] How many collisions must a neutron with initial energy of 1 MeV make to slow down to $\frac{1}{40}$ eV in (a) atomic hydrogen, (b) atomic deuterium, (c) graphite, and (d) uranium-238? (Inelastic collisions can be ignored in this energy range.)

1-11. Consider an unstable particle with internal energy E_0 moving with velocity $\mathbf{V}$ in the Lab frame. It undergoes *spontaneous dissociation* into two fragments with masses m_1 and m_2 and internal energies E_1 and E_2. The velocities of the fragments make angles θ_1 and θ_2 with respect to $\mathbf{V}$ in the Lab frame. Show that the relation between θ_1 and θ_2 is

$$\frac{m_2}{m_1}\sin^2\theta_2 + \frac{m_1}{m_2}\sin^2\theta_1 - 2\sin\theta_1\sin\theta_2\cos(\theta_1+\theta_2) = \frac{2\Delta E}{(m_1+m_2)V^2}\sin^2(\theta_1+\theta_2), \tag{1-10-6}$$

where $\Delta E = E_0 - E_1 - E_2$. (Solutions to Problems 1-11 to 1-14 appear in Chapter IV of the book by Landau and Lifshitz.[3])

1-12. In Problem 1-11, consider one of the fragments, and let $\mathbf{v}_L$ and $\mathbf{v}_C$ be its velocities in the Lab and CM frames. The angle between $\mathbf{v}_L$ and $\mathbf{V}$ is labeled θ here. Show that the angular distribution of the fragments in the Lab frame is

$$\frac{1}{2}\sin\theta\, d\theta\left[2\frac{V}{v_C}\cos\theta + \frac{1+(V^2/v_C^2)\cos 2\theta}{[1-(V^2/v_C^2)\sin^2\theta]^{1/2}}\right] \quad (0\leqslant\theta\leqslant\pi) \quad \text{if } v_C > V \tag{1-10-7}$$

and

$$\sin\theta\, d\theta\frac{1+(V^2/v_C^2)\cos 2\theta}{[1-(V^2/v_C^2)\sin^2\theta]^{1/2}} \quad (0\leqslant\theta\leqslant\theta_{\max}) \quad \text{if } v_C < V. \tag{1-10-8}$$

1-13. In Problem 1-11, assume that $v_{2C} > v_{1C}$ applies for the speeds of the fragments in the CM frame, and show that the range of possible values of the angle between the velocities of the fragments ($\theta_t = \theta_1 + \theta_2$) is

$$0 < \theta_t < \pi \qquad \text{if } v_{1C} < V < v_{2C}; \tag{1-10-9}$$

$$\pi - \theta_0 < \theta_t < \pi \qquad \text{if } V < v_{1C}; \tag{1-10-10}$$

$$0 < \theta_t < \theta_0 \qquad \text{if } V > v_{2C}, \tag{1-10-11}$$

where

$$\sin\theta_0 = \frac{V(v_{1C} + v_{2C})}{V^2 + v_{1C}v_{2C}}. \tag{1-10-12}$$

1-14. For the scattering problem discussed in Section 1-5, show that the cross section as a function of the energy lost by the projectile (ϵ) is

$$d\sigma = \pi D^2 \frac{d\epsilon}{\epsilon_{\max}}. \tag{1-10-13}$$

1-15. Define the CM momentum p_{CM} by the relation $E_{\text{CM}} = p_{\text{CM}}^2/2M_r$ and show that $p_{\text{CM}} = (2mE_{\text{Lab}})^{1/2}/(1 + \gamma)$, where $\gamma = (1 - v^2/c^2)^{-1/2}$. Here M_r is the reduced mass, and m is the mass of the projectile.

1-16. **(a)** A particle of mass m_1 and momentum p_1 collides elastically with a particle of mass m_2 that is initially at rest. In a relativistic calculation, show that the maximum possible momentum of m_2 after the collision in the Lab frame is

$$\frac{2m_2(m_2 + E_1)|\mathbf{p}_1|}{(m_2 + E_1)^2 - |\mathbf{p}_1|^2}.$$

(b) Show that for a proton with momentum equal to its rest energy/c, colliding with a stationary electron, the maximum postcollision momentum of the electron is 1.4 MeV/c. (Chen,[15] p. 1.)

1-17. A projectile of rest mass m_1 and velocity v is adsorbed by a stationary target of rest mass m_2, and a composite structure is formed. Show that the rest mass of the final structure is $M = (m_1^2 + m_2^2 + 2\gamma_1 m_1 m_2)^{1/2}$ and that its velocity is

$$V = \frac{\gamma_1 m_1 v c}{\sqrt{M^2c^2 + \gamma_1 m_1^2 v^2}}.$$

(Chen,[15] p. 272.)

1-18. **(a)** Show that if protons are scattered elastically by deuterons, the maximum scattering angle in both the CM and Lab frames is 180°.

(b) Show that if deuterons are scattered by protons, the maximum scattering angles are 120° and 30°, respectively.

REFERENCES

1. E. W. McDaniel, *Atomic Collisions: Heavy Particle Projectiles*, Wiley, New York, to be published.
2. H. S. W. Massey, E. W. McDaniel, and B. Bederson (Series Eds.), *Applied Atomic Collision Physics*, Academic, New York, 5 vols: Vol. 1, H. S. W. Massey and D. R. Bates (Eds.), *Atmospheric Physics and Chemistry* (1982); Vol. 2, C. F. Barnett and M. F. A. Harrison (Eds.), *Controlled Fusion* (1984); Vol. 3, E. W. McDaniel and W. L. Nighan (Eds.), *Gas Lasers* (1982); Vol. 4, S. Datz (Ed.), *Condensed Matter* (1983); Vol. 5, H. S. W. Massey, E. W. McDaniel, and B. Bederson (Eds.), *Special Topics* (1982).
3. L. D. Landau and E. M. Lifshitz, *Mechanics*, Addison-Wesley, Reading, Mass. (1960). R. D. Sard, *Relativistic Mechanics*, Benjamin, New York (1970), Chaps. 3 and 4. R. Hagedorn, *Relativistic Kinematics*, Benjamin, Menlo Park, Calif. (1973), Chaps. 3–6. E. Byckling and K. Kajantie, *Particle Kinematics*, Wiley, New York (1973), Chaps. 2–4. Particle Data Group, *Rev. Mod. Phys.* **56**, S38–S41 (1984). G. L. Catchen, J. Husain, and R. N. Zare, *J. Chem. Phys.* **69**, 1737 (1978). A. M. Baldwin, V. I. Goldanskii, and I. L. Rozenthal, *Kinematics of Nuclear Reactions*, Oxford University Press, London (1961).
4. D. R. Bates, *Phys. Rev. A* **34**, 1878 (1986).
5. R. D. Present, *The Kinetic Theory of Gases*, McGraw-Hill, New York (1958), pp. 140, 146, 219. J. H. Ferziger and H. G. Kaper, *Mathematical Theory of Transport Processes in Gases*, North-Holland, Amsterdam (1972).
6. W. D. Niven (Ed.), *The Scientific Papers of James Clark Maxwell*, Vol. II, Dover, New York (1952), pp. 26–78.
7. J. O. Hirschfelder, C. F. Curtiss, and R. B. Bird, *Molecular Theory of Gases and Liquids*, Wiley, New York (1964), pp. 31–35. See also T. Kihara, *Intermolecular Forces*, Wiley, New York (1978); R. B. Bernstein (Ed.), *Atom-Molecule Collision Theory*, Plenum, New York (1979); E. A. Mason, "Determination of Intermolecular Potentials," in Vol. 5, of Ref. 2.
8. P. Langevin, *Ann. Chim. Phys.* **28**, 317–321 (1903).
9. E. A. Mason and E. W. McDaniel, *Transport Properties of Ions in Gases*, Wiley, New York (1988).
10. S. Chapman and T. G. Cowling, *The Mathematical Theory of Non-uniform Gases*, 3rd ed., Cambridge University Press, London (1970). See also Refs. 5 and 7.
11. F. Brouillard and J. W. McGowan (Eds.), *Physics of Ion-Ion and Electron-Ion Collisions*, Plenum, New York (1983). See the chapters by F. Brouillard and W. Claeys, D. H. Crandall, E. Salzborn, and K. T. Dolder for discussions of the kinematics of intersecting beams.

12. F. T. Smith, *J. Chem. Phys.* **36**, 248 (1962); *Disc. Faraday Soc.* **33**, 183 (1962). J. B. Hasted, *Physics of Atomic Collisions*, 2nd ed., American Elsevier, New York (1972). M. L. Goldberger and K. M. Watson, *Collision Theory*, Wiley, New York (1964).

13. L. D. Landau and E. M. Lifshitz, Ref. 3. M. S. Child, *Molecular Collision Theory*, Academic, New York (1974). E. Segrè, *Nuclei and Particles*, 2nd ed., Benjamin-Cummings, Menlo Park, Calif. (1977).

14. S. Glasstone and M. C. Edlund, *The Elements of Nuclear Reactor Theory*, Van Nostrand, Princeton, N.J. (1952). E. Fermi, *Nuclear Physics*, University of Chicago, Chicago (1950).

15. M. Chen, *Berkeley Physics Problems with Solutions*, Prentice-Hall, Englewood Cliffs, N.J. (1974).

2

BACKGROUND INFORMATION FROM THE KINETIC THEORY OF GASES

We now present for future reference some results from the kinetic theory of gases. Emphasis is placed on those topics that are of special interest in the study of ionized gases. Derivations are given for only a few of the results quoted: for derivations of the remainder and for further discussion of their physical significance the reader is referred to the standard texts on kinetic theory.[1] Particular attention is directed to the treatises by Ferziger and Kaper,[2] Chapman and Cowling,[3] and Hirschfelder et al.[4] for rigorous discussions of the nonequilibrium properties of gases.

2-1. THERMODYNAMIC EQUILIBRIUM AND THE SCOPE OF KINETIC THEORY

A gas is in thermodynamic equilibrium if and only if it is in mechanical, chemical, and thermal equilibrium. The first requirement necessitates a balance of mechanical forces within the gas and between the gas and its surroundings. (In the special case in which no external forces are applied, the pressure of the gas must be uniform.) The requirement of chemical equilibrium is met when a system in mechanical equilibrium does not tend to undergo a spontaneous change of internal structure; for example, by a chemical reaction or a diffusive mass transfer. The thermal equilibrium requirement implies that the gas must have a uniform temperature and that this temperature must be equal to that of the surroundings.

The kinetic theory treats gases in both equilibrium and nonequilibrium conditions. In the first instance the major concerns are the equilibrium distribution of molecular velocities and energies, molecular mean free paths and collision frequencies, effusive molecular flow, equations of state, and the electric and magnetic properties of gases. In the study of a gas that is not in thermodynamic equilibrium interest is centered about its transport properties, that is, those properties that determine the behavior of the gas as regards diffusion, thermal and electrical conduction, and viscosity. The last-mentioned are termed *transport phenomena*, and they are investigated theoretically by what is referred to as transport theory.

2-2. THE MOLECULAR VELOCITY AND ENERGY DISTRIBUTIONS OF A GAS IN THERMODYNAMIC EQUILIBRIUM

Molecules of a gas in equilibrium move about in random fashion, the trajectory of each molecule consisting of a series of line segments consecutively joined together at the points in space at which the molecule suffers collisions. The average length of the line segments is the mean free path of the molecule in the gas. The line segments are essentially straight, except near the collision sites, where the interaction between the collision partners is appreciable. The curvature of the paths due to the earth's gravitational attraction is completely negligible on the laboratory scale. (The average radius of curvature for molecules of nitrogen at room temperature is of the order of 10^7 cm, for example.)

A. A Single Gas with No Applied Field

The speeds of the molecules of a gas in thermodynamic equilibrium are distributed according to *Maxwell's distribution*, which may be expressed in the form

$$f(v)\,dv = N_t\left(\frac{2}{\pi}\right)^{1/2}\left(\frac{m}{kT}\right)^{3/2} v^2 e^{-mv^2/2kT}\,dv. \tag{2-2-1}$$

Here $f(v)\,dv$ is the number of molecules with speeds between v and $v + dv$ cm/sec, m is the mass of one of the molecules in grams, T is the absolute temperature in K, and $k = 1.3806 \times 10^{-16}$ erg/K is Boltzmann's constant. Evidently, if N_t is the total number of molecules in the gas,

$$\int_0^\infty f(v)\,dv = N_t. \tag{2-2-2}$$

The function $f(v)$ is plotted in Fig. 2-2-1; $\bar{v}$ and v_R, the mean and root-mean-square speeds, respectively, are easily calculated in terms of moments of the

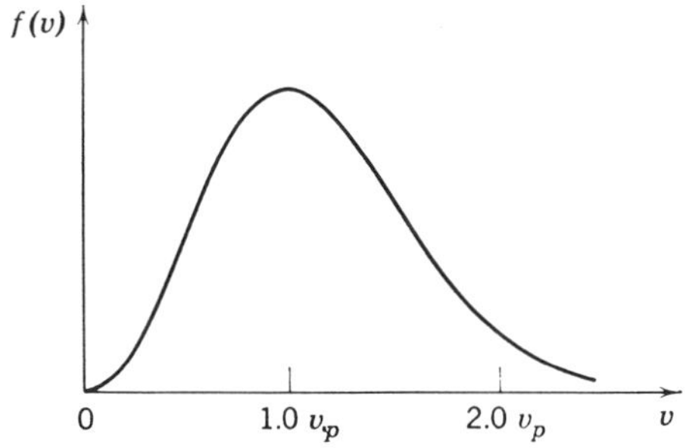

Figure 2-2-1. The Maxwellian distribution function for the speeds of the molecules of a gas in thermodynamic equilibrium.

distribution function:

$$\bar{v} = \frac{1}{N_t}\int_0^\infty v f(v)\, dv = \left(\frac{8kT}{\pi m}\right)^{1/2}, \tag{2-2-3}$$

$$v_R = \left[\frac{1}{N_t}\int_0^\infty v^2 f(v)\, dv\right]^{1/2} = \left(\frac{3kT}{m}\right)^{1/2}. \tag{2-2-4}$$

The most probable velocity of the distribution v_p is obtained by setting $df(v)/dv$ equal to zero and solving for v:

$$v_p = \left(\frac{2kT}{m}\right)^{1/2}. \tag{2-2-5}$$

It follows that

$$v_p/\bar{v}/v_R = 1/1.1284/1.2248. \tag{2-2-6}$$

Table 2-2-1 lists the mean velocity for a number of gases at 15°C. For comparison purposes, the muzzle velocity of a modern army rifle is about 8×10^4 cm/sec, whereas the vertical escape velocity from the earth is 1.1×10^6 cm/sec.

At 15°C the translational kinetic energies per molecule corresponding, respectively, to v_p and v_R are

$$\frac{m}{2} v_p^2 = kT = 0.025 \text{ eV} \tag{2-2-7}$$

and

$$\frac{m}{2} v_R^2 = \tfrac{3}{2}kT = 0.037 \text{ eV}. \tag{2-2-8}$$

Table 2-2-1 Values of the Mean Velocity $\bar{v}$, the Molecular Diameter D; the Mean Free Path λ, and the Collision Frequency ν, Calculated from the Kinetic Theory for a Number of Common Gases

Gas	Molecular Weight	$\bar{v}$ (10^3 cm/sec at 15°C)	D (10^{-8} cm)	λ (10^{-6} cm at 15°C, 760 torr)	ν (10^9 collisions/sec at 15°C, 760 torr)
H_2	2.016	174.0	2.74	11.77	14.8
He	4.002	123.5	2.18	18.62	6.6
CH_4	16.03	61.8	4.14	5.16	12.0
NH_3	17.03	59.8	4.43	4.51	13.3
H_2O	18.02	58.2	4.60	4.18	13.9
Ne	20.18	55.0	2.59	13.22	4.2
N_2	28.02	46.7	3.75	6.28	7.4
C_2H_4	28.03	46.7	4.95	3.61	12.9
C_2H_6	30.05	45.1	5.30	3.15	14.3
O_2	32.00	43.7	3.61	6.79	6.4
HCl	36.46	40.9	4.46	4.44	9.2
Ar	39.94	39.1	3.64	6.66	5.9
CO_2	44.00	37.2	4.59	4.19	8.8
Kr	82.9	27.1	4.16	5.12	5.3
Xe	130.2	21.7	4.85	3.76	5.8
Electron	5.49×10^{-4}	10.5×10^3			

Source: Kennard.[1]

The fraction of the molecules whose speeds exceed a certain value v is

$$\frac{1}{N_t}\int_v^\infty f(v)\,dv = 1 + \left(\frac{2m}{\pi kT}\right)^{1/2} v e^{-mv^2/2kT} - \Phi\left[\left(\frac{m}{2kT}\right)^{1/2} v\right], \tag{2-2-9}$$

where

$$\Phi(x) = \frac{2}{(\pi)^{1/2}}\int_0^x e^{-x^2}\,dx \tag{2-2-10}$$

is the error, or probability, integral, which is tabulated in books of mathematical tables. The sharpness of the Maxwellian distribution is indicated by the fact that only 12.55% of the molecules have $v > 1.5\bar{v}$, 1.70% have $v > 2\bar{v}$, and 0.01% have $v > 3\bar{v}$.

Knowing $f(v)$, we may easily calculate the distribution function for the molecular energies:

$$f(E)\,dE = N_t \frac{2}{(\pi)^{1/2}(kT)^{3/2}} e^{-E/kT} \sqrt{E}\,dE; \tag{2-2-11}$$

$f(E)\,dE$ is the number of molecules with kinetic energy in the range E to $E + dE$.

The distribution function for the component of velocity along one direction is also of interest. The number of molecules whose X components of velocity lie between v_x and $v_x + dv_x$ is

$$f_x\,dv_x = N_t \left(\frac{m}{2\pi kT}\right)^{1/2} e^{-mv_x^2/2kT}\,dv_x. \tag{2-2-12}$$

The function f_x has the shape of an error curve. It can be shown that the root mean square of any Cartesian component of the velocity equals Newton's (isothermal) value for the speed of sound in the gas:

$$(v_i)_R = \left(\frac{p}{\rho}\right)^{1/2}. \tag{2-2-13}$$

Here p is the pressure and ρ is the density. The right-hand side of (2-2-13) should be multiplied by $(\gamma)^{1/2}$, where γ is the ratio of the specific heats, to obtain the correct speed of sound.

Each component of the velocity is distributed independently of the other components. Therefore, if $p(\mathbf{v})\,dv_x\,dv_y\,dv_z \equiv p(v_x, v_y, v_z)\,dv_x\,dv_y\,dv_z$ is the probability that a given molecule has velocity components between v_x and $v_x + dv_x$, v_y and $v_y + dv_y$, and v_x and $v_z + dv_z$, it follows that

$$p(\mathbf{v})\,dv_x\,dv_y\,dv_z = \frac{1}{N_t^3} f_x f_y f_z\,dv_x\,dv_y\,dv_z. \tag{2-2-14}$$

B. Mixtures of Gases

In a mixture of gases that is in complete equilibrium, each of the species has a Maxwellian distribution characterized by the common temperature and the value of the molecular mass for that species, just as if the other constituents of the mixture were not present. We shall show that in a binary mixture of gases at the same temperature the mean value of the relative velocity of two unlike molecules $\bar{v}_0$ is given by

$$\bar{v}_0 = (\bar{v}_1^2 + \bar{v}_2^2)^{1/2}, \tag{2-2-15}$$

where $\bar{v}_1$ and $\bar{v}_2$ are the mean velocities of the two species. Evidently, if the two species are identical,

$$\bar{v}_0 = \sqrt{2}\,\bar{v}. \tag{2-2-16}$$

The proof of (2-2-15) which follows is taken from Present.[5] The mean value of the velocities of the type 1 molecules relative to those of type 2 is obtained by averaging v_0 over the distribution functions of both types of molecules. Thus

$$\bar{v}_0 = \iiint dv_{1x}\,dv_{1y}\,dv_{1z} \iiint dv_{2x}\,dv_{2y}\,dv_{2z} p_1(\mathbf{v}_1)p_2(\mathbf{v}_2)v_0, \tag{2-2-17}$$

where

$$v_0 = [(v_{1x} - v_{2x})^2 + (v_{1y} - v_{2y})^2 + (v_{1z} - v_{2z})^2]^{1/2}. \tag{2-2-18}$$

Using (2-2-12) and (2-2-14) in (2-2-17), we have

$$\bar{v}_0 = \left(\frac{m_1}{2\pi kT}\right)^{3/2}\left(\frac{m_2}{2\pi kT}\right)^{3/2}\int_{-\infty}^{\infty} dv_{1x} \cdots \int_{-\infty}^{\infty} dv_{2z} v_0 \exp\left(-\frac{m_1 v_1^2 + m_2 v_2^2}{2kT}\right). \tag{2-2-19}$$

The computation is simplified if we transform from $\mathbf{v}_1$ and $\mathbf{v}_2$ to the relative velocity $\mathbf{v}_0$ and the velocity of the center of mass in the Lab system $\mathbf{V}_{\mathrm{CM}}$. These velocities are given by the equations

$$\mathbf{v}_0 = \mathbf{v}_1 - \mathbf{v}_2, \qquad \mathbf{V}_{\mathrm{CM}} = \frac{m_1\mathbf{v}_1 + m_2\mathbf{v}_2}{m_1 + m_2}. \tag{2-2-20}$$

If we solve (2-2-20) in component form, we obtain

$$\begin{aligned} v_{1x} &= V_{\mathrm{CM}x} + \frac{m_2}{m_1 + m_2} v_{0x}, \\ v_{2x} &= V_{\mathrm{CM}x} - \frac{m_1}{m_1 + m_2} v_{0x}, \end{aligned} \tag{2-2-21}$$

with similar equations for the y and z components. Therefore,

$$\tfrac{1}{2}m_1 v_1^2 + \tfrac{1}{2}m_2 v_2^2 = \tfrac{1}{2}(m_1 + m_2)V_{\mathrm{CM}}^2 + \tfrac{1}{2}M_r v_0^2, \tag{2-2-22}$$

where $M_r = m_1 m_2/(m_1 + m_2)$ is the reduced mass of the pair of particles. Equation (2-2-22) expresses the fact that the total kinetic energy of the particles is equal to the kinetic energy of the total mass, moving with the velocity of the center of mass in the Lab system, plus the kinetic energy of relative motion (cf. Section 1-2).

The volume elements in the two new velocity spaces are given by the equations

$$\begin{aligned} dV_{CMx}\, dV_{CMy}\, dV_{CMz} &= V_{CM}^2\, dV_{CM} \sin\theta_{CM}\, d\theta_{CM}\, d\varphi_{CM}, \\ dv_{0x}\, dv_{0y}\, dv_{0z} &= v_0^2\, dv_0 \sin\theta_0\, d\theta_0\, d\varphi_0. \end{aligned} \qquad (2\text{-}2\text{-}23)$$

When (2-2-22) is substituted into the exponent in the integral in (2-2-19), the integrand is independent of the angles θ_{CM}, φ_{CM}, θ_0, and φ_0. Integration over all angles gives

$$\bar{v}_0 = (2\pi kT)^{-3}(m_1 m_2)^{3/2}(4\pi)^2 I_2\left(\frac{m_1 + m_2}{2kT}\right) I_3\left(\frac{M_r}{2kT}\right), \qquad (2\text{-}2\text{-}24)$$

where*

$$\begin{aligned} I_2\left(\frac{m_1 + m_2}{2kT}\right) &= \int_0^\infty \exp\left[-\frac{(m_1 + m_2)V_{CM}^2}{2kT}\right] V_{CM}^2\, dV_{CM} \\ &= \left(\frac{\pi}{2}\right)^{1/2}\left(\frac{kT}{m_1 + m_2}\right)^{3/2} \end{aligned}$$

and

$$I_3\left(\frac{M_r}{2kT}\right) = \int_0^\infty e^{-M_r v_0^2/2kT} v_0^3\, dv_0 = 2\left(\frac{kT}{M_r}\right)^2.$$

Equation (2-2-24) reduces to the simple form

$$\bar{v}_0 - \left(\frac{8kT}{\pi M_r}\right)^{1/2} = \left(\frac{8kT}{\pi m_1} + \frac{8kT}{\pi m_2}\right)^{1/2} = (\bar{v}_1^2 + \bar{v}_2^2)^{1/2}, \qquad (2\text{-}2\text{-}15)$$

which is the desired result.

C. A Gas in a Force Field

The energy distribution function given in (2-2-11) does not apply for a gas in an external field of force. If the force is describable by a scalar potential, the proper distribution function is obtained by inclusion of the *Boltzmann factor*, $e^{-V/kT}$, in (2-2-11). Here V is the potential energy per molecule and is a function of position. Thus the number of molecules per cm^3 with kinetic energy between E

*See Present[1] (pp. 266–267) for a discussion of the relationship of I_2 and I_3 to other integrals occurring frequently in kinetic theory.

and $E + dE$ is given by

$$N(E)\,dE = N_0 \frac{2}{(\pi)^{1/2}(kT)^{3/2}} e^{-(E+V)/kT} \sqrt{E}\,dE, \tag{2-2-25}$$

where N_0 is the total number density at a point at which $V = 0$. In this case the molecules are not uniformly distributed in configuration space. The variation of the number density with position is implicitly given by the equation

$$N = N_0 e^{-V/kT}. \tag{2-2-26}$$

The temperature is uniform throughout the gas, and the *Maxwell-Boltzmann distribution* of energies (2-2-25) corresponding to this temperature holds at each point.

2-3. THE MOLECULAR DIAMETER, MEAN FREE PATH, AND COLLISION FREQUENCY

A. Equations Derived from the Elastic Sphere Model

Consider a gas of number density N molecules/cm^3. If the molecules are assumed to be elastic spheres of diameter D whose velocities are distributed according to the Maxwellian law, it can be shown[6] that the number of collisions per cm^3-sec is $(\pi/\sqrt{2})N^2D^2\bar{v}$. The free path is usually defined as the distance that a molecule travels between successive collisions. Each collision marks the end of two free paths (one for each of the two collision partners), and the total number of free paths executed per cm^3-sec is $\sqrt{2}\,\pi N^2 D^2 \bar{v}$. The total length of all these free paths is $N\bar{v}$, and the average length of a free path is

$$\lambda = \frac{1}{\sqrt{2}\,\pi N D^2}, \tag{2-3-1}$$

which is designated as *Maxwell's mean free path*. This mean free path is the mean of all the paths traced out in unit time. An alternative definition leads to the *Tait mean free path*, which is the mean of all the paths being described at a given instant or the mean of all the distances traversed from a given instant to the next collision. The result of the Tait averaging procedure differs from (2-3-1) only in the substitution of the numerical factor 0.677 for $1/\sqrt{2}$, a difference of some 4%.

Note that (1-4-14) provides a value for the mean free path of a single molecule based on the assumption that all other molecules of the gas are stationary:

$$\lambda_{\text{stationary}} = \frac{1}{Q_s} = \frac{1}{Nq_s} = \frac{1}{\pi N D^2}. \tag{2-3-2}$$

From the cross section point of view, the moving molecule is assumed to be a point and each of the stationary molecules is assumed to have a radius equal to the true molecular diameter D.

In a mixture of two gases the mean free path for molecules of type 1 is

$$\lambda_1 = \left[\pi N_1 D_{11}^2 (2)^{1/2} + \pi N_2 D_{12}^2 \left(1 + \frac{m_1}{m_2}\right)^{1/2}\right]^{-1}, \tag{2-3-3}$$

where D_{ij} is the distance between centers of molecules of type i and j when they are in contact; that is,

$$D_{11} = D_1,$$
$$D_{12} = \tfrac{1}{2}(D_1 + D_2).$$

If the particles of type 1 are light and small compared with those of type 2, (2-3-3) reduces to

$$\lambda_1 = \frac{4}{\pi N_2 D_2^2}. \tag{2-3-4}$$

This equation gives the elastic sphere model prediction for the mean free path of electrons moving about in equilibrium with a gas. For reasons to be explained in Chapter 3, the elastic sphere model is unrealistic in the case of electrons and gives very poor agreement with experiment—a quantum mechanical treatment is required. The elastic sphere prediction is, however, quite good for atoms and molecules, and order-of-magnitude agreement with experiment is provided for ions in gases that are not highly ionized. With these gases consideration of the polarization attraction of the molecules by the ions (discussed in Section 1-8) leads to a reduction of the elastic sphere value for λ by about a factor of 5 and gives realistic results.

B. Variation of the Mean Free Path and Collision Frequency with Density and Temperature

The discussion in Section 2-3-A shows that the average collision frequency for a single molecule of a gas with a Maxwellian distribution is

$$\nu = \sqrt{2}\,\pi N D^2 \bar{v} \tag{2-3-5}$$

and that the mean free path is

$$\lambda = \frac{1}{\sqrt{2}\,\pi N D^2}. \tag{2-3-1}$$

Thus, according to the elastic sphere model, the collision frequency varies directly with the density of the gas and the square root of the absolute temperature, variation with temperature arising from the proportionality of ν to $\bar{v}$. The mean free path, on the other hand, varies only with the density and is inversely proportional to it.

Models more realistic than the elastic sphere model do, however, predict a variation of the mean free path with temperature. The Sutherland model (cf. Section 1-7) gives an expression of the form[7]

$$\lambda_S = \frac{\lambda}{1 + C/T}. \tag{2-3-6}$$

Here λ is the mean free path corresponding to the elastic sphere model (2-3-1) and C is a positive constant which has a different value for each gas. The reduction in the mean free path indicated by (2-3-6) is produced by the attractive field assumed in the Sutherland model.

If we let $\lambda(v)$ denote the average distance traveled by a single molecule of a particular speed v between successive collisions with molecules having a Maxwellian distribution of velocities, we find that $\lambda(v)$ is a function of v which increases monotonically from a value of 0 at $v = 0$ to a value $\sqrt{2}\,\lambda$ at $v = \infty$. This is made to appear reasonable by the following considerations: if the molecule is stationary, it travels zero distance before it is struck by another molecule. On the other hand, if the velocity of the single molecule approaches ∞, the other molecules may be considered stationary, and (2-3-2) applies. In this case the mean free path of the single molecule is greater than the value in (2-3-1) by the factor $\sqrt{2}$. The calculation of $\lambda(v)$ as a function of v is quite involved. It is presented in detail by Jeans.[8]

C. Numerical Results for Molecules in Pure Gases

The published values for molecular diameters and mean free paths are derived mainly from studies of viscosity, heat conductivity, diffusion rates, and equations of state. Loeb[9] has discussed in a very thorough manner all the methods of determining these quantities. None of the models used in kinetic theory provides an exact description of any physical phenomenon, and the values deduced for the molecular diameters and the mean free paths depend to some extent on the method of determination. Molecular diameters, mean free paths, and collision frequencies for a number of gases at 15°C are tabulated in Table 2-2-1. The values of λ were calculated by Kennard[10] from viscosity data, and (2-3-1) was then used to obtain D. The collision frequency ν was calculated from

$$\nu = \frac{\bar{v}}{\lambda}. \tag{2-3-7}$$

The discussion in Section 1-4 shows that the probability that a molecule will travel at least a distance x through the gas between collisions is $e^{-x/\lambda}$. Thus free paths appreciably greater than λ are extremely rare—only one molecule in 148 will travel as far as 5λ between collisions and only 1 in 22,027 as far as 10λ. It should be noted that the distribution of free paths does not peak near some mean value, as does the distribution of the molecular velocities, but rather the number of free paths longer than a certain value is a decreasing exponential function of the distance.

D. The Mean Free Path of a Charged Particle in a Highly Ionized Gas

The concept of the mean free path developed in Section 2-3-C does not apply to a charged particle moving through a gas that contains a high concentration of ions and electrons. The reason is that the motion of the particle is influenced, in this case, more by the cumulative effect of many small-angle Coulomb scatterings than by close impacts of the type considered so far.*

2-4. THE MEAN DISTANCE BETWEEN MOLECULES IN A GAS

The mean distance $\bar{d}$ between the molecules of a gas is given by

$$\bar{d} = \frac{1}{N^{1/3}}, \tag{2-4-1}$$

where N is the number density. The relative values of the mean free path, the mean spacing, and the molecular diameter for a gas at various pressures are of some interest, and representative values are tabulated below. The molecular diameter is taken to be 3×10^{-8} cm. The mean free path is assumed to be 10^{-5} cm at 760 torr pressure and to vary inversely with the pressure, the temperature being held constant. A temperature of 15°C is assumed throughout.

1. At $p = 760$ torr, $\bar{d} \approx 3.3 \times 10^{-7}$ cm and $\lambda = 10^{-5}$ cm;

$$\lambda/\bar{d}/D \approx 300/10/1.$$

2. At $p = 1$ torr, $\bar{d} \approx 3.0 \times 10^{-6}$ cm and $\lambda = 8 \times 10^{-3}$ cm;

$$\lambda/\bar{d}/D \approx 2 \times 10^5/100/1.$$

*Actually, we may speak of a sort of mean free path for a charged particle in a highly ionized gas, but reference is made in this case to the average depth of penetration required before multiple scattering through some large angle, usually taken to be 90° (see Appendix I).

3. At $p = 10^{-4}$ torr, $\bar{d} \approx 6.6 \times 10^{-5}$ cm and $\lambda \approx 80$ cm;

$$\lambda/\bar{d}/D \approx 3 \times 10^9/2000/1.$$

It will be noticed that at a pressure of 10^{-4} torr the mean free path is comparable with the dimensions of typical laboratory apparatus.

2-5. THE RELAXATION TIME OF A GAS*

Consider a gas in which the velocity distribution is initially non-Maxwellian. Maxwell[11] calculated the rate at which the gas must approach the steady-state condition. The result[12] is that deviations from the Maxwellian distribution are damped out exponentially, with a time constant equal to $1/\Gamma N$, where N is the number density of the gas and Γ is a constant that depends only on the structure of the molecules. This characteristic time is called the relaxation time, and it also measures the rate at which inequalities in the pressure and shear stresses must subside.

Γ is related to the coefficient of viscosity η and the hydrostatic pressure p by the equation

$$\Gamma = \frac{p}{\eta N}. \tag{2-5-1}$$

Numerical substitution into (2-5-1) shows that the relaxation time is comparable in magnitude with the mean free time for collisions in the gas, which is about 10^{-8} sec at standard conditions.

From the physical point of view the reason for the rapid approach to equilibrium is the efficient transfer of kinetic energy from one molecule to another in a collision. It was shown in Section 1-5 that a molecule transfers, on the average, half its excess energy to an identical collision partner in a single collision. Electrons, on the other hand, approach equilibrium much more slowly, since, on the average, an electron can transfer only a fraction $2m/M$ of its excess energy to a molecule in an elastic collision where m and M are the masses of the electron and molecule, respectively. (When an electric field is present the electrons in a gas approach an equilibrium distribution, but it is not the Maxwell-Boltzmann distribution.)

2-6. THE NUMBER OF MOLECULES CROSSING UNIT AREA IN A GAS PER SECOND

Let us now calculate the number of molecules of a gas in thermodynamic equilibrium that cross a surface element of unit area in the gas per unit time. Consider a small plane element of surface, dS, at an arbitrary location within the

*For a good general discussion of this subject, see J. D. Lambert, "Relaxation in Gases," in *Atomic and Molecular Processes*, D. R. Bates (Ed.), Academic, New York (1962); *Vibrational and Rotational Relaxation in Gases*, Oxford University Press, Oxford (1977).

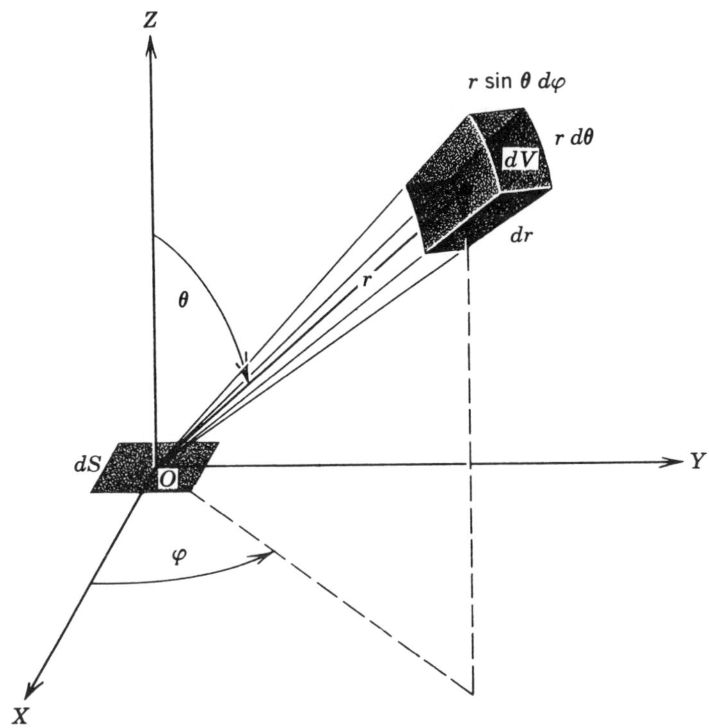

Figure 2-6-1. The coordinate system used for calculation of the rate of scattering of molecules from volume dV through surface dS.

gas and examine an element of volume, dV, located arbitrarily with respect to dS, as shown in Fig. 2-6-1. If the number density is N, there are $N\,dV$ molecules within dV. The volume of dV may be written as $r^2\,dr \sin\theta\,d\theta\,d\varphi$. From the form of the distribution function in (2-2-1) it is apparent that the number of molecules within dV with speeds between v and $v + dv$ is

$$n = N\left(\frac{2}{\pi}\right)^{1/2}\left(\frac{m}{kT}\right)^{3/2} v^2 e^{-mv^2/2kT}\,dv\,r^2\,dr \sin\theta\,d\theta\,d\varphi.$$

The pressure is assumed to be uniform, so N is independent of position. If the collision frequency is denoted by ν, the number of molecules leaving dV per second with speeds between v and $v + dv$ is $\nu n = \bar{v}n/\lambda$, since $\nu = \bar{v}/\lambda$. In this number, however, only those molecules scattered within the solid angle subtended at dV by dS will be headed toward dS. This fraction is given by $\cos\theta\,dS/4\pi r^2$. In this group of molecules only a fraction equal to $e^{-r/\lambda}$ will reach dS, the remainder undergoing further scattering. Thus the number leaving dV per second with speeds between v and $v + dv$ and passing directly through dS is given by

$$\frac{\bar{v}n}{\lambda}\frac{\cos\theta\,dS}{4\pi r^2}\,e^{-r/\lambda}.$$

The total number of molecules passing through dS per second from the half-space containing dV is obtained by integrating r from 0 to ∞, θ from 0 to $\pi/2$, φ from 0 to 2π, and v from 0 to ∞. The result of this multiple integration is $(N/4)\bar{v}\,dS$. An equal number of molecules will cross dS per second from the other side, with the result that *the total number of molecules passing through an imaginary surface of unit area in the gas per second is* $(N/2)\bar{v}$.

In probe theory one is required to know the number of particles that strike the unit surface area of an object located in the gas. We see that the kinetic theory provides a value of $(N/4)\bar{v}$ for this quantity.

From the foregoing discussion it is apparent that the rate of effusion of molecules through a small hole in the vessel containing the gas is $(N/4)\bar{v}$ molecules/cm^2-sec if the pressure outside the vessel is zero. Evidently, this result is valid only if the hole is very small compared with the mean free path, for only under this condition is the distribution of the velocities only slightly disturbed by the effusion of the gas. As the size of the hole is increased, we eventually pass into the regime of hydrodynamic streaming.

2-7. EQUATIONS OF STATE

Many equations of state[13] have been proposed to describe the interrelationship of the pressure, volume, and temperature of gases. Some of the most commonly used equations are listed below. The pressure, p, is measured in dyne/cm^2, V is the volume in cm^3 of one mole of the gas, and T is the absolute temperature in kelvin, and $R = 8.31 \times 10^7$ erg/mole-deg is the universal gas constant.

1. $pV = RT$ *Ideal gas law*

Gases deviate appreciably from ideal behavior only at pressures of about 1 atm, or higher, and for this reason the ideal gas law is nearly always the equation of state used in plasma applications.

2. $\left(p + \frac{a}{V^2}\right)(V - b) = RT$ *van der Waals equation*

The constant a arises from attractive forces which the molecules exert on one another when they are reasonably close together but somewhat more than a molecular diameter apart. The constant b, on the other hand, arises from shorter-range repulsive forces which come into play at a separation distance of about one molecular diameter.

3. $\left(p + \frac{a}{TV^2}\right)(V - b) = RT$ *Berthelot equation*
4. $p(V - b) = RT\,e^{-a/RTV}$ *Dieterici equation*

5. $p = \frac{RT}{V^2}\left(1 - \frac{c}{VT^3}\right)\left(V + B_0 - \frac{bB_0}{V}\right) - \frac{A_0}{V^2}\left(1 - \frac{a}{V}\right)$

Beattie-Bridgeman equation

By suitably choosing the five disposable constants in this equation, close agreement with experiment is obtained for many gases, including all the common gases, over a wide range of pressure and temperature.

6. $pV = RT + \frac{B}{V} + \frac{C}{V^2} + \cdots$ *Virial equation of state*

The quantities $B, C, \ldots$ in this equation are functions of temperature called the second, third, ... virial coefficients.* The coefficients $B, C, \ldots$ express the magnitude of the deviations from ideal gas behavior in terms of the forces between the molecules—hence the use of the term *virial.* (The Latin word for forces is *vires.*)

With the exception of the ideal gas law, all the equations listed here indicate the *critical point*, at which the vapor and liquid phases are identical. Let the parameters of the system at the critical point be denoted by p_c, V_c, T_c and the "reduced coordinates" be defined as $p_r = p/p_c$, $V_r = V/V_c$, $T_r = T/T_c$. If the equations of state are now expressed in terms of reduced coordinates, a functional form that is the same for all substances will be obtained. (This statement holds only for two-constant equations of state. It is not true for the Beattie-Bridgeman equation, for instance, which has five disposable constants.)

2-8. DIFFUSION, VISCOSITY, AND THERMAL CONDUCTIVITY

Ordinary diffusion is a process in which there is a net spatial transport of a substance due to a gradient in the relative concentration of that substance. The diffusive flow takes place in the direction opposite that of this gradient, and the flow rate is directly proportional to the magnitude of the gradient. The constant of proportionality is called the diffusion coefficient. Diffusive flow is separate and distinct from the flow resulting from a nonuniformity of the total pressure, and entirely different techniques are used to analyze them.

There are two kinds of ordinary diffusion. *Mutual diffusion* occurs in a mixture of gases of uniform total pressure but nonuniform composition, and it continues until the inequalities in composition have been eliminated by interdiffusion of the components. In the degenerate situation, in which the

*Oddly enough, pV, not RT, is designated the first virial coefficient [see E. A. Mason and H. W. Schamp, "The Virial Equation of State," in *Encyclopedia of Physical Chemistry and Chemical Physics*, Pergamon, Oxford (1964)].

components of the mixture are identical, we have the special case of *self-diffusion*. This phenomenon appears when selected molecules of a homogeneous gas diffuse through the remainder of the gas, and it may be closely approximated experimentally by the use of isotopic tracers. (Analysis of the gas is generally accomplished by mass spectrometric techniques if the isotopes are stable and by counting techniques if one or more of the isotopes are radioactive.)

A basically different type of diffusion, called *thermal*,[14] occurs when there is a temperature gradient rather than a composition* gradient in the gas. Thermal diffusion differs from ordinary diffusion in that it will always produce a state of nonuniform composition instead of leading to uniformity of composition. Thermal diffusion does, however, bring about mutual diffusion, as shown in the following example: Consider a closed tube containing an initially uniform mixture of two gases. If a temperature differential is established between the ends of the tube, thermal diffusion will occur, the molecules of the heavier gas diffusing toward the cooler end of the tube as the lighter gas diffuses toward the hotter end.* As a nonuniformity of composition begins to be established by thermal diffusion, mutual diffusion sets in and tends to restore the initial uniformity. A steady state is eventually attained in which there is a gradual change in the composition along the tube, transport by mutual diffusion just balancing that produced by thermal diffusion.

The effect of thermal diffusion was predicted independently by Enskog in 1911 and Chapman in 1916 in their developments of the rigorous kinetic theory. Thermal diffusion was first demonstrated experimentally in 1916 by Dootson[3] and has been put to practical use in the separation of isotopes.

Viscosity is a process involving the transport of momentum through a moving medium because of a gradient in the velocity of that medium. The shearing stress, which is related to the transport of the momentum of flow in the direction of the velocity gradient, is proportional to the magnitude of the velocity gradient. Viscosity is a manifestation of an internal friction.

Thermal conductivity is the transport of thermal energy through a medium due to a thermal gradient in the medium. The rate of heat flow is directly proportional to the magnitude of the temperature gradient.

It is evident that all these phenomena are closely related. Each involves the transport of some physical property of the gas. The rate of transport is directly proportional to the magnitude of the gradient that produces it, and the transport is oppositely directed to that gradient. Diffusion, viscosity, and thermal conductivity are known collectively as transport phenomena, and it is not surprising that the mathematical treatments of all three are quite similar. It

*The word "composition" rather than "concentration" is used in this discussion because "concentration" is frequently applied in the sense of mass per unit volume, which corresponds to the total number density in our nomenclature. A temperature gradient in a *pure* gas will produce a gradient in the concentration (i.e., the density will be less at the higher temperature) but not in the composition (the mole fraction is always unity).

*This is usually, but not always, true. In ionized gases, for instance, reversals occur. Reversals in ordinary gases are also known.

is worth noting that classical theory gives sufficiently accurate results in calculations of transport coefficients except for the lightest gases at low temperatures, for which quantum theory is required (see Section 3-10).

Diffusion is the most important of the transport phenomena for our purposes, and we shall now derive an expression for the diffusion coefficient, using the so-called mean free path approach.

2-9. DERIVATION OF THE DIFFUSION COEFFICIENT BY THE MEAN FREE PATH METHOD

Let us return to the discussion in Section 2-6 in which we calculated the number of molecules passing per unit time through unit area in a gas of uniform pressure. Now we shall assume that the pressure is nonuniform, and specifically, that a very small density gradient exists in the Z direction, the temperature being constant throughout the gas. The macroscopic flow of the gas will be negligible, but a diffusive type of flow will occur. Calculation of the net number of molecules passing through unit area in the X-Y plane per second will provide the diffusion coefficient characterizing the flow.

Let us expand the number density N in a Taylor series about the origin:

$$N(z) = N(0) + z\left(\frac{dN}{dz}\right)_o + \frac{z^2}{2}\left(\frac{d^2N}{dz^2}\right)_o + \cdots. \tag{2-9-1}$$

The subscripts denote that the derivatives are evaluated at the origin, which is, of course, an arbitrary point within the gas. The assumption that the gradient of N is small is meant to imply that the contribution to N of terms of higher than second order is negligible within a distance equal to a mean free path, λ. The discussion in Section 2-6 indicates the method of calculating the number of molecules passing downward through unit area in the X-Y plane per second. The original expression applies here if N, which is now a function of position, is taken inside the integral. A similar expression, in which the limits of θ are $\pi/2$ and π and $\cos\theta = -|\cos\theta|$, will give the number of molecules passing upward along the Z axis per unit area per second. Subtraction of the second expression from the first gives the desired result. J, the net downward transport of molecules per unit area per second, is

$$\begin{aligned} J &= \frac{\bar{v}}{4\pi\lambda}\int_0^\infty dr\, e^{-r/\lambda}N(z)\int_0^\pi \sin\theta\cos\theta\, d\theta \int_0^{2\pi} d\varphi \\ &= \frac{\bar{v}}{2\lambda}\int_0^\infty dr\, e^{-r/\lambda}N(r\cos\theta)\int_0^\pi \sin\theta\cos\theta\, d\theta \\ &= \frac{\bar{v}}{2\lambda}\int_0^\infty dr\, e^{-r/\lambda}N(r\xi)\int_{-1}^1 \xi\, d\xi, \end{aligned} \tag{2-9-2}$$

where $\xi = \cos\theta$. If the expansion for N given in (2-9-1) is substituted into (2-9-2), we obtain

$$J = \frac{\bar{v}}{2\lambda}\left[N(0)\int_0^\infty e^{-r/\lambda}\,dr\int_{-1}^{1}\xi\,d\xi + \left(\frac{dN}{dz}\right)_o\int_0^\infty re^{-r/\lambda}\,dr\int_{-1}^{1}\xi^2\,d\xi + \frac{1}{2}\left(\frac{d^2N}{dz^2}\right)_o\int_0^\infty r^2e^{-r/\lambda}\,dr\int_{-1}^{1}\xi^3\,d\xi\right]. \quad (2\text{-}9\text{-}3)$$

The first and third terms of (2-9-3) vanish when the integration over ξ is performed. The first term indeed must vanish for physical reasons, since there can be no net transport in a uniform gas. Evaluation of the second term shows that

$$J = \frac{\bar{v}\lambda}{3}\left(\frac{dN}{dz}\right)_o. \quad (2\text{-}9\text{-}4)$$

Thus, if the density increases as z increases ($dN/dz > 0$), J, as we have defined it, is positive, and there is a net flow of molecules downward along the $-Z$ axis.

The generalization to three dimensions is obvious. In this case the vector particle current density $\mathbf{J}$ is given in terms of the number density gradient by the equation

$$\mathbf{J} = -\frac{\bar{v}\lambda}{3}\nabla N. \quad (2\text{-}9\text{-}5)$$

The gradient of the number density is a vector that points in the direction of most rapidly increasing number density and whose magnitude equals the spatial rate of change of N in that direction. $\mathbf{J}$ is a vector pointing in the flow direction which expresses the net number of molecules flowing through unit area normal to the flow direction per second.

Equation (2-9-5) is known as *Fick's law of diffusion.* The constant of proportionality between $\mathbf{J}$ and ∇N is the diffusion coefficient $\mathscr{D}$. Thus, according to the mean free path approach used here,

$$\mathscr{D} = \frac{\bar{v}\lambda}{3}. \quad (2\text{-}9\text{-}6)$$

Equation (2-9-5) evidently applies to the case of self-diffusion. It also applies to the mutual diffusion of two dissimilar gases if the relative concentration of one of the gases is vanishingly small. Another interesting application is to the diffusion of slow neutrons through bulk matter. Here the neutrons are assumed to have a Maxwellian distribution characterized by the temperature of the medium through which they are diffusing and the nuclei of the medium are taken to be at rest.

It will be noted that the mean free path appears naturally in this approach to the diffusion problem. Similar treatments of viscosity and thermal conductivity also give results in which λ appears. For this reason these phenomena are sometimes called mean free path phenomena. However, the term is not entirely appropriate because, as we shall see, rigorous kinetic theory provides results in which the mean free path does not appear naturally.

The mean free path calculation of the molecular self-diffusion coefficient $\mathscr{D} = \mathscr{D}_{11}$ gives fair agreement with experiment. (The correct numerical coefficient is $\frac{1}{2}$ instead of $\frac{1}{3}$.) A similar calculation of the coefficient of mutual diffusion, $\mathscr{D}_{12}$, fails badly.[15] Rigorous kinetic theory, discussed in the next section, should be used for gas mixtures.

2-10. RIGOROUS EXPRESSIONS FOR THE DIFFUSION COEFFICIENT

The first rigorous calculation of the mutual diffusion coefficient was made by Maxwell[16] in 1867 for molecules interacting through a repulsive r^{-4} potential. His solution was exact, an accomplishment made possible by the special properties of the r^{-4} potential discussed in Section 1-8. Maxwell's result is

$$\mathscr{D}_{12} = \frac{kT}{NM_r v_0 q_D(v_0)} = \frac{0.376kT}{N(M_r c)^{1/2}}, \tag{2-10-1}$$

where N is the total number density $(N_1 + N_2)$, M_r the reduced mass of species 1 and 2, v_0 the relative velocity, $q_D(v_0)$ the diffusion cross section, and c the constant relating the force between the molecules to their separation distance:

$$f = cr^{-5}. \tag{2-10-2}$$

In 1872, Stefan[17] published a calculation of the mutual diffusion coefficient for hard elastic spheres, but it was not until 1905 that a general and rigorous calculation of $\mathscr{D}_{12}$ was made by Langevin.[18] Langevin's result, expressed in terms of the standard Ω collision integral of the refined theory subsequently developed by Chapman and Enskog,[19] is

$$\mathscr{D}_{12} = \frac{3kT}{16M_r N\Omega_D}, \tag{2-10-3}$$

where

$$\Omega_D = \left(\frac{kT}{2\pi M_r}\right)^{1/2} \int_0^\infty e^{-g^2} q_D(v_0) g^5 \, dg \tag{2-10-4}$$

and

$$g = \left(\frac{M_r v_0^2}{2kT}\right)^{1/2}. \tag{2-10-5}$$

In the approximation used in the derivation of (2-10-3), $\mathscr{D}_{12}$ is independent of the proportions of the gas mixture, depending only on N, not on N_1 and N_2 individually. Equation (2-10-3) is identical with the first approximation provided by the Chapman-Enskog theory, which, however, predicts a slight dependence on the composition of the gas mixture in higher approximation. The higher-order corrections to (2-10-3) are small and are usually ignored.[19]

Equation (2-10-3) may easily be evaluated for the elastic sphere model. In this case q_D becomes πD_{12}^2, where $D_{12} = \frac{1}{2}(D_1 + D_2)$, D_1 and D_2 being the molecular diameters of the two species; q_D is thus independent of the velocity, and the mutual diffusion coefficient is

$$\mathscr{D}_{12} = \frac{3}{8}\left(\frac{\pi kT}{2M_r}\right)^{1/2}(N\pi D_{12}^2)^{-1}. \tag{2-10-6}$$

In any of the foregoing expressions for $\mathscr{D}_{12}$ the self-diffusion coefficient $\mathscr{D}_{11}$ may be obtained by replacing the reduced mass M_r by half the molecular mass of the single species present.

It is important to note that the classical description of the diffusion process fails when the temperature is very low (cf. Section 3-10) or when more than one interaction potential applies, as in the diffusion of excited atoms through a gas of similar unexcited atoms and in the self-diffusion of atoms which do not have a closed shell configuration. The effect is to produce very large diffusion cross sections and small diffusion coefficients.[20] In such cases quantum theory must be used to calculate the diffusion cross section, which can then be used in the classical kinetic theory equations to obtain $\mathscr{D}_{12}$.*

We shall postpone further discussion of diffusion until we reach the chapter on ion and electron swarms in our companion volume. There the emphasis is on the drift velocities and geometrical distributions of electrons and ions diffusing through gases. For additional information concerning the theoretical and experimental evaluation of diffusion coefficients, the reader is referred to Chapman and Cowling,[3] Hirschfelder et al.,[4] Dalgarno,[20] and Waldmann.[21]

We now turn to a few of the fundamental topics that play an important role in rigorous treatments of neutral and ionized gases. Chapman and Cowling have presented a survey of the development of the kinetic theory, and a very detailed discussion of the theory is to be found in their book[3] and in that of Hirschfelder et al.[4] Reference should also be made to the authoritative review of the advanced theory by Grad.[22]

*For explanations of the effects that arise when more than one potential applies, see Note 2-3 in Section 2-16, and Sections 9-7-D-2 and 9-7-D-3 of E. W. McDaniel, *Atomic Collisions: Heavy Particle Projectiles*, Wiley, New York, to be published.

2-11. PHASE SPACE AND THE LIOUVILLE THEOREM[23]

Consider a dynamical system having s degrees of freedom. The motion of this system can be described in terms of s generalized coordinates $q_1, q_2, \ldots, q_s$ and the corresponding s generalized momenta $p_1, p_2, \ldots, p_s$. A set of values of these $2s$ variables may be regarded as the coordinates of a point in a space of $2s$ dimensions, called a *phase space*, in which the generalized coordinates and momenta serve as Cartesian coordinates. A given set of initial conditions is specified by a given single point, and as the system executes some particular motion the point representing the system in phase space traces out a trajectory in that space.

Now consider a huge ensemble of noninteracting replicas of this dynamical system to be started out simultaneously with different initial conditions, that is, at different points in phase space.* The state of this entire collection of systems at any time may be described by a set of points in phase space, and the histories of these systems may be represented by the motions of the various "phase points" through phase space. If the number of systems is large enough, the set of points may be represented by a continuous distribution function, $\rho(q_1, \ldots, q_s, p_1, \ldots, p_s)$, called the density in phase space. *Liouville's theorem* states that the density of such points about any one point remains constant with time. A proof of this theorem follows.

At a given instant of time the number of points in phase space representing systems which have coordinates and momenta lying in the range $\Delta q_1 \cdots \Delta q_s \, \Delta p_1 \cdots \Delta p_s$ at some fixed point $q_1, \ldots, q_s, p_1, \ldots, p_s$ is

$$\rho \, \Delta q_1 \cdots \Delta q_s \, \Delta p_1 \cdots \Delta p_s = \rho \, \Delta V,$$

where ΔV is an element of volume in the $2s$-dimensional phase space. As time progresses, points will pass through each of the two sides of ΔV which are perpendicular to the q_1 axis at a rate equal to $(\rho \dot{q}_1)\Delta q_2 \cdots \Delta p_s$, where $(\rho \dot{q}_1)$ is evaluated at that side of ΔV being considered. A similar flow occurs through the opposite side, and the net outward flow from ΔV through the sides perpendicular to q_1 is given by

$$\frac{\partial(\rho \dot{q}_1)}{\partial q_1} \Delta q_1 \, \Delta q_2 \cdots \Delta p_s$$

if higher-order terms are neglected. Now evaluate the outward flow through all remaining pairs of sides of the volume element ΔV corresponding to the remaining coordinates $q_2 \cdots p_s$. If we add all of these net outward flow rates, equate the sum to the net loss of points within ΔV, and pass to the limit of zero Δ,

*Alternatively, we could imagine the same system to be started repeatedly with different initial conditions and let time be the time elapsed after starting the system in each case.

we obtain the differential equation for ρ:

$$-\frac{\partial \rho}{\partial t} = \sum_{i=1}^{s} \left[\frac{\partial(\rho \dot{q}_i)}{\partial q_i} + \frac{\partial(\rho \dot{p}_i)}{\partial p_i} \right].$$

Here the symbol of partial differentiation with respect to time indicates that we are fixing our attention on a given stationary point in the phase space. Expanding the derivatives, we have

$$\frac{\partial \rho}{\partial t} + \sum_{i=1}^{s} \left[\rho \left(\frac{\partial \dot{q}_i}{\partial q_i} + \frac{\partial \dot{p}_i}{\partial p_i} \right) + \dot{q}_i \frac{\partial \rho}{\partial q_i} + \dot{p}_i \frac{\partial \rho}{\partial p_i} \right] = 0.$$

Hamilton's equations,

$$\dot{q}_i = \frac{\partial H}{\partial p_i}, \qquad \dot{p}_i = -\frac{\partial H}{\partial q_i} \qquad (i = 1, 2, \ldots, s), \tag{2-11-1}$$

where H is the Hamiltonian, show that the first term in the square brackets vanishes. Thus we obtain the result

$$\frac{\partial \rho}{\partial t} + \sum_{i=1}^{s} \left(\dot{q}_i \frac{\partial \rho}{\partial q_i} + \dot{p}_i \frac{\partial \rho}{\partial p_i} \right) = 0. \tag{2-11-2}$$

This equation expresses the fact that the total time rate of change of ρ along a trajectory in phase space is zero. The first term represents the time rate of change of ρ at a fixed point in phase space and is referred to as the local rate of change. The second term designates the time rate of change produced by flow, or streaming, in configuration space. The last term represents the change produced by streaming in momentum space.

Equation (2-11-2) is Liouville's theorem. It has the form of a continuity equation for the flow of points through phase space,

$$\frac{\partial}{\partial t}(\text{density}) + \nabla \cdot (\text{density flow}) = \text{source} \tag{2-11-3}$$

for the special case in which the source term is zero.* The theorem may be concisely written if use is made of the mobile derivative, that is, the derivative following the motion of a point. The mobile derivative in this space is defined in

*Here there is no source term in the continuity equation because the systems that compose the ensemble are incapable of interaction. The phase points, each of which represents one system of the ensemble, do not collide with one another: There is no mechanism for creation or destruction of the points, and none is abruptly transferred from one region of phase space to another by an impact with another phase point.

terms of the velocity $\mathbf{v}$, the applied force $\mathbf{f}$, and the gradient operators in s-dimensional configuration and momentum spaces, ∇_r and ∇_p, as follows:

$$\frac{D}{Dt} = \frac{\partial}{\partial t} + \mathbf{v} \cdot \nabla_r + \mathbf{f} \cdot \nabla_p. \tag{2-11-4}$$

We see that Liouville's theorem may now be expressed simply as

$$\frac{D\rho}{Dt} = 0. \tag{2-11-5}$$

This theorem is significant in kinetic theory and statistical mechanics. It shows that the volume in phase space occupied by a collection of points is constant in time—there is no tendency of the points to crowd into any particular region of phase space. The points may thus be regarded as composing an incompressible fluid. Another statement of the theorem is the following: If at some instant a point is known to lie within an infinitesimal volume of phase space, it will lie in a corresponding volume of the same size after a time t. The shape of the corresponding volume may be quite different, however.

The $2s$-dimensional phase space described in this section is not the only one that proves useful, although it does offer special advantages.[23] In the following section we utilize a phase space of six dimensions—three corresponding to spatial coordinates and three to the associated velocity components. A system of N particles is represented in this space by a collection of N points. Pierce[24] considers the motion of an electron beam through an arbitrary combination of applied electric and magnetic fields and shows that Liouville's theorem applies for the six-dimensional electron number density, provided that the fields do not depend on the positions of the electrons; that is, in six-dimensional phase space the electron density in the vicinity of a given electron does not change as the electron moves along the beam.

2-12. THE BOLTZMANN EQUATION

The rigorous kinetic theory of gases is based on the knowledge of the distribution function $f_i(\mathbf{r}, \mathbf{v}, t)$ for each of the various species of particle present: f_i is the number density for the ith species in six-dimensional phase space. Specifically, $f_i(\mathbf{r}, \mathbf{v}, t)\, d^3r\, d^3v$ is the number of particles of type i which, at time t, lie within the volume element $d^3\mathbf{r}$ centered at $\mathbf{r}$ in configuration space and which have velocities within the volume element d^3v centered at $\mathbf{v}$ in velocity space. The distribution function satisfies the Boltzmann integrodifferential equation. In certain cases f_i reduces to the Maxwell-Boltzmann distribution function.

The *Boltzmann equation*, which is the starting point for most rigorous calculations in kinetic theory, describes the effect of applied forces and collisions on the distribution function. It has the form of a continuity equation (2-11-3), as does the Liouville equation (2-11-2). However, the source term is no longer zero, for we now consider the interaction of particles through collisions with members of their own species and those of other species as well. Let $\mathbf{a}$ represent the acceleration produced by continuous or average forces. (If, for example, the species of particle under consideration is ionic, this acceleration may be produced by an externally applied electric field.) The effects of collisions and microscopic fluctuations are considered separately in the source, or collision, term. Thus the Boltzmann equation has the form

$$\frac{\partial f_i}{\partial t} + \mathbf{v} \cdot \nabla_r f_i + \mathbf{a} \cdot \nabla_v f_i = \left(\frac{\delta f_i}{\delta t} \right)_{\text{collisions}} \tag{2-12-1}$$

where ∇_r is now the gradient operator in three-dimensional configuration space and ∇_v is the gradient operator in the corresponding three-dimensional velocity space.

The collision term is really the sum of n integrals, where n is the number of different species present in the gas. Each of these integrals, designated by B with appropriate subscripts, gives the rate of change of the distribution function for species i caused by collisions with particles of species j as we move along the trajectory of particles i. The collision integral B_{ij} is evidently obtained by calculating the difference between the rates of scattering of particles i into and out of the volume element d^3v_i, due to collisions with particles j. The discussion of reaction rates in Section 1-4 shows that the collision integral has the following form:

$$B_{ij}\, d^3v_i = \iint f_j(\mathbf{v}_j')f_i(\mathbf{v}_i')v_0' I_s(\Theta, v_0')\, d\Omega_{\text{CM}}\, d^3v_j'\, d^3v_i'$$
$$- \iint f_j(\mathbf{v}_j)f_i(\mathbf{v}_i)v_0 I_s(\Theta, v_0)\, d\Omega_{\text{CM}}\, d^3v_j\, d^3v_i. \tag{2-12-2}$$

$f_j(\mathbf{v}_j')\, d^3v_j'$ and $f_i(\mathbf{v}_i')\, d^3v_i'$ are the numbers of particles in position to be scattered into velocity elements d^3v_j and d^3v_i, whereas $f_j(\mathbf{v}_j)\, d^3v_j$ and $f_i(\mathbf{v}_i)\, d^3v_i$ give the numbers in position to be scattered out of d^3v_j and d^3v_i:v_0 is the relative velocity and $I_s(\Theta, v_0)\, d\Omega_{\text{CM}}$ is the CM differential scattering cross section; $d\Omega_{\text{CM}} = \sin\Theta\, d\Theta\, d\varphi$ is the CM element of solid angle. The integration is to be performed over all scattering angles and all velocities of the scatterers j. The collision integral is very difficult to evaluate, in general, except for the case of a constant mean free time, $v_0 I_s(\Theta, v_0) = \text{constant}$.

Recognizing the fact that the distribution function defines a particle current in six-dimensional phase space, we may write the Boltzmann equation in the following physically meaningful form:

$$\frac{\partial f_i}{\partial t} + \nabla_r \cdot \boldsymbol{\gamma}_i + \nabla_v \cdot \mathbf{g}_i = \sum_{j=1}^{n} B_{ij}. \tag{2-12-3}$$

Here $\boldsymbol{\gamma}_i$ is the component of the particle current in configuration space

$$\boldsymbol{\gamma}_i = \mathbf{v} f_i \tag{2-12-4}$$

and $\mathbf{g}_i$ is the component in velocity space

$$\mathbf{g}_i = \mathbf{a} f_i. \tag{2-12-5}$$

The distribution function is seldom, if ever, measured, but various velocity moments can be determined experimentally. These moments are:

particle number density (scalar)
$$N = \int f(\mathbf{r}, \mathbf{v}, t)\, d^3v \tag{2-12-6}$$

average energy (scalar)
$$u = \frac{m}{2N} \int v^2 f\, d^3v \tag{2-12-7}$$

drift velocity (vector)
$$\mathbf{v}_d = \frac{1}{N} \int \mathbf{v} f\, d^3v \tag{2-12-8}$$

kinetic pressure (tensor)
$$\mathscr{p} = m \int (\mathbf{v} - \mathbf{v}_d)(\mathbf{v} - \mathbf{v}_d) f\, d^3v \tag{2-12-9}$$

heat flow (tensor)
$$\mathscr{H} = m \int (\mathbf{v} - \mathbf{v}_d)(\mathbf{v} - \mathbf{v}_d)(\mathbf{v} - \mathbf{v}_d) f\, d^3v \tag{2-12-10}$$

2-13. METHODS OF SOLUTION OF THE BOLTZMANN EQUATION*

The Boltzmann equation is quite intractable mathematically, and recourse must be made to expansion methods for its solution, except in the singular case of a constant mean free path. One of the most useful techniques is the *Chapman-Enskog method.* It applies to cases in which the system under consideration is close to thermal equilibrium, and it involves an expansion of the distribution function in powers of the perturbation parameter α, which produced the departure from equilibrium:

$$f = f^0 + \alpha f^1 + \alpha^2 f^2 + \cdots. \tag{2-13-1}$$

This procedure linearizes the Boltzmann equation. The perturbed distribution function is then expanded in Sonine polynomials, and the integral equation is reduced to an infinite set of algebraic equations. This method is highly

*This section is based largely on the masterly article by W. P. Allis, "Motions of Ions and Electrons," in *Handbuch der Physik*, Vol. XXI, Springer-Verlag, Berlin (1956), pp. 383–444.

developed and well described in the literature.[25] Unfortunately, the Chapman-Enskog method is not very useful for charged particles in an electric field, since the series does not converge well in this case.

Another perturbation method is also useful, wherein the total distribution function is taken to be the sum of the function f_0, characterizing the steady-state, unperturbed system, and a small contribution f_1 caused by a perturbation:

$$f(\mathbf{r}, \mathbf{v}, t) = f_0(\mathbf{r}, \mathbf{v}, t) + f_1(\mathbf{r}, \mathbf{v}, t). \tag{2-13-2}$$

Here $f_0(\mathbf{r}, \mathbf{v}, t)$ is not necessarily Maxwellian, but it is assumed to be known or deducible from physical considerations. If (2-13-2) is substituted into the Boltzmann equation, an equation for f_1 results. If this equation is linearized by neglecting second-order terms, the following equation is obtained:

$$\frac{Df_1}{Dt} + \mathbf{a}_1 \cdot \boldsymbol{\nabla}_v f_0 = -\nu f_1. \tag{2-13-3}$$

Df_1/Dt is the total time derivative of f_1 in six-dimensional phase space and $\mathbf{a}_1$ is the acceleration due to the perturbation. The right-hand side is really an integral whose integrand is linear in the perturbed distribution function. This collision integral is approximated by the term on the right-hand side of (2-13-3). The reciprocal of the effective collision frequency ν plays the role of a relaxation time—if the perturbation is removed, f_1 decays as $e^{-\nu t}$. There are numerous ways to solve (2-13-3). One, which is particularly satisfactory for light particles moving in a medium composed of heavy particles, consists of expanding the distribution function in spherical harmonics in velocity space. This technique is frequently used for electrons in gases and for neutrons in bulk matter of high atomic weight.

2-14. LIMITATIONS OF THE BOLTZMANN THEORY

Implicit in the use of the Boltzmann equation is the assumption that the particles of the gas move freely except "during" binary collisions whose durations are short compared with the mean free time. The Boltzmann equation is really well suited only for the treatment of problems involving relatively low density gases composed of uncharged molecules which are essentially spherically symmetric. Progress in the development of a kinetic theory for dense gases and for polyatomic molecules is described by Hirschfelder et al.[4]

Several alternative approaches to the Boltzmann theory are applied in the study of highly ionized gases. One approach makes use of the *Fokker-Planck equation*, which phrases the problem in terms of many simultaneous long-range interactions between charged particles in the gas rather than in terms of successive short-range interactions. The Fokker-Planck method has been discussed by a number of authors, some of whom are listed in Ref. 26. Another approach is afforded by the *Vlasov*, or *correlation-less Boltzmann, equation.*[27] In this case the collision integrals are completely neglected in the Boltzmann equation, and the result is used with Maxwell's equations for the electromagnetic field to form a set of equations that approximately describes a highly ionized gas. Still another attack on the problem employs *magnetohydrodynamics*, or hydromagnetics, which utilizes the equations of hydrodynamics together with Maxwell's equations and treats the ionized gas as a fluid.

2-15. QUANTUM STATISTICS[28]

The Maxwell-Boltzmann statistics used thus far in this chapter represent the classical approximation to the quantum statistics that rigorously apply, and it is necessary to mention the latter here, although in this book we use quantum statistics mainly in the theoretical description of individual scattering events (Section 3-18). Let us proceed in the context of ideal gases, whose particles have negligible mutual interaction.

In the case of *Maxwell-Boltzmann* (M-B) statistics, the particles are considered to be distinguishable, and any number of particles can be in any single-particle state. There are no symmetry requirements on the wave function of the system when any two particles are interchanged.

In fact, particles obey either *Bose-Einstein* (B-E) or *Fermi-Dirac* (F-D) statistics, depending on whether they have integral or half-integral spin. Thus "bosons" have spins of $0, \hbar, 2\hbar, 3\hbar, \ldots$, whereas "fermions" have spins of $\frac{1}{2}\hbar, \frac{3}{2}\hbar, \frac{5}{2}\hbar, \ldots$. The total wave function of a Bose-Einstein system must be symmetric with respect to the interchange of any two particles, whereas a Fermi-Dirac system must have an antisymmetric wave function. In the B-E case, there is no restriction on the number of particles that can occupy a given single-particle state, but in F-D statistics, there is no state for the whole system in which two or more particles can occupy the same single-particle state. The distribution functions for the various kinds of statistics are:

$$\bar{n}_s = \frac{1}{e^{(E_s - \mu)/kT} - 1} \qquad \text{(B-E)} \qquad (2\text{-}15\text{-}1)$$

$$\bar{n}_s = \frac{1}{e^{(E_s - \mu)/kT} + 1} \qquad \text{(F-D)} \qquad (2\text{-}15\text{-}2)$$

$$\bar{n}_s = N \frac{e^{-E_s/kT}}{\sum e^{-E_r/kT}} \qquad \text{(M-B).} \qquad (2\text{-}15\text{-}3)$$

Here N is the total number of particles, s labels the possible quantum states of a single particle, E_s is the energy of a particle in state s, n_s represents the number of particles in state s, and μ is the chemical potential per particle. In the *classical limit* of sufficiently low density or high temperature, both the B-E and F-D distributions reduce to the MB distribution (Problem 2-22).

2-16. NOTES

2-1. Jacobian of Velocity Transformation. In Section 2-2-B we transformed from the velocities v_{1x} and v_{2x} to the velocities v_{0x} and V_{CMx}, and so on. The Jacobian, J, equals unity for the "x transformation," and for the y and z transformations as well. Hence the volume element in the new Cartesian velocity space equals the volume element in the original Cartesian velocity space. A good discussion of Jacobian determinants is provided by W. Kaplan, *Advanced Calculus*, Addison-Wesley, Cambridge, Massachusetts (1952).

2-2. Phase Space. The phase space with $2s$ dimensions introduced at the beginning of Section 2-11 is called a Γ space (Γ for "gas"). The other phase space, with six dimensions, mentioned at the end of Section 2-11 is called a μ space (μ for "molecule").

2-3. Diffusion of Excited Atoms. In Section 2-10 it was stated that excited atoms diffusing through a gas of similar unexcited atoms can have very large cross sections. The reason is that an excited atom and one in the ground state can interact in either of two ways, depending on the symmetry of the electronic wave function with respect to the interchange of the nuclei. Denote the interaction energy for the symmetric state by $V^+(r)$ and for the antisymmetric state by $V^-(r)$, where r is the nuclear separation. The difference in these energies appears in the difference in the phase shifts, $\eta_l^+ - \eta_l^-$ (Section 3-15), which in turn appears in the cross section for excitation transfer, a new scattering process that is not possible for unexcited atoms. The interaction $V^+ - V^-$ can have a very long range and thereby can lead to a large cross section, greater than $10^{-13}\,\text{cm}^2$ in some cases. The excitation energy switches from atom to atom, in a fashion reminiscent of electron transfer in resonant charge transfer. The theories used to describe the two processes are similar.

In the case of the self-diffusion of atoms that do not have a closed-shell configuration, a similar situation arises. We must consider the spatial quantization of the atoms and the different ways in which their angular momenta can couple. The result is usually an interaction energy of long range, with a correspondingly large cross section.

2-4. Kinetic Theory of Polyatomic Gases. A landmark in this subject is C. S. Wang Chang, G. E. Uhlenbeck, and J. deBoer, "The Heat Conductivity and Viscosity of Polyatomic Gases," in J. deBoer and G. E. Uhlenbeck (Eds.), *Studies in Statistical Mechanics*, Vol. 2, Wiley, New York (1964), pp. 241–268. This article is an updated and expanded version of C. S. Wang Chang and G. E. Uhlenbeck, Report CM-681, University of Michigan, Engineering Research Institute (1951).

2-5. Tensor Notation. In Cartesian tensor notation, (2-12-9) and (2-12-10) take the form

$$\not{p}_{ij} = m \int (v_i - v_{di})(v_j - v_{dj}) f \, d^3v, \tag{2-16-1}$$

$$\mathscr{H}_{ijk} = m \int (v_i - v_{di})(v_j - v_{dj})(v_k - v_{dk}) f \, d^3v. \tag{2-16-2}$$

See Appendix A of Ferziger and Kaper.[2]

2-6. *Similarities between Neutron Physics and Atomic Collisions.* The production of fast fission neutrons, their slowing down in matter, and neutron multiplication in fission reactions have analogs in phenomena that occur in electron-beam-pumped gas lasers and in electron and ion drift tubes. For this reason, several problems on neutron physics are included at the end of this chapter.

2-17. PROBLEMS

Solutions for most of these problems appear in the references indicated.

2-1. Assuming a Maxwell-Boltzmann distribution, compute the temperature at which $(\overline{v^2})^{1/2}$ is equal to the escape velocity from the earth for (a) nitrogen and (b) helium. *Ans.:* 1.43×10^5 K for N_2; 2.04×10^4 K for He.

2-2. Show that

$$(\overline{v_x^2})^{1/2} = (\overline{v_y^2})^{1/2} = (\overline{v_z^2})^{1/2} = \left(\frac{p}{\rho}\right)^{1/2},$$

where p is the pressure and ρ is the density of a gas. This is the equation for the velocity of sound as derived by Newton. Why does it not apply to real gases?

2-3. A thin disk of area S moves in a gas of very low pressure with a velocity v that is small compared to the average molecular velocity. The Maxwell distribution is assumed to apply. Prove that for this motion a force of magnitude

$$F = 4\sqrt{\frac{2mkT}{\pi}}\, nSv$$

is required, where m is the mass of a molecule and n is the number density.

2-4. Prove that for an ideal gas at temperature T, the average impulse that one molecule gives to the walls of a containing vessel is $\bar{p} = \sqrt{2\pi mkT}$ and that the average square impulse is $\overline{p^2} = 8mkT$.

2-5. A large cylinder contains a gas with a M-B distribution at temperature T. In the wall of this cylinder is an aperture, small compared with the mean free path. Find the mean energy of the molecules that pass through the aperture. *Ans.:* $2kT$.

2-6. Two vessels of equal volume V contain the same gas at different pressures, p_1 and p_2 ($p_1 > p_2$). The temperatures are the same. There is a perforation of area S in the (thin) wall between the vessels. The pressures are so low that the mean free path of the molecules is large compared to the dimensions of the hole. Prove that the pressure difference $\Delta p = p_1 - p_2$ drops exponentially with the time:

$$(\Delta p)_t = (\Delta p)_{t=0} e^{-\alpha t},$$

where $\alpha = \bar{c}S/2V$ and $\bar{c}$ is the average molecular velocity.

2-7. Two vessels of equal volume V are joined by a *small* hole and contain some of the same gas. The temperatures are maintained externally at T_1 and T_2. Show that for equilibrium

$$\frac{p_1}{p_2} = \frac{\sqrt{T_1}}{\sqrt{T_2}}.$$

2-8. Assume that the earth's atmosphere is isothermal and that the gravitation field is homogeneous. Without using (2-2-26), show that the gas pressure at altitude z is $p(z) = p(0)e^{-mgz/kT}$, where we further assume that the atmosphere has only a single component with molecular mass m. Show that if we take $m = 29$ amu, then $mg/kT = 1.26 \times 10^{-4}$ per meter.

2-9. With the assumptions made in Problem 2-8, show that at an altitude of 170 km, the pressure is 5.6×10^{-10} atm, and the number density is 1.52×10^{16} molecules/m^3.

2-10. In a single-component, isothermal atmosphere, the molecular number density varies with altitude z according to the equation $n = n_0 e^{-z/H}$, where H is the scale height. Assume a collision cross section σ and calculate the altitude h at which a particle shot upward with very high speed has a probability 0.5 of escaping without a collision. *Ans.:* $h = H \ln(n_0 \sigma H/\ln 2)$, where n_0 is the number density at the surface.

2-11. **(a)** Consider N_2 filling a container 2.5 m high and having a temperature of 290 K. Show that the pressure at the top is 0.03% lower than that at the bottom.

(b) Repeat for H_2. *Ans.:* 0.002%.

2-12. Consider a N_2 molecule traveling horizontally at some instant with an energy corresponding to a temperature of 290 K. Show that it will drop a distance of 1.9×10^{-19} m while traveling 10^{-7} m.

2-13. **(a)** Show that the energy difference between the ground state and the first excited state for the H atom is 10.19 eV.

(b) Consider an ensemble of 10^{20} H atoms in equilibrium at 3300 K. Show that 2.32×10^4 atoms are in the first excited state.

(c) Show that they radiate energy at the rate of 3.78×10^{-6} W.

2-14. A molecule is in a long-lived metastable state (mean life 10 days). Show that the probability that the molecule will decay during the fifth day is 0.064.

2-15. A large number of atoms are excited to a long-lived excited state from which they decay at an average rate of 10^4 per second. What is the probability of observing 9700 or fewer disintegrations in a 1-sec interval? *Ans.:* 0.0015. (Chen,[29] p. 286.)

2-16. The mean-square displacement of a particle undergoing Brownian motion in a gas is inversely proportional to the viscosity of the gas. How does the mean-square displacement change if the pressure is doubled?

2-17. Consider a spherical bulb of 10 cm radius that is maintained at 300 K except for a 1-cm^2 area held at 77 K (liquid nitrogen temperature). At $t = 0$, the bulb contains water vapor at 0.1 torr. Show that about 4 sec is required for the cryogenic pumping action to reduce the pressure to 10^{-6} torr if it is assumed that every water molecule hitting the cold spot sticks to it. (Reif,[26] p. 260.)

2-18. N particles in an ensemble characterized by a Boltzmann distribution are distributed among three states of energy: $E = 0$, $E = kT$, and $E = 2kT$. Show that if the total energy is $1000kT$, $N \approx 2400$. (Chen,[29] p. 185.)

2-19. Consider the slowing down of neutrons in hydrogen, assuming the n-p scattering to be purely elastic and isotropic in the CM frame.

(a) If the initial energy is E_0, show that the average neutron energy is $(\frac{1}{2})^n E_0$ after n collisions.

(b) Show that after n collisions, the energy distribution is

$$p_n(E)\,dE = \frac{[\log(E_0/E)]^{n-1}}{(n-1)!}\,dE.$$

(c) If the high-energy neutrons (energy E_0) are produced at the rate q per cm^3-sec, find the steady-state neutron flux as a function of energy (and the total and absorption cross sections, assumed to be known functions of energy). (Cronin et al.,[30] p. 57; Segrè,[31] p. 621.)

2-20. A neutron of initial energy E_0 cascades down to energy E by making elastic collisions in hydrogen. Show that the probability $P_n(E)$ that it made n collisions in slowing down is

$$P_n(E) = \frac{[\log(E_0/E)]^{n-1}}{(n-1)!} \frac{E}{E_0}.$$

(Segrè,[31] p. 676.)

2-21. Consider a reactor containing a homogeneous mixture of uranium and carbon. Neutrons are produced at high energies by fission of the uranium, and then slowed down to thermal energies by the carbon moderator. The thermal neutrons have an average velocity v and mean free path $\lambda = 10$ cm. The average number of collisions that a thermal neutron makes before capture is $N = 100$, and an average of $k = 1.04$ fission neutrons are produced in each capture.

(a) Show that the time rate of increase in the neutron number density n is described by the differential equation

$$\frac{dn}{dt} = nv\left(\frac{k-1}{\lambda N} - \frac{A}{4V}\right),$$

where V is the volume and A the surface area of the reactor.

(b) Show that if the reactor is cubical, the length of a side required for criticality is $L = 3.75 \times 10^4$ cm. (Cronin et al.,[30] p. 57.)

2-22. Consider an ideal point-mass gas that obeys either F-D or B-E statistics. Then, if $f(E)\,dE$ is the fraction of the total number of particles having energies in the range E to $E + dE$, we have

$$f(E)\,dE = \frac{2}{\sqrt{\pi}} \frac{1}{(kT)^{3/2}} \frac{DT^{3/2}}{n} \frac{E^{1/2}\,dE}{\dfrac{1}{A}e^{E/kT} \pm 1},$$

where n is the gas number density and

$$D = \frac{(2\pi mk)^{3/2}}{h^3}.$$

Show that the M-B distribution is a good approximation for *all* energies if the following condition is satisfied:

$$\frac{DT^{3/2}}{n} \gg 1.$$

Prove that this condition is equivalent to

$$\left(\frac{2\pi}{3}\right)^{3/2}\left(\frac{\bar{d}}{\lambda_s}\right)^3 \gg 1,$$

where λ_s is the de Broglie wavelength corresponding to the root-mean-square momentum and $\bar{d}$ is the mean spacing between atoms. Discuss the physical significance of this condition.

REFERENCES

1. E. H. Kennard, *Kinetic Theory of Gases*, McGraw-Hill, New York (1938). J. H. Jeans, *An Introduction to the Kinetic Theory of Gases*, Cambridge University Press, Cambridge (1940). R. D. Present, *Kinetic Theory of Gases*, McGraw-Hill, New York (1958). L. B. Loeb, *Kinetic Theory of Gases*, 3rd ed., Dover, New York (1961). T. G. Cowling, *Molecules in Motion*, Harper, New York (1960). J. H. Jeans, *The Dynamical Theory of Gases*, Dover, New York (1954). (This book was first published by the Cambridge University Press in 1925.)
2. J. H. Ferziger and H. G. Kaper, *Mathematical Theory of Transport Processes in Gases*, North-Holland, Amsterdam (1972).
3. S. Chapman and T. G. Cowling, *The Mathematical Theory of Non-uniform Gases*, 2nd ed., Cambridge University Press, London (1952).
4. J. O. Hirschfelder, C. F. Curtiss, and R. B. Bird, *Molecular Theory of Gases and Liquids*, Wiley, New York (1964).
5. R. D. Present, Ref. 1, pp. 78–79.
6. J. H. Jeans, *An Introduction to the Kinetic Theory of Gases*, Ref. 1.
7. E. H. Kennard, Ref. 1, pp. 154–157.
8. J. H. Jeans, Ref. 6, pp. 138–141.
9. L. B. Loeb, Ref. 1, Appendix I.
10. E. H. Kennard, Ref. 1, p. 149.
11. W. D. Niven (Ed.), *The Scientific Papers of James Clerk Maxwell*, Vol. II, Dover, New York (1952), pp. 26–78.
12. J. H. Jeans, Ref. 6, p. 233.
13. E. H. Kennard, Ref. 1, Chap. V. J. O. Hirschfelder, C. F. Curtiss, and R. B. Bird, Ref. 4.
14. R. D. Present, Ref. 1, Chap. 7.
15. R. D. Present, Ref. 1, pp. 49–51.
16. J. C. Maxwell, in Ref. 11.
17. J. Stefan, *Wien. Ber.* **65**, 323 (1872).
18. P. Langevin, *Ann. Chim. Phys.* **5**, 245 (1905). This paper is one of the classics in the literature on kinetic theory. It contains a lengthy calculation of the mobility of ions interacting with molecules through the polarization potential, as well as the general computation of the diffusion coefficient. Langevin's calculation is an excellent example of the application of the rigorous momentum transfer method of Maxwell to the solution of a difficult problem. A similar calculation of the diffusion coefficient appears in Chapter 8 of the kinetic theory text by Present,[1] who also gives a general discussion of the momentum transfer method in Chapter 11.

19. J. O. Hirschfelder, C. F. Curtiss, and R. B. Bird, Ref. 4, Chap. 8. S. Chapman and T. G. Cowling, Ref. 3, Chap. 9. A. Dalgarno and A. Williams, *Proc. Phys. Soc. (London) A* **72**, 274 (1958). E. A. Mason and H. W. Schamp, *Ann. Phys. (N.Y.)* **4**, 233 (1958).

20. A. Dalgarno, "Diffusion and Mobilities," in D. R. Bates (Ed.), *Atomic and Molecular Processes*, Academic, New York (1962). E. A. Mason and T. R. Marrero, *AAMP* **6**, 155 (1970). T. R. Marrero and E. A. Mason, *J. Phys. Chem. Ref. Data* **1**, 1 (1972). E. A. Mason and E. W. McDaniel, *The Transport Properties of Ions in Gases*, Wiley, New York (1988).

21. L. Waldmann, "Transporterscheinungen in Gasen von mittlerem Druck," in *Handbuch der Physik*, Vol. XII, Springer-Verlag, Berlin (1958), pp. 295–514.

22. H. Grad, "Principles of the Kinetic Theory of Gases," in *Handbuch der Physik*, Vol. XII, Springer-Verlag, Berlin (1958), pp. 205–294.

23. E. H. Kennard, Ref. 1, Chap. IX. R. C. Tolman, *The Principles of Statistical Mechanics*, Oxford University Press, Oxford (1938), Chap. III. T. L. Hill, *Statistical Mechanics*, McGraw-Hill, New York (1956), Chap. I. L. D. Landau and E. M. Lifshitz, *Statistical Physics*, Addison-Wesley, Reading, Mass. (1958), Chap. I. G. E. Uhlenbeck and G. W. Ford, *Lectures in Statistical Mechanics*, American Mathematical Society, Providence, R. I. (1963). C. L. Longmire, *Elementary Plasma Physics*, Interscience, New York (1963). K. Huang, *Statistical Mechanics*, Wiley, New York (1963).

24. J. R. Pierce, *Theory and Design of Electron Beams*, 2nd ed., Van Nostrand, Princeton, N.J. (1954), pp. 48–51. See also A. Arakengy, *Am. J. Phys.* **25**, 519 (1957).

25. S. Chapman and T. G. Cowling, Ref. 3. J. O. Hirschfelder, C. F. Curtiss, and R. B. Bird, Ref. 4, Pt. 2.

26. L. Spitzer, *Physics of Fully Ionized Gases*, 2nd ed., Interscience, New York (1962), Chap. 5. W. P. Allis, "Motions of Ions and Electrons," in *Handbuch der Physik*, Vol. XXI, Springer-Verlag, Berlin (1956), pp. 429–444. S. Chandrasekhar, *Plasma Physics*, University of Chicago Press, Chicago (1960), Chap. VII. J. G. Linhart, *Plasma Physics*, 2nd ed., Interscience, New York (1961), Chap. 6. J. L. Delcroix, *Introduction to the Theory of Ionized Gases*, Interscience, New York (1960), pp. 116–118. A. N. Kaufman, "Plasma Transport Theory," in C. de Witt and J. F. Detoeuf (Eds.), *La Théorie des gaz neutres et ionisés*, Wiley, New York (1960), pp. 293–353. J. Enoch, *Phys. Fluids* **3**, 353 (1960). A. Simon, *Phys. Fluids* **4**, 691 (1961). C. L. Longmire, Ref. 23. D. R. Nicholson, *Introduction to Plasma Theory*, Wiley, New York (1983). W. M. Stacey, *Fusion Plasma Analysis*, Wiley, New York (1981). N. A. Krall and A. W. Trivelpiece, *Principles of Plasma Physics*, McGraw-Hill, New York (1973). F. Reif, *Fundamentals of Statistical and Thermal Physics*, McGraw-Hill, New York (1965). R. Kubo, M. Toda, and N. Hashitsume, *Statistical Physics II: Nonequilibrium Statistical Mechanics*, Springer-Verlag, Berlin (1985). H. Risken, *The Fokker-Planck Equation. Methods of Solution and Applications*, Springer-Verlag, New York (1984).

27. A. A. Vlasov, *Zh. Eksp. Teor. Fiz.* **8**, 291 (1938); see also U.S. Atomic Energy Commission translation 2729. N. Rostoker and M. N. Rosenbluth, *Phys. Fluids* **3**, 1 (1960). I. B. Bernstein, *Phys. Rev.* **109**, 10 (1958). J. E. Drummond (Ed.), *Plasma Physics*, McGraw-Hill, New York (1961). W. B. Thompson, *An Introduction to Plasma Physics*, Addison-Wesley, Reading, Mass. (1962). D. R. Nicholson, Ref. 26. W. M. Stacey, Ref. 26. N. A. Krall and A. W. Trivelpiece, Ref. 26. L. E. Reichl, *A Modern Course in Statistical Physics*, University of Texas Press, Austin (1980).

28. D. A. McQuarrie, *Statistical Mechanics*, Harper & Row, New York (1976). F. Reif, Ref. 26.
29. M. Chen, *Berkeley Physics Problems with Solutions*, Prentice-Hall, Englewood Cliffs, N.J. (1974).
30. J. A. Cronin, D. F. Greenberg, and V. L. Telegdi, *University of Chicago Graduate Problems in Physics with Solutions*, Addison-Wesley, Reading, Mass. (1967).
31. E. Segrè, *Nuclei and Particles*, 2nd ed., Benjamin-Cummings, Menlo Park, Calif. (1977).

3

COLLISION THEORY

In this chapter we discuss the theory of collisions of particles interacting through a central force,* that is, a force that acts along the line of centers of the particles and whose magnitude depends only on the distance of separation. Only binary nonrelativistic collisions are treated. The classical theory is presented first and is followed by a quantum mechanical treatment.

PART $\mathscr{A}$. THE TWO-BODY CENTRAL FORCE PROBLEM IN CLASSICAL MECHANICS

3-1. SEPARATION OF THE CENTER-OF-MASS MOTION FROM THE TOTAL MOTION

Consider two particles of masses m and M that interact with one another through a potential $V(r)$, which is a function only of the length of the vector $\mathbf{r}$ separating them. Let $\mathbf{R}_m$ and $\mathbf{R}_M$ denote the position vectors of the two particles in the Lab system of coordinates, so that $\mathbf{r} = \mathbf{R}_m - \mathbf{R}_M$. If no external forces are applied, Newton's second and third laws of motion for this system may be expressed as

$$\mathbf{f}_m = m\frac{d^2\mathbf{R}_m}{dt^2}, \tag{3-1-1}$$

$$\mathbf{f}_M = M\frac{d^2\mathbf{R}_M}{dt^2} = -\mathbf{f}_m. \tag{3-1-2}$$

*A reference on collisions in noncentral force fields, such as those described in Section 1-7, is J. O. Hirschfelder, C. F. Curtiss, and R. B. Bird, *Molecular Theory of Gases and Liquids*, Wiley, New York (1964). See also R. B. Bernstein (Ed.), *Atom-Molecule Collision Theory*, Plenum, New York (1979); G. Scoles (Ed.), *Atomic and Molecular Beam Methods*, Vol. I, Oxford University Press, New York (1988); R. D. Levine and R. B. Bernstein, *Molecular Reaction Dynamics and Chemical Reactivity*, Oxford University Press, New York (1987).

The forces $\mathbf{f}_m$ and $\mathbf{f}_M$, derivable from the potential $V(r)$, act on m and M, respectively. Now introduce the vector $\mathbf{R}$, which locates the center of mass of the two particles with respect to the origin of the Lab coordinates:

$$\mathbf{R} = \frac{m\mathbf{R}_m + M\mathbf{R}_M}{m + M} \tag{3-1-3}$$

(cf. Section 1-2). If we differentiate (3-1-3) twice with respect to the time, we obtain

$$\ddot{\mathbf{R}} = \frac{m\ddot{\mathbf{R}}_m + M\ddot{\mathbf{R}}_M}{m + M} = \frac{\mathbf{f}_m + \mathbf{f}_M}{m + M} = 0. \tag{3-1-4}$$

Thus we see that the center of mass travels with a uniform rectilinear velocity.

Now we multiply (3-1-1) by M and (3-1-2) by m and subtract the second from the first. The result is

$$\mathbf{f}_m = M_r \frac{d^2}{dt^2} (\mathbf{R}_m - \mathbf{R}_M), \tag{3-1-5}$$

where M_r is the reduced mass of the pair of particles, defined by the equation

$$M_r = \frac{mM}{m + M}. \tag{1-2-3}$$

Since $\mathbf{f}_m$ is in the direction of the vector $\mathbf{r} = \mathbf{R}_m - \mathbf{R}_M$, (3-1-5) for the relative motion has the same form as the equation of motion of a single particle of mass M_r in a central force field:

$$\mathbf{f}_m(r) = M_r \ddot{\mathbf{r}}. \tag{3-1-6}$$

3-2. REDUCTION TO A TWO-BODY PROBLEM IN TWO DIMENSIONS

Let us now take the cross product of the vector $(\mathbf{R}_m - \mathbf{R}_M)$ with both sides of (3-1-5). The left side will vanish because of the central force assumption. Therefore,

$$(\mathbf{R}_m - \mathbf{R}_M) \times \frac{d^2}{dt^2} (\mathbf{R}_m - \mathbf{R}_M) = 0$$

or

$$\frac{d}{dt}\left[(\mathbf{R}_m - \mathbf{R}_M) \times \frac{d}{dt} (\mathbf{R}_m - \mathbf{R}_M)\right] - \left[\frac{d}{dt} (\mathbf{R}_m - \mathbf{R}_M) \times \frac{d}{dt} (\mathbf{R}_m - \mathbf{R}_M)\right] = 0.$$

The second term on the left side of this equation obviously equals zero. If we integrate the first term with respect to the time, we obtain

$$(\mathbf{R}_m - \mathbf{R}_M) \times \frac{d}{dt}(\mathbf{R}_m - \mathbf{R}_M) = (\mathbf{R}_m - \mathbf{R}_M) \times (\mathbf{v}_m - \mathbf{v}_M) = \mathbf{r} \times \mathbf{v}_r = \mathbf{K}, \qquad (3\text{-}2\text{-}1)$$

where $\mathbf{v}_m$ and $\mathbf{v}_M$ are the velocities of m and M, respectively, in the Lab system, $\mathbf{v}_r$ is their relative velocity,* and $\mathbf{r}$ is the directed distance from M to m; $\mathbf{K}$ is a vector whose magnitude and direction are constant in time. Evidently, $\mathbf{K}$ is perpendicular to the plane defined by $\mathbf{r}$ and $\mathbf{v}_r$. The two particles and their center of mass thus remain in a plane perpendicular to the vector $\mathbf{K}$ throughout the collision. Since the collision takes place in a fixed plane containing the center of mass and the motion of the center of mass is known to be uniform, only a two-dimensional problem remains to be considered.

Figure 3-2-1 shows the collision in the plane perpendicular to $\mathbf{K}$. The impact parameter is b, and φ is the angle specifying the instantaneous orientation of the

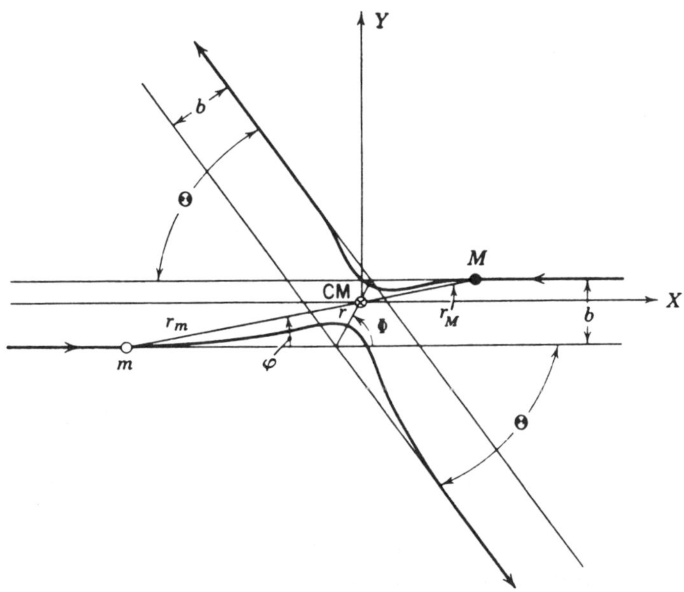

Figure 3-2-1. The mutual elastic scattering of particles m and M as viewed in the center-of-mass system, in which the motion is two-dimensional. The motion pictured here occurs in the plane of the paper, which is perpendicular to the vectors **K** and **J**. An interaction potential consisting of an attractive component and a shorter-range repulsion is assumed.

*The relative velocity of the collision partners at any arbitrary time is denoted by $\mathbf{v}_r$; $\mathbf{v}_0$ has been used to represent their relative velocity of approach at large r before their interaction becomes appreciable.

vector **r**. The particles are separated by a distance r, and r_m and r_M, the distances of m and M, respectively, from the center of mass, are in the ratio M/m. The value Φ is assumed by φ at the distance of closest approach, and Θ is the angle of scattering in the CM system. The CM trajectories of both m and M are symmetrical about the line of closest approach (the "line of apses") at the angle Φ. Proof of this statement is deferred until Section 3-5.

Let us now calculate the total kinetic energy of the particles $T_{\rm CM}$ in the center-of-mass system. Establish a rectangular Cartesian coordinate system X, Y, Z whose origin moves with the center of mass and whose Z axis points in the direction of the vector **K**. The X axis corresponds to $\varphi = 0$. It is evident that

$$T_{\rm CM} = \tfrac{1}{2}m(\dot{x}_m^2 + \dot{y}_m^2) + \tfrac{1}{2}M(\dot{x}_M^2 + \dot{y}_M^2). \tag{3-2-2}$$

The rectangular and polar coordinates are related by the equations

$$x_m = -\frac{M}{m + M} r\cos\varphi, \tag{3-2-3}$$

$$y_m = -\frac{M}{m + M} r\sin\varphi, \tag{3-2-4}$$

$$x_M = \frac{m}{m + M} r\cos\varphi, \tag{3-2-5}$$

$$y_M = \frac{m}{m + M} r\sin\varphi. \tag{3-2-6}$$

By substituting these relationships into (3-2-2), we may express the kinetic energy in the CM system as

$$T_{\rm CM} = \tfrac{1}{2}M_r(\dot{r}^2 + r^2\dot{\varphi}^2). \tag{3-2-7}$$

The angular momentum of the pair of particles about the center of mass is

$$\mathbf{J}_{\rm CM} = (\mathbf{r}_m \times \mathbf{p}_m) + (\mathbf{r}_M \times \mathbf{p}_M), \tag{3-2-8}$$

where $\mathbf{p}_m$ and $\mathbf{p}_M$ are the linear momenta of m and M, respectively, in the CM system. The component of $\mathbf{p}_m$ that is perpendicular to $\mathbf{r}_m$ is $mr_m\dot{\varphi}$, and a similar expression applies for M. Consequently,

$$J_{\rm CM} = r_m m r_m \dot{\varphi} + r_M M r_M \dot{\varphi} = M_r r^2 \dot{\varphi}. \tag{3-2-9}$$

Now the quantity $\mathbf{r} \times \mathbf{v}_r$ was shown in (3-2-1) to be a constant. The magnitude of this quantity equals $r(r\dot{\varphi}) = r^2\dot{\varphi}$, and the angular momentum is seen to be a constant of the motion, as is the total energy.

Let v_0 be the initial relative velocity of the particles, that is, the relative velocity outside the region in which the interaction becomes appreciable. At

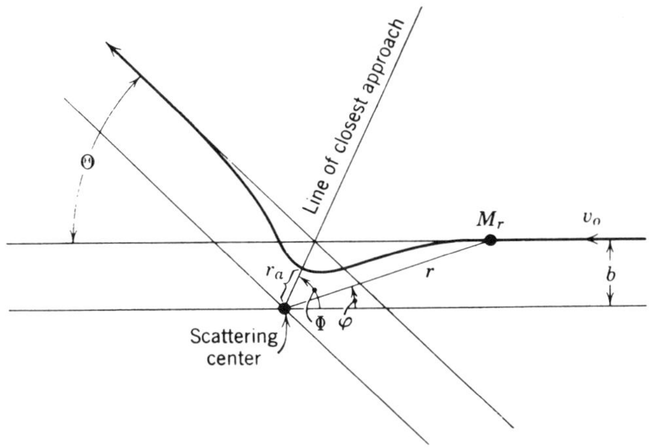

Figure 3-2-2. The one-body problem that is dynamically equivalent to the two-body problem illustrated in Fig. 3-2-1. The center-of-mass motion of two particles m and M interacting through a potential $V(r)$ can be obtained as follows: a single hypothetical particle of mass $M_r = mM/(m + M)$ is considered to be scattered by a center of force fixed in the CM frame. The potential describing this force field is $V(r)$, where r is the separation distance between m and M. The CM velocity of M_r is considered to be equal to v_0, where v_0 is the initial relative velocity of m and M. The hypothetical particle M_r is moving toward the stationary scattering center with impact parameter b. The instantaneous distance of M_r from the scattering center is r, and M_r is scattered through an angle Θ, whose value can be calculated by (3-4-5). This angle is also the scattering angle of both m and M in the CM frame.

some time before the collision occurs the angle $\varphi \to b/r$ and $\dot{\varphi} \to -b\dot{r}/r^2$. Then (3-2-7) and (3-2-9) show that the initial kinetic energy is $M_r v_0^2/2$ and the angular momentum is $M_r b v_0$ (cf. Section 1-2). Utilizing the constancy of the total energy and angular momentum of the system, we may then write, for any point along the trajectory,

$$E = \tfrac{1}{2}M_r v_0^2 = \tfrac{1}{2}M_r(\dot{r}^2 + r^2\dot{\varphi}^2) + V(r), \tag{3-2-10}$$

$$J = M_r b v_0 = M_r r^2 \dot{\varphi}. \tag{3-2-11}$$

Equations (3-2-10) and (3-2-11), which are the equations of motion in the CM system, furnish a complete description of the motion in terms of the interaction potential, the initial relative velocity, and the impact parameter. If we are given the potential, a particular trajectory may be specified by assigning values to the parameters v_0 and b or, alternatively, to the constants of motion (i.e., the total energy and the angular momentum).

Figure 3-2-2 describes the one-body problem that is dynamically equivalent to the two-body problem under discussion [cf. (3-1-5)]. We have already shown that the equivalent single-body motion is planar and can be drawn in two dimensions. The quantities b, r, φ, Φ, and Θ are given so that their significance in the one- and two-body problems may be compared.

3-3. REDUCTION TO THE EQUIVALENT ONE-BODY PROBLEM IN ONE DIMENSION

A further conceptual reduction of the problem is possible. If we substitute the value of $\dot{\varphi}$ from (3-2-11) into (3-2-10), we may write the following expression for r as a function of time:

$$\tfrac{1}{2}M_r v_0^2 = \tfrac{1}{2}M_r \dot{r}^2 + V_{\text{eff}}(r), \tag{3-3-1}$$

where $V_{\text{eff}}(r)$ is the "effective potential," defined by

$$V_{\text{eff}}(r) = V(r) + \frac{M_r v_0^2 b^2}{2r^2}. \tag{3-3-2}$$

Equation (3-3-1) contains no terms in φ. It can be regarded as describing the one-dimensional motion (along the r axis) of a particle of mass M_r with total energy $M_r v_0^2/2$ in an effective potential field described by $V_{\text{eff}}(r)$. The second term on the right in the expression for $V_{\text{eff}}(r)$ is equal to $J^2/2M_r r^2$ and represents the rotational kinetic energy of the system. Note that this term is a positive monotonic decreasing function of r. It can therefore be considered the source of a fictitious outwardly directed force, the "centrifugal force," and for this reason the second term on the right side of (3-3-2) is termed the "centrifugal potential." We use the symbol V_c to denote this quantity.

3-4. THE SCATTERING ANGLE IN THE CENTER-OF-MASS SYSTEM

Perhaps the most important feature of an elastic collision, from our point of view, is the angle of scattering in the center-of-mass system (see Fig. 3-2-1). Since the CM trajectories are symmetrical about the line of closest approach, the scattering angle Θ is related to the angle Φ by the equation

$$\Theta = \pi - 2\Phi. \tag{3-4-1}$$

The value, Φ, of the orientation angle φ, corresponding to the distance of closest approach, is easily calculated: dr/dt is given by (3-3-1) and $d\varphi/dt$ by (3-2-11). Thus we obtain the "orbit equation" by writing

$$\frac{dr}{d\varphi} = \frac{dr}{dt}\bigg/\frac{d\varphi}{dt} = \pm\frac{r^2}{b}\left[1 - \frac{V(r)}{M_r v_0^2/2} - \frac{b^2}{r^2}\right]^{1/2}. \tag{3-4-2}$$

The negative sign applies for the incoming branch of the trajectory and the positive sign for the outgoing branch. At the angle of closest approach Φ we have

$$1 - \frac{V(r_a)}{M_r v_0^2/2} - \frac{b^2}{r_a^2} = 0. \tag{3-4-3}$$

The largest real root of this equation is the distance of closest approach r_a.

We should note that a real solution of (3-4-3) does not always exist. If we assume an attractive potential of the form $V(r) \sim -r^{-n}$ and further assume that $n \geqslant 2$, there are values of the initial velocity and impact parameter for which no solution exists (see Section 3-6). The particle spirals inward until it is stopped by some repulsive force and then spirals back out. A solution always exists for a repulsive potential, however.

If a solution exists,

$$\Phi = -\int_{\infty}^{r_a} \frac{d\varphi}{dr}\,dr = -\int_{\infty}^{r_a} \frac{(b/r^2)\,dr}{\left[1 - \dfrac{V(r)}{M_r v_0^2/2} - \dfrac{b^2}{r^2}\right]^{1/2}}. \tag{3-4-4}$$

Equation (3-4-1) then shows Θ to be

$$\Theta(v_0, b) = \pi - 2b\int_{r_a}^{\infty} \frac{dr/r^2}{\left[1 - \dfrac{V(r)}{M_r v_0^2/2} - \dfrac{b^2}{r^2}\right]^{1/2}}. \tag{3-4-5}$$

Equation (3-4-5) is of particular importance to us, since the angle of scattering is the only aspect of an elastic collision that enters into the formulas for the transport coefficients.[1] The foregoing integral has been evaluated for several potentials used to describe neutral nonpolar gases, and the results applied to the calculation of transport properties, by various groups.[2]

The solution of (3-4-5) for integrable power-law potentials is treated by Whittaker[3] and Goldstein,[4] among others. If the potential is assumed to be of the form $V(r) \sim \pm r^{-n}$, a solution in terms of circular functions is obtainable for $n = +2$, $+1$, and -2. The integral exponents $n = -6, -4, -1, +3, +4$, and $+6$ lead to solutions in terms of elliptic integrals. The fractional exponents $n = +\frac{1}{2}, +\frac{3}{2}, -\frac{2}{3}, +\frac{2}{3}$, and $+\frac{4}{3}$ also lead to elliptic functions.

3-5. THE SYMMETRY OF THE TRAJECTORIES IN THE CENTER-OF-MASS SYSTEM

Having obtained the differential equation of the orbit in the CM system, we may now show that the trajectory of each of the particles is symmetrical about the line of closest approach, for which $\varphi = \Phi$. To do this, let us square both sides of (3-4-2) and make the substitution $\varphi = \alpha + \Phi$. The introduction of the angle α is equivalent to counterclockwise rotation of our polar coordinate system about the Z axis through the angle Φ, so that the line of closest approach now corresponds to $\alpha = 0$. The resulting equation is

$$\left(\frac{dr}{d\alpha}\right)^2 = \frac{r^4}{b^2}\left[1 - \frac{V(r)}{M_r v_0^2/2} - \frac{b^2}{r^2}\right]. \tag{3-5-1}$$

The trajectory of each particle will be symmetrical about the line of closest approach if reflection about this line produces no change. Mathematically, this reflection is accomplished by substitution of $-\alpha$ for α. Equation (3-5-1) is obviously invariant under this substitution. Further, the initial conditions $r = r(\alpha = 0)$ and $r' = (dr/d\alpha)_{\alpha=0} = 0$ are likewise unaffected by the substitution.

3-6. CLASSIFICATION OF THE ORBITS; ION-MOLECULE REACTIONS

A. Physical Basis

In Section 3-4 it was stated that a solution of (3-4-3) always exists for a repulsive potential, but not always for an attractive potential of the form $V(r) \sim -r^{-n}$ if $n \geqslant 2$. This statement is easily understood in terms of the effective one-dimensional motion and the centrifugal potential.

First let us consider motion in a typical repulsive potential field such as that described by the potential function $V(r)$ in Fig. 3-6-1. It has already been pointed out that if we wish to analyze the scattering problem in the single dimension r we must add to the real potential, $V(r)$, a fictitious centrifugal potential, $V_c(r)$, given

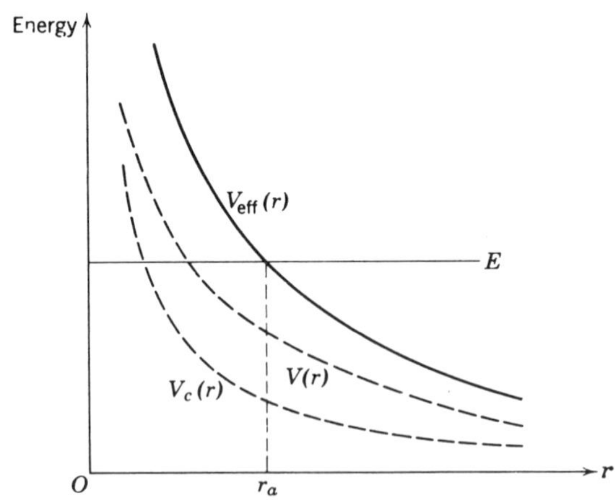

Figure 3-6-1. The potential functions used for analysis of elastic scattering in a typical repulsive potential field.

in terms of the angular momentum, J, of the system by the equation

$$V_c(r) = \frac{J^2}{2M_r r^2}. \tag{3-6-1}$$

This potential is plotted in Fig. 3-6-1 for some arbitrary but nonzero value of J. The total effective potential, $V_{\text{eff}}(r) = V(r) + V_c(r)$, is also shown in the figure. It is apparent that in this case the only significant effect of the centrifugal potential will be to supplement the real potential. If $V(r)$ decreases monotonically with increasing r, the effective potential will also, regardless of the value of the angular momentum. Hence in this one-dimensional formulation of the scattering problem a particle of total energy E will approach the scattering center and be reflected at the distance of closest approach, r_a, given by the intersection of the horizontal line at height E and the curve $V_{\text{eff}}(r)$. The radial velocity $\dot{r}$ at any point is given by the equation

$$E = \tfrac{1}{2}M_r\dot{r}^2 + V_{\text{eff}}(r), \tag{3-6-2}$$

whereas the angular motion must, of course, be such that the angular momentum is conserved throughout the scattering.

The shape of the effective potential curve is quite different if an attractive potential is assumed. However, if $V(r) \sim -r^{-n}$ with $n < 2$, the effective potential again decreases monotonically with increasing r, at sufficiently small r, and the one-dimensional motion is similar to that in the repulsive case in regard to reflection along the r axis at some finite distance of closest approach. (This situation is illustrated in Fig. 3-6-2, in which an attractive inverse-first-power potential is assumed.) Stable orbits* are possible only for $n < 2$, and for values of E lying in the well.

If, on the other hand, the attractive potential falls off with increasing r faster than r^{-2}, the attractive potential must dominate the centrifugal potential at small r, and an effective potential of the form shown in Fig. 3-6-3 results.† (An inverse-third-power attractive potential is assumed in this illustration.) Let us consider the motion for several different values of the total energy. A particle of energy $E_1 < E_0$ starting at large r and moving toward the center of attraction will evidently be reflected at $r = r_a$ by the potential barrier shown. If, however, the total energy is $E_2 > E_0$, the particle will be able to pass over the potential barrier. It will experience a repulsion only for $r > r_p$ and will sense an attraction thereafter, as its radial distance from the scattering center decreases. When $E > E_0$, the particle actually passes through the center of attraction, which is assumed to be a mathematical point in the model under consideration here.[5] A

*A particle is said to be in a *stable orbit* if the application of a small perturbation produces a small bounded excursion from the original orbit.

†The limiting case in which $n = 2$ must be treated separately. Here the shape of $V_{\text{eff}}(r)$ is determined by the relative magnitudes of J and the coefficient of r^{-2} in the expression for $V(R)$.

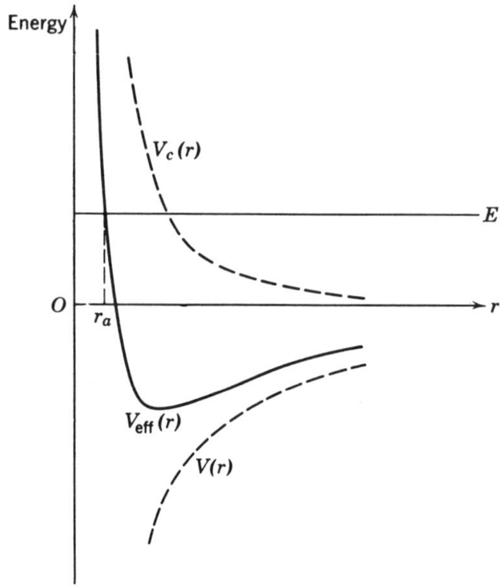

Figure 3-6-2. The potential functions used for analysis of elastic scattering in an attractive inverse-first-power potential field.

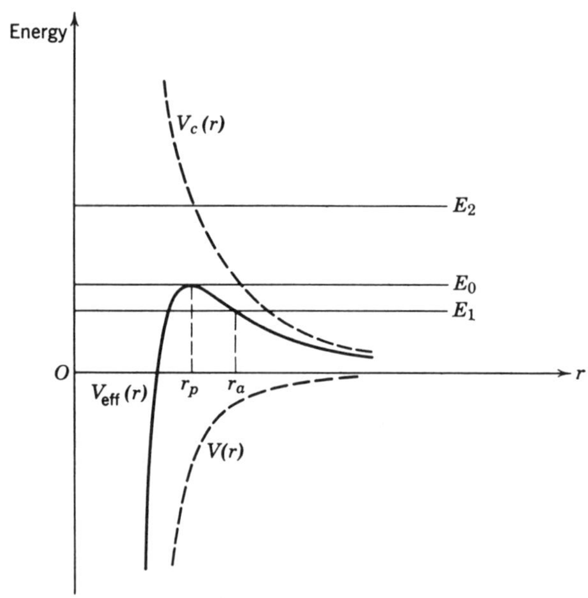

Figure 3-6-3. The potential functions used for analysis of elastic scattering in an attractive inverse-third-power potential field.

particularly interesting situation develops when the total energy just exceeds E_0, the value of $V_{\text{eff}}(r)$ at the peak of the potential curve. In this case the particle will spend a considerable time at a radial distance near the peak, where $\dot{r}$ is small by assumption, while spiraling inward toward the center. The particle is then said to "orbit" about the scattering center.* The angular motion speeds up as r decreases in order to conserve angular momentum, and a large number of revolutions may be made. (This is an unstable orbit, unlike those for which $n < 2$.) Under certain conditions the scattering angle may approach $-\infty$.[6] Figure 3-6-4 illustrates an orbiting collision for which Θ has a value between 6π and 7π. In this figure a long-range attractive potential and a short range, hard-core repulsive potential are assumed. A cusp appears in the trajectory when the particle is reflected from the hard core.

We see that the orbits may thus be meaningfully classified as *spiraling* or *nonspiraling*, and this distinction is frequently encountered in the literature. We shall illustrate the classification of orbits by discussing the important case of the attractive inverse-fourth-power polarization potential. The original treatment of this case by Langevin[7] appears in English translation in Ref. 8.

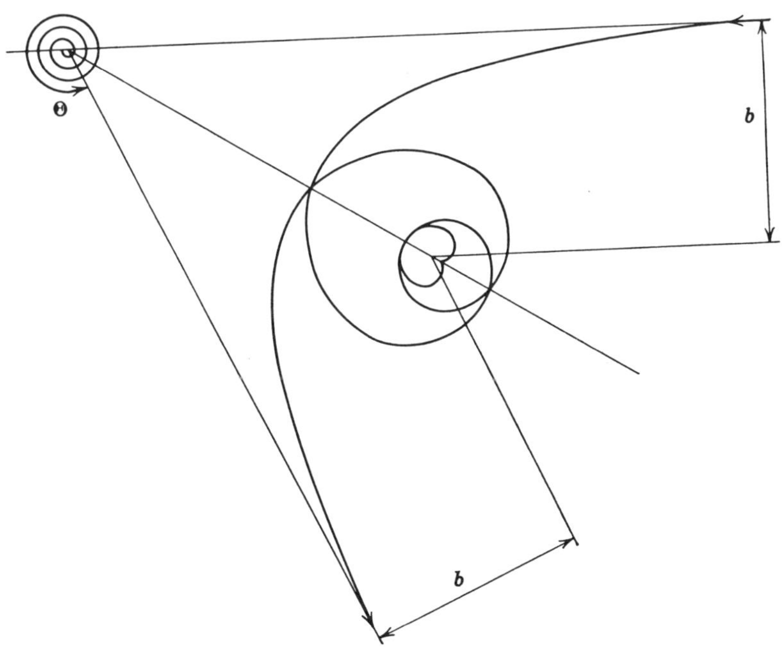

Figure 3-6-4. An orbiting collision for which the scattering angle Θ has a large value.

*This type of collision is sometimes referred to as a "sticky collision." As a rule, however, the term *sticky collision* means an impact in which the colliding particles temporarily form a complex structure with an appreciable amount of internal excitation energy. This structure then flies apart after a lifetime that is long compared to usual collision times.

B. Classification of the Orbits in the Polarization Potential Field

The inverse-fourth-power polarization potential has the form

$$V(r) = -\frac{\alpha e^2}{2r^4}, \tag{3-6-3}$$

where e is the ionic charge and α is the electric polarizability of the molecule. Two types of orbit are possible in this potential field. The orbits corresponding to large angular momentum resemble hyperbolas (Fig. 3-6-5a), whereas for small angular momentum an inward spiraling motion takes place until some repulsive force reverses the trend (Fig. 3-6-5b).

The scattering angle for the polarization potential may be expressed as

$$\Theta = \pi - 2\Phi = \pi - 2\int_0^{\rho_0} \frac{d\rho}{\sqrt{1 - \rho^2 + e^2\alpha\rho^4/M_r v_0^2 b^4}}, \tag{3-6-4}$$

in which $\rho = b/r$ and ρ_0 is the lower of the two positive roots of the polynomial in the denominator if such roots exist. If this polynomial has no real root, the integration goes from 0 to ∞. The existence of a real root is related to the nature of the orbit. If b is large enough for given v_0, a root exists, and an orbit of the type shown in Fig. 3-6-5a will result. For b less than a certain critical value b_0 no root exists, and the orbit is of the type presented in Fig. 3-6-5b. For b_0, the limiting value of b, the particles spiral into a circular orbit; b_0 is found by setting the discriminant of the square root in (3-6-4) equal to zero:

$$b_0 = \left(\frac{4e^2\alpha}{M_r v_0^2}\right)^{1/4}. \tag{3-6-5}$$

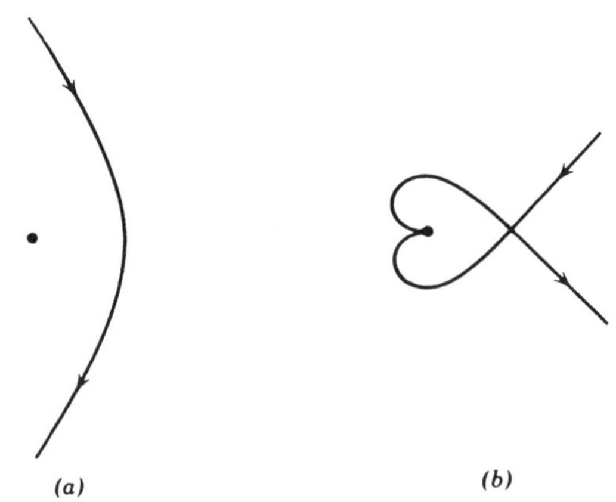

Figure 3-6-5. Types of orbit of a particle moving in the attractive inverse-fourth-power polarization potential field:[11] (a) large angular momentum, nonspiraling orbit; (b) small angular momentum, spiraling orbit.

Orbits for which $b < b_0$ pass through the origin if no repulsive core is present, whereas those orbits for which $b \geqslant b_0$ come no closer than $r_0 = b_0/\sqrt{2}$.* Several orbits of each type are accurately plotted in Fig. 3-6-6, which is a modified version of a drawing prepared by Langevin.[8] In this figure only the incoming branch of each spiraling trajectory is shown, the subsequent portion being omitted for clarity. The complete drawing appears in Fig. 4 of Appendix II of Ref. 8.

Now let us suppose that there exists a certain critical radius r_c such that a given type of reaction between the two particles under consideration is impossible if the distance of closest approach is greater than r_c and almost certain to occur if it is less than r_c. Then, if r_c lies between 0 and $b_0/\sqrt{2}$, all collisions for which $b < b_0$ must lead to this reaction. We may therefore assume that the cross section for the reaction is the same as the *cross section for orbiting collisions*, namely

$$q_0(v_0) = \pi b_0^2 = \frac{2\pi}{v_0}\left(\frac{e^2\alpha}{M_r}\right)^{1/2}. \tag{3-6-6}$$

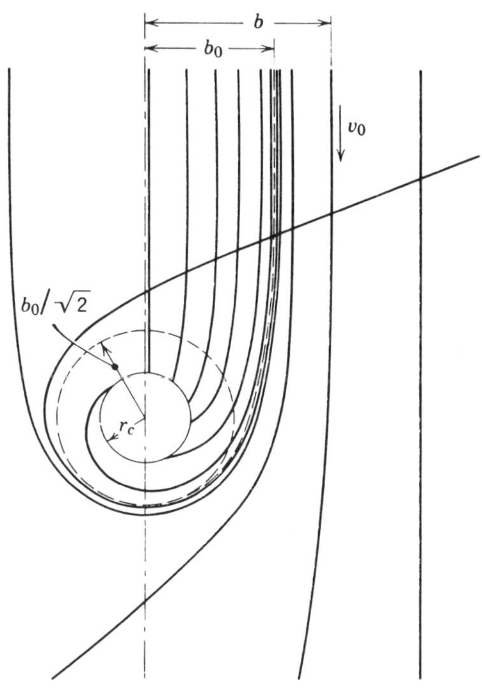

Figure 3-6-6. Representative trajectories for the inverse-fourth-power polarization potential as a function of the impact parameter b for a given relative velocity v_0. For clarity only the incoming branch of each spiraling trajectory is shown.

*For a short proof of this statement, see K. Yang and T. Ree, *J. Chem. Phys.* **35**, 588 (1961).

This type of analysis has been used in calculating the rates of ion-molecule reactions.[9] Since b_0 for the polarization potential varies inversely as the square root of v_0, there must be some collisions at the higher velocities for which it is not true that $r_c < b_0/\sqrt{2}$. Thus this model is not realistic when applied to the calculation of the reaction cross section at high energies—it predicts that this cross section should fall to zero, whereas it should actually drop to πr_c^2.

We may easily derive an alternative expression for the reaction cross section that is valid at high energies and holds for attractive or repulsive potentials of arbitrary form. Our starting point is the orbit equation (3-4-2). At the distance of closest approach r_a we have $dr/d\varphi = 0$. Thus (3-4-3), which is obtained from (3-4-2) by setting $dr/d\varphi = 0$, gives a relationship between the impact parameter and the distance of closest approach if we assume that the potential function and initial velocity are specified. Now let us suppose that an ion-molecule reaction is inevitable if r decreases to a value as small as r_c. The critical impact parameter b_0, which corresponds to the trajectory for which $dr/d\varphi = 0$ at $r = r_c$, is obtained by replacing r_a by r_c and b by b_0 in (3-4-3). Thus

$$1 - \frac{V(r_c)}{M_r v_0^2/2} - \frac{b_0^2}{r_c^2} = 0. \tag{3-6-7}$$

The reaction cross section is then equal to πb_0^2, as is apparent from Fig. 3-6-7, and we obtain

$$q_r(v_0) = \pi r_c^2 \left[1 - \frac{V(r_c)}{M_r v_0^2/2} \right]. \tag{3-6-8}$$

We now have the cross section for the reaction in terms of only one parameter, $V(r_c)$, if $V(r)$ is known. Equation (3-6-8) applies at all energies for attractive potentials that cannot produce orbiting and predicts a monotonic decrease from a high value at low energy toward a limiting value of πr_c^2 at high energies. For an attractive potential that can lead to orbiting, (3-6-8) is valid only at high energies. Equation (3-6-6) should be used for the polarization potential at low energies. Equation (3-6-8) applies in repulsive potentials only to energies above a certain threshold value, which is equal to the potential energy evaluated at the radius r_c. For repulsive potentials there is a hyperbolic increase in the reaction cross section from zero at the threshold toward a limiting value of πr_c^2 at high energies. This model was developed by Present and has been applied by him and by Mason and Vanderslice to the solution of specific problems.[10]

The cross section expressed by (3-6-6) is a good approximation to the diffusion cross section for the polarization potential. The reason[11] for this is that the final directions of the spiraling orbits are very nearly random (see Fig. 4, Appendix II, Ref. 8), whereas the deflections associated with the nonspiraling orbits are so small that they make a negligible contribution to an integral of the

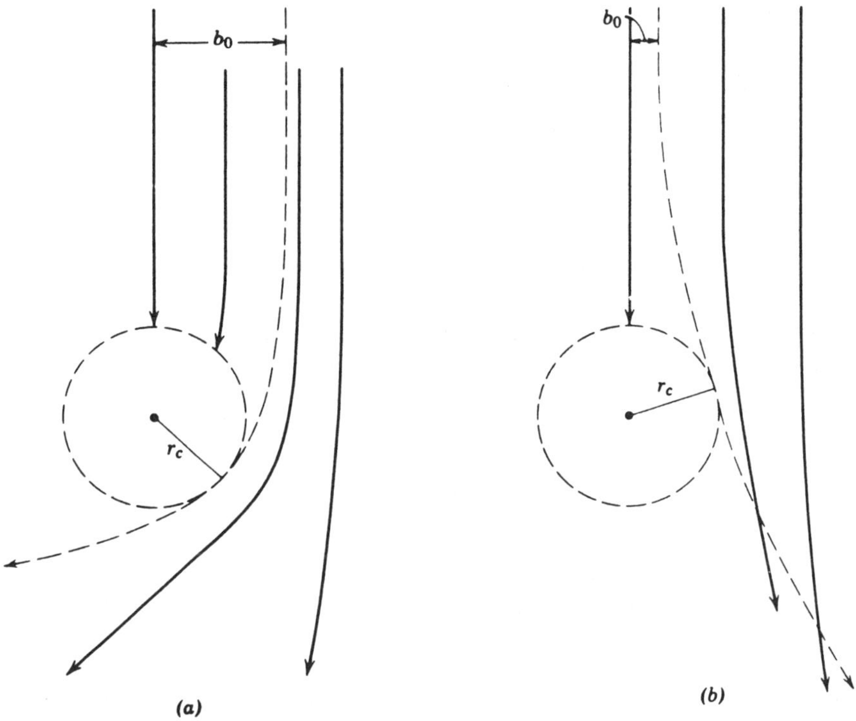

Figure 3-6-7. (*a*) Attractive potential; (*b*) repulsive potential. Trajectories for typical attractive and repulsive potentials corresponding to a given relative velocity of approach and various impact parameters. The interparticle separation distance decreases to a value less than the critical value r_c for any trajectory whose impact parameter is less than b_0. Each of these trajectories is assumed to lead to an ion-molecule reaction so that the reaction cross section is equal to πb_0^2.

type

$$q_D = \int (1 - \cos \Theta) I_s(\Theta)\, d\Omega_{\text{CM}}. \tag{1-6-1}$$

Vogt and Wannier[12] have also shown that the quantum mechanical description of the polarization potential is in many respects similar to the classical description.

3-7. EVALUATION OF THE SCATTERING CROSS SECTION

Let us now calculate the differential cross section $I_s(\Theta)\, d\Omega_{\text{CM}}$ for scattering into the element $d\Theta$ between the angles Θ and $\Theta + d\Theta$. The necessary and sufficient condition for scattering particles of initial relative velocity v_0 into the element of solid angle $d\Omega_{CM} = 2\pi \sin \Theta \, d\Theta$ is that the particles be incident in the annular impact ring of area $2\pi b \, db$ formed by circles of radius b and $b + db$, where b is related to Θ by (3-4-5). Hence

$$|2\pi b \, db| = |I_s(\Theta) 2\pi \sin \Theta \, d\Theta| \tag{3-7-1}$$

or

$$I_s(\Theta) \, d\Omega_{CM} = \left| \frac{b}{\sin \Theta} \frac{db}{d\Theta} \right| d\Omega_{CM}. \tag{3-7-2}$$

The differential cross section in the laboratory system, $I_s(\vartheta) \, d\Omega_{Lab}$, is given in terms of $I_s(\Theta) \, d\Omega_{CM}$ by (1-4-16), and the integral scattering cross section is obtained by integrating the differential cross section over the complete solid angle:

$$q_s = \int_0^{2\pi} \int_0^{\pi} I_s(\Theta) \, d\Omega_{CM} = \int_0^{2\pi} \int_0^{\pi} I_s(\vartheta) \, d\Omega_{Lab}. \tag{3-7-3}$$

These integrals diverge for potentials of infinite range, such as $V(r) \sim \pm r^{-n}$, because of a pole in the differential cross section in the forward direction, $\Theta = 0$. For potentials of this type, some deflection of the interacting particles occurs, no matter how large the impact parameter may be. Thus the effective cross sectional area of the particles appears to be infinite. This paradox is resolved in Section 3-11 by quantum mechanical considerations.

It can be seen from (3-7-2) that $I_s(\Theta)$ will become infinite if $\Theta = n\pi$ ($n = 0, 1, 2, \ldots$) for $b \neq 0$ because of the $\sin \Theta$ term in the denominator; it will also become infinite if $d\Theta/db = 0$. Both are possible for potentials that are attractive at large distances and repulsive at small distances. Both classical scattering phenomena have optical analogs* which become apparent in the quantum theory of scattering when the wave nature of the particles is explicitly introduced. The first effect is called a "glory" and the second, a "rainbow."[13]

3-8. COULOMB SCATTERING†

We now illustrate the use of the techniques developed above by deriving the differential cross section for *Coulomb scattering*. The potential that applies here is

$$V(r) = \frac{Zze^2}{r}. \tag{3-8-1}$$

*H. C. van de Hulst, *Light Scattering by Small Particles*, Wiley, New York (1957). R. Greenler, *Rainbows, Halos, and Glories*, Cambridge University Press, New York (1980).

†As the reader is doubtless aware, scattering by a Coulomb field is usually referred to as *Rutherford scattering* in nuclear physics. A clear and comprehensive treatment of Rutherford scattering appears in R. D. Evans, *The Atomic Nucleus*, McGraw-Hill, New York (1955), App. B.

It expresses the electrostatic interaction between particles of charge Ze and ze, separated by a distance r. Either charge may be positive or negative.

Again we introduce the variable $\rho = b/r$ and define a quantity α by the equation

$$\frac{V}{M_r v_0^2/2} = \alpha\rho, \tag{3-8-2}$$

so that

$$\alpha = \frac{2Zze^2}{M_r v_0^2 b}. \tag{3-8-3}$$

Equation (3-4-5) then becomes

$$\Theta = \pi - 2\int_0^{\rho_0} \frac{d\rho}{\sqrt{1 - \alpha\rho - \rho^2}}, \tag{3-8-4}$$

where ρ_0 is one of the roots of the equation

$$\rho^2 + \alpha\rho - 1 = 0. \tag{3-8-5}$$

The two roots of (3-8-5) are

$$\rho = -\frac{\alpha}{2} \pm \left(\frac{\alpha^2}{4} + 1\right)^{1/2}, \tag{3-8-6}$$

and since $\rho = b/r$ it is evident that the solution containing the positive square root must be chosen.

Integrating (3-8-4), we obtain

$$\Theta = \pi - 2\cos^{-1}\frac{\alpha}{\sqrt{\alpha^2 + 4}}, \tag{3-8-7}$$

so that

$$\frac{\alpha}{\sqrt{\alpha^2 + 4}} = \cos\left(\frac{\pi}{2} - \frac{\Theta}{2}\right) = \sin\frac{\Theta}{2}.$$

Thus

$$\alpha = 2\tan\frac{\Theta}{2} = \frac{2Zze^2}{M_r v_0^2 b}$$

and

$$b = \frac{Zze^2}{M_r v_0^2 \tan(\Theta/2)}. \tag{3-8-8}$$

The use of (3-7-2) then gives

$$I_s(\Theta)\, d\Omega_{\mathrm{CM}} = \left| \frac{b}{\sin\Theta} \frac{db}{d\Theta} \right| d\Omega_{\mathrm{CM}} = \frac{Z^2 z^2 e^4}{4M_r^2 v_0^4 \sin^4(\Theta/2)}\, d\Omega_{\mathrm{CM}}, \tag{3-8-9}$$

which is the desired result. It is interesting to note that precisely the same expression is obtained in the quantum mechanical treatment of Coulomb scattering (cf. Section 7-7). By dimensional analysis it can be shown[14] that if the interaction potential varies as $1/r^n$ the total scattering cross section varies as h^{2-2n}, where h is Planck's constant. Only for an inverse-first-power scattering potential does the cross section have no dependence on h, and therefore only in this case can the quantal description agree exactly with the classical description.

Because of the $\sin^4(\Theta/2)$ term in the denominator, the differential cross section expressed in (3-8-9) leads to infinite values for both the integral scattering and momentum transfer cross sections. The physical basis for the divergence of the integrals expressing these quantities is the infinite range of the Coulomb potential. This difficulty is circumvented in calculations of the properties of plasmas by the use of a "shielded" Coulomb potential, a stratagem that is not only mathematically essential but also physically realistic. This matter is discussed further in Section 3-17, in Appendix I, and in all books on plasmas. The shielded Coulomb potential also accurately represents the interaction between atoms colliding at high impact energies. The differential, momentum transfer, and integral collision cross sections for scattering of atomic projectiles by a Coulomb potential with exponential screening have been calculated by Everhart and his coworkers.[15] (See Section 9-13 of McDaniel.[1])

3-9. VARIATION OF THE DIFFERENTIAL SCATTERING CROSS SECTION WITH VELOCITY

It was stated in Section 1-8 that if the interaction potential between two colliding particles varies as $V(r) \sim r^{-n}$ the variation of the center-of-mass differential scattering cross section with velocity is given classically by the expression

$$I_s(\Theta) \sim v_0^{-4/n}, \tag{3-9-1}$$

where v_0 is the initial relative velocity of approach.* This statement is easily verified by dimensional considerations.

*The variation with scattering angle is also of interest. For a potential of the form $V(r) \sim r^{-n}$ classical theory predicts that $I_s(\Theta) \sim [\Theta^{(1+2/n)} \sin\Theta]^{-1}$ for small Θ [see E. H. Kennard, *Kinetic Theory of Gases*, McGraw-Hill, New York (1938), pp. 119–120.

Our starting point is (3-4-2), which now assumes the form

$$\frac{dr}{d\varphi} = \pm \frac{r^2}{b}\left(1 - \frac{k}{v_0^2 r^n} - \frac{b^2}{r^2}\right)^{1/2}. \tag{3-9-2}$$

All of the constants appearing in the middle term under the radical have been lumped into the single constant k. If b, v_0, and k are specified, the trajectories of the particles are determined. Let us now scale the velocity v_0 by a factor c and then scale the distances b and r appropriately to recover an orbit equation that is formally identical to the original equation (3-9-2). Evidently, if v_0 is scaled to give a new velocity $v_0^* = cv_0$, b and r must be replaced by b^* and r^*, defined by the equations $b^* = c^{-2/n}b$ and $r^* = c^{-2/n}r$. The new orbit equation is then

$$\frac{dr^*}{d\varphi} = \pm \frac{r^{*2}}{b^*}\left(1 - \frac{k}{v_0^{*2} r^{*n}} - \frac{b^{*2}}{r^{*2}}\right)^{1/2}, \tag{3-9-3}$$

and is seen to be formally identical to (3-9-2) so that the scattering angle Θ is the same as before. Since (3-7-2) shows that $I_s(\Theta)$ is proportional to b^2, it follows that

$$I_s(\Theta, cv_0) = c^{-4/n} I_s(\Theta, v_0). \tag{3-9-4}$$

Our proof is thus completed. Other proofs of the same result are given by Present and by Yang and Ree.[16]

As pointed out in Section 1-8, (3-9-1) yields the result that with elastic spheres, for which $n = \infty$, the scattering cross section is independent of the velocity. Hence the elastic sphere model is one of constant mean free path. On the other hand, $n = 4$ for particles interacting through the point-charge-induced dipole polarization potential, and here the differential scattering cross section varies inversely with the velocity. The collision frequency, which is proportional to $I_s(\Theta)v_0$, is then independent of v_0, and we have a constant mean free time situation. According to (3-9-1), the differential cross section is inversely proportional to v_0^4 for Coulomb scattering, in agreement with the result derived in Section 3-8.

As we shall see in Section 3-11, classical theory does not properly describe scattering at small angles in the CM system, and thus (3-9-1) cannot be used to obtain the velocity dependence of the integral scattering cross section q_s.* Quantum mechanics must be used for this purpose. Massey and Mohr[17] have obtained an approximate quantal solution for an interaction potential of the form $V(r) = -Cr^{-n}$. Their result is

$$q_s = B\left(\frac{C}{v_0}\right)^{2/(n-1)}, \tag{3-9-5}$$

*It does, however, give the correct dependence of the diffusion cross section on v_0 because the small-angle contributions are suppressed by the weighting factor $(1 - \cos\Theta)$ in the integral expressing q_D.

where B is a known constant. This equation is derived in Note 9-1 of McDaniel,[1] where a slightly more accurate result obtained by Landau and Lifshitz[17] is also discussed.

PART $\mathscr{B}$. THE QUANTUM THEORY OF ELASTIC SCATTERING

We now turn to the quantal view of elastic scattering. As in Part $\mathscr{A}$ of this chapter, we shall consider only nonrelativistic scattering in a spherically symmetric, time-independent potential field. [Scattering by a potential $V = V(r)$ is called "potential scattering."] After suitable preliminary discussion, we shall formulate the problem of elastic scattering in quantum mechanical terms and then develop methods for calculating the cross section for the case in which the collision partners are dissimilar. The mutual scattering of identical particles is then considered. The material presented here is based largely on treatments of the subject by Mott and Massey and by Burhop.[18] Other excellent sources of information are listed in Ref. 19.

3-10. THE INADEQUACY OF THE CLASSICAL TREATMENT OF SCATTERING

We may invoke the Heisenberg uncertainty principle to show how the classical theory fails, in general, to give an accurate description of collision processes. This principle applies to any pair of canonically conjugate variables* and states that the product of the uncertainties in the knowledge of the two variables (measured simultaneously) must be at least as great as Planck's constant, h, divided by 2π, that is, $\hbar = h/2\pi = 1.054 \times 10^{-27}$ erg-sec. Accordingly, if the pair of variables is chosen to be the X component of position and the X component of linear momentum, the following relationship holds:

$$\Delta x \cdot \Delta p_x \gtrsim \hbar. \tag{3-10-1}$$

Let us see what restrictions are thereby imposed on our knowledge of the dynamics of particles moving about in thermal equilibrium with a gas at room temperature. If we first take an argon atom and allow the uncertainty in its position along the X axis to be 10^{-8} cm, a distance approximately equal to an atomic diameter (see Table 2-2-1), we find that the X component of velocity

*Two dynamical variables are said to be *canonically conjugate* to one another if they satisfy a conjugate pair of Hamilton's "canonical equations of motion" (2-11-1). A given rectangular Cartesian spatial coordinate and the corresponding component of linear momentum evidently constitute such a pair of variables. The total energy E and the time t are also canonically conjugate; in the theory of relativity E is one component of the momentum four-vector and t plays the role of the corresponding spatial coordinate.

cannot be determined with a precision greater than $\Delta v_x \approx 1.59 \times 10^3$ cm/sec. A similar calculation for a hydrogen molecule yields $\Delta v_x \gtrsim 31.8 \times 10^3$ cm/sec, whereas an uncertainty of $\Delta x = 10^{-8}$ cm for an electron entails a minimum uncertainty in v_x of about 1.16×10^8 cm/sec. The mean thermal velocities of these particles at 15°C are 39.1×10^3, 174.0×10^3, and 1.05×10^7 cm/sec, respectively. It is apparent that the dynamics of a collision at thermal velocities are definitely hazy for an argon atom and much more so for a hydrogen molecule; for an electron the classical concept of a well-defined orbit is meaningless. At lower temperatures the classical treatment fails even more. We see that classical theory has some degree of validity in the analysis of collisions under "gas kinetic conditions" only for heavy particles or at high temperatures.

3-11. SCATTERING BY A CENTER OF FORCE OF INFINITE RANGE*

We have already remarked that an infinite value for q_s is predicted classically for scattering by any center of force of infinite range. The reason for the divergence of the cross section integral in this case is that some deflection, though perhaps small, is expected regardless of how large the impact parameter may be, and the sum of all of the contributions to the integral must therefore be infinite. In practice, the value for q_s deduced from a beam type of experiment (see Chapter 4) would then depend on the angular resolving power of the apparatus, that is, its ability to detect very small deflections of the particles from the beam. The concept of an integral scattering cross section would therefore be meaningless. Consideration of the uncertainty principle shows, however, that a finite value for q_s is expected provided the potential describing the interaction between projectile and target falls off sufficiently rapidly at large r. If the apparatus has a certain minimum resolving power, this finite value for q_s can, in principle, be determined experimentally. Let us explore this matter more fully.

In a quantal analysis of scattering the projectile is represented by a wave packet which spreads as it travels through space.[20] The quantal and classical descriptions of the scattering event are basically in accord if the following two conditions are met: (1) the de Broglie wavelength of the projectile must be small compared with the distance of closest approach to the target, and (2) the deflection of the projectile must not be obscured by the spread of the wave packet.

Consider the wave packet representing a projectile traveling with velocity v_0 in the Z direction toward a fixed target located at the origin of coordinates. The packet is assumed to be moving so that it will pass at distance y from the target if undeflected. The uncertainty principle states that the width of the wave packet, Δy, and the uncertainty in the transverse component of the projectile velocity,

*This discussion is based on Section 1 of the chapter by Burhop that appears in the book edited by Bates, Ref. 18.

Δv_y, are related by the equation

$$m\,\Delta y \cdot \Delta v_y \gtrsim \hbar, \tag{3-11-1}$$

where m is the mass of the projectile. The classical orbit concept will evidently be valid only if

$$y \gg \Delta y \tag{3-11-2}$$

or, according to (3-11-1), only if

$$y \gg \frac{\hbar}{m\,\Delta v_y}. \tag{3-11-3}$$

[Note that this inequality is not an explicit statement of condition (1), since the projectile wavelength is h/mv_0, not $\hbar/m\,\Delta v_y$.] The angle of spreading of the wave packet is approximately $\Delta v_y/v_0$, so that condition (2) is satisfied only if

$$\vartheta \gg \frac{\Delta v_y}{v_0}, \tag{3-11-4}$$

where ϑ is the angle of deflection.

If we eliminate Δv_y, we see that the collision may be described classically when*

$$\vartheta y \gg \frac{\hbar}{mv_0}. \tag{3-11-5}$$

The angle of deflection is given classically by $\vartheta \approx F_y\tau/mv_0$, where F_y is the Y component of the force on the projectile and τ is the "duration of the collision," which may be taken to be equal to y/v_0. If we put $F_y = |\partial V/\partial r|\,y/r$, we obtain $\vartheta y \approx |\partial V/\partial r|\,y^3/rmv_0^2$. Thus ϑy goes to zero as y goes to infinity (or ϑ goes to zero) if V falls off more rapidly than $1/r$ at large r. For such potentials the classical description is seen to fail at sufficiently small angles for any projectile wavelength h/mv_0. Deflections through angles smaller than some minimum angle $\vartheta_{\min} \approx \hbar/mv_0 y$ are not observable, and since it is these deflections that make q_s infinite classically, the quantum mechanical value for q_s may remain finite.† Improved resolution in the experimental apparatus does not lead to an indefinite increase in the measured value of the integral cross section.

*For further discussion of the criterion for the validity of the classical description of scattering, see Section 3-24.

†Actually, the integral cross section is finite only if $V(r)$ approaches zero more rapidly than $1/r^2$ at large r. The phase shifts (see Section 3-15) are finite if $V(r) \to 0$ faster than $1/r$, whereas the differential cross section at $\Theta = 0$ is bounded only if $V(r) \to 0$ faster than $1/r^3$. For proof of these statements, see E. H. S. Burhop (in Bates, Ref. 18, pp. 326–327).

A problem of special interest is the elastic scattering of an electron in the static field of an atom. To first approximation, if the perturbation of the atom by the electron is ignored, the mutual potential energy at a distance r from the center of the atom is

$$V(r) = -\frac{Ze^2}{r} + 4\pi e\left[\frac{1}{r}\int_0^r \rho(r')r'^2\,dr' + \int_r^\infty \rho(r')r'\,dr'\right], \qquad (3\text{-}11\text{-}6)$$

where Ze is the nuclear charge and ρ is the charge density of atomic electrons. This energy is usually expressed as

$$V(r) = \frac{-Z_p e^2}{r}, \qquad (3\text{-}11\text{-}7)$$

where Z_p is the effective nuclear charge. In general, Z_p falls off exponentially with r, for large r, and a finite value for q_s results if the problem is treated quantum mechanically.

3-12. REQUIREMENTS ON ANGULAR RESOLVING POWER FOR MEASUREMENTS OF THE INTEGRAL ELASTIC SCATTERING CROSS SECTION

Massey and Burhop[21] stress the point that care must be taken to ensure sufficient resolution for meaningful cross section measurement. To obtain 1% accuracy in the measured cross section for electrons elastically scattered against molecules, the apparatus must be able to sense deflections of as little as 11° for 1-eV electrons, 6.5° for 10-eV electrons, 2.3° for 100-eV electrons, 0.85° for 1000-eV electrons, and 0.2° for 10,000-eV electrons. These data apply to almost any type of target molecules.

The requirements are much more severe for atomic projectiles. Let ϑ_0 be the minimum angle of deflection to be counted as a collision if the error in the integral cross section is not to exceed 10%. For He projectiles on He targets ϑ_0 is calculated as 3.6° for an energy of 0.0255 eV, 2.0° for 0.0862 eV, and 0.59° for 1 eV. For argon atoms on argon the corresponding minimum angles are 0.70, 0.30, and 0.11°, respectively. (Here the angles and energies are measured in the laboratory system.) The resolving power required for integral cross section measurements increases roughly as the relative velocity of the collision partners, and for collisions of 100-eV protons with helium atoms deflections of less than 7 minutes of arc must be detectable in order that the measurement may be accurate within 10%.[21]

We are now in position to develop quantal expressions for the scattering cross section. The first step in this direction will be to show how the motion of the center of mass of the interacting particles may be separated from the total motion in the quantum mechanical treatment, as in the classical discussion.

3-13. SEPARATION OF THE CENTER-OF-MASS MOTION FROM THE TOTAL MOTION

The description of a pair of particles undergoing mutual scattering may be given quantum mechanically in terms of the *time-dependent* wave function $\Psi(\mathbf{R}_m, \mathbf{R}_M, t)$, where $\mathbf{R}_m = (X_m, Y_m, Z_m)$ and $\mathbf{R}_M = (X_M, Y_M, Z_M)$ are the laboratory position vectors of the particles (of mass m and M, respectively) and t is the time. The physical significance of this wave function is that $|\Psi(\mathbf{R}_m, \mathbf{R}_M, t)|^2\, d\tau_m\, d\tau_M$ is the probability that at time t particle m will be inside the volume element $d\tau_m$ at $\mathbf{R}_m$ and M will be within the volume element $d\tau_M$ at $\mathbf{R}_M$. This wave function obeys the time-dependent Schrödinger wave equation

$$-i\hbar\frac{\partial\Psi}{\partial t} = \frac{\hbar^2}{2m}\nabla_m^2\Psi + \frac{\hbar^2}{2M}\nabla_M^2\Psi - V(\mathbf{R}_m, \mathbf{R}_M)\Psi, \tag{3-13-1}$$

where

$$\nabla_m^2 = \frac{\partial^2}{\partial X_m^2} + \frac{\partial^2}{\partial Y_m^2} + \frac{\partial^2}{\partial Z_m^2}, \tag{3-13-2}$$

$$\nabla_M^2 = \frac{\partial^2}{\partial X_M^2} + \frac{\partial^2}{\partial Y_M^2} + \frac{\partial^2}{\partial Z_M^2}, \tag{3-13-3}$$

and $V(\mathbf{R}_m, \mathbf{R}_M)$ is the mutual potential energy of the system when m is at $\mathbf{R}_m$ and M is at $\mathbf{R}_M$. Because of the central force assumption, $V = V(|\mathbf{R}_m - \mathbf{R}_M|)$.

Now let us put

$$(m + M)\mathbf{R} = m\mathbf{R}_m + M\mathbf{R}_M \tag{3-13-4}$$

and

$$\mathbf{r} = \mathbf{R}_m - \mathbf{R}_M, \tag{3-13-5}$$

where $\mathbf{R} = (X, Y, Z)$ is the position vector of the center of mass of the two particles in the laboratory system and $\mathbf{r} = (x, y, z)$ is the vector joining the particles and pointing from M toward m. Then, since

$$\frac{\partial}{\partial X_m} = \frac{\partial}{\partial x} + \frac{m}{m + M}\frac{\partial}{\partial X}$$

and

$$\frac{\partial}{\partial X_M} = -\frac{\partial}{\partial x} + \frac{M}{m + M}\frac{\partial}{\partial X}$$

with similar relationships for the other coordinates and second derivatives, the operator

$$\frac{\hbar^2}{2m}\nabla_m^2 + \frac{\hbar^2}{2M}\nabla_M^2$$

transforms into the operator

$$\frac{\hbar^2}{2M_t}\nabla_R^2 + \frac{\hbar^2}{2M_r}\nabla_r^2,$$

where $M_t = m + M$ is the total mass and M_r is the reduced mass. Upon this transformation, the wave equation (3-13-1) becomes

$$-i\hbar\frac{\partial\Psi}{\partial t} = \frac{\hbar^2}{2M_t}\nabla_R^2\Psi + \frac{\hbar^2}{2M_r}\nabla_r^2\Psi - V(r)\Psi, \tag{3-13-6}$$

where $\Psi = \Psi(\mathbf{r}, \mathbf{R}, \mathbf{t})$.

We may now separate variables and obtain a solution of the form

$$\Psi(\mathbf{r}, \mathbf{R}, t) = F(\mathbf{r}, t)G(\mathbf{R}, t). \tag{3-13-7}$$

If we substitute (3-13-7) into (3-13-6), we obtain the equations

$$-i\hbar\frac{\partial F}{\partial t} = \frac{\hbar^2}{2M_r}\nabla^2 F - V(r)F + CF, \tag{3-13-8}$$

$$-i\hbar\frac{\partial G}{\partial t} = \frac{\hbar^2}{2M_t}\nabla^2 G - CG, \tag{3-13-9}$$

where C is a constant. Then if we put

$$F(\mathbf{r}, t) = f(\mathbf{r}, t)e^{iCt/\hbar} \tag{3-13-10}$$

and

$$G(\mathbf{R}, t) = g(\mathbf{R}, t)e^{-iCt/\hbar}, \tag{3-13-11}$$

the following equations result:

$$-i\hbar\frac{\partial f(\mathbf{r}, t)}{\partial t} = \frac{\hbar^2}{2M_r}\nabla^2 f(\mathbf{r}, t) - V(r)f(\mathbf{r}, t), \tag{3-13-12}$$

$$-i\hbar\frac{\partial g(\mathbf{R}, t)}{\partial t} = \frac{\hbar^2}{2M_t}\nabla^2 g(\mathbf{R}, t). \tag{3-13-13}$$

Thus we see that the wave function may be expressed as

$$\Psi(\mathbf{r}, \mathbf{R}, t) = f(\mathbf{r}, t)g(\mathbf{R}, t), \tag{3-13-14}$$

where f and g satisfy (3-13-12) and (3-13-13).

Equation (3-13-12) is the wave equation that describes the behavior of the interacting particles in the center-of-mass system. As in the classical case, this motion may be regarded as that of a particle of mass M_r about a fixed scattering center. Equation (3-13-13) describes the motion of the center of mass of the particles in terms of the motion of a free particle of mass M_t in the laboratory system.

3-14. THE QUANTUM MECHANICAL FORMULATION OF THE SCATTERING PROBLEM

From the foregoing discussion we see that we are free to formulate the scattering problem in terms of the behavior of hypothetical particles of mass M_r interacting with a fixed center of force. In the classical case we were able to determine the exact trajectory of each projectile in terms of its initial velocity v_0 and impact parameter b (or, alternatively, in terms of its total energy E and angular momentum J). According to the general principles of quantum mechanics, however, the projectile orbit is not precisely defined, and we are able to calculate only the average behavior of a large number of projectiles in their interaction with the target.

A. The Incident Beam and Its Wave Function

Let us imagine an infinitely wide beam of structureless projectiles approaching along the $-Z$ axis a structureless scattering center fixed at the origin of coordinates. A spherically symmetric force field is assumed, so that the potential energy of interaction V is a function only of the separation distance r. The incident beam is to be homogeneous and monoenergetic, and the particles in the beam are to be dissimilar to the target particle representing the scattering center. (Identical projectiles and targets require special handling in quantum mechanics and are treated separately in Section 3-18.) Furthermore, the beam intensity is to be uniform across any plane normal to the Z axis and is to be constant in time. To each projectile is ascribed a fictitious mass $M_r = mM/(m + M)$, where m and M are the true masses of the projectile and target, respectively. As is well known, the quantal representation of such a beam is a plane de Broglie wave of constant amplitude traveling in the $+Z$ direction. Let us suppose that the current density of the beam is A particles/cm^2-sec, so that A particles pass through unit area normal to the beam direction per second. Some of these projectiles will experience scattering by the center of force, the number of scattered particles being directly proportional to the integral elastic scattering cross section

$$q_s = \int_0^{\pi} \int_0^{2\pi} I_s(\Theta)\, d\Omega_{\text{CM}} = 2\pi \int_0^{\pi} I_s(\Theta) \sin\Theta\, d\Theta. \tag{3-14-1}$$

We wish to determine the differential cross section $I_s(\Theta)\, d\Omega_{\text{CM}}$, appearing in (3-14-1), as a function of angle and energy, and then obtain q_s as a function of energy by integration.

For quantitative expression of the intensity of scattering in any direction (Θ, φ) let us refer to Fig. 3-14-1. In this figure dS represents a small surface at a large distance r, in the direction (Θ, φ), from the origin. It is assumed to be normal to the radius vector. The solid angle subtended at the origin by dS is $d\Omega_{\text{CM}}$. From our previous discussion of the significance of the differential cross section [cf. (1-4-1)] we see that $AI_s(\Theta)\, d\Omega_{\text{CM}}$ equals the number of projectiles scattered through dS per second.

The plane wave representing the incident beam of particles is described by a wave function $Ce^{i(\kappa z - \omega t)}$, where C is the amplitude, κ the wave number, and ω the angular frequency of the wave; κ is given in terms of the de Broglie wavelength λ, the particle velocity v_0, or the initial kinetic energy T_0 by the

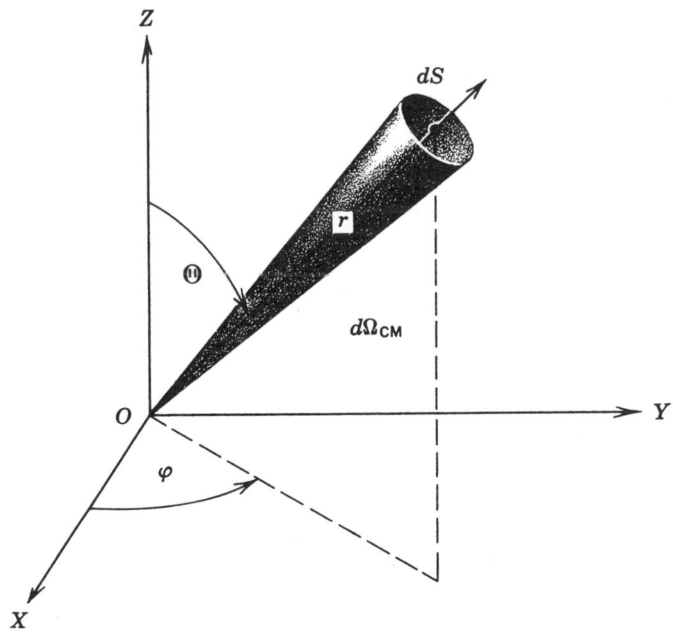

Figure 3-14-1. The spherical polar coordinate system used in discussion of the elastic scattering problem. The base of the cone is the surface dS referred to in the text.

equation

$$\kappa = \frac{2\pi}{\lambda} = \frac{M_r v_0}{\hbar} = \frac{\sqrt{2M_r T_0}}{\hbar} \tag{3-14-2}$$

and ω is given in terms of the frequency ν, or the total energy E, by the equation

$$\omega = 2\pi\nu = \frac{E}{\hbar}. \tag{3-14-3}$$

The scattering problem we have formulated is a steady-state problem, and the energy of the incident de Broglie wave is unchanged in the process of elastic scattering. Consequently, the term $e^{-i\omega t}$ is common to all the wave functions we shall write down, so that we may drop this common time factor and indicate only the spatial dependence of the wave functions.* The *time-independent* wave function for the incoming plane wave is denoted by $\psi_{\text{inc}}(r, \Theta) = Ce^{i\kappa z}$. Note that dependence of ψ_{inc} on φ is not implied, since all aspects of the problem are independent of the azimuth angle.

We have taken the current density of the incoming beam to be A projectiles/cm^2-sec. The current density may also be expressed in terms of C, the amplitude of the wave function, and thus we may obtain the relationship between A and C. We know that the average number of particles expected to be present in unit volume at a given location in the incoming beam is given by $\psi_{\text{inc}}^*\psi_{\text{inc}} = |\psi_{\text{inc}}|^2 = C^2$, where ψ_{inc}^* is the complex conjugate of ψ_{inc}. Each of these particles is traveling toward the origin along the $-Z$ axis with velocity v_0. If we now imagine a right circular cylinder of length v_0 and unit cross-sectional area to be laid along the $-Z$ axis, it is evident that this cylinder will contain $C^2 v_0$ projectiles at any instant and that within the next second just these projectiles, and no more, will pass through the right-hand end of the cylinder. The current density is then, by definition, equal to $C^2 v_0$, and A and C are thus related by the equation

$$C^2 = \frac{A}{v_0}. \tag{3-14-4}$$

B. The Scattered and Total Wave Functions

The total wave function will consist of a component associated with the incoming plane wave and another representing that portion of the incident wave which experiences scattering. The scattered wave must have the form of an outgoing spherical wave whose amplitude decreases, for large r, as $1/r$, in order

*The formal separation of the space and time dependence may be accomplished by the method used in Section 3-20.

that the radial current density may fall off as the inverse square of the distance from the center of force and that the number of scattered particles may thus be conserved. The wave function for the scattered wave will then have the asymptotic form

$$\psi_{\text{scatt}} \approx \frac{C}{r} f(\Theta) e^{i\kappa r}, \tag{3-14-5}$$

where $f(\Theta)$ is usually called the *scattered amplitude*. We shall now establish the relation of $f(\Theta)$ to $I_s(\Theta)$, the differential elastic scattering cross section per unit solid angle. Equation (3-14-5) shows that the radial current density is equal to $v_0|C|^2|f(\Theta)|^2/r^2$ for large r, and thus the number of projectiles scattered per second through the surface element dS at (Θ, φ) in Fig. 3-14-1 is

$$\frac{v_0|C|^2}{r^2} |f(\Theta)|^2 \, dS = v_0|C|^2 |f(\Theta)|^2 \, d\Omega_{\text{CM}} = A|f(\Theta)|^2 \, d\Omega_{\text{CM}}.$$

The number of particles scattered into $d\Omega_{\text{CM}}$ per unit time is also known to equal $AI_s(\Theta) \, d\Omega_{\text{CM}}$. It then follows that

$$I_s(\Theta) = |f(\Theta)|^2; \tag{3-14-6}$$

hence it is not necessary to know $\psi(r, \Theta)$ completely to determine the differential or integral cross section—we need know only the asymptotic form of the scattered wave function. (In the foregoing discussion only the scattered wave function was used to calculate the number of particles passing through dS per second. The incident beam is assumed to be collimated in such a way that its unscattered component is prevented from reaching dS, which also is assumed to be at a great distance from the scattering center. Such collimation is always provided in experimental apparatus.)

To obtain $f(\Theta)$, and thus the cross sections, we must solve the time-independent wave equation for the motion of particles of mass M_r and positive total energy ($E > 0$) in the potential field of the fixed scattering center. The solution is required to have the asymptotic form

$$\psi = \psi_{\text{inc}} + \psi_{\text{scatt}} \approx e^{i\kappa z} + \frac{e^{i\kappa r}}{r} f(\Theta), \tag{3-14-7}$$

where we have set the amplitude of the incoming wave equal to unity for convenience. The wave number κ was expressed in terms of the initial kinetic energy by (3-14-2). However, T_0 equals the total energy because the mutual potential energy of the particles is zero at very large r, and we may write

$$\kappa = \frac{\sqrt{2M_r E}}{\hbar}. \tag{3-14-8}$$

3-15. SOLUTION OF THE WAVE EQUATION BY THE METHOD OF PARTIAL WAVES

The time-independent wave equation is

$$\nabla^2\psi + \frac{2M_r}{\hbar^2}[E - V(r)]\psi = 0 \tag{3-15-1}$$

where, in spherical coordinates, the Laplacian operator is

$$\nabla^2 = \frac{1}{r^2}\frac{\partial}{\partial r}\left(r^2\frac{\partial}{\partial r}\right) + \frac{1}{r^2 \sin\Theta}\frac{\partial}{\partial\Theta}\left(\sin\Theta\frac{\partial}{\partial\Theta}\right) + \frac{1}{r^2\sin^2\Theta}\frac{\partial^2}{\partial\varphi^2}. \tag{3-15-2}$$

The last term in this expression vanishes in the present problem because of the lack of dependence of any of our functions on φ. Let us put

$$U(r) = \frac{2M_r V(r)}{\hbar^2}, \tag{3-15-3}$$

so that the wave equation may be written in the convenient form

$$\nabla^2\psi + [\kappa^2 - U(r)]\psi = 0. \tag{3-15-4}$$

We see a solution of (3-15-4) that is everywhere bounded, continuous, and single-valued and that has the asymptotic form indicated in (3-14-7).

A. Separation of the Wave Equation

Since by assumption V depends only on r, the wave equation may be separated into two equations, one of which contains only the variable r and the other only Θ. (In the more general case in which there is dependence on the azimuthal angle we would have a third equation in φ.) To perform this separation, let us write

$$\psi(r, \Theta) = L(r)Y(\Theta) \tag{3-15-5}$$

and substitute this expression into (3-15-4). The result may be written in the form

$$\frac{1}{L}\frac{d}{dr}\left(r^2\frac{dL}{dr}\right) + r^2[\kappa^2 - U(r)] = -\frac{1}{Y}\left[\frac{1}{\sin\Theta}\frac{d}{d\Theta}\left(\sin\Theta\frac{dY}{d\Theta}\right)\right]. \tag{3-15-6}$$

Note that the left side of (3-15-6) depends only on r and the right side, only on Θ. If both sides are to be equal for all values of r and Θ, both must be equal to some constant, which we shall denote as $l(l+1)$. Thus

$$\frac{1}{r^2}\frac{d}{dr}\left(r^2\frac{dL}{dr}\right)+\left[\kappa^2-U(r)-\frac{l(l+1)}{r^2}\right]L=0 \tag{3-15-7}$$

and

$$\frac{1}{\sin\Theta}\frac{d}{d\Theta}\left(\sin\Theta\frac{dY}{d\Theta}\right)+l(l+1)Y=0. \tag{3-15-8}$$

Equation 3-15-8 is a special case of Legendre's equation. Because it is a second-order equation it has two linearly independent solutions, each of which may be expressed as a power series in $\cos\Theta$. Both solutions become infinite for $\Theta = 0$ unless l is zero or a positive integer, and therefore the only physically acceptable solutions of the wave equation are those corresponding to $l = 0, 1, 2, 3, \ldots$. As we shall soon see, l is a measure of the angular momentum of the projectile about the fixed scattering center. For this reason l is called the *angular momentum quantum number*. The first few acceptable solutions of (3-15-8) are

$$P_0(\cos\Theta) = 1,$$

$$P_1(\cos\Theta) = \cos\Theta,$$

$$P_2(\cos\Theta) = \tfrac{1}{2}(3\cos^2\Theta - 1),$$

$$P_3(\cos\Theta) = \tfrac{1}{2}(5\cos^3\Theta - 3\cos\Theta).$$

The functions $P_l(\cos\Theta)$ are the well-known Legendre polynomials.

B. Expansion of the Wave Functions in Partial Waves

We now see that the desired wave function may be written in the form

$$\psi(r,\Theta)=\sum_{l=0}^{\infty}A_lP_l(\cos\Theta)L_l(r), \tag{3-15-9}$$

where the A_l are arbitrary constants and the $L_l(r)$ are solutions of (3-15-7) for particular values of l. The terms in the foregoing sum are known as *partial waves*. Now let us assume that $U(r)$ falls off faster than $1/r$ at large r and that if $U(r)$ has a pole at the origin it is not of higher order than $1/r$. Then there are two independent solutions of (3-15-7), one of which is finite at the origin, the other being infinite there.[22] The constants A_l must be chosen so that (3-15-9) represents the sum of an incoming plane wave and an outgoing spherical scattered wave. The requirement that the solution be bounded everywhere necessitates the choice of that solution, $L_l(r)$, of (3-15-7) that is finite at the origin.

We can simplify the radial wave equation by introducing the functions $G_l(r)$ defined by the equation

$$L_l(r) = \frac{G_l(r)}{r}. \tag{3-15-10}$$

The $G_l(r)$ satisfy the equation

$$\frac{d^2 G_l(r)}{dr^2} + \left[\kappa^2 - U(r) - \frac{l(l+1)}{r^2}\right] G_l(r) = 0. \tag{3-15-11}$$

The last two terms in the square brackets in (3-15-11) tend to zero at large r; hence we would expect the asymptotic form of any solution of (3-15-11) to be $\sin(\kappa r + \epsilon_l)$, where ϵ_l is a constant. To see whether this expectation is correct, set

$$G_l(r) = u_l(r) e^{i\kappa r}. \tag{3-15-12}$$

Substitution of (3-15-12) into (3-15-11) gives the following expression for $u_l(r)$:

$$\frac{d^2 u_l}{dr^2} + 2i\kappa \frac{du_l}{dr} - \left[U(r) + \frac{l(l+1)}{r^2}\right] u_l = 0. \tag{3-15-13}$$

For large r, $u_l(r)$ will be nearly constant, and d^2u_l/dr^2 will be much smaller than $\kappa(du_l/dr)$. Neglecting d^2u_l/dr^2, we can integrate (3-15-13) and obtain

$$2i\kappa \log u_l = \int^r \left[U(r) + \frac{l(l+1)}{r^2}\right] dr + \text{constant}.$$

The integral converges if and only if $U(r)$ falls off faster than $1/r$ for large r, and $u_l(r)$ thus approaches a constant value as r approaches infinity for a potential of this form. This observation justifies the assumption made that $u_l(r)$ is slowly varying. Accordingly, by expressing the exponential in (3-15-12) in trigonometric form we see that for potentials that fall off faster than r^{-1} for large r, $G_l(r)$ has the asymptotic form

$$G_l(r) \approx \sin(\kappa r + \epsilon_l), \tag{3-15-14}$$

where ϵ_l is a constant.

The solution of (3-15-7) that is finite at the origin will then have the asymptotic form

$$L_l(r) \approx \frac{1}{\kappa r} \sin\left(\kappa r - \frac{l\pi}{2} + \eta_l\right), \tag{3-15-15}$$

where η_l is a constant for given κ and $U(r)$ called the lth-*order phase shift*; η_l is the phase shift of the lth partial wave due to the action of the scattering potential. The term $-l\pi/2$ is inserted into (3-15-15) so that η_l will be zero if $U(r)$ is zero.

We must now determine the constants A_l in the expansion (3-15-9) so that the total wave function ψ will have the form indicated in (3-14-7). To do this, we require the expansion of the incoming plane wave $e^{i\kappa z}$ in partial waves:[23]

$$\psi_{\text{inc}} = e^{i\kappa z} = e^{i\kappa r \cos\Theta} = \sum_{l=0}^{\infty} (2l+1) i^l P_l(\cos\Theta) j_l(\kappa r). \qquad \text{(3-15-16)}$$

Here $j_l(\kappa r)$ is the spherical Bessel function which is defined in terms of the ordinary Bessel function of half-integral order by the equation

$$j_l(\kappa r) \equiv \left(\frac{\pi}{2\kappa r}\right)^{1/2} J_{l+1/2}(\kappa r). \qquad \text{(3-15-17)}$$

The values of the first few spherical Bessel functions are

$$j_0 = \frac{\sin \kappa r}{\kappa r}, \qquad \text{(3-15-18)}$$

$$j_1 = \frac{\sin \kappa r}{(\kappa r)^2} - \frac{\cos \kappa r}{\kappa r}, \qquad \text{(3-15-19)}$$

$$j_2 = \left[\frac{3}{(\kappa r)^3} - \frac{1}{\kappa r}\right] \sin \kappa r - \frac{3}{(\kappa r)^2} \cos \kappa r \qquad \text{(3-15-20)}$$

and the recurrence relation among these functions is

$$j_{l+1}(\kappa r) = \frac{2l+1}{\kappa r} j_l(\kappa r) - j_{l-1}(\kappa r). \qquad \text{(3-15-21)}$$

The asymptotic values of the spherical Bessel functions for large κr are

$$j_l(\kappa r) \approx \frac{1}{\kappa r} \sin\left(\kappa r - \frac{l\pi}{2}\right), \qquad \text{(3-15-22)}$$

whereas for small κr,

$$j_l(\kappa r) \approx \frac{(\kappa r)^l}{1 \times 3 \times 5 \cdots (2l+1)}. \qquad \text{(3-15-23)}$$

The asymptotic form of the incident plane wave for large r is then seen to be

$$\psi_{\text{inc}} = e^{i\kappa z} \approx \sum_{l=0}^{\infty} (2l + 1)i^l P_l(\cos \Theta) \frac{\sin(\kappa r - l\pi/2)}{\kappa r}. \tag{3-15-24}$$

If we subtract this expression from the asymptotic expression for the total wave function

$$\psi \approx \sum_{l=0}^{\infty} A_l P_l(\cos \Theta) \frac{\sin(\kappa r - l\pi/2 + \eta_l)}{\kappa r}, \tag{3-15-25}$$

we obtain for the scattered wave

$$\psi_{\text{scatt}} = \psi - \psi_{\text{inc}} \approx \sum_{l=0}^{\infty} \frac{1}{\kappa r} P_l(\cos \Theta) \left[A_l \sin\left(\kappa r - \frac{l\pi}{2} + \eta_l\right) - (2l + 1)i^l \sin\left(\kappa r - \frac{l\pi}{2}\right)\right]. \tag{3-15-26}$$

This wave function must represent an outgoing spherical wave. When (3-15-26) is expressed in terms of exponential functions by use of the equation $\sin x = (e^{ix} - e^{-ix})/2i$, the term in $e^{-i\kappa r}/r$ which appears must vanish, since it would represent an incoming spherical wave. This requirement determines the values of the constants A_l. If we express the square bracket in (3-15-26) in exponential form, we obtain

$$\frac{1}{2i} e^{i(\kappa r - l\pi/2)} [A_l e^{i\eta_l} - i^l(2l + 1)] - \frac{1}{2i} e^{-i(\kappa r - l\pi/2)} [A_l e^{-i\eta_l} - i^l(2l + 1)].$$

Evidently, it is then required that

$$A_l = (2l + 1)i^l e^{i\eta_l}, \tag{3-15-27}$$

and the total wave function is seen to be

$$\psi(r, \Theta) = \sum_{l=0}^{\infty} (2l + 1)i^l e^{i\eta_l} L_l(r) P_l(\cos \Theta). \tag{3-15-28}$$

C. The Scattering Cross Section

According to (3-14-7), the scattered wave function is $(e^{i\kappa r}/r)f(\Theta)$. The right side of (3-15-26) can have this form only if

$$f(\Theta) = \frac{1}{2i\kappa} \sum_{l=0}^{\infty} (2l + 1)(e^{2i\eta_l} - 1) P_l(\cos \Theta). \tag{3-15-29}$$

Note that $f(\Theta)$ is complex, having the form

$$f(\Theta) = A + iB, \tag{3-15-30}$$

where

$$A = \frac{1}{2\kappa} \sum_{l=0}^{\infty} (2l + 1) \sin 2\eta_l P_l(\cos \Theta) \tag{3-15-31}$$

$$B = \frac{1}{2\kappa} \sum_{l=0}^{\infty} (2l + 1)(1 - \cos 2\eta_l) P_l(\cos \Theta) \tag{3-15-32}$$

and i is the imaginary operator. Thus we see that

$$I_s(\Theta) = |f(\Theta)|^2 = A^2 + B^2 = \frac{1}{\kappa^2} \left| \sum_{l=0}^{\infty} (2l + 1) e^{i\eta_l} \sin \eta_l P_l(\cos \Theta) \right|^2. \tag{3-15-33}$$

Integration of $I_s(\Theta)$ over the complete solid angle gives for the integral elastic scattering cross section

$$q_s = \frac{4\pi}{\kappa^2} \sum_{l=0}^{\infty} (2l + 1) \sin^2 \eta_l. \tag{3-15-34}$$

To obtain (3-15-34), we use the well-known relationship

$$\int_0^\pi P_m(\cos \Theta) P_n(\cos \Theta) \sin \Theta \, d\Theta = \frac{2}{2m + 1} \delta_{mn}, \tag{3-15-35}$$

where δ_{mn}, the Kronecker delta, equals unity if $m = n$ and zero if $m \neq n$.

The integral cross section is also directly related to $f(0)$, the scattered amplitude in the forward direction $\Theta = 0$. Since $P_l(\cos 0) = 1$ for all l,

$$f(0) = \frac{1}{2i\kappa} \sum_{l=0}^{\infty} (2l + 1)(e^{2i\eta_l} - 1).$$

Thus

$$f(0) - f^*(0) = \frac{1}{2i\kappa} \sum_{l=0}^{\infty} (2l + 1)(e^{2i\eta_l} - 1) - \frac{1}{2i\kappa} \sum_{l=0}^{\infty} (2l + 1)(e^{-2i\eta_l} - 1)$$

$$= \frac{i\kappa q_s}{2\pi},$$

so that

$$q_s = \frac{2\pi}{i\kappa} [f(0) - f^*(0)] = \frac{4\pi}{\kappa} \mathscr{I}[f(0)], \tag{3-15-36}$$

where the symbol $\mathscr{I}$ denotes the imaginary component. This result is known as the *optical theorem*, a more general form of which (that includes inelastic contributions to the scattering) appears in Note 3-3.

It must be emphasized that the expressions for the differential and integral cross sections developed here apply only to collisions between dissimilar particles. The expressions applicable to collisions between identical particles are developed in Section 3-18. Furthermore, it must be remembered that (3-15-29) for the scattered amplitude is valid only for potentials that decrease faster than $1/r$ for large r.

The method of partial waves had its origin in Rayleigh's analysis of the scattering of sound waves by spherical obstacles.[24] The partial wave technique was first applied to atomic collisions by Faxén and Holtsmark.[25]

D. The Relation between the Classical Impact Parameter and the Partial Waves

In the classical treatment of scattering of projectiles from a beam incident on a fixed target particle the projectiles approach the scattering center with a random distribution of impact parameters. In a spherically symmetric force field all projectiles within the annular impact ring of inner radius b and outer radius $b + db$ are scattered between angles Θ and $\Theta + d\Theta$. The scattering angle is calculated as a function of the impact parameter (3-4-5), and the statistical angular scattering distribution is then determined by (3-7-2).

Classically, the angular momentum of the system may be written in terms of the impact parameter as

$$J = M_r v_0 b. \qquad (3\text{-}2\text{-}11)$$

The angular momentum might also be expressed as

$$J = L\hbar, \qquad (3\text{-}15\text{-}37)$$

where L may be zero or have *any* positive value. The impact parameter may thus be expressed in terms of the angular momentum by the equation

$$b = \frac{L\hbar}{M_r v_0} = L\lambda\!\!\!^{-}, \qquad (3\text{-}15\text{-}38)$$

where $\lambda\!\!\!^{-}$ denotes the de Broglie wavelength of the projectile divided by 2π. We may now visualize the incident beam as being divided into shells by cylinders that are coaxial with the line corresponding to a head-on trajectory. The radii of these cylinders are to be $l\lambda\!\!\!^{-} = 0, \lambda\!\!\!^{-}, 2\lambda\!\!\!^{-}, 3\lambda\!\!\!^{-}, \ldots$, so that the inner radius of the lth shell is $l\lambda\!\!\!^{-}$ and its outer radius is $(l + 1)\lambda\!\!\!^{-}$. Classically, all projectiles whose impact parameters lie between $l\lambda\!\!\!^{-}$ and $(l + 1)\lambda\!\!\!^{-}$ would approach the target within the lth shell, and the angular momentum of these particles would lie in the range $l\hbar$ to $(l + 1)\hbar$.

In quantum mechanics, however, the angular momentum is quantized, and it is characterized by the number $l = 0, 1, 2, 3, \ldots$, where l is now to be regarded as a quantum number. Furthermore, a group of projectiles of angular momentum quantum number l does not approach the target with a well-defined impact parameter but rather must be viewed quantally as moving mostly in the lth shell, as defined above. According to the general principles of quantum mechanics, the angular momentum of these particles is of *magnitude* $\sqrt{l(l+1)}\hbar$.* Now let us examine the expansion of the incoming wave in terms of the partial waves:

$$e^{i\kappa z} = \sum_{l=0}^{\infty} (2l+1) i^l P_l(\cos\Theta) j_l(\kappa r). \tag{3-15-16}$$

The first few spherical Bessel functions that constitute the radial part of this expansion are plotted in Fig. 3-15-1. From this figure it is seen that the first and largest maximum of $j_l(\kappa r)$ occurs near $\kappa r \equiv r/\lambda\!\!\!^{-} \approx 1.5l$ and consequently, most of

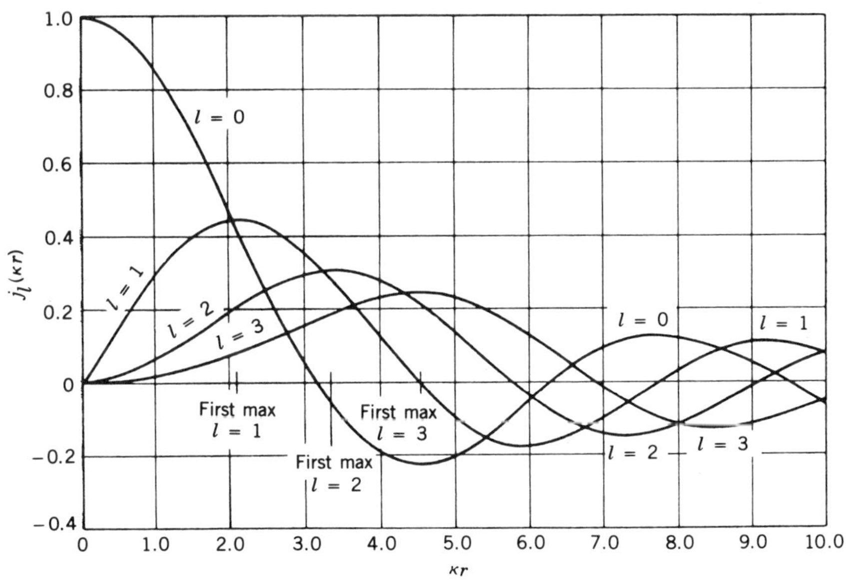

Figure 3-15-1. The first four spherical Bessel functions.

*We may show that $J = \sqrt{l(l+1)}\hbar$ from equations that we have already derived. Equation (3-6-1) expresses the classical centrifugal potential in terms of the angular momentum as $J^2/2M_r r^2$. This quantity represents the rotational kinetic energy of the system. The term $l(l+1)/r^2$ appearing in the radial wave equation (3-15-7) has the same dimensions as $U(r)$ and thus must represent an energy when multiplied by the factor $\hbar^2/2M_r$. The energy it represents is also evidently the rotational kinetic energy of the system. Thus, if the two terms representing this energy are equated, it follows that $J = \sqrt{l(l+1)}\hbar$.

the projectiles whose angular momentum quantum number is l will be found somewhere within the shell whose radii are $l\lambda\!\!\!^{-}$ and $(l + 1)\lambda\!\!\!^{-}$.

Spectroscopic notation is used to designate the various values of l. Thus $l = 0$ collisions are called "s-wave collisions," $l = 1$ collisions are "p-wave collisions," $l = 2$ collisions are "d-wave collisions," and so on. All wave functions associated with the s wave are independent of angle, and s-wave scattering is consequently spherically symmetric about the scattering center.

It is interesting to note that the cross-sectional areas of the shells into which the incoming beam is divided are given by

$$\pi(l + 1)^2\lambda\!\!\!^{-2} - \pi l^2\lambda\!\!\!^{-2} = \pi(2l + 1)\lambda\!\!\!^{-2}.$$

This fact accounts for the weighting factor $2l + 1$ appearing in the terms of the expansion (3-15-16). The area $\pi(2l + 1)\lambda\!\!\!^{-2}$ represents the upper limit on the absorption cross section* for particles in the lth partial wave, since no more particles can be removed from the beam than were in it originally. Equation (3-15-34) indicates, however, that the maximum cross section for elastic scattering of projectiles in the lth partial wave is $4\pi(2l + 1)\lambda\!\!\!^{-2}$. This apparently paradoxical result is due to the diffraction of the de Broglie waves representing the beam.[26]

E. The Physical Significance of the Phase Shifts as Illustrated by s-Wave Scattering by a Spherical Potential Well

We may illustrate the origin and significance of the phase shifts by solving the simple problem of the s-wave scattering of projectiles by a spherical potential well. Such a well is illustrated in Fig. 3-15-2 and described by the equations

$$V(r) = \begin{cases} -V_0 & (r < D) \\ 0 & (r > D). \end{cases} \tag{3-15-39}$$

If the radius D of the well is small compared with the rationalized de Broglie wavelength $\lambda\!\!\!^{-}$ of the incident projectiles, few projectiles with nonzero angular momentum can reach the edge of the potential well. Consequently, only s-wave collisions are effective in the scattering, and we need consider only the first partial wave, corresponding to $l = 0$. There are many physical problems in which s-wave scattering predominates, and the problem treated here is one of real significance.

If we consider only the s wave and assume its amplitude to be unity, the wave function for the incident plane wave, is, by (3-15-16),

$$\psi_{\text{inc}} = e^{i\kappa z} \approx \frac{\sin \kappa r}{\kappa r}, \tag{3-15-40}$$

*An example of the absorption of particles from a beam is the attachment by gas molecules of electrons in a beam passing through the gas.

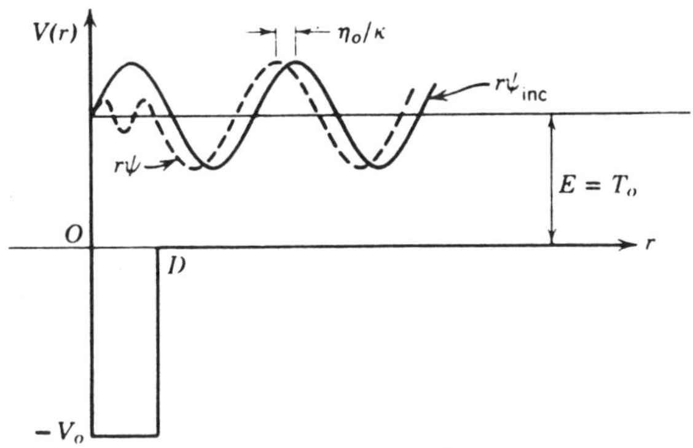

Figure 3-15-2. Scattering of *s* wave by a spherical potential well of radius D and depth V_0. The incident and total wave functions are multiplied by r and plotted along the horizontal axis corresponding to $V(r) = E = T_0$; $V(r)$ is the potential energy of the projectile, E is its total energy, and T_0 is its kinetic energy when $r > D$, that is, when the projectile is outside the well.

where κ, the wave number of the incident wave, is given in terms of the initial kinetic energy T_0 by the equation

$$\kappa = \left(\frac{2M_r T_0}{\hbar^2}\right)^{1/2}. \tag{3-14-8}$$

The total wave function is the sum of the wave function for the incident wave and that for the outgoing spherical scattered wave:

$$\psi = e^{i\kappa z} + f_0 \frac{e^{i\kappa r}}{r}. \tag{3-15-41}$$

Here f_0 is the amplitude of the scattered *s* wave, given by (3-15-29) to be

$$f_0 = \frac{1}{2i\kappa}(e^{2i\eta_0} - 1), \tag{3-15-42}$$

where η_0 is the phase shift for the $l = 0$ wave. Using (3-15-25) and (3-15-27), we find the asymptotic form of the total wave function to be

$$\psi \approx \frac{e^{i\eta_0}}{\kappa r} \sin(\kappa r + \eta_0), \tag{3-15-43}$$

and the integral elastic scattering cross section, according to (3-15-34), is

$$q_s = \frac{4\pi}{\kappa^2} \sin^2 \eta_0. \tag{3-15-44}$$

We must now evaluate the phase shift for the s wave in terms of the potential. Note that (3-15-11) for the modified radial wave function $G(r)$ must be satisfied both inside and outside the well. Thus

$$\frac{d^2G}{dr^2} + [\kappa^{*2} - U(r)]G = 0, \tag{3-15-45}$$

where κ^* is the wave number appropriate for the region under discussion. Inside the well the solution for $G(r)$ that vanishes at the origin is

$$G(r) = A \sin \kappa_i r, \tag{3-15-46}$$

so

$$\psi_{\text{inside}} = \frac{G(r)}{r} = \frac{A}{r} \sin \kappa_i r, \tag{3-15-47}$$

where κ_i, the wave number for $r < D$, is given by

$$\kappa_i = \left[\frac{2M_r(E - V)}{\hbar^2}\right]^{1/2} = \left[\frac{2M_r(E + V_0)}{\hbar^2}\right]^{1/2}. \tag{3-15-48}$$

The wave function (3-15-43) outside the well must now be matched to the function (3-15-47) inside the well. The standard boundary conditions are to be applied at the edge; that is, ψ and $d\psi/dr$ must be continuous at $r = D$. Then

$$\frac{e^{i\eta_0}}{\kappa} \sin(\kappa D + \eta_0) = A \sin \kappa_i D, \tag{3-15-49}$$

$$e^{i\eta_0} \cos(\kappa D + \eta_0) = A\kappa_i \cos \kappa_i D. \tag{3-15-50}$$

Dividing the first of these equations by the second, we eliminate A and obtain

$$\tan(\kappa D + \eta_0) = \frac{\kappa}{\kappa_i} \tan \kappa_i D \tag{3-15-51}$$

or

$$\eta_0 = -\kappa D + \tan^{-1}\left(\frac{\kappa}{\kappa_i} \tan \kappa_i D\right). \tag{3-15-52}$$

Equation (3-15-52) gives the phase shift in terms of the radius of the well and the wave numbers inside and outside the well; η_0 may also be expressed in terms of the total energy of the projectile and the depth and radius of the well, since

$$E = \frac{\kappa^2 \hbar^2}{2M_r} \tag{3-15-53}$$

and

$$V_0 = \frac{(\kappa_i^2 - \kappa^2)\hbar^2}{2M_r}. \tag{3-15-54}$$

Equation (3-15-52) shows that in general the phase shift η_0 goes to zero as the velocity of the incident particle goes to zero. The integral scattering cross section given by (3-15-44) then generally has a finite limit for very slow projectiles:

$$q_s \xrightarrow[\kappa \to 0]{} 4\pi D^2 \left(\frac{\tan \kappa_0 D}{\kappa_0 D} - 1 \right)^2, \tag{3-15-55}$$

where

$$\kappa_0^2 \equiv \kappa_i^2 - \kappa^2 = \frac{2M_r V_0}{\hbar^2}. \tag{3-15-56}$$

The following exceptional cases should be noted, however. If $\kappa_0 D$ happens to equal $\tan \kappa_0 D$, the limit in (3-15-55) vanishes as the velocity goes to zero. In addition, for $\kappa_0 D = \pi/2, 3\pi/2, 5\pi/2, \ldots$, the phase shift does not go to zero with κ, and the integral scattering cross section goes to infinity as the velocity approaches zero. The infinite values for q_s which arise in the latter cases are associated with the existence of allowed energy levels within the potential well.[27]

The incident and total wave functions, multiplied by r, are shown in Fig. 3-15-2 plotted against r. The curve labeled by $r\psi_{\text{inc}}$ represents the incident wave outside the range of the interaction and its uniform extension into the well, assuming that the potential is removed. Note that the amplitude and wavelength of the solid curve are constant. The dashed curve, on the other hand, represents the actual wave function, multiplied by r, as determined by the action of the potential well. Within the well the amplitude of this curve is generally smaller than that of the unmodified curve, and its wavelength, $\lambda_i = 2\pi/\kappa_i$, is always shorter because the kinetic energy of the projectile increases when it enters the well. It is apparent that outside the well there must be a difference in phase between the modified and unmodified waves. The phase difference between the two waves outside the well is equal to the phase shift η_0, which may be represented in Fig. 3-15-2 as κ times the distance between corresponding crests of the modified and unmodified waves. For attractive potentials, such as assumed in the present example, the phase shift is positive, and the incoming wave may be regarded as *drawn in* toward the scattering center. Repulsive potentials, however, lead to a negative η_0, and the incident wave may then be thought of as *pushed out* from the scattering center.

The scattering that occurs when the amplitude of the wave function inside the potential well is very small compared with that outside the well is called *potential scattering*. It usually appears when a wave of short wavelength inside the well is to be joined to one of long wavelength outside. However, there may be certain values for the kinetic energy of the incident projectile for which the derivative of the wave function is almost zero at the edge of the well, and then the interior and exterior wave functions may be joined with substantially equal amplitudes. The scattering in this case is termed *resonance scattering*, and the cross section assumes a large value. The resonance effect is very sharp and can usually be ignored in the analysis of atomic collisions. Figure 3-15-3 shows wave functions schematically for the cases of pure potential scattering, scattering near a resonance, and pure resonance scattering.[28]

F. Calculation of the Phase Shifts

Equation (3-15-29) for the scattered amplitude $f(\Theta)$ was derived on the assumption that the scattering potential $V(r)$ tends to zero faster than $1/r$. It was not necessary to assume that $V(r)$ would become negligible outside some finite radius D, but the method of partial waves is most easily applied and most useful when such a radius exists. We have pointed out that the first and largest maximum of the radial wave function $j_l(\kappa r)$ occurs near $r = 1.5l\lambda\!\!\!^{-}$, and for r much smaller than this value $j_l(\kappa r)$ is small and varies about as r^l, according to (3-15-23). If the radius beyond which the potential is negligible is small compared with $l\lambda\!\!\!^{-}$, the lth partial wave function will be small in the region in which the potential is significant, and the lth partial wave will not be appreciably affected by the

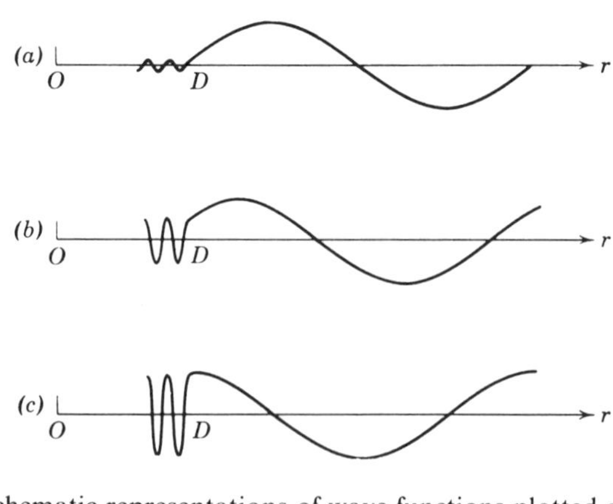

Figure 3-15-3. Schematic representations of wave functions plotted against the distance from the center of a potential well. Curve (a) corresponds to a projectile energy between resonances; curve (b), to an energy near resonance; and curve (c), to a resonance energy.

scatterer. This means that the phase shift associated with this wave, and consequently the contribution of this wave to the cross section, will be small. Therefore, we see that only those partial waves associated with values of l extending from zero to some maximum value of the order of $D/\lambda\!\!\!^{-} = D\kappa$ need be considered in the calculation of cross sections. Since κ is proportional to the square root of the incident energy, it then follows that fewer phase-shift calculations are required for low-energy projectiles and that the method of partial waves is most useful for low energies.

Several methods are available for calculation of the phase shifts. Clearly, by solving the differential equation (3-15-11) for $G_l(r)$, subject to the appropriate boundary conditions and comparing the solution to that when $V(r) = 0$, we can always find η_l numerically. In addition, several approximate methods are frequently used. In particular, we cite the *Born* and *Jeffreys* approximations to the phase shifts. The Born is useful when the phase shift is small and the Jeffreys when η_l is not small. Variational methods have also proved valuable. All of the methods mentioned here are treated in detail by Bransden,[29] Bates,[30] and Joachain,[31] and in fact in most books on atomic collision theory. It should be pointed out that the Jeffreys approximation is also referred to as the Wentzel-Kramers-Brillouin (W-K-B), the classical, the phase integral, and the asymptotic approximation.

3-16. THE BORN APPROXIMATION

An approximate method, due to Born,[32] for the calculation of cross sections is of great importance in the study of atomic collisions. This method is valid only under certain conditions but has the virtue of permitting the cross sections to be calculated with much less labor than is required by the exact method of partial waves. We now derive the Born expression for the scattered amplitude. The reader is referred to references cited in the last section for discussion of the Born approximation to the phase shifts.

The basic assumption in the derivation is that the effect of the scattering potential is small, so that the interaction between the particles may be treated as a perturbation. A sufficient, but not necessary, condition, therefore, is that $V(r)$ be much smaller than E, the total energy of the projectiles.* This condition is evidently always obeyed for sufficiently fast projectiles. Our starting point is the wave equation

$$\nabla^2\psi + [\kappa^2 - U(r)]\psi = 0 \qquad (3\text{-}15\text{-}4)$$

and we seek a solution having the asymptotic form, for large r,

$$\psi \approx e^{i\kappa z} + \frac{e^{i\kappa r}}{r} f(\Theta). \qquad (3\text{-}14\text{-}7)$$

*A thorough discussion of the conditions for validity of the Born approximation is given by Bransden.[19]

Mott and Massey[33] show that the most general bounded solution of the equation

$$\nabla^2\psi + \kappa^2\psi = F(x, y, z) \tag{3-16-1}$$

is

$$\psi = G(x, y, z) - \frac{1}{4\pi}\int \frac{e^{i\kappa|\mathbf{r}-\mathbf{r}'|}}{|\mathbf{r}-\mathbf{r}'|} F(\mathbf{r}')\, d\tau', \tag{3-16-2}$$

where G is the general solution of the equation

$$\nabla^2 G + \kappa^2 G = 0. \tag{3-16-3}$$

Therefore, the general solution of the wave equation (3-15-4) is

$$\psi = G - \frac{1}{4\pi}\int \frac{e^{i\kappa|\mathbf{r}-\mathbf{r}'|}}{|\mathbf{r}-\mathbf{r}'|} U(r')\psi(\mathbf{r}')\, d\tau'. \tag{3-16-4}$$

In order that this solution may have the proper asymptotic form (3-14-7), we must set $G = e^{i\kappa z}$.

Now let us assume that the scattering potential falls off sufficiently rapidly at large r' that there is an asymptotic region in which r is large in comparison with those values of r' for which there is a significant contribution to the integrand. If we let $\mathbf{n}$ represent a unit vector in the direction of $\mathbf{r}$ (see Fig. 3-16-1), we have

$$|\mathbf{r}-\mathbf{r}'| \approx r - \mathbf{n}\cdot\mathbf{r}' + \cdots \tag{3-16-5}$$

and (3-16-4) then becomes

$$\psi \approx e^{i\kappa z} - \frac{e^{i\kappa r}}{r}\frac{1}{4\pi}\int e^{-i\kappa\mathbf{n}\cdot\mathbf{r}'} U(r')\psi(\mathbf{r}')\, d\tau'. \tag{3-16-6}$$

Under the assumption that the scattering is weak, we may suppose that each point in the scattering region has a wave of full strength incident upon it. Thus we may replace the wave function $\psi(\mathbf{r}')$ in (3-16-6) by the unperturbed wave function $e^{i\kappa z'}$. Comparison of (3-14-7) and (3-16-6) then shows that

$$f(\Theta) = -\frac{1}{4\pi}\int e^{i\kappa(\mathbf{n}_0-\mathbf{n})\cdot\mathbf{r}'} U(r')\, d\tau', \tag{3-16-7}$$

where $\mathbf{n}_0$ is a unit vector along the Z axis.

We may easily evaluate the integral in (3-16-7) by introducing the angles α and β, where α is the polar angle measured from the direction of the vector

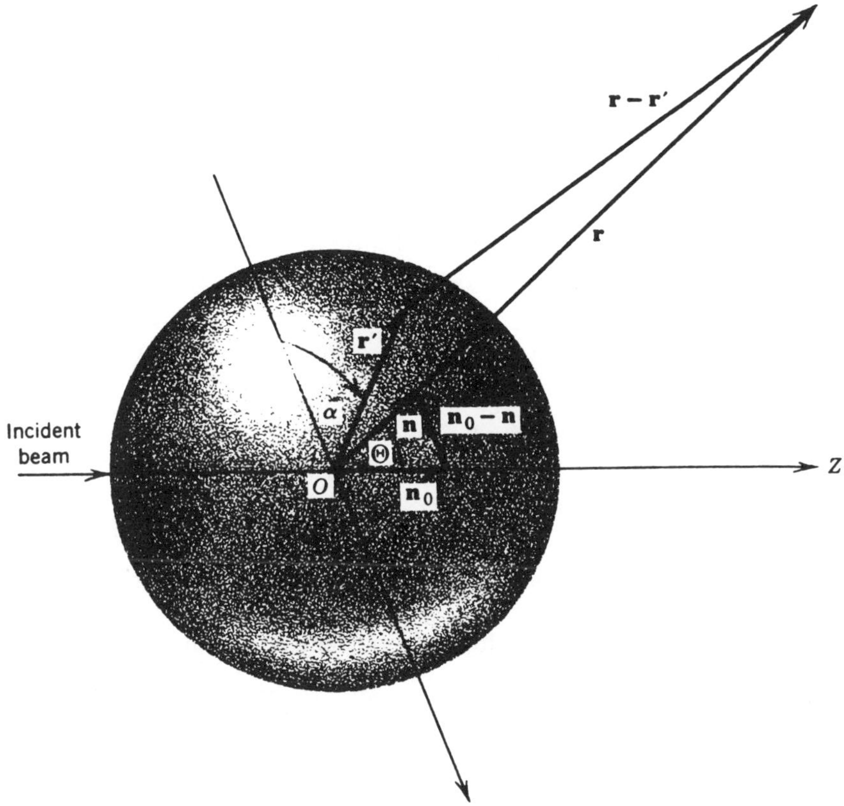

Figure 3-16-1. The geometric relationships among the vectors used in computing the elastic scattering of particles in the Born approximation. The scattering center is located at O and is effectively localized within the spherical region shown shaded: $\mathbf{r}$ is the vector from O to the point at which the amplitude of the scattered wave is to be computed; $\mathbf{r}'$ is the vector from O to a point contributing to the scattering; α is the polar angle between $\mathbf{r}'$ and the axis through O parallel to the vector $\mathbf{n}_0 - \mathbf{n}$. The corresponding azimuthal angle β is not shown.

$\mathbf{n}_0 - \mathbf{n}$ and β is the corresponding azimuthal angle. Then

$$f(\Theta) = -\frac{1}{4\pi}\int_0^{2\pi} d\beta \int_0^{\pi} \sin\alpha\, d\alpha \int_0^{\infty} e^{iKr'\cos\alpha} U(r') r'^2\, dr', \qquad (3\text{-}16\text{-}8)$$

where

$$K \equiv \kappa\,|\mathbf{n}_0 - \mathbf{n}| = 2\kappa \sin\frac{\Theta}{2}. \qquad (3\text{-}16\text{-}9)$$

K is the magnitude of the momentum transferred during the collision divided by

$\hbar$. If we perform the integrations over α and β and drop the dashes, we obtain

$$f(\Theta) = -\frac{2M_r}{\hbar^2}\int_0^\infty \frac{\sin Kr}{Kr} V(r) r^2 \, dr, \tag{3-16-10}$$

which is the *Born approximation for the scattered amplitude* $f(\Theta)$.

The approximation derived in this section is sometimes known as the *first Born approximation.* The method used here can be continued to higher approximations by an iterative procedure. For example, if

$$\psi = e^{i\kappa z} - \frac{1}{4\pi}\int \frac{e^{i\kappa|\mathbf{r}-\mathbf{r}'|}}{|\mathbf{r}-\mathbf{r}'|} U(r') e^{i\kappa z'} \, d\tau' \tag{3-16-11}$$

is substituted for ψ in the equation

$$f(\Theta) = -\frac{1}{4\pi}\int e^{-i\kappa\mathbf{n}\cdot\mathbf{r}'} U(r')\psi(\mathbf{r}') \, d\tau', \tag{3-16-12}$$

the *second Born approximation for the scattered amplitude* is obtained (Section 3-23-A).

Although the formalism developed in this chapter applies directly only to elastic scattering, both the Born approximation and the method of partial waves can be used in the study of inelastic collisions as well. Examples of these applications are discussed in Chapters 5 and 6.

3-17. COULOMB SCATTERING IN THE BORN APPROXIMATION

In Coulomb scattering, for which the potential is

$$V(r) = \frac{Zze^2}{r} \tag{3-17-1}$$

the integral in (3-16-10) becomes $\int_0^\infty \sin Kr \, dr$, which is mathematically meaningless. To avoid this difficulty, let us assume for the moment that the potential has the form

$$V(r) = \frac{Zze^2}{r} e^{-\gamma r}, \tag{3-17-2}$$

where γ is a positive quantity. Then the integral in (3-16-10) becomes

$$\int_0^\infty \sin Kr \, e^{-\gamma r} \, dr = \frac{K}{K^2 + \gamma^2}.$$

If we now let $\gamma \to 0$, this expression goes to $1/K$, and then $f(\Theta)$ becomes

$$f(\Theta) = -\frac{2Zze^2M_r}{\hbar^2K^2}.$$

Since $K = 2\kappa \sin(\Theta/2)$, the scattered amplitude can be written

$$f(\Theta) = -\frac{Zze^2M_r}{2\hbar^2\kappa^2}\frac{1}{\sin^2(\Theta/2)}.$$

Now $\hbar\kappa$ is the linear momentum $M_r v_0$, so that we finally obtain for the differential scattering cross section

$$I_s(\Theta)\,d\Omega_{\mathrm{CM}} = \frac{Z^2z^2e^4}{4M_r^2v_0^4\sin^4(\Theta/2)}\,d\Omega_{\mathrm{CM}}. \tag{3-17-3}$$

This equation, even though obtained by an approximation method, is exactly the same as the classical result (3-8-9). It is also in exact agreement with the result obtained in an exact quantum mechanical calculation,[34] as shown in Section 7-7.

In the foregoing derivation the inclusion of the $e^{-\gamma r}$ term to obtain the "screened Coulomb potential" in (3-17-2) was simply an *ad hoc* step of mathematical expediency. It is worth pointing out, however, that screened Coulomb potentials are accurate representations of the true potentials in certain types of physical problems. For example, (3-17-2) gives the mutual potential energy of two charged particles in a plasma if $1/\gamma$ is taken to be the debye length of the plasma.* In this case the exponential term takes into account the shielding effect of charged particles in the neighborhood of a given ion or electron in the plasma. In atomic physics, if $1/\gamma$ is taken to be of the order of atomic dimensions, (3-17-2) may represent the potential energy of an ion or electron in the vicinity of a neutral atom. Here the screening factor allows for the shielding of the nucleus of the atom by its orbital electrons.

3-18. THE SCATTERING OF IDENTICAL PARTICLES[35]

Up to this point no mention has been made of the symmetry effects that appear when the colliding particles are identical, and we must now modify the equations for the scattering cross section to take these effects into account. Their physical basis is the fact that it is impossible to distinguish experimentally between identical projectile and target particles, and the wave function describing the system must therefore satisfy certain symmetry properties with respect to

*See Appendix I.

the interchange of the coordinates of the two particles. The wave function, in fact, must be symmetric or antisymmetric with respect to the interchange of the particles, depending on whether their total (combined) spin is even or odd, that is, an even or odd multiple of $\hbar$.

In order to justify the statement concerning the symmetry character of the wave function, we reason as follows. Because the collision partners are indistinguishable by assumption, the states of the system obtained from each other by simply interchanging the particles must be completely equivalent physically. This means that the wave function describing the system can change only by an unimportant phase factor as a result of a single interchange. Let $\psi(1,2)$ represent the spatial part of the wave function of the system* before the interchange occurs. The numbers 1 and 2 refer to the two particles composing the system. The new wave function describing the system after the particles are interchanged must be given, according to what has been said, by the equation

$$\psi(2,1) = e^{ia}\psi(1,2),$$

where a is a real constant. If the interchange is repeated, we must recover the original wave function, that is,

$$\psi(1,2) = e^{2ia}\psi(1,2).$$

Thus $e^{2ia} = 1$ and $e^{ia} = \pm 1$, so that

$$\psi(2,1) = \pm\psi(1,2).$$

We see that the wave function must either be unchanged or simply undergo a change of sign if the two particles are interchanged. In the first case the wave function is said to be symmetric, whereas in the second it is antisymmetric.

An interchange of the positions of the particles is equivalent to a reversal of the direction of the radius vector joining them. In the center-of-mass coordinate system r remains unchanged in this reversal, whereas the angle Θ becomes $\pi - \Theta$ and z becomes $-z$, since $z = r\cos\Theta$. Therefore, if we assume that the interaction potential is a function only of r, the proper asymptotic expression for the spatial part of the wave function is now

$$\psi \approx e^{i\kappa z} \pm e^{-i\kappa z} + \frac{e^{i\kappa r}}{r}[f(\Theta) \pm f(\pi - \Theta)] \qquad (3\text{-}18\text{-}1)$$

(with the positive sign to be used to obtain the symmetric wave function and the

*When the spin is considered, the total wave function is expressed as a product of two functions, one of which depends only on the spatial coordinates and the other on the spin. The statement above concerning the symmetry or antisymmetry of the wave function refers to the spatial part of the total wave function.

negative sign, the antisymmetric wave function) instead of

$$\psi \approx e^{i\kappa z} + \frac{e^{i\kappa r}}{r} f(\Theta), \tag{3-14-7}$$

which is correct in the absence of symmetry. The wave function in (3-18-1) contains two incident plane waves of equal amplitude traveling in opposite directions, and the outgoing spherical wave takes into account the scattering of both waves. The differential cross section per unit solid angle is given by the square of the coefficient of the term $e^{i\kappa r}/r$ and is related to the probability that *either* of the particles will be scattered into the element of solid angle at Θ. Thus, if the total spin of the particles is even, the differential cross section per unit solid angle is

$$(I_s)_{\text{sym}} = |f(\Theta) + f(\pi - \Theta)|^2 \tag{3-18-2}$$

and for odd total spin

$$(I_s)_{\text{anti}} = |f(\Theta) - f(\pi - \Theta)|^2. \tag{3-18-3}$$

The appearance of the interference term $[f(\Theta)f^*(\pi - \Theta) \pm f^*(\Theta)f(\pi - \Theta)]$ is characteristic of quantum mechanics. This term does not appear in the classical treatment of scattering, in which the cross section per unit solid angle would be given simply by

$$(I_s)_{\text{class}} = |f(\Theta)|^2 + |f(\pi - \Theta)|^2. \tag{3-18-4}$$

In the foregoing discussion it was assumed that the total spin of the particles had some definite value, but we must as a rule consider a distribution among the possible spin states. To determine a cross section, we shall assume that all the spin states are equally probable. It can be shown that the total number of spin states is $(2s + 1)^2$ for two particles of spin s and that $s(2s + 1)$ of these states correspond to an even total spin and $(s + 1)(2s + 1)$ to an odd total spin, if s is half-integral. If s is integral, the weighting is reversed. Thus, if s is half-integral, the probability that the system will have even total spin is $s(2s + 1)/(2s + 1)^2 = s/(2s + 1)$, and the probability of odd total spin is $(s + 1)/(2s + 1)$. The differential cross section per unit solid angle is therefore

$$(I_s)_{\text{FD}} = \frac{s}{2s + 1}(I_s)_{\text{sym}} + \frac{s + 1}{2s + 1}(I_s)_{\text{anti}} \tag{3-18-5}$$

for half-integral s. By a similar analysis we can show that for integral s the appropriate expression is

$$(I_s)_{\text{BE}} = \frac{s}{2s + 1}(I_s)_{\text{anti}} + \frac{s + 1}{2s + 1}(I_s)_{\text{sym}}. \tag{3-18-6}$$

The subscripts FD and BE in (3-18-5) and (3-18-6) are abbreviations for *Fermi-Dirac* and *Bose-Einstein.* In relativistic quantum mechanics it is shown that particles with half-integral spin obey Fermi-Dirac statistics, hence (3-18-5) applies to the scattering of particles of this type, which are called fermions. Equation 3-18-6 applies to particles of integral spin, which obey Bose-Einstein statistics and are called bosons. Electrons, positrons, protons, neutrons, and nuclei of odd mass number are fermions, and nuclei of even mass number (such as deuterons and alpha particles) are bosons.

It is now of interest to write down the expressions for the cross section in terms of the phase shifts for particles obeying these types of statistics. For distinguishable particles we know that

$$I_s(\Theta) = \frac{1}{4\kappa^2}\left|\sum_{l=0}^{\infty}(2l+1)(e^{2i\eta_l}-1)P_l(\cos\Theta)\right|^2, \tag{3-18-7}$$

which is just the square of the expression given in (3-15-29) for the scattered amplitude. Two changes are required, however, when we write the corresponding expressions for identical particles. First we must multiply I_s by a factor of 2, since we cannot distinguish between the scattered waves representing the two collision partners. Second, assuming for the moment that the total spin has a definite even value, terms of odd l must be excluded from the sum because $P_l(\cos\Theta)$ is odd for all odd l. Thus for even total spin

$$(I_s)_{\text{sym}} = \frac{1}{2\kappa^2}\left|\sum_{\text{even } l}(2l+1)(e^{2i\eta_l}-1)P_l(\cos\Theta)\right|^2 \tag{3-18-8}$$

and similarly, for odd total spin

$$(I_s)_{\text{anti}} = \frac{1}{2\kappa^2}\left|\sum_{\text{odd } l}(2l+1)(e^{2i\eta_l}-1)P_l(\cos\Theta)\right|^2. \tag{3-18-9}$$

Equations (3-18-5) and (3-18-6) are to be used with these equations to obtain the scattering distributions when there is a mixture of spin states.

Examples of symmetry effects in the elastic scattering of identical particles are presented in Section 7-8 of this volume and in Sections 9-11, 9-12, and 9-13 of McDaniel.[1]

PART $\mathscr{C}$. THE QUANTUM THEORY OF INELASTIC COLLISIONS

The objective of this discussion is to indicate some of the basic concepts and methods employed in the quantum theory of inelastic collisions. Our treatment here is comparable in depth to the discussion of the quantum theory of elastic scattering, just concluded. Emphasis is laid on the fundamental aspects of the theory, and some of the discussion is only qualitative. The reader should consult the standard general references[18,19,36–38] for more detailed and comprehensive treatment of this vast and difficult subject. Quantal and classical approximation methods will be discussed in Part $\mathscr{D}$.

3-19. TIME-INDEPENDENT AND TIME-DEPENDENT DESCRIPTIONS OF SCATTERING

Either time-independent or time-dependent theory may be used to discuss atomic collision problems. Here we shall outline the basic approach followed in each of these formulations, although in succeeding sections of this chapter we shall use only the time-independent theory. In Chapter 8 time-dependent theory is applied to the discussion of photoionization.

The starting point in the *time-independent* quantum theory of atomic collisions is the time-independent Schrödinger wave equation for the system of colliding particles. In elastic scattering the scattered amplitude $f(\Theta)$ and differential cross section $I_s(\Theta)\,d\Omega_{CM}$ are obtained in terms of the phase shifts η_l (see Section 3-15-C) or the scattering matrix $\boldsymbol{S}$.* In the more general inelastic collisions the scattering is described by an integral equation, and the cross section can again be expressed in terms of the $\boldsymbol{S}$ matrix. In either case the total Hamiltonian H of the system is assumed to approach an unperturbed Hamiltonian H_0 when the collision partners are widely separated, and correspondingly the eigenstates of H are assumed to approach the eigenstates of H_0 when the particles are far apart. However, the perturbed eigenfunctions are not unique, and boundary conditions must be applied to obtain solutions for the specific problem at hand. Thus the "stationary-state method" described here involves the solution of a boundary-value problem. The method is not applicable if the Hamiltonian depends explicitly on the time.

In those problems for which the Hamiltonian is *time dependent* the system is described by the time-dependent wave equation

$$i\hbar\frac{\partial\Psi(\mathbf{r},t)}{\partial t} = H(\mathbf{r},t)\Psi(\mathbf{r},t), \tag{3-19-1}$$

where $H(\mathbf{r},t)$ is the Hamiltonian operator, $\Psi(\mathbf{r},t)$ is the time-dependent wave function of the system, and $\mathbf{r}$ denotes the coordinates of all the particles composing the system under consideration. (The symbol $\mathbf{r}$ is not meant to represent a vector here.) An example of a problem in which the Hamiltonian depends explicitly on the time is the calculation of the transition probabilities for an atom in an external, time-varying electric field. Now we have an initial-value problem, since in order to calculate the probability that the system will undergo

*The scattering matrix is used in Section 4-8-B. It is discussed in detail by Child,[13] Bransden,[19] Schiff,[19] Taylor,[19] Newton,[19] and Mott and Massey.[36]

a transition from one state to another we must solve (3-19-1) subject to the initial condition that Ψ has a specified value at some time $t = t_0$. Approximation methods must be used.[13,18,19,36,37] The most important of these, *Dirac's method of variation of constants*, may be applied if the Hamiltonian operator can be expressed in the form

$$H(\mathbf{r}, t) = H_0(\mathbf{r}) + V(\mathbf{r}, t) \tag{3-19-2}$$

and if V is small enough compared to H_0 to be treated as a small perturbation. A solution may be possible even if V is not small compared with H_0, provided that V changes fast enough, in which case the *sudden approximation* may be used, or slowly enough, in which case the *adiabatic approximation* is appropriate.*

Time-dependent methods may also be used in scattering problems even if the Hamiltonian does not explicitly depend on the time. Let us suppose that the collision partners are infinitely far apart at $t \to -\infty$ and that the system is in an initial state $\Psi_a(-\infty)$, which is an eigenstate of H_0, to which the Hamiltonian H tends as $t \to -\infty$. Then, as the time progresses in the positive direction and the particle separation decreases to some finite value, the interaction V becomes effective, and the wave function of the system evolves with time according to (3-19-1). After the encounter the interaction V again decreases to zero as the collision partners in their final states recede from one another toward infinite separation. At $t \to +\infty$, the system will be in some final state $\Psi_b(+\infty)$, and the problem is to find the relation between Ψ_a and Ψ_b. We see that time-dependent theory may be used in the analysis of scattering problems because even if V does not contain the time explicitly the interaction perturbation is transient in the sense that it starts from zero initially, rises to some finite value, and then falls to zero as the particles pull away from one another.

In order to calculate the transition probability and, subsequently, the collision cross section, it is necessary to introduce some artificial process of switching the interaction on and later switching it off. This switching may be accomplished abruptly by supposing the Hamiltonian to be

$$H = \begin{cases} H_0 & (-\infty < t < 0), \\ H_0 + V & (0 \leqslant t \leqslant \tau), \\ H_0 & (\tau < t < +\infty), \end{cases} \tag{3-19-3}$$

or it may be done continuously, or adiabatically, in a time-independent interaction potential by replacing $V(\mathbf{r})$ by $e^{-\epsilon|t|}V(\mathbf{r})$, in which $\tau = 1/\epsilon$ is a time constant that is long compared with the characteristic times of the system. In the second method the Hamiltonian $H = H_0 + e^{-\epsilon|t|}V$ approaches the unperturbed Hamiltonian H_0 as $t \to -\infty$ and as $t \to +\infty$ but assumes the form $H = H_0 + V$

*See Chapter 20 of D. Bohm, *Quantum Theory*, Prentice-Hall, Englewood Cliffs, N.J. (1951). Chapter 21 of Bohm's book is also useful in the comparison of time-dependent and time-independent formulations of scattering problems. See also L. S. Rodberg and R. M. Thaler.[19]

at $t = 0$. The parameter ϵ is eventually allowed to approach zero in the calculation so that the switching is done adiabatically. The final state $\Psi_b(+\infty)$ is, of course, related to the initial state $\Psi_a(-\infty)$ by (3-19-1) and may be expressed in terms of an operator S according to the equation

$$\Psi_b(+\infty) = S_{ab}\Psi_a(-\infty). \tag{3-19-4}$$

3-20. EXCITATION AND IONIZATION OF THE HYDROGEN ATOM BY ELECTRON IMPACT

We now proceed to discuss some of the elementary features of time-independent inelastic collision theory. From the physical standpoint perhaps the simplest problem we could treat is that of excitation* of the hydrogen atom by proton impact, since here the target has the simplest structure of any atomic system, whereas the projectile is regarded as a structureless point possessing mass and charge. Furthermore, the projectile is dissimilar to the atomic electron, and there are no electron symmetry effects. Only "direct scattering" is possible (i.e., there is no electron exchange). However, the problem of electron impact excitation of the hydrogen atom is simpler in another sense, provided we ignore symmetry and exchange effects, since in this problem there is only a single heavy particle, namely, the nucleus of the target atom. This fact permits us to ignore the motion of the nucleus completely, for we may consider it to be at rest in the laboratory frame initially and further consider it to have infinite mass so that it will not recoil in the collision. If, in addition, we neglect electron spin, we are left with a problem which is truly simple (formally), yet which illustrates the basic method of attack for a large class of inelastic collisions. Having developed the solution for this prototype problem, we may see how to generalize it to account for the finite mass of the collision partners, structural complications, electron exchange and other rearrangement phenomena, symmetry effects, and electron spin. However, we do not wish to leave the impression that it is an easy matter to solve the problems we shall formulate. It is difficult to make detailed calculations on even the simplest systems, and the equations to which we are led in the consideration of complex systems are highly intractable, if not completely insoluble. Here we shall follow closely the definitive treatment of Mott and Massey.[18]

A. Definition of the Excitation Cross Section

Let us consider, then, a beam of monoenergetic electrons moving along the $-Z$ axis toward the origin of the laboratory system of coordinates, at which point the nucleus of a hydrogen atom assumed to be in its ground state is located. The

*Here we use the term *excitation* in its broad sense to include ionization, which is, of course, the process of excitation to an unbound state in the continuum.

intensity of the incident beam is taken to be one electron per unit cross sectional area per second. The number of electrons scattered per second through an angle ϑ into the solid angle $d\Omega_{\text{Lab}}$ after exciting the target atom into its nth state has the dimensions of area and equals the *differential cross section for excitation*, $I_{0n}(\vartheta)\, d\Omega_{\text{Lab}}$ [cf. (1-4-2) and (1-4-3)]. The integral cross section for excitation to state n is then

$$q_{0n} = \int_0^{2\pi} \int_0^{\pi} I_{0n}(\vartheta) \sin \vartheta \, d\vartheta \, d\varphi. \tag{3-20-1}$$

B. Separation of the Time and Space Dependence of the Wave Function

Since the nucleus of the target atom is so much heavier than the projectile and orbital electrons, the nucleus may be considered to remain at rest throughout the collision, and no term corresponding to nuclear kinetic energy appears in the time-dependent wave equation, which may be written

$$i\hbar \frac{\partial \Psi(\mathbf{r}_b, \mathbf{r}_a, t)}{\partial t} = \left[-\frac{\hbar^2}{2m_e} \nabla_{r_b}^2 - \frac{\hbar^2}{2m_e} \nabla_{r_a}^2 + V(\mathbf{r}_b, \mathbf{r}_a) \right] \Psi(\mathbf{r}_b, \mathbf{r}_a, t). \tag{3-20-2}$$

Here m_e is the electron mass and V is the interaction potential energy operator, given in terms of the beam electron coordinates $\mathbf{r}_b$, the atomic electron coordinates $\mathbf{r}_a$ and the interelectron distance r_{ba} by the expression

$$V(\mathbf{r}_b, \mathbf{r}_a) = -\frac{e^2}{r_b} - \frac{e^2}{r_a} + \frac{e^2}{r_{ba}}. \tag{3-20-3}$$

Since the potential energy does not depend explicitly on the time, we may separate the wave function $\Psi(\mathbf{r}_b, \mathbf{r}_a, t)$ into the product of two functions, one of which depends only on the spatial coordinates and the other only on the time:

$$\Psi(\mathbf{r}_b, \mathbf{r}_a, t) = \psi(\mathbf{r}_b, \mathbf{r}_a)\Phi(t). \tag{3-20-4}$$

Equation (3-20-4) is a particular solution, but a general solution can be written as a sum of such separated solutions. If we substitute the right side of (3-20-4) into (3-20-2) and then divide each term of the resulting equation by the product $\psi\Phi$, we obtain

$$\frac{i\hbar}{\Phi} \frac{d\Phi}{dt} = \frac{1}{\psi} \left(-\frac{\hbar^2}{2m_e} \nabla_{r_b}^2 \psi - \frac{\hbar^2}{2m_e} \nabla_{r_a}^2 \psi + V\psi \right). \tag{3-20-5}$$

Since the left side of (3-20-5) depends only on the time and the right side only on the spatial coordinates, both sides must be equal to the same separation constant, which we denote E. We may readily integrate the resulting time equation to obtain the solution for Φ:

$$\Phi(t) = ce^{-iEt/\hbar}, \tag{3-20-6}$$

where c is an arbitrary constant. The equation for ψ is the time-independent wave equation

$$\left[-\frac{\hbar^2}{2m_e}\nabla_{r_b}^2 - \frac{\hbar^2}{2m_e}\nabla_{r_a}^2 + V(\mathbf{r}_b, \mathbf{r}_a)\right]\psi(\mathbf{r}_b, \mathbf{r}_a) = E\psi(\mathbf{r}_b, \mathbf{r}_a). \tag{3-20-7}$$

Since (3-20-7) is homogeneous in ψ, the constant c may be chosen to normalize ψ. Then a particular solution of the time-dependent wave equation is

$$\Psi(\mathbf{r}_b, \mathbf{r}_a, t) = \psi(\mathbf{r}_b, \mathbf{r}_a)e^{-iEt/\hbar}. \tag{3-20-8}$$

The quantum mechanical operator for the total energy of the system is $i\hbar(\partial/\partial t)$. If this operator is applied to the total wave function of the system given by (3-20-8), the following equation results:

$$i\hbar\frac{\partial\Psi}{\partial t} = E\Psi. \tag{3-20-9}$$

This is an eigenvalue equation; Ψ is an eigenfunction of the operator appearing on the left side, and the constant E on the right is the corresponding eigenvalue. Thus we see that E is equal to the total energy of the system. We may evaluate E by equating it to the sum of the internal energy of the target atom in its ground state and the kinetic energy of the incident electron, $T_{\text{Lab}} = m_e v_0^2/2$, when it is far from the target. An energy eigenfunction like the Ψ of (3-20-8) is said to represent a stationary state of the system, since the probability density $\Psi^*\Psi$ associated with this wave function is independent of time.

C. Calculation of the Excitation Cross Section

Let us now expand the wave function $\psi(\mathbf{r}_b, \mathbf{r}_a)$ in terms of the complete set of eigenfunctions of the unperturbed hydrogen atom $u_n(\mathbf{r}_a)$ and functions $F_n(\mathbf{r}_b)$ representing the projectile electrons:

$$\psi(\mathbf{r}_b, \mathbf{r}_a) = \left(\sum_n + \int\right) u_n(\mathbf{r}_a)F_n(\mathbf{r}_b) \equiv \underset{n}{\mathsf{S}}\, u_n(\mathbf{r}_a)F_n(\mathbf{r}_b). \tag{3-20-10}$$

The functions $u_n(\mathbf{r}_a)$ satisfy the equation

$$-\frac{\hbar^2}{2m_e}\nabla_{r_a}^2 u_n - \frac{e^2}{r_a}u_n = E_n u_n, \tag{3-20-11}$$

where E_n is the energy of the hydrogen atom in its unperturbed nth state. In (3-20-10) we sum over the states of the discrete spectrum and integrate over the states of the continuous spectrum. Thus the expression for the cross section which we shall obtain will apply for ionization of the target as well as for excitation to a bound state. By substituting (3-20-10) into (3-20-7) and using (3-20-11), we obtain

$$\underset{n}{\mathrm{S}}\, u_n(\mathbf{r}_a)\left[\frac{\hbar^2}{2m_e}\nabla_{r_b}^2 + (E - E_n)\right]F_n(\mathbf{r}_b) = \left(\frac{e^2}{r_{ba}} - \frac{e^2}{r_b}\right)\psi(\mathbf{r}_b, \mathbf{r}_a). \quad (3\text{-}20\text{-}12)$$

Now we multiply both sides of this equation by $u_n^*(\mathbf{r}_a)$ and integrate over the spatial coordinates of the atomic electron. Because of the orthonormality of the wave functions,

$$\left[\frac{\hbar^2}{2m_e}\nabla_{r_b}^2 + (E - E_n)\right]F_n(\mathbf{r}_b) = \int\left(\frac{e^2}{r_{ba}} - \frac{e^2}{r_b}\right)\psi(\mathbf{r}_b, \mathbf{r}_a)u_n^*(\mathbf{r}_a)\,d\mathbf{r}_a. \quad (3\text{-}20\text{-}13)$$

We note that the right side of (3-20-13) vanishes for large r_b, so that the functions $F_n(\mathbf{r}_b)$ become the wave functions for a free electron of energy $(E - E_n)$ at large r_b, where they satisfy the wave equation

$$\left[\nabla_{r_b}^2 + \frac{2m_e}{\hbar^2}(E - E_n)\right]F_n(\mathbf{r}_b) = 0. \quad (3\text{-}20\text{-}14)$$

The wavelength associated with this electron energy is $\lambda = 2\pi/\kappa_n$, where

$$\kappa_n^2 = \frac{2m_e(E - E_n)}{\hbar^2}. \quad (3\text{-}20\text{-}15)$$

We see that this wavelength is real only if $E > E_n$, that is, if the electron has enough energy to excite the nth state of the target atom. We shall consider only those values of n for which this condition is satisfied.

Since we have assumed the hydrogen atom to be in its ground state initially, the function $F_0(\mathbf{r}_b)$ must represent the sum of the incident electron wave and an elastically scattered electron wave. The other functions $F_n(\mathbf{r}_b)$ can represent only inelastically scattered waves. Consequently, the asymptotic forms of these wave functions are

$$F_0 \sim e^{i\kappa_0 z} + \frac{1}{r_b}e^{i\kappa_0 r_b}f_0(\vartheta) \quad (3\text{-}20\text{-}16)$$

and

$$F_n \sim \frac{1}{r_b}e^{i\kappa_n r_b}f_n(\vartheta) \qquad (n = 1, 2, 3, \ldots). \quad (3\text{-}20\text{-}17)$$

Equations (3-20-16) and (3-20-17) give the boundary conditions on the scattering. It follows from (3-20-17) that $r_b^{-2}|f_n(\vartheta)|^2$ is the number of electrons per unit volume at distance r_b from the target atom that have excited the atom to its nth state. The number of electrons that in one second cross unit area perpendicular to their direction of motion is proportional to $\kappa_n r_b^{-2}|f_n|^2$. Since the number of electrons in the incident beam crossing unit area per second is proportional to its wave number κ_0, the differential cross section for excitation of the atom from its ground state to the nth level is

$$I_{0n}(\vartheta)\, d\Omega_{\text{Lab}} = \frac{\kappa_n}{\kappa_0} |f_n(\vartheta)|^2\, d\Omega_{\text{Lab}} \tag{3-20-18}$$

(cf. Section 3-14). We see then that we need only the asymptotic form of the functions $F_n(\mathbf{r}_b)$ to obtain the excitation cross sections. However, the functions f_n cannot be calculated exactly, even though the wave functions for the unperturbed hydrogen atom are completely and exactly known. We must resort to approximate methods for their calculation. One of the most important and useful approximations is the *Born approximation*, which is valid at high impact velocities.

Before discussing the Born approximation, we shall rewrite (3-20-13) in a form that will prove more convenient and will also make clear the impossibility of obtaining an exact solution for the cross section. In doing this, we introduce the matrix that describes the interaction of the beam electron with the nucleus and orbital electron of the target atom. This matrix has elements

$$V_{mn} = \int u_n^*(\mathbf{r}_a)\left(\frac{e^2}{r_{bn}} - \frac{e^2}{r_b}\right) u_m(\mathbf{r}_a)\, d\mathbf{r}_a. \tag{3-20-19}$$

Then if we use (3-20-10) and (3-20-19) in (3-20-13) and introduce the wave number κ_n via (3-20-15), we obtain

$$\left(\nabla_{r_b}^2 + \kappa_n^2 - \frac{2m_e}{\hbar^2} V_{nn}\right) F_n(\mathbf{r}_b) = \frac{2m_e}{\hbar^2} \mathop{\mathrm{S}'}_{m} V_{mn} F_m(\mathbf{r}_b), \tag{3-20-20}$$

where the prime indicates that the term $m = n$ is to be omitted in the summation. Thus we see that we have an infinite set of coupled differential equations to solve in order to obtain the asymptotic form of the wave functions $F_n(\mathbf{r}_b)$. Clearly, we must use approximate methods, and the approximations consist of selecting only the most important matrix elements and putting all the others equal to zero. The inelastic scattering amplitude is exactly

$$f_n(\vartheta) = -\frac{1}{4\pi}\frac{2M_r}{\hbar^2} \int u_n^*(\mathbf{r}_a) e^{-i\boldsymbol{\kappa}_n \cdot \mathbf{r}_b} \left(\frac{e^2}{r_{ba}} - \frac{e^2}{r_b}\right) \psi(\mathbf{r}_b, \mathbf{r}_a)\, d\mathbf{r}_a\, d\mathbf{r}_b, \tag{3-20-21}$$

which is the generalization of (3-16-12) to inelastic scattering.

D. The Excitation Cross Section in the First Born Approximation[39]

The basic assumption in the Born approximation is that there is little interaction between the projectile and target. Specifically, we assume the following:

1. The incident wave is undistorted by the interaction, so that we may represent the beam electron by an undistorted plane wave moving in the direction defined by the unit vector $\mathbf{n}_0$ (usually taken to have the direction of the polar axis):

$$F_0(\mathbf{r}_b) = \exp i\kappa_0 \mathbf{n}_0 \cdot \mathbf{r}_b. \tag{3-20-22}$$

2. Excitation to any final state n comes as the result of a direct transition from the initial state, and intermediate states play no significant role. Since there is then no coupling with intermediate states, we put $V_{mn} = 0$ for $m \neq 0$, where 0 designates the initial state.
3. The potential energy of interaction between the scattered electron that produced the excitation and the atom in its final (nth) state is small, so that distortion of the scattered wave can be neglected. Since, according to the definition (3-20-19), V_{nn} is a measure of this interaction, we put $V_{nn} = 0$.

Thus we write the wave function for the system as

$$\psi(\mathbf{r}_b, \mathbf{r}_a) = \exp(i\kappa_0 \mathbf{n}_0 \cdot \mathbf{r}_b) u_0(\mathbf{r}_a). \tag{3-20-23}$$

Under these assumptions, the infinite set of equations (3-20-20) reduces to a single equation for excitation to state n:

$$(\nabla_{r_b}^2 + \kappa_n^2) F_n(\mathbf{r}_b) = \frac{2m_e}{\hbar^2} V_{0n} \exp i\kappa_0 \mathbf{n}_0 \cdot \mathbf{r}_b. \tag{3-20-24}$$

Now only one matrix element $V_{0n} = V_{n0}$ is involved in each calculation. We must solve (3-20-24) subject to the boundary conditions

$$F_n(\mathbf{r}_b) \sim \frac{1}{r_b} f_n(\vartheta) \exp i\kappa_n \mathbf{n}_n \cdot \mathbf{r}_b,$$

$$F_n(0) = 0. \tag{3-20-25}$$

As shown by Mott and Massey,[18] the solution of (3-20-24) with the correct asymptotic form is

$$F_n(\mathbf{r}) = \frac{m_e}{2\pi\hbar^2} \int \frac{e^{i\kappa_n|\mathbf{r}-\mathbf{r}_b|}}{|\mathbf{r}-\mathbf{r}_b|} \exp(i\kappa_0 \mathbf{n}_0 \cdot \mathbf{r}_b) V_{0n}\, d\mathbf{r}_b. \tag{3-20-26}$$

The asymptotic form of this solution is

$$F_n(\mathbf{r}) \sim \frac{m_e}{2\pi\hbar^2}\frac{1}{r}e^{i\kappa_n r}\int \exp\left[i(\kappa_0\mathbf{n}_0 - \kappa_n\mathbf{n})\cdot\mathbf{r}_b\right]V_{0n}\,d\mathbf{r}_b, \tag{3-20-27}$$

where $\mathbf{n}$ is a unit vector in the direction of the vector $\mathbf{r}$. Thus we finally obtain for the differential excitation cross section

$$\begin{aligned} I_{0n}(\vartheta)\,d\Omega_{\text{Lab}} &= \frac{\kappa_n}{\kappa_0}|f_n(\vartheta)|^2\,d\Omega_{\text{Lab}} \\ &= \frac{\kappa_n}{\kappa_0}\frac{m_e^2}{4\pi^2\hbar^4}\left|\int \exp\left[i(\kappa_0\mathbf{n}_0 - \kappa_n\mathbf{n})\cdot\mathbf{r}_b\right]V_{0n}\,d\mathbf{r}_b\right|^2 d\Omega_{\text{Lab}}. \end{aligned} \tag{3-20-28}$$

It is possible to extend this method by an iterative process and obtain the *second Born approximation* (cf. Section 3-23). However, this procedure is laborious, and it is better to start with more accurate initial approximations for the wave functions than to extend the Born approximation in order to achieve greater accuracy.

It is interesting to note that the cross section for elastic scattering can also be obtained in the Born approximation from the equations presented here. Let us suppose that the initial state and the final state are the same, and that no intermediate states are concerned. Then the matrix element appearing in (3-20-27) is, say,

$$V_{00} = \int u_0^*(\mathbf{r}_a)\left(\frac{e^2}{r_{ba}} - \frac{e^2}{r_b}\right)u_0(\mathbf{r}_a)\,d\mathbf{r}_a = V(\mathbf{r}'), \tag{3-20-29}$$

and we are led directly to (3-16-7), which we have already evaluated in Section 3-16.

3-21. GENERAL TWO-BODY COLLISIONS

Now we wish to generalize the results of the last section by extending our treatment to an arbitrary pair of colliding atomic or molecular structures and by removing the restriction that one of the structures can be considered to remain at rest in the laboratory frame. We shall describe the motion of the system of particles in terms of the motion of the center of mass of the entire system, the relative motion of the centers of mass of the two colliding structures, and the motion of the constituent particles of each structure relative to the center of mass of that structure. Since the motion of the center of mass of the complete system is inconsequential with respect to the calculation of cross sections, we shall ignore this contribution to the total motion.

To see how to generalize our previous discussion, we substitute the expression for the potential energy (3-20-3) into the wave equation (3-20-7) and expand the wave function as in (3-20-10). Then, if we substitute $E = E_0 + m_e v_0^2/2$, we obtain, using only the $n = 0$ term,

$$\left\{u_0(\mathbf{r}_a)\left[\frac{\hbar^2}{2m_e}\nabla_{r_b}^2 F_0(\mathbf{r}_b) + \frac{m_e v_0^2}{2} F_0(\mathbf{r}_b)\right] + F_0(\mathbf{r}_b)\left[\frac{\hbar^2}{2m_e}\nabla_{r_a}^2 u_0(\mathbf{r}_a) + \left(E_0 + \frac{e^2}{r_a}\right)u_0(\mathbf{r}_a)\right] + \left(\frac{e^2}{r_b} - \frac{e^2}{r_{ba}}\right)u_0(\mathbf{r}_a)F_0(\mathbf{r}_b)\right\} = 0, \tag{3-21-1}$$

where, as we have said, E_0 is the total energy of the target atom in its ground state. The first term in square brackets in (3-21-1) clearly corresponds to the unperturbed motion of the incident electron, the second term in square brackets, to the internal motion of the unperturbed target, and the final term, to the negative of the energy of interaction between the projectile and target. We must have a term similar to each of these terms in the wave equation for the generalized problem. (Furthermore, in this wave equation there should be a term corresponding to the motion of the center of mass of the entire system, but we have already justified its omission. This term did not appear in the generalization of the prototype problem, since in it the center of mass of the entire system was taken to be stationary in the laboratory frame.)

In the general problem, then, the first term above is to be replaced by one describing the relative motion of the two structures:

$$\left(\frac{\hbar^2}{2M_r}\nabla_r^2 + \frac{M_r v_0^2}{2}\right)F(\mathbf{r}). \tag{3-21-2}$$

Here $\mathbf{r}$ denotes the relative coordinates, that is, the spatial coordinates of the center of mass of one structure relative to that of the other; M_r represents the reduced mass of the two structures, and v_0 is their relative speed of approach when they are still widely separated, that is, before they begin to interact.

Next we need to consider the internal motions of the two collision partners. These motions are described by the equations

$$[H_A(\mathbf{r}_A) - E_A]u(\mathbf{r}_A) = 0$$

and

$$[H_B(\mathbf{r}_B) - E_B]v(\mathbf{r}_B) = 0, \tag{3-21-3}$$

where H_A and H_B are the Hamiltonians of the two unperturbed structures. These wave equations are satisfied by the sets of eigenfunctions and eigenvalues $[u_k(\mathbf{r}_A), E_A^k]$ and $[v_l(\mathbf{r}_B), E_B^l]$. To simplify the notation, we shall not distinguish these two sets of eigenfunctions, but we designate each pair of states of the two structures by a single index n. We may then write the composite wave function of the two structures $\xi_n(\mathbf{r}_A, \mathbf{r}_B)$ as the product of the two functions $u_k(\mathbf{r}_A)v_l(\mathbf{r}_B)$, and the corresponding energy eigenvalue E_n will be the sum $E_A^k + E_B^l$. The wave function ξ_n will satisfy the equation

$$[H_A(\mathbf{r}_A) + H_B(\mathbf{r}_B) - E_A - E_B]\xi = 0. \qquad (3\text{-}21\text{-}4)$$

If, finally, we generalize the interaction term by introducing the potential energy $V(\mathbf{r}, \mathbf{r}_A, \mathbf{r}_B)$, we obtain the complete time-independent wave equation for the entire system, except for the trivial motion of its center of mass:

$$\left[\frac{\hbar^2}{2M_r}\nabla_r^2 - H_A(\mathbf{r}_A) - H_B(\mathbf{r}_B) + \frac{M_r v_0^2}{2} + E_0 - V(\mathbf{r}, \mathbf{r}_A, \mathbf{r}_B)\right]\xi = 0. \qquad (3\text{-}21\text{-}5)$$

Proceeding then by the method of Section 3-20, we may show that the differential cross section in the center-of-mass coordinate frame for excitation from state p to state q of the combined system is, in the first Born approximation,

$$I_{pq}(\Theta)\, d\Omega_{CM} = \frac{M_r^2}{4\pi^2\hbar^4}\frac{\kappa_q}{\kappa_p} \times \left| \iiint V(\mathbf{r}, \mathbf{r}_A, \mathbf{r}_B) \exp\,[i(\kappa_p\mathbf{n}_0 - \kappa_q\mathbf{n})\cdot\mathbf{r}] \times \xi_q^*(\mathbf{r}_A, \mathbf{r}_B)\xi_p(\mathbf{r}_A, \mathbf{r}_B)\, d\mathbf{r}_A\, d\mathbf{r}_B\, d\mathbf{r} \right|^2 d\Omega_{CM}, \qquad (3\text{-}21\text{-}6)$$

where

$$\kappa_p = \frac{M_r v_0}{\hbar}$$

$$\kappa_q^2 = \frac{2M_r}{\hbar^2}\left(\frac{M_r v_0^2}{2} + E_p - E_q\right) \qquad (3\text{-}21\text{-}7)$$

and $\mathbf{v}_0$ is the initial velocity of approach of the colliding structures. The cross section in the coordinate system in which one of the structures is at rest initially may be obtained by applying (1-4-16).

3-22. REARRANGEMENT COLLISIONS

Electron exchange scattering, charge transfer and ion-molecule reactions, ion-atom interchange collisions, and other reactions in which there is a transfer of particles between the colliding structures require special consideration. In this section we shall show how the results of Sections 3-20 and 3-21 may be generalized to describe these *rearrangement collisions*.

A. Exchange Collisions between Electrons and Hydrogen Atoms

The simplest rearrangement collision is *exchange scattering* of a free electron by a hydrogen atom initially at rest in its ground state. In this case the incident electron is captured into the nth level of the atom, and the orbital electron is ejected into the continuum.

We have already derived the basic equations describing *direct scattering*, in which the incident electron produces excitation to state n but remains in the continuum after the impact. In doing so, we expanded the wave function of the system in the form

$$\psi(\mathbf{r}_b, \mathbf{r}_a) = \underset{n}{\mathsf{S}}\, u_n(\mathbf{r}_a)F_n(\mathbf{r}_b), \tag{3-20-10}$$

where $\mathbf{r}_b$ and $\mathbf{r}_a$ denote the laboratory coordinates of the beam and atomic electrons, respectively. As shown in (3-20-16), $F_0(\mathbf{r}_b)$ represents the sum of the incident wave and a scattered wave. If the height of the excited state n above the ground state $(E_n - E_0)$ is smaller than the initial kinetic energy of the incident electron (T_0), $F_n(\mathbf{r}_b)$ represents a scattered wave and is described, as in (3-20-17), by an exponential function with an imaginary exponent which corresponds to an undamped wave. If, on the other hand, $E_n - E_0 > T_0$, $F_n(\mathbf{r}_b)$ cannot correspond to an undamped scattered wave, hence must be described by an exponential with a negative real exponent:

$$F_n(\mathbf{r}_b) \sim r_b^{-1} e^{-k_n r_b} f_n(\vartheta) \qquad (E_n - E_0 > T_0), \tag{3-22-1}$$

where $k_n^2 = -\kappa_n^2$, and κ_n^2 is given by (3-20-15). These values of n relate to the possibility of capture of the incident electron and ejection of the atomic electron.

To obtain the cross section for exchange collisions, we expand the wave function of the system in the alternative form

$$\psi(\mathbf{r}_b, \mathbf{r}_a) = \underset{n}{\mathsf{S}}\, u_n(\mathbf{r}_b)G_n(\mathbf{r}_a). \tag{3-22-2}$$

Then, if we assume that the functions G_n have the asymptotic form

$$G_n(\mathbf{r}_a) \sim r_a^{-1} e^{i\kappa_n r_a} g_n(\vartheta), \tag{3-22-3}$$

the differential cross section for capture of the incident electron into the nth state and ejection of the atomic electron into the solid angle $d\Omega_{\text{Lab}}$ is

$$I_{0n}(\vartheta)\, d\Omega_{\text{Lab}} = \frac{\kappa_n}{\kappa_0} |g_n(\vartheta)|^2\, d\Omega_{\text{Lab}} \tag{3-22-4}$$

[cf. (3-20-18)]. Here we have treated the beam and atomic electrons as distinguishable, but this is permissible only if their spins are antiparallel. The method for obtaining the cross section for the general case in which the incident beam is unpolarized is discussed later in this section.

To calculate $G_n(\mathbf{r}_a)$, we substitute (3-22-3) into the wave equation (3-20-7), multiply by $u_n^*(\mathbf{r}_b)$, and integrate over the spatial coordinates of the beam electron to obtain

$$\left[\frac{\hbar^2}{2m_e}\nabla_{r_a}^2 + (E - E_n)\right] G_n(\mathbf{r}_a) = \int \left(\frac{e^2}{r_{ba}} - \frac{e^2}{r_a}\right) \psi(\mathbf{r}_b, \mathbf{r}_a) u_n^*(\mathbf{r}_b)\, d\mathbf{r}_b \tag{3-22-5}$$

[cf. (3-20-13)]. Equation (3-22-5) is exact, but we can obtain only approximate solutions to it. We assume some appropriate form for ψ on the right side of (3-22-5) and seek a solution of the form (3-22-3) by the methods used in Section 3-20.

To obtain a solution in the *first Born approximation*, we take

$$\psi(\mathbf{r}_b, \mathbf{r}_a) = \exp(i\kappa_0 \mathbf{n}_0 \cdot \mathbf{r}_b) u_0(\mathbf{r}_a) \tag{3-22-6}$$

on the right side of (3-22-5) and find (by the method described by Mott and Massey,[18] Chapter VI)

$$\begin{aligned} G_n(\mathbf{r}_a) \sim{} & \frac{m_e}{2\pi\hbar^2} \frac{1}{r_a} e^{i\kappa_n r_a} \\ & \times \iint \left(\frac{e^2}{r_a} - \frac{e^2}{r_{ba}}\right) u_n^*(\mathbf{r}_b) u_0(\mathbf{r}_a) \exp\left[i(\kappa_0 \mathbf{n}_0 \cdot \mathbf{r}_b - \kappa_n \mathbf{n} \cdot \mathbf{r}_a)\right] d\mathbf{r}_b\, d\mathbf{r}_a, \end{aligned} \tag{3-22-7}$$

where $\mathbf{n}$ is a unit vector in the direction of the ejected electron. The wave function (3-22-6) does not satisfy the orthogonality conditions

$$\begin{aligned} \int [\psi(\mathbf{r}_b, \mathbf{r}_a) - F_n(\mathbf{r}_b) u_n(\mathbf{r}_a)] u_n^*(\mathbf{r}_a)\, d\mathbf{r}_a &= 0, \\ \int [\psi(\mathbf{r}_b, \mathbf{r}_a) - G_n(\mathbf{r}_a) u_n(\mathbf{r}_b)] u_n^*(\mathbf{r}_b)\, d\mathbf{r}_b &= 0, \end{aligned} \tag{3-22-8}$$

which are satisfied by the true wave function, but the error is small for high impact velocity.

We must now take note that in (3-22-5) it is not really clear whether we should use the post interaction (e^2/r_a), which is the energy of interaction between the ejected atomic electron and the nucleus, or the prior interaction (e^2/r_b), which is that between the incident beam electron and the nucleus. This uncertainty causes no concern if the wave functions u_0 and u_n are exact solutions of the wave equation, since in this case the two interactions give the same result.[40] (Because of detailed balancing, we should be able to reverse the time in the wave equation and obtain the same cross section by viewing the collision in the reverse direction.) However, if only approximate wave equations are available, the two interaction terms do not give the same result, and it is uncertain whether we should use the prior or post interaction or some mean of the two.

The expression

$$g_n(\vartheta) = \frac{m_e}{2\pi\hbar^2} \iint \left(\frac{e^2}{r_b} - \frac{e^2}{r_{ba}}\right) u_n^*(\mathbf{r}_b) u_0(\mathbf{r}_a) \exp\left[i(\kappa_0 \mathbf{n}_0 \cdot \mathbf{r}_b - \kappa_n \mathbf{n} \cdot \mathbf{r}_a\right) d\mathbf{r}_b \, d\mathbf{r}_a \tag{3-22-9}$$

for the exchange amplitude was first used by Oppenheimer[41] and the approximation in which it appears is referred to as the *Born-Oppenheimer approximation.** Bates et al.[40] have studied the applicability of this approximation and have shown that generally it yields poor results for electron excitation of bound states. In the treatment of ionization, account must be taken of symmetry in the final state.†

B. The Effect of the Exclusion Principle on Exchange Scattering[42]

Since the incident and atomic electrons are identical particles obeying Fermi-Dirac statistics, we need to consider the effect of the Pauli principle on the scattering. To do this, we must introduce the electron spin into the wave function for the system and furthermore require that this wave function be antisymmetric (cf. Section 3-18). For scattering of electrons by hydrogen atoms we seek a solution of the form

$$\begin{aligned}\psi = \underset{n}{\mathsf{S}} \{&[u_n(\mathbf{r}_b) w_n^+(\mathbf{r}_a) + u_n(\mathbf{r}_a) w_n^+(\mathbf{r}_b)]\chi_s \\ &+ [u_n(\mathbf{r}_b) w_n^-(\mathbf{r}_a) - u_n(\mathbf{r}_a) w_n^-(\mathbf{r}_b)]\chi_t\},\end{aligned} \tag{3-22-10}$$

where u_n and $w_n^\pm$ are the wave functions of the bound and free electrons, respectively, and χ_s and χ_t are the singlet and triplet electron spin functions. The summation is carried out over all possible spatial and spin states. Then, solving the wave equation and putting in the statistical weighting factors described in Section 3-18, we find that the differential cross section for a beam of *unpolarized electrons* has the form[43]

$$I_{0n}(\vartheta) \, d\Omega_{\text{Lab}} = \frac{\kappa_n}{\kappa_0}\left(\frac{1}{4}|f_n + g_n|^2 + \frac{3}{4}|f_n - g_n|^2\right) d\Omega_{\text{Lab}}. \tag{3-22-11}$$

*This approximation is separate and distinct from the Born-Oppenheimer approximation used in the discussion of molecular energy levels (see Section 5-5).

†See R. Peterkop, *Proc. Phys. Soc.* (*London*) **77**, 1220 (1961); S. Geltman, M. R. H. Rudge, and M. J. Seaton, *Proc. Phys. Soc.* (*London*) **81**, 375 (1963); M. R. C. McDowell and J. H. Williamson, *Physics Letters* **4**, 159 (1963); R. K. Peterkop, *Theory of Ionization of Atoms by Electron Impact*, D. G. Hummer (Translation Ed.), Colorado Associated University Press, Boulder, Colo. (1977).

This expression gives the cross section for scattering *or* ejecting an electron into the element of solid angle $d\Omega_{\text{Lab}}$, with excitation of the hydrogen atom from its ground state to the nth level. The same approach indicated here is followed for electron scattering by other targets, with the form of the linear combination of f and g being determined by the number of electrons and the multiplicities of the states involved.

C. General Rearrangement Collisions

We now want to obtain the probability that two structures A and B, in states k and l, respectively, will collide, become rearranged, and produce structures C and D in states p and q, respectively. The results will be applicable, for example, to the treatment of charge transfer. We express the wave equation for the complete system (neglecting the motion of its center of mass) in the form that is most convenient for discussion of the final structures C and D:

$$\left[-\frac{\hbar^2}{2M_r'}\nabla_\rho'^2 + H_{\text{C}}(\mathbf{r}_{\text{C}}) + H_{\text{D}}(\mathbf{r}_{\text{D}}) + V(\mathbf{r}_{\text{C}}, \mathbf{r}_{\text{D}}, \boldsymbol{\rho}) - E\right]\zeta = 0. \tag{3-22-12}$$

Here $\boldsymbol{\rho}$ denotes the relative coordinates of the centers of mass of the *final* structures, whereas $\mathbf{r}_{\text{C}}$ and $\mathbf{r}_{\text{D}}$ refer to the internal coordinates of C and D with respect to their centers of mass; M_r' designates the reduced mass of the final system, $M_{\text{C}}M_{\text{D}}/(M_{\text{C}} + M_{\text{D}})$. H_{C} and H_{D} are the Hamiltonian operators of the internal motion of the structures C and D, and $V(\mathbf{r}_{\text{C}}, \mathbf{r}_{\text{D}}, \boldsymbol{\rho})$ is the energy of interaction between C and D. The similarity of (3-22-12) to (3-21-5) is obvious.

We shall distinguish a pair of stationary states of the initial structures A and B by the index n and write the corresponding composite wave function as $\xi_n(\mathbf{r}_{\text{A}}, \mathbf{r}_{\text{B}})$ and the energy as E_n. Similarly, a particular pair of stationary states of the final structures C and D will be distinguished by the index s, with composite wave function $\zeta_s(\mathbf{r}_{\text{C}}, \mathbf{r}_{\text{D}})$ and energy E_s; $\zeta_s(\mathbf{r}_{\text{C}}, \mathbf{r}_{\text{D}})$ is the product of the two wave functions, $x_i(\mathbf{r}_{\text{C}})$ and $y_j(\mathbf{r}_{\text{D}})$, of the separate structures C and D, and E_s is the sum of the corresponding energy values, E_{C}^i and E_{D}^j. By generalizing the results of Section 3-22-A, we obtain *in the Born approximation*

$$I_{ns}(\Theta)\,d\Omega_{\text{CM}} = \frac{M_r'^2}{4\pi^2\hbar^4}\frac{\kappa_s}{\kappa_n}\left|\iiint V(\mathbf{r}_{\text{C}}, \mathbf{r}_{\text{D}}, \boldsymbol{\rho}) \exp\left[i(\kappa_n\mathbf{n}_0\cdot\mathbf{r} - \kappa_s'\mathbf{n}\cdot\boldsymbol{\rho})\right] \times \xi_n(\mathbf{r}_{\text{A}}, \mathbf{r}_{\text{B}})\zeta_s(\mathbf{r}_{\text{C}}, \mathbf{r}_{\text{D}})\,d\mathbf{r}_{\text{A}}\,d\mathbf{r}_{\text{B}}\,d\boldsymbol{\rho}\right|^2 d\Omega_{\text{CM}} \tag{3-22-13}$$

as the differential cross section (in the relative coordinates $\boldsymbol{\rho}$) for the rearrangement reaction in which the sth state of the C–D system is excited from the nth state of the A–B system. Here $\kappa_n = M_r v_0/\hbar$ and $\kappa_s' = M_r' v_s/\hbar$, where $\mathbf{v}_0$ and $\mathbf{v}_s$ are the initial and final relative velocities, respectively. In (3-22-13) $\mathbf{r}$ denotes the relative coordinates of the centers of mass of the initial structures A and B.

PART $\mathscr{D}$. APPROXIMATION METHODS IN COLLISION THEORY*

3-23. QUANTUM MECHANICAL APPROXIMATIONS

The discussion of the preceding three sections indicates that it is impossible to obtain exact solutions for atomic collision cross sections. In most instances the requisite wave functions are not known with complete accuracy, and the approximate wave functions on which we must rely are frequently not orthogonal. Evidently, as the structural complexity of the colliding systems increases, the difficulty of obtaining good wave functions also increases and more complex reactions become possible. Calculations on molecular systems are particularly difficult. To make matters worse, the structure of the equations to be solved is such that approximate methods must be used even if the necessary wave functions are completely and exactly known, as in the case of the hydrogen atom. Some approximate methods are discussed in this section.

It should be pointed out that the range of validity of these approximations depends to a considerable degree on the masses of the colliding structures. In particular, collisions of free electrons differ significantly from heavy particle collisions because the masses of atomic systems are so much greater than the mass of an electron. In a collision between heavy particles the energy of relative motion exceeds the threshold energy for inelastic reactions down to much lower relative velocities than in the electron impact case. This fact is important because the relative velocity, rather than the energy, is the parameter that generally determines which of the prospective approximations might be acceptable, and in many instances, the lower the relative velocity, the more refined the approximation must be if it is to be satisfactory. Semiclassical and classical approximation techniques are treated in Section 3-24.

A. The Born Approximation; the Born Series[36,44]

Unless otherwise indicated, the term "Born approximation" refers to the *first Born approximation.* This simplest of all scattering approximations (vintage 1926) has already been discussed in Sections 3-16, 3-17, 3-20, 3-21, and 3-22. In it, the incident and scattered particles are treated as plane waves $e^{i\mathbf{K}_i\cdot\mathbf{r}}$ and $e^{i\mathbf{K}_f\cdot\mathbf{r}}$ that remain undistorted by the interaction. Only forward $i \rightarrow f$ transitions between the initial and final states i and f are considered, and for electron impact, electron exchange and spin effects are not taken into account. A single equation provides the solution to the scattering problem; no coupled equations are involved. The first Born approximation is valid for high-velocity impacts, but accurate cross sections can be obtained by its use only if accurate wave functions for the stationary states of the colliding structures are available. At low relative velocities, the first Born approximation generally overestimates the

*See Problem 3-23.

cross section, although it can be used at very low impact energies provided that the interaction energy is too shallow to accommodate a bound state of the interacting system.[45] The Born approximation is also valid for long-range scattering, and so is useful for low-energy collisions when the interaction has a long-range component (e.g., in electron-polar molecule collisions).

To show how we can obtain higher-order Born approximations, we shall develop the Born series, or expansion, displaying the results in modern Dirac notation, which makes the underlying physics more transparent. Consider the simplest situation, structureless projectiles being scattered by a static potential $U(\mathbf{r})$. We begin with a previous equation, (3-16-4). If we solve this integral equation by iteration, starting with the incident plane wave $\phi_{\boldsymbol{\kappa}_i}(\mathbf{r}) = e^{i\boldsymbol{\kappa}_i\cdot\mathbf{r}}$ as the zero-order approximation, we obtain

$$\begin{aligned}\psi_{\boldsymbol{\kappa}_i}(\mathbf{r}) = \phi_{\boldsymbol{\kappa}_i}(\mathbf{r}) &+ \int G_0^+(\kappa, \mathbf{r}, \mathbf{r}')U(\mathbf{r}')\phi_{\boldsymbol{\kappa}_i}(\mathbf{r}')\,d\mathbf{r}' \\ &+ \int G_0^+(\kappa, \mathbf{r}, \mathbf{r}')U(\mathbf{r}')G_0^+(\kappa, \mathbf{r}', \mathbf{r}'')U(\mathbf{r}'')\phi_{\boldsymbol{\kappa}_i}(\mathbf{r}'')\,d\mathbf{r}'\,d\mathbf{r}'' + \cdots.\end{aligned} \tag{3-23-1}$$

Writing this *Born series* as shown implies an assumption that the sequence of terms converges toward the exact wave function. The function G_0^+ is the "outgoing Green's function":

$$G_0^+(\kappa, \mathbf{r}, \mathbf{r}') = -\frac{1}{4\pi}\frac{e^{i\kappa|\mathbf{r}-\mathbf{r}'|}}{|\mathbf{r}-\mathbf{r}'|}. \tag{3-23-2}$$

The corresponding *Born series for the scattering amplitude* is

$$f = -\frac{1}{4\pi}\left\langle \phi_{\boldsymbol{\kappa}_f} \middle| U + UG_0^+U + UG_0^+UG_0^+U + \cdots \middle| \phi_{\boldsymbol{\kappa}_i} \right\rangle. \tag{3-23-3}$$

Here $\phi_{\boldsymbol{\kappa}_f}$ is the final plane wave function. The first term in this series is

$$f_{\mathrm{B1}} = -\frac{1}{4\pi}\left\langle \phi_{\boldsymbol{\kappa}_f} \middle| U \middle| \phi_{\boldsymbol{\kappa}_i} \right\rangle = -\frac{1}{4\pi}\int e^{i\boldsymbol{\kappa}\cdot\mathbf{r}}U(\mathbf{r})\,dr, \tag{3-23-4}$$

the *first Born approximation to the scattering amplitude*. The *second Born approximation* consists of the sum of the first two terms, and so on. The series that we have obtained here are perturbation expansions in powers of the potential.

In (3-23-4), $\boldsymbol{\kappa}$ is the momentum transferred during the collision, divided by $\hbar$. The middle form is the Fourier transform of the interaction with respect to $\boldsymbol{\kappa} = \boldsymbol{\kappa}_i - \boldsymbol{\kappa}_f$. The form at the right in (3-23-4) will be recognized as (3-16-7), which we worked out in detail earlier. The scattering amplitude in the first Born approximation also applies to inelastic collisions if a multiplicative factor κ_f/κ_i is introduced into these forms.

B. The Bethe-Born (Bethe) Approximation[36,46]

In 1930, H. A. Bethe proposed a modification of the Born approximation in which he made the additional assumption that the product of the momentum transfer and the range of interaction is small. Then the exponential term in the Born approximation (3-23-4) for the scattered amplitude can be expanded, and the integrations carried out term by term. A series of terms corresponding to the atomic transition moments is obtained (electric dipole, electric quadrupole, etc.). A Bethe-Born calculation of the cross sections for excitation and ionization of atoms is carried out in detail in Section 6-14.

C. The Coulomb-Born Approximation[36,47]

This approximation is useful in describing collisions of electrons and ions with target ions, in which the Coulomb interaction of the projectile with the target nucleus can be important. The Born plane wave functions are replaced by Coulomb wave functions corresponding to the nuclear charge. For highly charged ions, the long-range Coulomb interaction is dominant, and all other interactions can be treated as small perturbations. The Coulomb-Born approximation is never good for electron-neutral scattering. It is better, comparatively speaking, for electron-ion collisions than is the Born approximation for electron-neutral collisions.

D. Versions of the Born Approximation that Include Electron Exchange[36,48]

The first attempt to account for electron exchange was made by J. R. Oppenheimer in 1928. He made the same basic assumptions as used in the Born approximation, but his method of calculating the exchange amplitude (3-22-9) is much less satisfactory than the Born approximation for the direct scattering amplitude, mainly because in the *Born-Oppenheimer* (*B-O*) *approximation* the initial and final states are not orthogonal. The result is that the addition of any constant to the interaction potential gives a nonzero change in g. The B-O approximation tends to overestimate the cross sections to an excessive degree.*

An improvement was made by V. I. Ochkur, who suggested that (3-22-9) for g should be regarded as an expansion in inverse powers of the impact energy, with the result that g is expressed in terms of the direct amplitude f in (3-20-27). Then, only the first term should be retained because higher terms would cause the result to diverge at lower energies. The Ochkur approximation to the

*Bates has developed a formulation that is correct to first order and that includes the orthogonality correction: D. R. Bates, *Proc. R. Soc. London A* **247**, 294 (1958). In treatments of rearrangement scattering within the Born-Oppenheimer approximation, the boundary conditions can be satisfied only by including momentum translation factors. This problem was first discussed by D. R. Bates and R. McCarroll, *Proc. R. Soc. London A* **245**, 175 (1968). The literature on momentum translation factors is extensive. An especially useful modern reference is J. Delos, *Rev. Mod. Phys.* **53**, 287 (1981).

exchange integral is a good approximation only at impact energies high enough for the integral to be small. According to Massey and Burhop, any success that the *Born-Ochkur approximation* may have at low energies in improving the agreement between theory and experiment must be regarded as empirical.

In the *Born-exchange approximation*, the exchange scattering amplitude is related to the direct scattering amplitude by the equation (Massey and Burhop,[46] p. 465)

$$g(k', \kappa) = f(\kappa, k') = e^{i\delta(\kappa, k')} f_B(\kappa, k') \tag{3-23-5}$$

with

$$f(k', \kappa) = f_B(k', \kappa),$$

where f_B is the usual Born approximation to the direct scattered amplitude. Here $e^{i\delta}$ is an arbitrary phase factor; it is difficult to know which value of δ to choose. This method does not suffer from nonorthogonality difficulties, but it does suffer from all the shortcomings of the Born approximation.

E. The Quantal Impulse (Sudden) Approximation[36,49]

This approximation, introduced by E. Fermi in 1936, describes a many-body scattering problem in terms of known two-body scattering amplitudes. The underlying idea can be illustrated in terms of the scattering of a structureless projectile (1) by a hydrogen atom, the nucleus being labeled (2) and the orbital electron (3). The Hamiltonian for the system is

$$H = H_0 + V_{12} + V_{13} + V_{23}, \tag{3-23-6}$$

where H_0 is the sum of the kinetic energy operators. We now attack the problem by assuming that the collision of the projectile with the orbital electron takes place as if the electron were free except for having a momentum probability distribution that is determined by its interaction with the nucleus. Thus we are assuming that apart from determining the momentum distribution of the bound electron, V_{23} has little effect and can be ignored in the scattering calculation. Another statement of the basic assumption is that the collision between the projectile and the electron is sudden (i.e., the collision time is small compared with the characteristic period of the orbital motion of the electron). One would expect the impulse approximation to be useful when (1) the impact energy greatly exceeds the binding energy of the target particle being considered, and (2) the reduced wavelength of the relative motion of the projectile and target system is much smaller than the mean separations of the particles of the target.

F. The Distorted Wave Approximation[50]

The starting point for discussion of this approximation is the infinite set of coupled differential equations (3-20-20), derived for the excitation of hydrogen atoms by electron impact but capable of being generalized to more complex systems and other collisions. These equations may be solved in the Born approximation by representing the incident beam electron by an undistorted plane wave

$$F_0(\mathbf{r}_b) = \exp(i\kappa_0 \mathbf{n}_0 \cdot \mathbf{r}_b) \tag{3-23-7}$$

and taking the interaction matrix elements V_{nn} and V_{mn} to be zero except for $m = 0$, where 0 designates the initial state and n the final state of the system. As the relative impact velocity decreases, we try to improve this approximation by considering more terms on the right side of (3-20-20). The *distorted wave* (DW), or *distortion, approximation* ignores transitions through intermediate states (as does the Born) but takes account of the distortion of the incident and scattered waves by the static field of the target. The distortion is allowed for by retaining the matrix elements V_{nn} and V_{00}, and the transition occurs via V_{0n}; all other elements are set equal to zero. The infinite set of equations (3-20-20) then reduces to the pair of equations

$$\left(\nabla^2 + \kappa_0^2 - \frac{2m_e}{\hbar^2} V_{00}\right) F_0 = \frac{2m_e}{\hbar^2} V_{0n} F_n, \tag{3-23-8}$$

$$\left(\nabla^2 + \kappa_n^2 - \frac{2m_e}{\hbar^2} V_{nn}\right) F_n = \frac{2m_e}{\hbar^2} V_{0n} F_0. \tag{3-23-9}$$

The neglect of $V_{0n}F_n$ is valid when F_n is much smaller than F_0 (weak coupling). This approximation may not be justified, however, if the matrix element V_{0n} becomes large, producing strong coupling between (3-23-8) and (3-23-9). If weak coupling is assumed, the first equation may be solved to give a function with asymptotic form

$$F_0 \sim e^{i\kappa_0 z} + r^{-1} e^{i\kappa_0 r} f_0(\vartheta). \tag{3-23-10}$$

When this solution is put into the second equation, a solution is found for it with asymptotic form

$$F_n \sim r^{-1} e^{i\kappa_n r} f_n(\vartheta). \tag{3-23-11}$$

The function in (3-23-10) represents the incident particle. It corresponds to a distorted wave with the asymptotic form of a plane wave plus an outgoing spherical wave. The scattered particle is represented by a distorted wave function with the asymptotic form of an outgoing spherical wave.[18,36]

If the coupling constant V_{0n} is so large that the right side of (3-23-8) cannot be set to zero (strong coupling), we may reduce (3-20-20) to a set of simultaneous ordinary differential equations that can be solved with the aid of a computer.[36]

G. The Perturbed Stationary State and Perturbed Rotating Atom Approximations[36,51]

The number of matrix elements that are important in the description of the interaction between colliding systems generally increases rapidly as the relative impact velocity decreases, so that the use of the distorted wave or second Born approximation does not permit reliable calculations to be made at much lower velocities than does the first Born approximation. It is evident that slow collisions cannot be treated accurately by expanding the wave function of the system in terms of the eigenfunctions of the isolated target, but Mott[52] suggested that it might be legitimate to perform the expansion in terms of the eigenfunctions that would describe the quasi-molecule formed by the colliding structures if their relative position vector were momentarily fixed in space. This assumption is equivalent to what is frequently called the *perturbed stationary-state* (*PSS*) *approximation*. It allows for the gradual nature of the collision in the near-adiabatic region by treating the kinetic energy of relative motion as a perturbation.

The PSS approximation is rarely an adequate approximation except for symmetric resonance charge transfer.[37,53] For other reactions, at any given impact velocity, collisions at sufficiently small impact parameters generally are not nearly adiabatic, as assumed, because of the rapid rotation of the internuclear axis, and these close impacts normally give the dominant contribution to the calculated cross section at low energies. However, close encounters make only a small contribution to the cross section for symmetric resonance charge transfer at low velocities, because the probability of this reaction is high up to very large values of the impact parameter.[53]

Originally, it was mistakenly assumed that transitions are unlikely between states of the quasi-molecule formed by the slowly interacting bodies and that consequently the only matrix element that need be considered is the one joining the states of the quasi-molecule tending to the initial and final states of the target in the separated atoms limit. It was also widely believed that this formulation of the PSS approximation tends to the first Born approximation in the limit of weak interactions and high velocities, but this is not true because of the rotation of the internuclear axis during the collision.

As used, the PSS was a distorted wave approximation, but because coupling of the different n states was always strong, the two-state DWA was a poor approximation. The PSS approximation does adequately describe transitions between S states where there is no n coupling.

As Bates[53] points out, the PSS approximation is best understood by examining its true limit, the *perturbed rotating atom* (*PRA*) *approximation*,[54] in which the eigenfunctions and eigenenergies of the quasi-molecule are replaced by those of the target perturbed by the projectile, the same rotating frame of reference being used as in the PSS approximation. (Actually, the eigenenergies of the isolated target are used in practice, the error thereby introduced being small at high velocities.) At any given relative velocity the assumption that the interaction of the projectile causes the eigenfunctions to follow the rotation of the internuclear axis is invalid if the impact parameter b is less than some value b_1 or greater than some value b_2. For $b < b_1$ the rotation is too rapid, whereas for $b > b_2$ the interaction is too weak. As the impact velocity is increased, b_1 becomes larger and b_2 smaller, and at sufficiently high v_0 the assumption concerning the eigenfunctions is obviously invalid.

In the PRA approximation there is strong coupling between states differing only in magnetic quantum number, this coupling being a manifestation of the reluctance of the eigenfunctions to follow the rotation of the internuclear axis. When this coupling is allowed for, the modified PRA approximation tends to the first Born approximation in the weak interaction and high-velocity limits. Account must also be taken of coupling terms in the PSS approximation, particularly those terms that connect the initial and final states to states differing only in magnetic quantum number either in the combined atom limit or in the separated atoms limit. Because the need for retaining these terms was not initially realized, all PSS inelastic cross sections calculated before about 1960 are incorrect.[53] (Symmetric resonance charge transfer does not fall in this category.) The PSS and PRA approximations are first-order treatments; they are approximate applications of the molecular expansion method.

H. The Polarized Orbital Approximation[55]

As we have seen, one approach to the calculation of collision cross sections is to expand the scattering wave function in an appropriate set of basis functions and then truncate this set, so that only a finite set of coupled equations has to be solved. A different approach is provided for slow electron scattering by the *polarized orbital method.* It attempts to incorporate the essential physics of the scattering into the form of the wave function. The projectile, as it approaches the target, induces electric multipole moments in it, and these moments affect the motion of the projectile. In this method, only the effect of the dipole moment is acknowledged. This fact prompts the choice of a trial wave function for electron-atom collisions, of the form

$$\psi^t = (1 \pm P_{12})[\psi_0(\mathbf{r}_2) + \phi_0(\mathbf{r}_1, \mathbf{r}_2)]F_0(\mathbf{r}_1), \tag{3-23-12}$$

where $\psi_0(\mathbf{r}_2)$ is the original, unperturbed wave function of the target, $\phi_0(\mathbf{r}_1, \mathbf{r}_2)$ gives the polarization distortion of the target during the collision, and $F_0(\mathbf{r}_1)$ is the wavefunction of the projectile (1). P_{12} is the exchange operator that interchanges electrons 1 and 2. The function $\phi_0(\mathbf{r}_1, \mathbf{r}_2)$ can be calculated by perturbation theory using the electrostatic dipole perturbation, and then one can calculate an effective polarization potential in which the projectile is to move. We see that the polarized orbital approximation (3-23-12) takes into account both the effect of polarization and the effect of exchange in a relatively simple fashion.

If the distortion exchange terms in the polarized orbital analysis are set equal to zero, we obtain the *exchange-adiabatic approximation.* Without exchange, we have the *adiabatic approximation*, which merely states that the electrostatic potential describing the interaction between projectile and target varies so slowly that the target electrons can smoothly adjust to the disturbance. Mathematically, the kinetic energy of the projectile is to be neglected.

The *distorted wave-polarized orbital approximation* leads to equations similar to (3-23-2) and (3-23-3), except that the factors involving the potentials are modified to account for the polarization of the target, as in the polarized orbital approximation.

I. The Close-Coupling (Truncated Eigenfunction Expansion) Approximation[56]

Here the wave function for the system of projectile plus target is expanded in terms of the complete set of eigenfunctions ψ_0 (assumed known) of the target Hamiltonian. For a structureless projectile colliding with an N-electron atom, we could write

$$\psi(\mathbf{r}_1, \mathbf{r}_2) = A \sum_{\gamma} F_{\gamma}(\mathbf{r}_1)\psi_0(\gamma, \mathbf{r}_2), \qquad (3\text{-}23\text{-}13)$$

where $\mathbf{r}_1$ represents the spatial and spin coordinates of the projectile and $\mathbf{r}_2$ the coordinates of the atomic electrons. Any possible combination of the good quantum numbers belonging to the total system is denoted by γ. The symbol A implies antisymmetrization of the total wave function. The expansion coefficients $F_{\gamma}(\mathbf{r}_1)$ describe the radial motion of the projectile relative to the target in its various quantum states. In the *close-coupling approximation*, only a relatively small number (M) of terms are retained in (3-23-13), the number equaling the number of channels* considered. All the open channels and some closed channels should be included.

The radial partial wave scattering functions $F_i(r)$ in (3-23-13) satisfy a set of M coupled integrodifferential equations of the form

$$\left[\frac{d^2}{dr^2} + \kappa_i^2 - \frac{l_i(l_i+1)}{r^2}\frac{\hbar^2}{2m_e}\right]F_i(r) = \sum_{j=1}^{M} V_{ij}(r)F_j(r) + \int_0^{\infty} W_{ij}(r, r')F_j(r')\,dr', \qquad (3\text{-}23\text{-}14)$$

*When two composite systems collide, various outcomes of the collision may be possible: elastic scattering, excitation, ionization, dissociation, charge transfer, and so on. Each possible outcome, characterized by the number of particles plus their chemical identities and quantum states, is associated with a collision channel. A channel is said to be *open* if it is allowed by the conservation laws; otherwise, it is *closed*.

where V_{ij} is a direct electron-electron plus electron-nucleus potential, and W_{ij} is the exchange kernel. The r' refer to the exchanged electron. These "close-coupling equations" can give accurate cross sections provided that, in the spirit of perturbation theory for bound states, all of the target states lying close to the initial and final states in energy are included.

However, the slow convergence of the truncated expansions makes close-coupling calculations intractable at higher impact energies, at which more states are accessible and more angular momenta become important. Also, it is virtually impossible to apply the method to scattering from excited states, since close spacing of the excited levels requires that a large number of states be included in the expansion. Finally, the large number of angular momenta involved in heavy particle collisions makes the close-coupling method infeasible for this important class of collisions. The method is, however, effectively exact in the prediction of the positions and shapes of threshold effects and resonances in low-energy collisions of electrons with ground-state atoms.

As far as heavy particle collisions are concerned, there are two basic approaches: one, an expansion in products of atomic states, and the other, an expansion in molecular states. The atom expansion should be best at high velocities, the molecular at low velocities. By limiting the basis set or by omitting certain couplings, the various approximations can be obtained.

J. The Static Exchange Approximation[57]

When considering the elastic scattering of electrons, the use of only the first term in (3-23-13) is a good approximation provided that the coupling between the target ground state and its other bound and continuum states is weak. This *static exchange approximation* is tantamount to assuming that the target remains undistorted during the collision and that no virtual transitions to excited states take place. The wave functions for the *e-H* elastic scattering problem are written

$$\psi_{\pm}(\mathbf{r}_1, \mathbf{r}_2) = F_1^{\pm}(\mathbf{r}_1)\psi_1(\mathbf{r}_2) \pm F_1^{\pm}(\mathbf{r}_2)\psi_1(\mathbf{r}_1). \tag{3-23-15}$$

The spatial part of the wave function must be either symmetrical (ψ_+) or antisymmetrical (ψ_-) with respect to the interchange of the electrons. To these functions correspond the singlet spin state ($S = 0$) and the triplet spin state ($S = 1$), respectively. The functions F_1 satisfy the equation

$$(\nabla_1^2 + \kappa_1^2)F_1^{\pm}(\mathbf{r}_1) = 2V_{11}(\mathbf{r}_1)F_1^{\pm}(\mathbf{r}_1) \pm 2\int K_{11}(\mathbf{r}_1, \mathbf{r}_2)F_1^{\pm}(\mathbf{r}_2)\,d\mathbf{r}_2, \tag{3-23-16}$$

where

$$V_{11}(\mathbf{r}_1) = \left\langle \psi_1 \left| -\frac{e^2}{r_1} + \frac{e^2}{|\mathbf{r}_1 - \mathbf{r}_2|} \right| \psi_1 \right\rangle \tag{3-23-17}$$

is the static (direct) part of the effective potential and

$$K_{11}(\mathbf{r}_1, \mathbf{r}_2) = \psi_1^*(\mathbf{r}_1)\psi_1(\mathbf{r}_2)\left[\frac{e^2}{|\mathbf{r}_1 - \mathbf{r}_2|} - (E_{\text{total}} - 2E_1)\right] \tag{3-23-18}$$

is the exchange part. The *direct potential* V_{11} is the electrostatic energy associated with the projectile and the undistorted hydrogen atom, averaged over ψ_1, and K_{11} is a nonlocal *exchange potential* (see Section 4-8-A).

K. The Pseudostate Approximation[58]

Computational difficulties increase rapidly with the number of target eigenfunctions used in the close-coupling expansion; in practice, only a few bound states and continuum states are included. One partial remedy for this shortcoming is to add to the approximate wave function in the truncated version of (3-23-13) functions of the form $\Phi_n(\mathbf{r}_2)F_n(\mathbf{r}_1)$, where Φ_n is not a target eigenfunction but rather a function chosen to represent some appropriate average of bound and continuum states. The states thus introduced are fictitious, not real, and for this reason, they are called *pseudostates*. There is no unique way in which to choose the pseudostates to be used, although questions of normalization and orthogonality need to be addressed. The inclusion of pseudostates helps offset one of the effects of truncating the eigenfunction expansion, namely, the loss of flux from the open channels in the expansion to the remaining open channels. Incidentally, another effect of the truncation of the expansion is the loss of terms that contribute to the description of the distortion of the target. This defect is remedied by adding terms of the type used in the polarized orbital approximation. The pseudostates can introduce spurious thresholds and resonances, and these artifacts can lead to inaccuracies in computed cross sections.

L. The Correlation Approximation[59]

First, some background information, which is presented in much more detail by Weiss.[60] In the *independent particle model of the atom*, the state of the atom is described by a single configuration, each electron being assigned to some one-electron wave function, or orbital. The wave function for the entire atom is then taken to be an antisymmetrized product of orbitals, with the population of the orbitals determined by Bohr's Aufbau Principle. For a closed shell atom, the wave function is a single determinant formed from the product of the single-particle orbitals and spin functions. In general, the wave function for excited states and open shells is a linear combination of determinants that gives a pure *LS* state (i.e., one that is simultaneously an eigenfunction of the operators L^2 and S^2). The *Hartree-Fock approximation* is the variational formulation of the independent particle model of the atom. *Correlation energy* is the difference between the Hartree-Fock value of the total energy of the atom and the exact energy. The term *correlation* thus relates to the error in the Hartree-Fock model, which represents the effect of the electrons on each other in an average way without concern for the detailed fashion in which the electronic motions are correlated. Correlation effects may be calculated by the method of configuration interaction (CI) or the technique of superposition of configurations (SOC).*

Now back to collisions. When the initial or final states of the target are not well separated in energy from all other states, then the close-coupling approximation often converges only slowly. An effective stratagem may be to add a series of terms involving powers of the interelectron distance r_{ij} to a close-coupling expansion that already includes the states of crucial importance. Thus correlation effects are explicitly introduced into the wave function. This technique is the *correlation approximation.* The added terms usually involve square integrable $N + 1$ electron functions known as L^2 functions.

M. The Closure Approximation[61]

We can illustrate the essential feature of this approximation by considering a specific calculation as outlined by Walters.[61] He starts with the second Born term for the transition $0 \to f$:

$$f_{f0}^{B2} = -\frac{1}{8\pi^4}\sum_n \lim_{\eta\to 0^+} \int \frac{d\boldsymbol{\kappa}\langle \boldsymbol{\kappa}_f \psi_f | V | \boldsymbol{\kappa}\psi_n\rangle\langle \boldsymbol{\kappa}\psi_n | V | \boldsymbol{\kappa}_0 \psi_0\rangle}{\kappa_n^2 - \kappa^2 + i\eta}, \tag{3-23-19}$$

where atomic units are used. The sum over intermediate states ψ_n in (3-23-19) poses a serious problem because it covers all bound and all continuum states of the target. However, if the intermediate states are all assigned an average energy $\bar{\epsilon}$, κ_n^2 is replaced by

$$\overline{\kappa^2} = \kappa_0^2 + 2(\epsilon_0 - \bar{\epsilon}), \tag{3-23-20}$$

which is independent of n. Then all of the n dependence in (3-23-19) resides in the eigenfunctions ψ_n, and the completeness of the set of states can be utilized:

$$\sum_n |\psi_n\rangle\langle\psi_n| = 1. \tag{3-23-21}$$

The trivial evaluation of the sum then provides the approximation

$$\overline{f_{f0}^{B2}} = -\frac{1}{8\pi^4} \lim_{\eta\to 0^+} \int \frac{d\boldsymbol{\kappa}\langle \boldsymbol{\kappa}_f \psi_f | V | \boldsymbol{\kappa}\rangle\langle \boldsymbol{\kappa} | V | \boldsymbol{\kappa}_0 \psi_0\rangle}{\overline{\kappa^2} - \kappa^2 + i\eta}. \tag{3-23-22}$$

*Computer programs are available to compute full CI wave functions. See A. Hibbert, *Comp. Phys. Comm.* **9**, 142 (1975).

This *closure approximation* was first used by H. S. W. Massey and C. B. O. Mohr in 1934. Their choice of $\bar{\epsilon} = \epsilon_0$ led to serious problems; the approximation can be improved by evaluating explicitly a few of the terms with their exact energies and using an average energy for the remainder. This matter is discussed by Walters.[61]

N. The Quantum Defect Method[36,62]

Let us consider the scattering of an electron in a modified Coulomb field, one that has the asymptotic form $-Ze^2/r$ but that deviates from the Coulomb form at small r. There is an infinite discrete set of bound states for each value of l. We can classify these states in terms of the limit to which they pass if the field becomes purely Coulombic for all r. We can write for the energy of a state

$$E_{nl} = -\frac{2\pi^2 me^4 Z^2}{\nu_{nl}^2 h^2}, \tag{3-23-23}$$

where $\nu_{nl} = n - \mu_{nl}$. The quantity μ_{nl} is called the "quantum defect"; it varies slowly with energy and tends to zero in the limit of a pure Coulomb field. The quantum defect method for atomic collisions originated with M. J. Seaton in 1955. One of its useful aspects is that it provides a relation between the cross section for the lth partial wave and the quantum defect $\mu_l(\kappa^2)$, considered as a function of energy interpolating between bound-state positions and extrapolated to positive energies:

$$\cot \sigma_l(\kappa^2) = (1 - e^{\pi\beta/\kappa}) \cot \pi\mu_l(\kappa^2), \tag{3-23-24}$$

where

$$\frac{\beta^2}{4\nu_{nl}^2} = -\frac{8\pi^2 mE_n}{h^2}. \tag{3-23-25}$$

The quantum defect method has proved valuable in the calculation of cross sections for electron-positive ion collisions. In the limit $\kappa^2 \to 0$, (3-23-20) gives

$$\eta_l = \pi\mu_l, \tag{3-23-26}$$

so that if the bound-state energies are known from spectroscopic observations, we can obtain the phase shifts, and thus the cross section, with very little effort. Seaton[62] has developed a multichannel quantum defect method, which is a generalization of the single-channel method.

O. The *R*-Matrix Method[63]

The reactance (**R**) matrix is related to the scattering (**S**) matrix (Sections 3-19 and 4-8) by the equation

$$\boldsymbol{S} = (1 + i\boldsymbol{R})(1 - i\boldsymbol{R})^{-1}. \quad (3\text{-}23\text{-}27)$$

The **R** matrix is Hermitian, and if the potential describing the scattering is real, the elements R_{ij} are real and the matrix is symmetric. It is frequently advantageous to work with **R** rather than with **S** because the R_{ij} are real. Furthermore, any approximation to **R** that preserves the symmetry of the matrix ensures that **S** is unitary and hence that the number of particles in the system is conserved.

In this method, configuration space is divided into two regions. For the scattering of electrons from an atom, exchange can be neglected outside some radius r_0. Hence for $r > r_0$, the collision is described by coupled differential (rather than integrodifferential) equations that often have analytical solutions easily obtainable by numerical methods. The basic problem, then, is to calculate the **R** matrix in the internal region ($r < r_0$). The **S** matrix and cross section can then be obtained from the **R** matrix through the solution in the external region.

The **R**-matrix method was introduced by E. P. Wigner in 1946 in the context of nuclear physics. It has been subsequently developed by P. G. Burke, C. J. Noble, B. Schneider, and others for the study of collisions of electrons with atoms and molecules. This method is meeting with considerable success in a variety of types of electron-molecule collisions, including elastic scattering, vibrational excitation, electronic excitation, and dissociative attachment.

3-24. CLASSICAL AND SEMICLASSICAL APPROXIMATIONS

We begin with a rather general discussion of the validity of the classical approach that is based on the treatment by Percival.[64]

A. Conditions for Validity of Classical Methods

In quantum mechanics, a point in phase space has no significance because the uncertainty principle prohibits precise knowledge of both position and momentum simultaneously (Section 3-10). However, if many quantum states are involved, the concept of phase space is useful, as in statistical mechanics (Section 2-11). For consistency with the uncertainty principle, a single quantum state is represented not by a point but by a region of volume $(2\pi\hbar)^f$, where f is the number of degrees of freedom of the physical system. Accordingly, the volume in phase space associated with N quantum states of the system is $N(2\pi\hbar)^f$. To show the relevance of phase space to our present discussion, recall our discussion of the quantal close coupling method in Section 3-23-I. This method is useful if the number of coupled states is small and if $\lambda \gtrsim r_0$, where λ is the wavelength of the projectile and r_0 is the radius of the target. Computation difficulties are excessive otherwise.

We approach the *classical limit* when

$$\lambda \ll r_0. \tag{3-24-1}$$

Here the significant volume of phase space becomes large compared with $(2\pi\hbar)^f$, and the dynamical behavior of the system changes little over the volume $(2\pi\hbar)^f$. The wavelengths of the wave functions of the system are small compared to the characteristic distances of the system, and we are in the region of large quantum numbers. In this classical limit,

$$A \gg \hbar, \tag{3-24-2}$$

where A is the appropriate classical action for the corresponding classical problem. [Incidentally, if we use the Bohr-Sommerfeld action integrals and quantization conditions, it is easy to see that (3-24-2) corresponds to the regime of large quantum numbers. We may also note that in (3-24-1), if we identify λ with the wavelength of light and r_0 with the size of an obstacle, we have the condition for geometrical optics, in which diffraction effects are negligible.]

Conditions (3-24-1) and (3-24-2) are reasonably well satisfied under certain conditions for many kinds of electron–heavy particle collisions. They are especially well fulfilled for target atoms in highly excited states. For these "Rydberg" atoms, adjacent energy levels are closely spaced, and a large volume of phase space is important. The electric field at the outer edges of the atom varies only slowly with r, and the valence electrons are almost free. The atom is of enormous size, and its quantum numbers are large (states with $n \gtrsim 100$ are of great interest in astrophysical applications).

Classical and semiclassical methods are also appropriate for many kinds of heavy particle–heavy particle collisions. Indeed, the condition (3-24-1) shows that a classical treatment of the *relative* motion of two heavy collision partners is valid if the relative velocity exceeds 10^5 cm/sec, which corresponds to thermal motion at room temperature. (Of course, the application of classical methods to transitions in heavy particle collisions is another matter.)

B. Methods and Approximations

1. *The Semiclassical Impact Parameter Method*[65]

Many cross section calculations on systems for which the concept of a well-defined trajectory is valid for the relative motion have been made in the *impact parameter formulation.* The trajectory is taken to be rectilinear, and the impact parameter b plays the role of angular momentum through the relationship

$$M_r v_0 b = l\hbar. \tag{3-24-3}$$

Quantum mechanics is used to treat the electronic motion, so the impact parameter method is semiclassical. This method is most successful for high-energy, heavy particle collisions and low-energy transfers. An alternative approach is to integrate numerically the classical equations of motion to obtain the trajectory, or to apply classical perturbation theory to calculate it. The latter technique is applicable for large impact parameters and small energy transfers.

2. *The Classical Trajectory Monte Carlo Method*[66]

This method provides a means of evaluating collision cross sections. The equations governing the relative motion of the collision partners are integrated step by step on a computer for a large number of different impact parameters. The final states of the particles are determined, and the outcome of the collision is recorded. A Monte Carlo method is used for random selection of the impact parameter and the relevant target parameters (such as the position and momentum of the bound electrons). Thousands of collisions are studied, and cross sections are calculated from the relative probabilities of the various results. We have here the simulation of a scattering experiment on a computer, the initial conditions for each collision being chosen by using a sequence of random numbers to provide a statistical distribution that characterizes the initial physical conditions of an actual experiment.

3. *The Classical Impulse, Binary Encounter Approximation*[67]

Here we have the classical counterpart of the quantal impulse (sudden) approximation described in Section 3-23-E. In the present case we make the following assumptions:

1. The projectile follows a *classical trajectory*.
2. Initially, the target is in state i, with binding energy E_i. The binding forces are imagined to be switched off until after the collision, when in the case of excitation, for example, they are switched on again to produce a final state with binding energy E_f. This assumption is the *impulse approximation*.
3. During the collision, there are no interactions among the target electrons, and the interaction of each of these electrons with the projectile may be calculated independently. The separate cross sections are summed to obtain the total cross section. The *binary encounter approximation* is made here.

We are assuming that the probability of a given energy transfer ΔE is the same as if the projectile were passing through a gas of free electrons whose velocity distribution is the same as that of the classical atom. Classical binary encounter theory is most appropriate when the collisions are sudden (i.e., when the interaction time is short compared with the period of orbital motion of the target electron) and when the resulting ΔE is small.

The classical impulse, binary encounter approximation goes back to J. J. Thomson in 1912. It has been widely applied, especially to the case of the electronic and ionic projectiles impinging on neutral and ionized targets. Its success appears to be due mainly to two factors:

1. The correspondence principle[68] assures us that classical theory is accurate if the action integrals associated with the interactions obey (3-24-2), as is the case in many types of collisions.
2. The Coulomb potential has special properties of great importance here.[68] For distinguishable particles interacting through it, the classical and quantal scattering formulae are identical, and for indistinguishable particles, the formulas differ only because of interference between the direct and exchange terms. Furthermore, the microcanonical velocity distribution of a target electron is the same as the quantal velocity distribution.

We shall return to the classical impulse, binary encounter approximation in Section 6-13, where we discuss classical calculations of the cross sections for electron-impact excitation and ionization.

4. *The Eikonal Approximation*[69]

When the wavelength of the projectile is small compared with the distances over which the scattering potential changes appreciably, the concept of a classical trajectory acquires meaning. If r_0 is the range of the potential, the condition may be stated as

$$\kappa r_0 \gg 1. \tag{3-24-4}$$

This condition is the basis of semiclassical scattering approximations, which have been very useful for heavy particle, and also for electron, scattering. Further, if the energy of the projectile, E, is large compared with a typical value of the potential, V_0, so that

$$\frac{V_0}{E} \ll 1, \tag{3-24-5}$$

the eikonal approach to scattering problems becomes feasible, as we shall show here.

First let us consider the classical limit of the time-independent wave equation for the projectile-target relative motion in elastic scattering by a structureless potential $V(r)$:

$$\left[-\frac{\hbar^2}{2m}\nabla^2 + V(r) - E\right]\psi(\mathbf{r}) = 0. \tag{3-24-6}$$

We write

$$\psi(\mathbf{r}) = e^{iS(\mathbf{r})/\hbar} \tag{3-24-7}$$

and obtain

$$\frac{1}{2m}[-i\hbar\nabla^2 S + (\nabla S)^2] = E - V(r). \tag{3-24-8}$$

We go to the classical limit when $\nabla^2 S \ll (\nabla S)^2$, this condition being equivalent to the limit $h \to 0$. In this limit, $S = S_0(\mathbf{r})$, and

$$\frac{1}{2m}(\nabla S_0)^2 = E - V(r). \tag{3-24-9}$$

If $S_0(\mathbf{r})$ is taken to be Hamilton's characteristic function, (3-24-9) is the classical Hamilton-Jacobi equation. In optics, this equation is called the *eikonal equation.* By integrating (3-24-9), we can determine the trajectories that are normal to the surfaces $S_0(\mathbf{r}) =$ constant. Since the normals to the surfaces are parallel to ∇S, a formal solution of (3-24-9) is

$$S_0(\mathbf{r}) = \int ds\{2m[E - V(r)]\}^{1/2}, \tag{3-24-10}$$

where the integration is along a trajectory. If we substitute $S_0(\mathbf{r})$ into (3-24-7), we obtain the "eikonal wave function." The use of this approximate wave function in the integral equation for the scattering amplitude (3-16-12) is the basis of the *eikonal approximation.* This approximation is most accurate for high impact energies or for weak interactions. It represents the relative motion of the collision partners by a distorted plane wave, and may be regarded as a first-order correction to the Born approximation, which treats the relative motion as an undistorted plane wave. If the impact energy greatly exceeds the interaction energy and the internal energy of the system, the trajectory of the system will not deviate significantly from a straight-line path.

5. *Multichannel Eikonal Treatment*[70]

Generalization of the eikonal method to inelastic collisions can be carried out by writing the system wave function (3-23-13) as

$$\psi(\mathbf{r}, \mathbf{r}_1) = \sum_n A_n(\mathbf{r})\psi_n(\mathbf{r}_1)e^{iS_n(\mathbf{r})/\hbar}, \tag{3-24-11}$$

where the eikonal S_n for classical relative motion with local wave number $\kappa_n(\mathbf{r})$ in channel r under the static interaction $V_{nn}(\mathbf{r})$ is the solution of

$$(\nabla S_n)^2 = 2m[E - E_n - V_{nn}(\mathbf{r})] = \hbar^2\kappa_n^2(\mathbf{r}), \tag{3-24-12}$$

where E is the total energy of the system and E_n is the internal energy associated with the eigenfunction ψ_n. For a straight-line trajectory for relative motion ($\mathbf{r} = \mathbf{b}, z$) along the Z axis, the probability amplitudes satisfy the set of coupled equations

$$\frac{i\hbar^2}{m}\kappa_n\frac{\partial A_n}{\partial z} = \sum_{m \neq n} A_m(\mathbf{b}, z)V_{mn}(\mathbf{b}, z)e^{i(S_m - S_n)}, \tag{3-24-13}$$

where b is the impact parameter. Substitution of (3-24-13) into (3-20-21) yields

$$f_{in}(\vartheta) = -\frac{1}{4\pi}\frac{2m}{\hbar^2}\sum_m \langle e^{i\mathbf{k}_n\cdot\mathbf{r}}|V_{mn}(r)|A_m(\mathbf{r})e^{iS_m(\mathbf{r})}\rangle \tag{3-24-14}$$

for the scattering amplitude for $i \to n$ transitions as given by the multichannel eikonal treatment.

6. *The Glauber Approximation*[71]

In 1959, while seeking a high-energy scattering approximation more satisfactory than the first Born approximation, R. J. Glauber proposed an eikonal method in which all orders of the perturbation expansion were summed, with the leading term being the Born approximation. The Glauber approximation was introduced for nuclear problems and was not applied to atomic collisions until 1968. However, it has subsequently been used extensively in calculations on elastic and inelastic collisions of electrons with atoms and molecules, at intermediate and high energies. One of its useful features is that it satisfies the optical theorem, thereby being applicable in certain situations where the potential is too strong for the Born approximation to be valid.

The two main assumptions underlying the *Glauber approximation* are (1) all important atomic states ψ_n have the same energy, and (2) the scattering is confined to small angles in the forward direction. The first assumption is the closure assumption discussed in Section 3-23-M, which should be consulted at this point. The Glauber approximation played an important role in focusing attention on eikonal-type approximations in atomic collisions. Lately, the approximation in its original form has seen little use as its limitations have become better understood. These limitations include (1) the nonphysical singularity in the elastic forward-scattering amplitude, and (2) the alternating purely real and imaginary nature of the multiple scattering expansion.

7. *The Semiclassical S-Matrix Method*[72]

Semiclassical techniques in collision theory offer the possibility of retaining the computational simplicity of classical mechanics without sacrificing any essential quantal features. Of special importance is the *semiclassical* ***S****-matrix method*, which involves the analog of the quantal **S** matrix discussed in Sections 3-19 and 4-8-B. It has been applied with considerable success to elastic, inelastic, and reactive collisions between heavy particles at low energies. The development of the method began in the late 1960s and is due mainly to W. H. Miller and R. A. Marcus. Miller started from the semiclassical limit of the Feynman propagator[73] and followed along lines similar to those that establish the eikonal relation between physical and geometrical optics. Marcus utilized the semiclassical connection between the Schrödinger and Hamilton-Jacobi equations to obtain a generalized WKB form for the multidimensional wave function. This approach is an extension of the WKB treatment of motion in one dimension.

The semiclassical **S**-matrix method provides the ability to calculate accurate cross sections by integrating the classical equations of motion. The scattering amplitude may be represented as an integral over all possible phase-weighted classical trajectories that are relevant, with the phase expressed in terms of the classical action. The development of increasingly accurate methods of evaluating this integral has been the subject of much research. Here we have a semiclassical theory of scattering that combines exact* classical dynamics with the quantal principle of superposition. All quantum effects, arising as they do from the superposition of probability amplitudes, are contained at least qualitatively within the calculations. Such effects include interference, tunneling, resonances, selection rules, diffraction, and quantization itself.

We now reproduce Gentry's outline[72] of the application of the **S**-matrix method to the vibrational excitation of a diatomic molecule by an atom. As we have seen, the dynamical part of the problem is to be calculated by purely classical means. Use is made of only those so-called "root trajectories" that satisfy double-ended quantal boundary conditions for the initial and final states, and we associate with each root trajectory a phase that is the action integral computed along the trajectory. The results of each trajectory calculation are cast in terms of the action-angle variables, with the vibrational action n being the continuous classical analog of the vibrational quantum number v, and the conjugate angle variable q being the vibrational phase. Both n and q are functions of time during the collision, and they are evaluated numerically. Asymptotically, at $t \rightarrow \pm\infty$, n is a constant of the motion and $q \sim t$. For an initial relative kinetic energy T_i, we calculate the probability of a specified vibrational transition $v_i \rightarrow v_f$ by first identifying all of the classical trajectories for which both boundary conditions $n(-\infty) = v_1$ and $n(\infty) = v_f$ are satisfied. The first of these conditions ($n_i = v_i$) is the initial boundary condition for the

*Here common usage dictates that "exact" refer to numerical calculations that are accurate to within a few significant figures.

trajectory. The second condition ($n_f = v_f$) is obtained by searching the space of the initial phase

$$\bar{q}_i = q_i - R\left(\frac{m}{2T_i}\right)^{1/2} \tag{3-24-15}$$

for the relationship

$$n_f(\bar{q}_i) = v_f. \tag{3-24-16}$$

Here R is the separation distance between the atom and the center of mass of the molecule, and m is the reduced mass of the system. The elements of the **S**-matrix are then given by

$$S_{v_i v_f} = \sum\left[2\pi \frac{\partial n_f(\bar{q}_i, n_i)}{\partial \bar{q}_i}\right]^{-1/2} e^{(i/\hbar)\xi(\bar{q}_i, n_i)}, \tag{3-24-17}$$

where the phase ξ is given by the integral

$$\xi(\bar{q}_i, n_i) = -\int_{t_i}^{t_f} (mR\dot{R} + q\dot{n})\, dt. \tag{3-24-18}$$

The sum here is over all trajectories that satisfy (3-24-16). Finally, the transition probability is

$$P_{v_i v_f} = |S_{v_i v_f}|^2. \tag{3-24-19}$$

The expression for the **S** matrix in (3-24-17) is the "primitive" semiclassical result; it fails near an extremum in $n_f(\bar{q}_i, n_i)$. This failure, which is associated with the use of the stationary phase (saddle point) method in evaluating **S**, can often be avoided by using, instead, the "uniform approximation," which avoids a spurious singularity near an extremum. The coherent combination of **S**-matrix amplitudes is an important feature of the **S**-matrix method, as it allows the semiclassical description of interference phenomena, a capability not found in other semiclassical methods.

The semiclassical **S**-matrix method has another useful feature that emerges upon analytical continuation of the classical solutions (i.e., if complex values of the time steps, coordinates, momenta, and phases are considered). It is possible to find complex-valued root trajectories that satisfy real-valued boundary conditions even for transitions that are classically forbidden. (A transition is said to be classically forbidden if no real-valued root trajectories can be found for it.) Thus we can obtain information about quantum mechanical tunneling, elastic scattering beyond the rainbow angle, and similar phenomena that are inaccessible to other semiclassical methods. It is important to point out that here "forbidden" means that no solution of the classical equations leads to the transition in question; there is no implication that any conservation law is violated.

Apart from vibrational excitation in collinear collisions, the **S**-matrix theory becomes computationally infeasible for three-dimensional systems, even in comparison with full quantal methods.

8. *Other Approaches*

The discussion up to this point by no means covers all of the approaches and methods used in classical and semiclassical collision theory. The reader may refer to the books and reviews in Ref. 74 for material not covered here and for elaboration of the topics that we have sketched in this section.

3-25. NOTES

*3-1. The **T** Matrix.* We wish to recast (3-20-18) into a form that illustrates use of the **T** matrix. We generalize our previous result slightly and use the symbols i and j to denote arbitrary initial and final states, respectively, of the particles and of the system comprising them. We shall explicitly use the interaction potential V. Let Ψ_i be the exact solution of the Schrödinger equation having the asymptotic form

$$\Psi_i = e^{i\mathbf{K}_i\cdot\mathbf{r}}\psi_i + \text{outgoing waves.} \tag{3-25-1}$$

ψ_i is the wave function for the initial state of the target, and $e^{i\mathbf{K}_i\cdot\mathbf{r}}$ is the plane wave function of the projectile, which we shall take to be an electron (mass m). The final state of the system has the wave function

$$\Phi_f = e^{i\mathbf{K}_f\cdot\mathbf{r}}\psi_f, \tag{3-25-2}$$

where ψ_f and $e^{i\mathbf{K}_f\cdot\mathbf{r}}$ represent the final states of the target and electron. In Dirac notation, the normalization condition for the initial and final states of the target is

$$\langle\psi_i|\psi_j\rangle = \delta_{ij}. \tag{3-25-3}$$

The elements of the **T** matrix are defined as

$$T_{ji} = \langle\Phi_j|V|\Psi_i\rangle, \tag{3-25-4}$$

and the scattering amplitude [corresponding to f_n in (3-20-18)] is related to

T_{ji} by

$$f_{ji} = -\frac{m}{2\pi\hbar^2} T_{ji}. \tag{3-25-5}$$

Bransden proves this relationship on pages 139–144 of his book.[19]

We can make (3-25-5) plausible by recalling our treatment of elastic scattering in the Born approximation and examining (3-16-7) for the scattering amplitude. If we write out $U(r) \equiv 2M_r V/\hbar^2$ and replace M_r by m, we see that f in (3-16-7) can be written in the form

$$f = -\frac{m}{2\pi\hbar^2} \langle \Phi_j | V | \Psi_i \rangle, \tag{3-25-6}$$

where the matrix element is T_{ji}. In this restricted case, Ψ_i and Φ_j contain only the electron wave functions.

Using (3-25-5) we can now write (3-20-18) in the desired form:

$$I_{j \leftarrow i}(\vartheta)\, d\Omega_{\text{Lab}} = \frac{\kappa_j}{\kappa_i} \frac{m^2}{4\pi^2\hbar^4} |T_{ji}|^2\, d\Omega_{\text{Lab}}. \tag{3-25-7}$$

This result gives the differential cross section for the transition $i \to j$ in terms of the modulus squared of the corresponding element of the **T** matrix. It applies to elastic and inelastic collisions. Properties of the **T** matrix are discussed by Taylor,[19] Bransden,[19] and Mott and Massey.[36]

3-2. *Scattering by a Complex Potential.* We can account phenomenologically for the possibility of inelastic or reactive scattering of projectiles out of the elastic channel by using a *complex* (or "*optical*") *potential*:

$$V = V_R - iV_I. \tag{3-25-8}$$

The occurrence of nonelastic collisions is regarded as absorption from the projectile beam and is described by the imaginary part of the potential, as shown by the continuity equation

$$\nabla \cdot \mathbf{J} = -\frac{2}{\hbar} V_I |\Psi|^2. \tag{3-25-9}$$

Let us take the case of a central complex potential. The partial wave analysis of Section 3-15 may be generalized to cover the present situation if we introduce complex phase shifts (η_l). We obtain for the lth partial elastic cross section

$$q_l^{\text{el}} = \frac{\pi}{\kappa^2}(2l+1)|\beta_l e^{2i\text{Re}\cdot\eta_l} - 1|^2 \leqslant \frac{4\pi(2l+1)}{\kappa^2} \tag{3-25-10}$$

and for the integral elastic cross section

$$q_{\text{int}}^{\text{el}} = \sum_{l=0}^{\infty} q_l^{\text{el}} = \frac{\pi}{\kappa^2} \sum_{l=0}^{\infty} (2l + 1)|\beta_l e^{2i\text{Re}\cdot\eta_l} - 1|^2, \tag{3-25-11}$$

where

$$\beta_l(\kappa) = e^{-2\text{Im}\cdot\eta_l} \tag{3-25-12}$$

is the absorption factor. The integral absorption cross section is given as

$$q_{\text{int}}^{\text{abs}} = \sum_{l=0}^{\infty} q^{\text{abs}} = \frac{\pi}{\kappa^2} \sum_{l=0}^{\infty} (2l + 1)(1 - \beta_l^2), \tag{3-25-13}$$

where

$$q_l^{\text{abs}} = \frac{\pi}{\kappa^2}(2l + 1)(1 - \beta_l^2) \leqslant \frac{\pi(2l + 1)}{\kappa^2}. \tag{3-25-14}$$

Note that pure elastic scattering in the lth partial wave corresponds to $l = 1$ and $q^{\text{abs}} = 0$. There is a lack of symmetry here, in that nonelastic scattering ($0 \leqslant \beta_l < 1$) is always accompanied by elastic collisions. Finally, it may be shown that the total cross section $q_{\text{tot}} \equiv q_{\text{int}}^{\text{el}} + q_{\text{int}}^{\text{abs}}$ obeys the *generalized optical theorem*

$$q_{\text{tot}} = \frac{4\pi}{\kappa} \text{Im} \cdot f_{\text{el}}(0). \tag{3-25-15}$$

[This equation may be compared with (3-15-36), which was derived on the assumption of only elastic scattering.] Equation (3-25-15) relates the complete cross section for all types of scattering events to the imaginary part of the *elastic* scattering amplitude in the forward direction. The results discussed here are proved by Dalgarno[74] and by Bransden and Joachain.[44] Interesting applications to nuclear physics are discussed by many authors.*

3-3. Levinson's Theorem. In 1949, N. Levinson proved an important relationship between the phase shift and the number of bound states of the target. For *local* potentials (i.e., those depending only on r), *Levinson's theorem* states that

$$\eta_l(E = 0) - \eta_l(E = \infty) = n_a \frac{\pi}{2} + n_b \pi, \tag{3-25-16}$$

*See, for example, E. Segrè, *Nuclei and Particles*, Second Edition, Benjamin/Cummings, Reading, Mass. (1977).

where n_b is the number of bound states for angular momentum l, and $n_a = 1$ or 0 depending on the existence or nonexistence of an s-wave bound state at zero energy. The lth partial wave phase shift is denoted by η_l, which is evaluated here at zero energy and at infinite energy. For a finite potential, η_l vanishes at infinite projectile energy, so η_l passes through $(n_b - 1)\pi$, $(n_b - 2)\pi, \ldots, \pi$ as the impact energy increases. The partial cross section q_l goes through zero at $(n_b - 1)$ values of E.

In the event of a *nonlocal potential* (Section 4-8), there may be bound states that are normalized and that lie in the continuum. Equation (3-25-16) is invalidated by such states and must be modified. P. Swan has formulated a *modified Levinson's theorem* of the form

$$\eta_l(E = 0) - \eta_l(E = \infty) = (n_b + n_p)\pi. \tag{3-25-17}$$

Here n_b is the number of bound states and n_p is the number of families of degenerate solutions that vanish at infinity. These solutions are related to bound states that could exist if they were not excluded by the Pauli principle.

Levinson's theorem is discussed at greater length by Bransden,[19] Dalgarno,[74] and Bransden and Joachain.[44]

3-4. *Pseudopotentials.*[75,76] The optical spectra and reactions of atoms and molecules are quantitatively determined by their valence electrons. However, the core electrons affect the wave functions for the valence electrons because the potential acting on a valence electron when it is near the nucleus depends on the amount of screening of the nuclear charge by the core electrons. Also, the wave function of a valence electron must be orthogonal to all of the core electron orbitals; this requirement causes severe problems.

In 1935, H. Hellman and P. Gombas proposed the use of pseudopotentials to alleviate these difficulties. They reasoned that the main effect of the orthogonality requirement is that the valence electrons are excluded from the core, and that a short range repulsive potential produces a similar effect. The potentials that control the motion of the valence electrons can be modified empirically to achieve this result. These modified potentials (*pseudopotentials*) permit the use of empirical information to perform calculations with greater precision than obtainable in other approaches. The pseudopotential concept has proved useful in electron scattering calculations, in studies of electron affinities and interatomic potentials, and in problems in quantum chemistry and solid-state physics.

As pointed out by Dalgarno,[76] the pseudopotential is a nonlocal, energy-dependent, and angular momentum-dependent potential. Its main value in atomic physics is for systems with more than one electron; the potential for such a system can be pieced together from one-electron potentials that have been calculated previously.

3-5. *Good Quantum Numbers.* A quantum number that corresponds to an observable quantity Q is said to be a "good quantum number" if Q is conserved, that is, if its expectation value $\langle Q \rangle$ is a constant of the motion, independent of time. The total time derivative of $\langle Q \rangle$ is

$$\frac{d}{dt}\langle Q \rangle = \frac{i}{\hbar}\langle [\hat{H}, \hat{Q}] \rangle + \left\langle \frac{\partial \hat{Q}}{\partial t} \right\rangle.$$

Hence, to have a good quantum number, the operator $\hat{Q}$ corresponding to Q must not depend explicitly on the time, and it must commute with the Hamiltonian: $[\hat{Q}, \hat{H}] = 0$. The commutation of $\hat{Q}$ with $\hat{H}$ means that both the observable corresponding to $\hat{Q}$, namely Q, and that corresponding to $\hat{H}$, namely the energy E, can be precisely specified at the same time (i.e., that they both have definite values which will surely be equal to the values obtained in an "ideal measurement" of the quantities). We can find a set of functions ψ_n that span the space of the problem at hand and that are simultaneously eigenfunctions of both $\hat{H}$ and $\hat{Q}$:

$$\hat{H}\psi_n = E_n\psi_n, \qquad \hat{Q}\psi_n = q_n\psi_n.$$

The real numbers E_n and q_n are the eigenvalues associated with the eigenfunctions, which are orthonormal. The ψ_n are energy eigenfunctions (time independent). If the system is in state n, its total time-dependent wave function is $\Psi_n = \psi_n \exp[-(i/\hbar)E_n t]$. Such a state of definite energy is called a *stationary state.*

The quantum number corresponding to $\hat{Q}$ will be good if $\hat{Q}$ has a diagonal matrix in the representation used (i.e., when computed with respect to the functions ψ_n). In this case, $\langle \psi_m | \hat{Q} | \psi_n \rangle = q_m \delta_{mn}$. Here $\hat{Q}$ acting on ψ_n gives a real number times ψ_n and does not mix the eigenfunctions. Remember that ψ_n is orthogonal to ψ_m if $m \neq n$. In general, if a function ϕ is not an eigenfunction of an operator $\hat{A}$, then $\hat{A}$ acting on ϕ will not give ϕ times a real constant but rather mixes in other functions and produces a linear combination of them so that the system is not in a well-defined state corresponding to a single member of the set of functions. The problem of solving the Schrödinger time-independent wave equation with a given $\hat{H}$ may be considered as equivalent to finding the particular set of orthonormal functions that makes the matrix of $\hat{H}$ computed with respect to these functions diagonal. We can also speak of good quantum numbers associated with nonobservable entities such as the parity.

3-26. PROBLEMS

Solutions for most of these problems appear in the references indicated.

3-1. A particle is moving in the field of force described by

$$F_x = 2yz(1 - 6xyz), \; F_y = 2xz(1 - 6xyz), \; F_z = 2xy(1 - 6xyz).$$

Can a potential function $V(x, y, z)$ be defined? Is the total energy $E = T + V$ constant during the motion of a particle? Repeat the problem for the force field

$$F_x = y^2 + z^2 + 2(xy + yz + zx),\ F_y = x^2 + z^2 + 2(xy + yz + zx),$$
$$F_z = x^2 + y^2 + 2(xy + yz + zx).$$

(Chen,[77] p. 14.)

3-2. Consider scattering of particles by a repulsive force $f(r) = k/r^3$. Prove that the classical differential cross section is

$$\sigma(\theta)\, d\theta = \frac{k}{2E} \frac{(1 - x)\, dx}{x^2(2 - x)^2 \sin \pi x},$$

where x is the ratio θ/π and E is the energy. (Landau and Lifshitz,[78] p. 54.)

3-3. Show that the classical differential cross section for scattering by a rectangular potential well

$$V = \begin{cases} -V_0 & (r < a) \\ 0 & (r > a) \end{cases}$$

is

$$\sigma(\Theta) = \frac{n^2 a^2}{4 \cos(\Theta/2)} \frac{[n \cos(\Theta/2) - 1][n - \cos(\Theta/2)]}{[1 + n^2 - 2n \cos(\Theta/2)]^2}.$$

where $n = \sqrt{(E + V_0)/E}$, E being the energy of the incident particles. (Landau and Lifshitz,[78] pp. 54–55.)

3-4. Calculate classically the differential cross section for the potential

$$V(r) = \begin{cases} +\infty & (r < a) \\ -V_0 & (a < r < a_0) \\ 0 & (r > a_0). \end{cases}$$

This potential corresponds to a rigid core of radius a surrounded by an attractive square well that extends out to $r = a_0$. (Seshadri,[79] p. 244.)

3-5. Suppose that two particles interact through the repulsive Coulomb potential. Show by classical mechanics that in a head-on collision the distance of closest approach equals twice the value of the impact parameter for which the scattering angle is 90°. (Evans,[80] pp. 844–845.)

3-6. Show that the classical cross section for a particle to fall to the center of an attractive potential field $V(r) = -\beta/r^2$ is $\sigma = \pi\beta/E$, where E is the energy at infinity. Landau and Lifshitz[78] (p. 51) solve this problem and the corresponding one for $V(r) = -\beta/r^n$, where $n > 2$.

3-7. Particles of mass m_1 are attracted by a sphere of mass m_2 and radius R according to Newton's law of gravitation. Show that the classical collision cross section is $\sigma = \pi R^2(1 + 2\gamma m_2/Rv_\infty)^2$, where γ is the gravitational constant. (Landau and Lifshitz,[78] pp. 51–52.)

3-8. Assume that particles of energy E are scattered by a repulsive potential $V(r)$ such that $V(0) > E$ and $V(\infty) = 0$. Given the cross section as a function of scattering angle for a given energy E, deduce the functional form of $V(r)$. (Landau and Lifshitz,[78] pp. 52–53.)

3-9. Consider the elastic scattering of structureless particles from a rigid sphere:

$$V(r) = \begin{cases} +\infty & (r < a) \\ 0 & (r > a). \end{cases}$$

Calculate the phase shifts, and obtain the differential and integral cross sections in the low- and high-energy limits. (Rapp,[19] pp. 537–539; Mott and Massey,[36] pp. 38–40; Bransden and Joachain,[44] pp. 479–480.)

3-10. Evaluate the integral elastic cross section for s-wave (low-energy) scattering by a rectangular potential barrier:

$$V(r) = \begin{cases} V_0 \quad (>0) & (r < a) \\ 0 & (r > a). \end{cases}$$

(Mott and Massey,[36] pp. 37–38; Bransden and Joachain,[44] p. 497; Rapp,[19] pp. 540–541.)

3-11. Consider the scattering from the rectangular potential well

$$V(r) = \begin{cases} -V_0 & (r < a) \\ 0 & (r > a). \end{cases}$$

Find the phase shifts for all l and the corresponding partial cross sections. (Mott and Massey,[36] pp. 29–36; Rapp,[19] pp. 540–541; Bransden and Joachain,[44] pp. 477–479.)

3-12. Use the Born approximation to the phase shifts to calculate η_0 for scattering by (a) the screened Coulomb potential $V(r) = V_0 e^{-\gamma r}/r$, and (b) the polarization potential $V(r) = V_0/(r^2 + d^2)^2$, where d is a constant. (Bransden and Joachain,[44] p. 498.)

3-13. Calculate, in the Born approximation, the scattering amplitudes, and the differential and integral cross sections, for the following potentials: (a) the rectangular potential well, (b) the polarization potential, (c) the exponential potential $V(r) = V_0 e^{-\gamma r}$, and (d) the Gaussian potential

$$V(r) = V_0 e^{-\gamma^2 r^2}.$$

(Bransden and Joachain,[44] p. 498; ter Haar,[81] p. 58; Mott and Massey,[36] pp. 42–43; Landau and Lifshitz,[19] pp. 483–484.)

3-14. Consider the repulsive potential $V = A/r^2$.

(a) Find the scattering phase shifts and the differential cross section at small angles.

(b) Find the differential cross section in the Born approximation and according to classical scattering theory. Discuss the limits of applicability of the results. (ter Haar,[81] p. 58; Mott and Massey,[36] pp. 41–42.)

3-15. Calculate the differential cross section in the Born approximation for scattering by a delta-function potential. (ter Haar,[81] p. 59.)

3-16. Calculate the zeroth-order phase shift and integral cross section for the scattering of low-energy projectiles by the potential $V(r) = \alpha/r^4$ $(\alpha > 0)$. (Landau and Lifshitz,[19] pp. 504–505.)

3-17. Calculate, in the Born approximation, the cross section for elastic scattering of an electron by a spherical nucleus with charge uniformly distributed through its volume. (Kogan and Galitskiy,[82] p. 23.)

3-18. Calculate, in the Born approximation, the cross section for the scattering of fast neutrons by a Coulomb field. (Kogan and Galitskiy,[82] p. 23.)

3-19. Visualize a homonuclear diatomic molecule as consisting of two identical spherically symmetric scattering centers separated by the vector $\mathbf{l}$. If the scattering amplitude is known for a projectile of a given kind colliding with one of the isolated "atoms," what is the scattering cross section for this molecule? Neglect multiple scattering between the two "atoms." (Park,[83] pp. 338–339.)

3-20. Particles are scattered by a simple cubic structure with lattice spacing d. The particle–lattice point interaction is

$$V = \frac{-2\pi d\hbar^2}{m} \sum_i \delta(\mathbf{r} - \mathbf{r}_i).$$

Calculate the scattering amplitude in the Born approximation, and show that scattering occurs in directions consistent with the Bragg law. (Cronin et al.,[84] p. 34.)

3-21. Two identical particles, which have spin but which interact only through a short-range central force, collide at low relative velocity. Calculate the quantal differential cross section. (ter Haar,[81] p. 61.)

3-22. Calculate the quantal differential cross section for the mutual elastic scattering of (a) two electrons, and (b) two alpha particles. (ter Haar,[81] p. 61.)

3-23. Listed below are 27 approximations used in atomic collision theory. The notations *e* and *HP* indicate their applicability to collisions in which *electrons* and/or *heavy particles* serve as projectiles. Justify these notations in each case.

1. The Born approximation; the Born series: e, also HP for certain phenomena, not including charge transfer
2. The Bethe-Born (Bethe) approximation: same as 1.
3. The Coulomb-Born approximation: electrons and ions colliding with ions
4. The Born-Oppenheimer approximation: e
5. The Born-Ochkur approximation: e
6. The Born-exchange approximation: e
7. The quantal impulse (sudden) approximation: e, HP
8. The distorted wave (DW) approximation: e, HP
9. The perturbed stationary-state (PSS) approximation: HP
10. The perturbed rotating atom (PRA) approximation: HP
11. The polarized orbital approximation: e
12. The exchange-adiabatic approximation: e
13. The adiabatic approximation: e
14. The distorted wave–polarized orbital approximation: e
15. The close-coupling (truncated eigenfunction expansion) approximation: e, HP
16. The static exchange approximation: e
17. The pseudostate approximation: e, HP
18. The correlation approximation: e
19. The closure approximation: electrons and protons
20. The quantum defect approximation: e
21. The ***R***-matrix method: e, perhaps HP in some cases
22. The semiclassical impact parameter method: HP
23. The classical trajectory Monte Carlo method: HP
24. The classical impulse, binary encounter approximation: e, HP
25. The eikonal approximation: e, HP
26. The Glauber approximation: e, HP
27. The semiclassical ***S***-Matrix method: HP

REFERENCES

1. J. O. Hirschfelder, C. F. Curtiss, and R. B. Bird, *Molecular Theory of Gases and Liquids*, Wiley, New York (1964), p. 50 and Sec. 7-4. J. H. Ferziger and H. G. Kaper, *Mathematical Theory of Transport Processes in Gases*, North-Holland, Amsterdam (1972). E. W. McDaniel, *Atomic Collisions: Heavy Particle Projectiles*, Wiley, New York, to be published.
2. J. O. Hirschfelder, C. F. Curtiss, and R. B. Bird, Ref. 1, Secs. 8-3, 8-4, and 8-5. See also L. A. Viehland, E. A. Mason, W. F. Morrison, and M. R. Flannery, *ADNDT* **16**, 495 (1975). H. Pauly, "Elastic Scattering Cross Sections I: Spherical Potentials," in R. B. Bernstein (Ed.), *Atom-Molecule Collision Theory*, Plenum, New York (1979), p. 111. T. Su and M. T. Bowers, "Classical Ion-Molecule Collision Theory," in M. T. Bowers (Ed.), *Gas Phase Ion Chemistry*, Vol. 1, Academic, New York (1979), p. 83.
3. E. T. Whittaker, *A Treatise on the Analytical Dynamics of Particles and Rigid Bodies*, 4th ed., Cambridge University Press, Cambridge (1937), also published by Dover, New York (1944), see Chap. IV.
4. H. Goldstein, *Classical Mechanics*, 2nd ed., Addison-Wesley, Reading, Mass. (1980), Chap. 3.
5. See T. Kihara, M. H. Taylor, and J. O. Hirschfelder, *Phys. Fluids* **3**, 715 (1960). The behavior of the particle at the origin depends on the details of the scattering model assumed. The model under discussion here is the "transparent core model" of Kihara, Taylor, and Hirschfelder. These authors also treat the "rigid core model," which is the limit $D \to 0$ of the Sutherland potential (1-7-4), and the "random scattering model," which produces random scattering of the trajectories of the particles reaching the origin. Failure to recognize the effects of different assumed boundary conditions at the origin has led to discrepancies in published values of collision integrals.
6. M. A. Eliason, D. E. Stogryn, and J. O. Hirschfelder, *Proc. Nat. Acad. Sci. USA* **42**, 546 (1956). H. M. Mott-Smith, *Phys. Fluids* **3**, 721 (1960).
7. P. Langevin, *Ann. Chim. Phys. Ser.* **8**, **5**, 245–288 (1905).
8. E. W. McDaniel, *Collision Phenomena in Ionized Gases*, Wiley, New York (1964).
9. E. W. McDaniel, V. Čermák, A. Dalgarno, E. E. Ferguson, and L. Friedman, *Ion-Molecule Reactions*, Wiley, New York (1970). M. T. Bowers (Ed.), *Gas Phase Ion Chemistry*, Academic, New York, Vols. 1 and 2 (1979), Vol. 3 (1984).
10. R. D. Present, *Proc. Nat. Acad. Sci. USA* **41**, 415 (1955); *Kinetic Theory of Gases*, McGraw-Hill, New York (1958), pp. 152–153. E. A. Mason and J. T. Vanderslice, *J. Chem. Phys.* **28**, 253, 1070 (1958).
11. G. H. Wannier, *Bell Syst. Tech. J.* **32**, 170 (1953).
12. E. Vogt and G. H. Wannier, *Phys. Rev.* **95**, 1190 (1954).
13. R. B. Bernstein, *Chemical Dynamics via Molecular Beam and Laser Techniques*, Oxford University Press, New York (1982). H. Pauly, Ref. 2. R. D. Levine and R. B. Bernstein, *Molecular Reaction Dynamics and Chemical Reactivity*, Oxford University Press, New York (1987). M. S. Child, *Molecular Collision Theory*, Academic, New York (1974).
14. E. J. Williams, *Rev. Mod. Phys.* **17**, 217 (1945).

15. E. Everhart, G. Stone, and R. J. Carbone, *Phys. Rev.* **99**, 1287 (1955). G. H. Lane and E. Everhart, *Phys. Rev.* **117**, 920 (1960).

16. R. D. Present, *Kinetic Theory of Gases*, McGraw-Hill, New York (1958), pp. 113–114. K. Yang and T. Ree, *J. Chem. Phys.* **35**, 588 (1961).

17. H. S. W. Massey and C. B. O. Mohr, *Proc. R. Soc. London A* **144**, 188 (1934). L. D. Landau and E. M. Lifshitz, *Quantum Mechanics, Non-relativistic Theory*, 2nd. ed., Addison-Wesley, Reading, Mass. (1965).

18. N. F. Mott and H. S. W. Massey, *The Theory of Atomic Collisions*, 2nd ed., Oxford University Press, Oxford (1952). E. H. S. Burhop, "Theory of Collisions," in D. R. Bates (Ed.), *Quantum Theory*, Vol. I, *Elements*, Academic, New York (1961). *Note:* The *third* edition of Mott and Massey is cited in Ref. 36.

19. B. H. Bransden, *Atomic Collision Theory*, 2nd ed., Benjamin-Cummings, Menlo Park, Calif. (1983). L. D. Landau and E. M. Lifshitz, *Quantum Mechanics, Non-relativistic Theory*, 2nd ed., Addison-Wesley, Reading, Mass (1965). A. Messiah, *Quantum Mechanics*, Interscience, New York, Vol. I (1961), Vol. II (1962). E. Merzbacher, *Quantum Mechanics*, 2nd ed., Wiley, New York (1970). T. Y. Wu and T. Ohmura, *Quantum Theory of Scattering*, Prentice-Hall, Englewood Cliffs, N.J. (1962). C. Cohen-Tannoudji, B. Diu, and F. Laloë, *Quantum Mechanics*, 2 vols., Wiley, New York (1977). D. Rapp, *Quantum Mechanics*, Holt, Rinehart and Winston, New York (1971). J. R. Taylor, *Scattering Theory*, Wiley, New York (1972). R. G. Newton, *Scattering Theory of Waves and Particles*, 2nd ed., Springer, New York (1982). C. J. Joachain, *Quantum Collision Theory*, North-Holland, Amsterdam (1975). A. S. Davydov, *Quantum Mechanics*, Pergamon, Oxford (1965). L. I. Schiff, *Quantum Mechanics*, 3rd ed., McGraw-Hill, New York (1968). L. S. Rodberg and R. M. Thaler, *Introduction to the Quantum Theory of Scattering*, Academic, New York (1967). J. J. Sakurai, *Advanced Quantum Mechanics*, Addison-Wesley, Reading, Mass (1967); *Modern Quantum Mechanics*, Benjamin/Cummings, Menlo Park, Cal. (1985).

20. N. F. Mott and H. S. W. Massey, Ref. 18, pp. 14–18. L. S. Rodberg and R. M. Thaler, Ref. 19.

21. H. S. W. Massey and E. H. S. Burhop, *Electronic and Ionic Impact Phenomena*, Oxford University Press, Oxford (1952).

22. N. F. Mott and H. S. W. Massey, Ref. 18, Chap. VI.

23. N. F. Mott and H. S. W. Massey, Ref. 18, pp. 20–22. B. H. Bransden, Ref. 19.

24. Lord Rayleigh, *The Theory of Sound*, Dover, New York (1945), published first by Macmillan, London (1894). Also see P. M. Morse, *Vibration and Sound*, 2nd ed., McGraw-Hill, New York (1948), Chap. VII.

25. H. Faxén and J. Holtsmark, *Z. Phys.* **45**, 307 (1927).

26. H. S. W. Massey and C. B. O. Mohr, *Proc. R. Soc. London A* **141**, 434 (1933). N. F. Mott and H. S. W. Massey, Ref. 18, pp. 38–40. J. M. Blatt and V. F. Weisskopf, *Theoretical Nuclear Physics*, Wiley, New York (1952), pp. 318–325. M. S. Child, Ref. 13. L. S. Rodberg and R. M. Thaler, Ref. 19.

27. N. F. Mott and H. S. W. Massey, Ref. 18, pp. 30–31. E. H. S. Burhop, Ref. 18, pp. 314–316.

28. J. M. Blatt and V. F. Weiskopf, Ref. 26, p. 382.

29. B. H. Bransden, Ref. 19.

30. D. R. Bates (Ed.), *Quantum Theory*, Vol. 1, Academic, New York (1961).

31. C. J. Joachain, Ref. 19.

32. M. Born, *Z. Phys.* **38**, 803 (1926).

33. N. F. Mott and H. S. W. Massey, Ref. 18, Chap. VI, Sec. 4.

34. N. F. Mott and H. S. W. Massey, Ref. 18, Chap. III. B. H. Bransden, Ref. 19.

35. L. D. Landau and E. M. Lifshitz, Ref. 19, Chap. IX. R. P. Feynman, R. B. Leighton, and M. Sands, *The Feynman Lectures on Physics*, Vol. 3, Addison-Wesley, Reading, Mass. (1965). E. H. S. Burhop, Ref. 18. A. S. Davydov, Ref. 19. L. I. Schiff, Ref. 19.

36. N. F. Mott and H. S. W. Massey, *The Theory of Atomic Collisions*, 3rd ed., Clarendon Press, Oxford (1965).

37. D. R. Bates, "Transitions," in D. R. Bates (Ed.), *Quantum Theory*, Vol. I, Academic, New York (1961).

38. M. R. C. McDowell and J. P. Coleman, *Introduction to the Theory of Ion-Atom Collisions*, North-Holland, Amsterdam (1970).

39. M. Born, *Z. Phys.* **37**, 863 (1926); **38**, 803 (1926).

40. D. R. Bates, A. Fundaminsky, and H. S. W. Massey, *Philos. Trans. R. Soc. London A* **243**, 93 (1950).

41. J. R. Oppenheimer, *Phys. Rev.* **32**, 361 (1928).

42. M. J. Seaton, "The Theory of Excitation and Ionization by Electron Impact," in D. R. Bates (Ed.), *Atomic and Molecular Processes*, Academic, New York (1962).

43. E. H. S. Burhop, Ref. 18.

44. B. H. Bransden, Ref. 19. J. R. Taylor, Ref. 19. B. H. Bransden and C. J. Joachain, *Physics of Atoms and Molecules*, Longman Group, Harlow, Essex, England (1983).

45. L. D. Landau and E. M. Lifschitz, Ref. 19.

46. H. A. Bethe, *Ann Phys. (Leipzig)* **5**, 325 (1930). H. S. W. Massey and E. H. S. Burhop, *Electronic and Ionic Impact Phenomena*, 2nd ed., Vol. I, Oxford University Press, London (1969).

47. B. H. Bransden, Ref. 19. H. S. W. Massey and H. B. Gilbody, *Electronic and Ionic Impact Phenomena*, 2nd ed., Vol. IV, Oxford Univerity Press, London (1974).

48. H. S. W. Massey and E. H. S. Burhop, Ref. 46. P. G. Burke, *Potential Scattering in Atomic Physics*, Plenum, New York (1977). B. H. Bransden, Ref. 19.

49. J. P. Coleman, *Case Stud. At. Collision Phys.* **1**, 99 (1969).M. R. C. McDowell and J. P. Coleman, *Introduction to the Theory of Ion-Atom Collisions*, North-Holland, Amsterdam (1970).

50. B. H. Bransden, Ref. 19. H. S. W. Massey and E. H. S. Burhop, Ref. 46. Y. Itikawa, *Phys. Rep.* **143**, 69 (1986).

51. H. S. W. Massey and H. B. Gilbody, Ref. 47.

52. N. F. Mott, *Proc. Cambridge Philos. Soc.* **27**, 553 (1931).

53. D. R. Bates, "Theoretical Treatment of Collisions between Atomic Systems," in D. R. Bates (Ed.), *Atomic and Molecular Processes*, Academic, New York (1962).

54. D. R. Bates, *Proc. R. Soc. London A* **240,** 437 (1957); **243**, 15 (1957). A. M. Arthurs and A. R. Holt, *Proc. Phys. Soc. London* **82**, 1073 (1963) have shown that the procedure of using rotating coordinates in scattering problems, which can be justified for exact calculations, breaks down in calculations based on the Born approximation.

55. B. H. Bransden, Ref. 19. H. S. W. Massey and E. H. S. Burhop, Ref. 46. R. J. Drachman and A. Temkin, *Case Stud. At. Collision Phys.* **2**, 399 (1972).

56. B. H. Bransden, Ref. 19. H. S. W. Massey and E. H. S. Burhop, Ref. 46.

57. P. G. Burke, Ref. 48. B. H. Bransden, Ref. 19. B. H. Bransden and C. J. Joachain, Ref. 44.

58. B. H. Bransden, Ref. 19. B. H. Bransden and M. R. C. McDowell, *Phys. Rep.* **30**, 207 (1977).

59. P. G. Burke and J. F. Williams, *Phys. Rep.* **34**, 325 (1977).

60. A. W. Weiss, *AAMP* **9**, 1 (1973).

61. H. R. J. Walters, *Phys. Rep.* **116**, 1 (1984). B. H. Bransden and M. R. C. McDowell, Ref. 58.

62. M. J. Seaton, *Rep. Prog. Phys.* **46**, 167 (1983). H. S. W. Massey and E. H. S. Burhop, Ref. 46.

63. B. H. Bransden, Ref. 19. P. G. Burke and C. J. Noble, *Phys. Rep.* (to be published) (1988). P. G. Burke, Ref. 48.

64. I. C. Percival, "Atomic Scattering Computations," in P. G. Burke and B. L. Moiseiwitsch (Eds.), *Atomic Processes and Applications*, North-Holland, Amsterdam (1976).

65. M. R. C. McDowell and J. P. Coleman, Ref. 49.

66. I. C. Percival and D. Richards, *AAMP* **11**, 2 (1975). B. H. Bransden and R. K. Janev, *AAMP* **19**, 2 (1983).

67. M. R. C. McDowell and J. P. Coleman, Ref. 49. I. C. Percival and D. Richards, Ref. 66. A. Burgess and I. C. Percival, *AAMP* **4**, 109 (1968). D. R. Bates and A. E. Kingston, *AAMP* **6**, 269 (1970). D. R. Bates, *Phys. Rep.* **35**, 305 (1978).

68. A. Norcliffe, *Case Stud. At. Phys.* **4**, 1 (1975).

69. L. I. Schiff, Ref. 19. D. Rapp, Ref. 19. B. H. Bransden, Ref. 19. P. G. Burke, Ref. 48. C. J. Joachain and F. Byron, *Phys. Rep.* **34**, 233 (1977).

70. M. R. Flannery and K. J. McCann, *J. Phys. B.* **7**, 2518 (1974).

71. B. H. Bransden and M. R. C. McDowell, Ref. 58. P. G. Burke, Ref. 48. H. R. J. Walters, Ref. 61. T. T. Gien, *Phys. Rep.* **160**, 123 (1988).

72. W. H. Miller, *Adv. Chem. Phys.* **25**, 69 (1974); **30**, 77 (1976). W. R. Gentry, "Vibrational Excitation II: Classical and Semiclassical Methods," in R. B. Bernstein (Ed.), *Atom-Molecule Collision Theory*, Plenum, New York (1979). M. S. Child, "The Semiclassical *S*-Matrix," in M. S. Child, *Molecular Collision Theory*, Academic, New York (1974).

73. R. P. Feynman and A. R. Hibbs, *Quantum Mechanics and Path Integrals*, McGraw-Hill, New York (1965).

74. M. V. Berry and K. E. Mount, *Rep. Prog. Phys.* **35**, 315 (1972). I. C. Percival, *Adv. Chem. Phys.* **36**, 1 (1977). K. W. Ford and J. A. Wheeler, *Ann. Phys.* (*N.Y.*) **7**, 259, 287 (1959). M. S. Child (Ed.), *Semiclassical Methods in Molecular Scattering and Spectroscopy*, Reidel, Boston (1980). B. C. Eu and M. L. Sink, *Semiclassical Theories of Molecular Scattering*, Springer-Verlag, Berlin (1983). M. R. Flannery, *Ann. Phys.* (*N.Y.*) **61**, 465 (1970); **79**, 480 (1973). W. H. Miller (Ed.), *Dynamics of Molecular Collisions*, Vols. 1 and 2, Plenum, New York (1976). M. S. Child, *AAMP* **14**, 225 (1978). A. S. Dickinson and D. Richards, *AAMP* **18**, 166 (1982). M. Baer (Ed.), *Theory of Chemical Reaction Dynamics*, 4 vols., CRC Press, Boca Raton, Fla. (1985). S. M. Younger and T. D. Märk, Semi-empirical and Semi-classical Approximations for

Electron Ionization," in T. D. Märk and G. H. Dunn (Eds.), *Electron Impact Ionization*, Springer-Verlag, Berlin (1985). M. S. Child, *Molecular Collision Theory*, Academic, New York (1974). A. Dalgarno, "Theory of Ion-Molecule Collisions," in E. W. McDaniel et al., Ref. 9.

75. J. N. Bardsley, *Case Stud. At. Phys.* **4**, 299 (1975).
76. A. Dalgarno, *At. Phy.* **4**, 325 (1975).
77. M. Chen, *Berkeley Physics Problems with Solutions*, Prentice-Hall, Englewood Cliffs, N.J. (1974).
78. L. D. Landau and E. M. Lifshitz, *Mechanics*, Addison-Wesley, Reading, Mass. (1960).
79. S. R. Seshadri, *Fundamentals of Plasma Physics*, American Elsevier, New York (1973).
80. R. D. Evans, *The Atomic Nucleus*, McGraw-Hill, New York (1955).
81. D. ter Haar (Ed.), *Problems in Quantum Mechanics*, 3rd ed., Pion, London (1975), also published by Academic, New York (1975).
82. V. I. Kogan and V. M. Galitskiy, *Problems in Quantum Mechanics*, Prentice-Hall, Englewood Cliffs, N.J. (1963).
83. D. Park, *Introduction to the Quantum Theory*, 2nd ed., McGraw-Hill, New York (1974).
84. J. A. Cronin, D. F. Greenberg, and V. L. Telegdi, *University of Chicago Graduate Problems in Physics with Solutions*, Addison-Wesley, Reading, Mass. (1967).

4

ELASTIC SCATTERING OF ELECTRONS BY NEUTRAL ATOMS AND MOLECULES

This chapter is devoted to experimental studies of elastic scattering of electrons by neutral structures, and to certain aspects of the associated collision theory. The elastic scattering of atoms, molecules, and ions is treated in Ref. 1. Although there are certain features common to the scattering of electrons and "heavy particles," the differences in their scattering behavior are so pronounced that different types of apparatus and techniques of theoretical analysis are required for their study. Hence division of our discussion of elastic scattering into two distinct parts is appropriate. Inelastic collisions of electrons with heavy particles are considered in Chapters 5 and 6, and in Chapter 7, which deals with various kinds of collisions of electrons with atomic and molecular ions.

We have already demonstrated in Section 1-5 that the mean fractional amount of energy lost by an electron in elastic collisions with molecules initially at rest is approximately

$$\Delta \approx \frac{2m}{M}, \tag{1-5-21}$$

where m and M are the electronic and molecular masses, respectively. Since the indicated mass ratio is always less than 10^{-3}, practical interest in the elastic scattering of electrons usually does not center about the attendant loss of energy. Rather, as a rule, it is the change in direction experienced by the scattered electrons that is the important aspect of the collision. The simple calculation in Section 1-5 indicates that the scattering should be isotropic in the Lab system as well as in the CM. This prediction, being based on the classical elastic sphere model, could hardly be expected to be correct. However, both experiment and more realistic calculations show that large-angle scattering is very pronounced, and this fact must be carefully considered in the design of apparatus utilizing electron beams. Large-angle scattering is particularly prominent at low energies but cannot be ignored even at high energies.*

Certain integrals of the angular differential cross section for elastic scattering are also of great interest. The unweighted integral over all angles [i.e., the *integral elastic scattering cross section* ($q_s \equiv \sigma_s$)], is evidently related to the total probability of elastic collision. A more important quantity is the *cross section for momentum transfer* ($q_D \equiv q_m \equiv \sigma_m$), which determines the rate of diffusion of electrons, the velocity at which they drift through a gas under the influence of an electric field, and the electrical conductivity of the gas. The momentum transfer cross section plays the dominant role in many applications of atomic collisions, not only for electrons but also for heavy particles.[2] This cross section was defined in Section 1-6.

A comment on the ambiguity in the term *total cross section* is in order here. In recent times it has become conventional to use it only as "the cross section for the totality of all possible kinds of collisions" ($q_t \equiv \sigma_t$), the sum of all (angle-integrated) cross sections for the various scattering channels. Unfortunately, the term "total elastic cross section" is still occasionally used for the integral of the differential elastic cross section over all angles ($q_s \equiv \sigma_s$)

We now begin a more detailed discussion of electron scattering. The range of electron energies of principal interest to us in this chapter (and throughout the remainder of the book as well) extends from zero up to about 1000 eV. However, collisions at much higher energies are also important and will be considered briefly.

4-1. THE INTEGRAL ELASTIC SCATTERING CROSS SECTION ($q_s \equiv \sigma_s$)†

The first studies of slow electron scattering were conducted by P. Lenard in 1903, but quantitative measurements were not made until 1921, when Ramsauer[3] introduced his method of determining electron cross sections. We shall discuss only studies performed from that date onward and concentrate on

*An interesting illustration of the large-angle scattering of electrons is furnished by cloud chamber photographs of β-ray tracks. Similar photographs of tracks made by alpha particles show that heavy particle scattering is confined, by contrast, almost entirely to extremely small angles about the forward direction. Examples of such photographs are presented by E. Rutherford, J. Chadwick, and C. D. Ellis, *Radiations from Radioactive Substances*, Cambridge University Press, Cambridge (1951), Plates III and VIII. Of course, the path of a particle shown in a cloud-chamber photograph is not determined solely by elastic scattering. Some of the deflections experienced by the particle are produced by inelastic events, but the angular scattering distributions of electrons and heavy particles differ in much the same way in inelastic scattering as they do in elastic collisions.

†Modern discussions of many of the experimental techniques described in this section and throughout the book appear in G. Scoles (Ed.), *Atomic and Molecular Beam Methods*, Vol. I, Oxford University Press, New York (1988). Among the topics covered there are beam sources, particle detection methods, mass selection, state selection, velocity selection, time-of-flight techniques, optical and spectroscopic methods, spin-polarized beams, and laser-excited beams.

much later work. The early measurements are treated by Brown,[4] and by Bederson and Kieffer.[5] Other sources of information are the books by Massey and Burhop,[6] Massey,[6] Hasted,[7] and Huxley and Crompton.[8] These books, as well as the chapter by Csanak et al.,[9] are valuable general references for our entire treatment of electron scattering. The book edited by Shimamura and Takayanagi[10] contains useful reviews of various aspects of electron-molecule collisions. Fano and Rau[11] provide much insight into various aspects of collision theory.

We shall subdivide our discussion of the cross section measurements according to the general type of experimental technique used. The first class of experiments to be considered involves the passage of a sharply defined, nearly monoenergetic beam of electrons through a gas and direct observation of the scattering that occurs. Then we shall treat the time-of-flight method, and finally, "crossed-beam" studies.

A. Direct Single-Beam Measurements

Ramsauer's method of electron cross section measurement is discussed in some detail here as an example of techniques employing single electron beams.

Ramsauer's scattering apparatus is shown schematically in Fig. 4-1-1. Electrons are produced photoelectrically at the zinc plate P and accelerated to the desired energy by a potential difference between P and the first slit S_1. A properly adjusted, uniform magnetic field perpendicular to the plane of the drawing causes some of these electrons to travel in circular paths through S_1 and the other collimating slits, S_2 to S_7, and into the Faraday cage F, if they do not undergo collisions in their passage through the apparatus. Any electrons

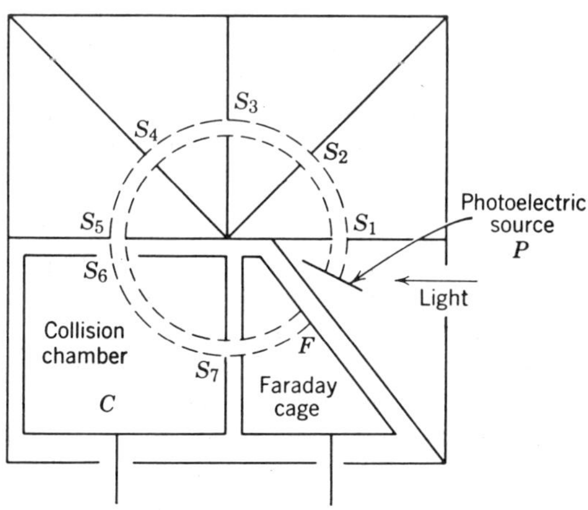

Figure 4-1-1. Ramsauer's apparatus for the measurement of electron collision cross sections (1921). (Adapted from Ramsauer.[3])

experiencing elastic scattering through angles greater than the angular aperture of the system will fail to reach F. Also, any electrons that suffer even slight reductions in their velocity as a result of inelastic encounters will thenceforth travel in orbits of smaller radius and be prevented from reaching F.

Now consider the beam of nearly monoenergetic electrons that remain unscattered before entry into the collision chamber C. Any of the electrons that experience collisions before passage through S_7 will be collected by C; the remainder is received by the Faraday cage F. It then follows that if the pressure and the arc length between S_6 and S_7 are known, the total collision cross section can be determined by measuring the currents to C and F and using the relationship in (1-4-9).

It is important in beam experiments that a given projectile avoid multiple collisions with target particles; otherwise, it might be deflected out of the beam in the first impact but scattered into the collector in a subsequent collision. Multiple collisions will occur to a negligible extent if the product of the target pressure and thickness is so low that the probability of even a single impact is small. Experimental checks are made to establish that the desired "thin target" conditions prevail by determining whether the collected currents vary linearly with the target gas pressure. In Ramsauer's apparatus rather high gas pressures (up to about 10^{-2} torr) were used, but the flight path of the electrons was short enough that only single collisions occurred in most cases. In the apparatus shown in Fig. 4-1-1 the mean diameter of the path of the unscattered electrons was 20 mm. The beam was 1 mm wide and 8 mm high.

It must be emphasized that the measured cross section will equal the integral elastic scattering cross section only if the electron energy is lower than the threshold energy for excitation of any of the levels of the target gas molecules. Otherwise, inelastic collisions make a contribution to the measured cross section. In Ramsauer's experiments it was impossible to separate the contributions of inelastic and elastic scattering. However, most of his work was done at energies low enough that inelastic collisions did not play a major role. In certain other types of apparatus to be discussed later in this chapter, the scattered electrons are analyzed in energy by retarding potential techniques or deflection in electric or magnetic fields, so that only elastically scattered electrons are detected.

It is obvious that since any apparatus has a finite resolving power the cross sections that are measured correspond to scattering through angles greater than some nonzero minimum angle and cannot refer to the true integral values. This is not so serious a matter in electron measurements as it is in the study of the scattering of heavy particles, since the scattering is not nearly so strongly peaked in the forward direction. It is serious for electron scattering at beam energies greater than about 100 eV and also at low energies in certain cases (e.g., near the Ramsauer minimum where the isotropic s wave vanishes).

Massey and Burhop[6] have collected data obtained in single-beam experiments on atoms and have averaged the results in some cases where more than one set of data is available. Their plots of total collision cross section, q_t, for the noble gases versus electron velocity in $(V^{1/2})$ are reproduced in Fig. 4-1-2. The

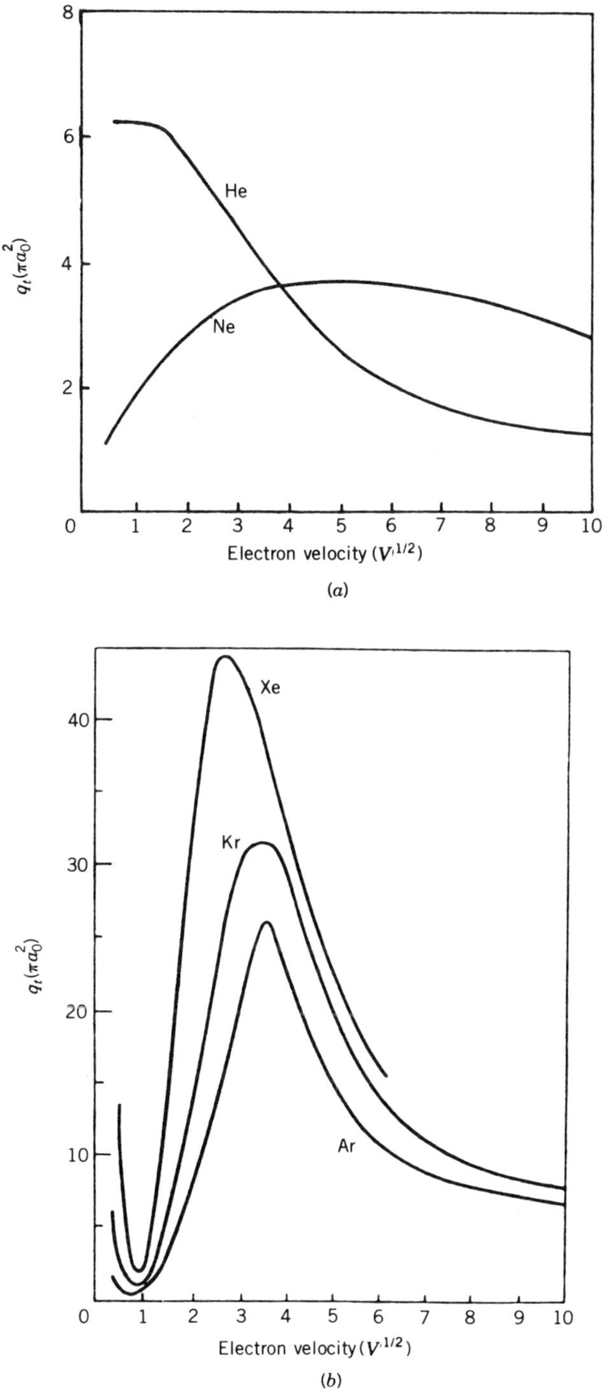

Figure 4-1-2. Total electron collision cross sections for atomic targets plotted as a function of the electron velocity in ($V^{1/2}$). (Adapted from Massey and Burhop.[6]) Modern, more accurate data on the Ramsauer minimum in Ar and Kr are presented in Figs. 4-1-5 and 4-3-7.

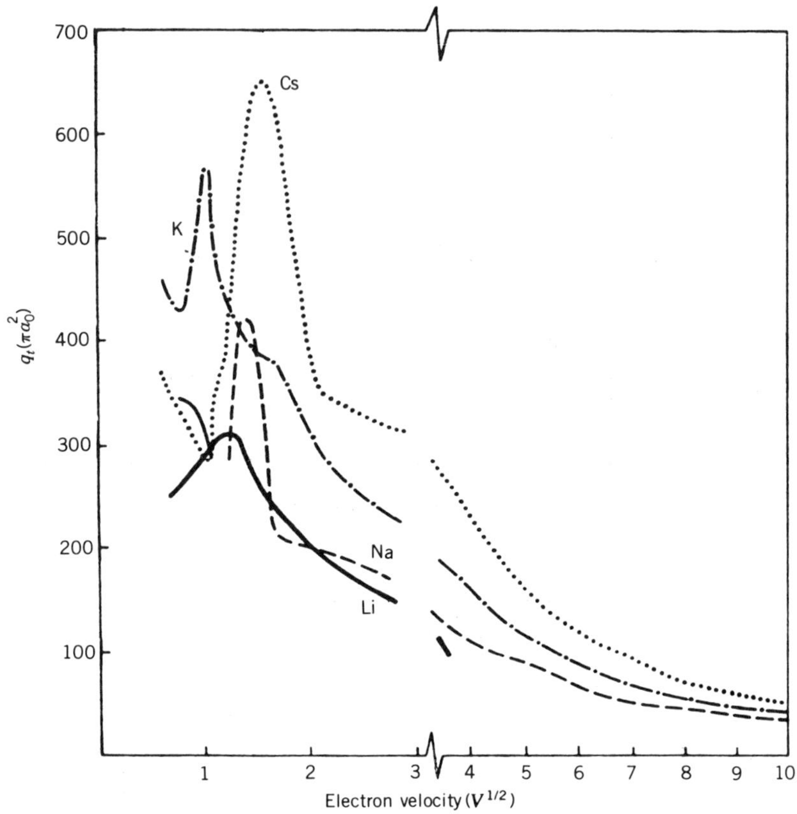

(c)

Fig. 4-1-2 *(Continued)*

electron energy at any point along the abscissa is obtained in eV by squaring the value of the abscissa at that point. If E represents the energy in eV, the electron velocity is given by the equation

$$v = 5.93 \times 10^7 (E)^{1/2} \qquad \text{cm/sec.} \tag{4-1-1}$$

Most of the data shown here were obtained by the Ramsauer method or some variant thereof.[5] An exception is the case of Li, for which the atomic beam recoil method (Section 4-1-B) was used in a modern experiment.[12] The differences between the curves and the data of the individual investigations were seldom as large as 10%, except at very low energies. This fact, coupled with the general agreement of these results with those obtained by other methods, indicates that the curves are reliable over most of the energy range.

Actually, the success of the Ramsauer technique must be counted as nothing short of astounding. In the entire field of atomic collisions, only a handful of experimental techniques developed before 1940 have stood the test of time and produced data that deserve serious attention at this date. Ramsauer's technique is the oldest of this select group. As testimony to Ramsauer's accomplishment, we point out that his technique has continued to be used in modified form up into modern times.[13]

The most striking feature of Fig. 4-1-2 is the rapid variation of q_t with electron velocity in most of the gases. Of particular interest are the curves in Fig. 4-1-2*b* for the heavier noble gases. It will be noticed that in xenon, krypton, and argon the cross section falls to a very low value in the vicinity of 1 eV. This marked transparency in a narrow range of energy, observed in certain other gases as well, was discovered by Ramsauer[3] and independently by J. S. Townsend and V. A. Bailey,[8] who used a swarm method (see Section 4-2-A). The *Ramsauer-Townsend effect* was in violent disagreement with the classical theory of scattering, which predicted a monotonic increase in the cross section with decreasing electron energy. Quantum theory was required for the explanation of the observed scattering (Section 4-8), and indeed, the necessity of explaining the scattering behavior of slow electrons provided a powerful impetus to the development of the quantum theory of atomic collisions.

The potential field of the outer electrons of the target particles is the major factor in determining the probability of collision. The polarizability of the molecules,[14] and thus the distortion of the potential field by the incident electrons, is also important. Consequently, there is close similarity within a chemically similar group of gases, but large differences are observed as we go from one group to another. It is particularly interesting to note the enormous difference in the cross sections for the noble gases and the alkalis. For example, at 2 eV, q_t is about 40 times as large for cesium as for xenon, which lies adjacent to cesium in the periodic table. The small values for the noble gases are due to the compact structure of the atoms, which have closed outer electron shells, whereas the huge values for the alkalis are a reflection of the diffuse structure of these atoms, each of which possesses one loosely bound valence electron. Interesting general trends may be noted among the monoatomic elements of Fig. 4-1-2: for electron energies of about 100 eV the collision probability is inversely proportional to the ionization potential and directly proportional to the polarizability of the atoms.* At high energies the cross section drops off about as $1/v$. This behavior is indicative that polarization attraction is the determining factor in the scattering at high energy.

A few general comments about beam methods for determining elastic scattering cross sections are in order. An advantage offered by these methods is the direct nature of the measurements they permit. However, the usual difficulties associated with the use of very low energy beams of charged particles (due principally to stray fields, contact potentials, mutual electrostatic repulsion, and the reflection of particles from surfaces) impose a lower limit on the electron

*See B. Fricke, *J. Chem. Phys.* **84**, 862 (1986) for an analysis of the correlation between the electric polarizability and the ionization potential of atoms.

beam energy. In beam experiments on elastic scattering extreme gas purity is usually not so essential as it is in certain other types of measurements, such as those of electron attachment cross sections, in which the impurities may have cross sections orders of magnitude larger than those of the main target gas. However, insulating surface layers of condensable impurities can cause trouble by acquiring electric charge, particularly at beam slits. It is therefore necessary to guard against the entry of pump oil vapor into the collision chamber by careful baffling and cold-trapping of the pumps, and a bakeable system that permits the elimination of layers which are formed in spite of all precautions is desirable. Oil-free turbomolecular pumps and closed-loop helium cryopumps offer obvious advantages and have largely replaced oil diffusion pumps in atomic collisions experiments. To obtain high conductivity, critical surfaces are frequently coated with gold, either by electroplating or by vacuum deposition. Bare molybdenum surfaces appear to be especially stable and have also recently become widely used. Good electrostatic and magnetic shielding of the collision volume is an obvious necessity. The only solution to the problem of mutual electrostatic repulsion is to operate with low beam currents and to make sure that the measured cross sections are independent of the beam current.

There are many vacuum- and gas-handling techniques that deserve careful study and that are discussed in Chapter 3 of the book by Hasted,[7] which is useful for its treatment of experimental techniques. Here we mention only two points. First, a static gas filling of the collision chamber should be avoided; a dynamic (flowing) arrangement should be used in order to sweep out impurities arising from outgassing of surfaces. Second, in about 1960, H. Ishii and K. Nakayama discovered that the McLeod gauge, which had been used in many experiments as an absolute manometer to measure gas target pressure, may give highly inaccurate readings because of a pumping effect. Since the gauge must be isolated from the collision chamber by a cold trap, mercury vapor is continually distilled from the gauge reservoir into the trap, sweeping gas molecules out of the gauge. The result is that the gas pressure is lower in the gauge than in the chamber. For the lightest gases, this "Ishii effect" gives errors of a few percent; for a heavy gas such as xenon, errors of about 50% can arise. The magnitude of the effect depends on the temperature of the mercury and on the radius of the connecting tube. Hasted[7] discusses techniques for minimizing the effect and for estimating the size of the effect in a given experimental arrangement. In modern experiments, the McLeod gauge has been almost entirely replaced by the capacitance manometer and the spinning-rotor gauge, which operates on the viscosity of the gas.

An additional factor that has played a significant role in the evolution of modern experimental atomic collisions research is the ever-increasing use of on-line computers to operate experiments. The technique can greatly increase the effective running time during a given period and result in better statistics in the data. It also usually results in more frequent and accurate checks being made on experimental variables during the course of an experiment.

We now move ahead more than 60 years and discuss two experiments that take advantage of recent improvements in experimental techniques. The first experiment is that of Wagenaar and de Heer[15] at the FOM Institute in Amsterdam. Their apparatus, which evolved to the form described below over the years 1979–1985, provides an interesting comparison with Ramsauer's.[3] Here the magnetic field of the Ramsauer method is eliminated and a linear, rather than a circular, geometry is employed. The target gas is contained at a pressure typically on the order of 1 Pa (7.5×10^{-3} torr) in a collision cell of adjustable length (0.2 to 10 mm) with entrance and exit orifices of 1-mm and 2-mm diameter, respectively. An electron gun and a Faraday cage with a 1 mm × 3 mm entrance are located fore and aft in separate differentially pumped regions. The distance from the exit plate of the collision cell to the current-measuring Faraday detector is 70 mm. Wagenaar and de Heer[15] measured the total collision cross section for the noble gases over the energy range 20 to 100 eV. Their results for He and Ne agreed satisfactorily with previous FOM results, but those for Ar, Kr, and Xe differed significantly. De Heer and his colleagues had previously used a much longer (42 mm) gas cell at lower pressures and experienced difficulty in measuring the pressure accurately. The 1985 FOM cross sections[15] are systematically higher than most of the previous results, including Ramsauer's. Wagenaar and de Heer attribute the difference to better suppression of small-angle scattering events in their detection of the transmitted electron beam. The large energy spread (0.35 eV) of their incident electron beam was not a matter of concern, but would be intolerable in certain other kinds of measurements to be described. Wagenaar et al.[16] have supplemented their total cross sections with small-angle elastic differential cross sections for the noble gases over the energy range 20 to 200 eV.

The second modern experiment to be discussed here was performed by Raith and his colleagues[17] at the University of Bielefeld. They used the time-of-flight (TOF) technique[18] for low-energy electron scattering studies at extremely high energy resolution. Their apparatus, shown in Fig. 4-1-3, is similar in geometry to the FOM apparatus[15] but utilizes TOF measurements for incident electron energy selection. According to Raith,[18] the TOF technique offers the advantage of speed, reliability, and long-time stability in high-energy resolution measurements, and the inherent sources of error are relatively obvious and easy to analyze. In the Bielefeld experiment, Ferch et al.[17] measured the total cross section of argon in the energy range 0.08 to 20 eV, with a "resolvable energy" increasing from about 0.003 to 2.9 eV as the electron energy is swept up through this range. The resolvable energy, ΔE, is the minimum separation required for two unknown structures of $q_t(E)$ in order that they be resolvable. Ferch et al. found the Ramsauer minimum for argon to be located at 0.345 eV, at which the resolvable energy is about 10 meV.

In the TOF apparatus shown schematically in Fig. 4-1-3, electrons from a thermionic cathode are accelerated to 200 eV to form a beam that is chopped in the gate by sweeping it across a 0.1-mm-wide slit at a frequency of 150 to

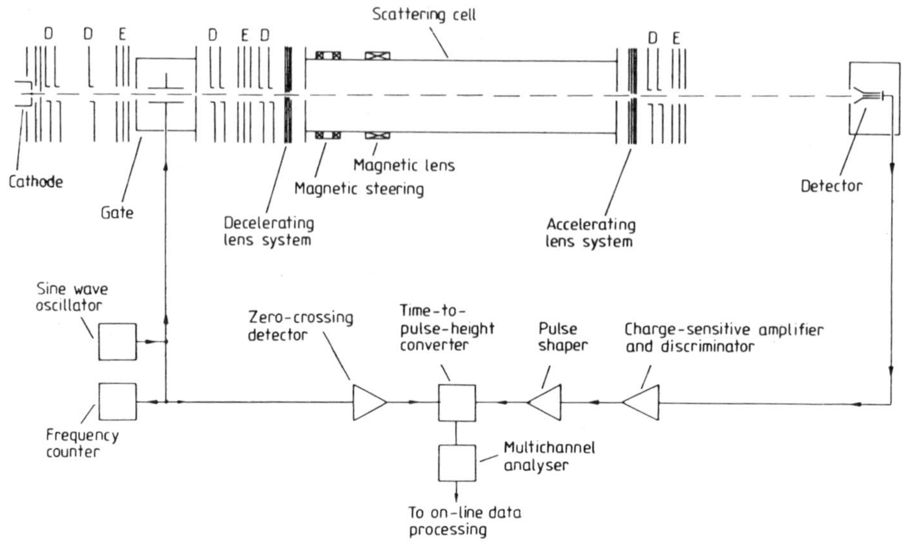

Figure 4-1-3. Layout of TOF spectrometer (not to scale) and electronics for data accumulation: *D*, deflector; *E*, einzel lens. (From Ferch et al.[17])

300 kHz. Bursts of electrons of about 7 ns width are thus produced. The 200-eV electrons in each burst are decelerated before entering the target cell, and those emerging from the cell are accelerated back to 200 eV for detection in a channel electron multiplier. Typically, only 0.1 of an electron per pulse reaches the detector. Several carefully designed electric and magnetic lenses are required for focusing and beam alignment to reduce the forward-scattering discrimination and to achieve a suitably high transmitted electron intensity. The flight time is determined individually for each detected electron by measuring the time between its detection and the following zero transition of the gate voltage ("inverted timing"). Flight-time data are accumulated for up to 30 min to form a flight-time spectrum, both with and without gas in the scattering cell (Fig. 4-1-4). The target gas usually produces an attenuation of the electron intensity by a factor between 0.7 and 0.2. Application of the equation

$$I = I_0 e^{-N q_t x} \tag{1-4-9}$$

for any given flight time gives the total cross section as a function of the flight time and hence as a function of *E*, the electron impact energy. The kinetic energy of each electron that reaches the detector is determined individually from the TOF measurements. Therefore, the energy resolution is independent of the energy distribution of the primary electron beam, which distribution limits only the energy range that can be covered in one TOF measurement. The electrons coming from the cathode have an energy spread of about 0.6 eV.

Figure 4-1-4. Example of a TOF measurement, here with CO_2.[17] (*a*) Time distribution $I_0(t)$ of the electrons detected without target gas by means of "inverted timing." The left and right wings of the spectrum correspond to electron energies of 0.007 and 0.5 eV, respectively. Note the pronounced structure in the initial distribution $I_0(t)$. (*b*) Time distribution $I(t)$ obtained with target gas. (*c*) Distribution of $\ln(I_0/I)$, a quantity that is proportional to the total cross section σ_t. Note that the structure of $I_0(t)$ does not lead to false structures in σ_t.

The data on argon obtained by Ferch et al. are shown as a solid curve extending from 0.08 to 20 eV in Fig. 4-1-5.[17] As indicated above, the energy resolution of the Bielefeld apparatus is extremely high at low electron energies, but poor at high energies. However, Ferch et al. extended their measurements up to 20 eV in order to make comparisons with other experimental and theoretical results. Since there is no strong variation of $q_t(E)$ between the Ramsauer minimum and 20 eV, the poor resolution at the higher energies is of no

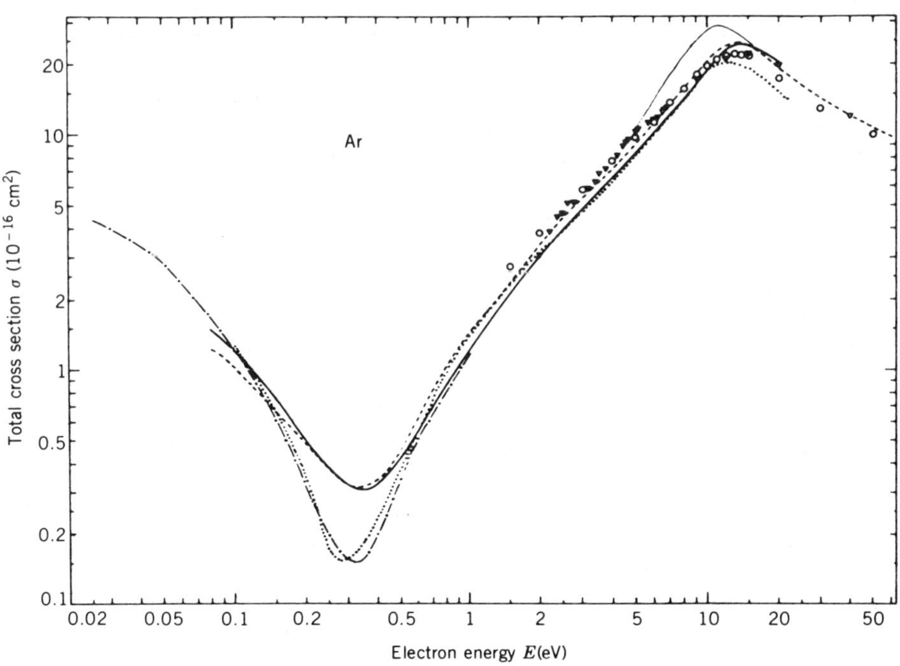

Figure 4-1-5. Total cross sections for scattering of electrons by argon atoms obtained at Bielefeld by TOF techniques (solid curve). The other data are discussed by Ferch et al.[17]

consequence. In closing this discussion of the Bielefeld experiment, it may be of some interest to point out that the great variation in the argon cross section as the electron beam energy is changed necessitates the use of a correspondingly wide gas pressure range in the measurements: 10^{-3} to 10^{-5} torr!

The TOF technique has also been used for important measurements at the University of Indiana[19] and at other laboratories where comparisons of electron and positron scattering were made (see Section 4-7).

B. Crossed-Beam Methods

The study of collisions between electrons and chemically unstable systems, such as hydrogen atoms, presents special problems and usually requires the use of crossed-beam techniques if direct measurements are to be made. Some of these techniques were introduced around 1930.[5] They were greatly extended and refined by Fite, Stebbings, and Brackmann,[6,7,20] in the late 1950s, and improvements (especially in vacuum techniques, detectors, and electronics) have continued to be made up to the present time. Crossed-beam techniques had already produced a sizable body of invaluable information by the early 1960s; now they are being used in many laboratories for various kinds of experiments and must be regarded as indispensable in atomic collision physics. In this section we first describe the crossed-beam apparatus used by Neynaber and his colleagues for electron scattering studies on atomic hydrogen, oxygen, and nitrogen. Then we shall describe the atomic beam recoil technique used by Bederson's group.

The technique used by Neynaber et al.[21] for measuring the scattering cross sections of atoms normally present in molecular form is the following. The number of electrons scattered from the intersection of a dc electron beam and a modulated (i.e., a mechanically interrupted) molecular beam is compared with the number scattered when the beam is partially dissociated. The molecular dissociation is accomplished in a radio-frequency discharge, and the degree of dissociation (typically about 30%) is measured with a mass spectrometer. From the scattering and dissociation data, the ratios of the atomic to molecular scattering cross sections are obtained at various energies. The absolute atomic values are calculated by multiplying these ratios by the molecular cross sections obtained by other investigators who used single-beam methods.

A schematic diagram of the apparatus is shown in Fig. 4-1-6. Not shown is the vacuum envelope, which is made of cadmium-plated mild steel in order to minimize stray magnetic fields. A neutral beam consisting entirely of molecules may be produced by the effusive flow of a molecular gas through a hole in the side of an unexcited discharge tube located in the first of three vacuum chambers. Alternatively, a partially dissociated beam may be obtained by striking a discharge within the source. The pure or mixed beam passes through a bulkhead slit into an intermediate chamber whose primary function is to provide vacuum isolation from the source. Here, also, any charged particles present in the beam are removed by a pair of oppositely charged deflector plates.

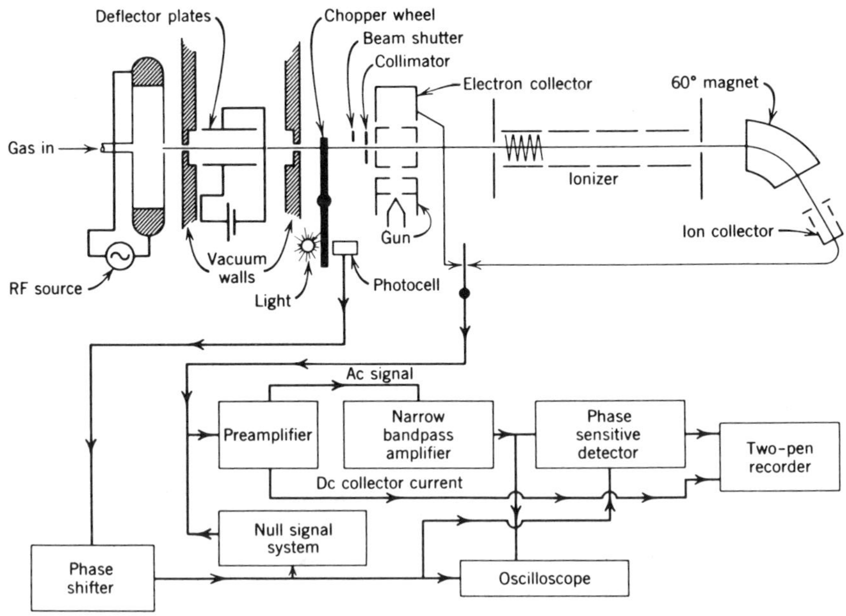

Figure 4-1-6. The crossed-beam apparatus used by Neynaber et al.[21] The scattering occurs in the electron gun. The distance from the RF source to the electron gun is 0.21 m. The electron beam diameter is 2.5 mm. The molecular beam is rectangular, typically 2.5×5.0 mm. Electron currents of several microamperes are used, and the electron detection sensitivity is about 10^{-13} A.

As the beam emerges into the scattering chamber, it is chopped by a slotted rotating wheel at 101.8 Hz, and it is further collimated by a slit just in front of the electron gun mount. It may be interrupted by a beam shutter, if desired. The neutral beam proceeds through the electron gun, where it intersects the electron beam and scatters electrons at the chopping frequency. The attenuation of the electron beam due to scattering by the neutral beam is measured as an ac signal. Finally the neutral beam enters a mass spectrometer, where it is partially ionized so that the degree of dissociation of the original beam may be determined.

Typical pressures in the scattering chamber are 6×10^{-7} and 2×10^{-7} torr with the gas flow on and off, respectively. The presence of background gas in the collision region dictates the use of a modulated neutral beam, since a pressure as low as 6×10^{-7} torr corresponds to a number density of about 2×10^{10} molecules/cm^3. If the neutral beam is run in the dc mode, it is impossible to distinguish between electrons scattered from the beam and from the background gas. With mechanical interruption of the beam, however, only scattering at the modulation frequency enters into the cross section calculation, and the dc contribution due to the background gas may be eliminated.

The energy of the electrons, as well as the energy spread, is determined by applying retarding potentials (V) to the collector and observing the collector currents (I). The energy spread (defined as the width at half-maximum of dI/dV versus V) is 0.45 eV at 11 eV and 0.33 eV at 2.3 eV.

The apparatus of Neynaber et al. represented the state of the crossed-beam art in 1960. In certain more modern instruments, electron energy selectors and analyzers have been added to provide good energy resolution. Analog detection has been largely replaced by digital detection schemes. (The apparatus described in the figures and references of Section 4-3 incorporate such improvements.) Also, the background pressure has usually been substantially reduced, so that beam chopping is not required in many experiments.

Bederson's group at New York University has developed and refined the *atomic-beam recoil method* for electron scattering studies in the energy range of about 1 to 25 eV.[12,22] When it is applied to the measurement of integral cross sections, an atomic (or molecular) beam is cross-fired by an electron beam, and the total scattering is observed by measuring the decrease in atomic-beam intensity in the forward direction, caused by momentum transfer during the scattering process. In the apparatus described in Ref. 22, neither the atomic nor the electron beam is usually modulated, but the apparatus is operated under electron beam "on" and "off" conditions to assess the amount of scattering. A combination of long geometry, good collimation, and velocity selection of the atomic beam is necessary to achieve good angular resolution. This apparatus[22] utilizes a 5-m-long vacuum envelope, a thermionic cathode for electron production, and a hexapole magnet for neutral beam focusing and velocity resolution. The hexapole also permits state selection in the neutral beam when it is desired. (The use of a hexapole for this purpose is described in Section 4-6-A.) The width of the electron energy distribution is about 0.5 eV, adequate for the experiments described in Refs.[22-24]. When alkali atoms are being studied, the detector is of the surface-ionization type. The principal limitation to the recoil method is the relatively poor detection efficiency for other kinds of neutral atoms. The main advantages of the method derive from the fact that the observations made after collisions are on the atom, not on the electron. Hence absolute cross sections can be measured without determining the number density of atoms in the neutral beam. Also by determining the atomic state before and after the collision, one can investigate spin-exchange collisions, or more generally, angular-momentum-changing collisions (see Section 4-6). Bederson's technique can be used to study both elastic and inelastic scattering, and to measure both integral and differential cross sections.

C. Data on the Integral Elastic Scattering Cross Section

We conclude our discussion of the integral elastic scattering cross section with Fig. 4-1-7, which gives experimental data on $q_t(E)$ for eight molecular gases. This figure was taken from the chapter by Csanak et al.,[9] which contains an extensive bibliography and data index on electron scattering by molecules. Actually, the data in Fig. 4-1-7 are "vibrationally elastic cross sections"; they contain contributions from rotational, but not vibrational (or electronic) excitation. The

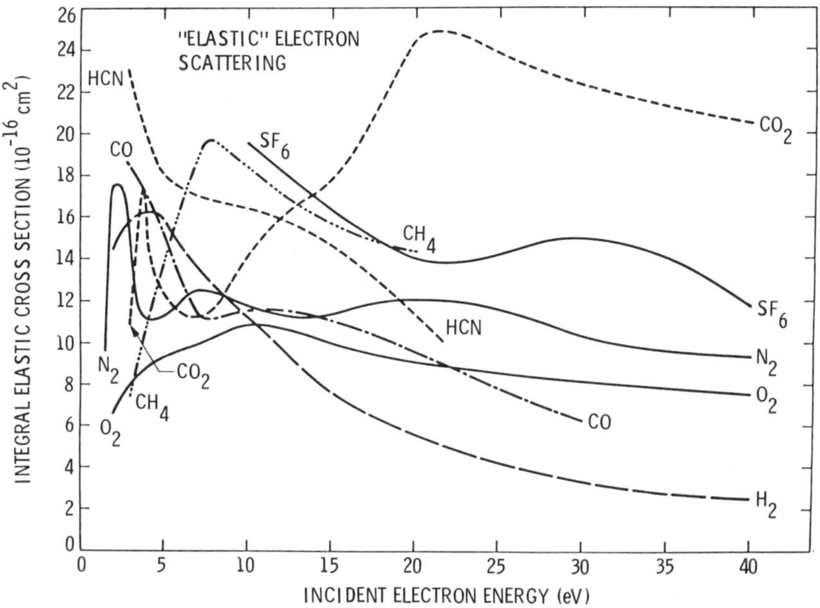

Figure 4-1-7. Integral vibrationally elastic scattering cross section as a function of incident electron energy for H_2, O_2, N_2, CO, CO_2, SF_6, HCN, and CH_4. (From Csanak et al.[9])

situation can be explained as follows: Rotational energy levels of molecules are typically spaced a few meV apart, not enough for electrons producing rotational excitation to be readily separated from those experiencing elastic scattering. On the other hand, typical vibrational and electronic level spacings are about 0.1 eV and several eV, respectively. Hence discrimination against electrons that produce vibrational and electronic excitation is a simple matter in experiments which detect the scattered electrons. It is impossible in transmission experiments that detect the unscattered electrons.

4-2. ELECTRON SWARM TECHNIQUES FOR DETERMINING THE MOMENTUM TRANSFER CROSS SECTION ($q_D \equiv q_m \equiv \sigma_m$)

A. The Drift Velocity Method

Rigorous transport theory shows that the momentum transfer cross section, q_D, may be determined as a function of the electron energy from measurements of the average velocity of electronic drift through a gas in a dc electric field. Such measurements are made in "drift tubes," examples of which will be described in

Ref. 1. Inside the drift region of the tube there exists a gas of number density N and a nominally uniform electric field E; the parameter E/N determines the drift velocity and average energy of the electrons in the gas for a given gas temperature. The drift velocity is measured as a function of this "energy parameter." The possibility of extracting cross section information from drift velocity data is based on the fact that at a given E/N, the steady-state drift velocity is achieved by a balance of the accelerations in the field direction *between* collisions of electrons with gas molecules and the decelerations that occur *during* collisions. Considerable demands are placed on the experimenter who desires to obtain sufficiently accurate drift velocity data to permit the extraction of reliable cross sections. The theoretical analysis of the data is also demanding, as we shall see. It is appropriate to defer further discussion of this important method until Ref. 1, where we shall be better equipped to describe it.

B. Microwave Methods

Several microwave techniques have been developed for determining the electron momentum transfer cross section, most of the early work having been done by Brown and his colleagues in the 1940s and 1950s at MIT. The various methods are summarized by Brown,[4] Huxley and Crompton,[8] Wharton,[25] and Gilardini.[26] Microwave experiments are useful because they can cover the energy range extending from thermal values up to several eV. The quantity they yield directly is the collision frequency for momentum transfer, ν_m, from which q_D may readily be obtained. But at energies up to several eV, q_D for electrons differs little from q_s, the integral elastic scattering cross section, so microwave data may meaningfully be compared with the results of other types of experiment. In some cases the lack of agreement among the low-energy results is pronounced. It is currently believed that some of the microwave data might be faulty because the electrons were not in thermal equilibrium, as had been assumed. The main reason that we discuss microwave methods here is that they have important application in studies of plasmas and ordinary ionized gases. It is important to make the connection between microscopic atomic collisions cross sections and macroscopic properties of plasmas and gases.

Let us now sketch one of the microwave methods—that used by Phelps et al.[27] Consider electrons oscillating in a gas of uniform pressure with a radian frequency ω under the influence of an electric field of intensity $E = E_0 e^{i\omega t}$. Because of collisions, the molecules may be considered to exert a continuous viscous damping force on the electrons. According to Newton's second law, the average motion of the electrons may be described by the equation

$$f = -cv_d - eE_0 e^{i\omega t} = \frac{d(mv_d)}{dt}, \tag{4-2-1}$$

where f denotes the total force on an electron, c is the damping constant, v_d the average velocity of the electrons,* and m the electronic mass. A physically realistic choice for the damping constant is $m\nu_m$, where ν_m is the collision frequency for momentum transfer, given in terms of the integral collision frequency ν and the center-of-mass scattering angle Θ by the equation

$$\nu_m = \overline{\nu(1 - \cos\Theta)}. \tag{4-2-2}$$

We shall assume here that ν_m is independent of v_d, so that the solution of (4-2-1) for v_d is

$$v_d = \frac{-e/m}{i\omega + \nu_m} E. \tag{4-2-3}$$

Now the electronic current density J_e is given by

$$J_e = -nev_d, \tag{4-2-4}$$

where n is the number density of the electrons; hence the complex conductivity is

$$\sigma_e = \sigma_r + i\sigma_i = \frac{J_e}{E} = \frac{ne^2/m}{i\omega + \nu_m}. \tag{4-2-5}$$

The real and imaginary components, σ_r and σ_i, respectively, of the complex conductivity σ are seen to be equal to

$$\sigma_r = \frac{ne^2}{m\omega} \frac{\nu_m/\omega}{(\nu_m/\omega)^2 + 1} \tag{4-2-6}$$

and

$$\sigma_i = \frac{-ne^2}{m\omega} \frac{1}{(\nu_m/\omega)^2 + 1}. \tag{4-2-7}$$

The ratio of the real and imaginary components is therefore

$$\left|\frac{\sigma_r}{\sigma_i}\right| = \frac{\nu_m}{\omega}. \tag{4-2-8}$$

It follows that in principle the collision frequency, and thus the cross section, for momentum transfer can be determined by measuring the complex conductivity

*v_d is called the "drift velocity" of the electrons and is given in terms of the velocity distribution function by (2-12-8).

in a microwave discharge if the pressure and applied frequency are known. The analysis presented here is a drastic oversimplification of the difficult analysis actually required in practice. Two sources of complication are the dependence of the collision frequency on the velocity and the fact that the velocity distribution of the electrons must be taken into account.

A block diagram of the apparatus used by Phelps et al.[27] to determine the conductivity ratio in (4-2-8) is shown in Fig. 4-2-1. A 10-cm pulsed magnetron is used periodically to break down the gas contained at a few torr pressure in the resonant cavity shown at the extreme right and thus provide a plasma whose conductivity is to be measured. The change in the cavity impedance due to the presence of the free electrons is measured with a continuous-wave tunable magnetron and a standing-wave detector, which is sensitive only for a period of a few microseconds during the afterglow.* The transient standing-wave detector consists of a calibrated waveguide-beyond-cutoff attenuator followed by a superheterodyne receiver whose local oscillator frequency is controlled by a delayed sweep from a single-sweep oscilloscope. The output of the receiver is observed on the oscilloscope, and the standing-wave ratio is determined by adjusting the calibrated attenuator to maintain constant amplitude at the output of the receiver. The delay circuit for the oscilloscope and the modulator for the pulsed magnetron are synchronized by a trigger generator operating at 60 Hz. A cavity wavemeter and a bolometer monitor the output of the continuous-wave magnetron. The power incident on the cavity is controlled by a calibrated variable attenuator. A useful evaluation of the various microwave methods and the data obtained by them is provided in the review by Bederson and Kieffer.[5]

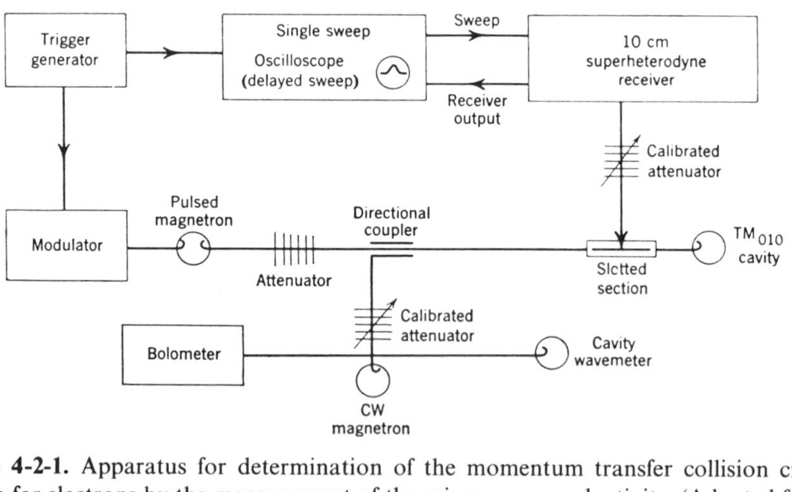

Figure 4-2-1. Apparatus for determination of the momentum transfer collision cross section for electrons by the measurement of the microwave conductivity. (Adapted from Phelps et al.[27])

*The *afterglow* is the period following the removal of the source of excitation during which the gas remains ionized to an appreciable extent.

4-3. MEASUREMENT OF THE ANGULAR DISTRIBUTION OF ELASTICALLY SCATTERED ELECTRONS; THE ANGULAR DIFFERENTIAL SCATTERING CROSS SECTION

$$\left[I_s(\vartheta, \varphi)\, d\Omega \equiv \frac{d\sigma(\vartheta, \varphi)}{d\Omega}\, d\Omega \right]$$

A. The Zone Apparatus of Ramsauer and Kollath

Ramsauer and Kollath[28] were among the first to study the angular dependence of the elastic scattering of electrons. Their "zone" apparatus, shown in Fig. 4-3-1, was used for measurements at low electron energies on several simple gases. The apparatus had an overall length of about 0.2 m, and pressures of the order of 10^{-3} torr were used in the scattering chamber. Electrons from an oxide-coated filament entered the collision volume at the center of the sphere defined by the curved metal plates, and the angular scattering distribution was determined by measuring the electron currents to each of the 11 plates. There was cylindrical symmetry about the axis of the electron beam so that the reduction of the experimental data could be effected in a straightforward manner.*

For any target, inelastic scattering eventually becomes significant as the electron beam energy is raised from initially low values, but the elastic contribution to the total scattering may be separated out if retarding potential techniques or some other form of energy analysis are employed to reject inelastically scattered electrons.

The zone apparatus proved not to be as successful as the original Ramsauer apparatus designed for integral cross section measurements (Fig. 4-1-1). The

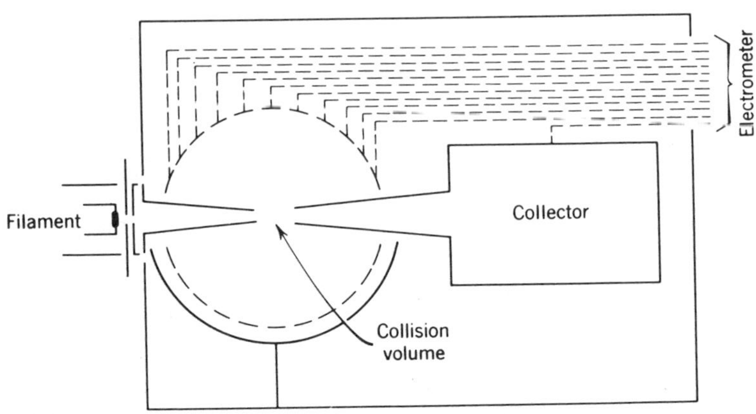

Figure 4-3-1. Ramsauer and Kollath's zone apparatus for the determination of angular scattering distributions (1932). (Adapted from Ramsauer and Kollath.[28])

*In none of the measurements of this section does the scattering distribution (or the angular differential cross section) depend on the azimuthal angle φ. The use of spatially oriented projectile electrons or target particles (or both) is required for such an effect (Section 4-6). Here we may write the angular differential cross section (integrated over φ) as $2\pi I_s(\vartheta) \sin \vartheta\, d$.

small number of electron collecting plates provided inadequate angular resolution to reveal the full angular structure in many measurements. Electron collection was incomplete and variable because of reflection from the plates. Stray fields and poor surface conductivity resulted in distorted electron trajectories and inordinately large scattered electron currents at small angles and low electron beam energies. However, the Ramsauer-Kollath data hold up surprisingly well in comparison with many of the modern results, and they have played an important role in the evolution of our understanding of electron scattering.

There are modern versions of the Ramsauer-Kollath technique that involve passing an energy-selected electron beam through a gas that is not confined to a collision cell and detecting the scattered electrons in a rotatable, collimated Faraday cup (see, e.g., Ref. 29). Such arrangements encounter a problem in the change of scattering volume with polar angle ϑ, and the determination of the corrections to be applied is not simple, especially at low beam energies.[30,31] Instruments of this type have been superseded by high-resolution crossed-beam electron spectrometers, to which we now turn.

B. Modern High-Resolution Electron Spectrometers

The apparatus of Ramsauer and Kollath was primitive by today's standards, but several decades were required after 1932 for the development of techniques that produced really significant improvements, especially those related to the narrowing of the energy and angular spreads of the electron beams. Such improvements are required if structure in the cross section as a function of incident electron energy and of scattering angle is to be fully (or at least highly) resolved. The interesting history of these developments is surveyed by Massey and Burhop.[6] Here we leap forward over the intermediate research to the year 1960 and cite a number of key papers published from that date to the present.[32-50] These papers contain descriptions of apparatus in which excellent electron energy resolution and satisfactory beam intensity were obtained by utilizing the energy dispersing and direction-focusing properties of electrostatic fields of various geometries* and by taking many precautions in the construction and operation of the instruments. Vastly improved angular resolution was achieved by using modern detectors of great sensitivity for the scattered electrons. An electron multiplier, for example, can detect electrons (postaccelerated to a suitably high energy) with nearly 100% efficiency in the pulse-counting mode, and digital electronics permit operation with pulse rates corresponding to currents as low as about 10^{-19} A, a factor of 10^4 lower than that detectable with the most sensitive current-measuring electrometer. Hence one may use an electron multiplier behind a very narrow slit to intercept and detect electrons scattered through an extremely small solid angle and thus obtain angular resolution of a high order.

*The principle of operation of the 127° selector/analyzer is treated in Note 4-1.

Each of the "electron spectrometers" referred to here was used to obtain data of special interest. Other modern examples of high-resolution collision apparatus are described in Chapters 5, 6, and 7. In the present context, it may be of interest to mention that there is a substantial literature on charged particle optics and lens design; Refs. 51–54 are especially useful.

During the late 1950s, Marmet and Kerwin[32] developed a 127° cylindrical electrostatic energy selector that provided an electron beam of approximately 60-meV energy width, which they later used for measurements of electron impact ionization of atoms and molecules. Their instrument was probably the first with an energy resolution better than 0.1 eV (100 meV). Some of its features were incorporated by Schulz[33] into the spectrometer shown in Fig. 4-3-2, which typifies the potent instruments in use in several laboratories during the early 1960s. It consists of a 127° cylindrical electrostatic energy selector separated from an identical analyzer by a collision chamber in which the electron beam is crossed by a dc atomic beam. A thoria-coated iridium filament emits electrons that pass into the selector, where they are deflected by the cylindrical grids and focused on the exit slit. The two grids in the selector have radii of 10 and 15 mm, respectively, and a suitably adjusted radial electrostatic field is established between them. Electrons elastically scattered at an angle of 72° are admitted into the analyzer and passed through a slit into an electron multiplier operated in the current-measuring mode. A vibrating reed electrometer operated at +2000 V is used to measure the current. A servomechanism brings the output to ground potential, and the signal is applied to the Y axis of an X-Y recorder. All electrodes near the electron beam are gold plated, and the whole system is baked at 420°C. The pressure in the collision chamber is varied between 5×10^{-4} and 5×10^{-3} torr, and the electron beam current ranges from 10^{-10} to 10^{-9} A. The

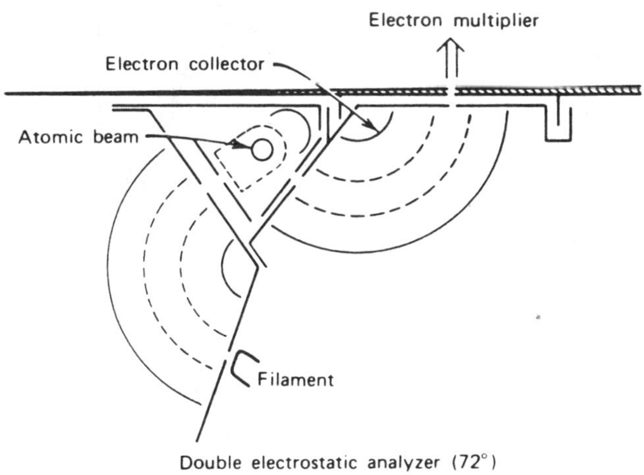

Figure 4-3-2. One of Schulz's 127° electrostatic spectrometers, set to collect electrons scattered at 72°. (Adapted from Schulz.[33])

spectrometer and electron multiplier regions are separately pumped by 300-liter/sec oil diffusion pumps with liquid nitrogen traps. The effective pressure in the atomic beam is about 100 times the background pressure, so beam modulation is not required. With no gas input, a background pressure of 10^{-9} torr is achieved. The overall energy resolution of the spectrometer is approximately 70 meV.

Over a long period centered about the 1960s, Schulz made important studies of elastic and inelastic collisions of electrons using electrostatic spectrometers of various types. With the instrument of Fig. 4-3-2, he discovered, near the end of 1962, the now-famous resonance in helium for elastically scattered electrons at an energy of 19.3 ± 0.1 eV, below the first excited (2^3S) state at 19.8 eV. This resonance appears as a sharp dip in the scattered electron signal as the energy is varied (see Fig. 4-3-3); in transmission experiments, the resonance appears as a sharp peak. Scattering resonances are related to structural properties of the

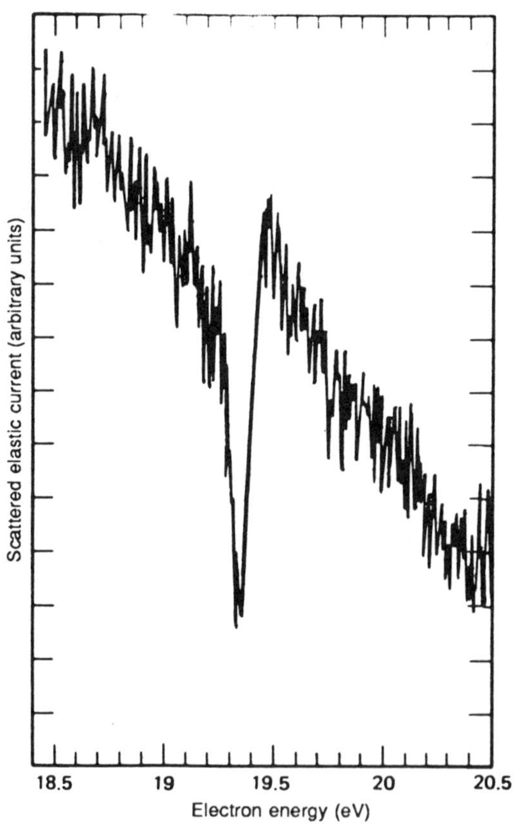

Figure 4-3-3. Schulz's X-Y recorder trace of the He resonance at 19.3 eV. The zero of the elastic cross section is suppressed. The decrease in the cross section at 19.3 eV is approximately 14%. (Adapted from Schulz.[33])

scattering complex (the negative ion in the case of electron-atom scattering), and a great deal has been learned about the structure of atoms, molecules, and ions from studies of resonances. A number of reviews of this work have been published.[55–60] Additional examples of resonances are given later in this section; theoretical aspects of resonance scattering are discussed in Section 4-8. The role of resonances in inelastic collisions is discussed in subsequent chapters.

Meanwhile, Simpson, at the U.S. National Bureau of Standards, had developed an electron spectrometer with 180° hemispherical focusing and an energy resolution of 50 meV.*[35] He and his colleagues discovered several resonances in 1963, and over the years made many important advances in the electron optics of spectrometer design.

In the 1970s, Williams made a series of measurements of elastic scattering of electrons from atomic hydrogen[61] and the noble gases[62] with the apparatus of Ref. 41, which has an energy resolution of 30 meV and an angular resolution of 2°. His work on the hydrogen atom was especially important because in this case the wave functions are completely and exactly known. Cross sections can be calculated with greater accuracy for this structure than for any other, and more useful comparisons can be made between theory and experiment.

The next spectrometer to be described is one of many constructed by Read and his colleagues at the University of Manchester (Fig. 4-3-4)[43] and used for elastic scattering investigations.[44] Elastic scattering spectra can be obtained with high resolution if *either* the incident or scattered electron beam has good energy definition. Only a high-resolution energy selector, of the 180° hemispherical electrostatic type with a 50-mm mean radius, is used here. For studies of inelastic collisions, a similar high-resolution analyzer can be attached. Thermionic electrons are used to form a beam with a radius of 0.25 mm, and electrons scattered from the target region are collected over a 7° acceptance angle by the low-resolution analyzer system, which can be rotated over an angular region of 30 to 118° with respect to the incident beam. The analyzer system consists of a three-cylinder lens of 10 mm diameter and a channel electron multiplier (CEM) operated as a pulse counter. For studies of elastic scattering on neon and argon (Fig. 4-3-5), only electrons having energies greater than 0.8 times the incident energy are passed to the detector. This feature provides discrimination against inelastic scattering. The hemispheres in the selector are made of spun molybdenum. All other materials used in the

*The most accurate method of assessing the energy resolution of an instrument is to compare the width and depth of a particular resonance as measured by it with later, presumably more accurate information about the resonance obtained by the use of other spectrometers. Computed data on instrumental broadening of the resonance profiles are also brought to bear. Read[57] used this method to deduce the overall resolution of the Simpson[35] and Schulz[34] instruments.

It is important to realize that in a scattering experiment, one must have a reasonable electron beam current and good stability, as well as suitably high energy resolution. With a given apparatus, it is possible to increase the energy resolution at the expense of the beam current, but this measure can be carried to the point of being counterproductive and can degrade the overall performance of the spectrometer.

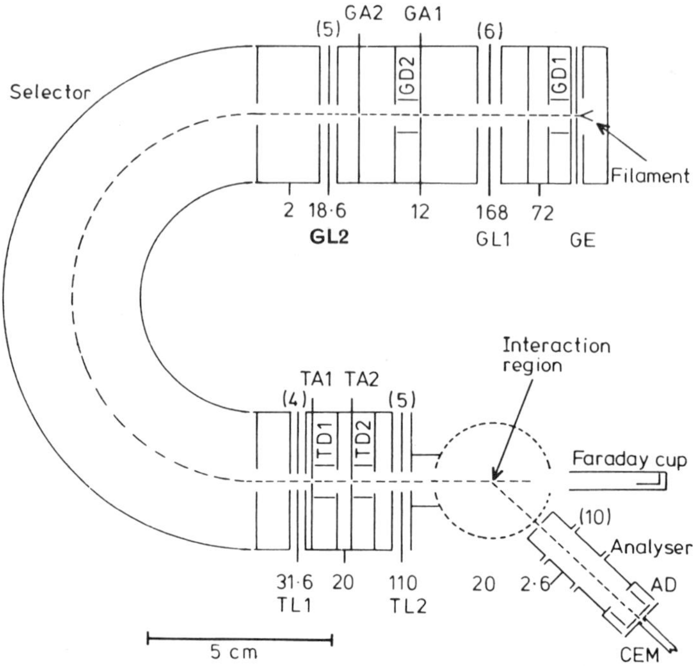

Figure 4-3-4. Schematic diagram of one of the Manchester spectrometers used for elastic scattering studies, approximately to scale.[43] The target gas beam enters perpendicular to the plane of the paper. GD1, GD2, TD1, TD2, and AD are electrostatic deflectors. The unbracketed numbers indicate typical applied voltages. The numbers in brackets above the lenses indicate diameters in millimeters.

construction of the spectrometer are also nonmagnetic and bakable to 300°C. An external pair of Helmholtz coils is used to cancel the earth's magnetic field to less than 10^{-6} T over the volume of the spectrometer. Additional shielding to reduce fluctuating magnetic fields is provided by a double layer of Mumetal surrounding the apparatus. Gas is introduced to the interaction region through a platinum-iridium tube of 0.5 mm internal diameter; the resulting crossed-beam arrangement provides reduced Doppler broadening. A liquid nitrogen–trapped 15-cm mercury diffusion pump maintains the background pressure lower than that in the target region by a factor of 100. The reader should consult the paper of Brunt et al.[43] for details of how they dealt with the many factors that can degrade the energy resolution of such an instrument. With considerable skill and care, they were able to record elastic scattering spectra containing features whose measured full-width at half-maximum (FWHM) is 12 meV. The Manchester group has also excelled in the development of control electronics and position-sensitive detectors incorporating microchannel plates[63] (see Section 5-9).

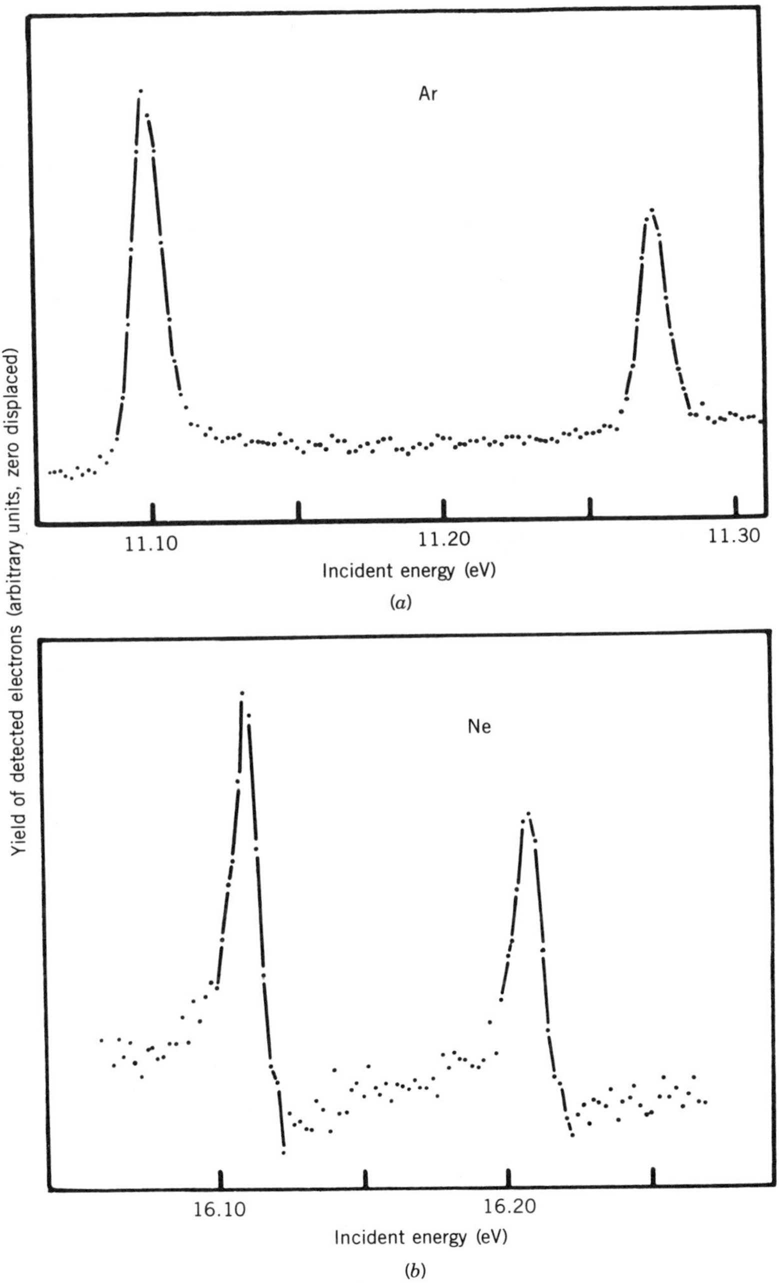

Figure 4-3-5. Elastic scattering spectra obtained by Brunt et al.[43] The energy width per channel is 1.99 meV. (*a*) Scattering through 113° by argon, $3p^5(^2P_{3/2,1/2})4s^2$ resonances. (*b*) Scattering through 100° by neon, $2p^5(^2P_{3/2,1/2})3s^2$ resonances.

More recently, another spectrometer with 12-meV resolution has been constructed by Linder's group at the University of Kaiserslautern[47] (see Fig. 4-3-6). It is built of stainless steel and ceramics. This instrument can operate down to 50-meV incident electron energy, and it allowed Weyhreter et al.[47] to map the Ramsauer minimum in differential scattering on Ar, Kr, and Xe for the first time at high resolution. Their data on krypton are presented in Fig. 4-3-7.

Ehrhardt's group at Kaiserslautern has played a major role in the development of apparatus for the study of various types of atomic collisions. One of their instruments[48] used for differential cross section measurements on elastic and excitation collisions[48,49] is shown in Fig. 4-3-8. This crossed-beam spectrometer utilizes a 127° cylindrical energy selector and a double 127° analyzer, the latter to improve measurements at small scattering angles and to reduce the widths of energy-loss peaks (Section 5-3). The electron detector assembly can be rotated around the gas beam from -30 to $+108°$. A primary electron beam current of about 10^{-9} A is used. To improve the surface properties of the apparatus, various components are heated to different temperatures in the range 360 to 470 K. The background pressure in the vacuum chamber is about 2×10^{-7} torr. Measurements can be made down to 50 meV primary beam energy. The overall energy resolution of this instrument is 22 meV (FWHM); some of Ehrhardt's instruments provide considerably higher resolution, although at lower beam current.

The optical systems of the electron gun and detector assembly can be tuned to nearly constant transmission over relative energy variations of more than a factor of 10. Sohn et al.[50] stress the necessity of controlling and monitoring the transmission functions of the gun and detector in order to avoid instrumental

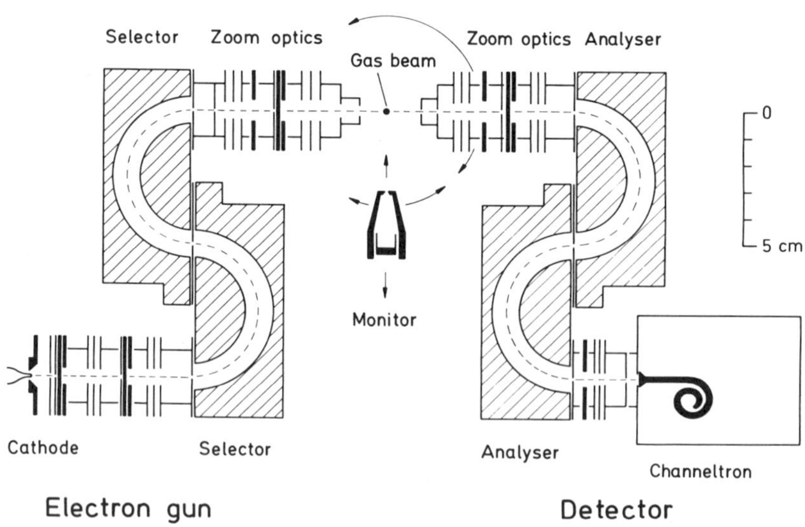

Figure 4-3-6. Electron spectrometer developed for differential scattering experiments at collision energies down to $E = 50$ meV. (Weyhreter et al.[47])

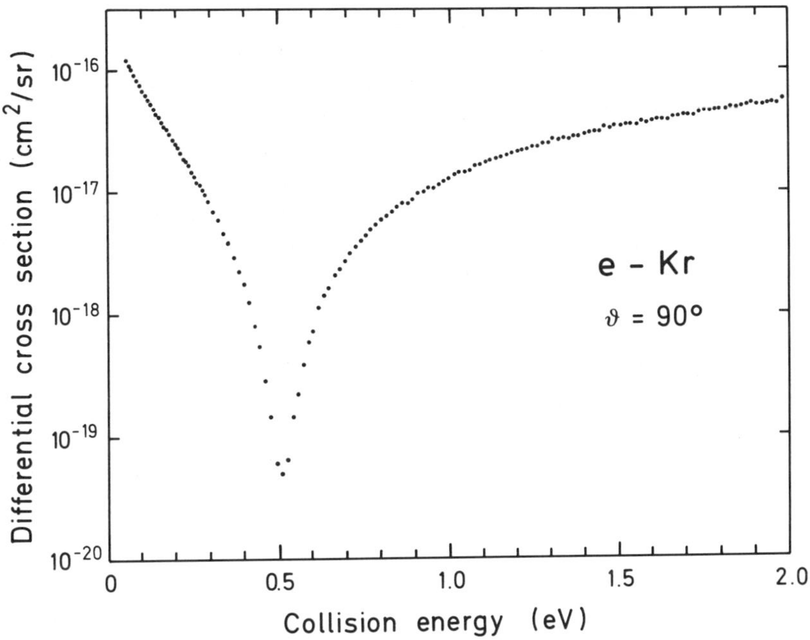

Figure 4-3-7. Energy dependence of the differential cross section for e-Kr scattering at a scattering angle of $\vartheta = 90°$. (Weyhreter et al.[47])

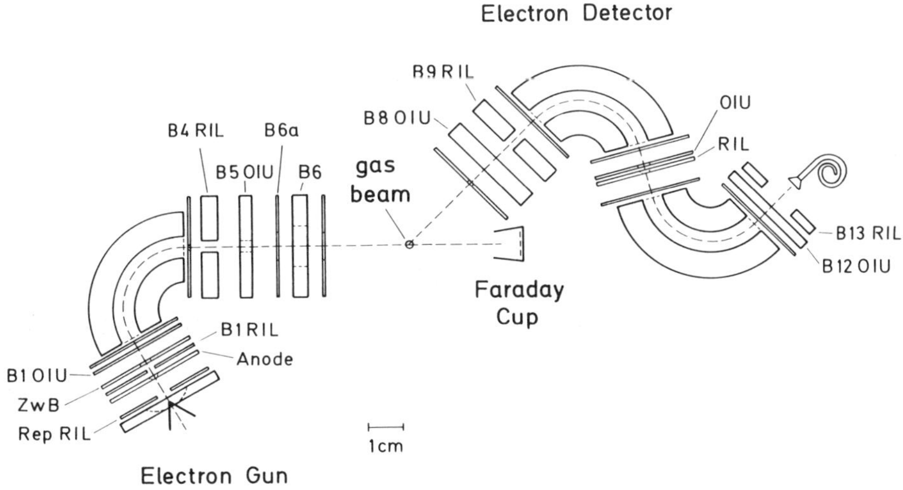

Figure 4-3-8. The crossed-beam electron spectrometer of Kochem et al.[48,49]

distortions of peak shapes and inaccuracies in cross-section measurements. The basic reason for the problem is that in order to change the electron beam energy, many of the potentials throughout the apparatus must be altered, and the focal lengths of the electron optical systems are thereby changed.

The ability to reach an electron beam energy as low as 50 meV (corresponding to a temperature of 389 K in an equilibrium situation) is important not only because it permits the close examination of electron scattering features at low energies, but also because it allows the study of inelastic processes at threshold, and below, where many important applications occur in plasmas and non-ionized gases. Such studies will be discussed in Chapter 5.

At the end of this discussion of modern spectrometers, which has emphasized so strongly the thrust toward higher and higher energy resolution, it is appropriate to make a direct comparison between some results obtained more than two decades ago with good resolution and those obtained much later (in 1981) with much higher resolution. Figure 4-3-9 shows the electron-helium

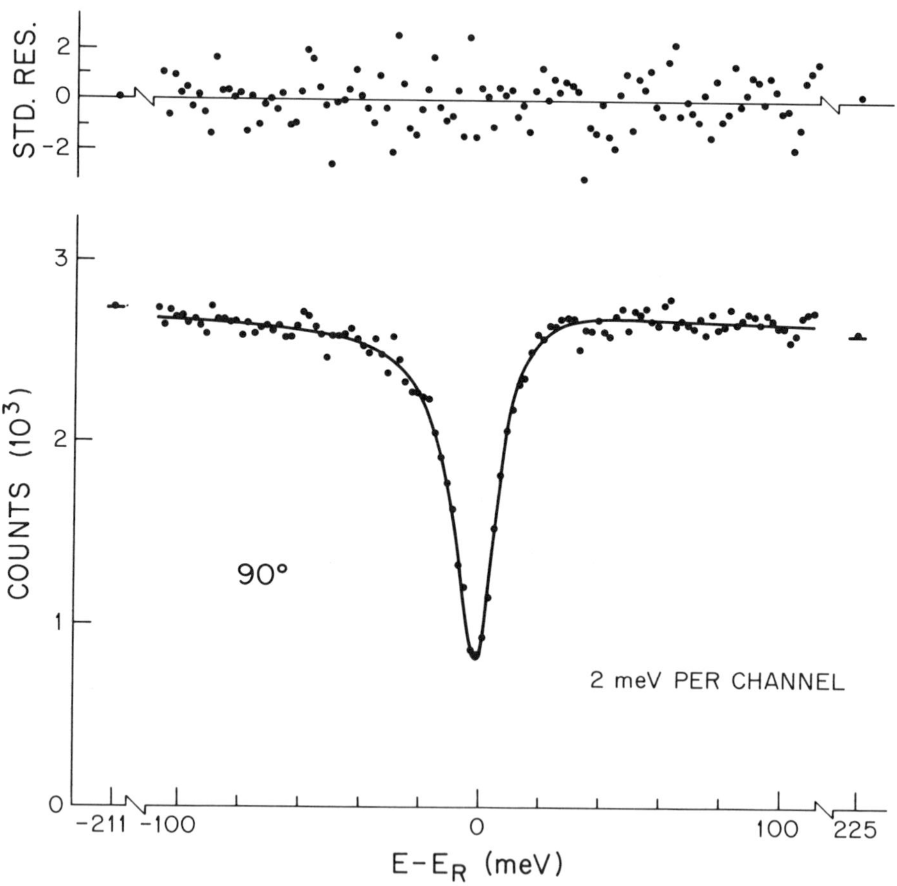

Figure 4-3-9. The helium resonance at 19.3 eV as mapped by Kennerly et al.[64]

elastic scattering resonance observed by Kennerly et al.,[64] with an instrumental energy width of about 5 meV. The resonance, as shown here for 90° scattering, appears as a nearly symmetrical dip of 73% below the ambient level, with a width of about 15 meV. Figure 4-3-9 is to be compared with Figs. 4-3-3, which shows the observation made by Schulz[33] in 1962 with an energy resolution of about 70 meV.

To achieve their high energy resolution, Kennerly et al.[64] use a photoionization source as an alternative to the usual practice of selecting a narrow slice of electron energy from a broad initial energy distribution. They cross a supersonic beam of metastable barium atoms with a He-Cd ultraviolet laser beam and obtain low-energy photoelectrons by the reaction

$$h\nu(3.815\,\text{eV}) + \text{Ba}^*(^1D_2) \rightarrow \text{Ba}^+ + \text{e}(17\,\text{meV}). \tag{4-3-1}$$

The photoelectrons are accelerated and focused to form the incident beam. The performance of this instrument, shown schematically in Fig. 4-3-10, is degraded somewhat, below the ideal expected, by space-charge effects, Doppler broadening, and surface potential drifts, but the net energy resolution is very impressive. The maximum electron beam currents that are obtainable, however, are so low ($10^{-12} - 10^{-11}$ A) that this method appears to be restricted to the study of strong resonances. It is worth noting that synchrotron radiation can be used as an alternative to the laser ultraviolet, with the advantage that it can be tuned continuously over a wider spectral range.

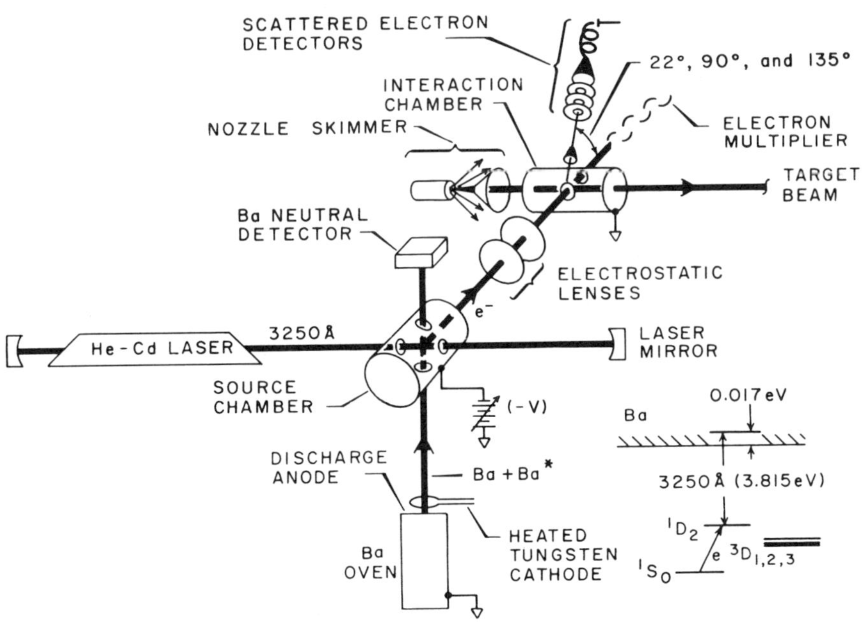

Figure 4-3-10. The apparatus of Kennerly et al.[64]

C. The Evaluation of the Angular Differential Cross Section

We must now address the difficult question of how differential scattering cross sections can be derived from measured angular distributions of elastic scattering. The direct analysis based on scattering and detection geometry is now generally considered to involve so much uncertainty as to be risky. The preferred approach at present is to use a phase shift analysis whose rationale can be understood by examination of equations developed in Chapter 3 during our discussion of the theory of elastic scattering. If spin effects are neglected, we can express the CM differential scattering cross section per unit solid angle in the form

$$I_s(\Theta) = |f(\Theta)|^2 = A^2 + B^2, \qquad \text{(3-15-33)}$$

where

$$A = \frac{1}{2\kappa} \sum_{l=0}^{\infty} (2l + 1) \sin 2\eta_l P_l(\cos \Theta), \qquad \text{(3-15-31)}$$

and

$$B = \frac{1}{2\kappa} \sum_{l=0}^{\infty} (2l + 1)(1 - \cos 2\eta_l) P_l(\cos \Theta). \qquad \text{(3-15-32)}$$

Here κ is the electron wave number, and η_l is the phase shift of the partial wave with angular momentum quantum number l. The CM scattering angle appears only in the form of the Legendre polynomials $P_l(\cos \Theta)$, so the angular dependence of each partial wave is known. At low electron energies, and below the excitation threshold in particular, only the first few partial waves (say up to order $l = L$) contribute appreciably, and the corresponding phase shifts are real. The higher-order ($l > L$) phase shifts cannot be neglected completely, but they are small and can be calculated by a simple scheme. The fact that the higher-order waves do not penetrate the inner structure of the target means that the long-range polarization potential $V(r) \sim -\alpha/r^4$ discussed in Section 1-8 dominates their scattering. For this potential the higher-order phase shifts can be calculated with sufficient accuracy by Born's approximation, with the result

$$\eta_l \approx \frac{\pi \alpha \kappa^2}{(2l - 1)(2l + 1)(2l + 3)} \quad (\text{for } l \geqslant 1), \qquad \text{(4-3-2)}$$

where α is the polarizability of the target. Then the partial wave contribution for $l > L$ can be summed analytically to give

$$f_B^L(\vartheta) = \pi \alpha \kappa \left[\frac{1}{3} - \frac{1}{2} \sin \frac{\vartheta}{2} - \sum_{i=1}^{L} A_i P_i(\cos \vartheta) \right], \qquad \text{(4-3-3)}$$

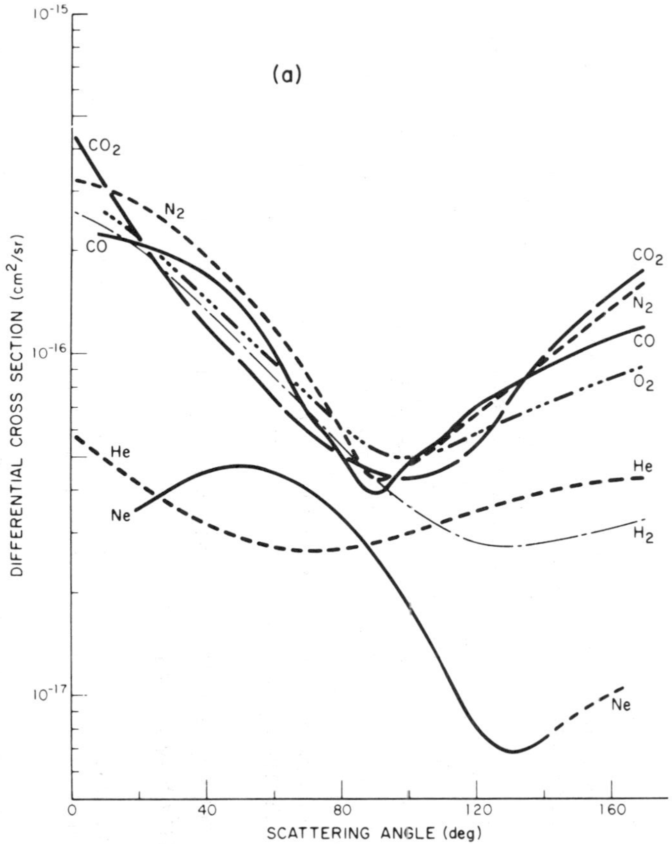

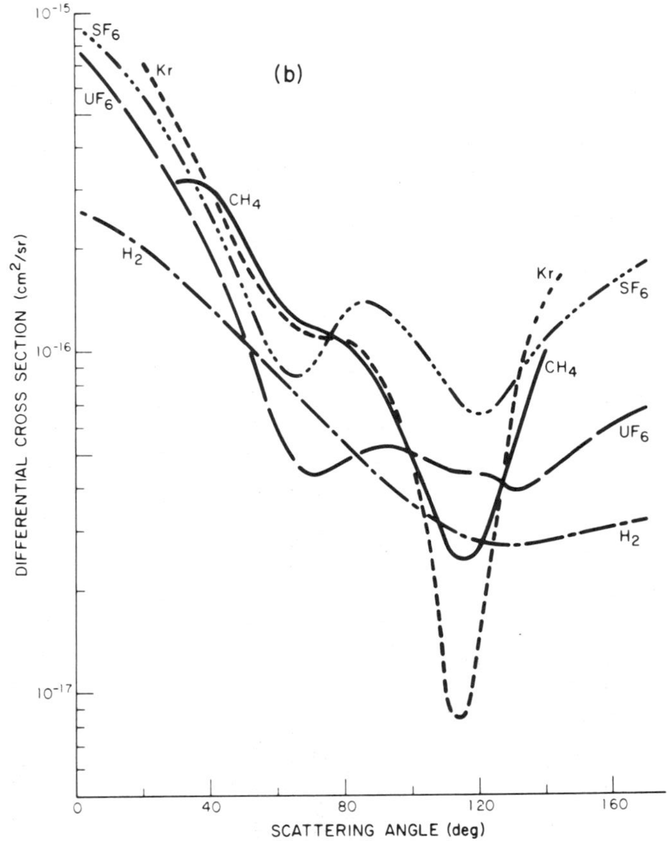

Figure 4-3-11. Experimental angular differential cross sections for the elastic scattering of electrons: (*a*) H_2, CO_2, N_2, CO, O_2 (10 eV) and He, Ne (12 eV); (*b*) Kr, H_2, CH_4, SF_6, UF_6 (10 eV). (From Csanak et al.[9]) For measurements on CO and N_2 at 20 to 100 eV, see J. C. Nickel, C. Mott, I. Kanik, and D. C. McCollum, *J. Phys. B* **21**, 1867 (1988).

where $A_i = [(2i - 1)(2i + 3)]^{-1}$. Here we have switched to the Lab scattering angle ϑ. The total scattering amplitude can consequently be expressed as a sum of terms of the usual form for $l = 0, 1, \ldots, L$ [cf. (3-15-29)] added to the Born scattering amplitude f_B^L:

$$f(\vartheta) = \frac{1}{2i\kappa}\left(\sum_{l=0}^{L} \{(2l+1)[\exp(2i\eta_l) - 1]P_l(\cos\vartheta)\} + 2i\kappa f_B^L(\vartheta)\right). \quad (4\text{-}3\text{-}4)$$

If trial values for the low-order phase shifts are put in (4-3-4), $I_s(\vartheta)$ can be calculated and compared with the experimental angular scattering distribution at the given electron energy. These phase shifts can then be varied until the best fit with experiment is obtained. Once a satisfactory set of phase shifts is determined, the integral elastic scattering cross section, as well, can be calculated by

$$q_s = \frac{4\pi}{\kappa^2}\sum_{l=0}^{\infty}(2l+1)\sin^2\eta_l. \quad (3\text{-}15\text{-}34)$$

It should be emphasized that the determination of phase shifts does not require absolute angular scattering distributions.[31,42] One may consult Refs. 31 and 42 for the modifications of the analysis required when the target has nonzero spin.

D. Data on the Angular Differential Cross Section*

A survey of data on differential elastic scattering from atoms and molecules is presented in Fig. 4-3-11. The most striking feature of these graphs of differential cross sections versus scattering angle is the appearance of distinct maxima and minima, which arise from the diffraction of the electron waves by the target structures. (This phenomenon was first observed by E. C. Bullard and H. S. W. Massey in 1932.[6]) In general, the complexity of the angular pattern increases with the atomic number of the scatterer in the case of atoms, or with other measures of the complexity of the target in the case of molecules. Differential cross sections are useful in their own right in certain applications. They are even more useful in relation to collision theory, as they provide much more sensitive tests of the theory than do total cross sections.

4-4. DATA ON THE MOMENTUM TRANSFER CROSS SECTION ($q_D \equiv q_m \equiv \sigma_m$)

The momentum transfer cross section q_D is defined in terms of the CM differential elastic scattering cross section, $I_s(\Theta)\,d\Omega_{CM}$, and the CM scattering

*Special attention should be directed to some recent data of very high quality. At Münster, absolute differential cross sections for the elastic scattering of electrons from Hg atoms have been reported by G. Holtkamp, K. Jost, F. J. Peitzmann, and J. Kessler, *J. Phys. B* **20**, 4543 (1987).

angle Θ, by the equation

$$q_D(v_0) = \int (1 - \cos \Theta) I_s(\Theta)\, d\Omega_{CM}. \qquad (1\text{-}6\text{-}1)$$

Accordingly, the momentum transfer cross section for electrons in a given gas may be obtained easily from measurements of the laboratory differential cross section for elastic scattering provided that they cover a sufficiently wide angular range. As is apparent from (1-6-1), q_D differs significantly from the integral collision cross section q_s only when the scattering is concentrated in the forward or the backward direction. At low relative velocities of approach, v_0, these two cross sections are nearly the same for electrons.

According to Section 4-2, the momentum transfer cross section can also be obtained as a function of electron energy by analysis of electron swarm experiments. The most important kind of measurement for this purpose is that of the electron drift velocity as a function of the energy parameter E/N (see Ref. 1).

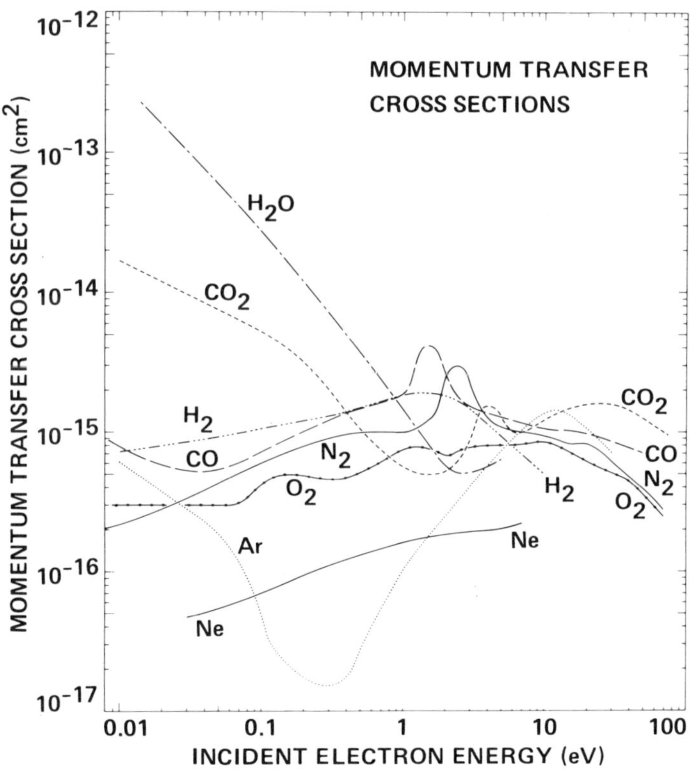

Figure 4-4-1. Experimental momentum transfer cross sections for electrons on Ne, Ar, H_2, N_2, O_2, CO, CO_2, and H_2O over the energy range 0.01 to 100 eV. (Csanak et al.[9])

If enough of the low-order phase shifts are known, q_D can be calculated from the equation

$$q_D(v_0) = \frac{4\pi}{\kappa^2} \sum_{l=0}^{\infty} (l+1) \sin^2 (\eta_l - \eta_{l+1}), \tag{4-4-1}$$

where the sum is truncated at $l = L$ (see Section 4-3). To derive (4-4-1), we start with $q_D = -2\pi \int_0^\pi I_s(\Theta)(1 - \cos \Theta)\, d(\cos \Theta)$, which is (1-6-1), make the substitution $\mu = \cos \Theta$, and obtain $I_s(\Theta) = |f(\Theta)|^2$ from (3-15-29). Thus

$$\begin{aligned}
q_D &= \frac{\pi}{2\kappa^2} \int_0^\pi \left| \sum_{l=0}^{\infty} (2l+1)(e^{2i\eta_l} - 1) P_l(\cos \Theta) \right|^2 (1 - \cos \Theta)\, d(\cos \Theta) \\
&= \frac{\pi}{2\kappa^2} \sum_{l=0}^{\infty} \sum_{m=0}^{\infty} (2l+1)(2m+1)(e^{2i\eta_l} - 1)(e^{-2i\eta_m} - 1) \int_{-1}^{+1} (1 - \mu) P_l(\mu) P_m(\mu)\, d\mu \\
&= \frac{\pi}{\kappa^2} \sum_{l=0}^{\infty} (2l+1) |e^{2i\eta_l} - 1|^2 - \frac{2\pi}{\kappa^2} \sum_{l=0}^{\infty} (l+1)(e^{2i\eta_l} - 1)(e^{-2i\eta_{l+1}} - 1) \\
&\qquad - \frac{2\pi}{\kappa^2} \sum_{l=0}^{\infty} l(e^{-2i\eta_{l-1}} - 1)(e^{2i\eta_l} - 1) \\
&= \frac{\pi}{\kappa^2} \sum_{l=0}^{\infty} (l+1)(e^{2i\eta_l} - 1)(e^{-2i\eta_l} - e^{-2i\eta_{l+1}}) \\
&\qquad + \frac{2\pi}{\kappa^2} \sum_{l=0}^{\infty} l(e^{2i\eta_l} - 1)(e^{-2i\eta_l} - e^{-2i\eta_{l-1}}) \\
&= \frac{\pi}{\kappa^2} \sum_{l=0}^{\infty} (l+1) |e^{2i\eta_l} - e^{2i\eta_{l+1}}|^2 = \frac{4\pi}{\kappa^2} \sum_{l=0}^{\infty} (l+1) \sin^2 (\eta_l - \eta_{l+1}).
\end{aligned}$$

Figure 4-4-1 presents a survey of momentum transfer data prepared by Csanak et al.[9] Additional data will be presented in Ref. 1.

4-5. THE ELASTIC SCATTERING OF HIGH-ENERGY (keV) ELECTRONS[65]

One of the goals in the study of electron collisions with atoms and molecules is to obtain information about static properties of the target. The complexity of the problem of extracting certain kinds of structural information from experimental data by comparison with the results of collision theory dictates that the electron energy be high enough to ensure adequacy of the first Born approximation. If the scattering experiments are conducted in the keV region, simple relations connect theory and experiment, and one has available a probe of the electron density of the target.

For elastic scattering of electrons from neutral atoms, the first Born approximation gives the angular differential cross section to be

$$\left(\frac{d\sigma}{d\Omega}\right)_{\text{el}} = \frac{4}{K^4}[Z - F(K)]^2, \tag{4-5-1}$$

where K is the electron-atom momentum transfer, $4/K^4$ is the Rutherford cross section in atomic units, Z is the atomic number, and $F(K)$ is the coherent X-ray form factor, which is a Fourier-Bessel transform of the radial electron density, $D(r)$.[65] Now define

$$\sigma(K) \equiv \frac{(d\sigma/d\Omega)_{el}}{(d\sigma/d\Omega)_R} - Z^2 = -2ZF(K) + F^2(K), \tag{4-5-2}$$

where $(d\sigma/d\Omega)_R$ is the Rutherford cross section, and use it to define the sum of the average of the electron-nuclear attractive potential energy and the classical Coulomb contribution to the electron-electron potential energy:

$$\begin{aligned}\bar{V}_{en} + \bar{V}^c_{ee} &= -2Z\int_0^\infty \frac{dr\, D(r)}{r} + \int d\mathbf{r}\rho(\mathbf{r})\int \frac{d\mathbf{r}'\rho(\mathbf{r}')}{|\mathbf{r} - \mathbf{r}'|} \\ &= -\frac{4Z}{\pi}\int_0^\infty dK\, F(K) + \frac{2}{\pi}\int_0^\infty dK\, F^2(K) = \frac{2}{\pi}\int_0^\infty dK\, \sigma(K). \end{aligned} \tag{4-5-3}$$

Note that (4-5-1) allows direct determination of $F(K)$ as

$$F(K) = Z - \sqrt{\left(\frac{d\sigma}{d\Omega}\right)_{\text{el}} \Big/ \left(\frac{d\sigma}{d\Omega}\right)_R} \tag{4-5-4}$$

and hence the separate determination of V_{en}. Then if V^c_{ee} is evaluated from (4-5-3), electron correlation effects can be studied by comparing $\sigma(K)$ and $F(K)$, derived from experiment, with the theoretical values for these quantities calculated with Hartree-Fock wave functions. For scattering on molecules, the analysis is more complex.[65]

Most of the early experiments involved measurements of the total scattered intensity (elastic plus inelastic), and the analysis of the data was quite complicated. Modern techniques, however, permit precise measurements of the elastic cross section separately, with a concomitant improvement in the conclusions that can be drawn.

4-6. STUDIES INVOLVING POLARIZED PROJECTILES AND TARGETS

A. Mott Scattering; Production and Analysis of Polarized Beams*

1. *Electrons*

A beam, or any other ensemble, of electrons is said to be *polarized* if the electron spins have a biased spatial orientation such that there exists a direction for which the two possible spin states are not equally populated. The degree of polarization, P_e, is defined by the equation

$$P_e = \frac{N(\uparrow) - N(\downarrow)}{N(\uparrow) + N(\downarrow)}, \tag{4-6-1}$$

where $N(\uparrow)$ is the number of measurements that yield the value $+\hbar/2$, and $N(\downarrow)$ the value $-\hbar/2$ for the component of the spin along the experimentally selected quantization axis. As an example, if we find 80% of the electrons with spin $+\hbar/2$ in the direction of the unit vector $\mathbf{n}$ and 20% with $-\hbar/2$, then $\mathbf{P}_e = 0.6\mathbf{n}$. We may also say that 60% of the beam is totally polarized and 40% is unpolarized (see Fig. 4-6-1).

A polarized beam of neutral atoms or molecules can be prepared by sending an unpolarized beam through a spin filter employing an inhomogeneous magnetic field. This technique was first used by Stern and Gerlach in 1921; an example involving a hexapole field is described in Section 4-6-A-2. In the case of an alkali atom beam, splitting into two beams occurs, with half the atoms having spin "up" and half spin "down." We can reject one of these beams and use the other polarized beam, with an overall intensity loss of only a factor of 2. Unfortunately, conventional spin filters do not work for free electrons. Using an argument due to N. Bohr, Kessler[66] and Mott and Massey[67] show how the Lorentz force acting on a free electron, combined with the uncertainty principle, negates the possibility of separating spin-up and spin-down electrons in an inhomogeneous magnetic field. Until recently, the known methods of producing polarized electrons provided beams of quite low intensity, and the use of

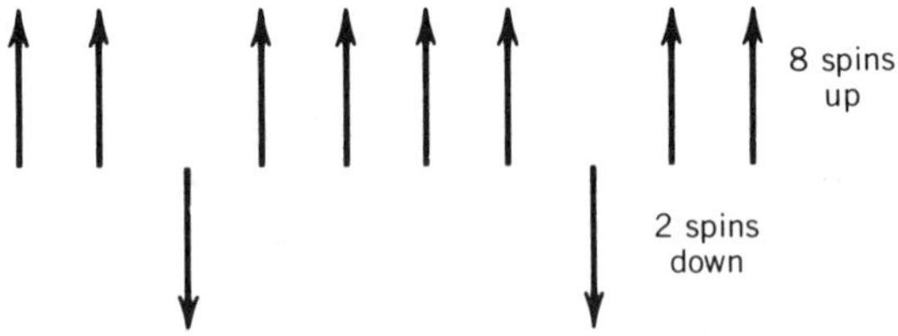

Figure 4-6-1. Partially polarized beam. (Based on Kessler.[66])

*This section is based on the book *Polarized Electrons*, by J. Kessler.[66]

polarization analyzers entailed intensity losses on the order of 10^4 to 10^5. Good atomic collision experiments involving both polarization and analysis of electrons became feasible only in the 1970s.

It has been known for many years that both electron polarization and analysis can be accomplished by *Mott scattering*, the wide-angle scattering of relativistic electrons by a heavy atom whose nucleus interacts strongly with the incident electron through spin-orbit coupling. The theory for this process was worked out through a solution of the Dirac relativistic wave equation by N. F. Mott in 1929.[66,67] Let us consider an electron moving with velocity $\mathbf{v}$ with respect to a scattering center that sets up an electrostatic field $\mathbf{E}$. In the rest frame of the electron there appears a magnetic induction

$$\mathbf{B} = -\frac{1}{c}\mathbf{v} \times \mathbf{E} = \frac{1}{mc}\mathbf{E} \times \mathbf{p}, \qquad (4\text{-}6\text{-}2)$$

where $\mathbf{p}$ and m are the electron momentum and mass, respectively. The electron spin is affected because the energy of the electron depends on its orientation with respect to $\mathbf{B}$; this "spin-orbit" contribution to the energy is given by

$$-\boldsymbol{\mu}\cdot\mathbf{B} = -\frac{e}{mc}\mathbf{s}\cdot\mathbf{B} = -\frac{e}{m^2c^2}\mathbf{s}\cdot[\mathbf{E} \times \mathbf{p}]. \qquad (4\text{-}6\text{-}3)$$

Here $\boldsymbol{\mu}$ is the magnetic dipole moment of the electron associated with its spin operator $\mathbf{s}$. If $\mathbf{E}$ is derived from a central field of potential energy $V(r)$, the spin-orbit energy can be written

$$\frac{1}{2m^2c^2}\frac{1}{r}\frac{dV}{dr}(\mathbf{s}\cdot\mathbf{l}), \qquad (4\text{-}6\text{-}4)$$

in which form it manifests its origin as an interaction between the spin of the incident electron and its orbital angular momentum, $\hbar\mathbf{l} = m\mathbf{r} \times \mathbf{p}$, about the nucleus.* Note that if $V(r)$ is a Coulomb potential, the spin-orbit energy falls off with increasing r more rapidly than $V(r)$ does, and hence may be neglected at moderately large distances from the nucleus.

Now we turn to Fig. 4-6-2. First look at the solid curve, which shows the potential energy assumed for the electron-nucleus interaction without spin-orbit coupling. Now imagine an electron approaching perpendicular to the drawing from above with spin up (↑), parallel to the drawing, on the right-hand side. Its

*A factor of 2 has been introduced into (4-6-4) to account for the "Thomas precession," a relativistic kinematic effect [see G. K. Woodgate, *Elementary Atomic Structure*, 2nd ed., Clarendon, Oxford (1980); R. D. Sard, *Relativistic Mechanics*, Benjamin, New York (1970)]. The pioneering work on spin dynamics done by L. H. Thomas during the mid-1920s is chronicled by J. D. Jackson, *Phys. Today* **40**, 34 (May 1987). Thomas developed relativistically covariant equations for the motion of a spin with any g factor in any arbitrary external electromagnetic field.

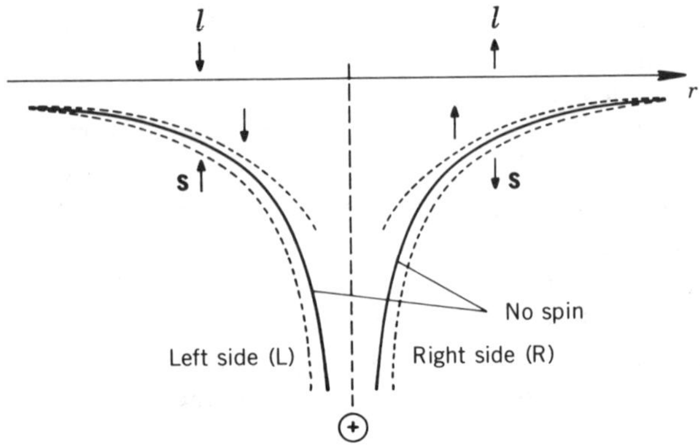

Figure 4-6-2. Potential curves with (dashed line) and without (solid line) spin-orbit coupling for incident electrons with spins up (↑) and down (↓).[66] The solid curves show the electrostatic interaction, V_{ES}, which in some cases is the electron-nucleus Coulomb interaction, to a good approximation. The dashed curves show the sum of V_{ES} and V_{SO}, the spin-orbit interaction. Note that at any given value of r, the slope of the lower dashed curve is greater than the slope of the upper dashed curve, so the force on the incident electron is greater for the lower curve ($F = -dV/dr$).

orbital angular momentum is represented by the arrow (↑), and the potential energy including the spin-orbit interaction is represented by the upper dashed curve. In this case, **s** is parallel to **l**. For an electron approaching on the same trajectory with spin down (↓), **s** and **l** are antiparallel, and the energy is lowered, to the lower dashed curve. If the electron approaches on the left-hand side of the nucleus as viewed from above, **l** is reversed (↓), and the two dashed curves on the left arise for the two orientations of **s** (↑↓). We are interested only in transversely polarized electron beams here; longitudinal polarization does not produce the effects we shall describe.

Figure 4-6-3 provides additional insight into this situation and also helps explain the left–right, or Mott, asymmetry in the scattering of a polarized beam.

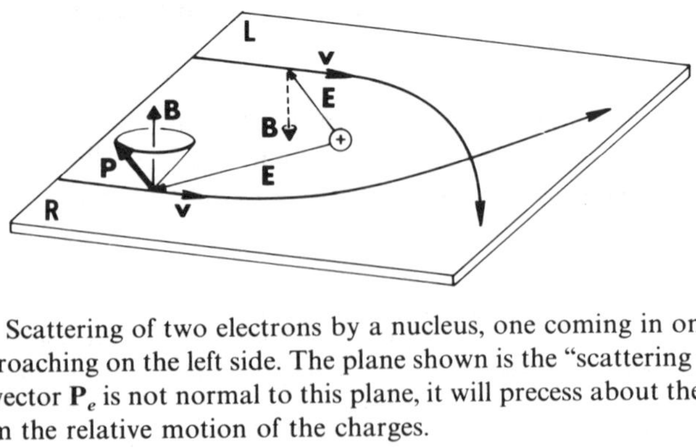

Figure 4-6-3. Scattering of two electrons by a nucleus, one coming in on the right side, the other approaching on the left side. The plane shown is the "scattering plane."[66] If the polarization vector $\mathbf{P}_e$ is not normal to this plane, it will precess about the magnetic field **B** arising from the relative motion of the charges.

First, if the polarization is complete and $\mathbf{P}_e$ points straight up out of the scattering plane, the e(↑) that pass the nucleus on the left side and are thus scattered toward the right are subjected to a stronger force than the e(↑) passing the nucleus on the right and being scattered toward the left. Hence the intensity of scattering toward the right is greater than that toward the left. This kind of analysis can be generalized to explain the left-right asymmetry observed also for complete downward polarization (↓) and for incomplete polarization. Second, Fig. 4-6-3 also shows that if the polarization $\mathbf{P}_e$ does not lie parallel or antiparallel to $\mathbf{B}$, the magnetic moment associated with $\mathbf{P}_e$ experiences a torque that causes $\mathbf{P}_e$ to precess. $\mathbf{P}_e$ retains its direction throughout the collision only if it is initially normal to the scattering plane.

Kessler[66] shows that for the simple situation of Fig. 4-6-4, the complex general result for the differential scattering cross section reduces to

$$\frac{d\sigma(\vartheta, \varphi)}{d\Omega} = (|f|^2 + |g|^2)[1 - S(\vartheta)\sin\varphi], \tag{4-6-5}$$

where f and g are the direct and exchange scattering amplitudes (discussed in Section 4-6-B) and $S(\vartheta)$ is the Sherman function, or analyzing power,

$$S(\vartheta) = i\,\frac{fg^* - f^*g}{|f|^2 + |g|^2}, \tag{4-6-6}$$

which is real. Two diametrically opposed points on the circle in Fig. 4-6-4 differ most in their values of $d\sigma/d\Omega$ when their φ values are 90° and 270°.

"Mott detectors" based on the left-right asymmetry of the scattering are used to measure the polarization of electron beams.[68] The same phenomenon can be used to produce a polarized beam from a beam of unpolarized electrons, the

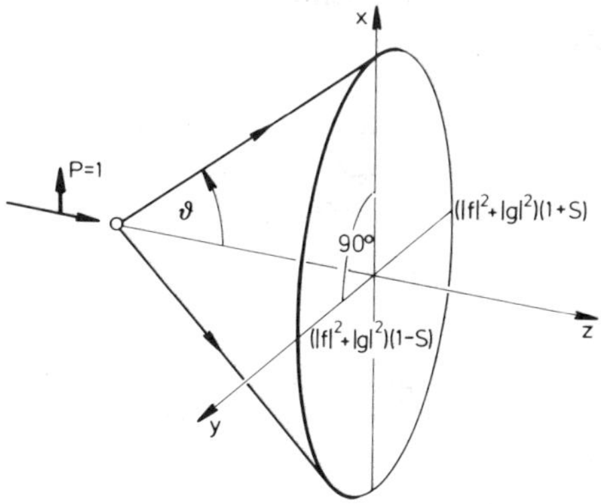

Figure 4-6-4. Left-right scattering asymmetry of an electron beam completely polarized in the x direction.[66]

magnitude of the effect again being determined, for a given electron energy and choice of target, by the Sherman function. If we consider an unpolarized incident beam in the geometry of Fig. 4-6-4, with equal numbers of e(↑) and e(↓), the beam scattered to the left at angle will have $N(\uparrow) \sim (1 + S)$ and $N(\downarrow) \sim (1 - S)$, whereas for the beam scattered at angle to the right, $N(\uparrow) \sim (1 - S)$ and $N(\downarrow) \sim (1 + S)$. Hence, in the "left-scattered" beam, $P_e = S$, and $P_e = -S$ in the "right-scattered" beam. However, even if the energy is optimized (at about 100 keV) and a good choice is made for the target (say, Au), Mott scattering is uncompetitive with other, recently discovered methods of polarized beam production for low-energy scattering experiments. According to the principles of electron optics, the deceleration of electrons from a high energy through four or five orders of magnitude to the desired level for the collision experiments brings about an unacceptable intensity loss. On the other hand, electron acceleration from the eV level up to 100 keV is favored, and high-energy Mott analyzers ("polarimeters") have been used successfully in many low-energy atomic collisions experiments.

Kessler[66] has discussed the various alternatives to Mott scattering for polarized beam production. For about a decade, starting in the late 1960s, the most satisfactory source in the present context was the one based on the "Fano effect,"[69] photoionization of an unpolarized cesium atomic beam by circularly polarized light with photon energy barely exceeding the photoionization threshold. The polarized electrons thus produced have the same helicity as that of the incident light. Electron currents of about 10 nA, with $P_e \sim 0.70$, are typical.

Detailed collision studies with polarized electrons became feasible only in the late 1970s with the development of the GaAs polarized electron source.[66,70,71] Now scattering experiments with polarized electrons are just as feasible as those with unpolarized electrons, although more expensive, and the accuracy is comparable, sometimes even better. The GaAs source,[72,73] first demonstrated by D. T. Pierce, F. Meier, and P. Zürcher in 1975, relies on the photoemission of polarized electrons from gallium arsenide when irradiated by circularly polarized light. The direction of the electron polarization is that of the light-beam axis, and the degree of polarization is about 25 to 45%. Currents of tens of μA are easily obtainable, and 1 mA can be achieved with special effort. The energy widths of the electron beams range from 30 meV to several hundred meV. Gallium arsenide sources are discussed at greater length in Note 4-2.

2. *Atoms and Molecules*

As pointed out earlier in this section, inhomogeneous magnetic fields have been used for decades for state selection in atomic and molecular beams. In recent years, hexapole magnets have been widely used for this purpose.[74] We shall now illustrate some of the techniques of state selection and beam handling by describing the method used by Fletcher et al.[75] to prepare a beam of H atoms in the $m_J = +\frac{1}{2}$ state for collisions with polarized electrons.

A tungsten oven is used to produce H atoms from H_2 molecules by thermal dissociation. A hexapole magnet (see Fig. 4-6-5) provides state selection of the ground-state H atoms entering it, the atoms with quantum number $m_J = -\frac{1}{2}$ being defocused, while those with $m_J = +\frac{1}{2}$ execute sinusoidal trajectories about the axis and are transmitted. The magnetic field in the hexapole is transverse to the axis of the H-atom beam and is a function of azimuthal position. Hence the direction of the spin of each of the $m_J = +\frac{1}{2}$ atoms depends on its position. To align the spins, a small solenoid is placed axially at the exit of the hexapole, and the spins of the atoms are adiabatically rotated into parallelism with their momenta by the 20-mT field of the solenoid. A small Helmholtz pair producing a 0.5-mT transverse field then induces adiabatic rotation of the atomic spins into a direction either parallel or antiparallel to the direction of the electron beam, which intersects the atomic beam at 90° about 90 cm from the H-atom source. After passing through the collision region, the hydrogen beam goes into a quadrupole mass analyzer, where it is ionized and its atomic and molecular fractions are determined. The atomic polarization, P_H, is about 0.99 in regions of high magnetic field, but drops to half of that value in regions of low field, such as the collision region, because of the coupling to the nuclear spin I. The low-field electronic polarization of a one-electron atom after perfect high-field state selection is given by $P = 1/(2I + 1)$. For ^1H, $I = \frac{1}{2}$ and $P = 0.5$; for ^{133}Cs, $I = \frac{7}{2}$ and $P = 0.125$.

B. The Complete Elastic Scattering Experiment; Scattering Asymmetries

1. *Spin Exchange Only, No Spin-Orbit Interaction (Spin-$\frac{1}{2}$, Low-Z Targets)*

In 1969, Bederson[76] published two short but influential papers exploring the question of how one can obtain the greatest possible amount of information in experiments involving elastic and inelastic electron collisions. He pointed out that the conventional measurements of integral and differential cross sections do

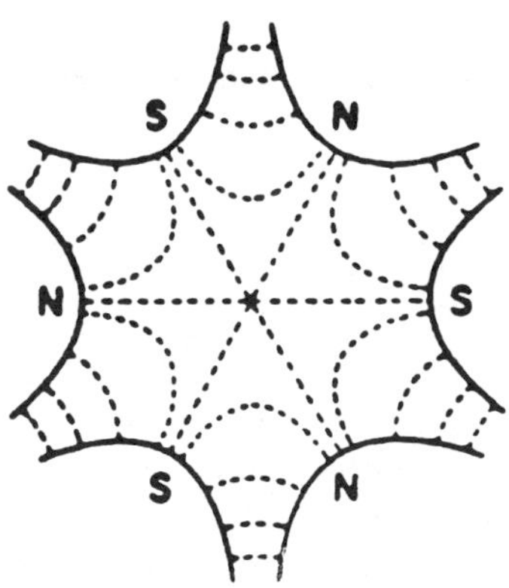

Figure 4-6-5. Field and potential lines for a magnetic field that varies as the square of the distance from the axis (field lines are dashed; potential lines are solid). The axis is through the geometrical center of the diagram perpendicular to the figure. The force on the atom acts radially. The hexapole field shown here is approximated by a permanent magnet with five collinear sections, each 83 mm long, and a total effective length of 457 mm, including the 11-mm spacing between adjacent sections. The magnet has a 3.2-mm-diameter gap and a pole-tip field strength of 0.85 T. (Drake.[74])

not usually provide results that can be directly and uniquely compared with the predictions of collision theory. The reason is that the spatial and spin orientations of the colliding particles usually influence the interaction potential, and without suitable precollision state selection and postcollision state analysis, the measurement averages over several different interaction potentials.

In his analysis of elastic scattering, Bederson[76] considers the case of greatest conceptual simplicity, that of an elastic collision with a "one-electron" atom, with no core angular momentum and with the valence electron in an *s* state. He also assumed no nuclear spin and no spin-orbit interaction between projectile and target atom (so that spin flips can occur only by spin exchange). In time-independent collision theory, the interaction between an electron and such an "ideal alkali atom" can be described by a *pair* of velocity-independent potentials, "singlet" and "triplet" according as the two involved electrons have total effective spin 0 or 1 $\hbar$, respectively, during the interaction. Associated with each of these two potentials is a scattering amplitude (singlet, S or triplet, T), that is, a spatial wave function with a magnitude and phase determined, relative to the unperturbed wave function, by the interaction potential. There are three independent quantities, observable at least in principle. They can be taken to be the magnitudes of S and T and the relative phase between them. Alternatively, we can use the direct (f) and exchange (g) amplitudes related to S and T by

$$S = f + g; \qquad T = f - g. \tag{4-6-7}$$

f and g are the amplitudes in the outgoing spherical wave functions for the direct and exchange scattering, respectively:

$$F(\mathbf{r})_{\text{direct}} \sim \frac{e^{ikr}}{r} f(\vartheta) \tag{4-6-8}$$

and

$$G(\mathbf{r})_{\text{exchange}} \sim \frac{e^{ikr}}{r} g(\vartheta). \tag{4-6-9}$$

If both the electrons and atoms are unpolarized, a measurement of the elastic differential cross section yields the average value

$$\frac{d\bar{\sigma}}{d\Omega} = \frac{1}{4}|f + g|^2 + \frac{3}{4}|f - g|^2 = \frac{1}{4}|S|^2 + \frac{3}{4}|T|^2. \tag{4-6-10}$$

To obtain this equation, put $n = 0$ in (3-22-11). The factors $\frac{1}{4}$ and $\frac{3}{4}$ are the statistical weights for the singlet and triplet states, respectively. This measurement yields only one relation between f and g at each energy and angle, so two additional independent measurements are required at each energy and angle to determine f and g (or S and T) experimentally. The possible measurements on

Table 4-6-1 Possible Elastic Collision Experiments Involving Electrons on One-Electron Atoms, for Various Combinations of Spin-Selected and Spin-Analyzed Beams[a]

Expt. No.	Type of Experiment[b]	Quantity Observed
I	$e(\uparrow\downarrow) + A(\uparrow\downarrow) \rightarrow e(\uparrow\downarrow) + A(\uparrow\downarrow)$	$\frac{3}{4}\lvert f - g\rvert^2 + \frac{1}{4}\lvert f + g\rvert^2$
IIa	$e(\uparrow\downarrow) + A(\uparrow) \rightarrow e(\uparrow) + A(\uparrow)$	$\frac{1}{2}\lvert f - g\rvert^2$
b	$\rightarrow e(\downarrow) + A(\uparrow)$	$\frac{1}{2}\lvert f\rvert^2$
c	$\rightarrow e(\uparrow) + A(\downarrow)$	$\frac{1}{2}\lvert g\rvert^2$
IIIa	$e(\uparrow) + A(\uparrow\downarrow) \rightarrow e(\uparrow) + A(\uparrow)$	$\frac{1}{2}\lvert f - g\rvert^2$
b	$\rightarrow e(\uparrow) + A(\downarrow)$	$\frac{1}{2}\lvert f\rvert^2$
c	$\rightarrow e(\downarrow) + A(\uparrow)$	$\frac{1}{2}\lvert g\rvert^2$
IV	$e(\uparrow) + A(\uparrow) \rightarrow e(\uparrow) + A(\uparrow)$	$\lvert f - g\rvert^2$
V	$e(\uparrow) + A(\downarrow) \rightarrow e(\uparrow) + A(\downarrow)$	$\lvert f\rvert^2$
VI	$e(\uparrow) + A(\downarrow) \rightarrow e(\downarrow) + A(\uparrow)$	$\lvert g\rvert^2$

[a]We assume no nuclear spin and no spin-orbit interaction between incident electron and target atom.

[b]e and A refer to electron and atom, respectively; $(\uparrow\downarrow)$ represents an unpolarized beam, and $(\uparrow)$ and $(\downarrow)$ represent polarized beams.

elastic scattering of an electron beam on a beam of one-electron atoms are tabulated in Table 4-6-1.

In each collision, we assign an arbitrary direction to the spin of the atom and consider the electron to be either parallel or antiparallel to it. In Table 4-6-1, on the left side of each "reaction equation," a single arrow such as $(\uparrow)$ or $(\downarrow)$ by the symbol e or A means deliberate polarization of one or both of the collision partners by some experimental beam preparation technique. On the right-hand side of each reaction equation, a single arrow by e or A may be appropriate as the result of experimental spin state analysis, but not necessarily so, depending on the experiment being performed.

Experiment I, with unpolarized electrons and atoms, has been discussed [see (4-6-10)]. Experiment II involves unpolarized electrons and a beam of polarized atoms. In the atomic recoil experiments performed by Bederson's group at New York University, postcollision spin analysis is performed on the atoms but not on the electrons. What is measured is IIc and IIa + IIb, which is equivalent to I − IIc, so the recoil measurement combined with I determines two out of the three parameters characterizing f and g.

To get a complete determination of f and g,* we must then use polarized electrons, one way now being to do Experiment II with postcollision spin-analyzed electrons. Then Experiment II gives IIa + IIc − IIb, or

$$\tfrac{1}{2}|f - g|^2 + \tfrac{1}{2}|g|^2 - \tfrac{1}{2}|f|^2.$$

*Although not pointed out by Bederson, a "complete experiment" requires a measurement of the rotation of the polarization vector, as in Sections 4-6-B-2 and 4-6-C (see Kessler,[66] p. 101).

This quantity gives $|f|^2$, when subtracted from I, so Experiment II with electron spin analysis, combined with Experiment I, also determines two out of the three parameters characterizing f and g. Hence Experiment II with recoil atom analysis, combined with Experiment II with postcollision electron analysis, completely determines f and g. Experiment III, involving a spin-polarized incident electron beam and an unpolarized atomic beam, provides an alternative to Experiment II and furnishes the same information.

Experiments IV, V, and VI utilize spin polarization in both incident electron and atom beams. The process in entry V is direct scattering, having amplitude f. In entry VI, the incident and atomic electrons are interchanged, the amplitude for the exchange scattering being g. In entry IV, the direct and exchange processes interfere. The indistinguishability of the electrons results in the triplet amplitude $T = f - g$, the minus sign being a consequence of Fermi-Dirac statistics.

Another approach to these considerations is to use the "scattering asymmetry," as done by the group at Yale University,[75] which made an extensive study of the scattering of polarized electrons by polarized hydrogen atoms to investigate the spin exchange in elastic and ionizing collisions. Their experiments were performed with the spins of the incident electrons and target electrons either antiparallel (↑↓) or parallel (↑↑), and they determined an elastic cross section asymmetry A defined by the equation

$$A = \frac{d\sigma(\uparrow\downarrow)/d\Omega - d\sigma(\uparrow\uparrow)/d\Omega}{d\sigma(\uparrow\downarrow)/d\Omega + d\sigma(\uparrow\uparrow)/d\Omega}. \tag{4-6-11}$$

In terms of f and g, or alternatively T and S, this expression for A reduces to

$$A = \frac{\mathrm{Re}\,(f^*g)}{d\bar{\sigma}/d\Omega} = \frac{|f||g|\cos\theta}{d\bar{\sigma}/d\Omega} = \frac{1-r}{1+3r}, \tag{4-6-12}$$

where θ is the relative phase of f and g, and $r = |T|^2/|S|^2$. Hence, if $d\bar{\sigma}/d\Omega$ is known, the measurement of A reveals the interference between the direct and exchange amplitudes. The experimental arrangement used here is illustrated schematically in Fig. 4-6-6 as drawing (1), along with all other arrangements possible subject to the restriction that only two selectors or analyzers are used. $\mathbf{P}_e$ and $\mathbf{P}'_e$ are the polarization vectors of the incident and scattered electrons, respectively, and $\mathbf{P}_a$ and $\mathbf{P}'_a$ are the polarization vectors for the incident and scattered H atoms. The two arrangements in each row of the figure provide the same information, so we need consider only three basic situations, and give only three asymmetries A, A', and A'' for rows 1, 2, and 3, respectively. A' and A'' can be expressed as

$$A' = |g|^2\left(\frac{d\bar{\sigma}}{d\Omega}\right)^{-1} - 1 \tag{4-6-13}$$

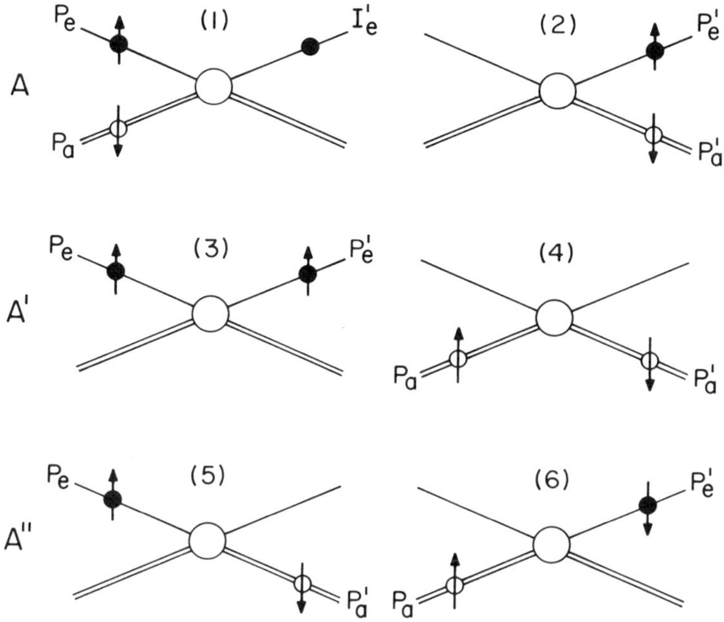

Figure 4-6-6. Possible experiments subject to the restriction that only two selectors or analyzers are used.[75] Particles are incident from the left with electrons denoted by single lines and atoms by double lines. The circles, solid for electrons and open for atoms, indicate measured beams with the measured quantities denoted by P_e and P_a for the electron and atom polarizations, respectively, in the incoming channels. Primes indicate quantities measured in the outgoing channels. Asymmetries measured in a given row are specified by A, A', and A'' in accordance with (4-6-12), (4-6-13), and (4-6-14).

and

$$A'' = |f|^2 \left(\frac{d\bar{\sigma}}{d\Omega}\right)^{1} - 1. \tag{4-6-14}$$

From (4-6-10) and the equations for A, A', and A'' we can show that the asymmetries are not linearly independent but are related by the equation

$$A - A' - A'' = 1. \tag{4-6-15}$$

Hence only two arrangements from any two rows in Fig. 4-6-6 are independent. Performance of any two of the corresponding independent experiments, together with a measurement of $d\bar{\sigma}/d\Omega$, does not suffice to determine the three scattering parameters $|f|$, $|g|$, and θ; $\cos\theta$, but not θ is determined, so the sign of θ is unknown. This ambiguity can be removed, however, by measuring the polarization rotation of either the electrons or the atoms induced by the collision.

In the Yale experiments, measurements were made of A for elastic scattering at 90° and the corresponding cross section asymmetry (A_I) for total ionization (i.e., with collection of all ejected electrons). A_I is given by the expression

$$A_I = \frac{\sigma_I(\uparrow\downarrow) - \sigma_I(\uparrow\uparrow)}{\sigma_I(\uparrow\downarrow) + \sigma_I(\uparrow\uparrow)}. \tag{4-6-16}$$

These ideas have been elaborated by several authors.[77]

2. *Spin-Orbit Interaction Only, No Exchange (Spin-Zero, High-Z Targets)*

We now consider the other limiting case, where electron exchange plays no role, and the spin-dependent effects are caused by spin-orbit interaction between the incident electron and the target atom. This process is described by two scattering amplitudes, $\mathscr{F}$ and $\mathscr{G}$, which contain the phases γ_1 and γ_2:

$$\mathscr{F} = |\mathscr{F}|e^{i\gamma_1} \quad \text{and} \quad \mathscr{G} = |\mathscr{G}|e^{i\gamma_2}. \tag{4-6-17}$$

$\mathscr{F}$ is governed mainly by the Coulomb interaction, and $\mathscr{G}$ describes the effects of the spin-orbit interaction. A complete experiment requires the determination of $|\mathscr{F}|$, $|\mathscr{G}|$, and $\gamma_1 - \gamma_2$. The conventional measurement of the differential cross section yields

$$\frac{d\sigma}{d\Omega} = |\mathscr{F}|^2 + |\mathscr{G}|^2. \tag{4-6-18}$$

Observing the polarization produced in scattering an unpolarized beam gives

$$P_e = i\frac{\mathscr{F}\mathscr{G}^* - \mathscr{F}^*\mathscr{G}}{|\mathscr{F}|^2 + |\mathscr{G}|^2} = -\frac{2|\mathscr{F}||\mathscr{G}|\sin(\gamma_1 - \gamma_2)}{|\mathscr{F}|^2 + |\mathscr{G}|^2}. \tag{4-6-19}$$

The package may be completed by scattering a polarized beam and observing the change of the polarization vector produced by the scattering. Such a measurement yields two new observables:

$$\mathscr{T} = \frac{|\mathscr{F}|^2 - |\mathscr{G}|^2}{|\mathscr{F}|^2 + |\mathscr{G}|^2} \tag{4-6-20}$$

and

$$\mathscr{U} = \frac{\mathscr{F}\mathscr{G}^* + \mathscr{F}^*\mathscr{G}}{|\mathscr{F}|^2 + |\mathscr{G}|^2} = \frac{2|\mathscr{F}||\mathscr{G}|\cos(\gamma_1 - \gamma_2)}{|\mathscr{F}|^2 + |\mathscr{G}|^2}. \tag{4-6-21}$$

$\mathscr{T}$ is related to the change in length of $\mathbf{P}_e$, $\mathscr{U}$ to the rotation of the $\mathbf{P}_e$ vector. The approach outlined here has been used by Kessler and his colleagues at the University of Münster in performing a "complete experiment" on the elastic scattering of electrons by mercury and xenon.[78]

C. Experiments

Many experiments have now been performed on various combinations of electron and neutral beams, with polarized and unpolarized electron beams, and state-slected neutral beams and otherwise. Much of this work is surveyed in reviews.[66,71] Among the laboratories that have been especially active in this experimentation are those in Bielefeld, City University of New York, Edinburgh, JILA, Mainz, Münster, the National Bureau of Standards, New York University, Stirling, and Yale. The neutral targets that have received the most attention are the H atom, the alkali atoms, the noble gases, and Hg. In our closing discussion, we shall concentrate on certain aspects of modern studies involving polarized incident electron beams.

In such experiments, $\mathbf{P}_e$ is known for the incident beam from measurements with a polarimeter, and the direction of $\mathbf{P}_e$ can be reversed, say, by a change in the sense of the circular light polarization in the case of the GaAs source. Some sort of asymmetry, such as A in (4-6-11), is always measured, with data being taken for both the original and reversed polarization directions at the same energy and same angle. The measurements thus involve ratios of differential cross sections at constant E and ϑ, so most of the systematic errors are eliminated.[71] This fact must be counted as a great advantage.

In studies with spin-polarized electron beams, a major goal has been the elucidation of the spin dependence in elastic scattering that arises from (1) spin exchange between the polarized electron beam and the target,[79] and from (2) the spin-orbit interaction.[80] The emphasis has been on the two limiting cases discussed in Section 4-6-B, where either (1) or (2), but not both, played a significant role in a given situation.* Both of these limiting cases can be described in terms of three parameters, and because of the availability of intense sources of polarized electrons, "complete experiments" can now be performed for (1) and (2). For increasingly more complex situations, the number of independent parameters soon becomes prohibitive. For example, in the case of a high-Z, spin-$\frac{1}{2}$ target (both spin exchange and spin-orbit interaction important), a complete determination of the collision matrix would require 18 different measurements at the same energy and scattering angle.[82] In such cases it might be concluded that the enormous effort required might be better directed to other kinds of measurements.

*The separation of the two effects may not be clean even with targets of rather low Z. In a recent study by McClelland et al.[81] of elastic scattering of polarized electrons by polarized Na atoms ($Z = 11$), both the exchange and spin-orbit interactions had to be taken into account to explain the experimental results.

We now turn to the work of Kessler's group on e-Hg, Xe scattering, performed with the apparatus shown in Fig. 4-6-7.[78] Polarized electrons are produced by light from the HeNe laser impinging on the GaAs cathode, and these electrons are focused into a beam traveling antiparallel to the laser beam. When the electrons enter the 90° deflector, they are sent vertically downward in the plane of the drawing, with the polarization direction that is indicated by the arrow that is perpendicular to the electron beam. The beam is deflected again by 90°, and then by 180°, before it crosses the atom beam (Hg or Xe) coming upward on the drawing. The electron source system (laser → GaAs cathode) is stationary, as is the scattered electron handling system (filter lens → Mott detector). The scattering angle, ϑ, that is observed is varied by rotating the 180° deflector and its associated elements.

The direction of the initial electron polarization, $\mathbf{P}_{ei}$, is unaffected by the passage of the electrons through the electrostatic deflectors and lenses. Both the transverse and longitudinal components of the polarization of the scattered beam $\mathbf{P}_{es}$, are determined from the left-right asymmetry in the two pairs of counters in the Mott detector. The transverse component can be measured directly. The longitudinal component cannot; it must be rotated into the transverse plane by the Wien filter[66] so that it can be measured (Section 4-6-A).

If measurements are made on all four observables in (4-6-18) to (4-6-21), then the elastic scattering processes have been completed, and a "complete experi-

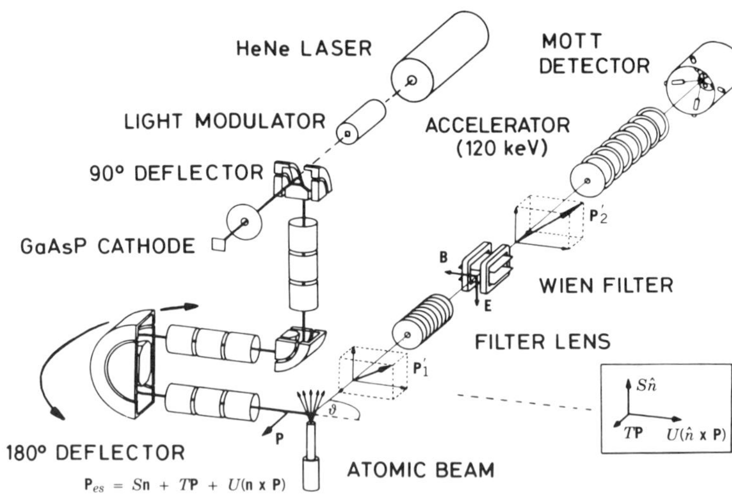

Figure 4-6-7. The apparatus used by the Münster group in studies of the elastic scattering of electrons by mercury and xenon atoms.[78] $\mathbf{P}_{ei}$ is the polarization of the electron beam incident on the atomic beam; $\mathbf{P}_{es}$ is the polarization of the scattered electron beam; $\mathbf{P}_{ef}$ is the final polarization of the electron beam after passage through the Wien filter. S is the Sherman function, and $\mathscr{T}$ and $\mathscr{U}$ are the parameters defined by (4-6-20) and (4-6-21). Electrons scattered at the angle ϑ are detected. $\mathbf{n}$ is the unit normal vector of the scattering plane.

ment" has been performed, for the E and ϑ specified. The Münster group has performed such measurements on Xe over the energy range 40 to 350 eV. Their values for the parameters P, T, and U versus ϑ at 100 and 50 eV are shown in Fig. 4-6-8. The agreement between theory and experiment is much better at the higher energy than at the lower, presumably because the complicating effect of the distortion of the atoms by the incident electrons is less severe at the higher energy.

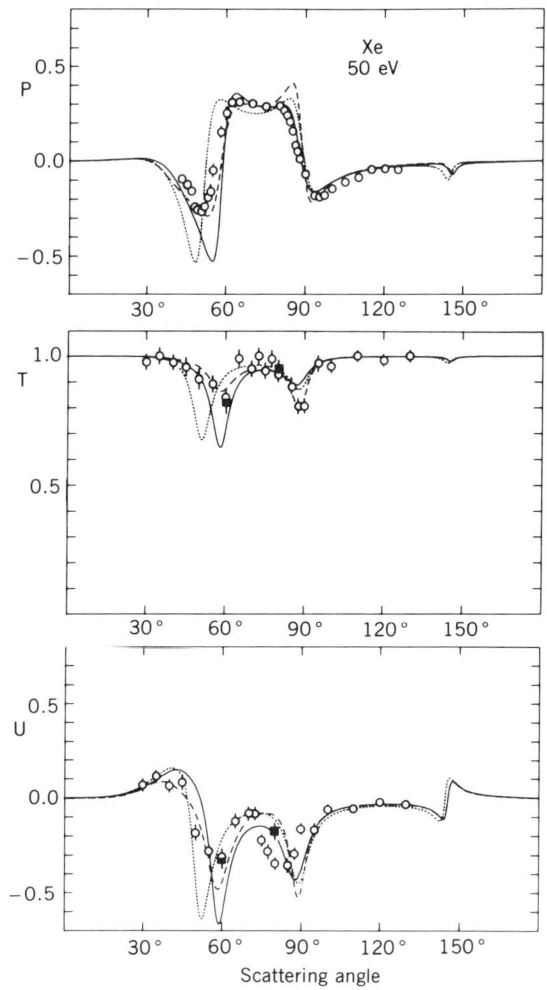

(a)

Figure 4-6-8. Some of the Münster experimental data (symbols) showing P, $\mathscr{T}$, and $\mathscr{U}$ versus scattering angle ϑ for electrons on Xe at energies of 100 and 50 eV.[78] Various theoretical results are indicated by lines.

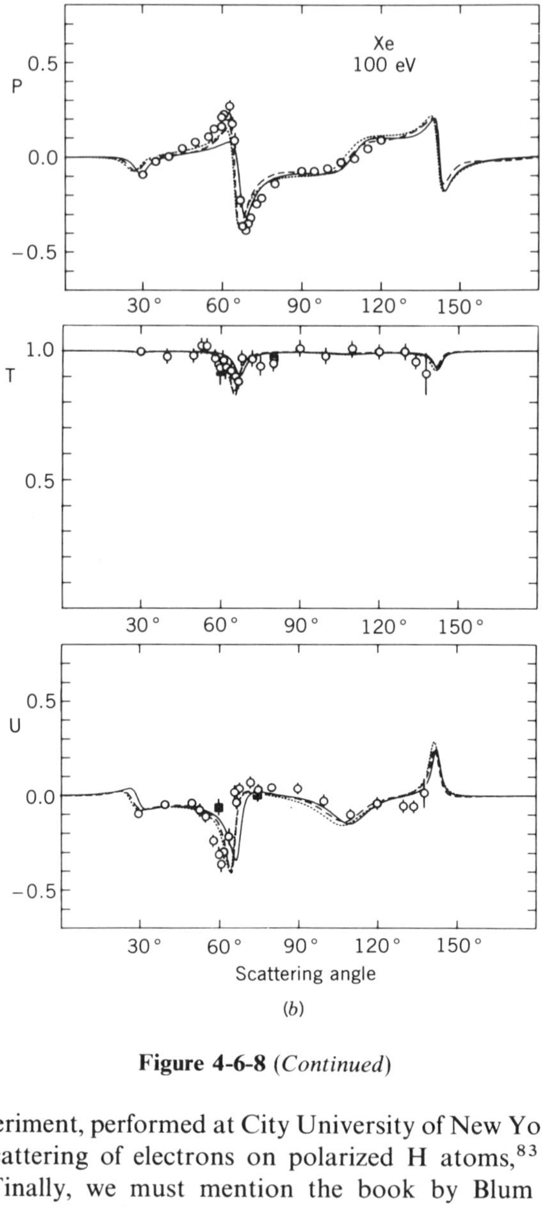

(*b*)

Figure 4-6-8 (*Continued*)

Another experiment, performed at City University of New York on the elastic and inelastic scattering of electrons on polarized H atoms,[83] is discussed in Section 6-11. Finally, we must mention the book by Blum et al.,[84] which provides a recent review of research with polarized electrons.

4-7. THE ELASTIC SCATTERING OF POSITRONS

The discovery of the positron in 1932 immediately raised questions about possible similarities and differences in the behavior of positrons (e^+) and their antiparticles (e^-) in various kinds of collisions. The masses and magnitudes of the electric charges of the positron and electron are the same, but the static interaction (associated with the Coulomb field of the undistorted target structure) is attractive for e^- and repulsive for e^+, whereas the induced-dipole polarization term is attractive for both. Reversal of the sign of the charge can make a marked difference in the convergence of approximations used in theoretical treatments of collision problems, changing a series of terms of one sign into an alternating series in which successive terms tend to cancel. Also, correlation effects are quite different with the two projectiles, partly because the Pauli exclusion principle does not operate against a positron impinging on the electronic cloud of the target. Exchange effects, arising from the indistinguishability of a projectile e^- from the electrons of the target, can occur only for incident electrons, whereas positronium formation (both real and virtual) necessarily involves positron projectiles.* Clearly, comparisons of e^+ and e^- behavior in both elastic and inelastic collisions against common targets can be very revealing and can provide stringent tests of approximations used in collision theory.

A formidable obstacle lay in the path of the experimentalist who wished to measure positron cross sections—the difficulty in producing suitably intense, collimated beams of low-energy positrons with a narrow energy spread. Depending on the apparatus and the measurement in mind, a flux on the order of 10^2 positrons per second is required for integral cross section studies, and 10^6 per second for angular differential measurements. Much higher fluxes are, in fact, desired. An important breakthrough was made in 1968 by D. E. Groce, D. G. Costello, J. W. McGowan, and D. F. Herring, who observed, while moderating fast positrons to low energies, that in a band of energies around 1 eV the number of positrons per unit energy interval was several orders of magnitude higher than expected. This finding promptly led to the development of scattering apparatus and the measurement of positron cross sections. The early work is surveyed by Massey, Burhop, and Gilbody[85] and by Griffith and Heyland.[86]

The first reliable data on positron collisions came in the form of cross sections for total scattering (all angles and all open channels) in about 1973. Subsequently, the experimental techniques matured very rapidly, partly because much could be borrowed from the technology of e^- experiments. In the early 1980s, cross sections for e^+ elastic differential scattering and e^+ inelastic processes, including positronium formation, began to appear. The more recent research is surveyed in Refs. 87–93.

*Positronium (Ps) is the bound state of the e^+–e^- combination; its ground state has a binding energy of 6.8 eV, half that of the H atom. It exists in two spin states: parapositronium with the e^+ and e^- spins opposed, and total spin $S = 0$; and orthopositronium with the e^+ and e^- spins parallel, and total spin $S = 1\hbar$. The lifetimes of the ground states for annihilation into two and three γ rays are 1.25×10^{-10} and 1.4×10^{-7} sec, respectively. Positronium formation is described by the equation $e^+ + X \rightarrow Ps + X^+$, and the threshold energy is the difference between the first ionization energy of the target and 6.8 eV. Virtual positronium formation involves a combination of e^+ and e^- whose existence is so brief that the structure cannot be detected. It may be described in terms of e^+–e^- correlations [see J. W. Humberston, *AAMP* **22**, 1 (1986); T. C. Griffith, *AAMP* **22**, 37 (1986)].

In most laboratories (e.g., University College London, Bielefeld, University of Texas–Arlington, and Swansea), commercially available accelerator-produced sources of ^{22}Na, of about 0.1 to 10 mCi strength, provide the primary, fast positrons according to the radioactive decay

$$^{22}_{11}\text{Na} \rightarrow ^{22}_{10}\text{Ne} + e^+ + \nu \qquad \text{(2.6 yr half-life, 0.54 MeV maximum } e^+ \text{ energy).}$$

A 1.28-MeV γ ray is emitted with negligible delay after the positron and can be detected by a plastic scintillator to provide a "start" pulse. At Wayne State University in Detroit, a Van de Graaff accelerator is used to produce 4.75-MeV protons that impinge on a boron target and form the short-lived positron emitter ^{11}C (20.3 min half-life, 0.97 MeV maximum e^+ energy) according to the nuclear reaction ^{11}B(p, n)^{11}C. In the case of the ^{22}Na sources, primary positrons enter "converters," or "moderators," which typically contain MgO or W, and which re-emit a small fraction of the incident e^+ spectrum in a narrow energy band at low energies. In the Detroit apparatus, boron is used both as the source and the converter, and a substantially lower energy (0.3 eV) and energy spread ($\Delta E = 0.1$ eV) can be achieved for the beam formed with the positrons emerging from the converter. The mechanism by which the low-energy positrons are produced in the converters is discussed in Refs. 89 and 90.

After emission from the converter, the slow positrons are accelerated to the desired energy, a beam is formed, and attenuation measurements are made to determine σ_T. Each of the basic experimental arrangements described in Section 4-1-A has been used, including the Ramsauer! The positrons can be detected at the end of their flight by counters that respond to the γ rays produced when the positrons are thermalized and annihilated in a solid target. More commonly, the e^+ are accelerated to a suitably high energy and detected in a pulse-counting electron multiplier. Several additional points should be made. First, annihilation will not be significant in the low-pressure gas filling the collision chamber. Second, γ rays and fast positrons from the source will cause trouble unless the beam positrons are guided along a curved path (say, by a longitudinal magnetic field) to prevent the detector from "seeing" the source, or unless a time-of-flight technique and electron multiplier detector are used. This kind of detector is nearly blind to γ rays, while the TOF method permits discrimination against prompt high-energy positrons from the source. The TOF method has been used at Bielefeld, University College London, and Arlington.

Charlton[91] has discussed the various systematic errors that can arise in total cross-section measurements, and summarized the experiments performed up to about the end of 1984. We shall present here a small sample of the excellent data now available on total positron scattering cross sections.

Figure 4-7-1 displays data obtained by the Detroit group for e^+ on the noble gases; the corresponding data for e^- at energies above a few eV were obtained on the same apparatus by adding a thermionic emitter of electrons. By making e^+ and e^- measurements on a single instrument, one can greatly reduce the effects of systematic errors in comparisons of the scattering behavior of the two

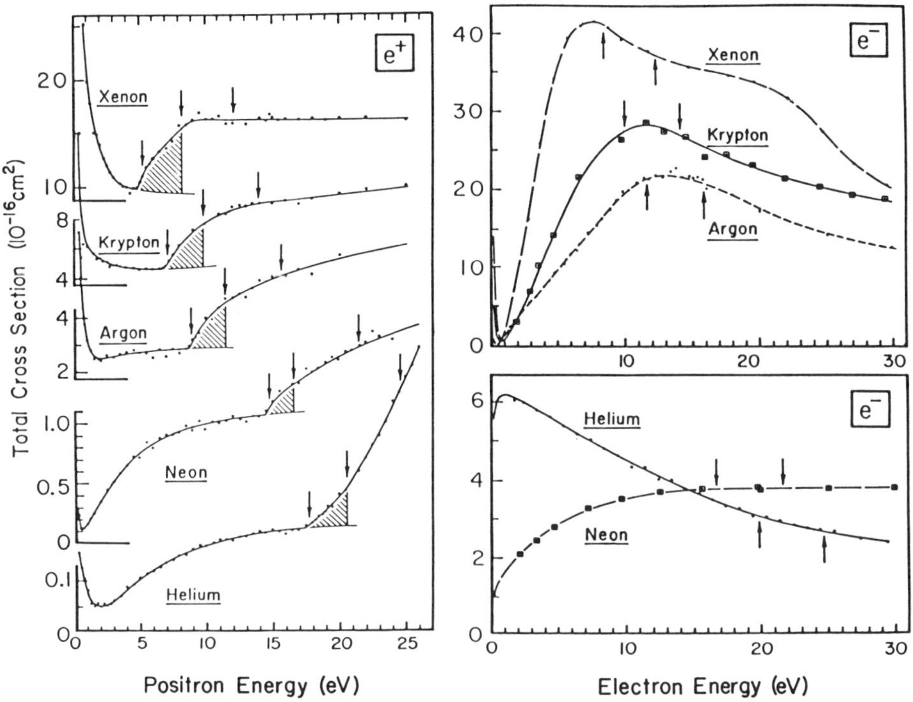

Figure 4-7-1. Total cross sections for e^+ and e^- projectiles on noble gas targets. (From Stein and Kauppila.[87])

particles. The lower-energy e^- data in Fig. 4-7-1 were taken from earlier work by Ramsauer and Kollath on Ne, Ar, Kr, and Xe and from the analysis of swarm experiments in the case of He. One may note a difference between e^+ and e^- scattering as regards the Ramsauer minimum. In the electron case, such a minimum occurs only in the heavier noble gases (Ar, Kr, and Xe); for positrons, a Ramsauer minimum is observed only for He, Ne, and possibly Ar. In the e^+ curves, the threshold energies for positronium formation, atomic excitation, and atomic ionization are indicated, respectively, by the three arrows in the order of increasing energy. At very low energies, only elastic scattering can occur; a dramatic increase in the total cross section is noticed in each of the noble gases at the threshold for positronium formation. If it is assumed that the elastic cross section varies smoothly as the energy is further increased, one would conclude that at higher energies, the inelastic cross sections become comparable to or greater than the elastic cross section. By contrast, the e^- curves show no marked increases in the total cross sections as the electron energy is increased through the threshold for excitation and ionization.

A few general comments about the various kinds of interactions between electrons (of both signs) and target structures are in order before we turn to the next figure.* The static interaction is important at all energies. The polarization attraction is significant only at sufficiently low impact energies that the target can polarize during the time of interaction. Electron exchange is important only at low energies. Excitation and ionization contribute to the scattering mainly in the energy region corresponding to the kinetic energies of the electrons bound in the target. Now we turn to Fig. 4-7-2, taken from the review by Raith[89] and adapted from results obtained by the Detroit group. For both N_2 and CO targets, the e^+ and e^- cross sections are merging as we go to very high energies. This behavior is consistent with the polarization and exchange interactions, becoming insignificant in comparison to the static interaction, which has the same magnitude for e^+ and e^- on a given target. At very low impact energies, the static and polarization interactions add in the case of e^- but tend to cancel for e^+, so that the elastic cross section for e^- can be much greater than for e^+.

Recent developments were surveyed at two conferences in 1987. At the Brighton ICPEAC, Sinapius[92] discussed a variety of experimental results, and astrophysical applications and positronium beams as well. Papers presented at the London Positron Conference[93] included a discussion of measurements of e^+ differential elastic scattering cross sections by Kauppila and Stein, and a

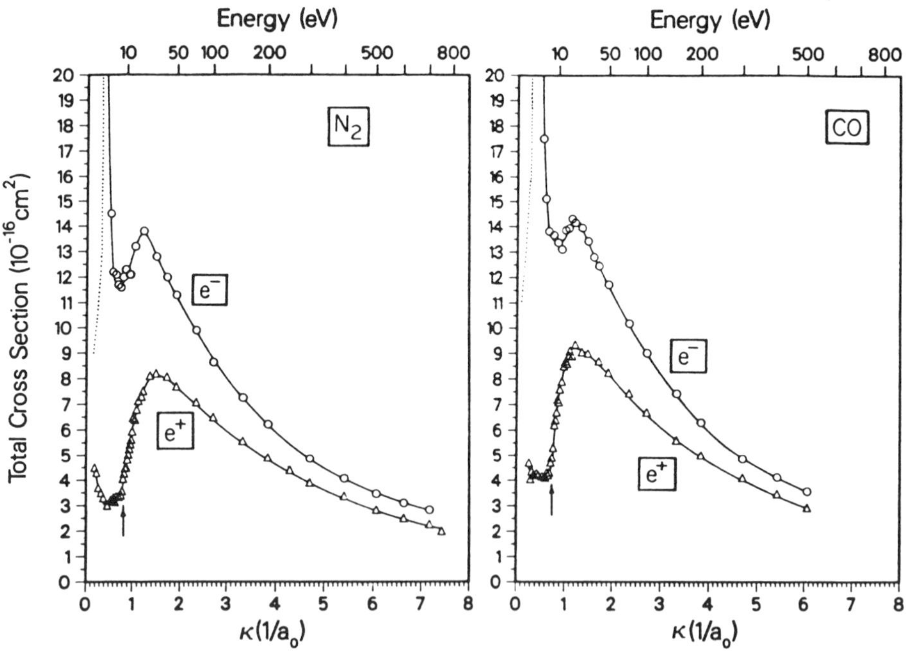

Figure 4-7-2. Total cross sections for e^+ and e^- projectiles on the isoelectronic target molecules N_2 and CO. (From Raith in Ref. 89, p. 7.)

*Additional material on electron-atom and electron-molecule interactions appears in Sections 4-8-4, 5-2, 5-3, and 5-4.

description of a high-intensity e^+ beam at the Brookhaven reactor by Weber et al. Positrons are emitted by high-specific-activity ^{64}Cu sources (870 Ci/g) produced in the reactor, which has a high thermal neutral flux. Fast-to-slow moderation can be performed in a self-moderation mode or with a transmission moderator. Up to 1.0×10^8 e^+ per second have been observed.

A great deal of effort has been directed toward the measurement of total cross sections, and recent data are presented in Ref. 94. Especially gratifying is the appearance of differential elastic cross sections, which are displayed in Ref. 95, as well as in Refs. 92 and 93. We shall return to positron collisions in Sections 6-15 and 6-16, where we discuss, among other topics, the partitioning of the total cross section into its components.

4-8. PHYSICAL BASIS AND THEORY OF SOME PHENOMENA OF SPECIAL INTEREST

Here we treat several specific topics and then conclude this section with some general comments concerning the theory of elastic scattering of electrons.

A. The Ramsauer-Townsend Effect

In Section 4-1 we discussed the Ramsauer-Townsend effect for electrons and illustrated it in Fig. 4-1-2*b* with data on electron scattering by Ar, Kr, and Xe. The following qualitative remarks show why this effect might be expected in the heavier noble gases. Owing to the compact structure of a noble gas atom, the force it exerts on an approaching electron is short range and strong at small r. Hence the potential field may be approximately represented by a deep, narrow potential well, such as that shown in Fig. 3-15-2. For electrons of very low energy the radius of the potential well is small compared with the electron wavelength, and only s-wave scattering will make an important contribution to the integral scattering. For certain bombarding energies the s wave function of the incident electron may be "drawn in" by the attractive potential to just the extent that a phase shift equal to an integral multiple of π results. Then, by (3-15-34), the s-wave contribution to the scattering cross section becomes zero, and since the higher wave contributions are known to be small, the integral cross section falls to a very low value. This effect may be regarded as a manifestation of the diffraction of the electron wave by the target atom.

The steepness of the wall of the potential well increases with the atomic number of the noble gas atom, so that the Ramsauer-Townsend effect for electrons is most pronounced with xenon. The effect is still strong in krypton and argon, but it is not observed in neon and helium (see Fig. 4-1-2*a*). However, the Ramsauer-Townsend effect does appear for certain other atoms and for some molecules as well.

Scattering by a repulsive potential cannot produce this sort of effect. With a repulsive potential, κD, the product of the wave number of the incident electron and the radius of the potential barrier, would have to be at least of the order of unity for the *s*-wave phase shift to be as large as $-180°$, and for a potential of this great a range higher-order waves would contribute to the scattering.

The first quantitative explanation of the Ramsauer-Townsend effect was provided in 1929 by J. Holtsmark, who made a phase shift analysis of the scattering of electrons by argon atoms.[5] He used a potential given by the Hartree method[96] initially, and discovered that a polarization correction was necessary to produce the effect. A great deal of attention has been given to the Ramsauer-Townsend effect since that time, and it is now well understood.[5,97–99] The basis of the explanation is the Pauli exclusion principle, which requires that the total wave function for the projectile-plus-target system be antisymmetric with respect to the interchange of the coordinates of the electrons. Additional *nonlocal*, or *exchange*, *potentials* are thereby introduced. They depend on the relative orientation of the spins of the electrons and thus produce spin-dependent effects even when the potential $V(r)$ is spin independent. These potentials describe the exchange that can occur between the projectile and a bound electron. The potentials are nonlocal, or energy dependent, because they involve an integral over the wave function of the scattered electron. The exclusion principle has the effect of producing a phase shift that decreases rapidly with increasing energy. This effect results from the repulsive character of the exchange potential, which acts against deep penetration of the projectile into closed-shell regions of the target. The exchange repulsion acting with the attractive polarization potential is essential for the Ramsauer-Townsend effect.

For incident electrons the mean interaction with an unpolarized atom is attractive, but it is repulsive for positrons. Nonetheless, the Ramsauer-Townsend effect is observed for positrons on He, Ne, and possibly Ar. This observation may be explained by recalling that the polarization interaction is attractive, independent of the sign of the charge. At low impact energies, this is the dominant interaction, and is the same for positrons as for electrons. Hence the Ramsauer-Townsend effect should not be ruled out on the basis of the static Coulomb interaction being repulsive for positrons.

B. Resonances

Up to now in our discussion of the theory of elastic scattering, we have been concerned primarily with the broad features of the variation of the cross section with impact energy. Different approaches are required to handle the rapid variations observed in scattering resonances. Here we resume our discussion of resonances, which were introduced in Section 3-15-E in the context of the spherical potential well and illustrated in Figs. 4-3-5 and 4-3-9.

The history of the observation and study of resonance effects in atomic physics has been surveyed by several authors.[5,56,100] It appears that the discovery of the Auger effect[5,96] by Pierre Auger in 1925 was probably the first observation of a phenomenon involving resonances. Here we shall concentrate on modern high-energy-resolution electron scattering studies, which began in the early 1960s. We shall attempt to explain the physical basis of resonances in elastic electron scattering, to classify these resonances, and to give some theoretical basis for understanding them. Resonances in inelastic electron collisions, and in photon and heavy particle processes, are discussed in later chapters.

1. *Electrons Incident on Neutral Atoms*

Schulz has written two important reviews on resonances in electron impact on neutral atoms and molecules.[55] Initially, following Schulz, we shall consider only electron impact on neutrals. The scattering features we examine here are known as *resonances*, *compound states*, and *temporary negative ions*. These three terms are synonyms in the present context. The scattering resonances occur at well-defined electron energies, and the incident electron attaches itself to the target atom or molecule for times longer than the normal time of passage through the target. Hence the projectile-target system at the resonance energy is called a compound state. Since the compound state has a negative charge, it is a temporary negative ion. The resonances can be regarded as nonstationary, short-lived states of the negative-ion system that decay by electron emission, not by photon emission as in the case of stationary states.

To deepen our understanding, we need to consider the "genealogy of resonances."[55] With malice aforethought, pick a positive ion of the target gas that has a particular, appropriate electron configuration, specified by the quantum numbers $(n_1l_1,\ n_2l_2,\ \ldots,\ n_{Z-1}l_{Z-1})$. Adding an electron to this "grandparent," we form an appropriate excited state (usually a Rydberg state) of the neutral atom. Then, when the projectile electron is added to this "parent" at the proper impact energy, we have the particular resonance we want to discuss.

Resonances that lie energetically below their parent are called *closed-channel*, or *Feshbach*, or *type I resonances*. The projectile electron virtually excites the target to a configuration that creates a potential well of sufficient depth to trap the electron briefly. The interaction between electron and excited neutral (parent) is strong enough to support a temporary bound state. Feshbach resonances usually lie approximately 0 to 0.5 eV below the parent. The resonance displayed in Fig. 4-3-9 is of the Feshbach type.

The other case is that of the *open-channel*, or *shape*, or *type II resonance*. Here the resonance lies energetically above the parent. The interaction between projectile electron and target is such that an intermediate state is formed without changing the configuration of the target atom. The potential has a penetrable barrier that traps the electron near the target. The barrier is formed by the angular momentum of the incident electron, so we expect *p*-, *d*-, *f*-..., wave resonances, but generally not *s*-wave resonances since in that case $l = 0$ and there is no barrier. The term "shape resonance" is appropriate because the shape of the potential is important in producing the resonance (see Fig. 4-8-1).

Shape resonances usually decay into their parents. Hence shape resonances generally have shorter lifetimes (perhaps as short as 10^{-15} sec) and greater

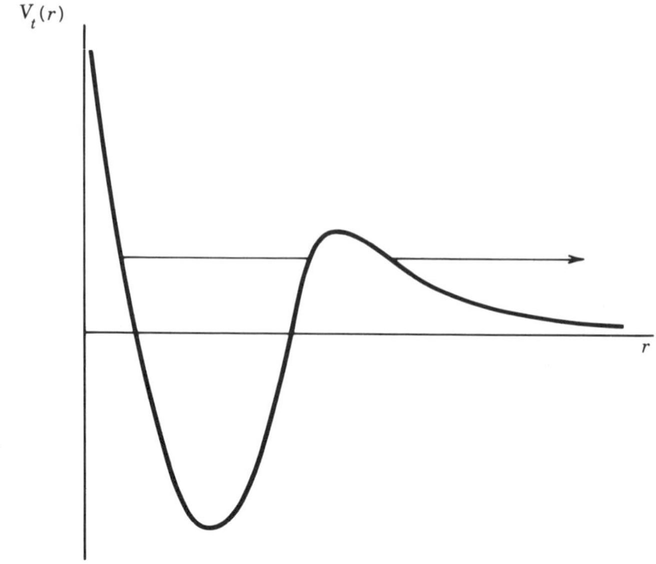

Figure 4-8-1. A shape resonance usually involves transitory capture of an incident electron to the ground-state atom. The addition of the repulsive centrifugal potential $l(l+1)\hbar^2/2M_r r^2$ to the electron-atom potential $V(r)$ can produce a total potential $V_t(r)$ that temporarily traps the electron. Quasi-stationary states exist at those electron impact energies for which constructive interference occurs between the various reflected waves.

widths (of the order of 1 eV) than do Feshbach resonances (frequently, 10^{-13} sec and 10 meV) from which the decay channel is closed (i.e., energetically forbidden). Indeed, the lifetimes of shape resonances can approach the usual time of transit of the electron through the atomic potential field. Such short lifetimes, and their concomitant broad energy widths (several eV), are inconsistent with the basic concept of an atomic resonance.

2. *Electrons Incident on Positive Ions*

A broader view of resonances than we have taken thus far would include electron impact on positive ions, with the formation of nonstationary states of neutral atoms, such as the $2s^2$ and $2s2p$ states of He. These "doubly excited" states have energies lying in the ionization continuum and decay by electron emission. They are excited states of the neutral atom, not a temporary negative ion, and for this reason were (artificially) excluded from the initial discussion. We must now consider such states and the associated scattering resonances. The easiest way to proceed is by examination of the energy-level diagram of the neutral helium atom. We arrange the energy states in a zero-order approximation in which we imagine the electron-electron Coulomb interaction to be turned off, the states being specified by products of wave functions for one-particle hydrogen-like states. In this approximation, the "singly excited" atom

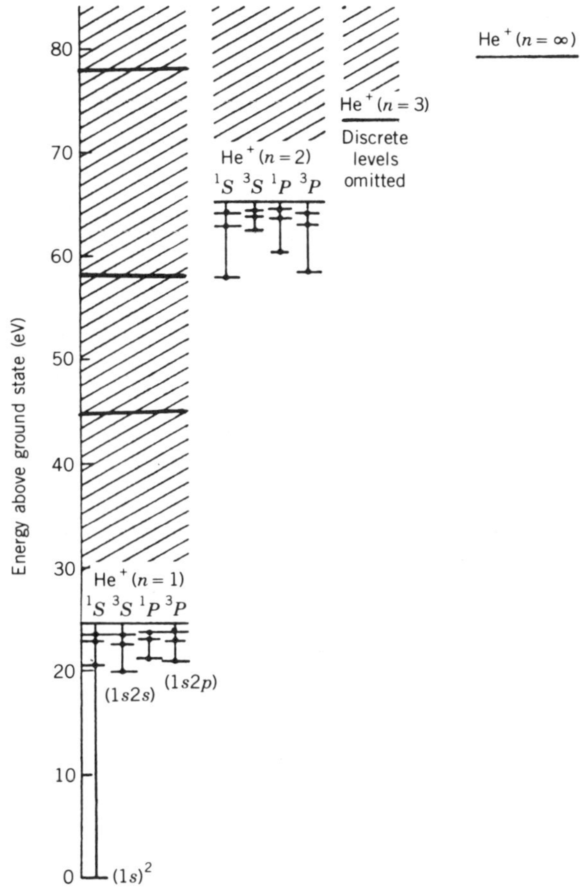

Figure 4-8-2. Energy-level diagram of the neutral He atom. The main purpose of this figure is to portray the states of the system composed of incident electron plus target He^+ ion and to illustrate the energy degeneracy between discrete doubly excited states and singly excited continuum states. The same kind of diagram would be appropriate for any two-electron system, although important differences in details arise. See pages 629–638 of Massey and Burhop[6] for a discussion of the H^- ion. (Based on Massey and Burhop,[6] Vol. 1.)

will have the set of configurations shown in the left-hand column of Fig. 4-8-2. Several of the infinite number of continuum states are indicated above the ionization limit. There is also a set of configurations for the doubly excited atom based on one electron being in the 2*s* orbital, and one in the 2*p* orbital, and so on, as shown in the other columns of Fig. 4-8-2. Any state of the real atom will be a superposition of such configurations. Often one configuration or a small number of them will be dominant. The important thing here is that each of the doubly excited states lies above the ionization threshold for the singly excited atom, and hence will coincide with a level lying in the continuum of the singly excited He atom.

We see that each bound doubly excited configuration is degenerate in energy with the singly excited continuum. Hence, near such a doubly excited configuration, the true wave function for the atom will be a linear combination of the two configurations, and the true state will not be bound. The atom will spend part of the time in the continuum state and must eventually break up. The lifetime of the doubly excited configuration is determined by the strength of the interaction with the relevant portion of the singly excited continuum (i.e., the portion overlapping it in energy).

Now let us look at the scattering of an electron by a He^+ ion. The initial system (electron plus ion) can be described as a singly excited state in the continuum of the He atom. If the total energy of this configuration lies close to that of a doubly excited configuration, the electron will spend some time in the doubly excited configuration before passing into the continuum state and escaping the ion. The phase shift of the projectile electron is altered because of the delay time associated with the capture of the electron by the ion. We would expect this effect to be manifested in the change of the elastic scattering cross section only over a narrow energy range because of the requirement of energy degeneracy.

It is important to realize that the resonance mechanism described here actually can operate with any two-electron system, such as the H^- ion formed by electron impact on the neutral H atom. The example used in the discussion above, that of an electron impinging on a He^+ ion, was chosen for its simplicity. The mechanism also works equally well for multielectron atomic systems, and for molecules.

3. *More Terminology: Single-Particle and Core-Excited Resonances*

Now it is necessary to introduce some additional terminology. We shall do so in the context of the scattering of an electron from a neutral target. A *single-particle resonance* (1 particle, 0 holes) is a shape resonance whose parent is the ground-state target, either an atom or a molecule. Attention is focused on a single "particle" (the incident electron), which is trapped within the penetrable potential barrier resulting from the combination of the static, polarization, exchange, and centrifugal interactions. None of the target electrons are promoted to higher orbitals, so no "holes" are created. Such resonances occur at low impact energies (usually, 0 to 4 eV) and have lifetimes of 10^{-15} to 10^{-10} sec, or longer.

A *core-excited resonance* has as the parent an excited state of the target atom or molecule. Usually, the resonance consists of a "hole" in one of the normally occupied orbitals, and of two "particles," which are two electrons in orbitals that are normally unoccupied. This is a (2-particle, 1-hole) resonance. Similarly, we might have a (3-particle, 2-hole) resonance, or, in principle, one with more particles and holes. Core-excited resonances can be of the Feshbach or of the shape type. Either valence or Rydberg excited states may serve as parents. Both of these types of states are singly excited states of neutral targets formed by promotion of a single electron into an orbital not filled in the ground state. When the principal quantum number of the electron does not change in the promotion, the excited state is a *valence* state. If the principal quantum number increases by one or more, the excited state is a *Rydberg* state.

4. *Perturbation Derivation of One-Level Breit-Wigner Formula for s-Wave Scattering*

Next we investigate these matters mathematically, using first-order time-independent perturbation theory and following the treatment of Massey and Burhop.[6] We make a number of simplifying assumptions: Only one discrete state and a single continuum are involved, only *s*-wave scattering occurs, the target is the hydrogen atom, and exchange effects are absent. Massey and Burhop also treat more complex situations in their Chapter 9.

We begin by writing down the complete time-independent wave equation (energy eigenvalue equation) for the system:

$$H\Psi \equiv (H_0 + H_1)\Psi = E\Psi, \tag{4-8-1}$$

where

$$H_0 = -\frac{\hbar^2}{2m}(\nabla_1^2 + \nabla_2^2) - e^2\left(\frac{1}{r_1} + \frac{1}{r_2}\right) \tag{4-8-2}$$

and

$$H_1 = \frac{e^2}{r_{12}}. \tag{4-8-3}$$

($\mathbf{r}_1$ and $\mathbf{r}_2$ are the coordinates of electrons 1 and 2, respectively, with respect to the proton.) As a crude approximation we take the wave functions of the doubly excited configuration to be the product

$$\phi(\mathbf{r}_1, \mathbf{r}_2) = \chi_a(\mathbf{r}_1)\chi_b(\mathbf{r}_2), \tag{4-8-4}$$

where χ_a and χ_b are the wave functions of the excited states of neutral hydrogen atoms with energies E_a and E_b. Then

$$H_0\phi = (E_a + E_b)\phi, \tag{4-8-5}$$

so we see that $(E_a + E_b)$ is the zero-order approximation to the energy of the configuration. The first-order approximation to the energy will be E_d, given by

$$E_d = E_a + E_b + \langle\phi|H_1|\phi\rangle. \tag{4-8-6}$$

Then we have

$$\langle \phi | H | \phi \rangle = E_d. \tag{4-8-7}$$

The wave function for the continuum configuration with energy E is written as

$$\psi_E(\mathbf{r}_1, \mathbf{r}_2) = \chi_0(r_2) F_0(E - E_0, \mathbf{r}_1), \tag{4-8-8}$$

where χ_0 is the wave function for the ground state of the H atom (energy E_0), and $F_0(\mathbf{r})$ is the wave function for motion of an electron in the mean static potential field V_{00} of the H atom with energy $E - E_0$. Since χ_0 is orthogonal to χ_a $(a \neq 0)$, ϕ and ψ_E are also orthogonal. Now we have

$$\left(\frac{\hbar^2}{2m} \nabla_1^2 + \frac{e^2}{r_2}\right) \chi_0 = -E_0 \chi_0 \tag{4-8-9}$$

and

$$\left[\frac{\hbar^2}{2m} \nabla_2^2 + V_{00}(r_1)\right] F_0 = -(E - E_0) F_0, \tag{4-8-10}$$

so

$$\int \chi_0(r_2) H \chi_0(r_2) F_0(\mathbf{r}_1)\, d\mathbf{r}_2 = E F_0(\mathbf{r}_1). \tag{4-8-11}$$

The function F_0 may be chosen to be real. The F_0 for all possible values of E form an orthogonal set; hence, from (4-8-11),

$$\begin{aligned} \langle \psi_{E''} | H | \psi_{E'} \rangle &= E' \int F_0(E'' - E_0, \mathbf{r}_1) F_0(E' - E_0, \mathbf{r}_1)\, d\mathbf{r}_1 \\ &= E' \delta(E'' - E'). \end{aligned} \tag{4-8-12}$$

Finally, if we use (4-8-5) and the fact that ψ_E and ϕ are orthogonal, we can easily show that

$$V_{E\phi} \equiv \langle \psi_E | H | \phi \rangle = \langle \psi_E | H_1 | \phi \rangle. \tag{4-8-13}$$

The symbol $V_{E\phi}$ is introduced to simplify the notation; $V_{E\phi}^2$ has dimensions of energy since the ψ_E are normalized per unit energy range. If H_1 goes to zero, $V_{E\phi}$ also vanishes.

We now expand the true wave function for the complete system for energy E in the form

$$\Psi_E = a\phi + \int b_{E'} \psi_{E'}\, dE', \tag{4-8-14}$$

where contributions from configurations with energies far from E are ignored. Ψ_E is meant to be an eigenfunction of the full Hamiltonian with energy E, so

$$\langle \phi | H | \Psi_E \rangle = E \langle \phi | \Psi_E \rangle = Ea. \tag{4-8-15}$$

Now, from (4-8-14),

$$H\Psi_E = aH\phi + \int b_{E'} H\psi_{E'}\, dE',$$

so

$$\begin{aligned}\langle \phi | H | \Psi_E \rangle &= a\langle \phi | H | \phi \rangle + \int b_{E'} \langle \phi | H | \psi_{E'} \rangle\, dE' \\ &= E_d a + \int b_{E'} V_{E'\phi}\, dE'.\end{aligned} \tag{4-8-16}$$

Then

$$E_d a + \int b_{E'} V_{E'\phi}\, dE' = Ea. \tag{4-8-17}$$

Equation (4-8-13) can be used, in similar fashion, to show that

$$V_{E'\phi} a + E' b_{E'} = E b_{E'}. \tag{4-8-18}$$

A mathematical difficulty arises in using (4-8-18) to relate $b_{E'}$ to a because $E - E'$ may equal zero. The problem may be resolved by writing the formal solution for $b_{E'}$ as

$$b_{E'} = \left[\frac{1}{E - E'} + z(E)\delta(E - E') \right] V_{E'\phi} a, \tag{4-8-19}$$

where, on substitution into (4-8-14) or (4-8-17), the principal part of the integral[101] over $E - E'$ is to be taken and $z(E)$ is to be determined as follows: Put (4-8-19) into (4-8-17) to get

$$E_d + \Delta E + z(E) V_{E'\phi}^2 = E, \tag{4-8-20}$$

where

$$\Delta E = P \int \frac{V_{E'\phi}^2}{E - E'}\, dE'. \tag{4-8-21}$$

Here P denotes "principal part of." Then we obtain

$$z(E) = \frac{E - E_d - \Delta E}{V_{E'\phi}^2}. \tag{4-8-22}$$

We now want to get the asymptotic form of Ψ_E for large r_1. In (4-8-14), $a\phi$ may be ignored, because the bound state wave functions vanish exponentially. Massey and Burhop[6] use (4-8-19) and (4-8-22) and the asymptotic form

$$\psi_E \sim \chi_0(r_2)[n(E)]^{1/2} r_1^{-1} \sin[\kappa(E) r_1 + \eta(E)] \tag{4-8-23}$$

for the continuum wave function, where $\kappa(E) = (2mE/\hbar^2)^{1/2}$ and

$$n(E) = \frac{1}{8\pi^3}\frac{d\kappa}{dE} = \frac{m}{2\pi\kappa h^2}. \tag{4-8-24}$$

They obtain, for the wave function for the complete system in the limit of $r_1 \to \infty$,

$$\Psi_E \sim a\chi_0(r_2) n^{1/2}(E)(\pi^2 + z^2)^{1/2} V_{E\phi} r_1^{-1} \sin[\kappa(E) r_1 + \eta(E) + \delta(E)], \tag{4-8-25}$$

where

$$\begin{aligned} \delta(E) &= -\arctan\frac{\pi}{z(E)} \\ &= -\arctan\frac{\pi(V_{E\phi})^2}{E - E_d - \Delta E}. \end{aligned} \tag{4-8-26}$$

We see that the interaction of the continuum state with the doubly excited discrete state introduces an additional phase shift, $\delta(E)$, into (4-8-25) with respect to (4-8-23). Equation (4-8-26) shows that $\delta(E)$ is small when the impact energy E is far from $(E_d + \Delta E)$, but that $\delta(E)$ changes by π as E is swept through a narrow energy range of width $\sim V_{E\phi}^2$ spanning the energy $(E_d + \Delta E)$. That is to say, we have a resonance in the elastic scattering of width $V_{E\phi}^2$ located at energy $(E_d + \Delta E)$, which is shifted from the energy E_d by the interaction between the continuum and discrete states.

If we now take only the first ($l = 0$) term in (3-15-29), evaluate $|f(\theta)|^2$ for this term, and integrate over the complete solid angle, we obtain the s-wave elastic cross section

$$q_{l=0}^{\text{el}} = \frac{\pi}{\kappa^2}|e^{2i(\eta+\delta)} - 1|^2 = \frac{\pi}{\kappa^2}|-e^{2i\delta} + 1 + e^{-2i\eta} - 1|^2. \tag{4-8-27}$$

We may also express this "partial cross section" as

$$q_{l=0}^{\mathrm{el}} = \frac{\pi}{\kappa^2}\left|\frac{-2i\pi V_{E\phi}^2}{E - E_d - \Delta E + i\pi V_{E\phi}^2} + e^{-2i\eta} - 1\right|^2, \tag{4-8-28}$$

where the term containing $V_{E\phi}$ is the contribution from the resonance and the remainder relates to scattering by the mean potential. If the resonance term dominates, we obtain the *Breit-Wigner one-level formula for s-wave scattering*:

$$q_{l=0}^{\mathrm{el}} = \frac{4\pi}{\kappa^2}\frac{\frac{1}{4}\Gamma^2}{(E - E_r)^2 + \frac{1}{4}\Gamma^2}, \tag{4-8-29}$$

where

$$\Gamma = 2\pi V_{E\phi}^2 \tag{4-8-30}$$

and

$$E_r = E_d + \Delta E. \tag{4-8-31}$$

The formula is designated "one-level" because only one discrete resonance state is considered. The cross section for s-wave scattering has a resonance peak of half-width Γ that is centered about E_r. Massey and Burhop[6] point out that in reality the shape of the cross section curve in the resonance region depends on the interference between resonance and mean potential terms, such as those in (4-8-28). They also evaluate the coefficient a of the discrete wave function in (4-8-14), thereby showing that the contribution from the discrete wave function ϕ is in fact not limited to the single energy E_d but rather spans a continuous range of energies with a maximum at E_r (close to E_d) and half-width $2\pi V_{E\phi}^2$. According to the uncertainty principle, this energy uncertainty would correspond to a finite, mean lifetime for the discrete state given by

$$\Delta t \simeq \frac{\hbar}{2\pi V_{E\phi}^2}. \tag{4-8-32}$$

Even with its drastic approximations, the first-order calculation presented here has the virtue of producing some important basic results and of bringing out the physical principles that are involved in resonance scattering. We now take a more general and abstract look at the problem, using the $\boldsymbol{S}$ matrix (see Section 3-19) in the complex κ plane. By letting κ assume complex values, we can study the effects of the poles and zeros of $S_l(\kappa)$ in the κ plane upon its behavior on the real axis.

5. *The Breit-Wigner Formula via the S Matrix in the Complex κ Plane*

Here we shall follow closely the discussion by Burke.[97,102] Other useful treatments of the **S** matrix in nonrelativistic scattering theory are given in Refs. 98, 99 and 103–110.

We begin with the asymptotic form of the partial radial wave function $L_l(r)$ that we obtained in Chapter 3 in our initial discussion of the partial wave analysis of elastic scattering:

$$L_l(r) \underset{r\to\infty}{\sim} \frac{1}{\kappa r} \sin\left(\kappa r - \frac{1}{2} l\pi + \eta_l\right). \tag{3-15-15}$$

This equation defines the energy-dependent, real phase shift η_l. We now introduce the functions $G_l(r) = rL_l(r)$ as in (3-15-10) and rewrite (3-15-15) in two equivalent alternative forms:

$$G_l(r) \underset{r\to\infty}{\sim} \cos\eta_l \left[\sin\left(\kappa r - \frac{1}{2} l\pi\right) + \cos\left(\kappa r - \frac{1}{2} l\pi\right) K_l\right] \tag{4-8-33}$$

and

$$G_l(r) \underset{r\to\infty}{\sim} \frac{e^{-i\eta_l}}{2i} [e^{-i\kappa r - (1/2)l\pi} - e^{i\kappa r - (1/2)l\pi} S_l]. \tag{4-8-34}$$

These two equations define the **K** matrix and the **S** matrix, which are related to the phase shifts by

$$K_l = \tan\eta_l \tag{4-8-35}$$

and

$$S_l = T_l + 1 = e^{2i\eta_l}, \tag{4-8-36}$$

where (4-8-36) also defines the **T** matrix. Since η_l is real, S_l is a complex number with modulus unity. Note that if $\eta_l = 0$ for all l (i.e., if there is no scattering), $S_l = 1$ and $T_l = 0$. We see that

$$S_l = \frac{1 + iK_l}{1 - iK_l}. \tag{4-8-37}$$

Here K_l, S_l, and T_l are 1×1 matrices. They generalize to higher-dimensional, nondiagonal matrices in treatments of inelastic scattering of electrons by complex atoms.

Now κ is to be regarded as a complex variable, so that $S(\kappa)$ is a function defined throughout the complex plane of κ. We refer again to Chapter 3 and the

lth-order radial wave equation:

$$\left[\frac{d^2}{dr^2} - \frac{l(l+1)}{r^2} - U(r) + \kappa^2\right]G_l(r) = 0. \tag{3-15-11}$$

Here

$$U(r) \equiv \frac{2M_r V(r)}{\hbar^2}. \tag{3-15-3}$$

We require that $U(r)$ be less singular than r^{-2} near the origin and that it vanish faster than r^{-3} at infinity. It is convenient to define two new solutions of (3-15-11) by the asymptotic forms

$$f_l(\kappa, r) \underset{r\to\infty}{\sim} \exp[-i(\kappa r - \tfrac{1}{2}l\pi)] \tag{4-8-38}$$

and

$$f_l(-\kappa, r) \underset{r\to\infty}{\sim} \exp[i(\kappa r + \tfrac{1}{2}l\pi)]. \tag{4-8-39}$$

Then the solution (4-8-34) can be written

$$G_l(r) = N[f_l(\kappa, r) + (-1)^{l+1} f_l(-\kappa, r) S_l(\kappa)] \tag{4-8-40}$$

for all r, N being a normalization factor. Next we define the Jost functions

$$f_l(\pm\kappa) = \lim_{r\to 0} \frac{(\kappa r)^l f_l(\pm\kappa, r)}{(2l+1)!!}, \tag{4-8-41}$$

which obey the equation

$$f_l(\kappa) + (-1)^{l+1} f_l(-\kappa) S_l(\kappa) = 0 \tag{4-8-42}$$

since $u_l(r)$ is zero at the origin. We see that

$$S_l(\kappa) = (-1)^l \frac{f_l(\kappa)}{f_l(-\kappa)}. \tag{4-8-43}$$

This equation can be used to determine the analytic properties of the $\boldsymbol{S}$ matrix from the simpler properties of the Jost functions. In this fashion Burke obtains the two symmetry relations

$$S_l(\kappa) S_l(-\kappa) = 1 \tag{4-8-44}$$

and

$$S_l(\kappa)S_l^*(\kappa^*) = 1, \tag{4-8-45}$$

or equivalently,

$$S_l(\kappa) = S_l^{-1}(\kappa) \tag{4-8-46}$$

and

$$S_l(\kappa) = S_l^*(-\kappa^*). \tag{4-8-47}$$

The following properties of the $\boldsymbol{S}$ matrix are then easily obtained:

1. The $\boldsymbol{S}$ matrix is unitary. This property, which is required for particle conservation, is expressed by the condition that $S^\dagger S = SS^\dagger = 1$, that is, that the Hermitian conjugate, $S^\dagger$, of the matrix equal its inverse. By definition, $(S^\dagger)_{ji} = S_{ij}$. When κ is real ($\kappa = \kappa^*$), one can then write, as before,

$$S_l(\kappa) = e^{2i\eta_l(\kappa)}, \tag{4-8-36}$$

   where η_l is a real phase shift.
2. The symmetry relations show that if the $\boldsymbol{S}$ matrix has a pole at κ, it also has a pole at $-\kappa^*$ and zeros at κ^* and $-\kappa$.

If we put $N = f_l(-\kappa)$ in (4-8-40), we obtain

$$G_l(r) = f_l(-\kappa)f_l(\kappa, r) - f_l(\kappa)f_l(-\kappa, r). \tag{4-8-48}$$

We see from (4-8-43) that a value of κ giving a pole in $S(\kappa)$ gives a zero in $f_l(-\kappa)$, and (4-8-48) then requires that

$$G_l^P(r) \underset{r\to\infty}{\sim} e^{i\kappa r}. \tag{4-8-49}$$

The state described here is called a Siegert resonance state. When κ lies in the upper half of the κ plane, $G_l^P(r)$ vanishes exponentially, so this wave function can be normalized. Furthermore, since the Hamiltonian for our problem is Hermitian, all normalizable wave functions correspond to energy eigenfunctions with real energy eigenvalues. Hence the corresponding value for κ^2 are real. (We shall use atomic units here, so that $\kappa^2 = E$.) Then, if a pole in the $\boldsymbol{S}$ matrix lies in the upper-half κ plane, it must lie on the imaginary κ axis, and $G_l^P(r)$ represents a bound state:

$$G^P(r) \underset{r\to\infty}{\sim} e^{-\alpha r} \qquad (\alpha = \text{real and positive}). \tag{4-8-50}$$

The wave function defined by (4-8-49) cannot be normalized if the pole in the **S** matrix lies in the lower-half κ plane, and we can no longer argue that the poles are confined to the imaginary axis. Figure 4-8-3 shows how the poles of the **S** matrix are distributed in a typical case. The properties of the **S** matrix itself are responsible for the symmetrical distribution of the poles about the imaginary κ axis. For realistic choices of $U(r)$ and l, only a few poles appear on the positive imaginary κ axis (bound-state spectrum), but an infinite number occur in the lower-half κ plane. As the strength of the potential is increased, the complex poles successively move onto the positive imaginary κ axis to give bound states.

Now let us examine an isolated pole in the **S** matrix that lies close enough to the real positive κ axis (the physical region) to produce observable resonance effects. We define its position by the equation

$$\kappa_r^2 = E_r - \tfrac{1}{2}i\Gamma, \tag{4-8-51}$$

where E_r will turn out to be the position of the resonance on the energy scale, and Γ the width of the resonance. Equation (4-8-45) shows that there is a zero in the **S** matrix at $E_r + \frac{1}{2}i\Gamma$. On pages 28–31 of his book, Geltman[99] shows how to obtain the desired result. He uses the definition

$$S_l(\kappa) = \frac{f_l(\kappa)}{f_l^*(\kappa^*)} \tag{4-8-52}$$

rather than (4-8-43), expands $f_l(\kappa)$ and $f_l^*(\kappa^*)$ in Taylor series about the values of κ and κ^* that correspond to the energies $E_r \pm \frac{1}{2}i\Gamma$, and retains only the linear

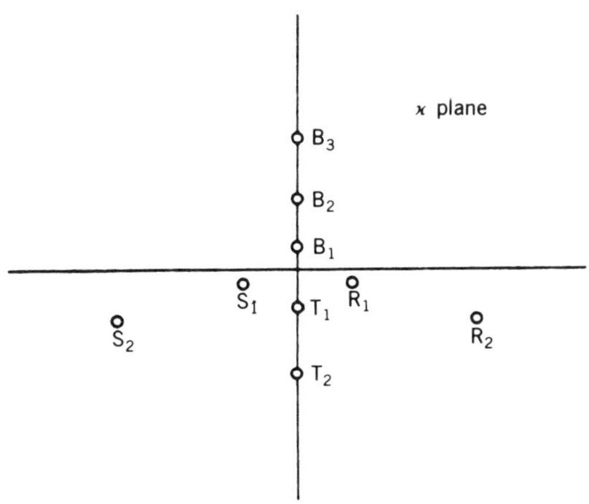

Figure 4-8-3. A possible distribution of poles in the single-channel **S** matrix. B_1, B_2, and B_3 correspond to bound states. R_1 and R_2 are resonance poles. The poles S_1 and S_2 are required by the symmetry properties of the **S** matrix. T_1 and T_2 are virtual state poles.[102]

terms. [Geltman shows that retention of only a linear term in an expansion of $f_l(-\kappa)$ in (4-8-43) would be incorrect.] We find that if all other singularities of S_l are remote, we can write

$$S_l(\kappa) = e^{2i\eta_b(\kappa)} \frac{E - E_r - \frac{1}{2}i\Gamma}{E - E_r + \frac{1}{2}i\Gamma} \tag{4-8-53}$$

for E close to E_r. This expression for S_l is unitary and has the required resonance structure. The "background phase shift" $\eta_b(\kappa)$ provides the slowly varying energy dependence produced by other, remote poles and zeros. Equation (4-8-53) is a general result that applies to all potentials of interest in electron scattering by atoms.

If we put $\eta_b(\kappa) = 0$, we again obtain the *Breit-Wigner one-level formula*, but now *for any partial wave* l:

$$q_l^{\text{el}} = \frac{4\pi}{\kappa^2}(2l + 1)\frac{\frac{1}{4}\Gamma^2}{(E - E_r)^2 + \frac{1}{4}\Gamma^2} \tag{4-8-54}$$

(see Fig. 4-8-4). In the general case, we define a total phase shift

$$\eta(\kappa) = \eta_b(\kappa) + \eta_r(\kappa), \tag{4-8-55}$$

where the resonance phase shift

$$\eta_r(\kappa) = \tan^{-1}\frac{\frac{1}{2}\Gamma}{E_r - E} \tag{4-8-56}$$

is obtained from (4-8-36) and (4-8-53). Then the partial cross section becomes

$$q_l^{\text{el}} = \frac{4\pi}{\kappa^2}(2l + 1)\sin^2\eta = \frac{4\pi}{\kappa^2}(2l + 1)(\sin^2\eta_b)\frac{(q + \epsilon)^2}{1 + \epsilon^2}. \tag{4-8-57}$$

Here ϵ is the reduced energy

$$\epsilon = \frac{E - E_r}{\frac{1}{2}\Gamma} \tag{4-8-58}$$

and

$$q = -\cot\eta_b \tag{4-8-59}$$

is the resonance shape parameter.

Figure 4-8-5 shows how the phase shift and partial elastic cross section vary with impact energy near a resonance. The total phase shift increases from the

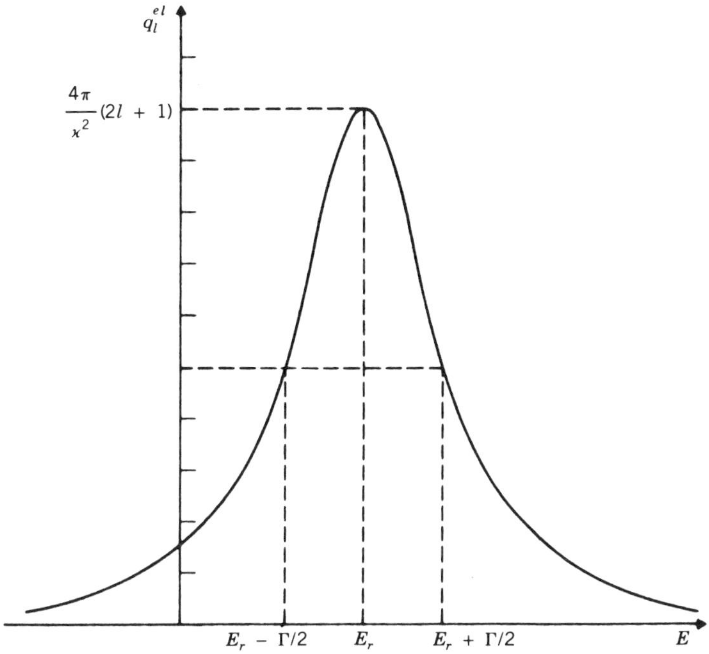

Figure 4-8-4. The partial (lth wave) elastic scattering cross section for electrons with incident energy E near an isolated resonance, according to (4-8-54).[96] Here we have a "pure Breit-Wigner resonance"; we assume that there is no contribution to the scattering amplitude from "hard-sphere" scattering in the lth wave, and no contributions from other nonresonant partial waves. The complex, many-body structure of the target is also neglected. This figure applies to both Feshbach and shape resonances.

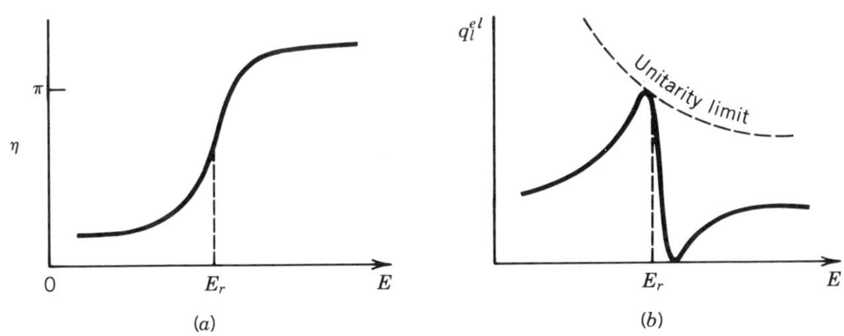

Figure 4-8-5. The variation of the total phase shift η and partial elastic cross section q_l^{el} near a resonance at energy E_r, according to (4-8-55), (4-8-56), and (4-8-57). Here, as opposed to Fig. 4-8-4, background effects are taken into account. (Adapted from Burke.[97])

value η_b at an energy well below resonance to the value $(\eta_b + \pi)$ well above resonance. (η_r is negligible at energies far from resonance.) Hence, if Γ is sufficiently small, the total phase shift will increase by about π in the vicinity of the resonance energy. There are two cases to consider: If η_b has a value between zero and $\pi/2$, as in Fig. 4-8-5, the partial cross section peaks and then decreases as E is raised through the resonance. However, if η_b lies between $\pi/2$ and π, the cross section first decreases through zero and then peaks as E is raised. Similar results are obtained if η_b lies outside the range zero to π. The dashed curve shows the maximum cross section that is allowed by the unitarity condition:

$$q_l^{\text{el}} = \frac{4\pi}{\kappa^2}(2l + 1). \tag{4-8-60}$$

It is of interest to note that the Breit-Wigner formula was first derived by G. Breit and E. P. Wigner in 1936 in connection with the radiative capture of slow neutrons by nuclei. Fermi[111] has also derived the one-level Breit-Wigner formula in the context of the (n, γ) reaction by applying time-*dependent* perturbation theory. We shall discuss inelastic electron scattering resonances, as well as resonances in photon and heavy particle collisions, in later chapters.

C. Elastic Scattering at Inelastic Thresholds; Cusps and Rounded Steps

We also wish to take note of another feature of elastic scattering that is manifested in unusual behavior of the cross section as a function of electron energy. Here we speak of discontinuous energy derivatives, called *cusps*, or *rounded steps*, that may appear at the onsets of new scattering channels. The differential cross section may either increase at energies below the threshold and then decrease above (a positive cusp), or do the opposite (a negative cusp). We may also have a discontinuity in the gradient, but with the same sign of slope on both sides of the threshold (an increasing, or decreasing, rounded step). Both positive cusps and decreasing rounded steps are possible in the integral elastic scattering cross section. Cusps and rounded steps also appear in inelastic and reactive scattering.

These effects may be explained by noting that the proximity of a threshold for a new process (scattering channel) is felt even at energies below the threshold at which transitions into the new channel are energetically possible. Virtual transitions can take place, according to $\Delta t \approx \hbar/\Delta E$, where ΔE is the energy deficit and Δt is the time over which energy conservation can be violated. The theory of cusps and rounded steps is discussed at length by Newton,[105] from whose book Fig. 4-8-6 is taken. In 1974, Cvejanović et al.[112] reported high-resolution measurements of the 2^3S cusp (and the 2^2S resonance) in elastic electron-helium scattering. Their observation of this cusp was the first made in electron collision studies.

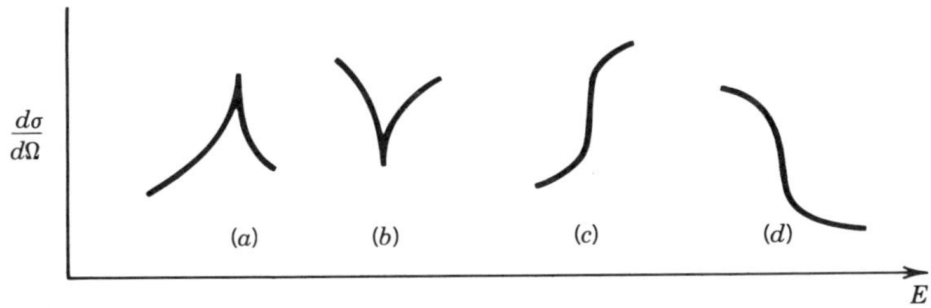

Figure 4-8-6. Illustration of "threshold anomalies" in elastic scattering. Cases (*a*) and (*b*) are cusps; (*c*) and (*d*) are rounded steps. (Adapted from Newton.[105])

D. Books and Reviews on the Theory of Elastic Scattering

The progress made during the last few decades on the computation of collision cross sections has been of great importance and indeed matches the progress made on the experimental side and in the area of basic theory. We have already referred to a number of works that deal with the theory of the elastic scattering of electrons.[5,6,9–11,60,67,69,70,73,79,84,96–99,102–110] For the most part these references are books that are mainly of tutorial value. We now add a few more references of that type, and many more that deal with the specific details of cross-section calculations. The latter references are especially useful in describing the application of the scattering approximations discussed in Chapter 3, and in assessing their utility and accuracy in calculations on various targets and at various impact energies. Reference T-17 is unique in that it relates to a journal concerned with the techniques of computer computations, *Computer Physics Communications.*

AUXILIARY THEORETICAL REFERENCES

T-1. M. A. Abdel-Raouf, "On the Variational Methods for Bound-State and Scattering Problems," Pts. 1 and 2, *Phys. Rep.* **84**, 163 (1982); **108**, 1 (1984).

T-2. E. A. G. Armour, "The Theory of Low-Energy Positron Collisions With Molecules," *Phys. Rep.* **169**, 1 (1988).

T-3. B. H. Bransden and M. R. C. McDowell, "Electron Scattering by Atoms at Intermediate Energies," Pts. I and II, *Phys. Rep.* **30**, 207 (1977); **46**, 249 (1978).

T-4. B. D. Buckley, P. G. Burke, and C. J. Noble, "Computational Methods for Low-Energy Electron-Molecule Collisions," in I. Shimamura and K. Takayanagi (Eds.), *Electron-Molecule Collisions*, Plenum, New York (1984).

T-5. A. Burgess and I. C. Percival, "Classical Theory of Atomic Scattering," *AAMP* **4**, 109 (1968).

T-6. P. G. Burke, "Atomic Processes," in T. R. Carson and M. J. Roberts (Eds.), *Atoms and Molecules in Astrophysics*, Academic, New York (1972), p. 1.

T-7. P. G. Burke, "*R* Matrix Theory of Atomic and Molecular Processes," in *At. Phys.* **5**, 293 (1977).

T-8. P. G. Burke, "Theory of Low-Energy Electron-Molecule Collisions," *AAMP* **15**, 471 (1979).

T-9. P. G. Burke, "Theory of Electron-Molecule Collisions," in R. C. Wooley (Ed.), *Quantum Dynamics of Molecules*, Plenum, New York (1980), p. 483.

T-10. P. G. Burke, "Electron and Photon Collisions with Molecules," in F. A. Gianturco (Ed.), *Atomic and Molecular Collision Theory*, Plenum, New York (1982), p. 69.

T-11. P. G. Burke and W. Eissner, "Low-Energy Electron Collisions with Complex Atoms and Ions," in P. G. Burke, W. B. Eissner, D. G. Hummer, and I. C. Percival (Eds.), *Atoms in Astrophysics*, Plenum, New York (1983) p. 1.

T-12. P. G. Burke and W. D. Robb, "The ***R***-Matrix Theory of Atomic Processes," *AAMP* **11**, 143 (1975).

T-13. P. G. Burke and J. F. Williams, "Electron Scattering by Atoms and Molecules," *Phys. Rep.* **34**, 325 (1977).

T-14. J. Callaway, "Scattering of Electrons by Atoms," *Adv. Phys.* **29**, 771 (1980).

T-15. J. Callaway, "The Variational Method in Atomic Scattering," *Phys. Rep.* **45**, 89 (1978).

T-16. J. P. Coleman, "The Impulse Approximation and Related Methods in the Theory of Atomic Collisions," *Case Stud. At. Collision Phys.* **1**, 99 (1969).

T-17. *Computer Physics Communications.* Volume **17** (1979) of this journal is especially relevant here.

T-18. A. S. Davydov, *Quantum Mechanics*, Pergamon, Oxford (1965).

T-19. Yu. N. Demkov, "New Results in General Theory of Atomic Collisions: Small Angle Classical and Semiclassical Forward and Backward Scattering," *ICPEAC XIII*, Berlin (1983), p. 171.

T-20. R. J. Drachman and A. Temkin, "Polarized Orbital Approximations," *Case Stud. At. Collision Phys.* **2**, 399 (1972).

T-21. G. F. Drukarev, *Collisions of Electrons with Atoms and Molecules*, Plenum, New York (1985).

T-22. K. W. Ford, D. L. Hill, M. Wakano, and J. A. Wheeler, "Quantum Effects near a Barrier Maximum," *Ann. Phys. (N.Y.)* **7**, 239 (1959).

T-23. K. W. Ford and J. A. Wheeler, "Semiclassical Description of Scattering," *Ann. Phys. (N.Y.)* **7**, 259 (1959).

T-24. K. W. Ford and J. A. Wheeler, "Application of Semiclassical Scattering Analysis," *Ann. Phys. (N.Y.)* **7**, 287 (1959).

T-25. F. A. Gianturco and A. Jain, "The Theory of Electron Scattering from Polyatomic Molecules," *Phys. Rep.* **143**, 347 (1986). F. A. Gianturco, in P. G. Burke and J. B. West (Eds.), *Electron-Molecule Scattering and Photoionization*, Plenum, New York (1988).

T-26. J. Hinze (Ed.), *Electron-Atom and Electron-Molecule Collisions*, Plenum, New York (1983).

T-27. Y. K. Ho, "The Complex Rotation Method and Its Application for Atomic Processes," *Phys. Rep.* **99**, 1 (1983).

T-28. C. J. Joachain, "Electron and Photon Collisions with Atoms and Ions," in F. A. Gianturco (Ed.), *Atomic and Molecular Collision Theory*, Plenum, New York (1982), p. 1.

T-29. C. J. Joachain and F. W. Byron, "Eikonal Theory of Electron- and Positron-Atom Collisions," *Phys. Rep.* **34**, 233 (1977).

T-30. N. F. Lane, "The State of the Theory," in S. C. Brown (Ed.), *Electron-Molecule Scattering*, Wiley, New York (1979), p. 147.

T-31. N. F. Lane, "The Theory of Electron-Molecule Collisions," *Rev. Mod. Phys.* **52**, 29 (1980).

T-32. N. F. Lane, "Fundamental Advances in the Theory of Electron-Molecule Collisions," *ICPEAC XIII*, Berlin (1983), p. 127.

T-33. B. L. Moiseiwitsch, "Relativistic Effects in Atomic Collisions Theory," *AAMP* **16**, 281 (1980).

T-34. B. L. Moiseiwitsch, "Scattering of Atomic Particles at Relativistic Energies," *Phys. Rep.* **118**, 133 (1985).

T-35. M. A. Morrison, "The Physics of Low-Energy Electron-Molecule Collisions: A Guide for the Perplexed and the Uninitiated," *Aust. J. Phys.* **36**, 239 (1983).

T-36. R. K. Nesbet, "Low-Energy Electron Scattering by Complex Atoms: Theory and Calculations," *AAMP* **13**, 315 (1977).

T-37. R. K. Nesbet, "Electronic Correlation in Electron-Atom Scattering," in H. Kleinpoppen and J. F. Williams (Eds.), *Coherence and Correlation in Atomic Collisions*, Plenum, New York (1980), p. 269.

T-38. R. K. Nesbet, *Variational Methods in Electron-Atom Scattering Theory*, Plenum, New York (1980).

T-39. R. K. Nesbet, "Variational Methods in Electron-Molecule Collisions," in J. Hinze (Ed.), *Electron-Atom and Electron-Molecule Collisions*, Plenum, New York (1983), p. 21.

T-40. D. W. Norcross and L. A. Collins, "Recent Developments in the Theory of Electron Scattering by Highly Polar Molecules," *AAMP* **18**, 341 (1982).

T-41. I. C. Percival and D. Richards, "The Theory of Collisions between Charged Particles and Highly Excited Atoms," *AAMP* **11**, 1 (1975).

T-42. T. Rescigno, V. McKoy, and B. Schneider (Eds.), *Electron-Molecule and Photon-Molecule Collisions*, Plenum, New York (1979).

T-43. B. I. Schneider, "The *R*-Matrix Theory of Electron-Molecule Scattering," in J. Hinze (Ed.), *Electron-Atom and Electron-Molecule Collisions*, Plenum, New York (1983), p. 121.

T-44. M. J. Seaton, "Quantum Defect Theory," *Rep. Prog. Phys.* **46**, 167 (1983).

T-45. K. Takayanagi, "Introduction to Electron-Molecule Collisions," in I. Shimamura and K. Takayanagi (Eds.), *Electron-Molecule Collisions*, Plenum, New York (1984).

T-46. H. R. J. Walters, "Perturbative Methods in Electron- and Positron-Atom Scattering," *Phys. Rep.* **116**, 1 (1984).

4-9. NOTES

4-1. The 127° Selector/Analyzer. The 127° cylindrical selector and analyzer (see, e.g., Fig. 4-3-2) are commonly used in electron scattering experiments. To understand their action, first consider two infinite, coaxial metal cylinders with potentials applied so as to create an inward radial electric force on an electron between the cylinders. The magnitude of the electric force varies as $1/r$. We use a cylindrical coordinate system, with the Z axis along the axis of the cylinders. An electron injected at radius R with $\mathbf{v} = (0, v_\theta, 0)$ will execute a circular orbit as a result of the balance of the electric and centrifugal forces if its energy and the field at radius R are such that $mR\omega^2 = eE(r = R)$, where $\omega = v/R$. Let other electrons with the same energy enter at the same point with velocities in the (r, θ) plane that make small angles with respect to the circular orbit. They experience a net restoring force pushing them toward this equilibrium orbit. For $r > R$, the inward electric force exceeds the outward centrifugal force; for $r < R$, the opposite is true. The displacement δr of each electron from the equilibrium orbit ($r = R$) obeys a differential equation of the harmonic oscillator type: $d^2\,\delta r/dt^2 = -2\omega^2\,\delta r$. We see that the electrons will oscillate about the equilibrium orbit with $\sqrt{2}$ times the angular frequency of the revolution around the central orbit. Hence the electrons cross the central orbit at points spaced $\pi/\sqrt{2}$ radians (127°17′) apart. Focusing in only one dimension is provided here, because the electric field is perpendicular to the axis of the cylinders and can furnish no restoring force in the axial direction. The operation of the 127° focusing sectors such as shown in Fig. 4-3-2 can be understood in terms of this discussion.

The hemispheric selector/analyzer employs two concentric hemispheres at different potentials; the electric field between them varies as $1/r^2$. Electrons enter at the center of the gap between the hemispheres and are focused at the other end of the gap after a deflection of 180°. The focusing is two-dimensional (see Klemperer and Barnett[53] and Farago[113]).

4-2. *Photoemission from Activated Gallium Arsenide.*[72,73] Thermionic emission has provided the traditional means of generating electron beams. The thermionic electrons are not spin polarized, and their energy distribution typically is 0.2 to 0.5 eV wide. Most modern experiments on atomic collisions that utilize thermionic electrons require that they be injected into a monochromator, which serves as a bandpass filter and provides a projectile beam of greatly reduced energy width.

If a crystal of the semiconductor GaAs is activated to negative electron affinity (NEA) and subjected to unpolarized laser light (photon energy about 1.6 eV), intense beams of unpolarized electrons may be drawn from the crystal with energy distributions as narrow as 31 meV FWHM (see Feigerle et al.[72]). The currents (up to several μA) are comparable to those achieved by coupling thermionic cathodes to electron monochromators and are at least 10 times as intense for energy distributions of comparable width.

If the incident laser light is circularly polarized, the photoelectrons emerge with a spin polarization typically in the range 25 to 45%. The quantum yield is usually 1 to 4%, and the electron current is about 20 μA per milliwatt of laser radiation. In scattering experiments at the U.S.

National Bureau of Standards (see Section 5-10-J-3), the following set of parameters has been used: $I_e = 1\ \mu\text{A}$, $P_e = 25\%$, $\Delta E = 200\ \text{meV}$, T (GaAs) $= 300\ \text{K}$. (Improved performance can be obtained by reducing the crystal temperature.) The spin polarization can easily be reversed by changing the handedness of the laser light. Hence the polarization can be modulated at a desired frequency by rotating the $\lambda/4$ plate that is interposed between laser and crystal (along with a linear polarizer) to produce circularly polarized light. The electron current remains constant as the polarization is modulated. Consequently, it is possible to use a polarization-modulated beam in a scattering experiment, measure the ac component of the scattering signal synchronous with the modulation, and directly obtain the spin-dependent signal.[73]

Photoemission from (NEA) GaAs can be understood as a three-step process: (1) absorption of a photon and creation of an electron-hole pair, (2) migration of the electron through the solid and toward the surface, and (3) passage through the surface layer and escape into the vacuum. Escape is possible only if the photon energy exceeds the work function of the solid, so that the electron can have an energy greater than the vacuum level.

In metals, electron-electron and electron-phonon collisions during step 2 degrade the electron energy and populate normally unoccupied states as low as the Fermi level, thereby rendering some of the electrons incapable of escape. Compensation of this effect by the use of photons with energy much higher than the work function produces undesirably broad energy distributions. On the other hand, in semiconductors such as GaAs, complete relaxation of the electron energy is blocked by the band gap, and if the photon energy is set slightly above the band gap (1.52 eV in GaAs), the energy distribution of the photoexcited electrons is the thermalized distribution of electrons in the conduction-band minimum. In normal GaAs, for example, these electrons do not have enough energy to escape the solid; the electron affinity is high, about 4 eV. However, the vacuum level of GaAs can be lowered below the bulk conduction band minimum by application of Cs and O_2 to produce the "negative electron affinity condition." The levels of the conduction-band minimum and the valence-band edge are shown as a function of position in Fig. 4-9-1, which, along with most of this discussion, is taken from Feigerle et al.[72] The (Cs + O) surface layer and two possible photoemission energy distribution curves (EDCs) are also shown in this figure for (NEA) GaAs. For small values of NEA, the vacuum level serves as a low-energy cutoff on the electron energy distribution and produces narrowed EDCs. The source operator can set the Cs/O balance on the surface to achieve the desired vacuum level (within limits). In addition to this control on the width of the energy distribution, the operator can also control the intensity of emitted electrons by varying the laser photon flux.

As discussed by Pierce and Meier,[72] the ability of the GaAs crystal to

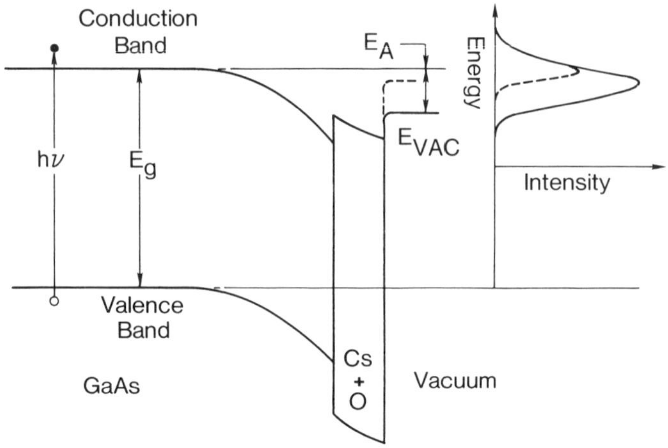

Figure 4-9-1. Variation of the energy levels of the conduction-band minimum and valence-band edge from the bulk (NEA) GaAs (left), through the band bending region and the (Cs + O) layer, and into the vacuum (right center). At the extreme right are shown two values for the vacuum level (solid and dashed lines), along with the two EDCs that would be produced for the resultant values of NEA. (From Feigerle et al.[72])

produce spin-polarized electrons depends on the spin-orbit splitting of the valence bands of GaAs near the center of the Brillouin zone (the Γ point). At the Γ point (the valence-band maximum) the degenerate p band is split into a fourfold degenerate $p_{3/2}$ level and twofold degenerate $p_{1/2}$ level, 0.34 eV lower in energy. The production of polarized electrons can be understood by examining transitions from the $p_{3/2}$ valence level (substates $m_j = +\frac{3}{2}, +\frac{1}{2}, -\frac{1}{2}, -\frac{3}{2}$) up to the $s_{1/2}$ level in the conduction band (substates $m_j = +\frac{1}{2}, -\frac{1}{2}$), 1.52 eV above the $p_{3/2}$ level at the Γ point. For circularly polarized light with positive helicity (σ^+), optical selection rules require that $\Delta m_j = +1$, and three times as many electrons go to the $m_j = -\frac{1}{2}$ state as to the $m_j = +\frac{1}{2}$ state. Consequently, for σ^+, we would expect from eq. (4-6-1) that $P_e = 50\%$, with the electron spins pointing preferentially in the direction opposed to the incoming laser light. If σ^- light were applied, the selection rule would be $\Delta m_j = -1$, and P_e would again have the value 50%, but the spin polarization would be along the incident light direction. The largest polarization is achieved for photon energies within 0.1 eV of the band-gap energy. In particular, one should avoid a photon energy high enough to excite electrons from the $p_{1/2}$ "split-off" valence band up to the $s_{1/2}$ conduction band; otherwise, the polarization goes to zero. Mechanisms that depolarize the electrons before they can escape from the crystal and thereby reduce the value of P_e for the beam below 50% are discussed by Pierce and Meier[72] and Pierce et al.[73]

4-10. PROBLEMS

Solutions for most of these problems appear in the references indicated.

4-1. It is possible in nuclear physics to measure certain kinds of cross sections with absolute accuracies considerably better than 1%, whereas in atomic physics, absolute accuracies of even several percent are seldom obtainable. While carefully avoiding any invidious speculation about the relative intelligence of atomic and nuclear physicists, offer some possible reasons for the difference in the levels of accuracy.

4-2. **(a)** From first principles of classical mechanics, derive the differential equation displayed in Note 4-1 that governs the oscillation of electrons about the equilibrium orbit in a 127° selector/analyzer.[53,113]

(b) Why is it not necessary to use quantum mechanics to analyze this device? (See Park,[114] p. 107.)

4-3. It is desired to transform the polarization of an electron beam from longitudinal to transverse, or vice versa. Show how this spin rotation may be accomplished in (a) a cylindrical electrostatic field, (b) a uniform magnetic field, and (c) a Wien filter. Include relativistic effects. (See Kessler,[66] pp. 230–232.)

4-4. **(a)** Show that the magnetic quadrupole lens can focus a beam of *charged* particles passing through it nearly parallel to its longitudinal (Z) axis, the plane in which the focusing occurs depending on the sign of the charge of the particles.

(b) Show that the magnetic hexapole lens (Fig. 4-6-5) is the simplest magnetic lens that can perform a similar function with *uncharged* particles having magnetic dipole moments polarized parallel or antiparallel to the X-axis. (Cronin et al.,[115] p. 20.)

4-5. In a Stern-Gerlach experiment, silver atoms emerge from a furnace at 1200°C, and a beam is produced by collimation. If the atoms then travel 1 m to a detector, show that the spot at the detector must have a diameter at least as large as about 2.8×10^{-4} cm.

4-6. Consider classically an electron (assumed to have no spin) moving with a speed $v = 10^4$ cm/sec at 45° to a constant, uniform magnetic field $B = 1$ T. Show that the path is a helix of radius 4.0×10^{-8} cm and pitch 2.5×10^{-7} cm. (Chen,[116] p. 113.)

4-7. Consider an electron moving in a magnetic field **B**. Show that the velocity component operators satisfy commutation relations of the form

$$\hat{v}_x\hat{v}_y - \hat{v}_y\hat{v}_x = \frac{ie\hbar}{\mu^2 c} B_z, \text{ etc.}$$

(ter Haar,[117] p. 38.)

4-8. Consider quantum mechanically an electron (assumed to have no spin) moving in a constant, uniform magnetic field **B**.

(a) Show that the energy levels of the electron are

$$E = \frac{eB\hbar}{mc}\left(n + \frac{1}{2}\right) + \frac{p_z^2}{2m}.$$

(Motion in the Z direction is not quantized.)

(b) Show that if the orbit is large, the magnetic flux through it is quantized. (Cronin et al.,[115] p. 31; Landau and Lifshitz,[109] p. 424; ter Haar,[117] p. 38.)

4-9. Repeat Problem 4-8 for the case of a uniform magnetic field with a uniform electric field perpendicular to it. (ter Haar,[117] p. 38.)

4-10. A particle of mass m, energy E, and angular momentum $l\hbar$ is captured in a spherical rectangular well of width a and depth V_0. Show that, according to the WKB approximation, the lifetime of the trapped state is approximately

$$\tau = \left(\frac{2ma^2}{E}\right)^{1/2} \left[\frac{\hbar^2 l(l+1)}{2ma^2 E}\right]^{\sqrt{l(l+1)}}.$$

[We neglect the angular momentum of the particle inside the well (Cronin et al.,[115] p. 34; ter Haar,[117] p. 36).] See pages 10–14 of ter Haar and pages 162–196 of Rapp[118] for many additional solved problems on tunneling.

4-11. Gamma rays are produced by the annihilation in flight of positrons against electrons. Show that in the direction of the line of flight, the γ-ray energy is $(E + m_0c^2 \pm pc)/2$, where E is the total energy of the positron, m_0 its rest mass, and p its momentum.

4-12. An e^-–e^+ pair undergoes annihilation into three γ rays when the particles are almost at rest. One of the γ rays is observed in the rest frame of the pair. What are the possible energies of this γ ray?

4-13. **(a)** The ground state of positronium is split into two hyperfine components. Show that the $^1S_0$ level lies lower than the $^3S_1$.

(b) Show that the binding energy of the ground state is 6.8 eV.

(c) A positronium atom at rest in the ground state decays by the emission of two γ rays. Show that each γ ray has an energy about equal to 0.5 MeV, and that the two rays are oppositely directed. (Chen,[116] p. 266.)

4-14. A positron and an electron are bound together in an atomic state with $l = 1$ orbital angular momentum. Show that the associated orbital magnetic moment is zero. (Chen,[116] p. 287.)

4-15. The electron and positron have magnetic moments of the same magnitude and g-factors of opposite sign. Show that the ground state of positronium (a ${}^1S_0$, ${}^3S_1$ doublet) cannot have a linear Zeeman effect by considering the total magnetic moment operator. (Cronin et al.,[115] p. 46.)

4-16. Negative muons are trapped in atomic Bohr orbits and eventually captured by protons in the nucleus in a process similar to K capture. Show that if various absorbing materials are considered, the muon capture rate should vary as Z^4, where Z is the atomic number of the material. (Chen,[116] p. 266.)

4-17. The negative muon has a mass 207 times that of an electron. If it is trapped by a proton to form a muonium atom: (a) What is the energy of a photon emitted when the atom decays from its first excited state to the ground state? (b) What is the radius of the first Bohr orbit? (c) What is the muon velocity in the nth circular Bohr orbit? (Chen,[116] p. 270.)

4-18. What are the eigenvalues of each of the four spin operators for an electron?

$$s_x,\ s_y,\ s_z \quad \text{and} \quad s^2 = s_x^2 + s_y^2 + s_z^2$$

4-19. Consider two particles that have Pauli spin matrices σ_1 and σ_2, respectively. Calculate and classify the eigenvalues of the Hamiltonian

$$H = A(\sigma_{1z} + \sigma_{2z}) + B\hat{\sigma}_1 \cdot \hat{\sigma}_2.$$

Do not take the exclusion principle into account. *Ans.:* For a singlet state, $E_1 = -3B$. For the triplet states, $E_3 = 2AS_z + B$, with $S_z = \pm 1, 0$. (Cronin et al.,[115] p. 30.)

4-20. Two distinguishable particles (each spin $\frac{1}{2}$) compose a system, their spins interacting according to $J\hat{\sigma}_1 \cdot \hat{\sigma}_2$, where J is a constant. Their magnetic moments are $\alpha\sigma_1$ and $\beta\sigma_2$. Find the exact energy eigenvalues of the system if a uniform magnetic field is applied. (Cronin et al.,[115] p. 30.)

4-21. An electron in a uniform, constant magnetic field $(0, 0, B)$ is shown to have its spin in the $+x$-direction at $t = 0$. For subsequent times, show quantum mechanically that the probability of finding the electron in the state $S_x = \frac{1}{2}$ is $\cos^2(\mu Bt)$, where μ is the magnetic moment of the electron. Similarly, show that for the $S_x = -\frac{1}{2}$ and $S_z = +\frac{1}{2}$ states, the probabilities are $\sin^2(\mu Bt)$ and $\frac{1}{2}$, respectively. (Cronin et al.,[115] p. 33.)

4-22. An atom has $J = \frac{1}{2}$ and is oriented in a uniform magnetic field such that $m_J = +\frac{1}{2}$. Suppose that the field is rotated through the angle $\phi = 60°$. Show that if a measurement is made immediately after the rotation, the probabilities of finding the atom with $m'_J = +\frac{1}{2}$ or $-\frac{1}{2}$ with respect to the rotated field are $\frac{3}{4}$ and $\frac{1}{4}$, respectively. (Cronin et al.,[115] p. 33; Feynman,[119] Chap. 6.)

4-23. Given the components of the spin function for particles of *spin 1* in a certain coordinate system, show that the components in a coordinate system arbitrarily rotated from the original one are

$$\begin{pmatrix} \psi'_1 \\ \psi'_0 \\ \psi'_{-1} \end{pmatrix} = \begin{pmatrix} e^{i(\psi+\phi)}\cos^2\dfrac{\theta}{2} & \dfrac{i}{\sqrt{2}}e^{i\psi}\sin\theta & -e^{i(\psi-\phi)}\sin^2\dfrac{\theta}{2} \\ \dfrac{i}{\sqrt{2}}e^{i\phi}\sin\theta & \cos\theta & \dfrac{i}{\sqrt{2}}e^{-i\phi}\sin\theta \\ -e^{-i(\psi-\phi)}\sin^2\dfrac{\theta}{2} & \dfrac{i}{\sqrt{2}}e^{-i\psi}\sin\theta & e^{-i(\psi+\phi)}\cos^2\dfrac{\theta}{2} \end{pmatrix} \begin{pmatrix} \psi_1 \\ \psi_0 \\ \psi_{-1} \end{pmatrix},$$

where θ, ψ, and ϕ are the Euler angles characterizing the rotation. (ter Haar,[117] p. 26.) Feynman[119] treats spin $\frac{1}{2}$ in Chapter 6, spin 1 in Section 12-6, and spin $\frac{3}{2}$ in Section 18-4.

4-24. Consider particles with total angular momentum quantum number $J = 1$ in a Stern-Gerlach experiment. The component of the angular momentum along an axis making an angle θ with the direction of the magnetic field in the apparatus has the well-defined value $M = +1$ when the particles enter the field. Show that the relative intensities of the emerging beam components are $w(+1) = \cos^4\theta/2$; $w(0) = \frac{1}{2}\sin^2\theta$; $w(-1) = \sin^4\theta/2$. Then, show that for $M = 0$, $w(+1) = \frac{1}{2}\sin^2\theta$; $w(0) = \cos^2\theta$; $w(-1) = \frac{1}{2}\sin^2\theta$; while, for $M = -1$, $w(+1) = \sin^4\theta/2$; $w(0) = \frac{1}{2}\sin^2\theta$; $w(-1) = \cos^4\theta/2$. (ter Haar,[117] p. 26.)

4-25. Suppose that two spin-$\frac{1}{2}$ particles interact through the magnetic dipole-dipole interaction

$$V = A\frac{(\hat{\sigma}_1\cdot\hat{\sigma}_2)r^2 - (\hat{\sigma}_1\cdot\mathbf{r})(\hat{\sigma}_2\cdot\mathbf{r})}{r^5}.$$

Let the spins be held at a fixed separation distance d. If at $t = 0$, one spin is parallel to $\mathbf{r}$ and the other spin antiparallel to $\mathbf{r}$, show that the time required for the first spin to become antiparallel to $\mathbf{r}$ and the second to become parallel is $\hbar d^3/4\pi A$. (ter Haar,[117] p. 29.) Chapters 4 and 7 of ter Haar's book contain many additional solved problems on spin systems and spin dynamics. Feynman's book[119] is highly recommended for a clear exposition of spin dynamics.

4-26. The Legendre polynomials constitute a complete, orthogonal set of functions that span the space of the polar scattering angle θ. Hence the CM differential cross section per unit solid angle for the mutual elastic scattering of two structureless particles may be represented as

$$\frac{d\sigma}{d\Omega} = AP_0(\cos\theta) + BP_1(\cos\theta) + CP_2(\cos\theta) + \cdots.$$

Use the Fourier trick to evaluate the expansion coefficients in terms of the phase shifts, and show that

$$A = \frac{\sigma_{\text{tot}}}{4\pi} = \frac{1}{\kappa^2} \sum_{l=0}^{\infty} (2l+1) \sin^2 \eta_l,$$

$$B = \frac{6}{\kappa^2} \sum_{l=0}^{\infty} (l+1) \sin \eta_l \sin \eta_{l+1} \cos(\eta_{l+1} - \eta_l),$$

$$C = \frac{5}{\kappa^2} \sum_{l=0}^{\infty} \left[\frac{l(l+1)(2l+1)}{(2l+1)(2l+3)} \sin^2 \eta_l + \frac{3(l+1)(l+2)}{2l+3} \sin \eta_l \right.$$
$$\left. \times \sin \eta_{l+2} \cos(\eta_{l+2} - \eta_l) \right].$$

(See p. Bransden and Joachain,[96] p. 497; ter Haar,[117] p. 58.)

4-27. By variational methods, show that any purely attractive one-dimensional potential has at least one bound state. (ter Haar,[117] p. 9.)

4-28. Calculate the Born cross section for elastic scattering of structureless particles by a screened Coulomb potential. Discuss the applicability of the Born approximation here. (ter Haar,[117] p. 59.)

4-29. Calculate in the Born approximation the differential and integral elastic cross sections for the scattering of fast electrons by ground-state hydrogen atoms. Ignore spin. *Ans*.: $\sigma = 7\pi/3\kappa^2$, where κ is the wave number of the electrons. (ter Haar, p. 59; Landau and Lifshitz,[109] p. 535.) Bethe and Jackiw[120] treat the elastic scattering of electrons at high and low energies, with and without spin effects in Chapters 13 to 16. Chapter 12 of Bransden and Joachain[96] is also a very useful reference.

4-30. Repeat Problem 4-29 for e-He scattering. *Ans*.: $\sigma = 28\pi/3\kappa^2$. (ter Haar,[117] p. 59.)

4-31. Consider particles impinging from the left upon a one-dimensional potential barrier of finite width and height. The time-independent wave function is $(Ae^{i\kappa x} + Be^{-i\kappa x})$ on the left and $(Ce^{i\kappa x} + De^{-i\kappa x})$ on the right. Show that if

$$\begin{pmatrix} C \\ B \end{pmatrix} = \begin{pmatrix} S_{11} & S_{12} \\ S_{21} & S_{22} \end{pmatrix} \begin{pmatrix} A \\ D \end{pmatrix},$$

then

$$|S_{11}|^2 + |S_{12}|^2 = 1$$
$$|S_{21}|^2 + |S_{22}|^2 = 1$$
$$S_{11} S_{12}^* + S_{21} S_{22}^* = 0,$$

so that the $\boldsymbol{S}$ matrix

$$\boldsymbol{S} = \begin{pmatrix} S_{11} & S_{12} \\ S_{21} & S_{22} \end{pmatrix}$$

is unitary. (Gasiorowicz,[104] p. 105.)

4-32. Specialize the potential barrier in Problem 4-31 to a rectangular barrier extending from $x = -a$ to $x = +a$, with $V = 0$ elsewhere. (a) Calculate the elements of $\boldsymbol{S}$ and show that the general conditions listed in Problem 4-31 hold. (b) Show also that $S(-\kappa) = S^{\dagger}(\kappa)$. (Gasiorowicz,[104] p. 106.) Additional derivations and problems concerning the $\boldsymbol{S}$ matrix appear in Chapter 11 of Segrè.[121]

REFERENCES

1. E. W. McDaniel, *Atomic Collisions: Heavy Particle Projectiles*, Wiley, New York, to be published.
2. H. S. W. Massey, E. W. McDaniel, and B. Bederson (Series Eds.), *Applied Atomic Collision Physics*, 5 vols, Academic, New York (1982–1984), see Vol. 1, Chap. 4; Vol. 5, Chaps. 3, 4, 5, 7, 14, 16.
3. C. Ramsauer, *Ann. Phys.* **64**, 513 (1921); **66**, 546 (1921).
4. S. C. Brown, *Basic Data of Plasma Physics*, Wiley, New York (1959).
5. B. Bederson and L. J. Kieffer, *Rev. Mod. Phys.* **43**, 601 (1971).
6. H. S. W. Massey and E. H. S. Burhop, *Electronic and Ionic Impact Phenomena*, 2nd ed., Vol. 1, Oxford University Press, London (1969); H. S. W. Massey, Vol. 2 (1969).
7. J. B. Hasted, *Physics of Atomic Collisions*, 2nd ed., American Elsevier, New York (1972).
8. L. G. H. Huxley and R. W. Crompton, *The Diffusion and Drift of Electrons in Gases*, Wiley, New York (1974).
9. G. Csanak, D. C. Cartwright, S. K. Srivastava, and S. Trajmar, "Elastic Scattering of Electrons by Molecules," in L. G. Christophorou (Ed.), *Electron-Molecule Interactions and Their Applications*, Vol. 1, Academic, New York (1984).
10. I. Shimamura and K. Takayanagi (Eds.), *Electron-Molecule Collisions*, Plenum, New York (1984).
11. U. Fano and A. R. P. Rau, *Atomic Collisions and Spectra*, Academic, New York (1986).
12. J. Perel, P. Englander, and B. Bederson, *Phys. Rev.* **128**, 1148 (1962).
13. D. E. Golden and H. W. Bandel, *Phys. Rev.* **138**, A-14 (1965). A. Salop and H. Nakano, *Phys. Rev. A* **2**, 127 (1970). G. Dalba, P. Fornasini, I. Lazizzera, G. Ranieri, and A. Zecca, *J. Phys. B* **12**, 3787 (1979).
14. T. M. Miller, "Atomic and Molecular Polarizabilities," in R. C. Weast (Ed.), *Handbook of Chemistry and Physics*, 67th ed, CRC Press, Orlando, Fla. (1986), pp. E-66 through E-75. See also T. M. Miller and B. Bederson, *AAMP* **25**, 37 (1988).

15. R. W. Wagenaar and F. J. de Heer, *J. Phys. B* **18**, 2021 (1985).
16. R. W. Wagenaar, A. de Boer, T. van Tubergen, J. Los, and F. J. de Heer, *J. Phys. B* **19**, 3121 (1986).
17. J. Ferch, B. Granitza, C. Masche, and W. Raith, *J. Phys. B* **18**, 967 (1985).
18. W. Raith, *AAMP* **12**, 281 (1976).
19. R. E. Kennerly and R. A. Bonham, *Phys. Rev. A* **17**, 1844 (1978). R. K. Jones, *Phys. Rev. A* **31**, 2898 (1985); *J. Chem. Phys.* **82**, 5424 (1985).
20. E. W. McDaniel, *Collision Phenomena in Ionized Gases*, Wiley, New York (1964).
21. R. H. Neynaber, L. L. Marino, E. W. Rothe, and S. M. Trujillo, O atoms: *Phys. Rev.* **123**, 148 (1961); H atoms: **124**, 135 (1961); N atoms: **129**, 2069 (1963).
22. B. Jaduszliwer, P. Weiss, A. Tino, and B. Bederson, *Phys. Rev. A* **30**, 1255 (1984).
23. B. Jaduszliwer, A. Tino, and B. Bederson, *Phys. Rev. A* **30**, 1269 (1984).
24. B. Jaduszliwer, G. F. Shen, J. L. Cai, and B. Bederson, *Phys. Rev. A* **31**, 1157 (1985).
25. C. B. Wharton, "Microwave Diagnostics for Controlled Fusion Research," in J. E. Drummond (Ed.), *Plasma Physics*, McGraw-Hill, New York (1961).
26. A. Gilardini, *Low Energy Electron Collisions in Gases*, Wiley, New York (1972).
27. A. V. Phelps, O. T. Fundingsland, and S. C. Brown, *Phys. Rev.* **84**, 559 (1951).
28. C. Ramsauer and R. Kollath, *Ann. Phys.* **12**, 529, 837 (1932).
29. J. R. Gibson and K. T. Dolder, *J. Phys. B* **2**, 741, 1180 (1969).
30. G. E. Chamberlain, S. R. Mielczarek, and C. E. Kuyatt, *Phys. Rev. A* **2**, 1905 (1970).
31. D. Andrick, *AAMP* **9**, 207 (1973).
32. P. Marmet and L. Kerwin, *Can. J. Phys.* **38**, 787 (1960).
33. G. J. Schulz, *Phys. Rev. Lett.* **10**, 104 (1963).
34. G. J. Schulz, *Phys. Rev.* **135**, A-988 (1964).
35. J. A. Simpson, *Rev. Sci. Instrum.* **35**, 1698 (1964).
36. C. E. Kuyatt and J. A. Simpson, *Rev. Sci. Instrum.* **38**, 103 (1967).
37. C. E. Kuyatt, *MEP* **7**, Pt. A, p. 1 (1968).
38. I. V. Hertel and K. J. Ross, *J. Phys. B* **1**, 697 (1968); **2**, 285, 484 (1969).
39. A. Stamatovic and G. J. Schulz, *Rev. Sci. Instrum.* **41**, 423 (1970).
40. M. J. W. Boness and G. J. Schulz, *Phys. Rev. A* **9**, 1969 (1974).
41. J. F. Williams and B. A. Willis, *J. Phys. B* **8**, 1641 (1975).
42. D. Andrick and A. Bitsch, *J. Phys. B* **8**, 393 (1975).
43. J. N. H. Brunt, F. H. Read, and G. C. King, *J. Phys. E* **10**, 134 (1977).
44. J. N. H. Brunt, G. C. King, and F. H. Read, *J. Phys. B* **10**, 1289 (1977).
45. D. F. Register, S. Trajmar, and M. K. Srivastava, *Phys. Rev. A* **21**, 1134 (1980).
46. D. F. C. Brewer, W. R. Newell, and A. C. H. Smith, *J. Phys. E* **13**, 114, 123 (1980). See also D. F. C. Brewer, W. R. Newell, S. F. W. Harper, and A. C. H. Smith, *J. Phys. B* **14**, 1749 (1981); W. R. Newell, D. F. C. Brewer, and A. C. H. Smith, *J. Phys. B* **14**, 3209 (1981).
47. M. Weyhreter, B. Barzick, A. Mann, and F. Linder, *Z. Phys. D* **7**, 333 (1988).
48. K.-H. Kochem, W. Sohn, K. Jung, H. Ehrhardt, and E. S. Chang, *J. Phys. B* **18**, 1253 (1985).

49. W. Sohn, K.-H. Kochem, K. Jung, H. Ehrhardt, and E. S. Chang, *J. Phys. B* **18**, 2049 (1985). K.-H. Kochem, W. Sohn, N. Hebel, K. Jung, and H. Ehrhardt, *J. Phys. B* **18**, 4455 (1985). W. Sohn, K.-H. Kochem, K.-M. Scheuerlein, K. Jung, and H. Ehrhardt, *J. Phys. B* **19**, 3625 (1986); **20**, 3217 (1987).
50. W. Sohn, K. Jung, and H. Ehrhardt, *J. Phys. B* **16**, 891 (1983).
51. F. H. Read, J. Comer, R. E. Imhof, J. N. H. Brunt, and E. Harting, *J. Electron Spectros. Relat. Pheno.* **4**, 293 (1974). F. H. Read, *Radiat. Eff.* **44**, 129 (1979). F. H. Read, *J. Phys. E* **16**, 636 (1983). E. Harting and F. H. Read, *Electrostatic Lenses*, Elsevier, Amsterdam (1976).
52. P. Grivet, *Electron Optics*, 2nd ed., Pergamon, Oxford (1972). P. W. Hawkes and E. Kasper, *Principles of Electron Optics* (2 vols.), Academic, New York (1988). *H. Wollnik, Optics of Charged Particles*, Academic, New York (1987).
53. O. Klemperer and M. E. Barnett, *Electron Optics*, 3rd ed., Cambridge University Press, Cambridge (1971).
54. Y. Ballu, *AEEP*, *Suppl.* **13-B**, 257 (1980).
55. G. J. Schulz, *Rev. Mod. Phys.* **45**, 378, 423 (1973).
56. D. E. Golden, *AAMP* **14**, 1 (1978).
57. F. H. Read, *Phys. Scr.* **27**, 103 (1983).
58. R. I. Hall and F. H. Read, "Molecular Spectroscopy by Electron Scattering," in Ref. 10.
59. J. B. Hasted and D. Mathur, "Electron-Molecule Resonances," in L. G. Christophorou (Ed.), Ref. 9.
60. A. Herzenberg, "Vibrational Excitation of Molecules by Slow Electrons," in Ref. 10; also in P. G. Burke and J. B. West (Eds.), *Electron-Molecule Scattering and Photoionization*, Plenum, New York (1988).
61. J. F. Williams, *J. Phys. B* **8**, 1683, 2191 (1975).
62. J. F. Williams and B. A. Willis, *J. Phys. B* **8**, 1670 (1975). J. F. Williams and A. Crowe, *J. Phys. B* **8**, 2233 (1975).
63. P. J. Hicks, S. Daviel, B. Wallbank, and J. Comer, *J. Phys. E* **13**, 713 (1980). T. A. York and J. Comer, *J. Phys. B* **17**, 2563 (1984). J. Comer, *ICPEAC XIII*, Berlin (1983), p. 63. Other important papers from other laboratories are R. W. Wijnaendts van Resandt and J. Los, *ICPEAC XI*, Kyoto (1979), p. 831; R. S. Gao, P. S. Gibner, J. H. Newman, K. A. Smith, and R. F. Stebbings, *Rev. Sci. Instrum.* **55**, 1756 (1984). A. Müller, N. Djuríc, G. H. Dunn, and D. S. Belíc, *Rev. Sci. Instrum.* **57**, 349 (1986).
64. R. E. Kennerly, R. J. Van Brunt, and A. C. Gallagher, *Phys. Rev. A* **23**, 2430 (1981).
65. R. A. Bonham, *ICPEAC XIII*, Berlin (1983), p. 73. R. A. Bonham and M. Fink, *High Energy Electron Scattering*, Van Nostrand Reinhold, New York (1974).
66. J. Kessler, *Polarized Electrons*, 2nd ed., Springer-Verlag, Berlin (1985). See also *ICPEAC XV*, Brighton (1987), p. 21.
67. N. F. Mott and H. S. W. Massey, *The Theory of Atomic Collisions*, 3rd ed., Clarendon, Oxford (1965).
68. G. D. Fletcher, T. J. Gay, and M. S. Lubell, *Phys. Rev. A* **34**, 911 (1986).
69. U. Fano, *Phys. Rev.* **178**, 131 (1969); **184**, 250 (1969). J. Kessler, *At. Phy.* **3**, 523 (1973). M. S. Lubell, *At. Phys.* **5**, 325 (1977).
70. W. Raith, *ICPEAC XIII*, Berlin (1983), p. 107.

71. J. Kessler, *Comments At. Mol. Phys.* **17**, 15 (1985).

72. D. T. Pierce, F. Meier, and P. Zürcher, *Appl. Phys. Lett.* **26**, 670 (1975); *Phys. Lett. A* **51**, 465 (1975). D. T. Pierce and F. Meier, *Phys. Rev. B* **13**, 5484 (1976). C. S. Feigerle, D. T. Pierce, A. Seiler, and R. J. Celotta, *Appl. Phys. Lett.* **44**, 866 (1984). H. J. Drouhin, C. Hermann, and G. Lampel, *Phys. Rev. B* **31**, 3859, 3872 (1985).

73. D. T. Pierce, R. J. Celotta, G. C. Wang, W. N. Unertl, A. Galejs, C. E. Kuyatt, and S. R. Mielczarek, *Rev. Sci. Instrum.* **51**, 478 (1980).

74. C. W. Drake, *MEP* **4**, Pt. B, p. 226 (1967). V. W. Hughes, R. L. Long, M. S. Lubell, M. Posner, and W. Raith, *Phys. Rev. A* **5**, 195 (1972).

75. G. D. Fletcher, M. J. Alguard, T. J. Gay, V. W. Hughes, P. F. Wainwright, M. S. Lubell, and W. Raith, *Phys. Rev. A* **31**, 2854 (1985).For a detailed description of a polarized atomic hydrogen beam, see also N. Chan, D. M. Crowe, M. S. Lubell, F. C. Tang, A. Vasilakis, F. J. Mulligan, and J. Slevin, *Z. Phys. D* **10**, 393 (1988).

76. B. Bederson, *Comments At. Mol. Phys.* **1**, 41, 65 (1969). For additional discussion of inelastic scattering with spin analysis, see K. Rubin, B. Bederson, M. Goldstein, and R. E. Collins, *Phys. Rev.* **182**, 201 (1969).

77. H. Kleinpoppen, *Phys. Rev. A* **3**, 2015 (1971). K. Blum and H. Kleinpoppen, *Phys. Rev. A* **9**, 1902 (1974). J. Byrne and P. S. Farago, *J. Phys. B* **4**, 954 (1971).

78. O. Berger and J. Kessler, *J. Phys. B* **19**, 3539 (1986). See also R. Möllenkamp, W. Wübker, O. Berger, K. Jost, and J. Kessler, *J. Phys. B* **17**, 1107 (1984).

79. G. Baum, M. Moede, W. Raith, and U. Sillmen, *Phys. Rev. Lett.* **57**, 1855 (1986).

80. F. Kaussen, H. Geesmann, G. F. Hanne, and J. Kessler, *J. Phys. B* **20**, 151 (1987).

81. J. J. McClelland, M. H. Kelley, and R. J. Celotta, *Phys. Rev. Lett.* **58**, 2198 (1987).

82. S. M. Khalid and H. Kleinpoppen, *Phys. Rev. A* **27**, 236 (1983).

83. D. M. Crowe, M. S. Lubell, F. J. Mulligan, J. Slevin, F. C. Tang, and A. Vasilakis, *ICPEAC XV*, Brighton (1987), Contributed Papers, p. 145. See also the paper by Lubell in Ref. 93.

84. H. J. Beyer, K. Blum, and R. Hippler (Eds.), *Coherence in Atomic Collision Physics*, Plenum, New York (1988).

85. H. S. W. Massey, E. H. S. Burhop, and H. B. Gilbody, *Electronic and Ionic Impact Phenomena*, Vol. 5, Oxford University Press, London (1974).

86. T. C. Griffith and G. R. Heyland, *Phys. Rep.* **39**, 169 (1978). T. C. Griffith, *AAMP* **15**, 135 (1979).

87. T. S. Stein and W. E. Kauppila, *AAMP* **18**, 53 (1982).

88. *Proc. Int. Conf. Positron Scattering Annihilation Gases*, York University, Toronto, Canada, July 1981, *Can. J. Phys.* **60**, Special Issue, 457–617 (1982).

89. J. W. Humberston and M. R. C. McDowell (Eds.), *Positron Scattering in Gases*, Plenum, New York (1984).

90. W. E. Kauppila, T. S. Stein, and J. M. Wadehra (Eds.), *Positron (Electron)-Gas Scattering*, World Scientific, Singapore (1986).

91. M. Charlton, *Rep. Prog. Phys.* **48**, 737 (1985).

92. G. Sinapius, *ICPEAC XV*, Brighton (1987), p. 73.

93. J. W. Humberston and E. A. G. Armour (Eds.), *Atomic Physics with Positrons*, Plenum, New York (1988). See papers by W. E. Kauppila and T. S. Stein and by M. Weber, K. G. Lynn, L. O. Roellig, A. P. Mills, and A. R. Moodenbaugh.

94. T. S. Stein, R. D. Gomez, Y. F. Hsieh, W. E. Kauppila, C. K. Kwan, and Y. J. Wan, *Phys. Rev. Lett.* **55**, 488 (1985); M. S. Dababneh et al., *Phys. Rev. A* **38**, 1207 (1988).
95. K. Floeder, P. Höner, W. Raith, A. Schwab, G. Sinapius, and G. Spicher, *Phys. Rev. Lett.* **60**, 2363 (1988). G. M. A. Hyder, M. S. Dababneh, Y. F. Hsieh, W. E. Kauppila, C. K. Kwan, M. Mahdavi-Hezaveh, and T. S. Stein, *Phys. Rev. Lett.* **57**, 2252 (1986).
96. B. H. Bransden and C. J. Joachain, *Physics of Atoms and Molecules*, Longman Group, Harlow, Essex, England (1983).
97. P. G. Burke, *Potential Scattering in Atomic Physics*, Plenum, New York (1977).
98. B. H. Bransden, *Atomic Collision Theory*, 2nd ed., Benjamin-Cummings, Menlo Park, Calif. (1983).
99. S. Geltman, *Topics in Atomic Collision Theory*, Academic, New York (1969).
100. M. A. Biondi, A. Herzenberg, and C. E. Kuyatt, *Phys. Today* **32**, 44 (Oct. 1979).
101. E. T. Whittaker and G. N. Watson, *A Course of Modern Analysis*, 4th ed., Cambridge University Press, Cambridge (1952).
102. P. G. Burke, *AAMP* **4**, 173 (1968).
103. P. Roman, *Advanced Quantum Theory*, Addison-Wesley, Reading, Mass. (1965).
104. S. Gasiorowicz, *Quantum Physics*, Wiley, New York (1974).
105. R. G. Newton, *Scattering Theory of Waves and Particles*, 2nd ed., Springer, New York (1982).
106. T.-Y. Wu and T. Ohmura, *Quantum Theory of Scattering*, Prentice-Hall, Englewood Cliffs, N.J. (1962).
107. J. R. Taylor, *Scattering Theory*, Wiley, New York (1972).
108. C. J. Joachain, *Quantum Collision Theory*, North-Holland, Amsterdam (1975).
109. L. D. Landau and E. M. Lifshitz, *Quantum Mechanics* (*Non-relativistic Theory*), 2nd ed., Pergamon, Oxford (1965).
110. A. Messiah, *Quantum Mechanics*, 2 vols., North-Holland, Amsterdam (1961).
111. E. Fermi, *Nuclear Physics*, University of Chicago, Chicago (1950).
112. S. Cvejanović, J. Comer, and F. H. Read, *J. Phys. B* **7**, 468 (1974).
113. P. S. Farago, *Free-Electron Physics*, Penguin Books, Harmondsworth, Middlesex, England (1970), pp. 70–75.
114. D. Park, *Introduction to the Quantum Theory*, 2nd ed., McGraw-Hill, New York (1974), pp. 102–107.
115. J. A. Cronin, D. F. Greenberg, and V. L. Telegdi, *University of Chicago Graduate Problems in Physics with Solutions*, Addison-Wesley, Reading, Mass. (1967).
116. M. Chen, *Berkeley Physics Problems with Solutions*, Prentice-Hall, Englewood Cliffs, N.J. (1974).
117. D. ter Haar (Ed.), *Problems in Quantum Mechanics*, 3rd ed., Pion, London (1975), also published by Academic, New York (1975).
118. D. Rapp, *Quantum Mechanics*, Holt, Reinhart and Winston, New York (1971).
119. R. P. Feynman, R. B. Leighton, and M. Sands, *The Feynman Lectures on Physics*, Vol. III, Addison-Wesley, Reading, Mass. (1964).
120. H. A. Bethe and R. W. Jackiw, *Intermediate Quantum Mechanics*, 2nd ed., Benjamin, New York (1968).
121. E. Segrè, *Nuclei and Particles*, 2nd ed., Benjamin-Cummings, Menlo Park, Calif. (1977).

5

EXCITATION AND DEEXCITATION OF NEUTRALS BY ELECTRON IMPACT

5-1. INTRODUCTION

An atom may be raised from an initial bound state to another bound state of higher electronic energy in a collision with an electron whose kinetic energy is sufficient to cause this transition. In electron impact on molecules, rotational and vibrational excitation may occur as well, with an increase in the energy of rotation and vibration of the molecule, respectively. Deexcitation of excited states is also possible in electron collisions, along with the more familiar deexcitation by the spontaneous or stimulated emission of radiation. Typically, for light molecules, adjacent rotational levels are spaced a few meV apart, so radiative transitions between rotational levels involve the emission or absorption of photons in the far-infrared and microwave regions of the spectrum. Vibrational levels are typically spaced about 0.1 eV apart, corresponding to infrared frequencies in radiative transitions. The spacing of low-lying electronic configurations of atoms or molecules is usually several eV, and decreases as the electronic energy increases. For our purposes it is appropriate to take several eV as a typical spacing. The photons emitted or absorbed in radiative electronic transitions are mostly in the visible and ultraviolet spectral regions. In many applications, a process of great interest is the loss of energy by electrons in collisions as they travel through a gas. Even energy loss through rotational excitation can be important. The loss per collision is only of the order of 1 meV, but the cross sections for pure rotational excitation can be large (greater than 10^{-16} cm^2).

Even in the simplest case of a diatomic molecule, the coupling of the rotational, vibrational, and electronic motions makes analysis of the structure and spectra of the molecule a formidable problem. However, if we look at the problem classically, we immediately see a mitigating factor—the frequencies of the three kinds of motion are so different that the motions may be considered separately in a zeroth-order approximation. This consideration is equally valid in a quantum mechanical treatment. We write down the zeroth-order solutions for the rotation and vibration of an isolated molecule (i.e., the solutions for the rigid rotor and harmonic oscillator, respectively) and apply second-order perturbation theory to describe the result of coupling the rotation and vibration of the nuclei. The electrons in the molecule come into play here as follows. Because the motion of the nuclei is so slow compared with that of the electrons, we may consider the nuclei to be frozen at some internuclear spacing and calculate the total electronic energy (kinetic plus potential) for this spacing. This calculation is then repeated for a large number of spacings, and the Coulomb repulsion energy between the nuclei is computed for each of these spacings. The sum of the total electronic energy and the nuclear repulsion energy may then be plotted as a function of the internuclear separation to provide a potential curve that governs the vibratory motion of the nuclei. A different potential curve is associated with each electronic state. For a given bound state (characterized by a potential curve with a minimum), the internuclear separation distance oscillates about its equilibrium value for that state. The instantaneous moment of inertia

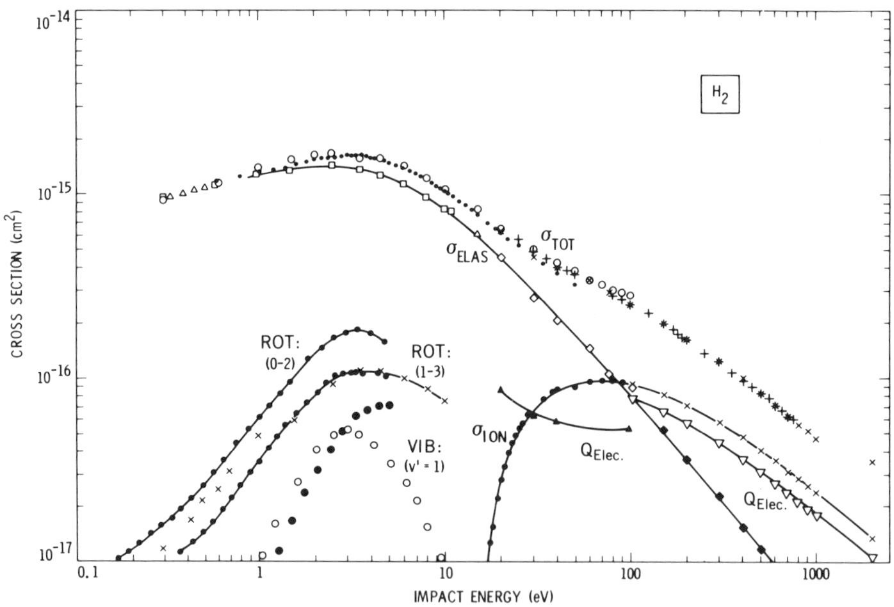

Figure 5-1-1. Cross sections for various types of electron collisions with H_2 molecules. *Elas.*, elastic scattering; *rot.*, rotational excitation; *vib.*, vibrational excitation; *elec.*, electronic excitation (above 100 eV, the estimated sum of all electronic excitation cross sections; below 100 eV, the sum of the cross sections measured for excitation of the six lowest electronic states); *ion.*, ionization; *tot.*, total cross section. (From Trajmar,[3] *ICPEAC XIV.*)

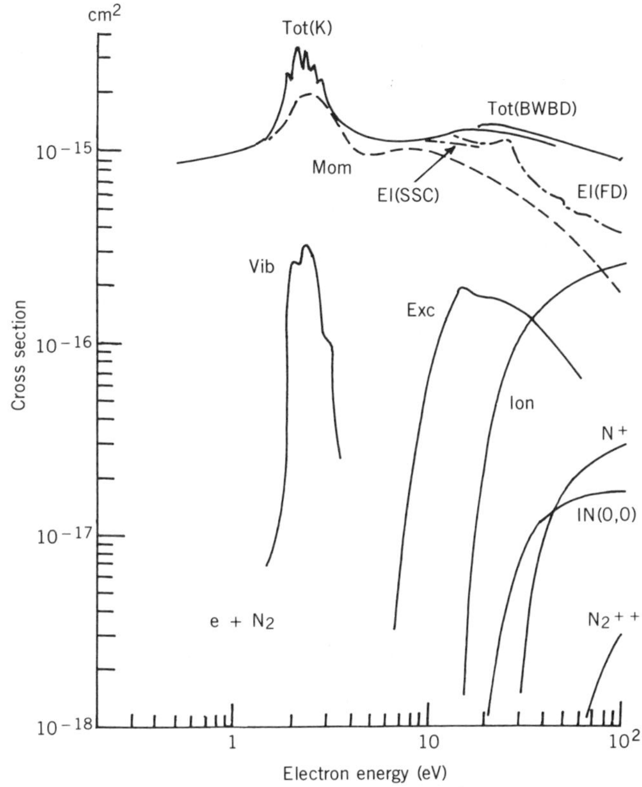

Figure 5-1-2. Cross sections for various types of electron collisions with N_2 molecules. *El.*, elastic scattering; *mom.*, momentum transfer; *vib.*, sum of vibrational excitation cross sections for ($v = 0 \to v'$); *exc.*, sum of 18 electronic excitation cross sections; *ion.*, total ionization; N^+, dissociative ionization; N_2^{2+}, double ionization; $1N$, cross section for emission of the first negative (0, 0) band; *tot.*, total cross section. Sources of these data are provided in Ref. 4, from which this figure is adapted. Some of the data are experimental, some theoretical. For additional data see Y. Itikawa et al., *J. Phys. Chem. Ref. Data* **15**, 985 (1986).

of the molecule, I, is determined by the instantaneous value of the separation distance and the masses of the nuclei.

The experimental and theoretical approaches to the study of rotational, vibrational, and electronic excitation are different enough to warrant division of our discussion along these lines, although in many collisions, changes in two, or even all three types of molecular motion occur simultaneously. Experimental studies of inelastic electron collisions date back to Lenard, only five years after the discovery of the electron by Thomson. However, measurements on excitation are beset by many intrinsic difficulties, and the field required many years to mature. Reliable cross sections for electronic and vibrational excitation were available by about 1965; the first directly measured cross sections for rotational excitation were published only in 1968. Since early in this century there has been a great demand for excitation cross sections because of their important role in astrophysics, atmospheric physics, gaseous electronics, plasma physics, and many other areas, and considerable reliance has been placed on theoretical calculations because of the paucity of experimental data of certain types. Applications of rotational cross sections are discussed in Refs. 1 and 2; the role of vibrational and electronic cross sections in natural and laboratory phenomena is described in Refs. 2 and 3.

At this early stage in our treatment it is useful to examine Figs. 5-1-1 and 5-1-2 in order to put the excitation processes into perspective. These figures display plots of cross sections for various kinds of electron collisions as functions of the electron energy. They show at a glance the energy ranges in which the various phenomena are likely to be significant. Before beginning our discussion of the separate modes of excitation, we should consider briefly the general question of the interaction of electrons with molecules.

5-2. MULTIPOLE MOMENTS AND ELECTRON-MOLECULE INTERACTIONS

In an atom there is a unique point that is natural to pick as the origin of coordinates in describing the internal motion and that gives the simplest expression for the electronic energy: namely, the nucleus of the atom. A molecule, on the other hand, has more than one nucleus and is "multicentered." Regardless of the choice of an origin of coordinates, the mathematical expression for the internal energy is inevitably complex because no single origin is the natural choice for expression of the energies of all the particles. As we shall see below, the interaction of the incident electron with the target molecule is usually anisotropic, and this fact also greatly complicates the analysis of electron-molecule collisions.

We approximate the general electron-molecule interaction by decomposing it into three parts, as we did in Section 4-7. The *static interaction* is the electrostatic interaction between the projectile and the target molecule under the assumption that the electron does not perturb the charge distribution of the molecule. The *polarization interaction* is associated with the distortion that, in fact, inevitably does occur. The *exchange interaction* can result in the exchange of the projectile electron and a target electron. This decomposition is not completely valid because parts of the interaction are of a dynamical nature and must be determined in the context of the collision problem being discussed.[4,5]

A. The Static Interaction

Let us consider a projectile electron approaching a molecule at a large distance with a low velocity. If the molecule has a net charge q, the projectile first feels the Coulomb potential, which varies as r^{-1}, where r is the electron-molecule separation distance. Closer in, if the molecule has a permanent electric dipole moment, the projectile feels the point charge–permanent dipole potential, which varies as r^{-2} at large r. On still closer approach, the projectile experiences the point charge–permanent quadrupole potential ($\sim r^{-3}$) if the target has a quadrupole moment. We need not consider still shorter-range permanent multipole interactions, such as the point charge–permanent octupole ($\sim r^{-4}$).

Let us now consider a localized charge distribution $\rho(\mathbf{r}')$ that is nonvanishing only inside a sphere of radius R around some origin. At any exterior point $\mathbf{r}$ the electric potential produced by this charge distribution can be obtained as an expansion in spherical harmonics[6] and expressed in rectangular coordinates as

$$\Phi(\mathbf{r}) = \frac{q}{r} + \frac{\boldsymbol{\mu}\cdot\mathbf{r}}{r^3} + \frac{1}{2}\sum_{i,j}^{3} Q_{ij}\frac{r_i r_j}{r^5} + \text{higher-order multipole terms.} \quad (5\text{-}2\text{-}1)$$

The scalar monopole moment of the charge distribution is q; the vector electric dipole moment is defined as

$$\boldsymbol{\mu} = \int \mathbf{r}'\rho(\mathbf{r}')\,d^3r', \quad (5\text{-}2\text{-}2)$$

while the 3×3 second-rank electric quadrupole moment tensor $\mathcal{Q}$ is expressed in Cartesian tensor form as

$$Q_{ij} = \int (3r_i'r_j' - r'^2\delta_{ij})\rho(\mathbf{r}')\,d^3r'. \quad (5\text{-}2\text{-}3)$$

Now we look at an electron interacting with a molecule (net charge Ne) and utilize the results of the spherical harmonic expansion sketched above. The starting point is

$$V_{\text{static}}(\mathbf{r}) = -e\int \frac{\rho(\mathbf{r}')}{|\mathbf{r}-\mathbf{r}'|}\,d\mathbf{r}', \quad (5\text{-}2\text{-}4)$$

which is the potential *energy*, $V = -e\Phi$, of the static interaction. The electron charge is $-e$. The charge distribution $\rho(\mathbf{r}')$ includes the continuous distribution of the electron cloud and the point charges of the nuclei (delta functions). The position vector of the incident electron is $\mathbf{r}$. We now specialize to the case of *linear* molecules, for which $\rho(\mathbf{r}')$ is axially symmetric. We take the origin of the vectors $\mathbf{r}$ and $\mathbf{r}'$ to be the molecular center of mass, and the Z axis to be along the molecular axis, which is described by the unit vector $\hat{R}$. Then for r large compared with the molecular dimensions, (5-2-1) gives

$$V_{\text{static}}(\mathbf{r}) \to -\frac{Ne^2}{r} - \frac{\mu e}{r^2}P_1(\hat{R}\cdot\hat{r}) - \frac{Qe}{r^3}P_2(\hat{R}\cdot\hat{r}) + \cdots, \quad (5\text{-}2\text{-}5)$$

where $\hat{r}$ is the unit vector along $\mathbf{r}$. The Legendre polynomials P_1 and P_2 are functions of the cosine of the angle χ between the molecular axis and the position vector of the electron, $\mathbf{r}$. Because of the axial symmetry, only one element of Q_{ij} survives in (5-2-5):

$$Q \equiv Q_{33} = \int (3z'^2 - r'^2)\rho(\mathbf{r}')\, d^3r', \tag{5-2-6}$$

which is the scalar quadrupole moment of the axially symmetric molecule.

If the molecule has a net charge, the monopole term dominates at large r. If the molecule is neutral and has a permanent dipole moment, the dipole term is important. A molecular quadrupole moment generally has a weak, but not necessarily negligible effect.

B. The Polarization Interaction

This contribution is due mainly to the point-charge-induced dipole attractive interaction. The electron induces an electric dipole in the molecule, and mutual attraction results. For large separation distance r, the electric field produced by the electron is essentially uniform over the molecular volume, and if the molecule is spherically symmetric, the polarization potential energy is

$$V_{\text{pol}}(\mathbf{r}) \to -\frac{\alpha e^2}{2r^4}, \tag{5-2-7}$$

where α is the scalar dipole polarizability of the molecule. The induced dipole moment will always point along the position vector $\mathbf{r}$ as the molecule rotates. The polarizability for most molecules, however, is a tensor of the second rank. It depends on the orientation of the molecule with respect to the polarizing electric field because the bound electrons can be displaced more easily along certain molecular axes than along others. For a linear molecule we have

$$V_{\text{pol}}(\mathbf{r}) \to -\frac{\alpha e^2}{2r^4} - \frac{\alpha' e^2}{2r^4} P_2(\hat{R}\cdot\hat{r}), \tag{5-2-8}$$

where α and α' are given in terms of the polarizabilities $\alpha_\parallel$ and $\alpha_\perp$ along the axis and perpendicular to the axis:

$$\alpha = \frac{\alpha_\parallel + 2\alpha_\perp}{3}, \qquad \alpha' = \frac{2}{3}(\alpha_\parallel - \alpha_\perp). \tag{5-2-9}$$

For small separation distances r and low impact velocities, the "asymptotic polarization potential" of (5-2-8) should be replaced by the "adiabatic polarization potential," which takes into account the nonuniformity of the electric field over the molecular volume. For large incident electron velocities, a velocity-dependent polarization potential should be used.[4]

C. The Exchange Interaction

Electron exchange comes into play only at small separation distances r. It is taken into account by antisymmetrization of the complete wave function (projectile plus molecule) with respect to the electrons. This step leads to a set of coupled integrodifferential equations which replace the usual coupled differential equations. It is here that a nonlocal interaction appears (cf. Section 4-8-A).

In this region of small r, the multipole expansion (5-2-5) breaks down, and the static interaction must be calculated by (5-2-4), which requires accurate molecular wave functions. The concept of the polarization interaction is also vitiated because the incident electron becomes indistinguishable from the target electrons and cannot be said to be the one that produces the polarization. It is more appropriate to speak of "correlation effects." These matters are discussed at length by Takayanagi[4] and Shimamura.[5]

5-3. ROTATIONAL EXCITATION AND DEEXCITATION OF NEUTRALS BY ELECTRON IMPACT

Electrons may produce rotational excitation by a direct mechanism in collisions with molecules, or by an indirect mechanism involving the formation of a resonance state and its subsequent decay (see Section 4-8-B and Refs. 7–9). Rotational excitation is of importance mainly in the energy range extending from the rotational thresholds at small fractions of 1 eV up to several tens of eV.

In directly exciting a molecule to a higher rotational state, an impinging electron must increase the total angular momentum $\mathbf{J}$ of the molecule by exerting a torque on it. A torque can be exerted only by a force with a component normal to the electron position vector $\mathbf{r}$, and since $\mathbf{F} = -\nabla V$, an angle-dependent electron-molecule interaction is required. Let us again specialize to the case of linear molecules and note that (5-2-5) and (5-2-8) contain terms involving the Legendre polynomials

$$P_1(\cos \chi) = \cos \chi; \qquad P_2(\cos \chi) = \tfrac{1}{2}(3 \cos^2 \chi - 1), \tag{5-3-1}$$

where χ is the angle between the molecular axis and the electron position vector. These terms in the electron-molecule potential energy can produce rotational excitation if certain conditions are fulfilled.

The cross section for direct rotational excitation is proportional to the square of the modulus of the transition matrix element connecting the specified initial and final quantum states (cf. Note 3-1). The electron impact selection rules are obtained by examination of these matrix elements.[4,5] Jung et al.[10] have prepared a table that gives the selection rules for ΔJ and cross sections for direct rotational transitions in collisions of electrons with various molecules (see the upper section of Table 5-3-1). The multipole moments considered here are the permanent dipole and quadrupole moments. As expected, the selection rules

Table 5-3-1 Selection Rules for ΔJ, and Total Cross Sections for Rotational Transitions Produced by Electron Impact, Ordered with Respect to the Most Relevant Interaction Mechanism[a]

Interaction	H_2	N_2	CO	H_2O
Nonresonant				
Weak quadrupole	Influence covered by resonance	$\Delta J = 0, \pm 2$ $\lesssim 1.5 \times 10^{-18}\ \text{cm}^2$		
Weak dipole			$\Delta J = \pm 1$ $\approx 1.7 \times 10^{-17}\ \text{cm}^2$	
Strong dipole				$\Delta J = \pm 1$ $2.5 \times 10^{-16}\ \text{cm}^2$
Shape resonance				
Attractive	$\Delta J = 0, \pm 2$ $0.8 \times 10^{-16}\ \text{cm}^2$	$\Delta J = 0, \pm 2, \pm 4$ $2.8 \times 10^{-16}\ \text{cm}^2$	$\Delta J = 0, \pm 1, \pm 2, \pm 3, \pm 4$ $\gtrsim 10^{-16}\ \text{cm}^2$	Influence covered by dipole
Repulsive	Influence not measurable			$\Delta J = 0, \pm 1$ $0.3 \times 10^{-16}\ \text{cm}^2$

Source: Jung et al.[10]

[a]More general discussions of electron-impact selection rules appear in Refs. 4 and 5, and in G. A. Natanson, *J. Phys. B* **18**, 4481 (1985). Note that H_2 and N_2, being homonuclear, have no permanent electric dipole moment.

differ for permanent dipole and quadrupole interactions. The permanent electric dipole interaction is especially important here because of the long range of the dipole field, which allows the incident electron to exert a torque on the molecule for a long time. The differential cross sections for dipole-induced rotational transitions are strongly peaked in the forward direction, and they rise sharply as the impact energy approaches zero.

Now we turn to resonance excitation. Recall from Section 4-8-B that at certain impact energies a projectile electron and a target molecule may form a compound state whose lifetime is significantly longer than the usual time of passage of the electron through the molecule. This long lifetime corresponds to an electron wave function with a large amplitude inside the target. Such resonances affect inelastic processes as well as elastic scattering. The compound state can decay with the ejection of an electron of undiminished CM kinetic energy; alternatively, the electron can lose kinetic energy by leaving the molecule in an excited, dissociated, or ionized state, or by producing a negative ion through dissociative attachment. "Superelastic" processes can also occur, in which the projectile electron leaves the collision site with increased kinetic energy that was gained by deexciting the original molecular quantum state. Resonances are important in applications because they can increase collision cross sections by one or two orders of magnitude.

When both resonance and direct processes are involved, so that the end result can be achieved either by the system passing through a resonance or by it not doing so, the resonance and direct scattering amplitudes interfere. Hence there is interference between the electron waves scattered by the two processes. Analysis of the angular and energy dependence of the cross sections in a given experiment can elucidate the scattering mechanisms that are involved. Sometimes, either a resonance or a direct process dominates the scattering; often, one or the other is not involved at all. In the case of electron collisions on H_2, a shape resonance involving the ${}^2\Sigma_u^+$ state of H_2^- strongly influences the scattering at impact energies of a few eV. This resonance is several eV wide and centered at 2 to 3 eV.[7] A Feshbach resonance is associated with the H_2^- state of ${}^2\Sigma_g^+$ symmetry at 11.3 eV.[8] The lower half of Table 5-3-1 relates to the influence of resonance formation in the rotational excitation of H_2, N_2, CO, and H_2O.

The spectrum of rotational energy levels for all molecules is characterized by small spacings.[11,12] Figure 5-3-1 displays low-lying vibrational-rotational levels (quantum numbers v and J) for a few diatomic molecules. The rotational energy is given approximately, in the rigid rotor approximation, by the equation

$$E_{\text{rot}} = \frac{\hbar^2}{2I} J(J+1). \tag{5-3-2}$$

Figure 5-3-1 suggests the difficulty of directly resolving the individual J levels in a collision experiment. This feat was accomplished first for the H_2 molecule, by Ehrhardt and Linder[13] in 1968, and was made possible by the uniquely large rotational spacings (about 50 meV) for the low-lying rotational states of H_2. For N_2, on the other hand, the spacing is about 1 meV.

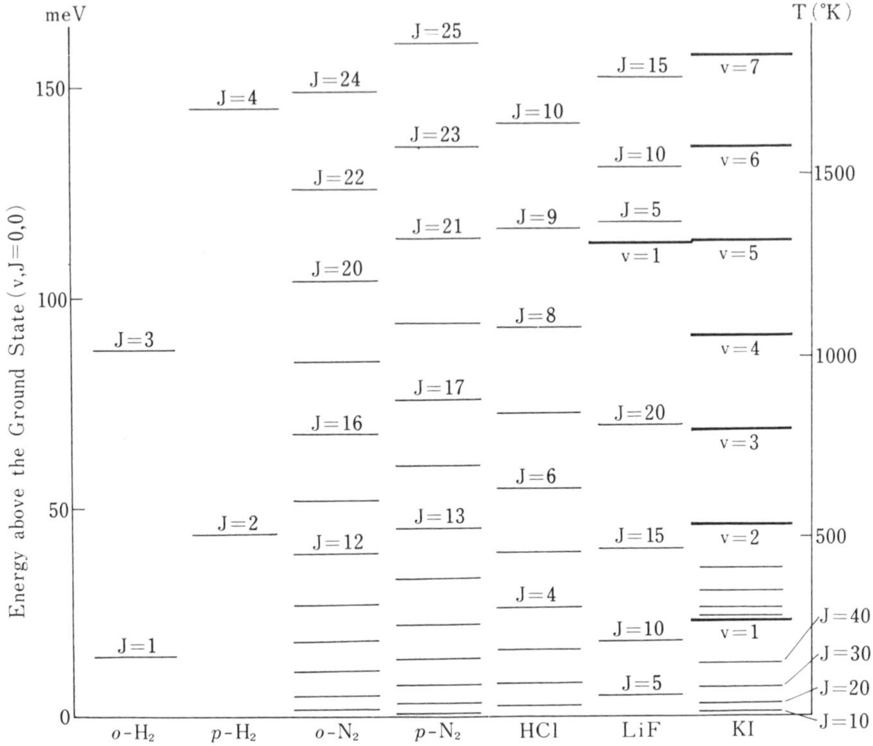

Figure 5-3-1. Vibrational-rotational levels (quantum numbers v, J) of a few diatomic molecules. The ($v = 1$, $J = 0$) level of H_2 lies 0.54 eV above the ground state (v, $J = 0, 0$). Rotational level spacings for H_2 are uniquely large, about $15J$ meV, where J is the quantum number for the upper level. For the ortho species of H_2 (o-H_2), the nuclear spins are parallel; for the para version (p-H_2), the nuclear spins are antiparallel. The effects of nuclear spin in various molecules are discussed in Refs. 11 and 12. (From Shimamura.[5])

The measurements by Ehrhardt and Linder[13] were followed by similar crossed-beam experiments of Linder and Schmidt,[14] who investigated elastic scattering, pure rotational excitation and deexcitation, pure vibrational excitation, and rotational-vibrational excitation of H_2 at electron impact energies of 0.3 eV to about 15 eV. Their cross sections for the rotational excitation $J = 1 \rightarrow 3$ appear in Fig. 5-1-1 in the region between 1 and 10 eV. The peak value of this cross section is about $1.1 \times 10^{-16}\ cm^2$ and is reached at an impact energy of about 4.0 eV.

Linder and Schmidt used a double 127° electrostatic electron spectrometer, similar to some of the instruments described in Section 4-3-B, with an energy resolution of about 30 meV. The first step in determining the rotational cross sections was to obtain the energy-loss spectrum at each of many scattering angles for various fixed electron impact energies. To accomplish this, the voltages on the scattered-electron energy analyzer were swept over a suitably wide range, in such a way that at each setting of the voltages only electrons scattered with a certain energy loss were transmitted to the detector. Figure 5-3-2 provides an example of an energy-loss spectrum for electrons on H_2 at 11.07 eV impact energy. (This spectrum was obtained by Joyez et al.[15] and was selected for this figure because it shows the rotational features more clearly than did the earlier work of Linder and Schmidt.[14]) In a spectrum such as that shown in Fig. 5-3-2, the peak locations are fixed by the energy levels of the molecule, and the areas under the peaks are related to the cross sections. The variation of the areas under the peaks with scattering angle provides relative values of the differential cross section, which can be used to determine the integral cross section for the process under consideration. Absolute values of both types of cross sections are obtained by normalization procedures, several of which are described in Refs. 2, 3, and 16.

Wong and Schulz[17] have also studied rotational excitation in H_2. They used the electron spectrometer described in Ref. 10 of Chapter 4 at an impact energy of 4.5 eV and an energy resolution of 18 to 22 meV. A typical counting rate when elastic scattering was being observed was 4×10^3 per second at a hydrogen

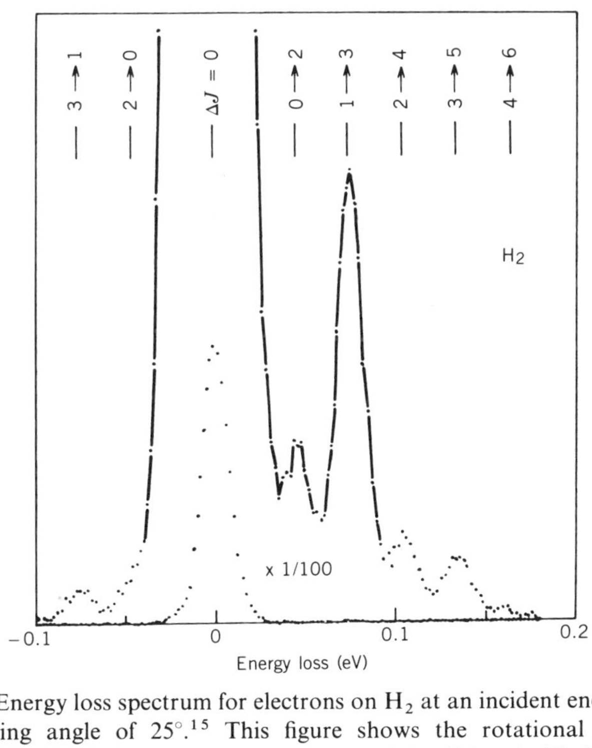

Figure 5-3-2. Energy loss spectrum for electrons on H_2 at an incident energy of 11.07 eV and a scattering angle of 25°.[15] This figure shows the rotational excitation and deexcitation of the ground vibronic state of H_2. The full width at half height of the peaks is 17 meV. Superelastic peaks appear at negative values of the energy loss.

background pressure of 2×10^{-5} torr. Wong and Schulz recorded energy-loss spectra at scattering angles of 20 to 100° for energy losses corresponding to $v = 0 \rightarrow 1, 2, 3$ and the associated rotational transitions. They reported absolute differential cross sections for rotational ($v = 0 \rightarrow 0$; $J = 1 \rightarrow 3$), vibrational ($v = 0 \rightarrow 1, 2, 3$; $\Delta J = 0$), and rotational-vibrational ($v = 0 \rightarrow 1, 2, 3$; $J = 1 \rightarrow 3$) excitations. Experimental errors in absolute magnitude for all of the inelastic cross sections were estimated to be about 10%, except for the transition to $v = 3$ (15%). The agreement of the cross sections for final vibrational states $v = 0$ and $v = 1$ agree with those of Linder and Schmidt[14] to within 5 to 10%, which is impressive in view of the fact that completely different methods of calibration were used in the two experiments. Figure 5-3-3 displays some of the Wong and Schulz data and indicates that the differential cross section for $J = 1 \rightarrow 3$ with $v = 0 \rightarrow 0, 1, 2, 3$ is isotropic. The ratio $R = \sigma(\Delta J = 0)/\sigma(J = 1 \rightarrow 3)$ for $v = 0 \rightarrow 1, 2, 3$ depends on the vibrational excitation.

Additional high-quality data come from Srivastava et al.,[18] who measured the differential cross section for pure rotational excitation ($J = 1 \rightarrow 3$) in H_2 at energies of 15 to 100 eV at 115°, 3 to 40 eV at 20°, and 40 eV at 10 to 135°. For intermediate impact energies and large scattering angles, the probability of pure

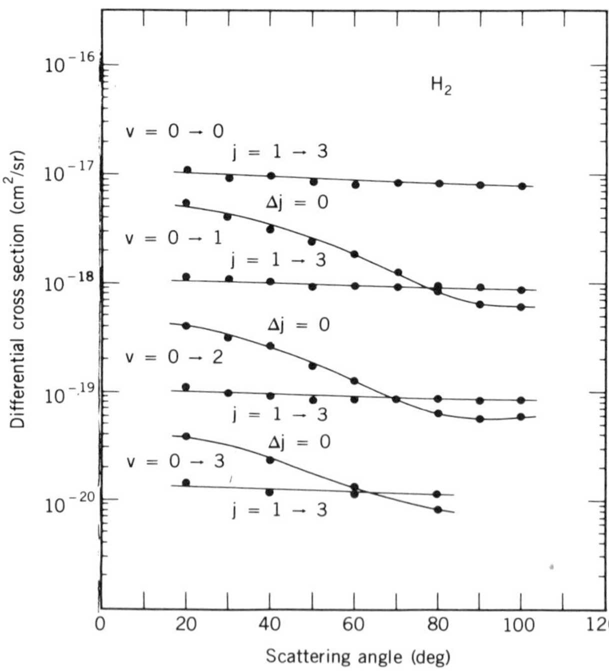

Figure 5-3-3. Angular dependence of the absolute differential cross sections for rotational ($v = 0 \rightarrow 0$; $J = 1 \rightarrow 3$), vibrational ($v = 0 \rightarrow 1, 2, 3$; $\Delta J = 0$), and rotational-vibrational ($v = 0 \rightarrow 1, 2, 3$; $J = 1 \rightarrow 3$) excitation of H_2.[17]

rotational excitation exceeds the probability of elastic scattering. The differential cross section is found to peak at about 4 eV for both large and small scattering angles, and to decrease slowly with increasing electron impact energy. Srivastava et al. also obtained good agreement with the results of Linder and Schmidt.[14]

More recently, Jung et al.[19] have measured the angular dependence of the rotational excitation of H_2 from $J_i = 1$ to $J_f = 3$ by electrons with energies of 0.6 eV and 0.2 eV. Comparison with various theories shows that the adiabatic nuclei rotation approximations are no longer valid when the impact energy is lowered into this region.

For molecules such as N_2, CO, and H_2O, about 100 rotational levels may contribute to a vibrational transition within an energy interval of about 80 meV. Jung et al.[10] studied these molecules at impact energies of 0.5 to 6 eV and an energy resolution of 10 to 18 meV in a crossed-beam electron spectrometer. They observed broadened peaks without substructure in the electron energy-loss spectrum. However, by making use of theoretical work of Read and Andrick[20] and Itikawa,[21] Jung et al.[10] were able to analyze the line shapes and extract cross sections for individual rotational transitions.* Müller et al.[22] have obtained cross sections for rotational excitation in vibrationally elastic collisions of electrons with methane (CH_4). Their electron beam energies ranged from 0.5 to 10 eV. They determined absolute cross sections from the broadened energy-loss peaks by an analysis due to Shimamura.[23]

Using Shimamura's deconvolution procedure, which assumes the validity of the adiabatic nuclei approximation,[23] Rädle et al.[24] have determined angular-differential rotational excitation cross sections in the elastic and first vibrational inelastic channels over the impact energy range 0.35 to 10 eV.

Also at Kaiserslautern, Zeigler et al.[25] have measured state-to-state angular-differential cross sections for rotationally inelastic e-Na_2 collisions over the impact energy range 150 to 300 eV. They observed, for the first time, large transition probabilities for $0 \leqslant \Delta J \leqslant 30$ excitations in backward scattering. The pronounced rotational rainbows that were seen are expected to be general features in high-energy electron-molecule scattering. For theoretical work on these rotational rainbows, see the paper by Korsch et al.[26] Recall that the rainbow in elastic scattering was introduced in Section 3-7.

A few cross sections for rotational excitation have also been derived by analyzing electron drift velocity experiments in terms of the Boltzmann transport theory. The cross sections labeled Rot $(0 \rightarrow 2)$ and Rot $(1 \rightarrow 3)$ below the electron energy of 0.5 eV in Fig. 5-1-1 were obtained by this method. Unfortunately, the implementation of this method is involved, and it is complicated by the presence of resonances and overlapping reactions.

We see that few experimental cross sections are available for rotational

*The first experimental study of resonant rotational excitation of N_2 was reported in 1973 by J. Comer and M. Harrison, *J. Phys. B* **6**, L-70 (1973). Broadening of the energy loss spectra was observed, although rotational cross sections were not obtained.

excitation. The data published up to about the end of 1983 are discussed in Refs. 2, 4, and 5, which also provide good treatments of the theory and exhaustive bibliographies. Shimamura[5] discusses scaling laws, which can be used to obtain cross sections for vibrational excitation of one type of molecule from data on another type, or which can provide the cross section at one energy if it is known for another energy for a given molecule. These scaling laws are of limited, not general, validity. Systematics in the data for vibrationally elastic scattering are covered by Csanak et al.[2] (It may be recalled from Section 4-1-C that most modern experimental data on electron-molecule scattering at low energies include contributions from rotational, but not vibrational, excitation, and that the term "vibrationally elastic" refers to such data.)

The rotational deexcitation of molecules by electron impact is also an important phenomenon and is intimately related to the excitation process. In fact, the cross section σ_{ij} for the deexcitation collision $j \rightarrow i$ is related to the cross section σ_{ji} for the inverse excitation process $i \rightarrow j$ by the principle of detailed balance, which is derived in Section 8-2. This principle requires that

$$g_j E_j \sigma_{ij} = g_i E_i \sigma_{ji}. \tag{5-3-3}$$

Here i and j denote the quantum states of the target molecule, g_i and g_j the statistical weights of the states, and E_i and E_j the kinetic energies of the projectile electron at large distances from the molecule.

Other quantities of interest that involve various kinds of inelastic collisions (not merely rotational excitation and deexcitation) are discussed by Takayanagi:[4] (1) the average electron energy loss per collision, (2) the stopping power of a target, (3) the stopping cross section, and (4) the momentum transfer cross section ($q_D \equiv \sigma_m$). We first encountered the momentum transfer cross section in (1-6-1), in a form that applied only to elastic collisions. Equation (1-6-1) may be generalized to an inelastic process of type n by writing

$$q_D^n = \int \left(1 - \frac{\kappa_f}{\kappa_i} \cos \Theta\right) I_n(\Theta)\, d\Omega_{cm}, \tag{5-3-4}$$

where I_n is the differential cross section per unit solid angle for the process n and κ_i and κ_f are the wave numbers of the incident and scattered electrons, respectively.

5-4. VIBRATIONAL EXCITATION AND DEEXCITATION OF NEUTRALS BY ELECTRON IMPACT

Vibrational excitation of a molecule by electron impact may occur by a direct mechanism if the projectile-target interaction depends on the vibrational coordinates of the molecule.[4,27] For a diatomic molecule, the interaction potential energy may be expanded as

$$V(\mathbf{r}, \mathbf{R}) = V(\mathbf{r}, \hat{R}, R = R_e) + (R - R_e)\frac{\partial V(\mathbf{r}, \mathbf{R})}{\partial R}\bigg]_{R=R_e} + \text{higher-order terms,} \tag{5-4-1}$$

where $\mathbf{r}$ is the projectile electron position vector relative to the molecular center of mass, $\mathbf{R}$ the internuclear spacing vector, and R_e the equilibrium internuclear spacing.* The first term on the right-hand side of (5-4-1) is for a rigid rotator. The second term contains the vibrational coordinate $(R - R_e)$, which represents the stretching or compression of the molecule. Excitation of the molecule involves increasing the amplitude of the vibratory motion, so the incident electron may couple to the vibration of the molecule through the factor $(R - R_e)$, higher powers of which appear in the higher-order terms. Hence the second and higher terms can cause vibrational transitions. If we consider only the second term of (5-4-1) and use harmonic oscillator wave functions for the vibrator, (3-25-7) gives for the differential cross section per unit solid angle

$$I_{j\leftarrow i}(\vartheta) = \frac{\kappa_j}{\kappa_i}\frac{m^2}{4\pi^2\hbar^4}|T_{ji}|^2 \tag{5-4-2}$$

with the transition matrix element

$$T_{ji} = \langle\Phi_j|V|\Psi_i\rangle = 0 \text{ unless } v_j - v_i = \pm 1. \tag{5-4-3}$$

Thus we have the approximate selection rule $\Delta v = \pm 1$ for upward and downward vibrational transitions. More realistic selection rules are obtained by using more appropriate wave functions to account for the anharmonicity of the vibration, and sometimes by using higher-order terms in (5-4-1). Cross sections for direct vibrational excitation usually exhibit a smooth, slowly varying energy dependence and contribute a "background" at a level of about 10^{-18} to 10^{-17} cm^2, typically, for small nonpolar molecules.

Vibrational excitation may also proceed indirectly by a resonance process in which the projectile electron is temporarily captured in a compound, negative ion state that decays into a final state consisting of the electron scattered with reduced energy and the target molecule in a vibrationally excited level[26] (cf. Section 5-3). Resonances can enhance all inelastic cross sections (not only those for vibrational excitation) by orders of magnitude. The energy dependence of the cross sections associated with resonances often shows isolated peaks or oscillatory structure. The resonance effects are frequently large and important, but are limited to certain ranges of impact energy, outside of which the background, direct mechanism is the only important possibility, except in special cases. At low energies, typical cross sections for vibrational excitation by resonance mechanisms are 10^{-16} cm^2.

*It is essential to realize that (5-4-1) does not represent the energy of the isolated molecule, for which the term linear in the vibrational coordinate equals zero.

A third mechanism for vibrational excitation, one that is a special type of resonance, has also been proposed. This mechanism involves the formation of a *virtual state*,[28] that is, a state of small positive energy which produces a large enhancement of the projectile electron wave function at the site of the molecular potential well without actually trapping the electron in a negative energy state.[26] The virtual state mechanism was proposed to explain vibrational excitation of molecules with large permanent dipole moments.[29] For such molecules, sharp threshold peaks appear in the differential cross sections for vibrational excitation, the widths being of the order of 0.1 eV and the peaks occurring only about 0.1 eV above the vibrational thresholds. For HCl and HF, Rohr and Linder[29] reported isotropic scattering in the vicinity of the peaks,* where the integral cross section becomes extremely large (4×10^{-15} cm^2 for $v = 0 \rightarrow 1$ in measurements on HBr). In 1987, Knoth and his colleagues[30] determined absolute rovibrational excitation cross sections for HCl and HF over the energy range 0.15 to 5.5 eV and the angular range 15 to 135°, with an energy inhomogeneity of about 20 meV. They observed strong simultaneous rotational and vibrational ($\Delta v = 1$) transitions at the threshold peaks with nonisotropic angular distributions. (For $v = 0 \rightarrow 1$, the differential cross sections for $J = 0 \rightarrow 0, 1, 2, 3$, and 4 decreased at small ϑ as $\vartheta \rightarrow 0$.) This angular dependence calls into question the validity of the virtual state model for threshold structures. The results of Knoth et al.[30] indicate that not only the *s* wave, but higher partial waves as well, contribute significantly to the scattering at the vibrational threshold. Threshold structures also appear for some nonpolar molecules, especially those with large polarizabilities, in addition to those with permanent electric dipole moments. In fact, about half of all of the molecules that have been studied exhibit threshold structures.[30] Additional experiments, more precise and providing more detail, are required to establish clearly the mechanism underlying the threshold peaks.

Most modern experimental studies of vibrational excitation employ high-resolution electron spectrometers of the types described in Section 4-3-B. The procedures followed are the same as in the studies of rotational excitation that were discussed in Section 5-3. The range of electron impact energies that are of greatest interest extends from threshold up to about 30 eV. Over most of this range, an incident electron that is not trapped in a compound state moves fast enough to pass the target molecule in a time short compared with a rotational period (10^{-12} sec, typically), so that the orientation of the molecule may be regarded as fixed in the laboratory during the collision. Only a very slow electron, with energy less than 0.1 eV, fails in this regard and presents special difficulties. The velocity of an electron with 0.1 eV energy is 1.88×10^5 m/sec.

Energy-loss spectra, of the type shown in Fig. 5-3-2 but emphasizing vibrational rather than rotational transitions, are displayed in Figs. 5-4-1 and 5-

*By contrast, vibrationally elastic scattering of electrons on polar molecules characteristically is strongly peaked in the forward direction, and the cross sections rise sharply as the impact energy approaches zero. Recall that this type of scattering includes both elastic scattering and rotational excitation.

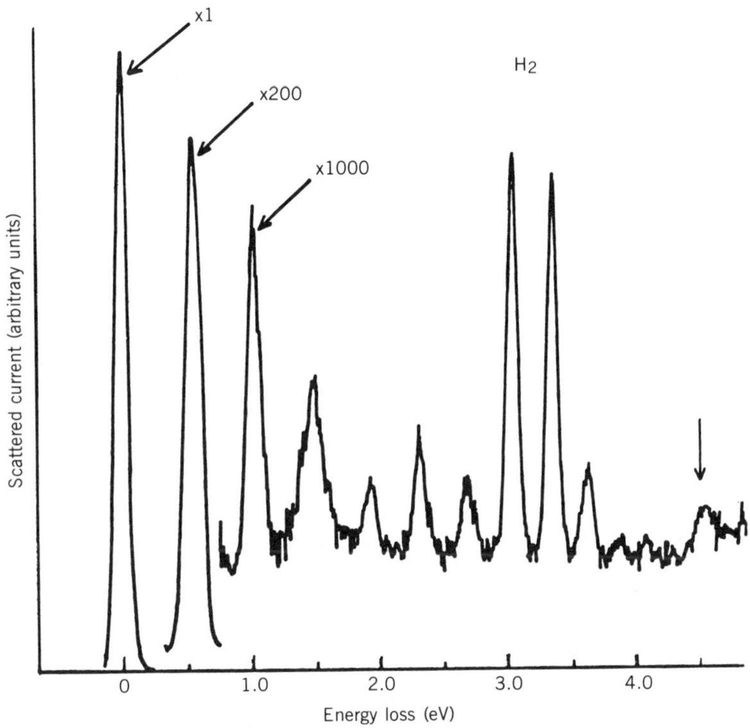

Figure 5-4-1. Energy-loss spectrum for electrons scattered from H_2 at an incident energy of 12.47 eV and scattering angle of 70°.[31] The dissociation limit is indicated by the vertical arrow. The excitation of vibrational levels of the ground electronic state from $v = 0$ to $v = 1, 2, \ldots, 11$ is in evidence.

4-2. The data for Fig. 5-4-1 were taken by Comer and Read[31] with an incident electron energy of 12.47 eV and a scattering angle of 70°. Excitation of vibrational levels of the ground electronic state from $v = 0$ all of the way up to $v = 11$ is in evidence. At 12.47 eV impact energy, a vibrational level of a resonance electronic state of H_2^- is excited; it decays preferentially into the vibrational levels $v = 7$ and $v = 8$ of the ground electronic state and thereby enhances the cross section for excitation of these levels.

Figure 5-4-2 shows energy-loss spectra obtained by Kochem et al.[32] in their studies of acetylene (C_2H_2). This linear, nonpolar molecule has five fundamental vibrational modes, with frequencies ν_n:

ν_1 symmetric C—H stretching:	Raman active	$D = 0$
ν_2 symmetric C≡C stretching:	Raman active	$D = 0$
ν_3 asymmetric stretching:	infrared active	$D > 0$
ν_4 symmetric bending:	Raman active	$D = 0$
ν_5 asymmetric bending:	infrared active	$D > 0$

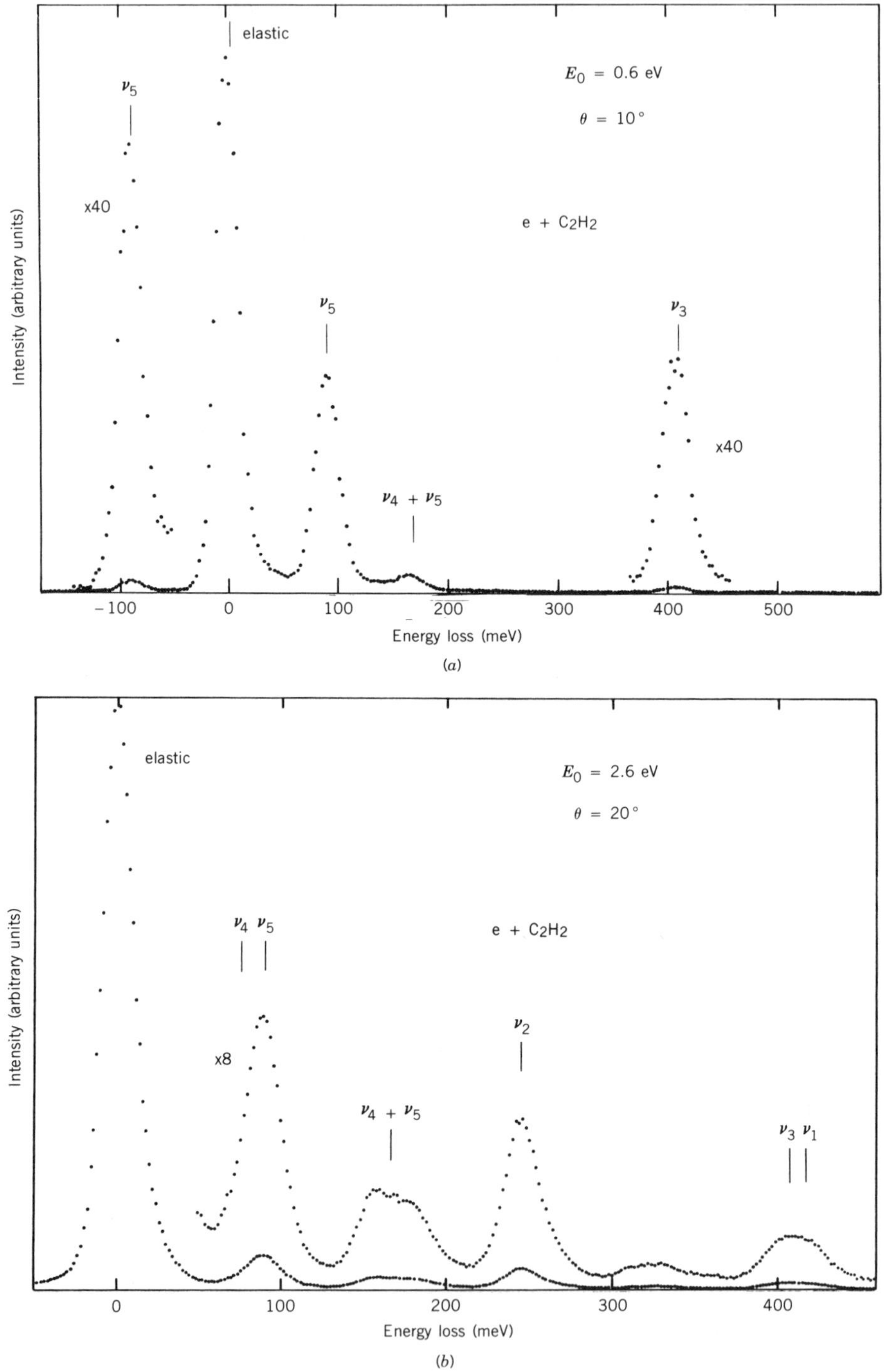

Figure 5-4-2. Vibrational energy-loss spectra for electrons scattered from C_2H_2 at incident energies of 0.6 eV (in the threshold region) and 2.6 eV (the resonance energy).[32] A superelastic peak is shown at −91 meV in (*a*). The peak at 164 meV is for the combination mode ($\nu_4 + \nu_5$). The ν_2 mode is evident in (*b*), but not in (*a*).

Here D is the transition dipole moment, defined as the magnitude of the vector moment:

$$\mathbf{D} = \langle j | \sum q_k \mathbf{r}'_k | i \rangle. \tag{5-4-4}$$

A given mode j is IR active if the transition dipole moment connecting it to the initial state i is nonzero, so that an electric dipole radiative transition of first order in the electric field intensity $\mathbf{E}$ can take place. If D is zero, the transition is of second order in E, as in Raman scattering, because the field must induce a dipole moment in the molecule as well as cause the transition. In electron impact on C_2H_2, only the infrared active modes of vibration are strongly excited in the threshold region, and the excitation proceeds by the direct dipole interaction; but the Raman active modes, as well, are stimulated in the resonance region. For the Raman active modes, the $^2\Pi_g$ shape resonance at 2.6 eV, with a width greater than 1 eV, underlies their excitation at this energy, whereas the IR active modes here are excited by both the resonance and the dipole interaction acting coherently. The ν_1 and ν_4 vibrations overlap with the stronger ν_3 and ν_5 modes, respectively, within the energy resolution of the experiment, and the excitation of these modes was obscured to some extent. However, Kochem et al.[32] were able to obtain absolute differential and integral cross sections for elastic scattering and for excitation of the ν_2, ν_3, and ν_5 vibrational modes from threshold to 3.6 eV. The angular range from 5 to 108° was covered.

The electron spectrometer used by Ehrhardt's group for this study of C_2H_2 has been employed in a similar study of the weakly polar molecule CO.[33] The contributions of both direct and resonance scattering in different ranges of energy and scattering angle were assessed for vibrationally elastic collisions and for vibrational excitation. For C_2H_2 and CO, no sharp, high threshold peaks were observed as had been seen with the strongly polar molecules HF, HCl, HBr, H_2O, and H_2S,[29] nor was the scattering in C_2H_2 and CO isotropic at threshold and slightly above.

Low-energy measurements have also been made by Ehrhardt's group on CO_2,[34] which in its ground state, is a linear molecule of $^1\Sigma_g^+$ symmetry and has no permanent dipole moment. Of the three fundamental modes of vibration, the symmetric-stretch mode (100)* with excitation energy of 172 meV does not change the molecular symmetry. The bending mode (010) has Π_u symmetry and an excitation energy of 83 meV, whereas the asymmetric-stretch mode has Σ_u symmetry and an excitation energy of 291 meV. Because of the change of molecular symmetry, the last two excitations are associated with time-dependent dipole moments and therefore are IR-active modes. The studies by Kochem et al.[34] indicate that the excitation of the two IR-active modes in the threshold region is determined by the direct interaction with the dipole

*For a discussion of this notation, which arises in the normal-mode analysis of molecular vibration, see Chapter 8 of J. I. Steinfeld, *Molecules and Radiation*, 2nd ed. MIT Press, Cambridge, Mass. (1985). Steinfeld also treats IR and Raman activity and many other subjects in molecular spectroscopy.

potential, whereas the excitation of the (100) mode is dominated by s-wave scattering that is strongly influenced by a virtual-state pole of the $\boldsymbol{S}$ matrix located close to the origin.

Now let us discuss how resonances are designated in the simplest case, that of electrons impinging upon neutral diatomic molecules. Such resonances can be represented as temporary negative ions with well-defined configuration and symmetry. For example, there are five Feshbach resonance series of NO^-. Series (a) has the configuration $(nl_1\lambda_1, nl_2\lambda_2) = (3s\sigma)^2$ and the resonance symmetry $^1\Sigma^+$; series (b) has $(3s\sigma, 3p\pi)$ and $^3\Pi$, and so on.[8] Under favorable conditions,[7] the angular distribution of inelastically scattered electrons is determined uniquely by the symmetries of the resonant state, the initial state, and the final state, so measurements of angular distributions are invaluable in working out the classification of resonances. Usually, only one partial wave is expected to participate in resonance scattering, and in many cases its identity may be determined from angular distribution measurements and angular momentum conservation. For example, the dominant partial wave in a given case might be determined to be the $d\pi$ wave, where the first (Latin) letter designates the angular momentum of the partial wave at infinity (here $l = 2$), while the second (Greek) letter refers to the component of this angular momentum along the internuclear axis of the molecule as the incident electron comes close to it. (In the corresponding case of scattering by an atom, a simpler designation would apply, and we could refer to the resonance as a *d-wave resonance*.)

We have already presented the integral cross section for excitation of the $v = 0 \rightarrow 1$ and $v = 0 \rightarrow 2$ transitions in H_2 as a function of impact energy in Fig. 5-1-1. We now take a global look at vibrational excitation of N_2. This molecule, like H_2, has no permanent electric dipole moment, so resonance excitation should dominate over the direct mechanism. Figure 5-4-3 provides a global view of resonance vibrational excitation of N_2 in the energy range 1 to 30 eV. Shown are differential cross sections for $v = 0 \rightarrow 1$ and $v = 0 \rightarrow 2$ at a scattering angle of 90° that were obtained by Wong and Schulz and discussed in a review by Schulz.[35] To understand this figure, we recall that at a resonance energy, the projectile electron is trapped in the vicinity of the molecule and closely interacts with it for an abnormally long time. During this trapping, the electron distorts the molecule more efficiently than is usually the case, with a resulting enhancement of the vibrational cross section. It is reasonable to expect that the details of the excitation process, and the observed cross section, are strongly influenced by the lifetime of the resonance for a given target.

When the lifetime of the resonance, τ, is much smaller than a typical vibrational period (10^{-14} sec), the electron is attached to the molecule to form some intermediate state, which decays with the emission of the electron before the molecule can vibrate. (The mechanism here is called the *impulse limit*.) The molecule is left in a vibrationally excited level. The intermediate state has a large energy width Γ, according to the uncertainty principle ($\Gamma\tau \approx \hbar$), so the vibrational cross section shows a broad hump in its energy dependence at this resonance. The humps in Fig. 5-4-3 near 8 and 14 eV are believed to be of this type.

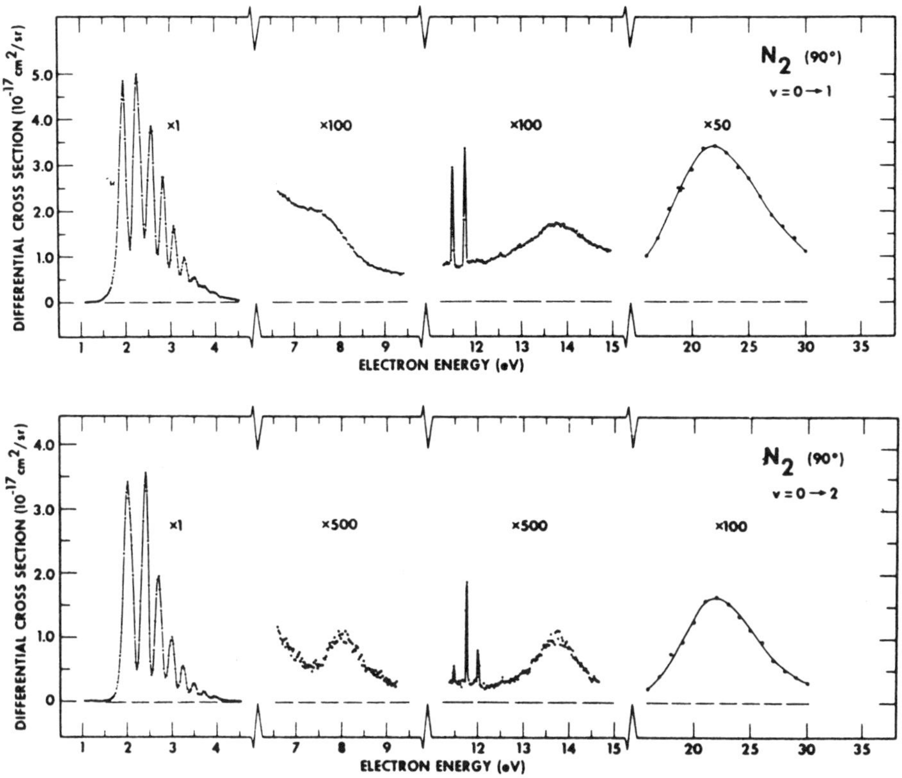

Figure 5-4-3. A global view of vibrational excitation of N_2 by electron impact in the energy range 1 to 30 eV.[35] The excitation is from $v = 0$ to $v = 1$ in the top part of the figure, and from $v = 0$ to $v = 2$ in the lower part. Excitation of N_2 vibrational levels up to $v = 17$ has been observed in energy-loss experiments by M. Allan, *J. Phys. B* **18**, 4511 (1985). Here the electron beam energy ranged up to 5 eV [see also *ICPEAC XV*, Brighton (1987), p. 93].

If the resonance lifetime (i.e., the lifetime of the computed state) becomes comparable with a typical vibrational period, the molecule can vibrate once or a few times before the compound state decays. Herzenberg[36] has proposed a "boomerang model" that successfully explains the excitation in such cases. We consider a diatomic molecule (N_2) and work in CM coordinates so that the motions of the two nuclei are replaced by the motion of an equivalent single particle, represented by the "nuclear wave packet" in what follows. In Fig. 5-4-4, the incident electron is captured to form an intermediate N_2^- state, and the nuclear wave packet propagates across the N_2^- potential well as the molecule stretches. The wave packet is reflected, and we assume that it is completely annihilated by autodetachment when it reaches the end of the first stretching-compression cycle. (It has been shown that only a single pair of counter-propagating wave packets need be considered.[36]) The superposition of the waves produces a standing wave whose node positions depend weakly on the

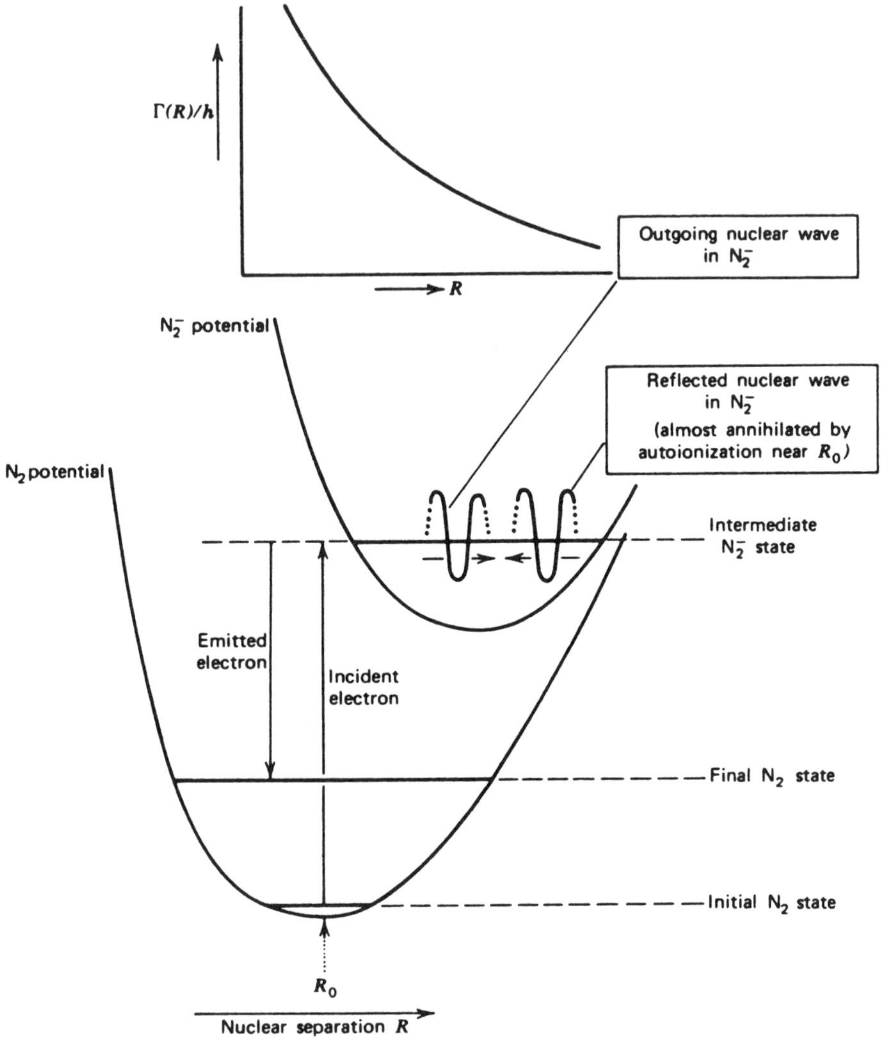

Figure 5-4-4. Illustration of the "boomerang" model for the vibrational excitation of the N_2 molecule by electron impact.[35] The magnitude and R dependence of the energy width Γ of the intermediate state (shown in the drawing at the top) are assumed to be such that only a single outgoing and reflected nuclear wave are important.

incident electron energy. Calculation of the overlap integral between the standing wave and the vibrational states of the neutral molecule shows that the vibrational cross sections have an oscillatory energy dependence, with peaks occurring at different impact energies, depending on the final states of excitation. A boomerang state is shown in Fig. 5-4-3 near 2.3 eV; it has a lifetime of about 10^{-14} sec.

If the lifetime of the compound state increases to the point where many molecular vibrations can take place, the compound state has time to develop a well-defined structure of vibrational levels before it decays. These levels decay to various vibrational levels of the neutral molecule with the emission of the attached electron. The vibrational cross sections, then, show sharp spikes corresponding to the energies of the vibrational levels of the compound state. The sharp, isolated peaks in the energy range 11.4 to 12 eV in Fig. 5-4-3 are of this type.

Finally, "ultrabroad" peaks, with widths of about 10 eV, are sometimes observed (see the 22-eV region in Fig. 5-4-3). Such peaks could probably not arise from a single resonance; it is more likely that they result from a number of closely spaced, overlapping resonances.

There is now a large quantity of accurate data on vibrational excitation by electron impact. Schulz gave an extremely useful survey of the diatomic data in 1973,[7] and his 1976 review[35] updates his previous treatment. Trajmar and Cartwright[3] summarize the vibrational excitation cross section measurements made since about 1970, while Herzenberg[36] provides an indispensable introduction to the theory. More general treatments that cover many aspects of electron-molecule collision theory at an introductory level have been prepared by Lane and Morrison and are highly recommended.[37]

Many good swarm measurements have been made on electrons in molecular gases, and some useful vibrational cross sections have been deduced from the results. However, difficulties arise when the cross sections exhibit structure, and only limited use of the swarm transport data has been possible to date.

Cross sections for vibrational deexcitation can be obtained from those for vibrational excitation by use of the principle of detailed balance (see Section 5-3).

5-5. DISSOCIATION OF NEUTRAL MOLECULES INTO NEUTRAL FRAGMENTS

Several types of dissociative events produced by electron impact on molecules and molecular ions are important and will be treated in this book. For simplicity, we enumerate them here in terms of electrons colliding with diatomic molecules and ions:

1. $e + XY \rightarrow e + X + Y$	Dissociation of molecule into neutral fragments in the electronic ground state
2. $e + XY \rightarrow e + X + Y^*$	Dissociation of molecule into neutral, electronically excited fragments
3. $e + XY \rightarrow 2e + X + Y^+$	Dissociative ionization of molecule (Chapter 6)
4. $e + XY \rightarrow X + Y^-$	Dissociative electron attachment to molecule (Ref. 38)
5. $e + XY^+ \rightarrow e + X + Y^+$	Dissociation of molecular ion (Chapter 7)
6. $e + XY^+ \rightarrow X + Y$	Dissociative electron-ion recombination (Ref. 38)

In all but the first process listed here, product atoms not shown as excited (*) may be in fact excited. If the molecular target is polyatomic, dissociation may result in rotationally and vibrationally excited molecular fragments. Here we discuss in detail only the first two processes; the remaining reactions are treated in Chapters 6 and 7 or in Ref. 38. As explained below, dissociation into neutral fragments may be viewed as vibrational excitation of the molecule into the dissociation continuum—hence its placement at this point in our discussion.

A. Molecular Potential Curves; Dissociation and Predissociation

We recall from Section 5-1 that in a rigorous treatment of molecular structure, it is necessary to include the effects of interaction among the three classes of internal motion in a molecule: electronic, vibrational, and rotational. The frequencies of these motions differ by orders of magnitude, however, so their interaction is relatively weak. Thus in first approximation (the *Born-Oppenheimer approximation**), the nuclei may be regarded as stationary and the energy of a particular electronic state calculated for different relative positions of

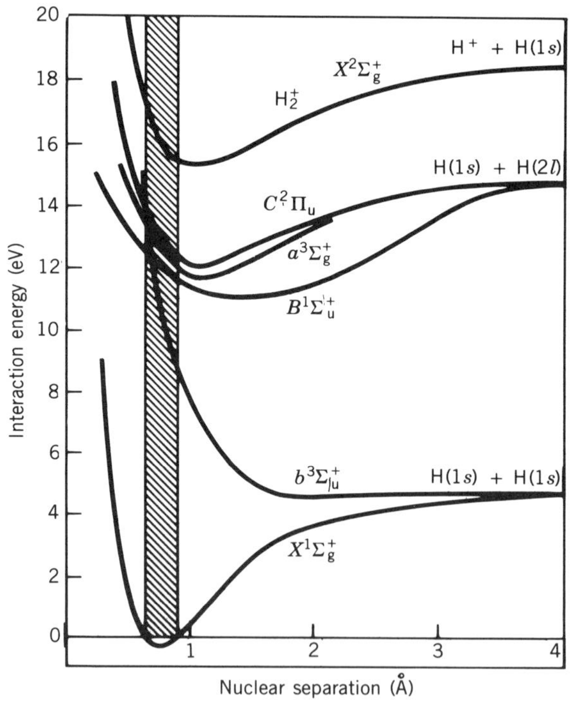

(*a*)

*This approximation is unrelated to the Born-Oppenheimer approximation that is discussed in Section 3-22 in connection with electron exchange scattering.

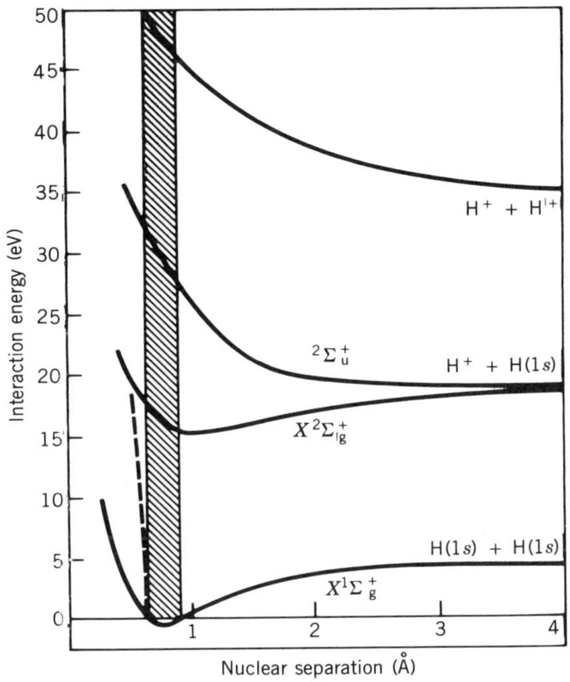

(b)

Figure 5-5-1. Potential energy curves for electronic states of H_2 and H_2^+: (*a*) within 20 eV of the ground state; (*b*) for higher-energy states. The united atom and separated atom limits[11,12,39] of a few states are given below.

United Atom Limit	Molecular State	Separated Atom Limit
He(1^1S)($1s\sigma$)^2	$X^1\Sigma_g^+$	$(\sigma_g 1s)^2$
He(2^1P) $1s\sigma$ $2p\sigma$	$B^1\Sigma_u^+$	$\sigma_g 1s\pi_u 2s$
He(2^1P) $1s\sigma$ $2p\pi$	$C^1\Sigma_u$	$\sigma_g 1s\pi_u 2p$
He(2^1S) $1s\sigma$ $2s\sigma$	$E^1\Sigma_g^+$	$\sigma_g 1s\sigma_g 2s$
He(2^3S) $1s\sigma$ $2s\sigma$	$a^3\Sigma_g^+$	$\sigma_g 1s\sigma_g 2s$
He(2^3P) $1s\sigma$ $2p\sigma$	$b^3\Sigma_u^+$	$\sigma_g 1s\sigma_u 1s$

The notation used here is explained by Sharp.[39] In describing polyatomic molecules, we must use potential energy surfaces rather than curves. (Adapted from Massey[39].)

the nuclei. The vibrational levels corresponding to this electronic state are then calculated by using this potential curve, and the rotational motion is introduced as a splitting of the vibrational levels.

Figure 5-5-1, based on Massey[39] and Sharp,[39] shows a few of the many known potential energy curves for various electronic states of the H_2 molecule and H_2^+ molecular ion. One class of potential curve, corresponding to an attractive state, possesses a minimum. Here there is a discrete set of levels associated with nuclear motion extending up to the top of the potential well and a continuum above. On the other hand, if the potential curve has no minimum, there are no discrete nuclear levels at all, and the state is repulsive. Whenever the nuclear vibrational level lies in the continuum, the molecule dissociates. The vertical distance from the lowest vibrational level near the bottom of the well to the horizontal asymptote corresponding to infinite separation of the nuclei is equal to the negative potential energy of the atoms when the molecule is formed adiabatically from atoms initially at rest with infinite separation. (The dissociation limit is the logical zero of potential energy.) If an amount of energy equal to the minimum dissociation energy is supplied internally to the molecule when it is in its lowest vibrational state, it dissociates into two atoms, each of which has zero kinetic energy.

In Fig. 5-5-1 the shaded area indicates the normal vibrational range of the nuclei of the H_2 molecule in the ground electronic state ($^1\Sigma_g$). To dissociate the ground-state molecule into two H atoms without electronic excitation of the molecule requires about 4.4 eV. This separation may be effected in thermal dissociation.* However, electron impact dissociation or photodissociation cannot occur by a change only of the vibrational state without a simultaneous electronic excitation. According to the *Franck-Condon principle*, which governs the latter types of dissociation, the nuclei do not have time to change their spacing appreciably during electronic excitation of the molecule, and an electronic transition corresponds to a *vertical* transition between potential curves on the energy diagram. Thus a minimum energy of about 8.9 eV is required to excite the H_2 molecule from its ground electronic and vibrational state to the first excited electronic state $b^3\Sigma_u^+$. This state is repulsive, and a molecule excited to it quickly dissociates into two H atoms, each with about 2.25 eV. If higher electronic states of the H_2 molecule are dissociated, the products include excited and ionized H atoms, as indicated at the right side of the curves of Fig. 5-5-1*a*. Excitation to the stable electronic states above $b^3\Sigma_u^+$ will be followed by the emission of ultraviolet light, as radiative transitions occur back to the ground state. Stable H_2^+ ions may be produced by the impact of electrons with energy greater than the vertical ionization energy 15.4 eV. About 18.0 eV is required to produce H^+ ions from H_2, and as we have indicated at least 8.9 eV is required to dissociate the molecule by electron impact.

When we consider an electron colliding with a molecule whose nuclei each outweigh the projectile electron by a factor of at least 2×10^3, we might wonder how the projectile can achieve the "impedance match" required to drive the

*If a gas is heated, molecules are driven to higher vibrational states by gas kinetic impacts, and some of the molecules may be excited by successive collisions to vibrational levels high enough to produce dissociation. No selection rules apply in the dissociation. There is no difficulty in satisfying the requirements of energy and momentum conservation because at least three particles are involved in the process—the "projectile" molecule and the dissociation fragments of the "target."

nuclei apart to infinity, even if it has the requisite kinetic energy. The answer lies in the discussion above. According to the Born-Oppenheimer principle, the motion of the nuclei is determined by the potential provided by the eigenvalues of the electronic Hamiltonian. Consequently, if the electronic structure of the molecule is altered, the nuclear motion may be drastically affected, so vibrational excitation or dissociation may result from an electronic transition produced in electron scattering, either of the direct or exchange type.

At this point, it is also necessary to discuss the phenomenon of *predissociation.* Let us refer to Fig. 5-5-2, which shows the potential energy curves and vibrational level diagrams for two electronically excited molecular states, A and B. We shall suppose that molecules in the ground electronic state, which is not shown in the drawing, are irradiated with light and that transitions up to the state A are possible. If state B were not present, we would observe a sequence of absorption bands corresponding to transitions from the ground state to all vibrational levels of state A and beyond the dissociation limit of A, into the dissociation continuum. Now we consider the effect of the second electronically excited state B, whose dissociation energy in this example is assumed to be less than that of state A, as shown in Fig. 5-5-2. Suppose that a molecule is in state A but vibrationally excited to the extent that the total internal energy exceeds the dissociation limit of the molecule in state B. Then, if certain selection rules are obeyed, a nonradiative transition may occur into the dissociation continuum of B, with the molecule dissociating before the dissociation limit of state A is reached. According to the Franck-Condon principle, such transitions from A into the continuum of B occur with high probability only in the neighborhood of the intersection of the potential curves of A and B. The possibility of deexcitation by predissociation reduces the lifetime of the discrete states of A and, by the uncertainty principle, thereby increases the width of these energy levels. This fact accounts for the diffuse appearance of the dissociation bands. A phenomenon analogous to predissociation occurs in atoms and involves a radiationless transition into the ionization continuum. Here reference is made to preionization, or autoionization, which is discussed in Sections 6-6 and 7-1.

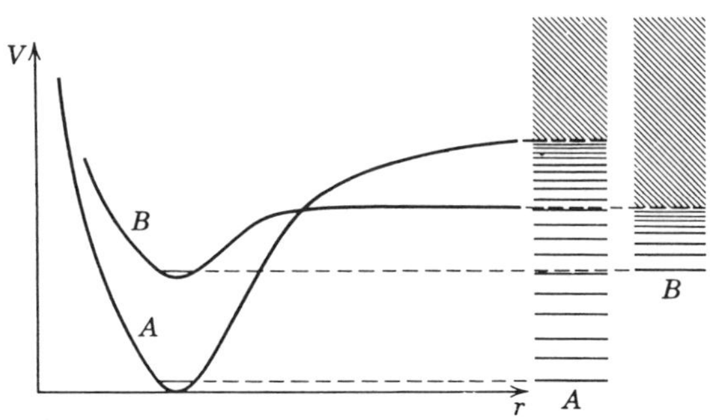

Figure 5-5-2. Potential energy curves and corresponding vibration term sequences leading to predissociation of a diatomic molecule.

The dissociation of molecules into neutral fragments is an important energy loss mechanism for electrons passing through gases. Also, this kind of dissociation can provide a strong source of photons, as well as fast and electronically excited particles. Such particles have high reactivity in general and may undergo reactions not possible otherwise. Zipf[40] has discussed the role played by electron-impact dissociation in planetary atmospheres, nuclear explosions, electrical discharges, gas lasers, and other areas. His review, and the one by Compton and Bardsley,[41] should be consulted by the reader who desires a deeper treatment of the subject than is provided here.

B. Dissociation into Ground-State, Neutral Fragments

The measurement of the cross sections for dissociation into neutral, electronic ground-state fragments is especially challenging, because such fragments are difficult to detect. It is also difficult to subtract out of the signal the contribution from neutral fragments left in electronically excited states. Still another matter is that of probing the final rotational and vibrational distributions of dissociation products that are molecular rather than atomic. Laser-induced fluorescence may provide the means of obtaining such data.

The most direct method of measuring the cross section for dissociation into ground-state neutrals is to remove the products selectively and measure the pressure change in a closed gas system. A known amount of gas is introduced into an evacuated tube that contains an electron source, a pressure gauge, and a freshly evaporated "getter" that will effectively trap the dissociation products but not the target gas molecules. If the source produces n_e electrons per second that travel a distance L through the gas, which occupies a volume V at a pressure P, the cross section for dissociation is given by the equation

$$\sigma = \frac{1}{n_e L} \frac{V}{P} \frac{dP}{dt}. \tag{5-5-1}$$

The getter is assumed to absorb gas emitted from the walls and all types of dissociation fragments, including ions. Any fragments not trapped by the getter are assumed to be absorbed on the walls. The spurious contribution to the pressure decrease caused by molecular decomposition at the hot filament in the electron source can be assessed by turning the electron beam off. Other problems, not so easily dealt with, which severely limit the number of gases to which this method can be applied include[41] (1) the absorption of the target gas on the walls of the tube (especially if the gas is polar), (2) the effect of the electron beam on the absorption and desorption of molecules on the walls, and (3) recombination of fragments in the gas phase to form molecules of the target gas. This method dates back to the 1920s. More recently it has been used by Corrigan[42] on H_2 and by Winters[43] on N_2, CH_4, and C_2H_6. Chung and Lin's theoretical cross section[44] for the process

$$e + H_2 \rightarrow H(1s) + H(1s) + e$$

is shown in Fig. 5-5-3, along with Corrigan's cross section as adjusted to enable his results to be compared with theory. The cross section for dissociation of H_2 is dominated by direct excitation to the antibonding $b^3\Sigma_u^+$ state (see Fig. 5-5-1*a*), although other dissociation mechanisms, such as the excitation of the $a^3\Sigma_g^+$ state and its decay to the $b^3\Sigma_u^+$ state, contribute.[41]

Another method of dealing with ground-state neutral fragments is to use a second electron beam to ionize them and a mass spectrometer to detect the ionized fragments. This "dual-chamber" method was introduced by Beck and Osberghaus[45] and has been used by various groups.[41] It has produced much useful information about the kinds of neutrals that are formed, but no accurate cross sections because of problems in assessing collection and detection efficiencies and in dealing with anisotropic distributions of fragments.

C. Dissociation into Electronically Excited, Neutral Fragments

Zipf[40] has pointed out that the recent development of new kinds of photomultiplier tubes, low-scattered-light holographic gratings, and computer-supported data acquisition techniques (especially the latter) has resulted in great advances in the study of dissociative excitation of gases by electron impact. It is now possible to measure cross sections as small as 10^{-24} cm^2 at a spectral

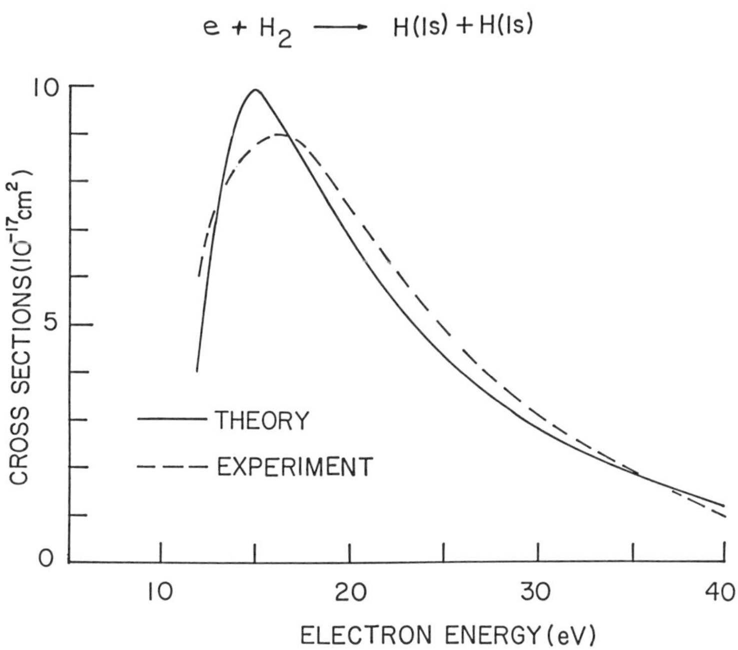

Figure 5-5-3. Chung and Lin's theoretical cross section for the dissociative reaction $e + H_2 \rightarrow H(1s) + H(1s) + e$, along with the experimental values of Corrigan[42] as adjusted by Chung and Lin[44] to subtract out the cross section for production of excited states.

resolution of less than 0.05 Å, and extensive data have been obtained for many of the diatomic and polyatomic molecules of greatest planetary and astrophysical interest over the wavelength range 500 to 13,000 Å.[40]

A simple case that we should consider first is that in which the projectile excites a molecule to a discrete state from which either predissociation, autoionization, or radiative relaxation occurs. The distribution of the total cross section into the partial cross sections associated with the various output channels is determined by the physical properties of the excited state into which the target molecule is driven by the projectile electron. Under the simple circumstances envisioned here, it may be possible to obtain quantitative data on the radiative lifetime of this excited state, the branching ratio for its decay into the various open channels, and potential surfaces relevant to the processes that are involved. Measurements and interpretation of data are greatly complicated if the medium is *optically thick*, which will be the case if the gas occupies a large volume and/or its number density is high. Then, photons emitted by relaxing molecules may be trapped in the gas rather than immediately escaping from it, collisional deexcitation becomes important, and the chemistry of the gas is drastically altered.

Many studies of dissociation into neutral, electronically excited fragments have involved detecting photons emitted by the decaying fragments and deducing cross sections for photon emission as a function of wavelength. In some cases, it is possible to get cross sections for dissociative excitation to the particular levels of the target molecule that are involved. Optical techniques of the type to be discussed in Section 5-7-B are available for fragments with low principal quantum number, n. Balmer emission from H atoms produced in the dissociation of H_2 has been studied for $n = 3$ to 11 decaying to $n = 2$.[41] "High-Rydberg atoms" ($n \gtrsim 15$) have long radiative lifetimes and usually are detected by the following methods:[41]

1. Collisional ionization: $A^{**} + M \nearrow A^+ + M + e$; $A^{**} + M \searrow A^+ + M^-$
2. Penning ionization: $A^{**} + M \rightarrow A + M^+ + e$
3. Resonance ionization near a metal surface
4. Ionization in a strong electric field (see Note 6-2)

In method 1, the CM kinetic energy of the collision pair (A^{**}, M) supplies the internal energy for ionization, and the method is feasible because so little energy is required to ionize a high-Rydberg atom (only 34 meV for the H atom in the $n = 20$ state). Method 2 will work if the excitation energy of A^{**} exceeds the ionization energy of M. Method 3 depends on the specific structure of the atoms and of the surface. The final method, (4), not only provides absolute cross sections but also an approximate knowledge of the n and l distribution of the high-Rydberg states. Methods 1 to 3 detect certain metastables as well as high-Rydberg atoms, whereas method 4 is more selective and detects only the latter species. A quantitative comparison of lifetimes may be useful: Metastable states have lifetimes of about 10^{-4} sec or longer. For high-Rydberg atoms, the radiative lifetime for a state with fixed l is proportional to n^3; for $n = 20$, $\tau \approx 10^{-5}$ to 10^{-4} sec.

Even when it is possible to study dissociation by measuring the radiation emitted by excited fragments, various factors make it difficult to interpret the data in terms of single, known processes and to obtain accurate cross sections. For example, H atoms with a given value of n can exist in n sublevels with different values of l and slightly different energies. (This "fine structure" is caused by relativistic effects.[46]) The sublevels can be mixed by the Stark effect produced by external electric fields, or by the motional electric field felt by the atom if it moves through an external magnetic field. A new set of levels will be produced, with slightly altered energies, and the lifetime of the excited H atoms can be drastically changed. Even weak electric fields (a few V/cm) can produce this effect, especially for atoms in high-n states and fragments produced with appreciable kinetic energy in the dissociation process.[41] An additional extremely serious problem arises from the difficulty in extracting the true cross section for the formation of excited states in a given state from the measured "emission cross section" (i.e., the effective cross section for emission of photons of a given wavelength). These and related problems are discussed in Section 5-6.

Another useful technique involves time-of-flight (TOF) measurements to obtain the kinetic energy distributions of excited dissociation fragments. Both metastable and high-Rydberg states have sufficiently long lifetimes to make this kind of study possible, by standard techniques similar to those discussed in Section 4-1. (The TOF techniques used here are restricted to states with lifetimes greater than about 10^{-6} sec. The internal energies of these states are usually higher than about 5 eV.) TOF studies have provided much useful information on radiative lifetimes and repulsive potential energy curves.[40]* Zipf[40] and Compton and Bardsley[41] have provided admirable surveys of the experimental and theoretical work done on dissociation of molecules into neutral fragments.

5-6. THE SIMILARITY OF A FAST ELECTRON TO A PHOTON; VIRTUAL QUANTA†

At this point it is desirable to interrupt our discussion of the various modes of excitation of neutrals by electron impact and make an excursion into classical electromagnetic theory. Our main objective will be to show that a fast charged particle moving in a straight line with constant velocity **v** produces an

*For other applications of TOF techniques to excitation studies of atoms and molecules, see N. J. Mason and W. R. Newell, *J. Phys. B* **19**, L-203, L-587 (1986); **20**, 1357, 3913 (1987).

†This section depends heavily on Chapters 11, 13, and 15 of the book by Jackson.[6]

electromagnetic field at a point off its trajectory that resembles the field of a pulse of plane-polarized radiation propagating in the same direction, and that in the limit $\beta \equiv v/c \to 1$, the fields are indistinguishable. This result will prove useful when we resume our discussion of excitation in Section 5-7.

Suppose that we have a set of rectangular Cartesian axes X, Y, Z embedded in an inertial frame of reference K, and an observer located at point P on the Y axis with coordinates $(0, b, 0)$, as shown in Fig. 5-6-1. A charge q is moving with constant speed v along the X axis. We introduce a second Cartesian coordinate system with axes X', Y', Z' parallel to X, Y, Z, respectively. The $X'Y'Z'$ axes are embedded in a second frame of reference K' that is inertial because K is. The origins, O and O', of the two coordinate systems are made to coincide at $t = t' = 0$, where t and t' are the times measured by an observer fixed in XYZ and $X'Y'Z'$, respectively. The four coordinates of an event (e.g., the explosion of a firecracker at time t) transform in this simple case according to the special Lorentz transformation (SLT):[47]

$$x' = \gamma(x - \beta ct), \tag{5-6-1a}$$

$$y' = y, \tag{5-6-1b}$$

$$z' = z, \tag{5-6-1c}$$

$$ct' = \gamma(ct - \beta x), \tag{5-6-1d}$$

where $\gamma \equiv (1 - v^2/c^2)^{-1/2}$. The four coordinates of a given event as measured by the observer fixed in K are x, y, z, t. The observer fixed in K' associates the numbers x', y', z', t' with the same event. The space and time components of any four-vector $\tilde{A}$ transform in the same way as do the components of the event four-vector: $\tilde{x} = (x, y, z, ct)$.

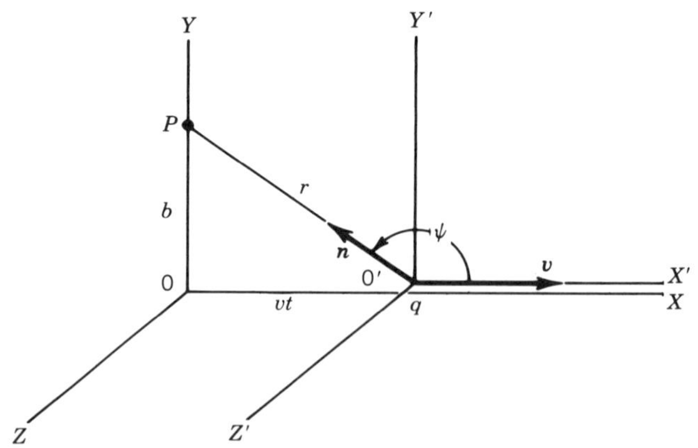

Figure 5-6-1. The coordinate systems used to calculate the electric and magnetic fields at the point P produced by a charge q moving with constant speed along the X-axis. The $X'Y'Z'$ axes are fixed in the rest frame of q.

The electromagnetic field is described by a second-rank tensor $F^{\alpha\beta}$ whose elements transform according to

$$F'^{\alpha\beta} = \frac{\partial x'^{\alpha}}{\partial x^{\gamma}} \frac{\partial x'^{\beta}}{\partial x^{\delta}} F^{\gamma\delta}. \tag{5-6-2}$$

For the SLT considered here, the electric field **E** and magnetic field **B** therefore transform as

$$\begin{aligned} E'_x &= E_x, & B'_x &= B_x, \\ E'_y &= \gamma(E_y - \beta B_z), & B'_y &= \gamma(B_y + \beta E_z), \\ E'_z &= \gamma(E_z + \beta B_y), & B'_z &= \gamma(B_z - \beta E_y). \end{aligned} \tag{5-6-3}$$

In the rest frame of the charge, K', the fields $\mathbf{E}'$ and $\mathbf{B}'$ at the observation point have components

$$\begin{aligned} E'_x &= -\frac{qvt'}{r'^3}, & E'_y &= \frac{qb}{r'^3}, & E'_z &= 0, \\ B'_x &= 0, & B'_y &= 0, & B'_z &= 0. \end{aligned} \tag{5-6-4}$$

In the K' frame, the point P is separated from O' by the distance $r' = [b^2 + (vt')^2]^{1/2}$, but in the K-frame coordinates, this separation distance is $[b^2 + (\gamma vt)^2]^{1/2}$. Here the only coordinate that needed transformation was $t' = \gamma t$, since $x = 0$ for the point P in the K frame. We see that in terms of the unprimed coordinates of K, the nonvanishing field components are

$$E'_x = -\frac{q\gamma vt}{(b^2 + \gamma^2 v^2 t^2)^{3/2}}, \qquad E'_y = \frac{qb}{(b^2 + \gamma^2 v^2 t^2)^{3/2}}. \tag{5-6-5}$$

If we use the inverse transformation to (5-6-3), obtained by interchanging primed and unprimed quantities and letting $\beta \to -\beta$, we get

$$\begin{aligned} E_x &= E'_x = -\frac{q\gamma vt}{(b^2 + \gamma^2 v^2 t^2)^{3/2}}, & B_x &= B'_x = 0, \\ E_y &= \gamma E'_y = \frac{\gamma qb}{(b^2 + \gamma^2 v^2 t^2)^{3/2}}, & B_y &= B'_y = 0, \\ E_z &= E'_z = 0, & B_z &= \gamma\beta E'_y = \beta E_y. \end{aligned} \tag{5-6-6}$$

Here we have the electric and magnetic fields at the observation point P fixed in K, in terms of the unprimed quantities of the frame K. We see that there is a magnetic induction $\mathbf{B} = (0, 0, B_z)$ that approaches E_y in magnitude as $\beta \to 1$. The E_y and E_x components are plotted in Fig. 5-6-2 for very small v ($\beta \approx 0, \gamma \approx 1$)

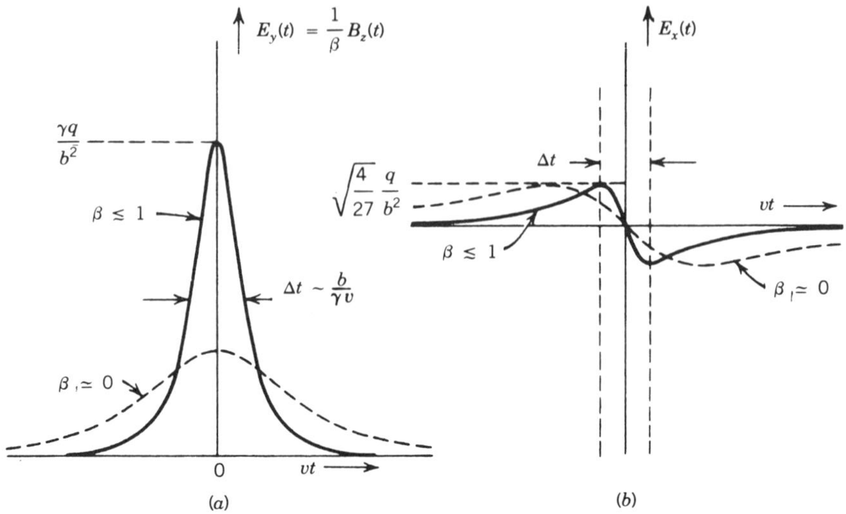

Figure 5-6-2. The transverse and longitudinal **E** fields (E_y and E_x) and the transverse B field (B_z) produced at the point P by the charge q moving in the X-direction. Both the slow charge ($\beta \approx 0$) and fast charge ($\beta \lesssim 1$) situations are shown. (Adapted from Jackson.[6])

and for high v ($\beta \lesssim 1, \gamma \gg 1$). For $\gamma \gg 1$, the peak transverse electric field, $E_y(t = 0)$, becomes equal to γ times its nonrelativistic value, while the "duration" of the field at P ($\Delta t \approx b/\gamma v$) decreases by the same factor, γ. Hence the integral of E_y times v is independent of velocity. For $\beta \to 1$ (highly relativistic motion), the observer at P sees **E** and **B** fields that are nearly transverse, mutually perpendicular, and equal in magnitude. These fields are what would be produced by a pulse of plane-polarized electromagnetic radiation traveling in the $+X$ direction. The longitudinal field E_x shown in Fig. 5-6-2 is not contained in such an electromagnetic pulse. However, it varies rapidly and symmetrically from positive to negative, and its time integral is zero. Hence an observing apparatus with any sort of realistic time constant will not detect the E_x field, so its presence does not invalidate our comparison of the field of a fast-moving charge to an electromagnetic pulse.

From (5-6-6) we see that $E_x/E_y = -vt/b$, so the electric field at P is directed along the unit vector **n**, which points radially from the present position of the charge to P. The influence of q at P comes from the "retarded position" of q, determined by the trajectory of the charge and the speed of light. However, here the motion of q is specified, and the retarded position is given uniquely and simply in terms of the *present* position. In the general case, the relation of the retarded position of the charge to the present position is not known, and one must use the Liénard-Wiechert potentials, which give the fields in terms of the retarded potentials and velocities.[6]

It is instructive to express $\mathbf{E}_P$ in the form

$$\mathbf{E}_P = \frac{q\mathbf{r}}{r^3\gamma^2(1 - \beta^2 \sin^2 \psi)^{3/2}}. \tag{5-6-7}$$

An observer at P sees an $\mathbf{E}$ vector at any instant of time that points from the instantaneous present position of q. For given r, E_P depends on ψ. If the charge is stationary, its electric field can be represented by Fig. 5-6-3*a*. (A different drawing, with equally spaced radial lines of force emanating from q, would more dramatically show the isotropic nature of the field.) Now, if this figure is put in motion with the constant speed v, it would be compressed longitudinally by the Lorentz contraction, and the black particles composing the lines on the paper

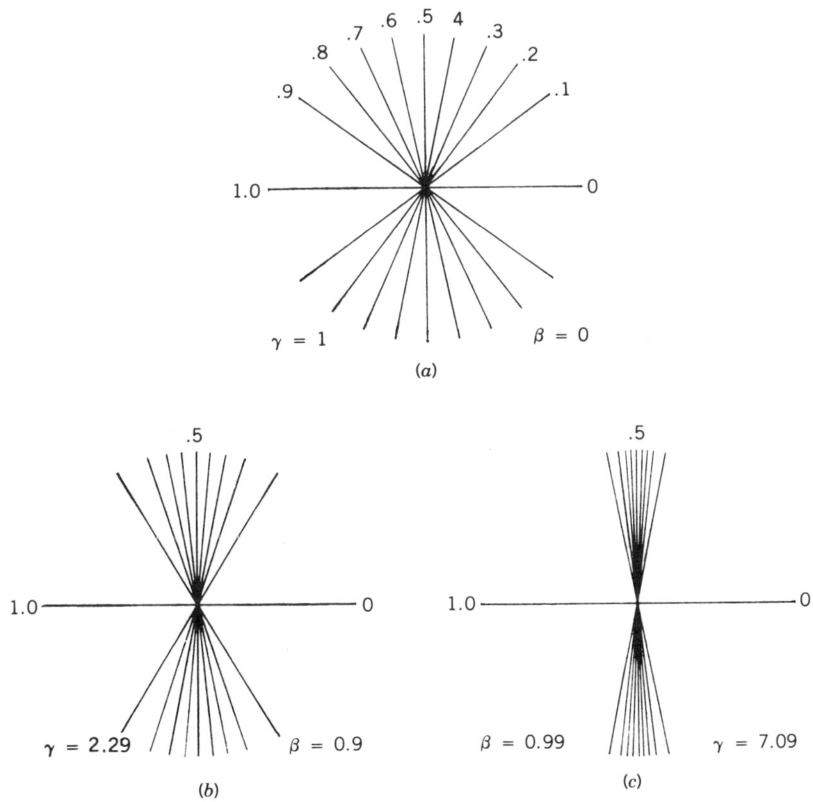

Figure 5-6-3. The concentration of the electric field of a moving charge into the equatorial plane, as described by (5-6-7). At all speeds of the charge, the field has an inverse-square dependence, but it is isotropic only at very low speeds ($\beta \to 0$). The decimals define the proportion of the total electric flux lying within a circular cone whose axis is the direction of motion of the charge. [Adapted from R. B. Leighton, *Principles of Modern Physics*, McGraw-Hill, New York (1959).]

would appear at different places. The picture witnessed by the observer P in K still represents the **E** field lines produced by q. Contraction compresses them at the sides and expands them ahead and behind in such a manner as to give the correct density for the lines of force as in (*b*) and (*c*) of Fig. 5-6-3.[48]

We can now find the Fourier transform of the **E** field described in (5-6-6) by using

$$\mathbf{E}(\omega) = \frac{1}{\sqrt{2\pi}} \int_{-\infty}^{\infty} \mathbf{E}(t) e^{i\omega t}\, dt. \tag{5-6-8}$$

Given the components $E_y(t)$ and $E_x(t)$, we can obtain from a table of Fourier transforms the results

$$E_y(\omega) = \frac{q}{bv}\left(\frac{2}{\pi}\right)^{1/2}\left(\frac{\omega b}{\gamma v} K_1\left[\frac{\omega b}{\gamma v}\right]\right), \tag{5-6-9}$$

$$E_x(\omega) = -i\frac{q}{\gamma v b}\left(\frac{2}{\pi}\right)^{1/2}\left(\frac{\omega b}{\gamma v} K_0\left[\frac{\omega b}{\gamma v}\right]\right). \tag{5-6-10}$$

Here K_1 and K_0 are modified Bessel functions of order unity and zero, respectively (see pages 107–108 of Jackson's book[6]). The electromagnetic field at P produced by the passing charge is a pulse in the time domain that corresponds to a certain distribution of plane waves traveling in the X direction, whose frequencies range from $-\infty$ to $+\infty$ and whose amplitudes are given by (5-6-8). This statement can be justified by examining the inverse Fourier transform

$$\mathbf{E}(t) = \frac{1}{\sqrt{2\pi}} \int_{-\infty}^{\infty} \mathbf{E}(\omega) e^{-i\omega t}\, d\omega. \tag{5-6-11}$$

A similar analysis is often used in discussions of quantum mechanical wave packets.[49]

The *frequency spectrum* of the equivalent pulse of plane-polarized radiation incident on the observer P in the X direction can easily be calculated for a relativistic charge. It is defined as the energy per unit area per unit frequency interval and is given by

$$\frac{dI_x}{d\omega}(\omega, b) = \frac{c}{4\pi}\left[|E_y(\omega)|^2 + |B_z(\omega)|^2\right] = \frac{c}{2\pi}|E_y(\omega)|^2 \qquad (\beta \approx 1), \tag{5-6-12}$$

where $E_y(\omega)$ is given by (5-6-9). The electric field also has an X component, $E_x(t)$, at P, but the corresponding magnetic field component, $B_z(t) \approx E_x(t)$, that would be required to produce a pulse of plane polarized radiation in the Y direction is missing. However, if we artificially supply the missing B_z component, we can calculate the frequency spectrum of an equivalent pulse incident on P in the Y

direction:

$$\frac{dI_y}{d\omega}(\omega, b) = \frac{c}{2\pi}|E_x(\omega)|^2. \tag{5-6-13}$$

The two frequency spectra are shown in Fig. 5-6-4. Because of its small amplitude, the pulse traveling in the Y direction is unimportant.

To obtain the energy per unit frequency interval in the equivalent radiation field generated by the moving charge, we integrate the sum of the frequency spectra over the appropriate range of impact parameters:

$$\frac{dI}{d\omega}(\omega) = 2\pi \int_{b_{\min}}^{\infty} \left[\frac{dI_x}{d\omega}(\omega, b) + \frac{dI_y}{d\omega}(\omega, b)\right] b\, db. \tag{5-6-14}$$

This integral diverges unless a cutoff is imposed at small impact parameters. The value of $b_{\min}$ to be used in a given application is chosen such that for $b > b_{\min}$, the **E** and **B** fields of the moving charge appear to generate at P plane-polarized pulses of electromagnetic radiation. Furthermore, for $b < b_{\min}$, the effects of the fields at P must be negligible or able to be accounted for in a different fashion. For bremsstrahlung (Section 7-9), $b_{\min}$ is determined through the uncertainty relation $b \approx \hbar/\Delta p$, where Δp is the momentum transfer. The general result for the energy per unit frequency interval is

$$\frac{dI}{d\omega}(\omega) = \frac{2}{\pi}\frac{q^2}{c}\left(\frac{c}{v}\right)^2 \left\{\chi K_0(\chi)K_1(\chi) - \frac{v^2}{2c^2}\chi^2[K_1^2(\chi) - K_0^2(\chi)]\right\} \tag{5-6-15}$$

(where $\chi = \omega b_{\min}/\gamma v$) and is illustrated in Fig. 5-6-5. It is important to note that the low frequencies make the main contribution to the energy spectrum.

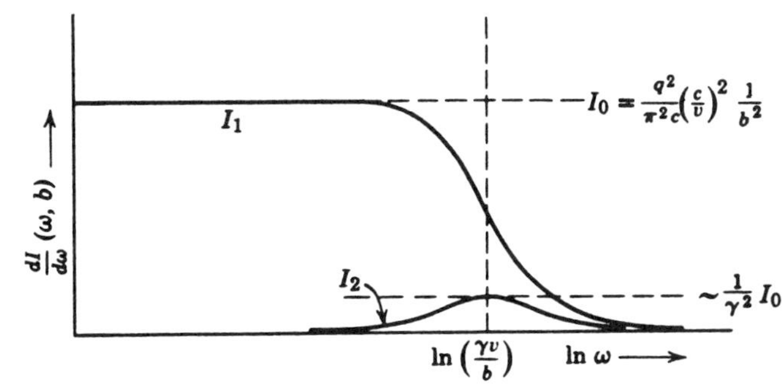

Figure 5-6-4. The frequency spectra of the two equivalent pulses of radiation. The upper curve, for the pulse traveling in the X direction, has the constant value I_0 at low ω. The lower curve, for the pulse in the Y direction, has a much smaller amplitude for $\beta \approx 1$. (Taken from Jackson.[6])

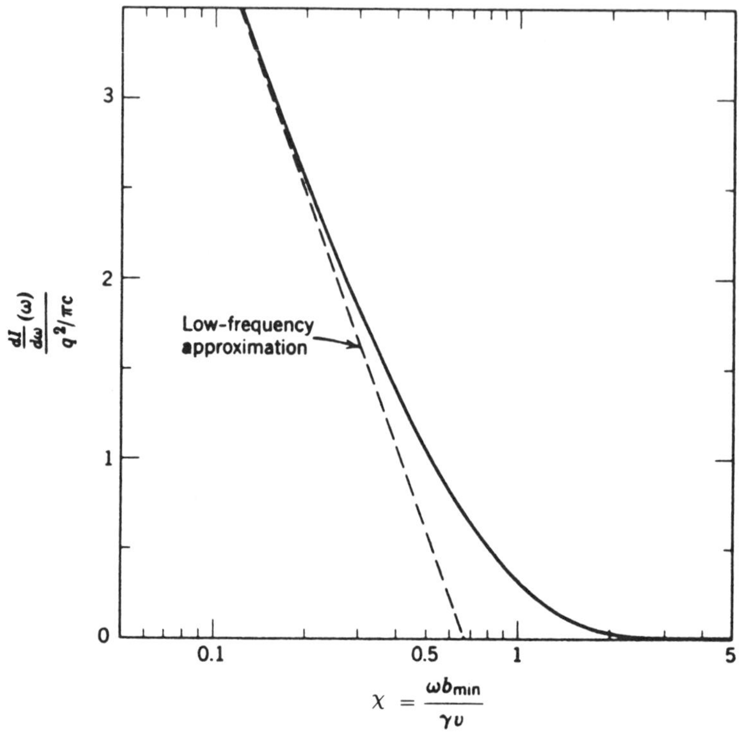

Figure 5-6-5. The frequency spectrum of virtual quanta for a relativistic charge. The energy per unit frequency $dI(\omega)/d\omega$ in units of $q^2/\pi c$ is plotted against the frequency in units of $\gamma v/b_{\min}$ (i.e., against $\chi = \omega b_{\min}/\gamma v$). The solid curve is a plot of (5-6-15). The dashed curve represents the low-frequency approximation obtained by putting $\chi \ll 1$ in (5-6-15). (Adapted from Jackson.[6])

We can now invoke the photon concept to describe the radiation pulse and calculate the number spectrum of *virtual quanta* $N(\hbar\omega)$ from the relation

$$\frac{dI}{d\omega}(\omega)\,d\omega = \hbar\omega N(\hbar\omega)\,d(\hbar\omega). \tag{5-6-16}$$

In the low-frequency limit ($\chi \ll 1$) of (5-6-15), the number of virtual quanta per unit energy interval is

$$N(\hbar\omega) \simeq \frac{2}{\pi}\frac{q^2}{\hbar c}\left(\frac{c}{v}\right)^2\frac{1}{\hbar\omega}\left(\ln\frac{1.123\gamma v}{\omega b_{\min}} - \frac{v^2}{2c^2}\right). \tag{5-6-17}$$

These considerations led C. F. von Weizsäcker and E. J. Williams in 1934 to develop independently the *method of virtual quanta*, which has been applied to such processes as the production of bremsstrahlung, collisional ionization of atoms in distant collisions, electron disintegration of nuclei, and production of pions in electron-nuclear collisions. (The basic idea in the method was used by E. Fermi in 1924 to relate the energy loss by ionization to the absorption of X-rays by atoms.) In a two-body collision, the lighter of the collision partners is designated the "struck particle," and the heavier the "incident particle." This choice is made because the lighter particle has the greater radiation scattering power. The spectrum of virtual quanta emitted by the "incident particle" is calculated, and the effects of the quanta on the "struck particle" are determined. These effects may involve either scattering or absorption. In effect, the method of virtual quanta allows the replacement of a charged particle interaction by a photon interaction. For example, in collisional ionization, the corresponding radiative process is the photoejection of atomic electrons by virtual quanta. Underlying the method is the assumption that the effects of the different frequency components of the effective radiation pulse add incoherently. The method of virtual quanta will appear again in Section 7-9 when we discuss bremsstrahlung. The classic reference to the method is the article by Williams.[50] The excellent discussion by Jackson[6] is also highly recommended.

5-7. ELECTRONIC EXCITATION AND DEEXCITATION OF NEUTRALS BY ELECTRON IMPACT

Electron collisions with neutral atoms and molecules may produce excitation by altering the electronic structure of the target. Ionization may also occur if the electron impact energy exceeds the threshold for ejection of a bound electron into the continuum. At intermediate and high energies, the greatest contribution to the total electron collision cross section comes from these processes. In "direct" events, the charge-charge Coulomb interaction is directly responsible for the transition—no special requirements such as those for rotational and vibrational excitation stated in Sections 5-3 and 5-4 are imposed. Indeed, recall from Section 3-20-D that use of the Coulomb interaction energy in the transition matrix element suffices to produce electronic excitation (and ionization) when the transitions are not forbidden by selection rules. We shall say more about selection rules in Section 5-7-G. Resonance mechanisms do not appear to be important in electronic excitation.[3] Core-excited shape resonances do significantly increase cross sections but over an energy range narrow compared to that over which direct excitation is important.

Accurate absolute cross sections for electronic excitation are notoriously difficult to measure, but there is a compensatory advantage associated with such measurements—the detection signals are usually unique to the state of the system to which the excitation occurs. Not only is more detailed information obtainable; it may also be used for "diagnostic" purposes in plasma physics and many other fields. The uniqueness of the signal resulting from excitation to a state from which optical radiation to a lower state is permitted is particularly valuable; observation of the radiation emitted by the solar corona, for example, allows estimates to be made of the chemical composition and atomic concentrations of the corona if the relevant excitation cross sections and transition probabilities are known.

The role of electronic excitation in natural and laboratory phenomena is discussed in Refs. 1, 3, and 40. Cross section data are presented by Trajmar and Cartwright[3] and by Zipf;[40] the period that they cover extends from about 1965 to the end of 1982. Heddle and Gallagher[51] have surveyed the measurements on atoms and positive atomic ions up to about the end of 1986. We have already presented some data on H_2 and N_2 in Figs. 5-1-1 and 5-2-2.

A. Energy-Loss Spectra

As with rotational and vibrational excitation, an important source of information on electronic excitation is the energy-loss spectrum, which in modern studies[3,8,16] is most often obtained with high-energy-resolution electron spectrometers such as described in Section 4-3-B. A beam of nearly monoenergetic electrons is passed through the target, and inelastically scattered electrons are collected. The energy losses are characteristic of the excitation process, and the energy spectrum of the scattered electrons provides information on these processes. This method is applicable to metastable as well as ordinary excited states. An advantage offered here is the ease of signal detection, for only a current measurement is required. An example of an energy-loss spectrum obtained with a modern spectrometer is shown in Fig. 5-7-1.[52]

It is possible to obtain differential cross sections for electronic excitation from energy-loss measurements,[3,16] as is the case for rotational and vibrational excitation (Sections 5-3 and 5-4). In fact, Müller-Fiedler et al. at Kaiserslautern have used an electron spectrometer to obtain energy-loss spectra and differential cross sections for inelastic scattering of electrons from *metastable* He(2^3S) atoms.[53] They studied transitions to the individual states 2^3P, 3^3S, 3^3P, 3^3D and to the sum of the $n = 4$ triplet states of helium. In the crossed-beam arrangement that was used, a metastable helium beam, with density of about 6×10^7 atoms/cm^3, was produced in a differentially pumped cold cathode dc discharge source. The incident electron energy was varied over the range 15 to 30 eV, and measurements were made on electrons scattered at angles between 10 and 40°. The cross sections are very large at small angles, but decrease by almost a factor of 100 as the scattering angle increases to the maximum value studied.

Retarding-potential-difference and trapped-electron techniques played an important role in excitation studies before the use of the high-resolution electron spectrometer became widespread, but they never produced any absolute excitation cross-section data. These techniques, associated especially with R. E. Fox and G. J. Schulz, are discussed in the books by McDaniel[54] and Hasted.[55]

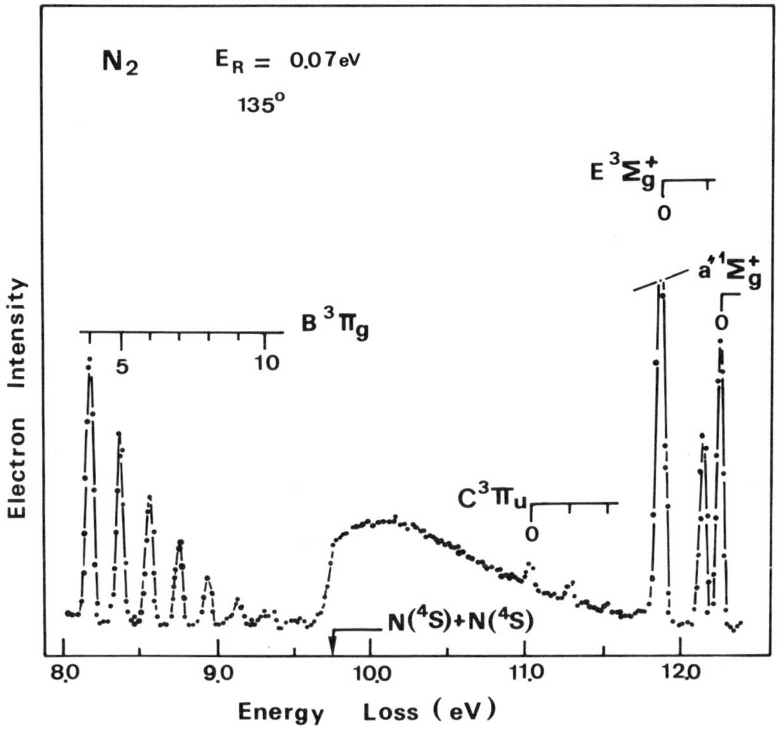

Figure 5-7-1. Energy-loss spectrum obtained by Mazeau et al.[52] for electrons of variable energy incident on N_2. The apparatus is set to receive electrons with an energy of 70 meV that are scattered at 135°. The incident electron energy equals the energy loss plus 70 meV.

B. Excitation Cross Section Measurements Involving Detection of Radiation

This method, which has its roots in the pioneering research of J. Franck and G. Hertz in 1913–1914, is very powerful, but the interpretation of the raw data it provides and accurate evaluation of excitation cross sections are difficult. In such measurements, a beam of electrons of known and variable energy is passed through a gas, and atoms or molecules are raised to electronic states of higher energy. Measurements are made of the wavelengths and intensity of the optical radiation resulting from spontaneous radiative transitions in deexcitation to lower levels, and "emission functions" or "emission cross sections" are calculated as functions of the electron impact energy. If it proves possible to account properly for various complicating effects that we shall discuss, true "excitation cross sections" may be obtained. This method has low sensitivity because the detector intercepts only a small fraction of the photons produced and because not all of the photons arriving at the detector are registered.

Such determinations are beset with difficulties in absolute radiometry, gas density measurement, electron beam handling and current measurement, radiation trapping, excitation transfer, cascading, and radiation polarization.[55–60] Published experimental data on a given excitation cross section generally have about the same dependence on the electron energy but seldom agree to within a factor of 3 in absolute magnitude. Van Zyl et al.[61] have performed a praiseworthy set of experiments designed to provide the best data possible on the electron-impact excitation of n^1S levels of He and thus provide "benchmarks" for future measurements. Their data are sufficiently accurate, and the uncertainties well examined, that other workers should be able to use the results in calibrating their own instruments and assessing the reliability of their techniques. The apparatus used by Van Zyl et al.[61] is shown in Fig. 5-7-2. An electron gun produces a beam of 2 to 200 μA that passes through a collision chamber containing helium, typically at about 8×10^{-4} torr pressure. (In this chamber the profile of the beam can be measured with a slit system that can be scanned across the beam.) The emerging beam is collected in a deep Faraday cup. Light emitted in spontaneous decay of the excited helium atoms along a short segment of the electron beam path (about 0.32 mm long) at the center of the chamber is collected by an achromat lens at the entrance to a calibrated spectroradiometer, in which it is mechanically chopped, dispersed by grating optics and an interference filter, and allowed to impinge on a cooled photo-

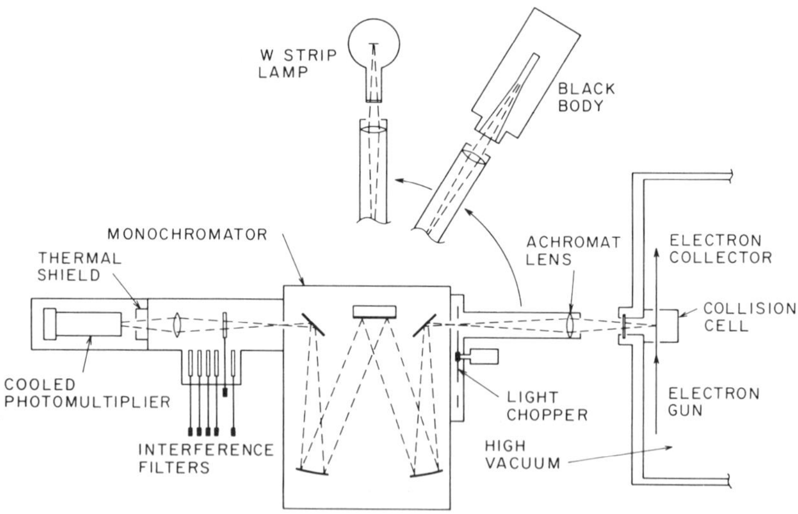

Figure 5-7-2. The apparatus used by Van Zyl et al.[61] to obtain benchmark cross sections for electron-impact excitation of He from measured $n^1S \rightarrow 2^1P$ emission cross sections ($n = 3$ to 6). The detected radiation had wavelengths ranging from 728.1 to 416.9 nm, corresponding to photon energies of 1.703 to 2.974 eV. The monochromator shown is of the Czerny-Turner grating type.

multiplier. The multiplier pulses are counted by two scalars, which are gated alternately in synchronism with the chopper wheel to separate the signal from the dark count. The spectroradiometer can be rotated to view a copper-melting-point blackbody or the filament of a tungsten strip lamp, both used as calibration standards.

Van Zyl et al.[61] measured absolute total emission cross sections for the excitation of ground-state He giving radiation from the $6^1S \rightarrow 2^1P$, $5^1S \rightarrow 2^1P$, $4^1S \rightarrow 2^1P$, and $3^1S \rightarrow 2^1P$ transitions. The electron impact energies were 50, 100, 500, 1000, and 2000 eV. The greatest accuracy was achieved at 500 eV, where the mean uncertainty was only $\pm 3.5\%$ "high confidence level" (HCL). The emission cross sections were adjusted to allow for branching and for cascading from higher levels and thus provide the level-excitation cross sections displayed in Table 5-7-1. The 2000-eV data agree well with the predictions of the Born approximation, in fact to within 2.5%.

Van Zyl et al. chose the indicated transitions for study because some of the subtle experimental problems associated with the radiation detection method should have a relatively small effect here, and hence attention could be focused on other problems associated with the basic parameters in these experiments.* Special mention should be made of the dynamic expansion technique that was used for target density measurements in the pressure range 10^{-3} to 10^{-4} torr, with an uncertainty of $\pm 1.2\%$ "high confidence level." The reader should consult Ref. 61 for details of the helium density measurements, the electron-beam handling techniques, and above all, the spectroradiometer calibration. Radiation cascading and the polarization of impact radiation are discussed in Sections 5-7-C and 5-7-D.

E. C. Zipf and R. W. McLaughlin[40] have described an apparatus developed for absolute emission cross-section measurements over the range 500 Å to 1.3 μm (13,000 Å). The apparatus employs both a 1-m, normal-incidence vacuum monochromator (for the region 500 to 1800 Å) and a $\frac{3}{4}$-m Czerny-

Table 5-7-1 Level-Excitation Cross Sections for Electrons on Ground-State Helium Atoms, Obtained by Van Zyl et al.[61]

Electron Energy (eV)	$\sigma(10^{-20}\ \text{cm}^2)$			
	3^1S	4^1S	5^1S	6^1S
50	33.7	13.3	6.42	3.40
100	23.3	9.06	4.74	2.23
500	8.45	3.26	1.58	0.849
1000	4.63	1.76	0.862	0.479
2000	2.45	0.906	0.447	0.241

*For example, the upper states are *S* states, so the radiation is isotropic.

Turner monochromator (for the range 1800 Å to 1.2 μm). Zipf and McLaughlin stress the importance of keeping stray electric fields and pump oil contamination to an extremely low level. By using efficient photon counters and sophisticated computer-assisted data acquisition techniques, they are able to measure excitation cross sections as small as 10^{-24} cm^2.

Some interesting and important work has also recently been done on electron impact leading to the emission of vacuum ultraviolet radiation (VUV), which covers the spectral range 30 to 150 nm, or 41.3 to 8.27 eV photon energy. The apparatus and techniques that have been developed and the data that have been obtained have considerable significance in plasma diagnostics, astrophysics, and various kinds of laboratory studies.

A group at the North Carolina State University has constructed a cross-section apparatus whose spectrometer-detector was calibrated with the continuum radiation from the storage ring of the electron synchrotron at the National Bureau of Standards in Gaithersburg, Maryland.[62] The electrons orbiting at some well-defined energy (here 282 MeV) provide copious radiation in the VUV region, and the radiation flux can be accurately calculated as a function of wavelength and direction.[6] (In the present case, the peak in the flux occurs at 20 nm.) The North Carolina apparatus contains an electron gun that can supply electrons at any energy in the range 30 to 500 eV. It has been used to measure, by the radiation detection method, cross sections for the emission of Lyman-α radiation following the dissociative excitation of H_2, and for the emission of the 92.0- and 93.2-nm lines from Ar II and the 53.7- and 58.4-nm He lines.[63] Previous accurate data for He,[64] which agree to within $\pm 2\%$ with the Born approximation at 350 eV, provide a good check on the North Carolina He cross sections. The overall accuracy of the He measurements is estimated to be about $\pm 5\%$. A major goal in this work has been to develop a primary, absolute light intensity standard for the VUV region suitable for calibration of VUV spectrometers. This standard apparatus utilizes the line radiation emitted as an electron beam of specified energy passes a known distance through a gas. If the emission cross section is accurately known, the flux of radiation can be calculated from the electron current and the target gas density, and the spectrometer can be calibrated from its response to the synchrotron flux. The apparatus employs a Minuteman 302-VM Seya-Namioka spectrometer with a Jobin-Yvon holographic grating and an EMI 9642/4B venetian blind photomultiplier.

Similar research has been conducted by McConkey's group at the University of Windsor in Ontario.[65] Of special significance is their accurate measurement of the cross section for excitation of Lyman-α radiation by electron impact on H_2 at 100 eV.[65]

The method described in this section cannot be applied to the measurement of cross sections for the excitation of neutrals to metastable levels because of the long lifetimes of these levels (typically above 10^{-4} sec). The metastables would move too far from the site of their production before they could radiate. However, such cross sections may be obtained by firing a beam of electrons through a gas and determining the rate of production of metastables by one of the following methods. Measurements can be made of the number of secondary electrons ejected when the metastables collide with a metal electrode within the gas volume. Alternatively, certain metastable states may be "quenched" by passage through an electric or magnetic field that produces a superposition of nearly degenerate states, some of which can radiate to lower states by the electric dipole process. The resulting radiation is detected by standard means. The quenching action is usually accomplished by electric fields of about 50 kV/m or magnetic fields of about 20 mT. The concentration of metastables may also be deduced from measurements of selective light absorption or by observations of anomalous dispersion. Both techniques offer low sensitivity. Experiments utilizing all of these techniques are discussed by McDaniel[54] and Hasted.[55]*

Special mention should be made here of J. F. Williams' measurement of absolute cross sections for excitation of the 2*S* and 2*P* states of atomic hydrogen.[66] Both the energy-loss and radiation detection methods were used. An electric field was employed to quench the 2*S* state. More recently, Warner et al.[67] have measured elastic and inelastic ($n = 2, 3$) differential cross sections of atomic hydrogen at higher energy resolution (typically, 20 meV).

We must now discuss several subjects that have an important bearing on the measurement of cross sections by the radiation detection method.

C. Cascading of Radiation

1. *Simple Model for Equilibrium with No Cascading*

Consider Fig. 5-7-3, adapted from the review by Heddle and Keesing.[57] Let N_e and N_g be the number densities of electrons and ground-state atoms, respectively, and v their relative velocity. Then, if the cross section for excitation of the atoms from state g to state j is q_j, the volume rate of production of atoms in

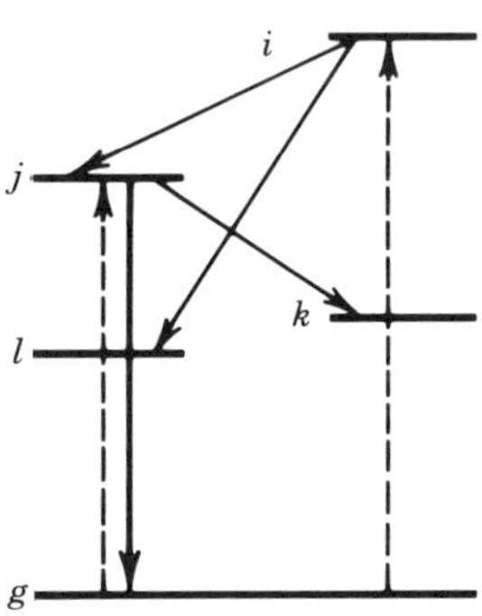

Figure 5-7-3. A schematic energy-level diagram used to illustrate the branching and cascading of radiation. The dashed lines indicate electron-impact-induced transitions; the solid lines radiative transitions. "Resonance radiation" is emitted in the deexcitation process $j \rightarrow g$, where g is the ground state of the structure. (Adapted from Heddle and Kessing.[57])

*For a recent review, see I. I. Fabrikant, O. B. Shpenik, A. V. Snegursky, and A. N. Zavilopulo, *Phys. Rep.* **159**, 1 (1988).

the state j is $N_g N_e v q_j$. If the electrons are in a beam with cross-sectional area S and current I, this production rate can be expressed as $N_g I q_j/eS$. The state j is radiatively depopulated at the volume rate $N_j A_j$, where $A_j = \sum_k A_{jk}$ and A_{jk} is the Einstein A coefficient (Chapter 8), the probability of a spontaneous transition from state j to the lower state k per atom per unit time. Usually, only the radiation from a single, given transition is observed. If that is the case here, we are interested in the rate of emission of photons in the single transition $j \to k$ per unit length of the electron beam: $\phi_{jk} = N_j A_{jk} S$. If no other processes are involved in the population or depopulation of state j, we obtain the equilibrium condition

$$N_g q_j \frac{I}{eS} = N_j A_j = \phi_{jk} \frac{A_j}{A_{jk} S}. \tag{5-7-1}$$

It then follows that

$$q_j = \frac{A_j}{A_{jk}} \frac{\phi_{jk}}{N_g I/e} \to q_j'. \tag{5-7-2}$$

We call this cross section the "apparent cross section" and designate it by the symbol q_j' in order to emphasize the dubious nature of the assumptions on which it is based.

2. *Effects of Cascading*

We now acknowledge that the electron beam may excite atoms to states i above j and that radiative transitions $i \to j$ may occur. These *cascade transitions* populate the state j at the rate $\sum_i \phi_{ij} = \sum_i N_i A_{ij} S$ per unit length of the electron beam. The ϕ_{ij} are required for determination of the cascade contribution. Usually, not all of the important ϕ_{ij} can be measured readily, but it may be possible to obtain them indirectly. For example, suppose that a state i lies close to j in energy, so that the transition connecting them has a wavelength too long for easy measurement. There is usually a transition such as $i \to l$ with shorter wavelength $\lambda_{il} \approx \lambda_{jk}$ that is easily measurable. Then ϕ_{il} can be determined, and used to provide ϕ_{ij} through the relation

$$\phi_{ij} = \phi_{il} \frac{A_{ij}}{A_{il}}. \tag{5-7-3}$$

With the effect of cascading considered, (5-7-1) should now contain the new volume production rate

$$N_g q_j \frac{I}{eS} + \sum_i \frac{\phi_{il}}{S} \frac{A_{ij}}{A_{il}}, \tag{5-7-4}$$

and its inclusion leads to the cross section

$$q_j = \left(\phi_{jk}\frac{A_j}{A_{jk}} - \sum_i \phi_{il}\frac{A_{ij}}{A_{il}}\right)\left(N_g\frac{I}{e}\right)^{-1}, \tag{5-7-5}$$

which we shall write as

$$q_j = q_j' - q_j^*. \tag{5-7-6}$$

Here $q_j^* = \sum_i q_{ij}^*$, and

$$q_{ij}^* = \frac{A_{ij}}{A_{il}}\frac{\phi_{il}}{N_g I/e}. \tag{5-7-7}$$

We define, by analogy with q_j',

$$q_i' = \frac{A_i}{A_{il}}\frac{\phi_{il}}{N_g I/e}, \tag{5-7-8}$$

and obtain

$$q_{ij}^* = q_i'\frac{A_{il}}{A_i}. \tag{5-7-9}$$

Hence, if we are given the branching ratio A_{ij}/A_i, measurement of the apparent cross section q_i' will provide the cascade correction to q_j'. It is unnecessary to measure the true cross section q_i to obtain the true cross section q_j, because the cascading transitions downward to state i are unimportant in this context. The analysis presented here is based on the discussion by Heddle and Keesing.[57]

D. Polarization of the Emitted Radiation

In an excitation study that involves the detection of radiation emitted from states excited by an electron beam, measurements are made of the intensity of radiation into a small solid angle $d\omega$ making an angle θ (usually 90°) with the beam. Comparison with standard light sources such as mentioned in Section 5-7-B will permit the evaluation of the differential emission cross section $q_{jk}(\theta)$ for exciting the radiative transition $j \rightarrow k$. To obtain the total emission cross section q_{jk}, we must take into account the variation of $q_{jk}(\theta)$ with θ. We assume the emitted radiation to be of the electric dipole type (Chapter 8), so the angular dependence of $q_{jk}(\theta)$ will have the form

$$q_{jk}(\theta) = a_{jk}(1 - P_{jk}\cos^2\theta), \tag{5-7-10}$$

where P_{jk} and a_{jk} are independent of θ but dependent on the electron energy.

Again following Heddle and Kessing,[57] we work in terms of ϕ_{jk}, the total rate of photon emission in the transition $j \to k$ per unit length of electron beam, and of $\phi_{jk}(\theta)$, the corresponding rate of emission into the solid angle $d\omega$. Then, from the measurements, we have to determine

$$\phi_{jk} = \int \phi_{jk}(\theta)\, d\omega. \tag{5-7-11}$$

It is useful to define $\phi_{jk}(\|)$ and $\phi_{jk}(\perp)$ as the photon fluxes emitted into unit solid angle perpendicular to the electron beam with electric vectors parallel and perpendicular to the electron beam, respectively. The photon flux into $d\omega$ at θ is then

$$\phi_{jk}(\theta) = \phi_{jk}(\|) \sin^2 \theta + \phi_{jk}(\perp)(1 + \cos^2 \theta), \tag{5-7-12}$$

and substitution into (5-7-11) gives

$$\phi_{jk} = \frac{8\pi}{3} [\phi_{jk}(\|) + 2\phi_{jk}(\perp)]. \tag{5-7-13}$$

The radiation emitted in any direction is polarized, and it is useful to define the polarization factor P_{jk} by the equation

$$P_{jk} = \frac{\phi_{jk}(\|) - \phi_{jk}(\perp)}{\phi_{jk}(\|) + \phi_{jk}(\perp)}. \tag{5-7-14}$$

[This factor is consistent with the factor P_{jk} introduced in (5-7-10).] Finally, we obtain

$$\phi_{jk}(\theta) = \frac{\phi_{jk}}{4\pi} \frac{1 - P_{jk} \cos^2 \theta}{1 - P_{jk}/3}, \tag{5-7-15}$$

which shows that measurements made at an angle θ_M such that $\cos^2 \theta_M = 1/3$ (i.e., $\theta_M = 54.7^\circ$) give information on the total photon flux independent of the state of polarization of the radiation. Note that (5-7-15) has the same form as (5-7-10).

We can define two new excitation cross sections, $q_j(\|)$ and $q_j(\perp)$, as in (5-7-2) but with the total flux replaced by the fluxes of polarized radiation $4\pi\phi_{jk}(\|)$ and $4\pi\phi_{jk}(\perp)$. Then we introduce a cross section

$$q_j(90) = q_j(\|) + q_j(\perp) \tag{5-7-16}$$

that depends on the total photon flux radiated perpendicular to the electron beam. The excitation cross section q_j is, however,

$$q_j = \frac{2}{3}[q_j(\|) + 2q_j(\perp)], \tag{5-7-17}$$

as must be the case for consistency with (5-7-13). The cross sections $q_j(\|)$ and $q_j(\perp)$ are important because they are related to the cross sections for excitation of the various Zeeman levels of the state j.[68] Usually, experimental cross section data are presented as $q_j(90)$, although the cross sections may actually be some weighted sum of $q_j(\|)$ and $q_j(\perp)$ unless allowance has been made for the polarization sensitivity of the apparatus.

For a given transition, the percentage polarization of the radiation emitted at 90° is $p = 100P$, and the percentage polarization for radiation at angle θ is

$$p(\theta) = \frac{p \sin^2 \theta}{1 - p \cos^2 \theta/100}. \tag{5-7-18}$$

In practice, the errors introduced by failure to account for polarization properly are seldom greater than about 10%, and are sometimes much less. However, as in spectroscopy,[46] knowledge of the polarization provides additional information about the process that is operative.

E. Excitation Transfer and the Imprisonment of Resonance Radiation

We shall now briefly discuss two pressure-dependent effects that can be significant in excitation studies involving the measurement of radiation. The first arises because a target particle excited by the electron beam may transfer some or all of its internal energy to another particle in a collision rather than disposing of this energy by emitting a photon. There is competition between the two deexcitation mechanisms, and at sufficiently high target pressure, collision excitation transfer can become important. Let us assume here that the transferred energy goes into electron excitation of the struck particle, although excitation of other modes, including translational, may be possible. The effect of the excitation transfer is to produce a spurious redistribution of the excitation energy provided by the electron beam, and a false spectrum of radiation emitted by the excited states of the target. If this process is not properly dealt with, some cross sections will appear to be too large and others too small. Another problem is that the radiation emitted by particles excited by the energy transfer will be unpolarized because there is no reference direction for this radiation.

The other pressure-dependent effect arises if the state to which excitation is being studied (j in Fig. 5-7-3) can return to the ground state (g) by the emission of electric dipole radiation. For "allowed" transitions of this type, the energy of the emitted photon equals the energy lost by the electron producing the excitation, and we speak of *resonance radiation*. Unless the target is optically thin, this radiation can excite another ground state particle to the state j, and this particle may emit a resonance photon inside or outside the volume of the electron beam. This process of emission and absorption of resonance photons may be repeated many times if the gas pressure is high, and the terms *imprisonment*, or *trapping*, *of resonance radiation* have been applied to it. R. G. Fowler[58] has suggested that *blockading* might be more appropriate because although there is always a net outward flux of radiation, this flux may be considered diminished by the absorption and reradiation of the photons. We note that here, as in the case of excitation transfer, the secondary radiation is unpolarized. These effects can be quite important, even at low pressures. They are discussed in detail by Heddle[58] and Massey and Burhop,[59] who give references to the important work by T. Holstein and A. V. Phelps in analyzing the effects of resonance radiation. Experimental checks are essential. It must be demonstrated that a linear relationship exists between the observed spectral line intensity and the target pressure. A similar check to show that the intensity depends linearly on the electron beam current will ensure that electron-impact excitation of target particles that have already been excited is not occurring to a significant extent. Finally, we mention the errors that can arise because of the changes in the state of polarization of the radiation at diffraction gratings or prism surfaces in monochromators.[55] These effects can be measured, and corrections can be applied.

F. Oscillator Strengths, Optical and Generalized

The concept of the oscillator strength arose in the late nineteenth century during attempts to explain the electrical and magnetic properties of matter. The classical model that was used envisioned electrons residing at equilibrium positions in atoms and responding elastically to weak disturbances.[69] For example,[46] when subjected to electromagnetic radiation, the electrons would execute forced oscillations, the amplitude and phase of which would depend on the frequency ν of the radiation, the resonant frequency ν_{ij} of the oscillator, and the damping constant γ (supposed to be small). The equation of motion for such an oscillator is

$$\ddot{x} + \gamma\dot{x} + (2\pi\nu_{ij})^2 x = -\frac{eE_0}{m}e^{i2\pi\nu t} \tag{5-7-19}$$

in one dimension. If radiation of intensity I_0 is incident on a uniform medium composed of these oscillators, the intensity is reduced to a value $I(d)$ at a depth, d, in the medium, where

$$I(d) = I_0 e^{-k_{ij}d} \tag{5-7-20}$$

[cf. (1-4-9)]. Solution of (5-7-19) provides the *absorption coefficient*:

$$k_{ij} = \frac{e^2}{4\pi mc} N_{ij} \frac{\gamma}{(\nu_{ij} - \nu)^2 + (\gamma/4\pi)^2}, \tag{5-7-21}$$

where N_{ij} is the number of oscillators per unit volume.[46] To make a realistic model, we replace these oscillators with N one-electron atoms per unit volume, each of which can absorb radiation when in state i and make transitions to many states j. The *oscillator strength*, f_{ij}, is then defined by the equations

$$Nf_{ij} = N_{ij} \quad \text{or} \quad f_{ij} = \frac{N_{ij}}{N}, \tag{5-7-22}$$

where

$$\sum_j f_{ij} = \sum_j \frac{N_{ij}}{N} = 1. \tag{5-7-23}$$

Only a fraction f_{ij} of the absorbing power of each oscillator is available for each channel $i \to j$. The total absorption coefficient is then

$$k_i = \sum_j k_{ij} = \frac{e^2 N}{4\pi mc} \sum_j f_{ij} \frac{\gamma}{(\nu_{ij} - \nu)^2 + (\gamma/4\pi)^2}. \tag{5-7-24}$$

Here we have a sum of Lorentz curves of absorption versus frequency. The frequency width of each absorption line at half the peak absorption is $\gamma/2\pi$.

The oscillator strength appears in classical treatments of such properties as the dielectric constant, the polarizability of atoms, and optical constants. Of course, the electrons in an atom do not have fixed equilibrium positions but, instead, a statistical distribution of positions given quantum mechanically by $|\Psi|^2$, which is normally time independent but which is distorted by disturbances. This statistical distribution can undergo free oscillations at a number of frequencies much larger than the number of electrons in the atom. To each frequency in this spectrum there corresponds a fraction of the integrated probability density which is usually represented by an irrational number rather than by an integer, and this number is the oscillator strength of the atom at that frequency. In the continuous spectrum, a spectral density of oscillator strength, $df/d\nu$, is used (Fano and Cooper[69]).

In quantum mechanics, the oscillator strength is given by the equation

$$f_{ij} = \frac{8\pi^2 m\nu_{ij}}{3h} |\mathbf{r}_{ij}|^2 = \frac{2m\hbar\omega_{ij}}{3\hbar^2} |\mathbf{r}_{ij}|^2, \tag{5-7-25}$$

where $\mathbf{r}_{ij}$ is the electric dipole matrix element connecting the states i and j.[59] The Einstein A_{ji} and B_{ij} coefficients (Chapter 8) are related to the corresponding oscillator strength by

$$A_{ji} = \frac{8\pi^2 \nu_{ij}^2 e^2}{mc^3} f_{ij} \tag{5-7-26}$$

and

$$B_{ij} = \frac{\pi e^2}{mh\nu_{ij}} f_{ij}. \tag{5-7-27}$$

In generalizing to treat transitions in which one of the states lies in the continuum, we define a differential oscillator strength

$$\frac{df}{d\nu} = \frac{8\pi^2 m^3 v\nu}{3h^3} \int |\mathbf{r}_{i\nu}|^2 \, d\omega, \tag{5-7-28}$$

where the continuum wave function is used in calculating $\mathbf{r}_{i\nu}$. This wave function is normalized to the number of states per unit frequency range per unit volume, which is $(m^2 v/h^2)\, d\omega$, where $d\omega$ is the element of solid angle around the asymptotic direction (ϑ, φ) of the electron in the continuum state.[59]

The *optical oscillator strength*, as the quantity f_{ij} in (5-7-25) is frequently called, is very useful in treating the emission and absorption of radiation, especially because of the sum rules, such as (5-7-23),* that it obeys.[46,59] This same quantity also appears in the classical calculation of the energy loss per unit length of a fast heavy particle passing through matter (see Jackson,[6] p. 626). However, for discussions of collisions, the *generalized oscillator strength*, introduced by H. A. Bethe[70] in 1930 and defined as

$$f_{0n}(K) = \frac{2m(E_n - E_0)}{K^2\hbar^2} |\epsilon_{0n}(K)|^2, \tag{5-7-30}$$

is much more useful. Here K is the momentum change experienced by the projectile electron in the collision (in units of $\hbar$), E_0 and E_n are the energies of the ground and excited states, respectively, and $\epsilon_{0n}(K)$ is the matrix element

*If we consider an N-electron atom and take continuum transitions into account, we obtain the more general sum rule

$$\sum_j f_{ij} + \int \frac{df_{iE}}{dE} \, dE = N \tag{5-7-29}$$

instead of (5-7-23).

$$\epsilon_{0n}(K) = \int \psi_0 e^{iKz} \psi_n^* \, d\mathbf{r} \tag{5-7-31}$$

(see Section 6-14-A). A power series expansion shows that when $K \to 0$, $f_{0n}(K)$ reduces to $f_{0n}(0)$, the optical oscillator strength defined in (5-7-25).[71]

As indicated in Problem 6-22, the Born approximation to the differential cross section per unit solid angle for excitation of a discrete level n from the ground state 0 is

$$I_{0n} = \frac{4m^2}{\hbar^4} \frac{\kappa_n}{\kappa_0} \frac{e^4}{K^4} |\epsilon_{0n}(K)|^2. \tag{5-7-32}$$

We can rewrite this expression in terms of the generalized oscillator strength as follows:

$$I_{0n}(K) = \frac{2me^4 \kappa_n}{K^2 \hbar^2 \kappa_0 (E_n - E_0)} f_{0n}. \tag{5-7-33}$$

Evidently, when the first Born approximation is valid, we can derive $f_{0n}(K)$ from measured values of $I_{0n}(K)$, obtaining a result independent of the electron impact energy and scattering angle, because $f_{0n}(K)$ is a function of K only. Indeed, establishing this independence constitutes an important test of the validity of the Born approximation for the calculation under discussion. Then, extrapolation of the values of $f_{0n}(K)$ to $K = 0$ should provide the optical oscillator strength $f_{0n}(0)$, which can be compared with spectroscopic values. In the case of optically forbidden transitions, $f_{0n}(0) = 0$, and $f_{0n}(K)$ should extrapolate to zero.

If the Born approximation is valid in a given situation, (5-7-32) relates the differential excitation cross section for some specified electron impact energy and scattering angle to the corresponding cross section at any other combination of impact energy and angle for which K^2 has the same value. Measurement of $I_{0n}(K)$ over a range of scattering angles at one impact energy can provide $f_{0n}(K)$ over a range of K^2, and then $I_{0n}(K)$ can be calculated at other energies and scattering angles that correspond to the range of K^2 covered.

The Born approximation is expected to fail as the electron impact energy is lowered sufficiently, and then (5-7-33) is no longer valid. However, an *effective generalized oscillator strength* (or *effective strength*) can be defined so that its use in (5-7-33) will result in the equation still being correct:

$$F_{0n}(E, K^2) = \frac{\hbar^2}{2me^4} \frac{\kappa_0}{\kappa_n} K^2 (E_n - E_0) I_{0n}(K). \tag{5-7-34}$$

The effective strength $F_{0n}(E, K^2)$ depends on the electron impact energy as well as on K^2, so values of F_{0n} at one energy do not imply anything about excitation cross sections at other energies. However, if measurements are made at low energies where the Born approximation is not valid, and then extended upward in energy, F_{0n} will eventually merge with $f_{0n}(K)$ and become independent of energy. Thus measurements of F_{0n} as a function of energy may be useful in investigating the breakdown of the Born approximation at low energies.

Lassettre's group, which is credited with much of the fruitful utilization of the concept of oscillator strengths,[72] has shown that $F_{0n}(E, K^2)$ also approaches $f_{0n}(0)$ as $K \to 0$,[73] this result being known as the "limit theorem." Hence, in the zero-momentum-transfer limit, optical selection rules should hold for electron impact excitation. However, there is a problem in the application of the limit theorem. It involves a nonphysical limit, in that even for a scattering angle of zero degrees the momentum transfer in an inelastic collision is not zero, so the extrapolation to $K = 0$ is not uniquely determined.[3]

The most important point to emerge from this section is that at sufficiently high impact energies ($\gtrsim 300$ eV, say) electrons scattered through small angles, with small values of K, excite atoms and molecules in much the same way as do photons. In fact, in the limit $K \to 0$, the spectra produced by the two types of excitation should be identical. The important practical consequence is that electrons may be used to generate optical excitation and ionization cross sections, and this equivalence of electrons and photons is especially valuable in the extreme UV and X-ray regions, where photon sources may not be available or may be expensive, as in the case of synchrotrons. The conclusions that we have drawn here are consistent with the demonstrated fast electron-photon equivalence in Section 5-6.

G. Selection Rules; Systematics of Electronic Excitation

It is useful now to reexamine Fig. 5-6-2, which shows the electric and magnetic fields produced by an electric charge q moving with constant velocity $\mathbf{v}$ with respect to an observation point P. Let the charge q be an electron, and suppose that a target molecule is centered at P. If the impact parameter of the electron trajectory, b, is much larger than the dimensions of the molecule, the electric field at the molecule will be approximately uniform over the structure at any given instant of time during the passage of the electron. As in the case of photon impact (Chapter 8), the electric field under these conditions will interact much more strongly with the transition electric dipole moment of the molecule than with its other multipole moments, and the transitions that are excited are mainly of the electric dipole type. These transitions are the *optically allowed* transitions in photoabsorption when the radiation wavelength, λ, is much greater than the dimensions of the molecule. The optical *electric dipole selection rules* apply.

When the impact parameter, b, is comparable with, but still larger than, the molecular dimensions, the molecule experiences a nonuniform electric field at any instant as the electron passes. The nonuniform components of the E field can interact with the transition electric quadrupole moment of the molecule and excite optically forbidden transitions. (For small b, electric octupole transitions may also be possible, although with lower probability.) Equation (5-7-32) shows that the differential cross sections for optically allowed transitions are proportional to $1/K^2$ for small K, and that they decrease rapidly as θ, and therefore K, increases. We thus see that optically allowed transitions tend to be strongly peaked in the forward ($\theta = 0$) direction. The forbidden transitions are enhanced, relative to the allowed, at scattering angles appreciably greater than zero, and they are favored at all scattering angles when the electron impact energy is low. All of these matters are treated in detail by Hall and Read in Section 3 of their review.[8]

Hall and Read[8] also discuss collisions involving very small impact parameters, in which the projectile electron comes close enough to the target for electron exchange to become possible (Sections 4-6-B, 4-8-A, and 5-2-C). This process is favored when the velocity of the incident electron is close to that of the valence electrons of the molecule, that is, when the impact energy is near the threshold energy for excitation of the state being considered. One possibility is the *spin-exchange reaction*: Suppose that a divalent molecule is initially in a singlet state ($S = 0$), with its two valence electrons having spins opposed ($m_1 = \frac{1}{2}$, $m_2 = -\frac{1}{2}$). If the projectile electron has spin up ($m_3 = \frac{1}{2}$) and exchanges roles with m_2, the molecule will be left in a triplet state ($S = 1$) with ($m_1 = \frac{1}{2}$, $m_3 = \frac{1}{2}$), while the emerging electron has spin down ($m_2 = -\frac{1}{2}$). The mechanism described here is the only effective one for producing a singlet-triplet transition. Spin-exchange processes are not peaked in the forward direction. Detection of these transitions is facilitated by using low impact energies and studying large-angle events.

We see that optical selection rules are not obeyed at low electron impact energies, and transitions that are forbidden in the electric dipole approximation because of symmetry or spin considerations are now allowed. Such transitions are permitted because the electron-molecule interaction, according to (5-2-5), contains the various multipole terms, and it contains electron exchange terms as well. Many more possibilities are available for nonzero transition matrix elements than in the case of photon impact. In fact, many electronic states of molecules have been discovered through electron impact excitation utilizing slow electrons and high-angle observations.[3] At low electron energies (within a few eV of threshold), dipole-forbidden transitions actually dominate the energy-loss spectra in most cases. However, the largest cross sections for electronic excitation are for optically allowed processes at intermediate electron energies (20 to 200 eV) and for small scattering angles. Such cross sections typically increase slowly with increasing impact energy from threshold up to a peak at about 10 times the threshold energy and then decrease gradually. Trajmar and Cartwright[3] provide a good discussion of selection rules for electronic excitation and give useful references to the subject. Representative papers from the group at University College London are cited in Refs. 74–76. They deal with both allowed and forbidden transitions. Reference 76 concerns excitation from the $a^1\Delta_g$ ($v = 0$) state of O_2 to $^1\Pi_u$ ($v' = 0, 1, \ldots, 7$); a Morse potential energy curve is determined for the $^1\Pi_u$ state from the relative intensities of tne energy-loss spectra.

H. Excitation by High-Energy Electrons (Energies above a Few Hundred eV)

At high energies both outer-shell and inner-shell excitation are important.[39,77,78] (Ionization by fast electrons is considered in Chapter 6.) We shall not attempt a broad coverage of this subject, and in fact we shall omit several interesting topics, such as electron diffraction in gases, that are treated by Massey[39] and Bonham and Fink.[77]

At present, in the study of *outer-shell* processes, attention is centered on the determination of accurate generalized oscillator strengths and Bethe surfaces.* Energy-loss spectra are mapped at small scattering angles and low energy losses (usually less than 50 eV). These studies closely resemble the photoabsorption techniques and can provide accurate optical oscillator strengths. Information about the momentum distribution of target electrons can be derived, as well as details concerning rotationally and vibrationally excited states, and valence and Rydberg states.

Important measurements have been made at electron impact energies in the range 20 to 35 keV, where energy resolutions of a few meV can be obtained! Here the incident beam is generated at high energy, slowed down to a few eV for passage through a Wien filter (which serves as an energy selector), and then brought back to high energy for collision with the target gas. The electrons scattered at angles less than 10^{-4} rad, say, are slowed down for analysis in another Wien filter, which disperses them laterally. These electrons are then reaccelerated and detected photographically. Details of these techniques are presented in publications by Geiger and his colleagues in Kaiserslautern.[80]

The reviews by Hall and Read[8] and Trajmar and Register[16] give references to other important work done by van der Wiel's group in Amsterdam, and by Kuyatt, Mielczarek, Celotta, and their colleagues at the National Bureau of Standards in Gaithersburg.

In *inner-shell* excitation, one or more inner-shell electrons may be excited to a valence or Rydberg state. The atomic or molecular state thus produced generally will decay within about 10^{-14} sec by the Auger process, in which one electron is ejected into the continuum while another fills the inner-shell vacancy (see Sections 6-6 and 6-7). It is often possible to obtain much higher energy resolution in energy-loss inner-shell spectra than in the corresponding photoabsorption spectra.[8] Inner-shell processes comprise many interesting and complex phenomena, and they are the subject of periodic international conferences that deal with electron, photon, and heavy particle impact. We shall not discuss these processes in depth, but merely give a few references to work of special interest. King et al.[81] made the first observation of resonances associated with inner-shell

*A *Bethe surface* is a three-dimensional plot of $df(K, E')/dE'$ as a function of E' and of $\ln K^2$, where E' is the electron energy loss, K is the momentum transfer, and $df(K, E')/dE'$ is the density of the generalized oscillator strength per unit range of E' (Inokuti;[69] Ulsh et al.[79]). Such a plot contains all the information about inelastic charged particle scattering for a particular target within the framework of the first Born approximation.

states. Shaw and his colleagues[82] made the first application of the "equivalent-core model" of inner-shell excitation, in which the outer electrons of the excited atom Z move in a mean field similar to that of the outer electrons of the atom $(Z + 1)$, so that the term values of these two atoms are similar. In Ref. 83, the first use of the electron energy-loss technique to study forbidden transitions was reported (a spin-forbidden excitation in N_2 and a parity-forbidden transition in Ar). Shaw et al.[84] have designed an electron impact spectrometer for use in inner-shell studies and obtained with it high-resolution (typically 60 meV) energy-loss spectra for CO at energies ranging from 200 to 1500 eV.

I. Excitation near Threshold

Here we refer to excitation of atoms and molecules by electrons that retain less than a few eV of their incident energy after producing the excitation in question. Measurements of such processes can provide useful spectroscopic information[8] and tests of the Wannier theory for near-threshold excitation.[85] High-energy resolution is necessary, and handling the scattered electrons, which may have energies in the low-meV region, is difficult. A "penetrating field" technique developed by Cvejanović and Read[86] has proved useful in this regard. A weak electric field is allowed to penetrate through two windows into a target region that is otherwise well shielded. A saddle point in the electric potential is produced at the point where the electron and target beams cross, and the electric field is able to focus low-energy scattered electrons over a wide range of scattering angles and send them out through the windows for detection. Higher-energy electrons are rejected in the collection process.

King et al.[87] in 1975 observed structures in the cross sections for excitation of He, Ne, and Ar near autoionizing energies of the atom. A quantum-mechanical model, called the "shake-down model," was advanced to explain these structures. It is based on the postcollision interaction between the scattered and ejected electrons (see Section 6-6).

An atom-detection technique has been used by Hammond et al.[88] to study near-threshold excitation to high-n states of Ar, Kr, Xe, H_2, N_2, and CO. A crossed-beam arrangement is used. Target particles excited to Rydberg states with sufficiently high values of n (16 to 46) are metastable, and their long lifetime permits them to travel to a field-ionization detector composed of parallel wire meshes. The ions formed in it are then accelerated and detected in a channel electron multiplier.

J. Cross Sections for Electronic Deexcitation

As in the case of rotational and vibrational excitation, cross sections for electronic excitation by electron impact may be used to obtain the corresponding cross sections for the inverse process through use of the principle of detailed balance (see Sections 5-3 and 5-4).

5-8. MEASUREMENTS OF LIFETIMES BY COINCIDENCE TECHNIQUES

Suppose that we have an ensemble of $N(t)$ atoms in some particular excited state j at time t, and that these atoms decay by the spontaneous emission of radiation to lower-energy states k. Let $A_j = \sum_k A_{jk}$, where A_{jk} is the Einstein A coefficient defined as the probability of decay to a particular state k per atom per unit time. It follows that

$$\frac{dN(t)}{dt} = A_j N(t), \tag{5-8-1}$$

and that

$$N(t) = N(0)e^{-t/\tau}, \tag{5-8-2}$$

where $N(0)$ is the number of atoms in state j at $t = 0$, and τ is a constant, called the *lifetime* of state j, given by

$$\tau = \frac{1}{A_j}. \tag{5-8-3}$$

Values of the lifetimes of some low-lying states of the hydrogen atom are given in Table 5-8-1. A strong dependence on the quantum numbers n and l is observed, but there is no dependence on m_l unless an external field exists and provides a direction of orientation. Highly excited states of atoms and molecules generally have much longer lifetimes than states of low internal energy. The lifetimes of hydrogenic ions with nuclear charge Z are given in terms of the hydrogen atom lifetime by

$$\tau(Z) = \frac{1}{Z^4}\,\tau(Z = 1) \tag{5-8-4}$$

Various techniques have been used to measure lifetimes of excited states of atoms and molecules,[55,57] but an electron-photon coincidence method devel-

Table 5-8-1 Lifetimes of a Few Low-Lying States of the H Atom, in Units of 10^{-8} sec Except for the 2*s* Metastable State

State	2*s*	3*s*	4*s*	2*p*	3*p*	4*p*	3*d*	4*d*	4*f*
Lifetime	$\frac{1}{7}$ sec	16	23	0.16	0.54	1.24	1.56	3.65	7.3

Source: Adapted from Bransden and Joachain.[12]

oped by Imhof and Read[89] offers distinct advantages, especially in the elimination of complications arising from cascading. Molecules in a thermal energy beam are excited by a continuous, monoenergetic electron beam whose energy is well above the threshold of the state being studied. Electrons that are scattered inelastically into a small element of solid angle in the forward direction enter an energy selector, and only those electrons that have lost the same amount of energy corresponding to excitation of this state are transmitted to an electron counter. Photons emitted in the deexcitation of this state are selected with a filter and detected with a photomultiplier. A pulse from the electron counter indicates the time of excitation of a molecule, and the corresponding photon pulse marks the demise of the excited state. The time separation of these two pulses equals the lifetime of the state of that molecule plus the constant-delay-time difference between the electron and photon counting channels. A statistical distribution built up over many events permits a determination of the mean life of the state. A delayed electron-photon coincidence technique is used for the time separation measurements. The energy of the incident electrons is chosen to maximize the yield of the electrons scattered with the energy loss corresponding to the excited state being studied. This technique can be used to measure lifetimes of vibrational levels of excited states of molecules in their decay to lower electronic states.[90] A recent paper describing lifetime measurements of VUV emitting levels in He, Ne, Ar, and CO is cited in Ref. 91. The electron-photon delayed coincidence technique has also been used to study the polarization of radiation from atoms excited by electron impact.[92]

A photon-photon delayed coincidence technique has been employed at the University of Manchester to measure molecular lifetimes. A beam of electrons excites various levels of the molecule, including some above the state whose lifetime is to be determined. A cascade of photons results during the successive stages of deexcitation, and the lifetime that is desired can be obtained from observations of delayed coincidences between photons emitted in transitions to and from the state in question.[93] A useful discussion of the difficulties inherent in the various kinds of coincidence measurements in atomic physics appears in a review by Trajmar and Register.[16]

5-9. POSITION-SENSITIVE DETECTION IN ENERGY-LOSS SPECTROMETRY

Several types of electron spectrometers were described in Section 4-3-B. Their use in electron energy-loss determinations and differential cross-section measurements was discussed in Sections 5-3, 5-4, and 5-7. We recall that an energy analyzer is used to disperse the scattered electrons and that only a small fraction of these electrons passes through a slit for detection. The resolving power of the selector is determined by the width of the slit—the narrower the slit, the better the resolution, but also, unfortunately, the more of the signal that is wasted.

Typically, only about 1% of the electrons in the dispersed image will negotiate the slit. Comer and his colleagues at the University of Manchester[94-100] have developed a position-sensitive detector technique that makes use of the whole image transmitted through the analyzer by simultaneously detecting electrons from the entire image and registering the position where each was detected. The Manchester technique utilizes a microchannel plate image intensifier to amplify the extended image composed of electrons arriving at the exit plane of the analyzer. (A microchannel plate is a two-dimensional array of many closely spaced and parallel single-channel electron multipliers, typically having diameters of about 20 μm and lengths of about 1 mm.[101]) The amplified electron image is converted into a photon signal by a phosphor screen. A linear charge-coupled device (CCD)[102] detects the light image and transforms it into an electrical signal, which is processed by a computer-controlled data acquisition system. (Alternative multidetection techniques that have been used elsewhere are discussed in Refs. 94 and 95.)

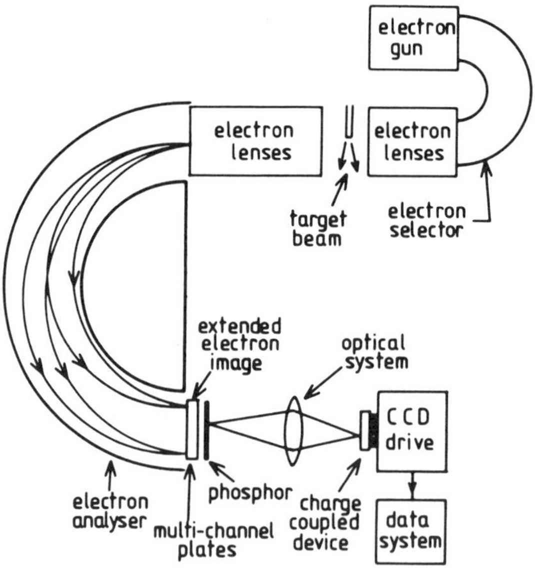

Figure 5-9-1. An electron spectrometer developed by Comer et al.[98,99] that utilizes position-sensitive detection of the scattered electrons.[95] The incident electron energy can be varied over the range 10 to 120 eV. Electrons scattered from the target beam into a narrow angular range are admitted into the analyzer; the angle at which scattering is observed can be varied from 0 to 90°. A 20-mm-long section of the image in the exit plane of the analyzer falls on the microchannel plate detector, corresponding to an energy-loss spectrum typically about 0.25 eV wide. By the data acquisition techniques described in the text, a 4-eV-wide continuous range of energy loss can be covered in a few hours.

We shall now discuss in more detail the Manchester instrument that is described in Refs. 98 and 99. The apparatus is shown in Fig. 5-9-1. An electron beam with variable and well-defined energy intersects a target beam, and the scattered electrons are dispersed in energy by a hemispherical electrostatic analyzer, which is made of molybdenum and is unusually large (0.19 m in mean diameter). The extended image in its exit plane is composed of electrons whose energy is linearly related to their distance from the entrance to the analyzer. The intensity of the image as a function of this distance gives the energy-loss spectrum. The electrons impinge on the microchannel plate multiplier, which provides an intensity gain of 10^6, and the electrons emerging from it are accelerated through 6 kV onto a phosphor screen. The photon image is transferred onto the CCD, which contains a linear array of 256 photosensitive elements that store information in the form of a charge distribution. The incident photons produce electron-hole pairs, and the electron component is allowed to accumulate under the "photogate" electrode, which is held positive for an integration period typically of the order of 2 msec. At the end of this period, the distribution of charges under the electrode corresponds to the pattern of light on the phosphor screen. The 256 batches of charge are then transferred in parallel into a pair of CCD analog shift registers that flank the array of photosensitive elements. The analog signals from the CCD are digitized and stored in the data acquisition system. While this 2-msec sample is being processed, the next 2-msec sample is being accumulated.

The multichannel plate detector spans an energy range corresponding to about 12% of the mean energy of the electrons passing through the analyzer. A wider range of energies can be scanned by varying the potential difference between the energy selector and analyzer, which amounts to varying the energy loss. This scanning is accomplished by a control processor, which steps the potential difference at the end of each complete readout cycle from the CCD by an amount equal to one channel width. Each step corresponds to a small shift in the energy-loss range covered by the image. For each step the data from the 256 segments of the detector are accumulated in the appropriate part of the final spectrum. Each channel of the detector is used to accumulate data for each part of the overall spectrum for an equal amount of time. In this way, any nonuniformity in gain of the detection channels is eliminated. Comer and his coworkers[94] have used this technique to obtain vibrational energy-loss spectra that are practically indistinguishable from spectra mapped by conventional techniques that require more than 100 times as long to acquire. Energy-loss spectra obtained by Comer et al. for N_2, O_2, CO, HCl, and Cl_2 are presented in Refs. 94–100. The apparatus described in Refs. 98 and 99 is differentially pumped so that corrosive gases may be used in the target beam without adversely affecting the stability of the instrument. An energy resolution of 18 meV (FWHM) was achieved in the measurements on Cl_2.[99] Reddish et al.[100] have also used a position-sensitive detection technique to map simultaneously excitation functions for several vibrational levels of N_2.

5-10. FINE DETAILS OF THE ELECTRON-IMPACT EXCITATION PROCESS

A. General Considerations

In Sections 4-3, 4-6, 5-3, 5-4, and 5-7, we saw how observations restricted to subsets of the incident and scattered electrons, of the target particles, and of the photons emitted following excitation provide information not accessible in total cross section measurements. Here we refer to the selective detection of electrons scattered in a certain direction and with specified energy loss, the spin selection of projectile electrons and spin analysis of scattered electrons, the state selection and analysis of target particles, and the selective detection of photons emitted in a narrow wavelength range and in a closely defined direction. In this section we show how this approach can be carried still further to gain deeper insight into the excitation process and to test the theory at a deeper level. Experimental studies based on these considerations and involving delayed electron-photon coincidence techniques are discussed in Section 5-10-J, together with related studies of electron scattering from laser-excited atoms.

We first need to set the stage for our detailed treatment of the fine details of the excitation process by discussing briefly some general considerations. We shall be concerned with collisions that may proceed from a given state through several different channels, each described by a different probability amplitude (f_k) for a transition to a different state (k). If we add these amplitudes and take the modulus square of the result, we obtain two kinds of terms: A term such as $|f_i|^2$ gives the cross section for the transition to state i, whereas one such as $f_i^* f_j$ is an interference term arising from the coherent nature of the excitation. ("Coherent" is one of several terms introduced here but not defined properly until Section 5-10-B.)

We are especially interested in excitation of atoms* to states of well-defined orbital angular momentum L, which have $(2L + 1)$ magnetic sublevels that are degenerate in energy in the absence of external fields. These sublevels cannot be separated in an electron energy spectrometer. Hence, in the traditional experiments that involve only detection of scattered electrons but not the photons as well, the measured angular differential cross section is an average over the degenerate magnetic sublevels.

Similarly, the traditional measurements on the photons but not on the scattered electrons as well provide useful, but limited, information. Measurements of the light intensity $I(\theta_v)$ and percentage polarization P as a function of the angle of emission can give the cross sections for excitation of the individual magnetic sublevels, but they are averaged over all the angles of scattering of the electrons (Section 5-7-D), and hence relate to the integral excitation cross section.

*We shall consider only atomic targets here, as molecules would present unnecessary complications at this stage. Sufficient unto the day are the evils thereof!

Only in the early 1970s did investigators begin to take advantage of the benefits that accrue from observations of both photons and scattered electrons in coincidence measurements. Such an arrangement has a lower degree of symmetry than does the traditional excitation experiment, with its complete axial symmetry. Owing to this fact and to the brevity of the collision time compared to the lifetimes of excited atomic states, *coherent excitation* is possible (Section 5-10-B), according to which the state of the excited atom is given, for example, by

$$|\psi\rangle = f_{+1}|1 \quad 1\rangle + f_0|1 \quad 0\rangle + f_{-1}|1 \quad -1\rangle \qquad (\text{at } t = 0) \qquad (5\text{-}10\text{-}1)$$

if we assume the excited atom to be in a ${}^1P$ state and denote the magnetic substates by $|1M\rangle$. If the electron-photon coincidence technique is used, *only electrons scattered in some selected direction and with a specified energy loss are detected*, and *only photons associated with these scattering events are observed.* The former restriction means that all the atomic states are prepared in the same way, corresponding classically to the same electron impact velocity, impact parameter, scattering angle, and energy loss. Hence each of the atoms whose radiation is detected was produced in the linear superposition of basis states in (5-10-1) immediately after the collision ($t = 0$). Each coefficient f_i in (5-10-1) is a complex scattering amplitude with a magnitude and phase that are well defined except for an overall phase factor. There exists a definite phase relationship among the basis states. Because of the coherent nature of the excitation, interference between these amplitudes can be measured in coincidence experiments. In special cases where the excited atoms are in a *pure state* (Section 5-10-C), coincidence experiments provide direct information on the magnitudes and relative phases of the scattering amplitudes. They generally measure two, or sometimes three, additional parameters that describe the excitation process beyond the differential cross section that is traditionally measured. It is in this sense that the electron-photon coincidence technique provides the close look at the excitation process that is so valuable. If, on the other hand, this technique is not used and the excitation process is *incoherent*, each excited atom is in a state of well-defined M, and the kind of information described above cannot be obtained.

In this section we concentrate on electron-photon coincidence studies that measure the angular distribution or the polarization of the light emitted by the excited atoms. We have already discussed other kinds of coincidence measurements in Section 5-8 and mentioned the noteworthy paper by King et al.[92] that reported in 1972 the first use of the e-$h\nu$ coincidence technique to study alignment in an atomic system.

Specifically, we shall see how coincidence studies and related observations provide detailed information on (1) properties of the target state excited in the collision (orientation and alignment parameters, multipole moments); (2) radiative properties of the excited target (Stokes' parameters of the emitted light, coherence effects, and quantum beats); and (3) parameters of the excitation collision (scattering amplitudes and phases) that are related to the properties of the excited states.

We shall not discuss the valuable contributions made to our understanding of the excitation process through the use of polarized electrons and ground-state atoms without e-*hv* coincidence. The reader may recall our discussion of these techniques for the study of elastic scattering of electrons in Section 4-6. Initially, we shall closely follow the definitive review by Blum and Kleinpoppen[103] and the book by Blum.[104] Then in Section 5-10-G we base our treatment on the book by Kessler.[105] This book and the review by Slevin[106] provide a very useful supplement to our treatment of experimental techniques in Section 5-10-J. Good general references are the books edited by Kleinpoppen and McDowell,[107] Kleinpoppen and Williams,[108] and Lutz et al.[109]

B. The Electron-Photon Coincidence Technique and Coherent Excitation[103,104]

In the traditional electron impact excitation experiments, a beam of electrons with energy E_0 and wave number $\boldsymbol{\kappa}_0$ passes through a target region, and *either* scattered electrons *or* photons emitted by excited target particles are detected. There is axial (azimuthal) symmetry about the incident electron beam. Now we describe an electron-photon coincidence technique that is used to provide a spatial restriction on electron-impact excitation events as follows. As in Fig. 5-10-1, we count only those electrons with energy E_1 that have been scattered

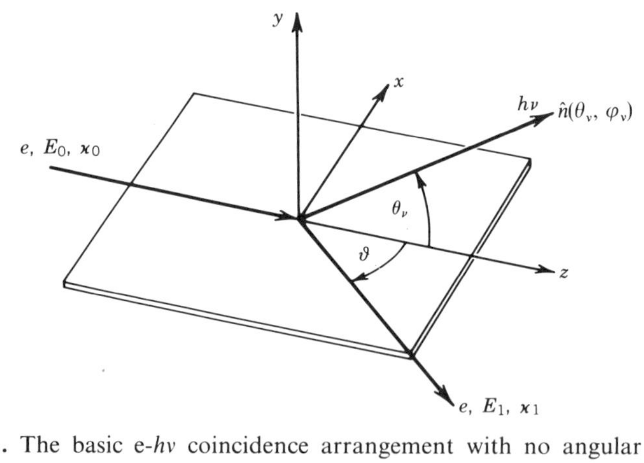

Figure 5-10-1. The basic e-*hv* coincidence arrangement with no angular momentum selection or analysis. The incident electron beam is along the Z axis. The direction chosen for observation of scattered electrons (that of the final electron momentum) defines, with the Z axis, the *scattering plane*. The X axis is placed in this plane, and the Y axis perpendicular to it, to form a right-hand coordinate system. Two kinds of measurements are made on the photons detected in delayed coincidence with the scattered electrons: the coincidence count rate as a function of the photon emission direction, and the polarization of the photons at various emission angles.

with a specified energy loss ($E_0 - E_1$) into the fixed direction indicated by the final wave number vector $\boldsymbol{\kappa}_1$. Photons are detected only if they are in delayed coincidence with these electrons and if they are emitted in a narrow cone $d\omega$ (θ_v, φ_v) in some selected direction specified by the unit vector $\hat{n}$. *Angular correlation* measurements are made by mapping the photon coincidence count as a function of emission direction.* Observations of the polarization of the photons (*polarization correlation*) provide an alternative approach for extracting information on the details of the excitation process.

The coincidence measurements described here involve only a selected subensemble of all the atoms excited by the electron beam, and this subensemble has special properties not characteristic of the total ensemble. Consequently, the radiation detected in delayed coincidence does not represent the radiation emitted by the full ensemble of excited particles. The arrangement sketched in Fig. 5-10-1 illustrates the basic "directional correlation" technique, in which neither polarized electrons nor state-selected targets are employed, and no spin analysis is performed on the photons or scattered electrons. Angular momentum selection and/or analysis imposes additional constraints and can provide additional information.

We now wish to examine some theoretical considerations relating to the experimental arrangement of Fig. 5-10-1. We shall assume that the excitation and radiative decay are independent in the sense that the scattered electrons have moved far from the target before the photons are emitted. Also assume that there is no spin-orbit interaction between the projectile and target, so that there are no explicit spin-dependent terms in the Hamiltonian describing the collision. Then the *L-S* coupling scheme will adequately describe the excitation, and changes in spin during the collision can result only from electron exchange.

For simplicity, consider an electron colliding with an unexcited atom in an S state. Use the following sets of quantum numbers to characterize the unperturbed and excited atom, the incident and scattered electron, and the initial and final states of total system:

$$\begin{array}{ll}
\text{initial atom} & \gamma_0 = n_0, L_0 = 0, M_0 = 0, S_0, M_{S0} \\
\text{incident electron} & \mathbf{p}_0, m_0 \\
\text{excited atom} & \gamma_1 = n_1, L_1 = L, M_1 = M, S_1, M_{S1} \\
\text{scattered electron} & \mathbf{p}_1, m_1 \\
\text{initial system} & \Gamma_0 = \gamma_0 \mathbf{p}_0 m_0 \\
\text{final system} & \Gamma_1 = \gamma_1 \mathbf{p}_1 m_1 .
\end{array} \tag{5-10-2}$$

$M \equiv M_L$, and M_S represent the projection of the orbital (L) and spin (S) angular momenta, respectively, of the atom along any axis selected for their measure-

*Angular correlation techniques have been employed in nuclear physics since shortly after World War II,[110] but were introduced into atomic collisions only in the 1970s.

ment; **p** is the electron linear momentum, and m gives the projection of the electron spin along the quantization axis. All necessary quantum numbers not explicitly shown for the atom are lumped in the symbol n.

We consider the transition $\Gamma_0 \rightarrow \Gamma_1$, whose scattering amplitude

$$f(\Gamma_1\Gamma_0) = f(\gamma_1\mathbf{p}_1m_1, \gamma_0\mathbf{p}_0m_0) \tag{5-10-3}$$

is normalized according to

$$|f(\Gamma_1\Gamma_0)|^2 = \sigma(\Gamma_1\Gamma_0), \tag{5-10-4}$$

where $\sigma(\Gamma_1\Gamma_0)$ is the angular differential cross section for the transition $\Gamma_0 \rightarrow \Gamma_1$. We are interested in both the magnitude and phase of the amplitude $f(\Gamma_1\Gamma_0)$. Assume here that the excited state has definite values of n_1, L_1, S_1. Then all possible experiments on the system can be analyzed in terms of bilinear combinations of the scattering amplitudes $f_{M'}f_M^* \equiv f(\Gamma_1'\Gamma_0)f(\Gamma_1\Gamma_0)^*$, where

$$\Gamma_1' = n_1, L, M', S_1, M_{S1}, \mathbf{p}_1, m_1, \tag{5-10-5}$$

differing from Γ_1 only by containing M' rather than M. Because the experimental arrangement here does not provide any spin selection or analysis, we are interested in $f_{M'}f_M^*$ averaged over the initial and final spin states:

$$\langle f_{M'}f_M^* \rangle \equiv \frac{1}{2(2S_0 + 1)} \sum_{m_0 M_{S_0} m_1 M_{S_1}} f(\Gamma_1'\Gamma_0)f(\Gamma_1\Gamma_0)^*. \tag{5-10-6}$$

We are concerned only with the dependence of the parameters $\langle f_{M'}f_M^* \rangle$ on M' and M, and we suppress the other quantities. Note that if we put $M' = M$, we obtain the spin-averaged differential cross section $\sigma(M)$ for the excitation of the magnetic substate M ($n_1, L, S_1, \mathbf{p}_1$ being fixed).

The element $\langle |f_M|^2 \rangle = \sigma(M)$ is proportional to the number of atoms of the selected subensemble of excited atoms that are in the magnetic sublevel L, M. All elements $\langle f_{M'}f_M^* \rangle$ with $M' \neq M$ vanish if the excitation is performed under conditions of axial symmetry (as is *not* the case here).[104] The corresponding subensemble is then, by definition, an *incoherent superposition* of magnetic substates; that is, it is a mixture of excited atoms, each of which is in a state of definite M, but with different atoms having different values of M.

If the excitation process has no axis of symmetry, the elements $\langle f_{M'}f_M^* \rangle$ with $M' \neq M$ will be nonzero in general. If at least one element with $M' \neq M$ is nonvanishing, the system is said to be a *coherent superposition* of magnetic sublevels. The parameters with $M' \neq M$ reflect the interference between the different magnetic substates, and they can be determined only in an experiment that lacks axial symmetry. (By contrast, the parameters with $M' = M$ can be derived by detecting the photons only, without any measurements on the scattered electrons.) Of course, no interference terms between different spin states are observed unless spin analysis is performed. In the case considered here the subensemble of excited atoms is an incoherent mixture of all spin states.

It should be emphasized that the set of parameters defined in (5-10-6) for all possible values of M and M' completely characterizes the subensemble of excited atoms that is under investigation.

C. The Density Matrix*

The parameters defined in (5-10-6) form a $(2L + 1) \times (2L + 1)$ *density matrix*, ρ,[104,105,111,112] the dimensionality being determined by the range of M' and M: $L, L-1, \ldots, -L$. For an ensemble of excited atoms with $L = 1$, the density matrix

$$\langle LM'|\rho|LM\rangle = \langle f_{M'} f_M^* \rangle \tag{5-10-7}$$

has the form

$$\rho = \begin{pmatrix} \langle |f_{+1}|^2 \rangle & \langle f_{+1} f_0^* \rangle & \langle f_{+1} f_{-1}^* \rangle \\ \langle f_0 f_{+1}^* \rangle & \langle |f_0|^2 \rangle & \langle f_0 f_{-1}^* \rangle \\ \langle f_{-1} f_{+1}^* \rangle & \langle f_{-1} f_0^* \rangle & \langle |f_{-1}|^2 \rangle \end{pmatrix}. \tag{5-10-8}$$

The differential cross sections $\sigma(M)$ for excitation of the various magnetic sublevels M are given by the diagonal elements $\langle LM|\rho|LM\rangle$. The trace of ρ gives the differential cross sections summed over M:

$$\operatorname{tr} \rho = \sum_M \sigma(M) = \sigma. \tag{5-10-9}$$

The number of independent measurements required for a complete determination of ρ equals the number of independent elements of the matrix. The diagonal elements are real, but the off-diagonal elements are complex, with two real components (magnitude and phase). The number of independent elements is reduced by the hermiticity condition

$$\langle LM'|\rho|LM\rangle = \langle LM|\rho|LM'\rangle^* \tag{5-10-10}$$

and by the symmetry requirement that the excited system must be invariant under reflection in the scattering plane.[103,104] In the case considered here, the spins are not observed in the scattering process, and it can be shown that spin

*No single wave function can describe an unpolarized beam because the function would have to contain information about the relative phases corresponding to the different spin orientations, and these phases cannot be defined for an unpolarized beam. The density matrix, on the other hand, is ideally suited to provide such a description, and it is introduced for that purpose. Its symbol and name reflect its similarity to the classical density in phase space $\rho(q, p)$ used in Section 2-11.

averaging leads to the relation

$$\langle LM'|\rho|LM\rangle = (-1)^{M'+M}\langle L - M'|\rho|L - M\rangle. \tag{5-10-11}$$

In the special case of singlet-singlet transitions ($S_0 = S_1 = 0$), no spin averaging is required, and

$$\langle LM'|\rho|LM\rangle = f_{M'}f_M, \tag{5-10-12}$$

with $f_M = f(n_1 LM\mathbf{p}_1, \mathbf{p}_0)$, where n_1, L, $\mathbf{p}_1$, $\mathbf{p}_0$ are fixed. Here it can be shown that

$$f_M = (-1)^M f_{-M} \tag{5-10-13}$$

and

$$\langle LM'|\rho|LM\rangle = (-1)^M\langle LM'|\rho|L - M\rangle. \tag{5-10-14}$$

For singlet-singlet transitions, it can also be shown that the single wave function

$$|\psi\rangle = \sum_M f_M|n_1 LM\rangle \tag{5-10-15}$$

characterizes the quantum state of the entire excited subensemble (i.e., all the atoms in it are in the same state $|\psi\rangle$ except for a phase factor[104]). Equation (5-10-15) shows that $|\psi\rangle$ contains $(2L + 1)$ complex amplitudes f_M, but because of the symmetry condition (invariance under reflection in the scattering plane), $|\psi\rangle$ is, in fact, completely specified by $(2L + 1)$ real parameters. The state $|\psi\rangle$ is called a *pure state*[104,105,111] because it is represented by a single wave function; it is a *completely coherent*[113] superposition of magnetic substates. Blum and Kleinpoppen[103] show that complete coherence is based on more stringent requirements than is coherence, which requires only that at least one off-diagonal element of ρ be nonzero.

In the more general case where the initial and/or final atomic spins are nonzero, we take conservation of total spin explicitly into account and write

$$\langle f_{M'}f_M^*\rangle = \frac{1}{2(2S_0 + 1)}\sum_S (2S + 1)f_{M'}^{(S)}f_M^{(S)*}. \tag{5-10-16}$$

where $S = S_0 \pm \frac{1}{2} = S_1 \pm \frac{1}{2}$ is the total "channel" spin, and $f_M^{(S)}$ is the amplitude for excitation of the state $|LMS\rangle$ in the channel with total spin S. If $S_0 = 0$, then $S = \frac{1}{2}$, and (5-10-16) reduces to the single term

$$\langle f_{M'}f_M^*\rangle = f_{M'}^{(1/2)}f_M^{(1/2)*}. \tag{5-10-17}$$

The condition of reflection invariance in the scattering plane requires that

$$f_M^{(1/2)} = (-1)^M f_{-M}^{(1/2)}. \tag{5-10-18}$$

Hence $2L + 1$ real parameters suffice to specify completely triplet levels excited from singlet states. In the event that both S_0 and S_1 are nonzero, there are two possible spin channels to consider: $S = S_0 - \frac{1}{2}$ and $S = S_0 + \frac{1}{2}$. Because $f_M^{(S)}$ depends only on the total spin, only the spin-averaged elements in (5-10-16) can be observed without spin analysis. For S_0 and S_1 not zero, the excited state cannot be specified by a single wave function; the subensemble of excited atoms must be described by a statistical mixture of wave functions or by their density matrix.

D. Excitation of *P* Levels; the Parameters σ, λ, and χ

In the excitation of ^1P states from a ^1S ground state, a single wave function, $|\psi\rangle = \sum_M f_M |n_1 LM\rangle$ in (5-10-15), characterizes the excited ensemble. Therefore, three parameters are sufficient to specify completely the excited system, and three independent measurements are required to determine the scattering amplitudes f_M. The set of three parameters σ, λ, χ evaluated for each electron impact energy and scattering angle proves to be convenient in this regard. Here σ is the total differential cross section given by (5-10-9). We choose f_0 to be real and define the relative phase χ between f_0 and f_1 by writing

$$f_1 = -f_{-1} = |f_1|e^{i\chi}. \tag{5-10-19}$$

Finally, we put

$$\lambda = \frac{|f_0|^2}{\sigma}. \tag{5-10-20}$$

Observation of the scattered electrons without the photons can provide σ; observation of the photons without the electrons can give λ integrated over all scattering angles. A determination of an off-diagonal element of ρ, say by an electron-photon coincidence measurement, is required to obtain the phase χ. The excitation of ^3P states from a ^1S ground state can be analyzed in the same fashion. If both S_0 and S_1 are nonzero, only the spin-averaged quantities $\langle f_{M'} f_M^* \rangle$ can be obtained unless spin analysis of the incident and final particles is performed.

E. Orientation and Alignment of Atomic Ensembles[103,104,114–118]

If no preferred axis exists in the excitation of the target atoms, *isotropic excitation* takes place. Measurement of the number of excited atoms in the various possible magnetic substates $M = L, L-1, \ldots, -L$ with respect to any arbitrary direction will reveal that these substates have been equally populated. Figure 5-10-2*a* illustrates this situation.

In the case where the excitation is produced by a beam of electrons and the scattered electrons are not observed, the axis of the electron beam is the preferred axis. We now have the situation pictured in Fig. 5-10-2*b*, where the Z axis is in the beam direction. Axial symmetry requires that ρ be diagonal in M:

$$\langle LM'|\rho|LM\rangle = \langle LM|\rho|LM\rangle\delta_{M'M}, \tag{5-10-21}$$

and we have an incoherent superposition of magnetic substates. The cross sections for excitation of the various substates will, in general, be unequal, so the populations will be unequal. If, as shown here, the ensemble is invariant under reversal of the symmetry axis, the population of substate $-M$ must be the same as for the substate $+M$; hence

$$\langle LM|\rho|LM\rangle = \langle L-M|\rho|L-M\rangle. \tag{5-10-22}$$

An ensemble with the symmetry property (5-10-22) is said to be *aligned*, and the mean value of the orbital angular momentum transferred to the atoms during the excitation is zero: $\langle \mathbf{L}\rangle = 0$.

An ensemble that is axially symmetric but not invariant under reversal of the symmetry axis is illustrated in Fig. 5-10-2*c*, where the populations for substates M and $-M$ are different. Here $\langle L_x\rangle = \langle L_y\rangle = 0$, but $\langle L_z\rangle \neq 0$. We say that this ensemble is *oriented* because its mean orbital angular momentum is not zero:

$$\langle \mathbf{L}\rangle \neq 0. \tag{5-10-23}$$

In experiments involving excitation by a beam and observation of the scattered electrons, there is a *plane* of symmetry, and invariance under reflection in this plane requires $\langle \mathbf{L}\rangle$ to be normal to the scattering plane.[104] Only $\langle L_y\rangle$

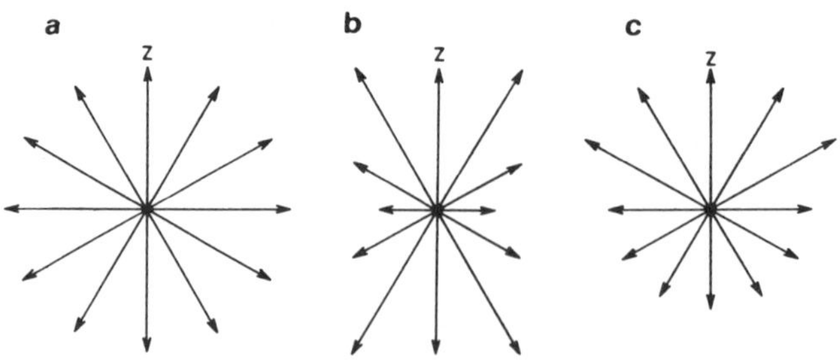

Figure 5-10-2. (*a*) Isotropic excitation; (*b*) Aligned axially symmetric system; (*c*) Oriented axially symmetric system. (From Blum and Kleinpoppen.[103]) Here $L = 2$, and the length of each vector is proportional to the population of the corresponding magnetic substate.

can be nonzero, and it can be shown that for ^1P excitation,

$$\sigma\langle L_y\rangle = -2\sqrt{\lambda(1-\lambda)}\sin\chi. \tag{5-10-24}$$

We see that the deviation of the phase χ from zero measures the degree of orientation of the ensemble.

Experimental studies of alignment and orientation in atomic collisions go back to the early 1970s; their objective is to elucidate the shape and dynamics of the electronic charge clouds of atoms that are excited in collisions. The information that can be obtained subjects the theory of collisions to a much more stringent test than does the cross section that has been measured traditionally. Indeed, in the most favorable cases, the complete quantum mechanical state of the system may be determined, so that a "perfect" or "complete" scattering experiment (cf. Section 4-6) is feasible. A significant advantage inherent in the alignment and orientation studies is that the results related to shape and dynamics can be expressed in terms of dimensionless parameters based on relative measurements; absolute measurements are not required.

Various choices of parameters have been used in these studies. For S- to -P excitation, four alignment and orientation parameters are required in the general case; they are functions of the scattering angle as well as the isotropic part of the differential cross section. A set of parameters that describe the charge cloud in an instructive fashion is the following: γ (the alignment angle of the charge cloud after the collision), P_l (the linear polarization of the charge cloud), ρ_{00} (the relative height of the charge cloud), and $L_\perp$ (the angular momentum transferred to the atom perpendicular to the scattering plane). Figure 5-10-3 will help in the visualization of these quantities. In the case of fully coherent excitation, the wave function is completely specified by γ and $L_\perp$. If, on the other hand, several processes that are distinguishable in principle contribute to the excitation, the excitation may not be coherent, as in the case of excitation of an $H(n^2P)$ level at low energies, where exchange effects give different amplitudes for singlet and triplet scattering. In this case, three parameters are needed: γ, $L_\perp$, and P_l. In the general case, for which reflection symmetry does not exist, all four parameters are required, as in the excitation of heavy noble gas atoms where spin-orbit effects during the collision are important. Spin-dependent phenomena in inelastic electron-atom collisions are treated in the book by Kessler[105] and in the reviews by Blum and Kleinpoppen[119] and Hanne.[120] The pioneering theoretical paper underlying the study of alignment and orientation is that of Macek and Jaecks.[114] Several recent publications treat both the theoretical and experimental work in the field in considerable detail.[115-118]

As we have pointed out, the e-$h\nu$ coincidence studies involve either a measurement of the angular distribution of the photons or an analysis of the polarization of the emitted light (a "coherence" measurement). An interesting and important technique has been developed by Hertel and Stoll[121] as an alternative to the coincidence measurements. They describe it as a time-reversed

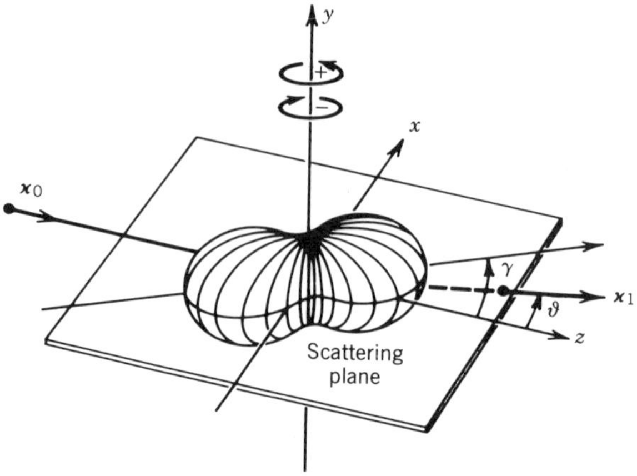

Figure 5-10-3. The angular part of the electron density of a P state produced by excitation of an atom originally in an S state. The length of the charge cloud is denoted by A, its width in the scattering plane by B, and its height perpendicular to this plane by H. The alignment angle γ is shown. The linear polarization is defined as $P_l = (A - B)/(A + B)$, and the relative height by $\rho_{00} = H/(A + B + H)$. The angular momentum transferred to the atom perpendicular to the scattering plane is $L_\perp$. (Adapted from Andersen et al.[115])

coincidence experiment in which laser light excites target atoms, a beam of monoenergetic electrons collisionally deexcites these atoms, and the electrons are superelastically scattered in the process. The scattering is analyzed as a function of the polarization of the laser light. The use of a laser for excitation provides an important advantage in terms of energy resolution; cross sections for hyperfine states separated by only 10^{-7} eV can be distinguished. The basic technique described here (and also in Section 5-10-J-3) can also be used to measure cross sections for excitation of the already excited target atoms. The theory of electron scattering by laser-excited atoms has been described by Macek and Hertel.[122]

F. State Multipoles

We now introduce a new set of parameters, which are linear combinations of the density matrix elements (5-10-7), and which provide an alternative and sometimes more convenient characterization of an ensemble of atoms. We define the *state multipoles*, or *statistical tensors*, $\langle T(L)^\dagger_{KQ}\rangle$ for systems in a state of definite orbital angular momentum L as

$$\langle T(L)^\dagger_{KQ}\rangle = \sum_{M'M} (-1)^{L-M}(LM', L - M|KQ)\langle LM'|\rho|LM\rangle, \quad (5\text{-}10\text{-}25)$$

where $(LM', L-M|KQ)$ is a Clebsch-Gordan (vector-coupling) coefficient.* For the case of electron-beam excitation with observation of the inelastically scattered electrons, we may use (5-10-7) to write

$$\langle T(L)^{\dagger}_{KQ} \rangle = \sum_{M'M} (-1)^{L-M}(LM', L-M|KQ)\langle f_M, f^*_M \rangle. \tag{5-10-26}$$

For an ensemble with given L, the number of multipoles is determined by the conditions

$$K \leqslant 2L, \qquad -K \leqslant Q \leqslant K. \tag{5-10-27}$$

The multipoles $\langle T(L)^{\dagger}_{KQ} \rangle$ contain the same information as ρ, and hence completely characterize the subensemble of excited atoms selected in the coincidence experiment. However, they are irreducible tensors[104] (of rank K, with components labeled by Q), and consequently, they transform more simply under rotation than do the components of ρ. (In fact, irreducible tensors transform in the same way that spherical harmonics do.) It is important to note that for axially symmetric systems, ρ is diagonal, and hence all multipoles with $Q \neq 0$ vanish.

1. *The Statistical Tensor of Rank Zero: The State Monopole*

The lowest rank tensor corresponds to a state *monopole*, and is given, according to our normalization scheme, by

$$\langle T(L)^{\dagger}_{00} \rangle = \frac{\sigma}{\sqrt{2L+1}}. \tag{5-10-28}$$

where σ is the total differential cross section in (5-10-9). For an ensemble such as that shown in Fig. 5-10-2*a*, with an isotropic orbital angular momentum distribution, the density matrix elements in (5-10-25) are constants independent of M, and the sum over M vanishes if $K \neq 0$. Hence the ensemble is characterized by the single parameter $\langle T(L)^{\dagger}_{00} \rangle$.

2. *The Statistical Tensor of Rank One: The Orientation Vector*

The tensor $\langle T(L)^{\dagger}_{1Q} \rangle$ is a vector (the *orientation vector*) whose three components $(Q = +1, 0, -1)$ are related to the components of the mean orbital angular momentum vector $\langle \mathbf{L} \rangle$ of the subensemble. Specifically,[103]

$$\langle T(L)^{\dagger}_{1Q} \rangle = \frac{\sqrt{3}\,\sigma}{\sqrt{L(L+1)(2L+1)}} \langle L_Q \rangle \qquad (Q = \pm 1, 0), \tag{5-10-29}$$

*The Clebsch-Gordan coefficients are the coefficients in the unitary transformations from one angular momentum representation to another.[117,123-126] where

$$\langle L_{+1}\rangle = -\frac{1}{\sqrt{2}}(\langle L_x\rangle + i\langle L_y\rangle),$$

$$\langle L_0\rangle = \langle L_z\rangle,$$

$$\langle L_{-1}\rangle = \frac{1}{\sqrt{2}}(\langle L_x\rangle - i\langle L_y\rangle) \tag{5-10-30}$$

are the spherical vector components of $\langle \mathbf{L}\rangle$.

The Wigner-Eckart theorem is used in obtaining (5-10-29). It requires that all irreducible tensor operators V_{KQ} satisfy the relation

$$\langle J'M'|V_{KQ}|JM\rangle = (-1)^{J'-M'}\begin{pmatrix} J' & K & J \\ -M' & Q & M \end{pmatrix}\langle J'||V_K||J\rangle. \tag{5-10-31}$$

The last factor on the right is a reduced matrix element: a scalar, independent of M', M, and Q. The factor immediately preceding it is a 3-j symbol, a number that reflects the geometry of the situation. The Wigner-Eckart theorem separates the purely geometrical part of the problem from the dynamical part. Irreducible tensors are usually discussed in the spherical coordinate representation, rather than in the Cartesian form, because of the simpler transformation properties in the former.

From our discussion of Fig. 5-10-2*c* we recall that an oriented, axially symmetric ensemble has $\langle L_x\rangle = \langle L_y\rangle = 0$ and $\langle L_z\rangle \neq 0$. It can be shown here that

$$\langle T(L)^\dagger_{10}\rangle = \frac{1}{\sqrt{L(L+1)}}\sum_M M\langle M|\rho|M\rangle, \tag{5-10-32}$$

where $\langle M|\rho|M\rangle$ is proportional to the population of the substate M. We see that $\langle T(L)^\dagger_{10}\rangle$ describes the excess of orbital angular momentum in the $+Z$ or $-Z$ direction. If the ensemble is not axially symmetric, $\langle T(L)^\dagger_{1,+1}\rangle$ or $\langle T(L)^\dagger_{1,-1}\rangle$ may be nonzero and produce orientation. A general definition of an *oriented system* is one for which at least one of the components of the orientation vector is nonzero. We should note that the orientation vector completely characterizes the orbital angular momentum distribution of an ensemble only if the particles have total angular momentum $J = \frac{1}{2}$. (Of course, electrons are included here.) In this case, the orientation vector is proportional to the spin polarization vector.

It is possible to show by invoking parity invariance that the orientation vector is a measure of the induced magnetic moment of the system.[103,104]

3. *The Statistical Tensor of Rank Two: The Alignment Tensor*

The next tensor in the hierarchy is of rank two and has five components ($K = 2$; $Q = \pm 2, \pm 1, 0$). We call $\langle T(L)^{\dagger}_{2Q}\rangle$ the alignment tensor because of the following considerations. If an ensemble is axially symmetric, as in Fig. 5-10-2*b*, $\langle T(L)^{\dagger}_{20}\rangle$ is the only nonzero component of this tensor, and also $\langle T(L)^{\dagger}_{10}\rangle = 0$ because the ensemble is aligned. A table of Clebsch-Gordan coefficients provides the result

$$\langle T(L)^{\dagger}_{20}\rangle = C(L) \sum_{M} (3M^2 - L(L+1))\langle LM|\rho|LM\rangle, \tag{5-10-33}$$

where

$$C(L) = \frac{\sqrt{5}}{\sqrt{L(L+1)(2L+1)(2L+3)(2L-1)}}. \tag{5-10-34}$$

This single nonzero component of $\langle T(L)^{\dagger}_{2Q}\rangle$ describes the orbital angular momentum distribution of a system like that in Fig. 5-10-2*b*. The fact that the magnetic quantum number appears as M^2 in (5-10-33) makes $\langle T(L)^{\dagger}_{20}\rangle \neq 0$ for an aligned system. For ensembles without axial symmetry, other components of $\langle T(L)^{\dagger}_{2Q}\rangle$ with $Q \neq 0$ will be nonzero. We say, in general, that an ensemble is aligned if at least one component of the tensor $\langle T(L)^{\dagger}_{2Q}\rangle$ is nonvanishing. The more general term *polarized ensemble* is applied if at least one multipole $\langle T(L)^{\dagger}_{KQ}\rangle$ with $K \neq 0$ is not zero.

The tensors $\langle T(L)^{\dagger}_{KQ}\rangle$ are related to the electric and magnetic multipoles of the atomic system. The Wigner-Eckart theorem can be used to show that the electric quadrupole tensor $\mathcal{Q}_{2Q}$ is proportional to the alignment tensor:

$$\mathcal{Q}_{2Q} = A(L)\bar{\mathcal{Q}}\langle T(L)^{\dagger}_{2Q}\rangle. \tag{5-10-35}$$

Here $A(L)$ is a numerical factor, and $\bar{Q}$ is the quantum mechanical quadrupole moment of the system:

$$\bar{Q} = \langle LL|Q_{20}|LL\rangle. \tag{5-10-36}$$

(compare with the classical definition in Section 5-2). The electric quadrupole tensor is partitioned into a dynamic factor $\bar{Q}$ and a geometric factor $\langle T(L)^{\dagger}_{KQ}\rangle$, which describe different attributes of the system.

The number of independent multipoles is restricted by the hermiticity condition

$$\langle T(L)^{\dagger}_{KQ}\rangle^* = (-1)^Q \langle T(L)^{\dagger}_{K-Q}\rangle \tag{5-10-37}$$

and by the symmetry considerations peculiar to the situation being considered. For example, a system with $L = 1$ that is invariant under reflection in a plane can be completely specified by the five parameters $\langle T(L)^{\dagger}_{00}\rangle$, $\langle T(L)^{\dagger}_{11}\rangle$, $\langle T(L)^{\dagger}_{22}\rangle$, $\langle T(L)^{\dagger}_{21}\rangle$, $\langle T(L)^{\dagger}_{20}\rangle$. However, only three of these parameters are independent for excitation of ^1P or ^3P levels from a ^1S state.

We have already expressed the orientation vector in terms of the spherical vector components of $\langle \mathbf{L}\rangle$ in (5-10-29). Similarly, if reflection invariance in the X-Z plane applies, the components of the alignment tensor can be written

$$\langle T(L)^{\dagger}_{20}\rangle = N_2 \frac{\sigma}{\sqrt{6}}(\langle 3L_z^2 - L^2\rangle),$$

$$\langle T(L)^{\dagger}_{21}\rangle = N_2 \frac{\sigma}{\sqrt{2}}\langle L_x L_z + L_z L_x\rangle,$$

$$\langle T(L)^{\dagger}_{22}\rangle = N_2 \frac{\sigma}{\sqrt{2}}\langle L_x^2 - L_y^2\rangle, \tag{5-10-38}$$

where

$$N_2 = \sqrt{30}[(2L+3)(2L+1)(L+1)L(2L-1)]^{-1/2} \tag{5-10-39}$$

is a normalization factor. An alternative set of parameters, introduced by Fano and Macek,[127] can be used to describe this situation: They define 0_{1-} to specify the orientation of the system, where

$$0_{1-} = \frac{\langle L_y\rangle}{L(L+1)}, \tag{5-10-40}$$

and A_{2+}, A_{1+} and A_0 to characterize the alignment:

$$A_{2+} = \frac{\langle L_x^2 - L_y^2\rangle}{L(L+1)}, \quad A_{1+} = \frac{\langle L_x L_z + L_z L_x\rangle}{L(L+1)}, \quad A_0 = \frac{\langle 3L_z^2 - L^2\rangle}{L(L+1)} \tag{5-10-41}$$

Blum and Kleinpoppen[103] display the explicit results for the monopole, orientation vector, and alignment tensor for the case of ^1P states:

$$\langle T(1)^{\dagger}_{00}\rangle = \frac{\sigma}{\sqrt{3}}, \qquad \langle T(1)^{\dagger}_{11}\rangle = -i\sigma\sqrt{\lambda(1-\lambda)}\sin\chi, \tag{5-10-42}$$

$$\langle T(1)^{\dagger}_{22}\rangle = \frac{1}{2}\sigma(\lambda - 1), \qquad \langle T(1)^{\dagger}_{21}\rangle = -\sigma\sqrt{\lambda(1-\lambda)}\cos\chi,$$

$$\langle T(1)^{\dagger}_{20}\rangle = \frac{1}{\sqrt{6}}\sigma(1 - 3\lambda).$$

The sign of χ can be determined only if the orientation vector is known. Similar relations hold for the excitation of ^3P states from a ^1S state.

G. The Polarization of Light Emitted by Atoms Excited by Electron Impact

Here we are interested in the polarization of the light produced by the impact of both unpolarized and polarized electron beams. The fact that the polarization characteristics depend on whether the photons are observed in coincidence with the scattered electrons indicates that coincidence measurements can reveal details of the excitation process that are not accessible if only the electrons or the photons are observed.

We shall use the geometry of Fig. 5-10-4, in which the incident electron beam is along the $+Z$ axis and the photons are detected only along the unit vector $\hat{n}$, oriented by the angles θ_v and φ_v. The unit vector $\hat{e}_1$ lies in the plane formed by $\hat{n}$ and the Z axis and is in the direction $(\theta_v + 90°, \varphi_v)$. The unit vector $\hat{e}_2$ is perpendicular to both $\hat{n}$ and $\hat{e}_1$; it is oriented by the angles $(90°, \varphi_v + 90°)$ and points in the direction of increasing φ_v. These three unit vectors span the helicity system of the emitted photons, $\hat{n}$ defining the quantization axis in this system. The polarization vector **P** of the photons has the direction of the electric field intensity **E** and is restricted to a plane perpendicular to $\hat{n}$ that is spanned by $\hat{e}_1$ and $\hat{e}_2$.

A set of four independent measurements must be made to determine the properties of a light beam completely, and four parameters are required for its complete specification. Before introducing such a set of parameters, we must define several quantities. $I(\alpha)$ will represent the light intensity transmitted through a linear-polarization filter with its transmission axis oriented at an angle α with respect to $\hat{e}_1$ in the $\hat{e}_1$–$\hat{e}_2$ plane. I^+ and I^- will denote the intensities

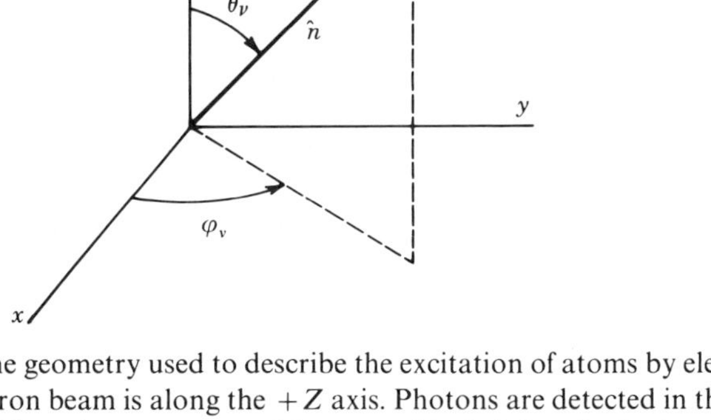

Figure 5-10-4. The geometry used to describe the excitation of atoms by electron impact. The incident electron beam is along the $+Z$ axis. Photons are detected in the direction of the unit vector $\hat{n}$. The unit vectors $\hat{e}_1$ and $\hat{e}_2$ are used in defining the Stokes parameters. $\hat{e}_1$ lies in the plane defined by $\hat{n}$ and the Z axis; $\hat{e}_2$ lies in the X-Y plane and makes an angle φ_v with respect to the Y axis (Adapted from Kessler.[105])

of the circularly polarized light* with positive and negative helicities, respectively, traveling in the direction of $\hat{n}$. Now we define the Stokes parameters (vintage 1852), which provide a convenient means of characterizing a beam of light:

1. The total intensity of the beam along $\hat{n}$:

$$I = I(0^\circ) + I(90^\circ) \tag{5-10-43}$$

2. The degree of linear polarization with respect to an axis oriented at 45° with respect to $\hat{e}_1$:

$$\eta_1 = \frac{I(45^\circ) - I(135^\circ)}{I} \tag{5-10-44}$$

3. The degree of circular polarization:

$$\eta_2 = \frac{I^+ - I^-}{I} \tag{5-10-45}$$

4. The degree of linear polarization with respect to $\hat{e}_1$:

$$\eta_3 = \frac{I(0^\circ) - I(90^\circ)}{I}. \tag{5-10-46}$$

All four of the Stokes parameters depend on θ_v and φ_v. For complete linear polarization in the $\hat{e}_1$ direction, $I = I(0^\circ)$, $\eta_1 = \eta_2 = 0$, and $\eta_3 = 1$. On the other hand, complete linear polarization in a direction of 45° with respect to $\hat{e}_1$ corresponds to $I = I(45^\circ)$, $\eta_1 = 1$, and $\eta_2 = \eta_3 = 0$.

There is a close relation between the polarization of the incident electron beam and that of the radiation emitted by the excited atoms. Table 5-10-1 shows this relation for the case where only the photons, not the scattered electrons, are detected. All the information presented here can be derived from considerations of symmetry. We assume that only electric dipole radiation is important in this context; results in Table 5-10-1 that cannot be extended to higher multipole transitions because of their different polarization characteristics are indicated by the notation "ed."

An important fact that emerges here is that a polarized electron beam can produce polarized radiation in cases where an unpolarized beam cannot. We should note that the symmetry arguments used to construct Table 5-10-1 merely

*In classical optics, the convention is that we look at the oncoming radiation and call the polarization "right-hand" if we see the **E** vector rotating clockwise, and "left-hand" if it appears to be rotating counterclockwise. The opposite convention is used in certain other fields of physics. Much confusion centers on this point. More will be said about helicity in connection with (5-10-50).

Table 5-10-1 Polarization of Light Emitted in Electron-Impact Excitation Experiments That Do Not Involve Electron-Photon Coincidence[a]

Electron Polarization (0 means unpolarized)	Position of Photon Detector along Direction	Light Polarization		
		η_1	η_2	η_3
0	y	0	0	$\neq 0$
Longitudinal (z)	y	0(ed)	0(ed)	$\neq 0$
Transverse (y)	y	$\neq 0$	$\neq 0$	$\neq 0$
0	x	0	0	$\neq 0$
Longitudinal (z)	x	0(ed)	0(ed)	$\neq 0$
Transverse (y)	x	0	0	$\neq 0$
0	z	0	0	0
Longitudinal (z)	z	0	$\neq 0$	0
Transverse (y)	z	0	0	0

Source: Kessler.[105]

[a]The Stokes parameters η_1 and η_3 relate to linear light polarization, η_2 to circular light polarization. The incident electron beam is in the $+Z$ direction. Transverse electron polarization is taken to be along the $+Y$ axis. The notation "ed" means that the result can be applied only to electric dipole transitions. Recent experimental and theoretical studies of relevance here are cited in Refs. 128 and 129, respectively.

show which kinds of light polarization can occur, not the magnitudes of the polarizations. The latter are determined by the dynamics of the excitation process in question. For example, η_1 can be nonzero only if the spin-orbit interaction plays a role.[105]

If now the photons are detected in coincidence with the scattered electrons, as in Fig. 5-10-1, the degree of symmetry is reduced, and the results of Table 5-10-2, apply. In this table, P_{ex}, P_{ey}, and P_{ez} are the Cartesian components of the electron polarization vector, $\mathbf{P}_e$. We see that most of the light observed in a coincidence experiment is polarized. If the photons are observed in some direction other than that of the X, Y, or Z axis, η_1, η_2, and η_3 are nonzero in general, and they all depend on all the components of the electron polarization. Hence coincidence experiments can provide more details concerning the dynamics of the excitation than can experiments without this feature.

The vectors $\hat{e}_1$ and $\hat{e}_2$ form a basis set for expression of any polarization vector $\hat{e}$:

$$\hat{e} = a_1\hat{e}_1 + a_2\hat{e}_2, \tag{5-10-47}$$

where $|a_1|^2 + |a_2|^2 = 1$. We can rewrite (5-10-47) in the form

$$\hat{e} = \hat{e}_1 \cos\beta + \hat{e}_2 e^{i\delta} \sin\beta, \tag{5-10-48}$$

Table 5-10-2 Polarization of Light Emitted in Electron-Impact Excitation Experiments That Involve Electron-Photon Coincidence[a]

Electron Polarization (0 means unpolarized)	Position of Photon Detector along Direction	Light Polarization η_1	η_2	η_3
0	y	$\neq 0$	$\neq 0$	$\neq 0$
$P_{ex}\hat{e}_x + P_{ez}\hat{e}_z$	y	$\neq 0$	$\neq 0$	$\neq 0$
$P_{ey}\hat{e}_y$	y	$\neq 0$*	$\neq 0$*	$\neq 0$*
0	x	0	0	$\neq 0$
$P_{ex}\hat{e}_x + P_{ez}\hat{e}_z$	x	$\neq 0$*	$\neq 0$*	$\neq 0$
$P_{ey}\hat{e}_y$	x	0	0	$\neq 0$*
0	z	0	0	0
$P_{ex}\hat{e}_x + P_{ez}\hat{e}_z$	z	0	$\neq 0$*	0
$P_{ey}\hat{e}_y$	z	0	0	0

Source: Kessler.[105]

[a]The scattering plane is the X-Z plane (see Fig. 5-10-1). The incident electron beam is in the $+Z$ direction. P_{ex}, P_{ey}, and P_{ez} are the Cartesian components of the electron polarization vector; $\hat{e}_x$, $\hat{e}_y$, and $\hat{e}_z$ are unit vectors. The asterisk (*) means that the value of η differs from that for unpolarized electrons.

where $\cos\beta = a_1$, $\sin\beta = a_2$, and δ is the phase that describes elliptical polarization. Then linear polarization is represented by

$$\hat{e}_{\text{linear}} = \hat{e}_1 \cos\beta + \hat{e}_2 \sin\beta \tag{5-10-49}$$

(with β being the angle between the polarization vector and $\hat{e}_1$) and circular polarization by

$$\hat{e}_{\text{circular}} = \hat{e}_{\pm 1} = \mp \frac{1}{\sqrt{2}}(\hat{e}_1 \pm i\hat{e}_2). \tag{5-10-50}$$

The helicity of the photon is indicated by the subscript ± 1. A photon is a spin-1 particle having zero rest mass that consequently can have its spin only parallel or antiparallel to the direction of the photon propagation. The state of helicity $+1$ refers to a photon with spin parallel to the propagation direction, helicity -1 to the opposite situation.

If a beam of light of intensity I and polarization given by (5-10-48) impinges upon a polarizer that transmits only light with polarization $\hat{e}$, the intensity of the transmitted light will be

$$I(\hat{e}) = \tfrac{1}{2}I(1 + \eta_3 \cos 2\beta + \eta_1 \sin 2\beta \cos\delta + \eta_2 \sin 2\beta \sin\delta). \tag{5-10-51}$$

The polarization state of the photons is completely characterized by the Stokes parameters, and these parameters can be used to express the results of any polarization experiment.

A beam that is linearly polarized at an angle α with respect to $\hat{e}_1$ will be reduced from intensity I to intensity $I(\hat{e})$ on passing through a linear polarizer, where

$$I(\hat{e}) = \tfrac{1}{2}I(1 + \eta_3 \cos 2\alpha + \eta_1 \sin 2\alpha). \tag{5-10-52}$$

On the other hand, if the incident beam is completely polarized along $\hat{e}_1$,

$$I(\hat{e}) = I \cos^2 \alpha, \tag{5-10-53}$$

which is the law of Malus. Finally, if the beam encounters a circular polarization filter,

$$I(\hat{e}) = \tfrac{1}{2}I(1 \pm \eta_2). \tag{5-10-54}$$

The degree of polarization of a beam of light is given by

$$P = \frac{\sqrt{(I\eta_1)^2 + (I\eta_2)^2 + (I\eta_3)^2}}{I} = \sqrt{\eta_1^2 + \eta_2^2 + \eta_3^2} \leqslant 1. \tag{5-10-55}$$

The value $P = 1$ corresponds to complete polarization, the beam being characterized by a single polarization vector $\hat{e}$. Only if $P = 1$ can we find a polarizer that will transmit the beam without loss; the beam in this case is in a *pure polarization state.* The light is completely coherent and arises from a transition between two pure atomic states. If $P < 1$, the beam is in an incoherent mixture of pure polarization states, each of which has a different polarization vector. We see that the value of P is connected with the degree of coherence of the excitation process. Measurement of P reveals whether the excited atoms are in a pure state ($P = 1$) or in a statistical mixture of such states ($P < 1$). The beam is unpolarized if $P = 0$.

The *photon density matrix* also proves useful here; it is given in the helicity representation as

$$\rho_{\lambda'\lambda} = \begin{pmatrix} \rho_{11} & \rho_{1-1} \\ \rho_{-11} & \rho_{-1-1} \end{pmatrix} = \frac{I}{2}\begin{pmatrix} 1 + \eta_2 & -\eta_3 + i\eta_1 \\ -\eta_3 - i\eta_1 & 1 - \eta_2 \end{pmatrix}. \tag{5-10-56}$$

The indices λ' and λ take on the values of the helicity: ± 1. The photon density matrix contains all the information about the angular distribution and polarization of the emitted radiation.

Another quantity that is frequently used to describe the radiation is the *coherence correlation factor*, μ, defined in terms of elements of the photon density matrix and related to the Stokes parameters as follows:

$$\mu = \frac{\rho_{12}}{\sqrt{\rho_{11}\rho_{22}}} = \frac{\eta_1 + i\eta_2}{\sqrt{1 - \eta_3^2}} = |\mu|e^{i\beta}. \tag{5-10-57}$$

where $|\mu|$ is the *degree of coherence* and β is the *effective phase difference*. If the light is completely coherent, $|\mu| = |P| = 1$.

H. Evaluation of Excitation and Radiation Parameters from Experimental Data on $n^1P \to 1^1S_0$ Transitions

Now let us examine the emission of light from n^1P excited atoms in their decay to the ground 1^1S_0 state.[106] The measured intensity of the light emitted into the solid angle $d\omega$ at (θ_v, φ_v) during the time interval $(t, t + dt)$ is

$$I(\hat{e}, \theta_v, \varphi_v) \sim |\langle {}^1S|\hat{e}\cdot\mathbf{r}|{}^1P\rangle|^2 e^{-\gamma t} \tag{5-10-58}$$

if it is assumed that the detector is sensitive only to polarization $\hat{e}$. Evaluation of the various electric dipole matrix elements in (5-10-58) leads to the following expressions for the Stokes parameters:

$$I = C\{\tfrac{1}{2}(1 - \lambda)(1 + \cos^2\theta_v - \sin^2\theta_v \cos 2\varphi_v) + \lambda \sin^2\theta_v + [\lambda(1 - \lambda)]^{1/2} \cos\chi \sin 2\theta_v \cos\varphi_v\}, \tag{5-10-59}$$

$$I\eta_1 = C\{-(1 - \lambda)\cos\theta_v \sin 2\varphi_v - 2[\lambda(1 - \lambda)]^{1/2} \cos\chi \sin\theta_v \sin\varphi_v\}, \tag{5-10-60}$$

$$I\eta_2 = C\{2[\lambda(1 - \lambda)]^{1/2} \sin\chi \sin\theta_v \sin\varphi_v\}, \tag{5-10-61}$$

$$I\eta_3 = -C\{\tfrac{1}{2}(1 - \lambda)[\sin^2\theta_v - (1 + \cos^2\theta_v)\cos 2\varphi_v] - \lambda \sin^2\theta_v - [\lambda(1 - \lambda)]^{1/2} \cos\chi \sin 2\theta_v \cos\varphi_v\}, \tag{5-10-62}$$

where

$$C = \frac{e^2(2\pi\nu)^4\, d\omega}{2\pi c^3 \hbar} |\langle {}^1S||\mathbf{r}||P\rangle|^2 \frac{\sigma}{3} e^{-\gamma t}. \tag{5-10-63}$$

Here γ is the decay constant for the transition, and σ, λ, and χ are the parameters defined in (5-10-9), (5-10-19), and (5-10-20). Equation (5-10-59), with (5-10-63), shows how the measurement of I at a fixed detector position can provide the decay constant, and hence the lifetime of the excited state.[89] In general, the time dependence of the intensity is of no direct interest. For determinations of λ and χ, the time-integrated total intensity

$$I(\hat{n}) = \int_0^\infty I(\hat{n}, t)\, dt \tag{5-10-64}$$

is what is needed, and $e^{-\gamma t}$ in (5-10-63) is replaced by $1/\gamma$ (a constant).

Blum and Kleinpoppen[103] express the Stokes parameters for singlet-singlet ($S = 0$, $S_f = 0$) transitions in terms of the multipoles $\langle T(L)^\dagger_{KQ} \rangle$. They consider the general case of arbitrary L, L_f and then specialize to $L = 1$, $L_f = 0$ (see Ref. 103, pp. 227–229). The Stokes parameters can also be expressed in terms of the alignment and orientation parameters.

As indicated by eq. (5-10-9), measurement of the total differential cross section yields the parameter σ. Two different kinds of measurements can be made to obtain λ and χ:

1. *Angular correlation studies* provide $I(\theta_v, \varphi_v)$, and a least-squares fit to (5-10-59) gives λ and $|\chi|$. The sign of χ cannot be determined because I depends on $\cos \chi$, not on χ itself.
2. *Polarization correlation studies* yield λ and χ. To see this, locate the photon detector at $\theta_v = \varphi_v = 90^\circ$, and rewrite eq. (5-10-52) in terms of β, the angle of a linear polarizer with respect to the incident electron beam axis:

$$I(\beta) = \tfrac{1}{2}I(1 + \eta_3 \cos 2\beta + \eta_1 \sin 2\beta). \qquad (5\text{-}10\text{-}65)$$

We now define the experimental polarization correlation parameter to be

$$P_C(\beta) = \frac{I(\beta) - I(\beta + \pi/2)}{I(\beta) + I(\beta + \pi/2)} \qquad (5\text{-}10\text{-}66)$$

and use the equations for the Stokes parameters to show that

$$\lambda = \tfrac{1}{2}[1 + P_C(0)] \qquad (5\text{-}10\text{-}67)$$

and

$$\cos \chi = \tfrac{1}{2}[\lambda(1 - \lambda)]^{1/2} P_C(\tfrac{3}{4}\pi). \qquad (5\text{-}10\text{-}68)$$

We see that again we can determine λ and $|\chi|$, but now a measurement of the circular polarization will provide the sign of χ, because

$$\eta_2 = \frac{I^+ - I^-}{I^+ + I^-} = +2[\lambda(1 - \lambda)]^{1/2} \sin \chi. \qquad (5\text{-}10\text{-}69)$$

This equation is obtained by dividing (5-10-61) by (5-10-59), with θ_v and φ_v put equal to 90°.

We see that the angular correlation and photon correlation measurements are equivalent except for the matter of the sign of χ. The choice between the two is largely based on practical considerations. For example, polarization measurements are difficult in the VUV region, and the angular correlation method is often selected here.

Once σ, λ, and χ have been determined, the multipoles can be obtained from (5-10-42). Then 0_{1-} is given by (5-10-24) and (5-10-40), and A_{2+}, A_{1+}, and A_0 by (5-10-38) and (5-10-41).

I. Other Important Topics

We shall not discuss in any detail several additional topics that are of considerable importance:

1. The observation of light emitted in nonsinglet transitions
2. The influence of fine and hyperfine interactions following the excitation of the atom
3. The excitation of states without definite L and parity
4. The excitation of heavy atoms, for which the spin-orbit interaction between the incident electron and the target atom is important
5. Quantum beats (fine structure beats, and beats between states of different L caused by coherent excitation of these states)
6. Studies of molecules

These subjects are treated in Refs. 103–106.

J. Experimental Techniques

Here we briefly describe a few representative experimental apparatus that illustrate some of the techniques used to probe the fine details of the electron-impact excitation process. The reader may consult Refs. 103–109 and 115–122 for more details, and for experimental data and comparisons with theory.

1. *Angular Correlation*

The first measurements of this type in atomic physics were made by Eminyan et al.[130] in Kleinpoppen's laboratory at the University of Stirling. The apparatus pictured in Fig. 5-10-5 was employed. A beam of atoms moving up out of the drawing is crossed by an energy-selected electron beam, and electrons scattered in a selected direction (ϑ, φ) with a specified energy loss are detected in a pulse-counting electron multiplier. Delayed coincidences are observed between inelastically scattered electrons and photons emitted by the excited atoms. The incident electron beam current is typically about 1 μA, and the beam diameter is about 1 mm, with an angular spread of 1° (FWHM). The electron and photon counters, along with the Faraday cup that collects the unscattered electrons, lie in a plane, and the electron analyzer/detector and the photon counter can be separately rotated about the collision region. The background pressure is about 2×10^{-8} torr.

The pulses from the electron and photon counters are amplified and sent into a constant-fraction timing discriminator. An electron pulse starts the ramp of a

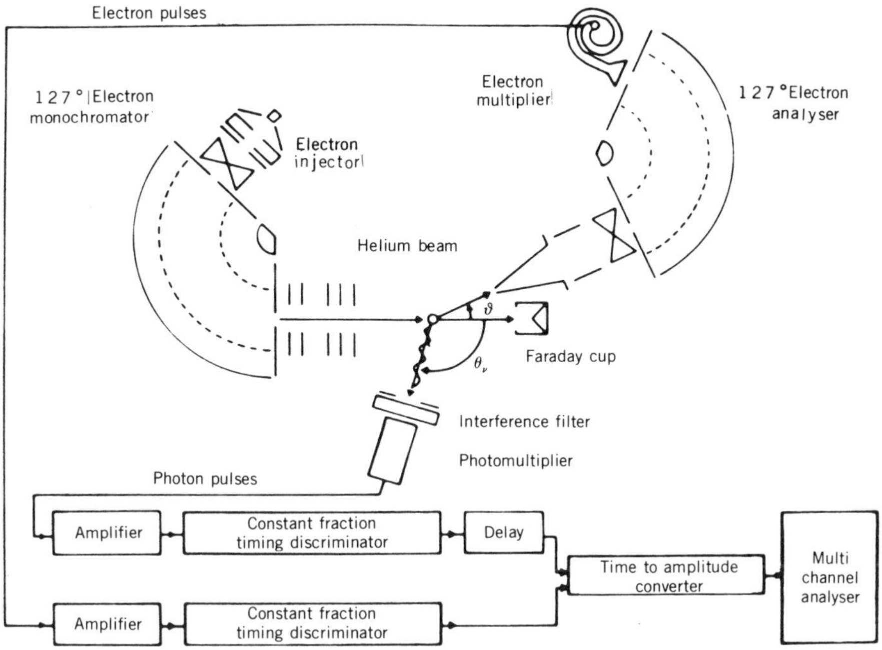

Figure 5-10-5. Schematic diagram of the e-$h\nu$ coincidence apparatus of Eminyan et al.[130] (not to scale). The plane of the drawing is the scattering plane. (From Eminyan et al.[130])

time-to-amplitude converter (TAC), and a photon pulse arrives later to stop the ramp. Thus the amplitude of the TAC output signal is proportional to the time difference between the start and stop pulses. A multichannel analyzer (MCA) records the resulting pulse height spectrum from the TAC. This technique has replaced the use of variable delay lines that were employed in the early angular correlation experiments in nuclear physics.[110]

Figure 5-10-6 shows a typical coincidence spectrum obtained for excitation of the 2^1P state of helium from the ground 1^1S_0 state by incident electrons of 80 eV. The electrons and photons that are associated with the same excitation events are correlated in time and produce the large peak. On the other hand, electron and photon counts from different excitation events are uncorrelated in time and merely provide a uniform background of random coincidences. With perfect instrumental time resolution, the peak would show a step-function rise followed by an exponential decay with the 2^1P lifetime (0.58 ns). The apparatus, in fact, has a time resolution of about 6 ns, and the peak shape in Fig. 5-10-6 results. The coincidence signal is the number of events under the peak corrected for the random coincidence background, and this number divided by the integration time T for the measurements gives the real coincidence signal rate.

The coincidence technique offers the important advantage of suppressing photon contributions from excited atoms produced by cascading rather than by

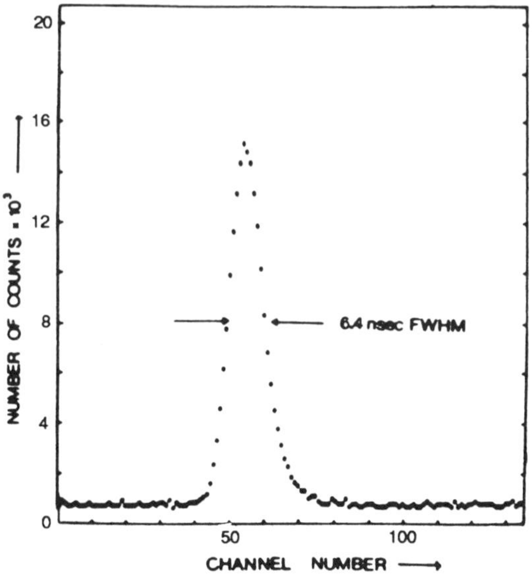

Figure 5-10-6. Delayed coincidence spectrum obtained by Eminyan et al.[130] for an electron energy of 80 eV, electron scattering angle $\vartheta = 16°$, and photon emission angle $\theta_v = 126°$. The accumulation time was about 12 hours. The channel width was 0.58 ns. The electron and photon count rates were 6 kHz and 21 kHz, respectively.

direct excitation. Provided that the bandpass of either the electron analyzer/detector or the photon detector is adequate to isolate the excited state in question, the only events correlated in time and contributing to the real coincidence signal are those associated with direct excitation.

Coincidence measurements are difficult, mainly because of the weak signals involved.[106] A basic experimental limitation arises because the photon detector subtends a solid angle of only about 1% of the total solid angle at the collision volume. At large electron scattering angles, the electron count rate may be of the order of 100 Hz, so that coincidence rates may be as low as 1 count per minute. This situation may be improved by using microchannel plates as position-sensitive detectors (Section 5-9) to register the emission of photons into many directions simultaneously. A second limitation arises from the fact that while the real coincidence rate is proportional to $\dot{N}$ (the rate at which the excited atoms are produced), the accidental coincidence rate is proportional to $\dot{N}^2$. For large $\dot{N}$, the accidental rate increases faster than the real rate, so there is a fundamental limitation on the enhancement of the real coincidence rate by merely increasing the rate of excitation events.

Eminyan et al.[130] deduced the ratio of the differential cross sections for exciting the degenerate magnetic sublevels of the 2^1P state and the relative phases of the corresponding excitation amplitudes. The electron scattering angle was varied between 16 and 40° and the incident electron energy between 40 and 200 eV.

Crowe et al.[131] have used similar techniques to study the $1^1S \rightarrow 3^1P$ excitation process in helium at electron impact energies of 34.6, 45.6, and 75.6 eV and electron scattering angles up to 35°. Their angular correlation function at 75.6 eV is shown in Fig. 5-10-7 along with data obtained by Eminyan et al. in 1975. None of the existing theoretical predictions were in satisfactory agreement with the values for λ, $|\chi|$, and the Fano-Macek orientation and alignment parameters obtained by Crowe et al.

Williams[132] also measured the angular correlation of Lyman-α photons resulting from the excitation of atomic hydrogen to the $2S$ and $2P$ levels by 54.4-eV electrons. The electron scattering angle was varied from 10 to 140°, and the angular correlation was measured for θ_v ranging from 10 to 140°. Again, no existing theory predicted values in agreement with Williams's excitation parameters, although theoretical differential cross sections for excitation to the $2S$ and $2P$ levels were consistent with his experimental results.

2. *Polarization Correlation*

First we discuss some recent experiments performed by McConkey and his colleagues at the University of Windsor, Ontario. Unpolarized incident electrons were used throughout in e-$h\nu$ coincidence studies. Becker et al.[133] studied the $X^1\Sigma_g^+ \rightarrow C^1\Pi_u \rightarrow X^1\Sigma_g^+$ excitation/deexcitation process in H_2 and

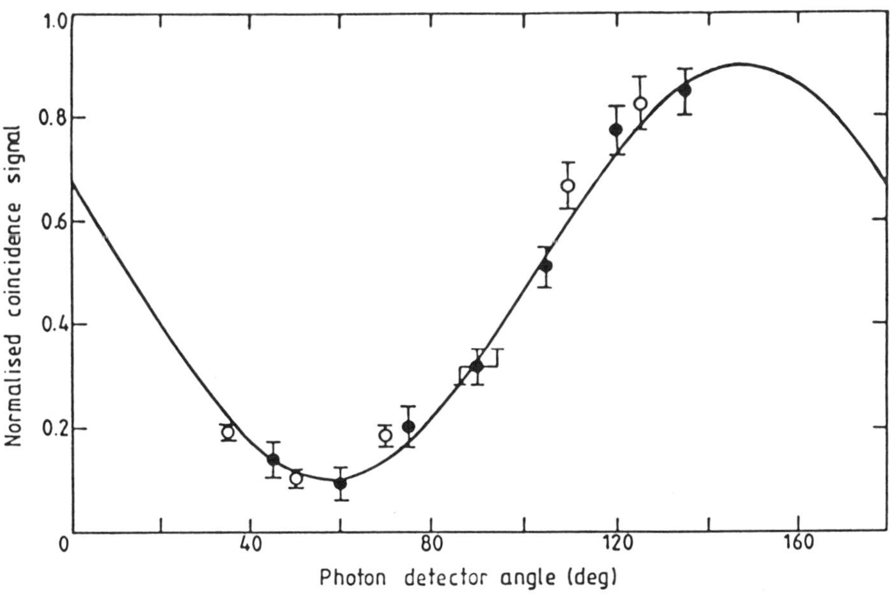

Figure 5-10-7. The 3^1P e-$h\nu$ angular correlation measured by Crowe et al.[131] in experiments on excitation of helium. The electron energy was 75.6 eV, and the electron scattering angle was 20°. The full curve is a least-squares fit of their data (solid circles) to (5-10-59). The open circles are data obtained by Eminyan et al. in 1975 at 80 eV.

D_2. They measured linear polarization correlations for integrated vibrational bands at electron scattering angles up to 25° and incident energies up to 80 eV. McConkey et al.[134] extended this study to measure the three normalized Stokes parameters characterizing the decay of the $d^3\Pi_u$ $(v_i = 0, N = 1)$ rotational level of H_2 at 25 eV of impact energy and a scattering angle of 60°. This experiment was the first application of the polarization correlation technique to molecules excited by electron impact to a specific rotational level. Finally, Khakoo et al.[135] used a VUV circular polarization analyzer developed by Westerveld et al.[136] to measure circular polarization correlations for the excitation of the 2^1P state of helium. Linear correlations were also measured, for electron energies of 50, 60, and 80 eV and scattering angles in the range 20 to 90°.

The first e-*hν* coincidence measurements with *polarized* electrons were made in 1983 by Kessler's group at Münster with the apparatus shown in Fig. 5-10-8.[137] Circularly polarized light from a krypton-ion laser ejects longitudinally polarized electrons from a GaAs photocathode. These electrons are deflected through 90°, and their polarization is rotated through 90° by two magnetic coils, so that $\mathbf{P}_e$ becomes transverse to the beam. The electrons then pass through a region of differential pumping, where they are again deflected through 90°. Finally, the electrons are focused onto a narrow section of an

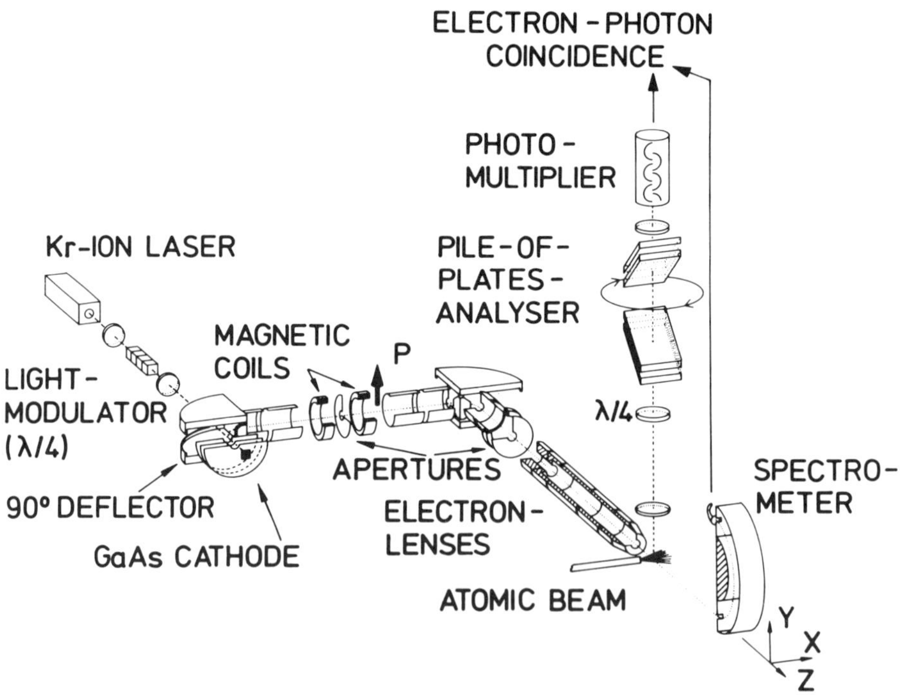

Figure 5-10-8. The e-*hν* apparatus used by Wolcke et al.[137] in Münster for the first atomic coincidence studies with polarized electrons.

atomic mercury beam. Some of these atoms are excited to the 6^3P_1 state, the electrons that produce the excitation losing 4.9 eV in the process. Electrons scattered in the forward direction with this energy loss are detected in coincidence with 254-nm photons emitted parallel to the electron polarization vector in the $6^3P_1 \rightarrow 6^1S_0$ deexcitation transition. The light analyzer consists of a collimating lens system, a $\lambda/4$ plate, a pile-of-plates analyzer, and a wavelength filter placed in front of the photomultiplier detector.

A $\lambda/4$ (quarter-wave) plate, such as the one shown in Fig. 5-10-8, is made of a transparent, anisotropic (birefringent or doubly refracting) material. There are two mutually perpendicular optical axes parallel to the flat surfaces of the plate. A linearly polarized light wave impinging normally on the plate will traverse it without a change in state of polarization if and only if the **E** vector is parallel to one of these axes. A wave polarized along one of the optical axes is propagated through the plate with a velocity different from that of a wave polarized along the other axis. A $\lambda/4$ plate has a thickness that just suffices to introduce a phase difference of $\lambda/4$ between two such waves. Thus a $\lambda/4$ plate can be used to convert linearly polarized light into circularly polarized light, and vice versa. In the Münster experiment, the $\lambda/4$ plate is used when circularly polarized light from the target region is studied, and its function is to produce linear polarization for analysis in the pile-of-plates. When linearly polarized light from the excited atoms is being studied, the $\lambda/4$ plate is moved out of the photon beam path.

The pile-of-plates is an analyzer of linear polarization. Its operation can be understood by examination of unpolarized light impinging at the Brewster angle upon a plate composed of some transparent, isotropic material. The reflected light is very nearly completely polarized, whereas the transmitted light has only a small degree of polarization. However, a stack of, say, 25 plates will have a large cumulative effect and produce very strong polarization in the transmitted beam (Problem 5-17). If the incident light approaches a plate at the Brewster angle, with its **E** vector in the plane of incidence ($\parallel$ polarization), the intensity of the reflected beam is zero, and the light is totally transmitted. If the **E** vector is normal to the plane of incidence ($\perp$ polarization), some of the light is reflected, the transmitted and reflected intensities being governed by the Fresnel relations. The pile-of-plates analyzer in Fig. 5-10-8 can be rotated azimuthally about the photon path, and the orientation for maximum count rate indicates the direction of linear polarization. The chevron arrangement eliminates the lateral deflection of the light beam produced by a single stack of plates. Quartz is chosen for the plate material because of its transparency to the 254-nm radiation emitted by the target, not because of its optical anisotropy, which is irrelevant here. The books by Rossi and Ditchburn are recommended for clear discussions of polarized light.[69]

Because the inelastically scattered electrons are observed in the forward direction ($\vartheta = 0$) the geometry of the analysis has the simple form shown in Fig. 5-10-9. Wolcke et al.[137] measured the Stokes parameters η_1, η_2, η_3 defined in (5-10-44), (5-10-45), and (5-10-46) over an impact energy range 7 to 15 eV. Here η_1

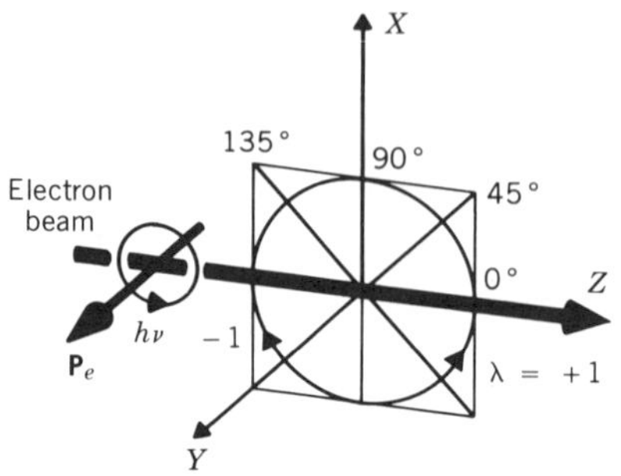

Figure 5-10-9. Geometry for the Münster e-$h\nu$ coincidence experiment with polarized electrons.[137] The inelastically scattered electrons are detected only along the Z axis (forward scattering). The photons are detected along the Y axis, which is the direction of the electron polarization, $\mathbf{P}_e$. The angles indicated are those that appear in the definition of the Stokes parameters for this simple geometry only. The numbers $\lambda = \pm 1$ refer to the helicities of the circularly polarized radiation.

and η_2 vanish identically if the electron beam is unpolarized, so these parameters are sensitive probes of spin-dependent interactions. In this experiment, reflection invariance of the electron-atom interaction with respect to the X-Z plane reduces the number of independent scattering amplitudes to two. Measurement of the Stokes parameters permits a complete determination of the moduli of these amplitudes and the phase between them if the total excitation cross section is known. The Münster group found η_1 and η_2 to be significantly different from zero over part of the energy range, thus confirming that spin-dependent interactions are important in this excitation process. These spin effects are associated with the breakdown of L-S coupling for the excited state, and the occurrence of electron exchange.

Finally, we must mention additional measurements of a different kind recently made on mercury atoms at Münster. Borgmann et al.[138] studied polarized electrons scattered inelastically from Hg atoms at energies ranging from the excitation threshold to 50 eV and scattering angles in the interval between 30 and 120°. In the excitation of the 6^3P_1, 6^3P_2, and 6^1P_1 target states, the left-right scattering asymmetry was measured as a function of scattering angle and impact energy. The conclusion drawn from the results is that the asymmetry is caused mainly by the combined effect of exchange and spin-orbit coupling of the Hg electrons.

3. *Superelastic Electron Scattering from Laser-Excited Targets*

In Section 5-10-E we briefly described the technique of superelastic scattering of electrons by laser-excited atoms, developed by Hertel and Stoll[121] in the 1970s. It is the inverse to the e-$h\nu$ coincidence technique, as can be seen from Fig. 5-10-10, in which both the directions of the arrows and the time sequence of events of drawing (*a*) are reversed to obtain drawing (*b*). One important aspect of the superelastic scattering technique is that when it can be applied, the count rates are typically several orders of magnitude higher than in the corresponding coincidence experiments. Also, the laser excitation feature permits the preparation of target atoms in various mixtures of angular momentum states by the use of different polarizations of the laser light. Hence it is possible to investigate the influence of target angular momentum on the course of the collision.

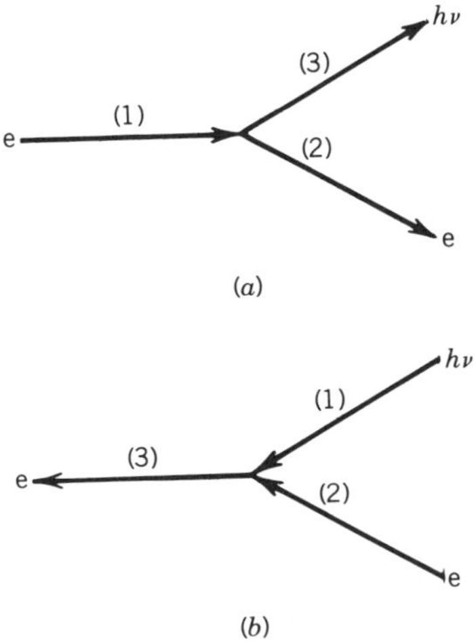

Figure 5-10-10. Comparison of (*a*) *electron-photon coincidence experiment* with (*b*) its inverse, *superelastic scattering of electrons by laser-excited atoms.* The numerals alongside the arrows indicate the time sequence. In (*b*), an electron scattered from the excited atom is detected only if it has fully deexcited the target atom and acquired the deexcitation energy in a superelastic collision.

McClelland et al.[139-141] at the National Bureau of Standards in Gaithersburg were the first to perform experiments of this type with spin-polarized electrons incident on state-selected atoms. In one experiment[139] they used linearly polarized laser light directed perpendicularly onto the scattering plane to provide sodium atoms in the $3P_{3/2}$ state. These excited atoms are not spin polarized. Electrons of 10 eV energy were fired at the target with their spins alternately parallel and antiparallel to the light beam. Surprisingly, spin asymmetries (cf. Section 4-6) were observed in the superelastic scattering. Data were obtained both as a function of the scattering angle with fixed laser polarization, and as a function of laser polarization direction at fixed scattering angle.

McClelland et al.[140] then studied the superelastic scattering of spin-polarized electrons separately from the spin-polarized $M_L = +1$ and $M_L = -1$ states of Na $3P_{3/2}$ atoms. The experimental apparatus was similar, but circularly polarized light of both helicities ($\lambda = +1$ and $\lambda = -1$) was used to produce the two M_L states. For each of these well-defined quantum states, the cross section for deexcitation to the $3S$ ground state is exactly the same as that for its excitation from the ground state. Measurements were made with the incident electron spin parallel to the target electron spin (pure triplet scattering), and with the spins opposed (equal contributions from singlet and triplet states). Hence the $M_L = +1$ and $M_L = -1$ scattering intensities can be separated into their singlet and triplet components. The spin symmetry

$$A = \frac{1}{P_e}\frac{I\uparrow\downarrow - I\uparrow\uparrow}{I\uparrow\downarrow + I\uparrow\uparrow} \tag{5-10-70}$$

was measured for each of these two values of M_L. (The $M_L = 0$ state has its angular momentum vector lying in the scattering plane and cannot be excited or deexcited by a projectile electron, whose angular momentum is normal to the scattering plane.) Here P_e is the polarization of the projectiles, and I is the scattering intensity for parallel ($\uparrow\uparrow$) and antiparallel ($\uparrow\downarrow$) electron spins. The asymmetry defined in (5-10-70) normalizes out various spin-averaged artifacts that may arise in the experiment. It can also be expressed in terms of two singlet (S_{+1}, S_{-1}) and two triplet (T_{+1}, T_{-1}) scattering amplitudes for excitation of the $M_L = +1$ and $M_L = -1$ states:

$$A_i = \frac{|S_i|^2 - |T_i|^2}{|S_i|^2 + 3|T_i|^2}, \tag{5-10-71}$$

where $i = +1$ or -1, according to the M_L value. Finally, if we substitute $S = (f + g)$ and $T = (f - g)$, we can express $A_{\pm}$ in terms of the direct $(f_{\pm 1})$ and exchange $(g_{\pm 1})$ amplitudes. In the experiment described here, four independent complex scattering amplitudes suffice to characterize the collision process completely. Measurements were made over the impact energy range 1 to 12 eV and at various scattering angles. The reader can refer to Refs. 139–141 for details on the variation of the spin asymmetries with energy and angle and for the new insight that the NBS results provide into the details of the inelastic scattering processes that were studied. It should be pointed out, as was done by McClelland et al.,[139] that the NBS experiments were preceded by a paper by Hanne et al.,[142] who discussed the "fine-structure effect" described in Ref. 139.

5-11. EXCITATION BY POSITRON IMPACT

As shown in Section 4-7, accurate cross sections are available for the total scattering of positrons, and for both integral and differential elastic scattering. Section 6-16 describes successful measurements of ionization cross sections for positron impact. The measurement of cross sections for *excitation* by positron impact presents especially severe difficulties, and progress here has lagged that for the other processes. Consequently, we shall not describe the activity in this area, but rather refer the reader to the review by Sinapius[143] and the proceedings of the Positron Conference held in London in 1987.[144]

5-12. BOOKS AND REVIEWS ON THE THEORY OF ELECTRON-IMPACT EXCITATION

Most of the books and reviews that are relevant here have already been cited in this chapter[3–5,8,26,30,37,39,41,59,70,104,105] or in Chapter 4 (see Section 4-8-D). However, special mention should be made of Takayanagi,[4] Shimamura,[5] and Compton and Bardsley,[41] who discuss recent theoretical work on electron-impact excitation. The book by Fano and Rau,[69] and the volumes edited by Blum et al.[145] and by Burke and West,[146] are also quite useful.

5-13. NOTES

5-1. Some Topics in Classical Mechanics.[147] Consider a rigid body composed of point masses. If it moves with one point stationary, its angular momentum about that point is

$$\mathbf{L} = \mathbf{I}\boldsymbol{\omega}, \tag{5-13-1}$$

where $\boldsymbol{\omega}$ is the angular velocity of the body, lying along the instantaneous axis of rotation. The quantity $\mathbf{I}$ may be designated equally well as a second-rank tensor or as a dyadic. It is called the moment of inertia tensor (or dyadic), or more succinctly, the inertia tensor (or dyadic). The x component of $\mathbf{L}$ may be written

$$L_x = I_{xx}\omega_x + I_{xy}\omega_y + I_{xz}\omega_z, \tag{5-13-2}$$

where $I_{xx} = m_i(r_i^2 - x_i^2)$, $I_{xy} = -m_i x_i y_i$, and so on.

The kinetic energy of rotation of the rigid body can be expressed as

$$T = \frac{\boldsymbol{\omega}\cdot\mathbf{L}}{2} = \frac{\boldsymbol{\omega}\cdot\mathbf{I}\cdot\boldsymbol{\omega}}{2}. \tag{5-13-3}$$

Small oscillations of a body, such as a molecule, that are capable of vibration are conveniently treated by a *normal-mode analysis.* A principal-axis transformation is made to diagonalize simultaneously the kinetic energy and potential energy matrices, and a set of *normal coordinates* is defined in the process. Each normal coordinate varies harmonically with time and corresponds to a pure vibration of the body with a single frequency. These basic vibrations (analogous to basis states in expansion of wave functions in quantum mechanics) are the *normal modes of vibration.* In each, all the particles vibrate with the same frequency and with the same phase (or exactly out of phase). The relative amplitudes of the particles are determined by the equations relating the original coordinates to the normal coordinates. Any physically realizable motion of the body can be constructed as a sum of the normal modes weighted with the appropriate amplitudes and phase factors. The normal modes of the C_2H_2 molecule are discussed in Section 5-4.

5-2. A Few Comments on Tensors. Tensors are objects that have physical meaning independent of any particular coordinate frame (i.e., the axes used to describe them). This fact accounts for their great importance in physics. A given tensor may be symmetrized and contracted to construct a new set of tensors that are simpler than the original. ("Contraction" means summing on one of the tensor indices. According to the Einstein convention, a repeated index implies summation.) Thus a second-rank tensor

T_{ij} (with nine independent components, in general) may be decomposed as follows:

$$\begin{aligned} T_{ij} &= [\tfrac{1}{2}(T_{ij} + T_{ji}) - \tfrac{1}{3}\delta_{ij}T_{kk}] + [\tfrac{1}{2}(T_{ij} - T_{ji})] + [\tfrac{1}{3}\delta_{ij}T_{kk}] \\ &= T_{ij}^{(2)} + T_{ij}^{(1)} + T_{ij}^{(0)}. \end{aligned} \tag{5-13-4}$$

The first part, $T_{ij}^{(2)}$, is symmetric and traceless and has five independent components. The second, $T_{ij}^{(1)}$, is antisymmetric and has three components. The third, $T_{ij}^{(0)} = T_{ii}$, is the trace, with one component. The process of breaking up a tensor into simpler types is called reduction of the tensor. A tensor incapable of such reduction is said to be *irreducible*; examples are scalars, vectors, and antisymmetric tensors of the second rank (see Chapter 3 of Good and Nelson[6] and Refs. 123–126). The state multipoles $\langle T(L)_{KQ}^{\dagger}\rangle$ introduced in Section 5-10-F are irreducible tensors. The dagger appearing in this symbol indicates it is the adjoint of another tensor that is not directly applicable in the discussion of multipoles (see Blum,[104] Chap. 4).

5-3. *Fast Fourier Spectroscopy.* If a molecular gas is subjected to an intense burst of electromagnetic radiation, and the total intensity of the light emitted by the excited molecules is recorded as a function of time, the signal has such a complicated form that little useful information can be extracted, as a rule. However, if the output is Fourier analyzed and displayed as a function of frequency, rather than of time, sharp peaks typically stand out, representing the quantum states excited by the burst. This technique of *fast Fourier spectroscopy* (see Steinfeld[147]) is one of the most spectacular applications of Fourier transforms (Section 5-6).

5-4. *Photon Spin States.* In Section 5-10-G we mentioned that the photon (spin 1, zero rest mass) must have its spin either parallel or antiparallel to its direction of motion. This matter is discussed by H. Frauenfelder and E. M. Henley, *Subatomic Physics*, Prentice-Hall, Englewood Cliffs, N.J. (1974), Sec. 5-5; R. P. Feynman, *Theory of Fundamental Processes*, Benjamin, New York (1962), Chap. 20; J. D. Bjorken and S. D. Drell, *Relativistic Quantum Fields*, McGraw-Hill, New York (1965), Sec. 14-5. A relativistic argument is required. The fact that the photon has only the two indicated spin states, when combined with the conservation laws for momentum, angular momentum, and energy, prohibits the spontaneous decay of a photon into two photons. However, decay into three photons is not disallowed.

5-5. *Virtual Particles.* The concept of virtual particles is invoked as an aid in intermediate calculations required in the theoretical treatment of certain physical problems. These particles are not real because they do not obey the relativistic relation between energy, momentum, and rest mass ($E = c\sqrt{p^2 + M_0^2c^2}$); hence they cannot be observed. *Virtual photons* are used in many kinds of calculations in quantum electrodynamics, the quantum field theory for electrons and photons. The electromagnetic force between two charged particles is "mediated" by the exchange of virtual photons between them. In elastic electron-electron scattering, the Coulomb force producing the deflection is envisioned in terms of a virtual photon emitted by one electron and absorbed by the other. During the interval between emission and absorption, the energy of the system is not conserved, but the energy discrepancy ΔE cannot be observed provided the virtual photon exists only for a time less than Δt, where $\Delta E\,\Delta t \approx \hbar$. (Similarly, the force between hadrons, say two neutrons, or a neutron and a proton, is mediated by virtual pions.) Virtual photons appear even in the discussion of isolated electrons. A "physical" electron does not always exist as a bare Dirac electron. Part of the time it is described as a Dirac electron plus a virtual photon, which it emits and reabsorbs. The electron also spends part of its time in more complicated states. The books cited in Note 5-4 are good references for this subject.

5-6. *Predissociation.* The theory of predissociation has some features in common with the theory of heavy particle collisions; for example, the concept of curve crossing appears in each (see L. D. Landau and E. M. Lifshitz, *Quantum Mechanics: Non-relativistic Theory*, 2nd ed., Pergamon, Oxford (1965), pp. 322–331; D. R. Bates, "Transitions" in Ref. 28, pp. 293–296).

5-7. *The Time Evolution of a Stationary State.* Consider, as an example, a *plane* rigid rotor whose time-independent wave functions $\psi_m(\varphi) = \psi_m(\varphi, t = 0)$ and energy levels E_m for the stationary states are

$$\psi_m(\varphi) = \frac{1}{\sqrt{2\pi}} e^{im\varphi}; \qquad E_m = \frac{\hbar^2 m^2}{2I}. \tag{5-13-5}$$

Here I is the moment of inertia, and m is the azimuthal (magnetic) quantum number: $m = 0, \pm 1, \pm 2, \ldots$. If the rotor is in one of these eigenstates, its time-dependent wave function is

$$\Psi_m(\varphi, t) = \psi_m(\varphi) e^{-iE_m t/\hbar}. \tag{5-13-6}$$

One may show that if $\psi(\varphi, t = 0) = A \sin^2 \varphi$, where A is a normalization constant, the wave function at time t is

$$\Psi(\varphi, t) = \frac{A}{2}[1 - (\cos 2\varphi) e^{-2i\hbar t/I}]. \tag{5-13-7}$$

Ter Haar treats this example on pages 8 and 118 of his book.[148] He also treats the time evolution of more complicated systems (wave packets) in which the wave function is a superposition of eigenfunctions, each with a different characteristic frequency $\omega_m = E_m/\hbar$ (see ter Haar, p. 17, Prob. 18).

5-14. PROBLEMS

Solutions for most of these problems appear in the references indicated. Problems on collision theory for excitation processes are grouped with those for ionization in Chapter 6.

5-1. Three points, each with mass m, have Cartesian coordinates $(d, 0, 0)$, $(0, d, 2d)$, and $(0, 2d, d)$, respectively. Show that the principal moments of inertia about the origin are $10md^2$, $10md^2$, and $2md^2$, and find a set of principal axes. (Goldstein,[147] p. 238.)

5-2. Consider a diatomic molecule with nuclei of masses m_1 and m_2 whose internal potential energy is $V(r) = -2V_0(1/\rho - 1/2\rho^2)$, where $\rho = r/d$. Here r is the internuclear spacing, and d is a characteristic length. Expand the effective potential energy in a power series, and show that for small oscillations, the quantum-mechanical energy levels are given by

$$E = -V_0 + \frac{L(L+1)\hbar^2}{2\mu d^2} + \hbar\omega_0\left(n + \frac{1}{2}\right) - \frac{3}{2}\,\frac{\hbar^3 L(L+1)(n+\frac{1}{2})}{\mu^2 d^4 \omega_0},$$

where μ is the reduced mass of the nuclei and $\omega_0 = \sqrt{2V_0/\mu d^2}$. (Cronin et al.,[149] p. 47.)

5-3. Given that the far-infrared spectrum of HBr consists of lines spaced $17\ \text{cm}^{-1}$ apart, show that the internuclear spacing of the molecule is 1.4×10^{-8} cm. (Cronin et al.,[149] p. 47.)

5-4. Given the dissociation energies of the H_2 and D_2 molecules to be 4.46 eV and 4.54 eV, respectively, show that the zero-point energy of H_2 is 0.27 eV. (Cronin et al.,[149] p. 47.)

5-5. Use the uncertainty principle and other semiquantitative arguments to estimate, for diatomic molecules, the relative order of magnitudes of the following quantities: (a) the spacings of the electronic, vibrational, and rotational energy levels; (b) the internuclear distance and the zero-point amplitude of the nuclear vibrations; and (c) the characteristic periods and velocities of the electronic and nuclear motion. (ter Haar,[148] p. 53.)

5-6. Show that the interaction energy of two magnetic dipoles of moments $\boldsymbol{\mu}_1$ and $\boldsymbol{\mu}_2$ is

$$\left[\frac{3(\boldsymbol{\mu}_1 \cdot \mathbf{r})(\boldsymbol{\mu}_2 \cdot \mathbf{r})}{r^2} - \boldsymbol{\mu}_1 \cdot \boldsymbol{\mu}_2\right]\frac{1}{r^3},$$

where $\mathbf{r}$ is the separation vector.

5-7. Consider a hydrogen atom in the $2p$ state. Show that the magnetic induction at the nucleus produced by the orbiting electron is $B = 0.43$ T.

5-8. Show that the expectation value of the electric quadrupole moment for a neutron-proton system in the $^3S_1$ state is zero.

5-9. Show in several ways that the quadrupole moment of a spherically symmetric charge distribution is zero. (Good and Nelson,[6] p. 114.)

5-10. An ellipsoid of revolution (semiaxes a, a, and c) is uniformly filled with a total amount of charge q. Locate the origin at its center and orient the Z axis along the axis of length $2c$. Show that the quadrupole moment tensor is

$$\mathcal{Q} = \begin{pmatrix} -\frac{q}{5}(c^2 - a^2) & 0 & 0 \\ 0 & -\frac{q}{5}(c^2 - a^2) & 0 \\ 0 & 0 & \frac{2q}{5}(c^2 - a^2) \end{pmatrix}.$$

(Good and Nelson,[6] p. 114.)

5-11. A charge distribution has the quadrupole moment tensor

$$\mathcal{Q} = \begin{pmatrix} 3 & 2 & 0 \\ 2 & 3 & 0 \\ 0 & 0 & -6 \end{pmatrix}.$$

Show that the quadrupole tensor, when referred to principal axes, has the form

$$\mathcal{Q} = \begin{pmatrix} 5 & 0 & 0 \\ 0 & 1 & 0 \\ 0 & 0 & -6 \end{pmatrix}.$$

(Good and Nelson,[6] p. 112.)

5-12. A localized charge distribution $\rho(\mathbf{r})$ is situated in an *external* electrostatic field $\mathbf{E}(\mathbf{r})$ described by a potential $\Phi(\mathbf{r})$ that varies slowly with position in the region occupied by the charge.

(a) Show that the electrostatic energy of the system is

$$W = q\Phi(0) - \boldsymbol{\mu} \cdot \mathbf{E}(0) - \frac{1}{6}\sum_i \sum_j Q_{ij} \frac{\partial E_j}{\partial x_i}(0) + \cdots,$$

where q, $\boldsymbol{\mu}$, and Q_{ij} are defined in Section 5-2.

(b) Show that the force on the charge distribution can be written

$$\mathbf{F} = q\mathbf{E}(0) + \{\nabla[\boldsymbol{\mu}\cdot\mathbf{E}(\mathbf{x})]\}_0 + \left\{\nabla\left[\frac{1}{6}\sum_{j,k} Q_{jk}\frac{\partial E_j}{\partial x_k}(\mathbf{x})\right]\right\}_0 + \cdots.$$

(c) Show that the torque on the charge distribution can be written, for any Cartesian component (say N_1), as

$$N_1 = [\boldsymbol{\mu} \times \mathbf{E}(0)]_1 + \frac{1}{3}\left[\frac{\partial}{\partial x_3}\left(\sum_j Q_{2j}E_j\right) - \frac{\partial}{\partial x_2}\left(\sum_j Q_{3j}E_j\right)\right]_0 + \cdots.$$

Other problems on multipoles: Problems 1 to 11 in Section 1, 1 to 5 in Section 8 of Good and Nelson;[6] Problems 1 to 7 in Chapter 4 of Jackson;[6] see also Chapter 2 of Nayfeh and Brussel.[6] Problems 57 to 60 on page 33 of ter Haar[148] deal with the electric quadrupole moment in quantum mechanics.

5-13. The vector potential **A** and the scalar potential V form a four-vector (A_x, A_y, A_z, iV). Consider a point charge q moving as in Fig. 5-6-1, and derive the fields at point P fixed in the Lab frame by making a Lorentz transformation of the four-vector from the rest frame of the charge to the Lab frame. (Chen,[150] p. 100.)

5-14. Solve the differential equation (5-7-19) for $x(t)$ by the Fourier transform method (see Jackson,[6] pp. 623–624). See Problem 19 on page 17 of ter Haar[148] for another method of solving for the motion of a driven oscillator in quantum mechanics.

5-15. Associated with the transition from the first excited state ($3p$) to the ground state ($3s$) of the sodium atom are an oscillator strength equal to 0.94 and a wavelength of 589 nm. Show that the radiative lifetime of the excited state is 17 ns. (Read,[151] p. 123.)

5-16. A transmission grating that is 5 cm wide is to be used to analyze, at normal incidence, the spectrum of sodium. Show that at least 982 grating lines are required to resolve the sodium D doublet at 5890 and 5896 Å in the first order. Show that for 982 lines, the angular separation of the two members of the doublet is 10^{-5} rad. (Cronin et al.,[149] p. 26.)

5-17. The *pile-of-plates polarizer* was invented by D. Arago in 1812; it is identical in construction to the analyzer shown in Fig. 5-10-8. Consider such a polarizer consisting of 25 plates made of fused silica (index of refraction equal to 1.46). Let unpolarized light be incident upon the first plate at the Brewster angle, and denote the polarization directions parallel and perpendicular to the plane of incidence by $\parallel$ and $\perp$, respectively. Show that at each of the 50 surfaces, the $\parallel$ component of the light is completely transmitted, whereas the intensity of the $\perp$ component suffers a fractional loss of 0.131 in transmission. Further show that if absorption in the plates is neglected, the intensity of the $\|$ component in the emerging light is greater than the $\perp$ intensity by a factor of 1,094. [E. Hecht and A. Zajac, *Optics*, Addison-Wesley, Reading, Mass. (1974).]

5-18. Laser light with $\lambda = 6000$ Å is focused by a telescope with a 2-m-diameter lens (or mirror) onto the surface of the moon, 400,000 km away. Ignoring atmospheric effects, show that the radius of the light pattern on the moon is approximately 150 m. (Chen,[150] p. 202.)

5-19. Show that a $\lambda/4$ plate can convert circular polarization into linear polarization, and vice versa.

5-20. The following optical devices are provided without being identified: two linear polarizers, a $\lambda/4$ plate, a $\lambda/2$ plate, and a circular polarizer. How could you identify each device with only a lamp and screen as auxiliary apparatus? How would you proceed if you had only one linear polarizer in the package instead of two? (Chen,[150] p. 188.)

5-21. A quartz crystal is cut with its optic axis parallel to the flat surfaces to make a $\lambda/4$ plate for light with $\lambda = 5829.90$ Å. Show that its thickness is 1.263×10^{-3} cm. (Chen,[150] p. 197.)

5-22. Show that for maximum polarization of light by reflection from a plane dielectric, the angle between the reflected and refracted rays is 90°. Show that the angle between the refracted ray and the normal to the dielectric surface is $\theta = \tan^{-1} n_1/n_2$, where n_1 and n_2 are the indices of refraction of the two dielectrics. (Chen,[150] p. 199.)

5-23. The wavelength range of visible light is 4000 to 7000 Å. Show that the corresponding frequency range is 7.5×10^{14} to 4.3×10^{14} Hz. (Chen,[150] p. 200; see also p. 188, Prob. III-3, and p. 191, Prob. IV-6.

5-24. A beam of electrons all of which have "spin-up" is in the pure state $|+\frac{1}{2}\rangle$, whereas a beam in the pure "spin-down" state is represented by the ket vector $|-\frac{1}{2}\rangle$. The electron polarization vector, $\mathbf{P}_e$, has components defined as expectation values of the Pauli spin matrices:

$$P_{ei} = \langle \sigma_i \rangle,$$

where $i = x, y, z$. We use the representation wherein

$$\sigma_x = \begin{pmatrix} 0 & 1 \\ 1 & 0 \end{pmatrix}, \quad \sigma_y = \begin{pmatrix} 0 & -i \\ i & 0 \end{pmatrix}, \quad \sigma_z = \begin{pmatrix} 1 & 0 \\ 0 & -1 \end{pmatrix},$$

and

$$\left|+\frac{1}{2}\right\rangle = \begin{pmatrix} 1 \\ 0 \end{pmatrix}, \quad \left|-\frac{1}{2}\right\rangle = \begin{pmatrix} 0 \\ 1 \end{pmatrix}.$$

Show that a beam in the pure state $|+\frac{1}{2}\rangle$ has $P_x = P_y = 0$, and $P_z = 1$, and a beam in the pure state $|-\frac{1}{2}\rangle$ has $P_x = P_y = 0$, and $P_z = -1$. (Blum,[104] pp. 3–5.)

5-25. Show that the Stokes parameters are related to the elements of the photon density matrix (5-10-56) by the following equations (Blum,[104] pp. 30–32):

$$\text{(a)}\ I = \rho_{11} + \rho_{-1,-1}, \qquad \text{(b)}\ I\eta_1 = -i(\rho_{1,-1} - \rho_{-1,1}),$$

$$\text{(c)}\ I\eta_2 = \rho_{11} - \rho_{-1,-1}, \qquad \text{(d)}\ I\eta_3 = -(\rho_{1,-1} + \rho_{-1,1}).$$

5-26. Show that the three density matrices displayed below are those for photons in pure states with linear polarization in the directions indicated:

$$\text{x direction:}\quad \rho = \frac{I}{2}\begin{pmatrix} 1 & -1 \\ -1 & 1 \end{pmatrix}, \qquad \text{y direction;}\quad \rho = \frac{I}{2}\begin{pmatrix} 1 & i \\ -i & 1 \end{pmatrix},$$

angle θ with respect to x-axis:

$$\rho = \frac{I}{2}\begin{pmatrix} 1 & -\cos 2\beta + i \sin 2\beta \\ -\cos 2\beta - i \sin 2\beta & 1 \end{pmatrix}.$$

Show further that circularly polarized light is represented by the density matrices

$$\rho = I\begin{pmatrix} 1 & 0 \\ 0 & 0 \end{pmatrix} \quad \text{and} \quad \rho = I\begin{pmatrix} 0 & 0 \\ 0 & 1 \end{pmatrix}$$

and identify the helicity associated with each matrix. (Blum,[104] pp. 32–34.)

5-27. Show that the density matrix for an electron can be written in the form

$$\rho = \tfrac{1}{2}[\overleftrightarrow{1} + (\mathbf{P}_e \cdot \boldsymbol{\sigma})],$$

and that if a magnetic field $\mathbf{B}$ is present, the equation of motion of $\mathbf{P}_e$ is

$$\frac{\partial \mathbf{P}_e}{\partial t} = \gamma(\mathbf{P}_e \times \mathbf{B}),$$

where γ is a constant you are asked to identify. (ter Haar,[148] p. 71; five other problems on the density matrix also are stated and solved in this book.)

5-28. Two stationary electrons are separated by a distance r. One emits a virtual photon that is absorbed by the other. Show that during the lifetime of the photon, the energy of the system is proportional to e^2/r, in accord with Coulomb's law. (Recall that the dimensionless fine structure constant is defined as $\alpha = e^2/c\hbar = 1/137$.)

5-29. Yukawa in 1935 suggested that the force between two nucleons could be visualized in terms of the exchange of virtual mesons between them. Let two nucleons be separated by a distance equal to the range of the nuclear force, r. A virtual meson in transit has an uncertainty in its position about equal to r. Make a reasonable assumption for its uncertainty in momentum, and obtain the relation $r \approx \hbar/mc$ between the range of the force and the mass, m, of the meson.

REFERENCES

1. H. S. W. Massey, E. W. McDaniel, and B. Bederson (Series Eds.), *Applied Atomic Collision Physics*, Academic, New York (1982–1984), 5 vol, see Vol. 1, Chaps. 4, 6, 8, 10, 11; Vol. 2, Chaps. 5, 7, 8; Vol. 3, Chaps. 1, 8–11, 14; Vol. 4, Chap. 3; Vol. 5, Chaps. 1, 2, 4, 6, 7, 14, 16.
2. G. Csanak, D. C. Cartwright, S. K. Srivastava, and S. Trajmar, "Elastic Scattering of Electrons by Molecules," in L. G. Christophorou (Ed.), *Electron-Molecule Interactions and Their Applications*, Vol. 1, Academic, New York (1984).
3. S. Trajmar and D. C. Cartwright, "Excitation of Molecules by Electron Impact," in Christophorou, Ref. 2. Updated by S. Trajmar, *ICPEAC XIV*, Stanford (1985), p. 77.
4. K. Takayanagi, "Introduction to Electron-Molecule Collisions," in I. Shimamura and K. Takayanagi (Eds.), *Electron-Molecule Collisions*, Plenum, New York (1984).
5. I. Shimamura, "Rotational Excitation of Molecules by Slow Electrons," in Shimamura and Takayanagi, Ref. 4. Updated in *ICPEAC XIV*, Stanford (1985), p. 93.
6. J. D. Jackson, *Classical Electrodynamics*, 2nd ed., Wiley, New York (1975). R. H. Good and T. J. Nelson, *Classical Theory of Electric and Magnetic Fields*, Academic, New York (1971). M. H. Nayfeh and M. K. Brussel, *Electricity and Magnetism*, Wiley, New York (1985).
7. G. J. Schulz, *Rev. Mod. Phys.* **45**, 423 (1973).
8. R. I. Hall and F. H. Read, "Molecular Spectroscopy by Electron Scattering," in Shimamura and Takayanagi, Ref. 4.
9. J. B. Hasted and D. Mathur, "Electron-Molecule Resonances," in Christophorou, Ref. 2.
10. K. Jung, Th. Antoni, R. Müller, K.-H. Kochem, and H. Ehrhardt, *J. Phys. B* **15**, 3535 (1982).
11. G. Herzberg, *Spectra of Diatomic Molecules*, Van Nostrand, Princeton, N.J. (1950).
12. B. H. Bransden and C. J. Joachain, *Physics of Atoms and Molecules*, Longman Group, Harlow, Essex, England (1983).
13. H. Ehrhardt and F. Linder, *Phys. Rev. Lett.* **21**, 419 (1968).
14. F. Linder and H. Schmidt, *Z. Naturforsch*, **26-a**, 1603 (1971).
15. G. Joyez, J. Comer, and F. H. Read, *J. Phys. B* **6**, 2427 (1973).
16. S. Trajmar and D. F. Register, "Experimental Techniques for Cross Section Measurements," in Shimamura and Takayanagi, Ref. 4.
17. S. F. Wong and G. J. Schulz, *Phys. Rev. Lett.* **32**, 1089 (1974).

18. S. K. Srivastava, R. I. Hall, S. Trajmar, and A. Chutjian, *Phys. Rev. A* **12**, 1399 (1975).

19. K. Jung, K.-M. Scheuerlein, W. Sohn, K.-H. Kochem, and H. Ehrhardt, *J. Phys. B* **20**, L-327 (1987).

20. F. N. Read and D. Andrick, *J. Phys B* **4**, 911 (1971). F. H. Read, *J. Phys. B* **5**, 255 (1972).

21. Y. Itikawa, *J. Phys. Soc. Jpn.* **32**, 217 (1972).

22. R. Müller, K. Jung, K.-H. Kochem, W. Sohn, and H. Ehrhardt, *J. Phys. B* **18**, 3971 (1985).

23. I. Shimamura, *Chem. Phys. Lett.* **73**, 328 (1980).

24. M. Rädle, G. Knoth, S. Klauer, K. Jung, and H. Ehrhardt, *ICPEAC XV*, Brighton (1987), Contributed Papers, p. 317.

25. G. Ziegler, M. Rädle, O. Pütz, K. Jung, H. Ehrhardt, and K. Bergmann, *Phys. Rev. Lett.* **58**, 2642 (1987). See also G. Ziegler, S. V. K. Kumar, P. Dittmann, and K. Bergmann, *Z. Phys. D* **10**, 247 (1988).

26. H. J. Korpsch, H. Kutz, and H. D. Meyer, *J. Phys. B* **20**, L-433 (1987).

27. A. Herzenberg, "Vibrational Excitation of Molecules by Slow Electrons," in Shimamura and Takayanagi, Ref. 4.

28. D. Bohm, *Quantum Theory*, Prentice-Hall, Englewood Cliffs, N. J. (1951). A. S. Davydov, *Quantum Mechanics*, Pergamon, Oxford (1965). D. R. Bates (Ed.), *Quantum Theory*, Vol. I, Academic, New York (1961).

29. K. Rohr and F. Linder, *J. Phys. B* **9**, 2521 (1976). G. Seng and F. Linder, *J. Phys. B* **9**, 2539 (1976). K. Rohr, *J. Phys. B* **11**, 1849, 4109 (1978).

30. G. Knoth, M. Rädle, H. Ehrhardt, and K. Jung, *Europhys. Lett.* **4**, 805 (1987). See also M. A. Morrison, *AAMP* **24**, 52 (1988) and G. Knoth, M. Rädle, M. Gote, H. Ehrhardt, and K. Jung, *J. Phys. B* **22**, 299 (1989). *Phys. Rev. Lett.* **62**, 1735 (1989).

31. J. Comer and F. H. Read, *J. Phys. B* **4**, 368 (1971).

32. K.-H. Kochem, W. Sohn, K. Jung, H. Ehrhardt, and E. S. Chang, *J. Phys. B* **18**, 1253 (1985).

33. W. Sohn, K.-H. Kochem, K. Jung, H. Ehrhardt, and E. S. Chang, *J. Phys. B* **18**, 2049 (1985).

34. K.-H. Kochem, W. Sohn, N. Hebel, K. Jung, and H. Ehrhardt, *J. Phys. B* **18**, 4455 (1985). For a rotational branch analysis of the excitation of the fundamental vibrational modes of CO_2 by slow electrons, see Th. Antoni, K. Jung, H. Ehrhardt, and E. S. Chang, *J. Phys. B* **19**, 1377 (1986). For related theoretical work, see H. Estrada and W. Domcks, *J. Phys. B* **18**, 4469 (1985). For measurements on the similar molecules OCS and CS_2, see W. Sohn, K. H. Kochem, K.-M. Scheuerlein, K. Jung, and H. Ehrhardt, *J. Phys. B* **20**, 3217 (1987).

35. G. J. Schulz, "A Review of Vibrational Excitation of Molecules by Electron Impact at Low Energies," in G. Bekefi (Ed.), *Principles of Laser Plasmas*, Wiley, New York (1976).

36. A. Herzenberg, *J. Phys. B* **1**, 548 (1968). D. T. Birtwistle and A. Herzenberg, *J. Phys. B* **4**, 53 (1971).

37. N. F. Lane, "The State of the Theory," in S. C. Brown (Ed.), *Electron-Molecule Scattering*, Wiley, New York (1979). M. A. Morrison, *Aust. J. Phys.* **36**, 239 (1983).

38. E. W. McDaniel, *Atomic Collisions: Heavy Particle Projectiles*, Wiley, New York, to be published.
39. H. S. W. Massey, *Electronic and Ionic Impact Phenomena*, 2nd ed., Vol. 2, Oxford University Press, London (1969). T. E. Sharp, *At. Data* **2**, 119 (1971).
40. E. C. Zipf, "Dissociation of Molecules by Electron Impact," in Christophorou, Ref. 2.
41. R. N. Compton and J. N. Bardsley, "Dissociation of Molecules by Slow Electrons," in Shimamura and Takayanagi, Ref. 4.
42. S. J. B. Corrigan, *J. Chem. Phys.* **43**, 4381 (1965).
43. H. F. Winters, *J. Chem. Phys.* **44**, 1472 (1966); **63**, 3462 (1975); *Chem. Phys.* **36**, 353 (1979).
44. S. Chung and C. C. Lin, *Phys. Rev. A* **17**, 1874 (1978).
45. D. Beck and O. Osberghaus, *Z. Phys.* **160**, 406 (1960).
46. G. K. Woodgate, *Elementary Atomic Structure*, 2nd ed., Clarendon, Oxford (1980).
47. R. D. Sard, *Relativistic Mechanics*, Benjamin, New York (1970).
48. R. P. Feynman, R. B. Leighton, and M. Sands, *The Feynman Lectures on Physics*, Vol. II, Addison-Wesley, Reading, Mass. (1964), Sec. 26-2.
49. R. L. Sproull and W. A. Phillips, *Modern Physics*, 3rd ed., Wiley, New York (1980).
50. E. J. Williams, *K. Dan. Vidensk. Selsk. Mat.-Fys. Medd.* **13**, No. 4 (1935).
51. D. W. O. Heddle, *Rev. Mod. Phys.* **61**, 221 (1989). For more details, see D. W. O. Heddle and J. W. Gallagher, *JILA Data Center Report*, to be published.
52. J. Mazeau, F. Gresteau, R. I. Hall, and A. Huetz, *J. Phys. B* **11**, L-557 (1978).
53. R. Müller-Fiedler, P. Schlemmer, K. Jung, H. Hotop, and H. Ehrhardt, *J. Phys. B* **17**, 259 (1984).
54. E. W. McDaniel, *Collision Phenomena in Ionized Gases*, Wiley, New York (1964).
55. J. B. Hasted, *Physics of Atomic Collisions*, 2nd ed., American Elsevier, New York (1972).
56. B. L. Moiseiwitsch and S. J. Smith, *Rev. Mod. Phys.* **40**, 238 (1968).
57. D. W. O. Heddle and R. G. W. Keesing, *AAMP* **4**, 267 (1968).
58. D. W. O. Heddle, *AAMP* **15**, 381 (1979); *MEP* **7**, Pt. A, 43 (1968).
59. H. S. W. Massey and E. H. S. Burhop, *Electronic and Ionic Impact Phenomena*, 2nd ed., Vol. 1, Oxford University Press, London (1969).
60. C. E. Kuyatt, *MEP* **7**, Pt. A, 1 (1968).
61. B. Van Zyl, G. H. Dunn, G. Chamberlain, and D. W. O. Heddle, *Phys. Rev. A* **22**, 1916 (1980).
62. A. McPherson, N. Rouze, W. B. Westerveld, and J. S. Risley, *Appl. Opt.* **25**, 298 (1986). A. McPherson, N. Rouze, D. Graves, W. B. Westerveld, and J. S. Risley, *SPIE*, **689**, 70 (1986). See also R. L. Kendrick, A. McPherson, N. Rouze, W. B. Westerveld, and J. S. Risley, *Appl. Opt.* **26**, 2029 (1987).
63. R. C. G. Ligtenberg, P. J. M. van der Burgt, W. B. Westerveld, and J. S. Risley, *ICPEAC XV*, Brighton (1987), Contributed Papers, pp. 179, 352. See also B. Van Zyl, M. W. Gealy, and H. Neumann, *Phys. Rev. A* **31**, 2922 (1985).
64. W. B. Westerveld, H. G. M. Heideman, and J. van Eck, *J. Phys. B* **12**, 115 (1979).
65. J. M. Woolsey, J. L. Forand, and J. W. McConkey, *J. Phys. B* **19**, L-493 (1986). J. W.

McConkey, P. Hammond, and M. A. Khakoo, *ICPEAC XV*, Brighton (1987), p. 105.

66. J. F. Williams, *J. Phys. B* **9**, 1519 (1976). See also J. F. Williams and B. A. Willis, *J. Phys. B* **8**, 1641 (1975) and J. F. Williams, *J. Phys. B* **21**, 2107 (1988).
67. C. D. Warner, G. C. King, P. Hammond, and J. Slevin, *J. Phys. B* **19**, 3297 (1986). P. M. Rutter, C. D. Warner, and G. C. King, *ICPEAC XV*, Brighton (1987), Abstracts of Contributed Papers, p. 221.
68. I. C. Percival and M. J. Seaton, *Philos. Trans. R. Soc. London A* **251**, 113 (1958). See also D. W. O. Heddle, *J. Phys. B* **16**, 275 (1983).
69. A. C. G. Mitchell and M. W. Zemansky, *Resonance Radiation and Excited Atoms*, Cambridge University Press, Cambridge (1934). B. Rossi, *Optics*, Addison-Wesley, Reading, Mass. (1957), Chap. 8. R. W. Ditchburn, *Light*, 2nd ed., Interscience, New York (1963), Chap. 15. U. Fano and J. W. Cooper, *Rev. Mod. Phys.* **40**, 441 (1968). M. Inokuti, *Rev. Mod. Phys.* **43**, 297 (1971). W. R. Newell, "Optical Oscillator Strengths by Electron Impact Spectroscopy," in W. Hanle and H. Kleinpoppen (Eds.), *Progress in Atomic Spectroscopy*, Pt. B, Plenum, New York (1978). M. Born and E. Wolf, *Principles of Optics*, 6th ed., Pergamon, Oxford (1980), Sec. 2-3. A. Corney, *Atomic and Laser Spectroscopy*, Clarendon, Oxford (1980). R. Loudon, *The Quantum Theory of Light*, 2nd ed., Clarendon, Oxford (1983). U. Fano and A. R. P. Rau, *Atomic Collisions and Spectra*, Academic, New York (1986).
70. H. A. Bethe, *Ann. Phys. (Leipzig)* **5**, 325 (1930).
71. E. N. Lassettre, *J. Chem. Phys.* **43**, 4479 (1965).
72. E. N. Lassettre and A. Skerbele, *MEP* **3**, Pt. B, 868 (1974). Also see ten papers by E. N. Lassettre and his colleagues on inelastic scattering of electrons, oscillator strengths, and cross sections in *J. Chem. Phys.* **40**, 1208–1276 (1964).
73. E. N. Lassettre, A. Skerbele, and M. A. Dillon, *J. Chem. Phys.* **50**, 1829 (1969).
74. T. Shuttleworth, D. E. Burgess, M. A. Hender, and A. C. H. Smith, *J. Phys. B* **12**, 3967 (1979).
75. M. A. Khakoo, W. R. Newell, and A. C. H. Smith, *J. Phys. B* **13**, 4263 (1980).
76. M. A. Khakoo, W. R. Newell, and A. C. H. Smith, *J. Phys. B* **16**, L-317 (1983).
77. R. A. Bonham and M. Fink, *High Energy Electron Scattering*, Van Nostrand Reinhold, New York (1974).
78. R. A. Bonham, *ICPEAC XIII*, Berlin (1983), p. 73. G. C. King, *ICPEAC XIV*, Stanford (1985), p. 215. H. R. J. Walters, *ICPEAC XV*, Brighton (1987), p. 147.
79. R. C. Ulsh, H. F. Wellenstein, and R. A. Bonham, *J. Chem. Phys.* **60**, 103 (1974).
80. H. Boersch, J. Geiger, and M. Topschowsky, *Phys. Lett.* **17**, 266 (1965). M. Essig and J. Geiger, *Proc. 7th Eur. Cong. Electr. Microsc.*, The Hague **3**, 64 (1980).
81. G. C. King, J. W. McConkey, and F. H. Read, *J. Phys. B* **10**, L-541 (1977). See also I. Harrison and G. C. King, *J. Phys. B* **19**, L-447 (1986).
82. D. A. Shaw, G. C. King, and F. H. Read, *J. Phys. B* **13**, L-723 (1980).
83. D. A. Shaw, G. C. King, F. H. Read, and D. Cvejanović, *J. Phys. B* **15**, 1785 (1982).
84. D. A. Shaw, G. C. King, D. Cvejanović and F. H. Read, *J. Phys. B* **17**, 2091 (1984).
85. F. H. Read, *J. Phys. B* **17**, 3965 (1984). P. Hammond, F. H. Read, S. Cvejanović, and G. C. King, *J. Phys. B* **18**, L-141 (1985).
86. S. Cvejanović and F. H. Read, *J. Phys. B* **7**, 1180 (1974). See also P. Hammond, G. C. King, J. Jureta, and F. H. Read, *J. Phys. B* **18**, 2057 (1985); **20**, 3547, 4255 (1987).

87. G. C. King, F. H. Read, and R. C. Bradford, *J. Phys. B* **8**, 2210 (1975). See also W. Sandner, *ICPEAC XV*, Brighton (1987), p. 117.
88. P. Hammond, F. H. Read, and G. C. King, *J. Phys. B* **17**, 2925 (1984). See also **21**, 3123 (1988).
89. R. E. Imhof and F. H. Read, *J. Phys. B* **4**, 450 (1971). See also I. Humphrey, J. F. Williams, and E. L. Heck, *J. Phys. B* **20**, 367 (1987).
90. A. J. Smith, R. E. Imhof, and F. H. Read, *J. Phys. B* **6**, 1333 (1973).
91. D. J. Chornay, G. C. King, and S. J. Buchman, *J. Phys. B* **17**, 3173 (1984).
92. G. C. King, A. Adams, and F. H. Read, *J. Phys. B* **5**, L-254 (1972).
93. K. A. Mohamed, G. C. King, and F. H. Read, *J. Electron Spectrosc. Relat. Pheno.* **12**, 229 (1977).
94. P. J. Hicks, S. Daviel, B. Wallbank, and J. Comer, *J. Phys. B* **13**, 713 (1980); *Nucl. Instrum. Methods* **195**, 323 (1982).
95. J. Comer, *ICPEAC XIII*, Berlin (1983), p. 63.
96. S. Daviel, B. Wallbank, J. Comer, and P. J. Hicks, *J. Phys. B* **15**, 1929 (1982).
97. B. Wallbank, S. Daviel, J. Comer, and P. J. Hicks, *J. Phys. B* **16**, 3065 (1983).
98. T. A. York and J. Comer, *J. Phys. B* **15**, 4629 (1982); **16**, 3627 (1983); **17**, 2563 (1984).
99. R. J. Stubbs, T. A. York, and J. Comer, *J. Phys. B* **18**, 3229 (1985).
100. T. Reddish, B. Wallbank, and J. Comer, *J. Phys. E* **17**, 100 (1984). Also see S. Hill, T. Reddish, T. York, and J. Comer, *ICPEAC XV*, Brighton (1987), Contributed Papers, p. 799.
101. B. Leskovar, *Phys. Today* **30**, 42 (Nov. 1977). M. Lampton, *Sci. Am.* **245**, 62 (Nov. 1981).
102. G. F. Amelio, *Sci. Am.* **230**, 22 (Feb. 1974).
103. K. Blum and H. Kleinpoppen, *Phys. Rep.* **52**, 203 (1979).
104. K. Blum, *Density Matrix Theory and Applications*, Plenum, New York (1981).
105. J. Kessler, *Polarized Electrons*, 2nd ed., Springer-Verlag, Berlin (1985). See also *ICPEAC XV*, Brighton (1987), p. 21.
106. J. Slevin, *Rep. Prog. Phys.* **47**, 461 (1984). See also *ICPEAC XIII*, Berlin (1983), p. 85.
107. H. Kleinpoppen and M. R. C. McDowell (Eds.), *Electron and Photon Interactions with Atoms*, Plenum, New York (1976).
108. H. Kleinpoppen and J. F. Williams (Eds.), *Coherence and Correlation in Atomic Collisions*, Plenum, New York (1980).
109. H. O. Lutz, J. S. Briggs, and H. Kleinpoppen (Eds.), *Fundamental Processes in Energetic Atomic Collisions*, Plenum, New York (1983), see pages 39, 519, 531, 611, and 621.
110. M. Deutsch, *Rep. Prog. Phys.* **14**, 196 (1951). K. Siegbahn (Ed.), *Alpha-, Beta-, and Gamma-Ray Spectroscopy*, 2 vols., North-Holland, Amsterdam (1965), see A. H. Wapstra, p. 539; P. R. Bell, p. 905; H. Frauenfelder, R. M. Steffen, S. R. de Groot, H. A. Tolhoek, and W. J. Huiskamp, p. 997.
111. U. Fano, *Rev. Mod. Phys.* **29**, 74 (1957).
112. C. Cohen-Tannoudji, B. Diu, and F. Laloë, *Quantum Mechanics*, 2 vols., Wiley, New York (1977). W. H. Louisell, *Quantum Statistical Properties of Radiation*, Wiley, New York (1973). H. A. Bethe and R. Jackiw, *Intermediate Quantum Mechanics*, 2nd ed., Benjamin, New York (1968). R. Loudon, *The Quantum Theory of Light*, 2nd ed.,

Clarendon, Oxford (1983). J. J. Sakurai, *Modern Quantum Mechanics*, Benjamin-Cummings, Menlo Park, Calif. (1985); *Advanced Quantum Mechanics*, Addison-Wesley, Reading, Mass. (1967).

113. H. Kleinpoppen, *AAMP* **15**, 423 (1979).
114. J. H. Macek and D. H. Jaecks, *Phys. Rev. A* **4**, 1288 (1971).
115. N. Andersen, J. W. Gallagher, and I. V. Hertel, *ICPEAC XIV*, Stanford (1985), p. 57.
116. I. V. Hertel, H. Schmidt, A. Bähring, and E. Meyer, *Rep. Prog. Phys.* **48**, 375 (1985).
117. R. N. Zare, *Angular Momentum*, Wiley, New York (1988).
118. N. Andersen, J. W. Gallagher, and I. V. Hertel, *Phys. Rep.* **165**, 1 (1988). This review is the first in a set of three to be published in *Physics Reports*.
119. K. Blum and H. Kleinpoppen, *AAMP* **19**, 187 (1983).
120. G. F. Hanne, *Phys. Rep.* **95**, 95 (1983).
121. I. V. Hertel and W. Stoll, *J. Phys. B* **7**, 570, 583 (1974); *AAMP* **13**, 113 (1977).
122. J. H. Macek and I. V. Hertel, *J. Phys. B* **7**, 2173 (1974).
123. M. E. Rose, *Elementary Theory of Angular Momentum*, Wiley, New York (1957).
124. A. R. Edmonds, *Angular Momentum in Quantum Mechanics*, 2nd ed., Princeton University, Princeton, N.J. (1960).
125. M. Tinkham, *Group Theory and Quantum Mechanics*, McGraw-Hill, New York (1964).
126. D. M. Brink and G. R. Satchler, *Angular Momentum*, Oxford University Press, London (1971).
127. U. Fano and J. M. Macek, *Rev. Mod. Phys.* **45**, 553 (1973).
128. A. Wolcke, K. Bartschat, K. Blum, H. Borgmann, G. F. Hanne, and J. Kessler, *J. Phys. B* **16**, 639 (1983). J. Goeke, G. F. Hanne, J. Kessler, and A. Wolcke, *Phys. Rev. Lett.* **51**, 2273 (1983).
129. K. Bartschat and K. Blum, *Z. Phys. A* **304**, 85 (1982). P. S. Farago and J. S. Wykes, *J. Phys. B* **2**, 747 (1969). J. Wykes, *J. Phys. B* **4**, L-91 (1971).
130. M. Eminyan, K. B. MacAdam, J. Slevin, and H. Kleinpoppen, *J. Phys. B* **7**, 1519 (1974).
131. A. Crowe, T. C. F. King, and J. F. Williams, *J. Phys. B* **14**, 1219 (1981). See also P. A. Neill and A. Crowe, *J. Phys. B* **21**, 1879 (1988); B. P. Donnelly, P. A. Neill, and A. Crowe, *J. Phys. B* **21**, L-321 (1988).
132. J. F. Williams, *J. Phys. B* **14**, 1197 (1981). See also J. F. Williams, *Aust. J. Phys.* **39**, 621 (1986) and J. F. Williams and E. L. Heck, *J. Phys. B* **21**, 1627 (1988).
133. K. Becker, H. W. Dassen, and J. W. McConkey, *J. Phys. B* **17**, 2535 (1984).
134. J. W. McConkey, S. Trajmar, J. C. Nickel, and R. McAdams, *J. Phys. B* **18**, L-207 (1985).
135. M. A. Khakoo, K. Becker, J. L. Forand, and J. W. McConkey, *J. Phys. B* **19**, L-209 (1986).
136. W. B. Westerveld, K. Becker, P. W. Zetner, J. J. Corr, and J. W. McConkey, *Appl. Opt.* **24**, 2256 (1985).
137. A. Wolcke, J. Goeke, G. F. Hanne, J. Kessler, W. Vollmer, K. Bartschat, and K. Blum, *Phys. Rev. Lett.* **52**, 1108 (1984). See also J. Goeke, G. F. Hanne, and J. Kessler, *ICPEAC XV*, Brighton (1987), p. 175; H. J. Beyer, K. Blum, H. A. Silim, M. C. Standage, and H. Kleinpoppen, *J. Phys. B* **21**, 2953 (1988).

138. H. Borgmann, J. Goeke, G. F. Hanne, J. Kessler, and A. Wolcke, *J. Phys. B* **20**, 1619 (1987). For a more recent and more elaborate experiment, see J. Goeke, G. F. Hanne, and J. Kessler, *Phys. Rev. Lett.* **61**, 58 (1988), in which e-$h\nu$ coincidence measurements are described.

139. J. J. McClelland, M. H. Kelley, and R. J. Celotta, *Phys. Rev. Lett.* **55**, 688 (1985).

140. J. J. McClelland, M. H. Kelley, and R. J. Celotta, *Phys. Rev. Lett.* **56**, 1362 (1986).

141. J. J. McClelland, M. H. Kelley, and R. J. Celotta, *J. Phys. B* **20**, L-385 (1987); also *ICPEAC XV*, Brighton (1987), Contributed Papers, p. 152. I. V. Hertel, M. H. Kelley, and J. J. McClelland, *Z. Phys. D* **6**, 163 (1987).

142. G. F. Hanne, Cz. Szymytkowski, and M. van der Wiel, *J. Phys. B* **15**, L-109 (1982).

143. G. Sinapius, *ICPEAC XV*, Brighton (1987), p. 73.

144. J. W. Humberston (Ed.), *Atomic Physics with Positrons*, Plenum, New York (1988).

145. H. J. Beyer, K. Blum, and R. Hippler (Eds.), *Coherence in Atomic Collision Physics*, Plenum, New York (1988).

146. P. G. Burke and J. B. West (Eds.), *Electron-Molecule Scattering and Photoionization*, Plenum, New York (1988).

147. H. Goldstein, *Classical Mechanics*, 2nd ed., Addison-Wesley, Reading, Mass. (1980). K. R. Symon, *Mechanics*, 3rd ed., Addison-Wesley, Reading, Mass. (1971). J. B. Marion, *Classical Dynamics of Particles and Systems*, 2nd ed., Academic, New York (1970). E. B. Wilson, J. C. Decius, and P. C. Cross, *Molecular Vibrations*, McGraw-Hill, New York (1955). J. I. Steinfeld, *Molecules and Radiation*, 2nd ed., MIT Press, Cambridge, Mass. (1985).

148. D. ter Haar (Ed.), *Problems in Quantum Mechanics*, 3rd ed., Pion, London (1975), also published by Academic, New York (1975). V. I. Kogan and V. M. Galitskiy, Problems in Quantum Mechanics, Prentice-Hall, Englewood Cliffs, N.J. (1963).

149. J. A. Cronin, D. F. Greenberg, and V. L. Telegdi, *University of Chicago Graduate Problems in Physics with Solutions*, Addison-Wesley, Reading, Mass. (1967).

150. M. Chen, *Berkeley Physics Problems with Solutions*, Prentice-Hall, Englewood Cliffs, N.J. (1974).

151. F. H. Read, *Electromagnetic Radiation*, Wiley, New York (1980).

6

IONIZATION OF NEUTRALS BY ELECTRON IMPACT

This chapter deals with the ionization of neutral atoms and molecules by "monoenergetic" electrons. (The ejection of bound electrons from *ions* by electron impact is treated in Chapter 7.) We are concerned with beam studies of targets sufficiently "thin" (see Section 4-1-A) that only "primary" ionization by the projectile electrons is significant. Electrical discharges and swarm experiments, in which "secondary" ionization by ejected electrons and electron multiplication can occur, are not covered; there is no discussion of the Townsend coefficients.[1] Similarly, thermal ionization[2] is not treated here. We shall spend considerable time on the measurement of integral ionization cross sections, which are of great importance in applications, but we shall spend more time on modern experiments that illuminate fine details of the ionization process and elucidate the underlying mechanisms. Most of our discussion of the theory of electron impact ionization is devoted to classical and semiclassical calculations, and to the Bethe-Born approximation, which was briefly mentioned in Section 3-23-B.

6-1. INTRODUCTION

The ejection of electrons from atoms and molecules by electron impact is a phenomenon of considerable practical interest.[3,4] Most ion sources used in atomic and nuclear physics, and in mass spectrometry, rely on this process. Electron impact is also responsible for the production of electrons in radio-frequency (RF), glow, and arc discharges. The action of various kinds of radiation detectors hinges on the ionization of the detector gas by primary or secondary electrons. Collisional ionization is also of great importance in astrophysics, planetary atmospheres, controlled fusion research, plasma physics, and gaseous electronics. In addition, experimental data on ionization are vital to the further development of the theory of atomic collisions, for they provide tests of the assumptions and approximations that are necessarily made in calculations. Not only are data required for ordinary stable atoms and molecules; many structures of practical and theoretical significance are ionic or are atoms that normally exist only in molecular form.

There is a vast literature on electron impact ionization. Useful books and reviews covering many aspects of the subject are cited in Refs. 4–12. We begin our discussion with the oldest part of the field, one especially important in applications, namely, the measurement of integral cross sections for ionization.

PART $\mathscr{A}$. THE INTEGRAL IONIZATION CROSS SECTION

The *integral cross section* refers to ionization irrespective of the direction of motion of the ionization products or of the angle of scattering of the electron projectile. In certain parts of our discussion, all physical processes that contribute to the ionization are counted; then we may speak of the *total, integral cross section.* (The various mechanisms of electron impact ionization are discussed in Section 6-5.)

6-2. GENERAL METHODS OF MEASUREMENT OF INTEGRAL IONIZATION CROSS SECTIONS[13,14]

About 100 years ago, research on electrical discharges and considerations of valence in chemistry stimulated speculation about the possibility of electron impact ionization.[12] Some 40 years elapsed before the first reliable measurements of cross sections were made. Notable among these pioneering measurements were those of J. T. Tate and P. T. Smith in the late 1920s and early 1930s, who obtained cross sections that were accepted as standards as late as the 1970s, and that appear to be more accurate than most of the other data published up to that time. Indeed, Tate and Smith made contributions of the same magnitude as those of their contemporary Ramsauer (see Section 4-1), and their work will figure prominently in the discussion that follows.

A. Single-Beam Experiments Involving the Total Collection of Residual Positive Ions; Apparent (Gross) Ionization Cross Sections

Most measurements of electron impact ionization cross sections have been made by passing a collimated and nearly monoenergetic beam of electrons through a gas or vapor and collecting essentially all of the positive ions formed in ionizing events. To obtain an accurate and meaningful value for the cross sections, we must use a "thin" target so that only a small fraction of the "primary" electrons will strike molecules of the target gas. Then an insignificant fraction of the projectiles will experience more than one collision, and an insignificant amount of secondary ionization will be produced by electrons ejected from molecules by the beam projectiles. If i_0 denotes the incident beam current; N, the number density of the gas target; x, the effective target thickness; and i^+, the collected positive ion current, the *apparent* (*or gross*) *ionization cross section* σ_i will be given by the equation

$$i^+ = i_0 N \sigma_i x. \tag{6-2-1}$$

[The similarity of this equation to (1-4-8) is obvious.] Since multiply charged ions may be produced and collected, the measured cross section is really the weighted sum

$$\sigma_i = \sigma^+ + 2\sigma^{2+} + 3\sigma^{3+} + \cdots, \tag{6-2-2}$$

where σ^{n+} is the sum of the cross sections for all events involving ejection of n electrons from a target molecule.

Although such an experiment provides no information about the charges and masses of the residual ions, σ_i does represent an unambiguous and useful cross section for the production of free electrons. Furthermore, measurements of this type can be made with reasonably good accuracy, and they directly yield absolute cross sections, because nondiscriminating detectors with very nearly 100% efficiency are used. Since e/m analysis of the residual ions is not attempted, all the ions produced inside the active region of the target gas are simply swept to a negatively charged collector plate located parallel to the electron beam and bounding the active volume. The active thickness of the target may be accurately defined by the use of guard electrodes. The positive ions produced by the electron beam have low energies, and a weak electric field suffices to sweep all the ions to the collector.

Actually, the apparent cross section will equal the true cross section at electron energies below the threshold for double ionization (which is usually several times the threshold for single ionization). Furthermore, with electron projectiles, single ionization from the outermost shell is usually much more probable than multiple ionization or inner-shell ionization, and σ_i is generally nearly equal to the cross section for single-electron ejection from the outer shell. (Inner-shell ionization is discussed in Section 6-7.)

The technique described here was used to great advantage in the early work of Tate and Smith, and later by D. Rapp and his colleagues (see Section 6-3-A).

B. Single-Beam Experiments Involving *e/m* Analysis of Residual Positive Ions; True (Partial) Ionization Cross Sections

If we wish to determine the cross section for production of ions of a given charge-to-mass ratio, e/m analysis is required. Ions may be drawn tranversely from a narrow section of the target gas and analyzed by one of several different mass spectrometric techniques. If careful attention is paid to the difficult question of discrimination in the ion production, collection, and detection processes, we may obtain the cross section for production of any particular ionic species relative to that of any other species produced in significant quantity (Section 6-3-B). The difficulty of accurately determining the effective volume of the target gas and the efficiency of the e/m analyzer for each ionization fragment has led to a concentration of effort in measuring relative cross sections. Absolute values for the "partial cross sections" may be obtained from these data by comparison with a cross section whose absolute value is already known. Since simple stable atoms and molecules are generally used as standards, the importance of accurate absolute cross sections for such structures is evident.

In certain types of studies, however, knowledge of absolute cross sections is not important. In recent years there has been considerable theoretical interest in the variation of the cross sections for production of single ionic species in the region near threshold (Section 6-8). Also, a great deal of work has been done to gain information concerning the quantum states in which the ions are formed. Evidently, e/m analysis is required for such investigations, but absolute cross sections are not.

C. Crossed-Beam Experiments on Unstable and Stable Targets

Crossed modulated beam techniques have been used since 1958 for cross section measurements on unstable targets. Most of these experiments have involved the intersection of a mechanically modulated *thermal* beam of neutral atoms with a dc beam of electrons of variable energy in apparatus somewhat similar to that shown in Fig. 4-1-6. A number of groups, including those of W. L. Fite and R. H. Neynaber, have made important ionization measurements with this slow neutral beam technique.[4,7–10] However, difficulties (mainly associated with cross modulation of the dc beam) appear when both beams consist of charged particles, and Dolder et al.[15] developed a technique that obviates these difficulties. Their technique will be described in detail in Section 7-2; it will be discussed briefly here because it served as the precursor of the important method implemented by M. F. A. Harrison et al. for measurements of the ionization of neutrals.

The first application of the Dolder-Harrison-Thonemann technique was to the study of ionization of He^+ ions by electrons, and was the first successful experiment with two intersecting beams of charged particles. The scheme they employed intersects a fast well-collimated beam of He^+ ions with an electron beam of variable energy and then resolves the ion beam with respect to its charge states after it passes through the intersection region. If the incident He^+ beam has an energy of several keV, the momentum imparted to any of the He^+ ions experiencing ionization is negligible compared with their initial forward momentum, so that the He^{2+} ions thus formed remain in the beam along with the He^+ ions that do not lose their remaining electrons. An important factor in the success of this technique appears to be the simultaneous modulation of both the ion and the electron beams and the ability to vary the relative phase of the modulation.

Recently, Harrison and his colleagues at Culham Laboratory[16–19] have developed a fast neutral beam technique for electron-impact ionization measurements that is similar in concept to that described above.[15] It involves generating a fast ion beam and obtaining the desired neutral beam by subjecting the ions to charge transfer. (The neutrals that are produced have almost exactly the same velocity as the ions in the original beam.) The neutral beam is then crossed by an electron beam of variable energy, and the resulting ions are deflected out of the neutral beam by a magnet, which also performs momentum analysis on the ions. Detection of the fast ions is then a simple matter. This technique has been applied by Harrison et al. to studies of the single ionization of H(2*S*);[16] He*;[17] C, O, N;[18] and ground state He.[19] It offers several important advantages over the slow neutral beam method. First, it provides a much purer beam of atoms. (A slow beam of H atoms, for example, that is produced in the conventional fashion by thermal dissociation of hydrogen gas or by dissociation in an RF discharge contains undissociated H_2 molecules; on the other hand, a fast beam that is entirely atomic can be produced by the technique described here.) Second, the flux of nonionized neutrals emerging from the collision region can be measured much more accurately if the neutrals are energetic; no normalization to other cross sections is required. The fast neutral technique has also been applied by Freund's group at the AT&T Bell Laboratories to the measurement of cross sections for multiple ionization.[20]

The latest technique for measuring integral ionization cross sections of unstable targets marks a return to thermal target beams, but with a significant difference. Gilbody's group[21] at Belfast passes short pulses of electrons through the beam at 90°, and sweeps out the slow-ion collision products with a transverse electric field for identification by their characteristic times of flight to a counter. Their H-atom data are presented in Section 6-4.

6-3. EXPERIMENTAL APPARATUS

A. Rapp's Apparatus for Measurement of Apparent (Gross) Ionization Cross Sections

A modernized version of the classic apparatus of Tate and Smith[22,23] was constructed by Rapp and his colleagues in the early 1960s and used for a large number of important measurements.[24–26] They made several improvements over the original design and experimental procedures of Tate and Smith (especially in pressure measurement, residual ion collection, and suppression of scattered electrons), and thereby were able to obtain significantly better data in some instances. For the most part, deviations from the original data were small.

The Rapp apparatus is sketched in Fig. 6-3-1. The entire system is bakeable and can be pumped down to 10^{-9} torr. An electron beam from a differentially pumped, indirectly heated cathode *K* is collimated by apertures and confined by a strong axial magnetic field (30 to 60 mT). The electrons are accelerated to the desired energy (1000 eV maximum) and passed through a chamber *C* containing the target gas at a pressure of about 5×10^{-5} torr. The emerging beam (current about 10^{-6} A) is collected on the plate ECP, which is maintained at a positive potential screened from the collision chamber by the region ECS. The current of positive ions (about 10^{-9} A) formed by the beam is collected on the upper plate IC (18.3 mm long), maintained at a negative potential relative to the parallel plate JC located symmetrically on the other side of the electron beam. To provide a uniform transverse field for collection of the ions, and to define accurately the length of the effective collision region, plates IC and JC are flanked by guard electrodes G and H, respectively, with G at the same potential as IC and H as JC. The potential difference between IC and JC need be only about 4 V to collect all ions produced from atomic gases, because such ions are formed with little kinetic energy. For molecular gases, however, Rapp et al. demonstrated that as much as 30 V is required in some of their studies to collect all the ions produced in dissociative ionization events. It appears that some of Tate and Smith's cross sections were several percent low because of their failure to use sufficiently high sweeping voltages.

The apparent ionization cross section σ_i is determined from (6-2-1). In a careful critique of experimental methods used in ionization experiments, Kieffer and Dunn[27] concluded that there are serious difficulties associated with the determination of *each* of the factors in this equation. Especially difficult is the accurate measurement of the target gas pressure, required for calculation of the number density in the collision chamber. In Section 4-1-A it was pointed out that many collision studies made before about 1961 were vitiated by the Ishii effect, which operates when cold traps are used with McLeod gauges for pressure measurements. This effect is not serious for H_2 and He, but it can lead to large

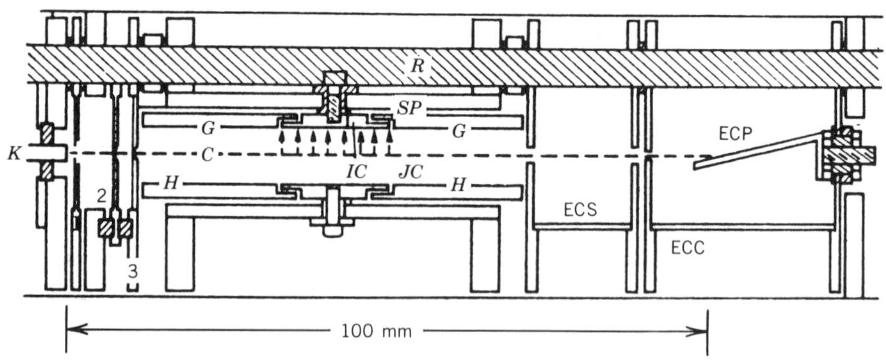

Figure 6-3-1. The apparatus of Rapp and his colleagues.[26] The electron beam passes from the cathode K at the left to the plate ECP at the right. Residual ions are swept upward by a transverse electric field. The ions collected on the plate IC are the ones involved in the calculations of the apparent ionization cross section.

errors for heavy gases. Rapp and Englander-Golden avoided this problem by using a McLeod gauge to measure absolute cross sections for H_2, and they obtained absolute cross sections for the other gases relative to H_2 by a new method.[26] McLeod gauge errors appear to account for most of the differences in absolute cross sections at the time of publication of Rapp's data.[26] The McLeod gauge is no longer used in collisions experiments; it has been replaced mainly by the capacitance manometer.[13]

Careful attention must be paid to the primary beam electrons as they traverse the collision chamber. The axial magnetic field, produced by a solenoid surrounding the ionization apparatus, confines the primary electrons to a well-defined beam. Scattering of the beam electrons by gas molecules would otherwise require the use of strong electric fields to prevent these electrons from reaching the ion collector IC, and such fields would cause a serious variation in the energy of the incident electrons along the beam path. Electrons scattered from the beam might also cause secondary electron emission from interior surfaces of the apparatus if magnetic confinement were not employed. Finally, the magnetic field prevents the beam from striking the rim of any of the apertures through which it passes.

The collimating magnetic field causes the beam electrons to travel in helices, the radii of which are determined by the electron speed, the angle at which each electron leaves the gun, and the collimation geometry. Accordingly, the path length of the primary electrons through the target gas is uncertain. Several analyses of this problem have been made,[13,14] but there has been little agreement on the magnitude of the effect in specific apparatus. However, the consensus now is that the "spiraling effect" leads only to small errors ($<1\%$) in experiments of the type described here.

B. Bleakney's Apparatus for Measurement of True (Partial) Ionization Cross Sections

The analyzer shown in Fig. 6-3-2 was employed by Bleakney[28] to measure the ionization potential of molecular hydrogen and by Bleakney and Smith[29] to obtain the cross section for double ionization of He atoms by single-electron impact. They used a water-cooled solenoid 1 m long and with a 125-mm bore that could produce a field of 150 mT continuously without iron in the magnetic circuit. With such a field it was possible to use a Dempster-type semicircular analyzer. Electrons from the filament F are accelerated between the first two diaphragms of the electron gun G and pass between plates A and B, as shown by the dashed lines. The electron beam is collected by the inclined plate P, which is held about 100 V positive with respect to the rest of the trap T to prevent the escape of secondary electrons. The positive ions formed along the magnetically confined electron path are accelerated through the two slits S_1 and S_2 by a small field between A and B and a stronger one between B and C. The ions then travel along circular arcs in the magnetic field. The radius of the circle along which a

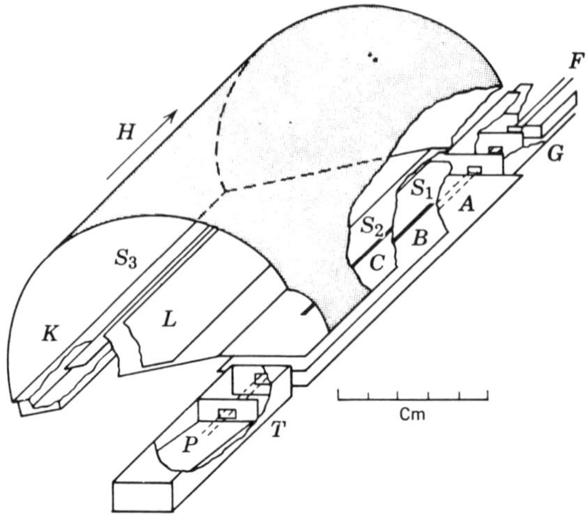

Figure 6-3-2. Bleakney's apparatus for appearance potential and partial ionization cross section measurements.[28,29]

given ion moves is determined by the equation

$$\frac{e}{m} = \frac{2V}{r^2 B^2} \tag{6-3-1}$$

where V is the potential difference through which the particular ion has fallen before it passes through S_2. The ions selected for detection pass through the third slit S_3 and fall on plate K, which is connected to an electrometer. Different ionic species are passed successively through S_3 by varying the accelerating potential V applied to S_2. The total ion current is measured by connecting plate L to the electrometer and setting V so that no ions pass through S_3. The entire apparatus within the vacuum tube is made of tantalum and tungsten with Pyrex insulators and can be baked out in a furnace mounted inside the solenoid. Apparatus somewhat similar to that shown in Fig. 6-3-2 was used by Tate and Smith[30] to measure the ionization potentials and probabilities of formation of singly and multiply charged ions in Na, K, Rb, Cs, Kr, and Xe at electron energies up to 700 eV.

All of the instruments[7–10,14] used for the early measurements of partial cross sections appear to have suffered badly from mass discrimination at various points extending from the ion production region to the detector, although the basic principles of operation of some of the instruments, such as Bleakney's, was sound. Even the *relative* cross sections obtained in the 1930s now appear to be accurate only to within 20 to 30%. The problems arising from mass discrimination are especially severe for molecular gases.[13,14,27] A discussion of these and other problems appears in pages 151–174 of Märk's review.[14] In recent years much progress has been made in understanding and reducing the effects that have degraded the accuracy of partial cross section measurements in the past,[5,14] and reliable data are accumulating at an impressive rate.

C. The Crossed Electron–Fast Neutral Beam Apparatus of Harrison and His Colleagues[16–19]

We shall describe this apparatus in the form in which it was used for measurements of the single ionization of ground-state helium atoms.[19] As shown in Fig. 6-3-3, helium ions are produced in a PIG ion source by electron bombardment in a hot cathode arc with an axial magnetic field, and He^+ ions are extracted at either 2 or 4 keV. The He^+ beam is focused with an einzel lens and deflected through a 60°-sector magnetic field. In then traversing a charge exchange chamber that is 57 mm long and contains helium gas, about 10% of the ions are neutralized, the remainder being subsequently removed from the beam by electrostatic deflection plates. The neutral helium atom beam then is intersected by an electron beam at 90° in the collision region. The He^+ ions formed therein are swept out of the neutral beam by a second 60°-sector magnetic field and detected with a calibrated electron multiplier. The non-ionized He atoms pass on to a vacuum thermopile whose sensitivity is regularly measured by directing onto it ion beams of known current and energy. Because of the thermal drift of the instrument, it proves inconvenient to make direct use of the thermopile output voltage during the actual cross section measurements. Instead, the current of secondary electrons ejected by the fast neutrals from a thin (2-μm-thick) aluminum target covering the thermopile is used to determine the neutral atom current,[19] and the secondary emission coefficient of this surface is frequently determined with the thermopile.

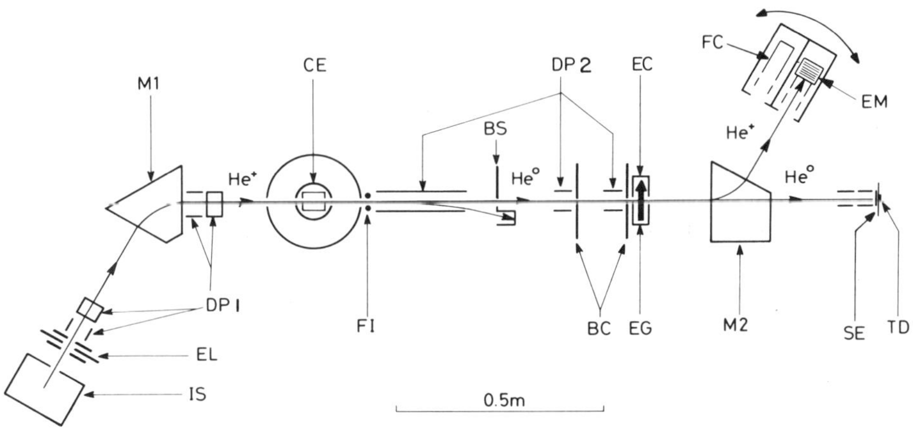

Figure 6-3-3. The crossed electron–fast neutral beam apparatus of Harrison and his colleagues.[19] IS, ion source (electron bombardment PIG type); EL, einzel lens; DP1, deflector plates to optimize path of ion beam; M1, ion selector magnet; CE, charge exchange cell containing helium gas; F1, field ionizer bars to ionize highly excited atoms; DP2, electrostatic deflector plates to remove ions from atom beam; BS, beam stop aperture; BC, beam collimators; EG, electron gun; EC, electron collector; M2, analyzer magnet; SE, secondary electron emission detector for fast helium atoms (aluminum target surface); TD, integral thermopile used for calibration of SE; EM, electron multiplier for detection of He^+ product ions; FC, Faraday cup for calibration of the electron multiplier.

This technique does not require a magnetic field to guide the electron beam or the use of a pressure gauge. The number density of the target atoms is calculated from their flux and velocity, and the collision volume and the beam density profiles can be accurately determined. However, it is difficult to produce a neutral beam that contains no metastables or long-lived ordinary excited states. Montague et al.[19] discuss the steps that are taken to reduce the excited component of the neutral beam. They also discuss the modulation of the electron beam performed in order to separate the real ionization signal from spurious background counts. Additional examples of the electron beam modulation appear in Chapter 7, where the method of calculating the cross sections from the experimental variables is explained.

The crossed electron–fast neutral beam method has also been used by other groups. Defrance and his colleagues in Louvain have studied the ionization of atomic hydrogen in the metastable $2S$ state.[31] Instead of using the standard technique of pulse modulating the beam intensities in order to separate the true signal from the spurious background,[16,21] Defrance et al. modulate the interaction by sweeping the electron beam periodically across the neutral beam, keeping the electron beam always parallel to its original direction. By so doing, they eliminate the necessity of mechanically scanning the beams to obtain their profiles and the beam overlap factor, and they can utilize an elegant method of measuring the background contribution. (These matters are discussed more fully in Section 7-1-C.) At the AT&T Bell Laboratories in Murray Hill, New Jersey, Wetzel et al.[20] have used the fast beam method to study the single and multiple ionization of the noble gas atoms, and Hayes et al.[20] similarly for the halogen atoms. They measured absolute partial cross sections for production of the separate charge states and calculated the cross sections from (6-2-2).

6-4. IONIZATION CROSS SECTION DATA

A large quantity of experimental data on integral ionization cross sections exists. Kieffer and Dunn[27] have compiled and evaluated absolute ionization cross sections obtained for atoms and diatomic molecules up to 1965. They also present selected relative cross sections for production of multiply charged ions and survey experimental techniques used for ionization measurements up to that date.

The Belfast Data Center[32,33] has reviewed the data up to 1986 for all stages of ionization of atoms ranging from hydrogen through nickel, and published a recommended set of cross sections and Maxwell-averaged rate coefficients over a wide range of temperature. Both experimental and theoretical data are considered. Analytical expressions are given for the recommended cross sections as functions of electron impact energy and for the rate coefficients as functions of temperature.

Tawara and Kato[34] have collected experimental data up to mid-1986 and published graphs of ionization cross sections for atoms and ions from hydrogen to uranium. They also include some theoretical data, and experimental cross sections for H_2, D_2, O_2 and N_2. Older integral cross-section data, obtained up to about 1970, are surveyed in Refs. 7–9 and by de Heer and Inokuti,[13] who also present much more recent information on atoms and molecules. Märk[5] has published a useful data index for molecules that extends up to about 1982. Partial cross section data have been collected and discussed recently by Märk.[5,14]

We begin our brief data survey with Fig. 6-4-1, which shows total, integral cross sections for the noble gases from near threshold up to 20 keV impact energy. The unlabeled data below 1 keV were obtained by Rapp and Englander-Golden[26] with the apparatus of Section 6-3-A; the high-energy data labeled SHWK come from Schram et al.[35]; the data marked S are those of Smith.[23] Recent results obtained at the AT&T Bell Laboratories[20] with the crossed electron–fast neutral beam method are in excellent agreement with the Rapp data for He, Ne, Ar, and Kr, as are the cross sections from other recent

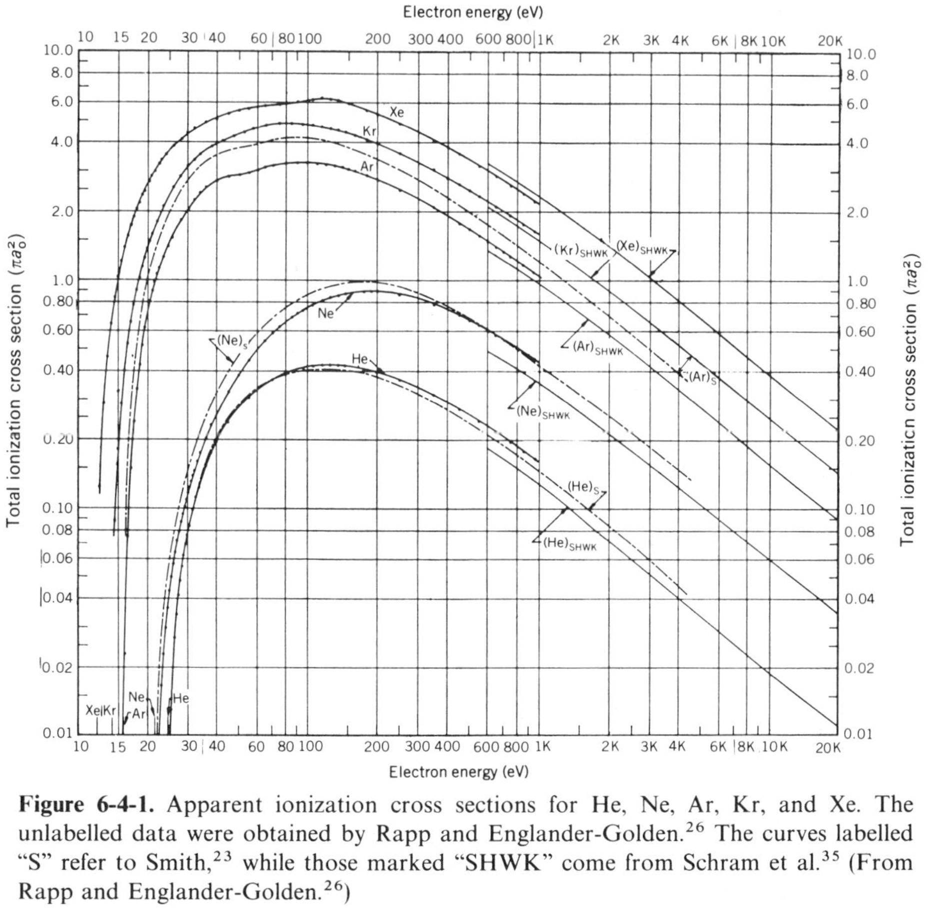

Figure 6-4-1. Apparent ionization cross sections for He, Ne, Ar, Kr, and Xe. The unlabelled data were obtained by Rapp and Englander-Golden.[26] The curves labelled "S" refer to Smith,[23] while those marked "SHWK" come from Schram et al.[35] (From Rapp and Englander-Golden.[26])

measurements.[19,36–38] Rapp's cross sections for Xe appear to be 10 to 15% too low, perhaps because of problems in accurately measuring the pressure of this heavy gas. Rapp's results, except for Xe, probably have an absolute accuracy of 6 to 7%. The results for He, Ne, Ar, and Kr labeled SHWK seem to be too low,[13] although they exhibit the correct energy dependence. Integral cross sections for H_2, O_2, N_2, CO, and NO obtained by Rapp and Englander-Golden[26] are presented in Fig. 6-4-2.

The results of three experiments[21,39,40] on the ionization of ground-state atomic hydrogen are shown in Fig. 6-4-3, along with several theoretical predictions discussed in Ref. 21. (The data near threshold agree well with the earlier low-energy results of McGowan and Clarke.[41]) Crossed electron-thermal beam methods were used in each case. The cross sections are relative, not absolute, because of the difficulty in measuring the number density of the neutral beam. Fite and Brackmann[39] normalized their data in two different ways. In one, they normalized their cross section at 500 eV to a Born calculation result. In the other, they normalized their data by using the absolute ionization cross sections for H_2 obtained by Tate and Smith.[22] In the Fite-Brackmann experiment, partial dissociation of H_2 gas is accomplished in a tungsten furnace at about 2000°C, and the emerging H/H_2 beam is collimated, mechanically chopped, and then intersected by the electron beam, which ionizes a small

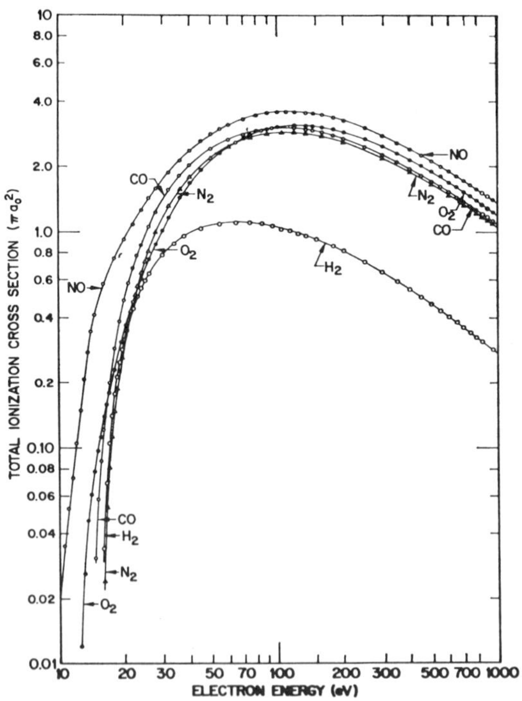

Figure 6-4-2. Apparent ionization cross sections for H_2, O_2, N_2, CO, and NO.[26]

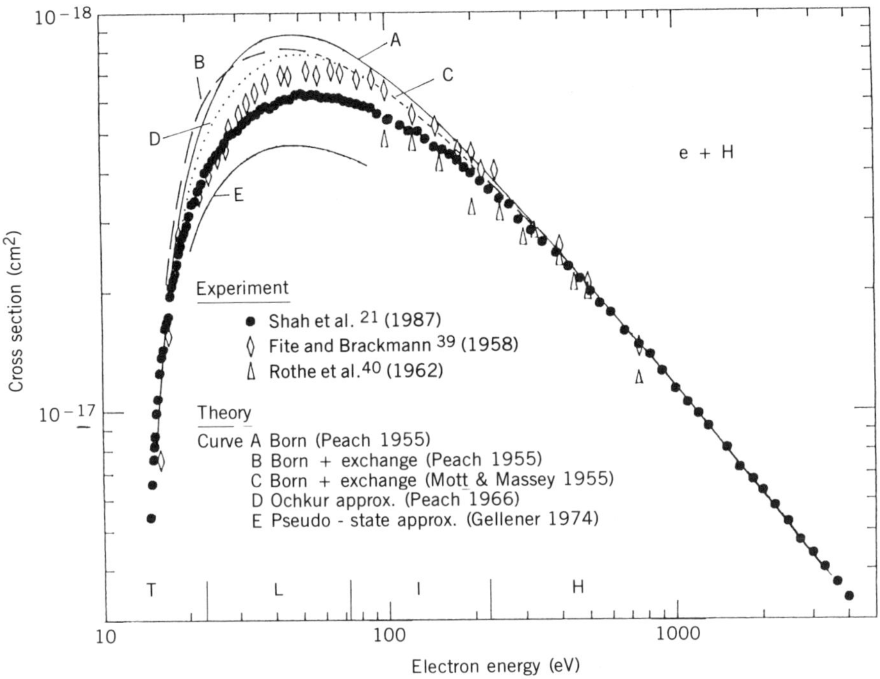

Figure 6-4-3. The integral ionization cross section for the H atom in its ground state.[21] Displayed are experimental data from the Belfast group,[21] Fite and Brackmann,[39] and Rothe et al.[40] References to the indicated theoretical results are provided in the paper by Shah et al.[21] The impact energy scale is divided into regions for which the collision dynamics and theoretical approaches are quite different, as in Fig. 1 of Ref. 12. The letters *T, L, I,* and *H* indicate the threshold, low-, intermediate-, and high-energy regimes, respectively.

fraction of each beam component. The mixed ion beam then passes through a magnetic sector mass spectrometer, and the H^+ and H_2^+ signals are recorded. By assuming that the H_2^+ signal (corrected for the amount of dissociation of the neutral beam) is governed by the Tate and Smith cross section for H_2, Fite and Brackmann could calculate cross sections for the ionization of H atoms.[7,39] Gilbody's group[21] normalized their data by comparison to known proton-impact cross sections.

Integral cross sections for the ionization of hydrogen atoms in the metastable 2*S* state are presented in Fig. 6-4-4. The crossed electron–fast neutral beam method was used in both of the indicated measurements.[16,31]

Figure 6-4-5 shows absolute partial cross sections for the single and multiple ionization of argon atoms by electron impact.[42] The weighted sum of these cross sections agrees to within several percent with the total, integral cross section reported by Rapp and Englander-Golden.[26]

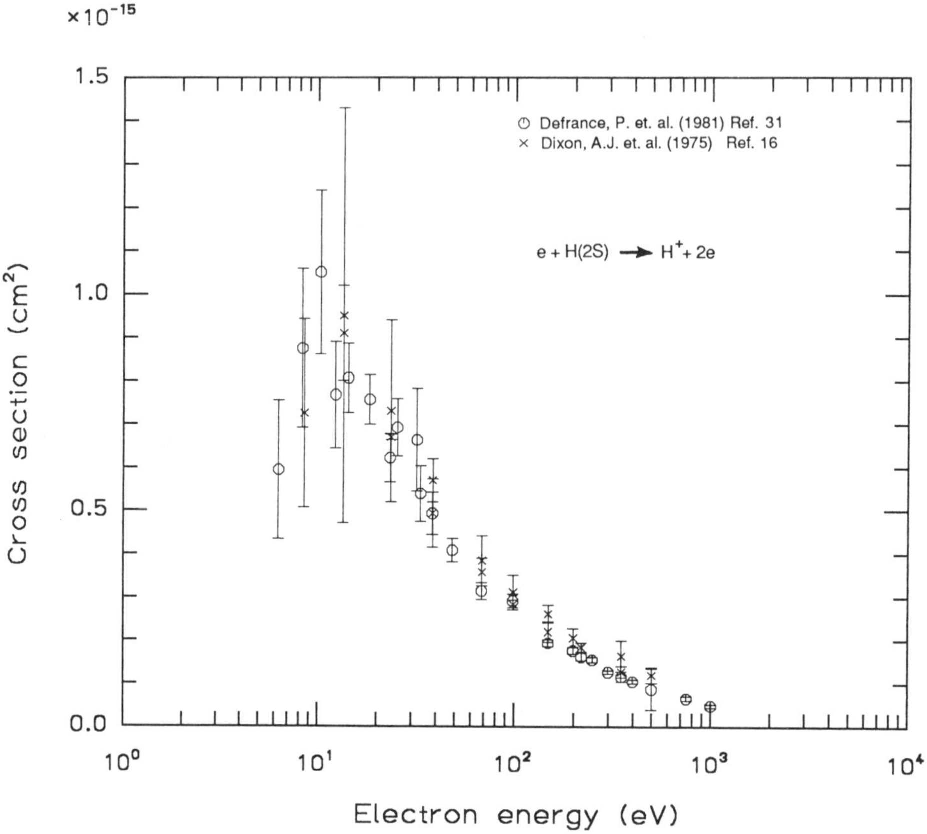

Figure 6-4-4. The integral ionization cross section for the H atom in its metastable 2*S* state.[34]

In Fig. 6-4-6, the bottom curve gives the relative cross section for single ionization of Ar, the middle curves show the contributions of the 3*d* and 4*p* autoionization processes (see Section 6-6), and the top curve gives the total ionization cross section.[43]

Partial cross sections for the ionization of ammonia by electron impact, as measured by Märk et al.,[44] are displayed in Fig. 6-4-7.

Figure 6-4-8 shows Tate and Smith's electron-impact ionization cross section for O_2[22] compared with data for ionization of O_2 by protons of the same velocity.[45] As shown in Section 6-14-C, the cross sections for ionization by electrons and by H^+ projectiles are expected to have the same value at the same velocity in the high-velocity limit. Approximate verification of this prediction has been established by McDaniel et al. for He, Ne, Ar, H_2, N_2, and CO, as well as for O_2.[10,45]

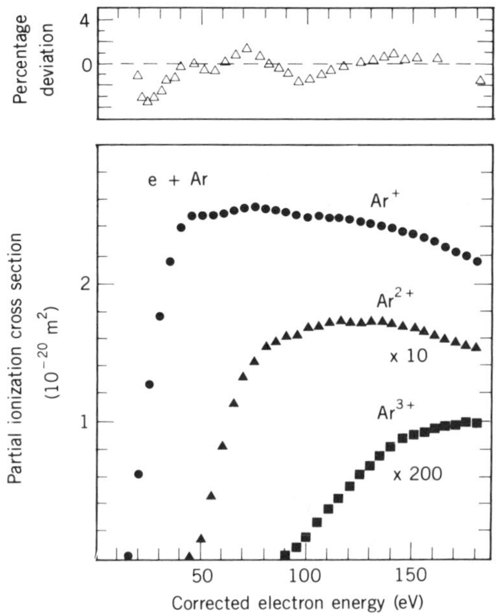

Figure 6-4-5. Absolute partial ionization cross sections for the Ar atom.[42] The panel at the top compares the weighted sum of the partial cross sections to the total, integral cross section measured by Rapp and Englander-Golden.[26] (From Märk and Dunn.[4])

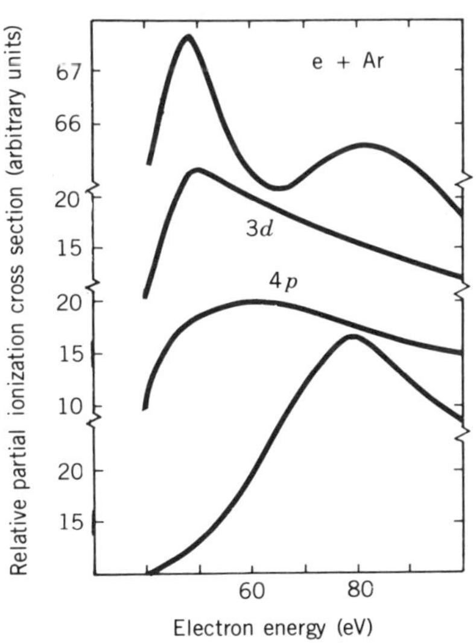

Figure 6-4-6. The relative cross section for single ionization of the Ar atom (bottom curve); contributions of the 3*d* and 4*p* autoionization processes (the two middle curves); the total, integral ionization cross section for Ar (top curve).[43] (From Märk and Dunn.[4])

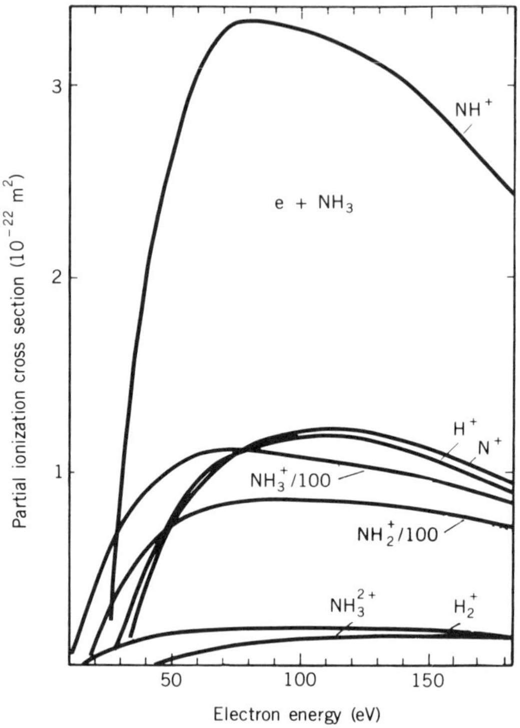

Figure 6-4-7. Partial ionization cross sections for the ammonia molecule, NH_3.[44] (From Märk and Dunn.[4])

PART ℬ. IONIZATION MECHANISMS: DETAILED STUDIES OF SPECIAL FEATURES OF IONIZATION

In Section 6-5 we shall enumerate and briefly describe the principal mechanisms by which electron impact ionization occurs. In subsequent sections we expand our discussion of two of them: autoionization and inner-shell ionization, and also describe several kinds of studies that provide insight into the ionization mechanisms.

6-5. OUTLINE OF PROCESSES BY WHICH ELECTRON IMPACT IONIZATION OCCURS[4,12]

Mechanisms for ionization of atoms by electron impact include:

1. $e + X \rightarrow X^+ + e + e$. *Single ionization of the outer shell, with the ion left in its ground state.* This process is direct (i.e., nonresonant) and is characterized by large impact parameters and small momentum transfers. It is

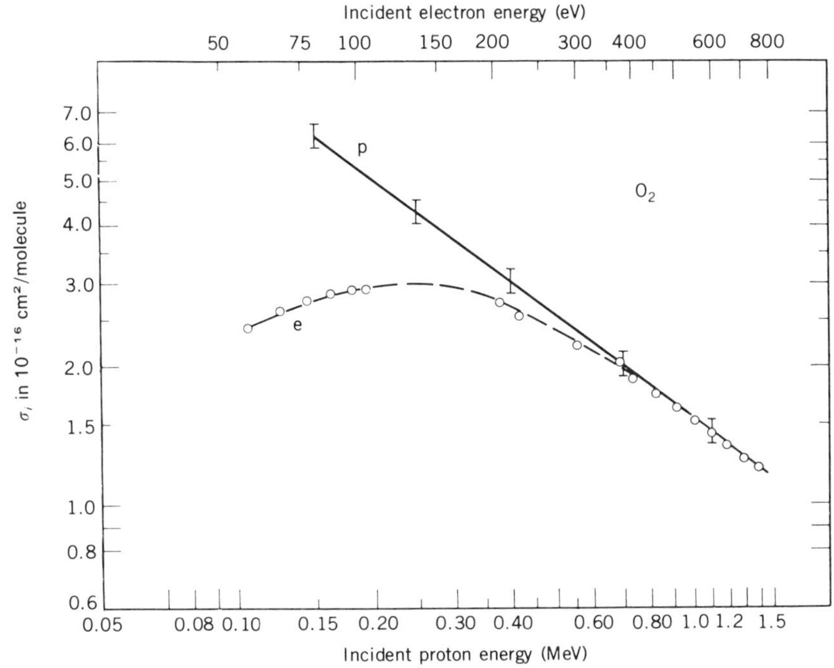

Figure 6-4-8. Comparison of experimental apparent ionization cross sections for protons and electrons of equal velocity incident on molecular oxygen, O_2.[45]

important in applications because of its large cross sections; for targets with only a few electrons, this mechanism accounts for about 90% of all ionizing events.[12] In the gentle collisions characterizing this process, only a single target electron is involved in the ionization event. The ejected electron and the scattered electron leave the collision site within about 10^{-16} sec of each other.[14] The cross section for this process rises almost linearly from threshold to a maximum at an impact energy typically about four or five times the threshold value. Then the cross section falls to a few percent of the peak value at impact energies of several keV. Figure 6-4-1 shows this kind of behavior, although the total cross sections plotted there contain contributions from other ionization mechanisms.

2. $e + X \rightarrow X^{+*} + e + e$. *Single outer-shell ionization, the ion being left in an excited state.* These events are associated with small impact parameters, and more than one atomic electron are involved simultaneously in the collision complex. Strong correlations usually exist among the electrons.
3. $e + X \rightarrow X^{++} + e + e + e$. *Double outer-shell ionization.* As in (2), the impact parameters are small, and the electron-electron correlation is usually strong.
4. $e + X \rightarrow X^{**} + e \rightarrow X^{+} + e + e$. *Autoionization, with double outer-shell excitation.* Here two outer-shell electrons are excited simultaneously, and

the discrete atomic state thus formed decays into a ground-state ion and a free electron. This process frequently contributes less than 1% to the total ionization cross section. Here also, collisions with small impact parameters are the important ones, and the electron correlation is strong. This mechanism involves a resonant process.

5. $e + X \rightarrow X^*(J) + e \rightarrow X^+(J') + e + e$. *Autoionization, with single outer-shell excitation.* The target atom is singly excited to a discrete state with angular momentum J, and a radiationless transition occurs to the ionization continuum of the atomic state with angular momentum J' and an ionization energy less than that for state J. This mechanism is important for heavy atoms at impact energies about 1 eV above the lowest ionization threshold.

When we consider *heavy atoms*, we find that several other mechanisms become important:

6. $e + X \rightarrow X^{n+} + (n+1)e$, $n > 2$. *Multiple outer-shell ionization.*
7. $e + X \rightarrow X^{+*} + e + e \rightarrow X^{+k} + (k+1)e$. *Inner-shell ionization followed by the emission of Auger electrons* in the reaction shown here (or alternatively by the emission of X-rays.)[46–48]
8. $e + X \rightarrow X^* + e \rightarrow X^{+k} + (k+1)e$. *Inner-shell excitation followed by autoionization.* As with 7 this mechanism can be viewed as a two-step process. It involves excitation of an inner-shell electron, followed by the decay of the highly excited state thus produced. This reaction is often called "excitation-autoionization".

Processes 7 and 8 are also very important in the case of heavy ionic targets. The cross sections for the single and multiple ionization of high-Z atoms and ions by inner-shell processes are generally very large, and consequently these processes are important in applications. The cross section for the excitation-autoionization reaction can be many times that for direct ionization.[49]

Counterparts of processes 1–8 also occur with molecular targets, which, however, can be ionized by additional mechanisms not available for atoms. Using diatomic molecules XY for simplicity in illustration, we can have

9. $e + XY \rightarrow X^+ + Y + e + e$. *Dissociative ionization.* This process was discussed in Section 5-5, and the reader is again directed to Fig. 5-5-1.
10. $e + XY \rightarrow XY^{+*} + e + e \rightarrow X^+ + Y + e + e$. *Fragmentation in a two-step process.*
11. $e + XY \rightarrow X^+ + Y^- + e$. *Ion-pair formation.* This process will be discussed later in connection with mechanisms for the formation of negative ions.

6-6. AUTOIONIZATION

Four mechanisms of autoionization have been listed in Section 6-5: processes 4, 5, 7, and 8. The first of these, autoionization with double outer-shell excitation, was discussed first in Section 4-8-B-2, in connection with electron scattering resonances. Atoms with several electrons have states that correspond to double excitation, that is, to two electrons being in higher than their normal orbits. If the energy of double excitation exceeds the ionization energy of the ground-state atom, the atom may be de-excited in a radiationless transition, wherein one of the excited electrons drops to a lower orbital and the resulting de-excitation energy is expended in ejecting the other excited electron from the atom, without the emission of radiation. This process, known as *autoionization, preionization*, or the *Auger effect*, occurs very quickly after the atom has been doubly excited. An example of this type of autoionization is presented in Fig. 6-6-1, where it is apparent that ionization of potassium can proceed by at least two independent processes. The direct ionization mechanism corresponds to a transition from the electronic ground-state configuration $3p^6 4s$ to the ionic ground state $3p^6$, which can also be reached by the process of autoionization through the $3p^5 4s^2$ excited state of the atom. The threshold for this second process is about 19 eV, at which energy the upper curve in Fig. 6-6-2 shows a sharp break. The ejection of the Auger electron is connected with the existence of various degrees of freedom among which energy may be distributed, and with the eventual concentration of this energy in a mode that leads irreversibly to a continuum state (auto-

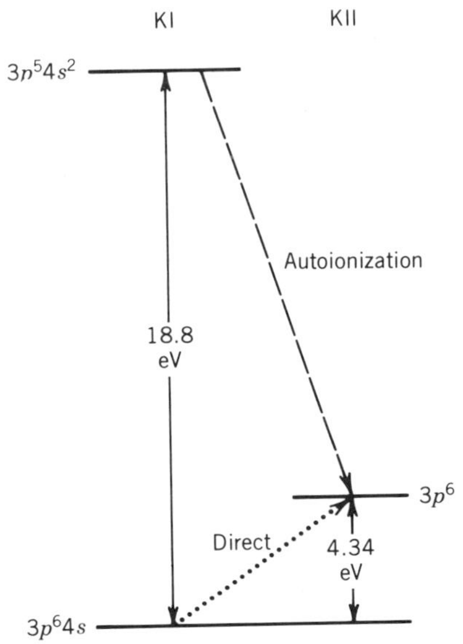

Figure 6-6-1. Illustration of transitions involved in direct, nonresonant ionization and autoionization of potassium.

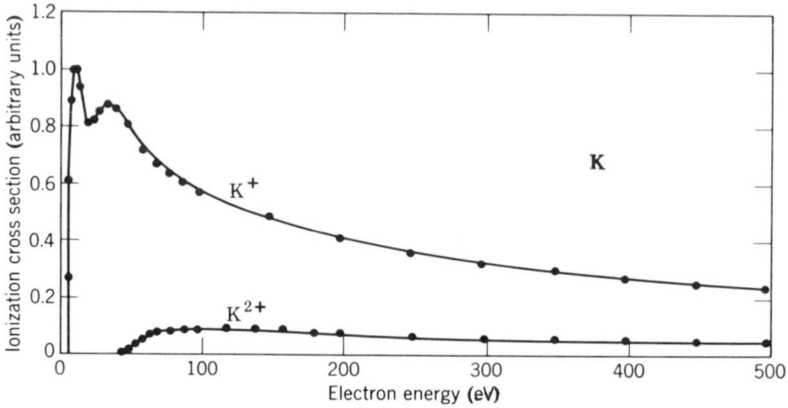

Figure 6-6-2. Relative cross sections for partial ionization of potassium by electron impact, obtained at *low electron energy resolution.* [From G. O. Brink, *Phys. Rev.* **127**, 1204 (1962).]

ionization, or predissociation, perhaps, in the case of a molecule). The lifetime against autoionization is generally much shorter than its radiative lifetime.

The example of potassium discussed above shows that the existence of autoionizing states can be inferred from structure on cross section curves; when the projectile energy is equal to the energy of an autoionizing state, a new mechanism for ionization becomes possible, and the cross section is enhanced in the vicinity of that energy. Evidence for autoionization is also provided when electrons are emitted with energy characteristic of the autoionizing level. Such studies of autoionizing states can be carried out by photon or heavy particle impact, as well as by electron bombardment.[46,50] Excitation by photons or high-energy electrons produces mainly optically allowed transitions, whereas heavy particles and low-energy electrons allow the observation of optically forbidden transitions (cf. Section 5-7-G).

Both ejected and scattered electrons have been studied by Comer's group at Manchester.[51] They use a crossed-beam electron spectrometer that permits detection of scattered or ejected electrons at angles in the range 0 to 100° with respect to the incident electron beam (see Section 4-3-B for descriptions of similar instruments). The instrumental energy width is about 45 meV (FWHM) for the ejected-electron spectra and 60 meV for the energy-loss spectra (see Section 5-7-A). The objective in studies of this type is to determine the energies and widths of the excited levels involved, and to measure the behavior of the line shape parameters as a function of the impact energy, angle of electron ejection, and final ion state. Additional papers from the Manchester laboratories are cited in Ref. 52. An ejected electron spectrum for Kr recorded by Baxter et al.[51] is displayed in Fig. 6-6-3. Obviously, much can be learned from a spectrum that is so rich in detail. Another example relevant to this discussion was presented in Fig. 6-4-6.

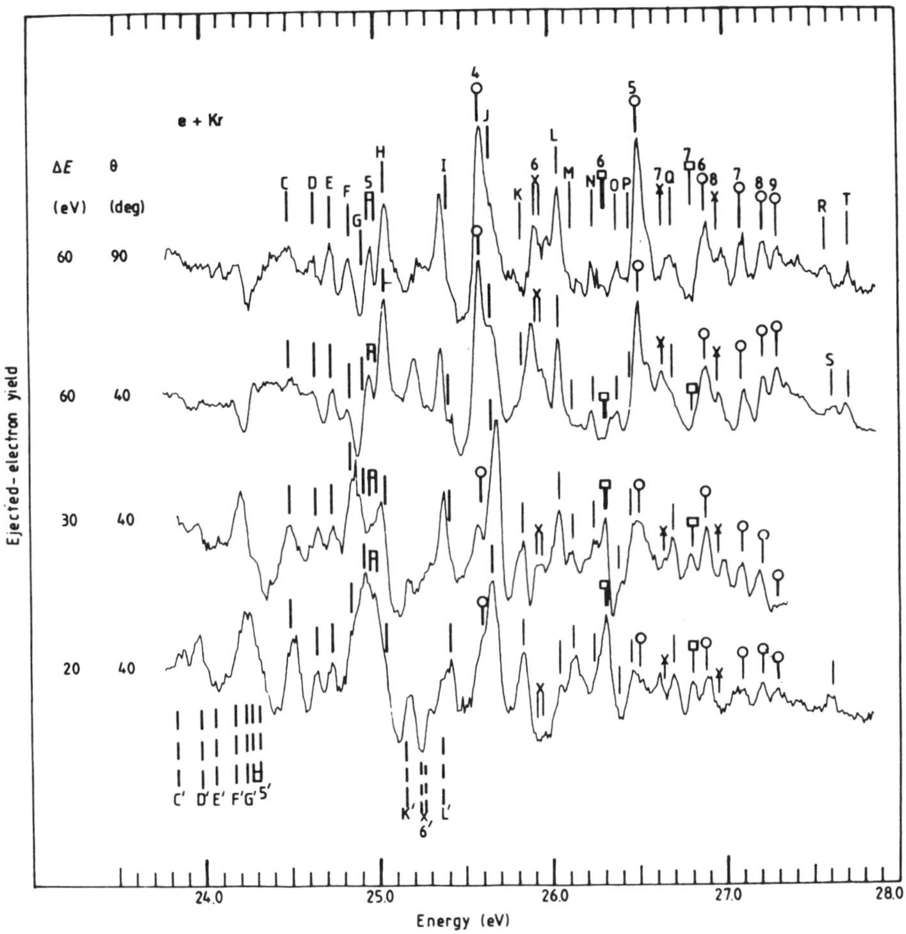

Figure 6-6-3. Ejected-electron spectra of krypton recorded at high electron energy resolution at various values of ΔE and θ, where ΔE is the energy difference between the projectile electron beam and the detected electron beam, and θ is the angle between the two beams.[51] The energy scale gives the energy of the autoionizing state where indicated by a full line. Features marked with a dashed line result from autoionization leaving the krypton ion in the $J = 1/2$ level; their energy is 0.666 eV greater than the indicated energy.

The final topic to be considered here is that of displaced electron energies. In 1966, Barker and Berry[53] reported experiments in which they excited autoionizing states of He in collisions with He^+ projectiles having energies of 1 to 4 keV, and measured the energy distribution of the ejected electrons. Although they could not resolve the electron energy peaks associated with individual autoionization peaks, their observations on the prominent composite peak at about 36 eV revealed an asymmetric energy profile and a shift of the ejected electron energies downward as the incident ion energy was decreased. Barker and Berry proposed a *postcollision interaction* (*PCI*) to explain these effects. The central idea is that the autoionizing atom decays before the Coulomb field of the receding He^+ becomes negligible. Hence the asymptotic energy of the ejected electron is reduced by the negative Coulomb energy; the electron energies are displaced. Hicks et al.[54] observed similar shifts in electron bombardment of He atoms near the thresholds for exciting autoionizing states, although the sign of the displacements was opposite to that in the ion impact experiments because of the opposite sign of the Coulomb energy. Read[55] has suggested an interpretation of these results in terms of a *shake-down model*, in which the decisive feature is the sudden perturbation produced when an electron is removed from the atom.

6-7. INNER-SHELL EXCITATION AND IONIZATION*

If an incident electron has sufficient energy, it can cause an inner-shell, or core, electron to be ejected from an atom, molecule, or ion. This process of inner-shell ionization[46-48] can be studied by observing the characteristic X-rays and/or Auger electrons that are subsequently emitted, or by analyzing the scattered electron energy-loss spectra. In the case of light targets, Auger emission is the usual decay mode, and the final state of the target is at least doubly charged (process 7 in Section 6-5). As we have seen, an important alternative to this mechanism is process 8, in which the initial step is inner-shell excitation, not ionization. For molecular targets, the dominant decay mode is dissociative ionization of the intermediate state.

Spectrometers have been developed at the University of Manchester that permit energy-loss measurements to be made at energy resolution of about 50 meV over an impact energy range of 50 to 1500 eV. At projectile energies above 200 eV, this energy resolution is superior to that obtainable in photoabsorption measurements.[47] An energy resolution of 50 meV is well matched to the study of inner-shell states having natural widths, Γ, of 50 to 200 meV. The lifetime, $\tau = \hbar/\Gamma$, corresponding to $\Gamma = 50$ meV is 1.32×10^{-14} sec. States with vacancies deep within the structure usually have natural widths exceeding 200 meV and can be studied more efficiently at lower energy resolution.

King et al.[56,57] have studied inner-shell excitation of Ar, Kr, and Xe at high resolution. Figure 6-7-1 shows an energy-loss spectrum for Ar at an impact energy of 1.5 keV and energy resolution of 65 meV. Argon has a ground-state configuration $1s^2 2s^2 2p^6 3s^2 3p^6$, and optically allowed transitions of a $2p$ electron to one of the orbitals $ns(n \geqslant 4)$ and $nd(n \geqslant 3)$ are observed. A weak parity-forbidden transition to the $4p$ orbital is also in evidence. Because the $2p$ hole can have $j = \frac{3}{2}$ or $j = \frac{1}{2}$, two Rydberg series are observed that converge to the $^2P_{3/2}$ and $^2P_{1/2}$ states of the ion. Analysis of the energies of the Rydberg series[58] provides accurate values of the $^2P_{3/2,1/2}$ ionization energies. The measurements

*This section overlaps and extends Section 5-7-H.

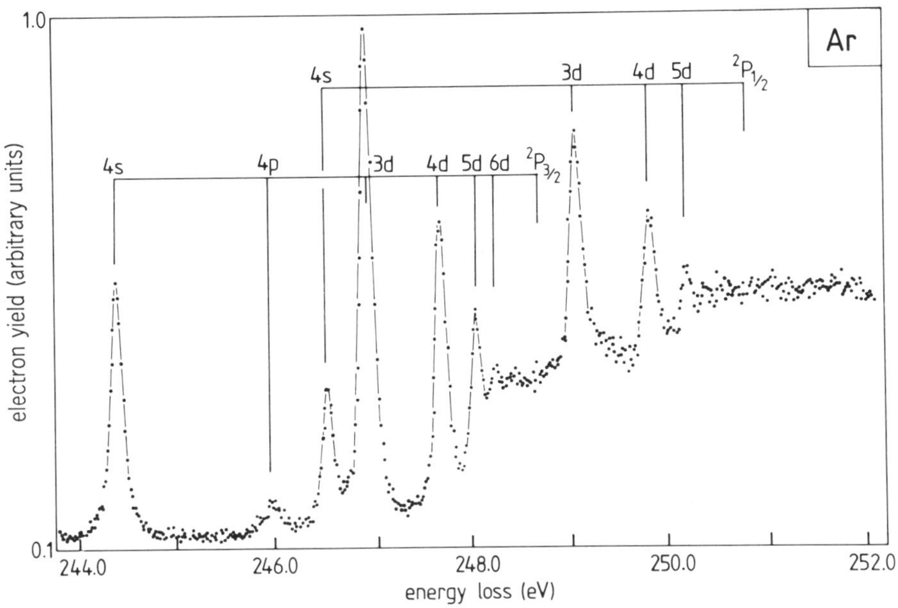

Figure 6-7-1. A high-energy-resolution spectrum of electron energy loss in argon, obtained at an incident energy of 1.5 keV and resolution of 65 meV.[57]

of King et al.[56] also permit thorough tests of the equivalent core model (Section 5-7-H). Finally, they provide accurate measurements of the widths and lifetimes of X-ray levels of Ar, Kr, and Xe.

In the case of molecules, additional information can be obtained in the form of vibrational spacings and intensities, from which potential curves can be deduced.[47] King et al.[59] obtained the high-resolution energy-loss spectrum for N_2 shown in Fig. 6-7-2 and revealed for the first time the vibrational structure of the $(1s)^{-1}(2p\pi)^1\Pi$ state of the nitrogen molecule. By analyzing the spectrum, they obtained the vibrational parameters ω and ωx, information about the mean internuclear separation R_0 in the inner-shell excited molecule, and the lifetime of the state.

Hippler et al.[60] have studied threshold behavior in inner-shell ionization of Ar and Xe. Crowe[61] has reviewed measurements of angular correlations between outgoing scattered electrons and Auger electrons from the decay of inner-shell ions produced by a directed beam of electrons. When the inner-shell vacancy is created, the initial ion state will, in general, be aligned with respect to the projectile beam, that is, there will be an unequal population of its magnetic sublevels (cf. Section 5-10-E). This alignment results in an anisotropic distribution of the Auger electrons and the X-rays, and the polarization of the emitted X-rays. The alignment arises from the ionization cross section having different values for different projections of the total angular momentum of the residual ion on the projectile beam direction, although the cross section is independent of

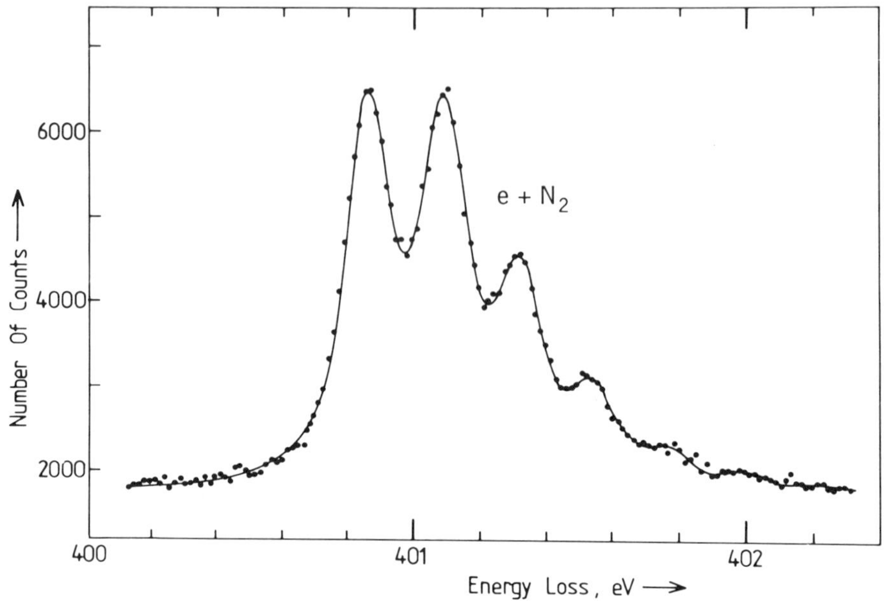

Figure 6-7-2. An energy-loss spectrum for N_2, showing the vibrational structure of the electronic state of the molecule formed by the promotion of an atomic 1*s* electron to the first unoccupied valence orbital ($2p\pi$).[59]

the sign of the projection. Study of the alignment of inner-shell vacancies in ions provides information on the process of ionization and on the wave functions of the inner-shell electrons, because calculations of the alignment are sensitive to the model and the approximations used.[62] The theory of inner-shell ionization has been reviewed by Powell[63] and by Burhop.[64]

6-8. THRESHOLD STUDIES*

Here the matter of concern is the variation of the "threshold ionization" cross section, σ_{th}, with projectile electron energy, E, in the close vicinity of the ionization threshold, E_{th}. The simplest process to be considered is single ionization of a neutral atom:

$$e + A \rightarrow A^+ + e + e. \tag{6-8-1}$$

Actually, the most meaningful energy here is

$$\Delta E \equiv E - E_{th}, \tag{6-8-2}$$

*For a recent review, see G. C. King, in P. G. Burke and J. B. West (Eds.), *Electron-Molecule Scattering and Photoionization*, Plenum, New York (1988).

which is the beam electron energy in excess of the minimum energy at which ionization can occur. We are interested in $\sigma_{th}(\Delta E)$ as $\Delta E \to 0$. Measurements of this quantity are extremely difficult because they require working as close to threshold as possible, but σ_{th} is rising from *zero* at E_{th}! High collection and detection efficiency is required to measure the weak signals associated with the low-energy electrons in the final state. The difficulty is compounded by the fact that one must use electron beams of narrow energy width ($\leqslant$40 meV, say) for faithful mapping of the cross section variation with energy.

Experimental threshold data are useful because they permit the threshold to be located by extrapolation, and because electrons produced in near-threshold ionization play a role in certain applications. These data are more important as tests of threshold ionization theories, although, thus far, few experimental studies have the requisite energy resolution and accuracy to provide definitive tests.

As pointed out by Read in his authoritative review,[65] the ionization threshold problem has received a great deal of attention, in part, because it is determined to a considerable degree by the interaction of three or more particles at distances so large that conventional quantum mechanical techniques are not feasible. Also, a high degree of correlation exists, and motion occurs in the vicinity of an unstable potential ridge (the Wannier ridge). The fact that threshold ionization is the simplest problem that contains all of these complications offers challenge to the theorist and hope that threshold studies will provide continuing insight into related, even more difficult, problems.

The first theoretical paper that is relevant to our discussion is Wigner's landmark contribution[66] in 1948, which dealt with processes in which the final state of the system consists of *two* particles escaping from one another, as in the photo-single-detachment reaction

$$h\nu + A^- \to A + e. \tag{6-8-3}$$

A few years later, Wannier[67] used arguments similar to Wigner's to show that the study of electron impact threshold ionization should be simpler than attempting to solve the "full" problem of ionization at higher energies. Near threshold, the value of ΔE is important only with regard to the *final* state. The initial-state wave function will change negligibly with small changes in ΔE, but the wave function of the final state will depend critically on ΔE because ΔE is the total energy of this state. The threshold behavior depends only on long-range interactions in the final state, so study of this behavior is especially suitable for probing long-range interactions and correlations between the outgoing electrons in process (6-8-1). The complicated physics in the "reaction zone" from which the electrons emerge in the final state is irrelevant here, although it must be investigated in the full problem. Hence the long-range interactions can be studied near threshold without their being obscured by the short-range effects. It also follows that the mechanism by which the final state is produced is unimportant, and the energy dependence of $\sigma_{th}(\Delta E)$ is insensitive to this mechanism. Accordingly, the near-threshold cross section for the photo-double-detachment process

$$h\nu + A^{-} \rightarrow A^{+} + e + e \tag{6-8-4}$$

has the same energy dependence as the electron-impact ionization process (6-8-1). Thus a valuable alternative exists for the study of threshold ionization processes, and it is possible to use optical techniques and take advantage of their high energy resolution.

We now move on to Wannier's classic paper on the theory of single ionization of neutral atoms and ions near threshold, published in 1953[67] and discussed recently by Read[65] and Rau.[68,69] The Wannier theory involves classical mechanical calculations coupled with statistical mechanical arguments. It predicts the threshold law

$$\sigma_{\text{th}} \sim (\Delta E)^{1.127} \tag{6-8-5}$$

for the processes (6-8-1) and (6-8-4); for electrons impinging on ions, the exponent has a value between 1.127 and unity, depending on the charge of the target ion.

The Wannier threshold law is most easily derived by considering a collinear system that consists of a stationary, point-like, positive ion of charge $+Ne$ and two outgoing electrons that initially move in opposite directions along the same straight line. The final state of the system corresponds to ionization if both electrons are free, or to excitation in the event that one of the electrons is bound to the ion in the final state and oscillates about it. Let us denote the position vectors of the electrons, measured from the ion, by $\mathbf{r}_1$ and $\mathbf{r}_2$. It can be shown that when ΔE is small, ionization occurs only when the ion and electrons start in, or close to, the symmetric state with $\mathbf{r}_1 = -\mathbf{r}_2 = \mathbf{r}$ and $\dot{\mathbf{r}}_1 = -\dot{\mathbf{r}}_2$. For this state, r space can be meaningfully divided into three regions. Close in to the ion, the problem is quantum mechanical; this "inner," or "reaction" zone extends out to a radius $r_{\text{min}} \sim$ (several Å). From this boundary out to a "critical radius" $r_c = 2e^2(N - 1/4)/E$, the potential energy predominates, and in this "Coulomb zone," the electron behavior is classical. Finally, in the "outer zone," $r > r_c$, the kinetic energy is dominant, and the electrons are considered to be free, classical particles. If spin-dependent effects are ignored, the behavior of the system in the classical zones can be investigated by integrating the classical equations of motion.

A set of universal trajectories can be derived, and it clearly shows the division of the collision events into ionization and excitation. When ΔE is small and positive, the trajectories start at the boundary $r = r_{\text{min}}$ with the same value of r_1 and with a uniform spacing in v_1. Hence they are uniformly distributed in phase space. Analysis of the subsequent change in trajectory spacing as the electrons move past the boundary $r = r_c$ provides the density of trajectories in the outer, or "asymptotic" region:

$$\rho \sim \frac{(\Delta E)^n}{\Delta v_{1i}}, \tag{6-8-6}$$

where Δv_{1i} is the initial velocity spacing at $r = r_{\min}$, and $n = 1.127$ for $N = 1$, $n = 1.056$ for $N = 2$, and $n = 1.036$ for $N = 3$, and so on. For a given initial phase space density, the cross section for ionization is proportional to $(\Delta E)^n$, so we have obtained the Wannier threshold law. Recall that N is the number of positive charges on the residual ion.

The general noncollinear motion of the two final-state electrons has also been analyzed, and the ergodic assumption that the initial phase space distribution of trajectories is uniform leads again to the Wannier threshold law.[65]

Cvejanović and Read[70] have made the most accurate measurement of the Wannier exponent n for single ionization by electron impact [cf. (6-8-1)]. They used their "penetrating field" technique, described in Section 5-7-I, to measure a partial yield as a function of the energy of electrons incident on helium atoms. The partial yield here is the yield of electrons with energies between 0 and E_m, where E_m is a constant, approximately 20 meV. Under their conditions, the measured yield varies as $(\Delta E)^{n-1}$. Analysis of their data, a sample of which appears in Fig. 6-8-1, gives $n = 1.131 \pm 0.019$, which is consistent with the Wannier value $n = 1.127$. The part of the spectrum in Fig. 6-8-1 that is immediately below the ionization threshold ($E_{\text{th}} = 24.58$ eV) is almost the mirror image of the part immediately above E_{th}. Thus an ionization cusp is in evidence (see Section 4-8-C).

A closer test of the Wannier threshold law equation (6-8-5) has been made by studying the photo-double-detachment process (6-8-4) for H^- ions. Data thus obtained by Donahue et al.[71] are displayed in Fig. 6-8-2. A complication in their experiment arises from spurious ionization following single-detachment events in which hydrogen atoms are left in highly excited states:

$$h\nu + H^- \rightarrow H^* + e. \tag{6-8-7}$$

Such atoms can be ionized when they pass through a magnetic field in Donahue's detector system. This field ionization (Note 6-2) occurs for excited states with principal quantum number equal to or exceeding 15, corresponding to binding energies less than B_m, where $B_m = 60$ meV. Because of this additional energy-dependent source of ionization, the electron yield should vary not as ΔE, but approximately as $(\Delta E + B_m)^n$.[65] Donahue et al.[71] used the form

$$\sigma \sim (\Delta E + E_0)^n \tag{6-8-8}$$

to fit their data, and obtained the best fit with $E_0 = (47 \pm 5)$ meV and $n = 1.15 \pm 0.04$. This value of n is consistent with the Wannier value $n = 1.127$.

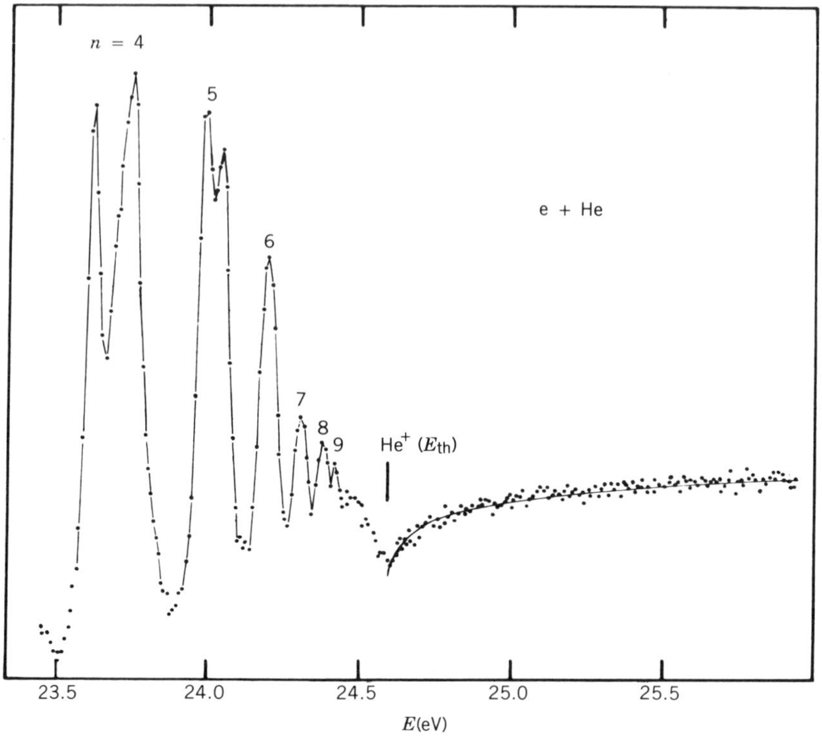

Figure 6-8-1. The yield of very low energy electrons produced in collisions of electrons with helium atoms, as a function of the incident energy E.[70] The curve drawn through the data points above the ionization threshold (E_{th}) has the energy dependence $(\Delta E)^{0.127}$, where $\Delta E = E - E_{th}$.

Other electron impact and photodetachment threshold measurements are also discussed by Read.[65]

Other topics important in the study of threshold ionization and discussed by Read[65] are:

1. Threshold behavior under the assumption that the residual ion acts as a screened charge of magnitude $eN(r)$, where r is measured from the nucleus.
2. Multiple ionization.[72]
3. Threshold behavior when the total angular momentum of the system is not zero, as in the Wannier treatment and certain others.
4. Differential threshold ionization; the sharing of the energy between the two electrons; the relative directions and separate angular momenta of these electrons.
5. Near-threshold excitation processes.

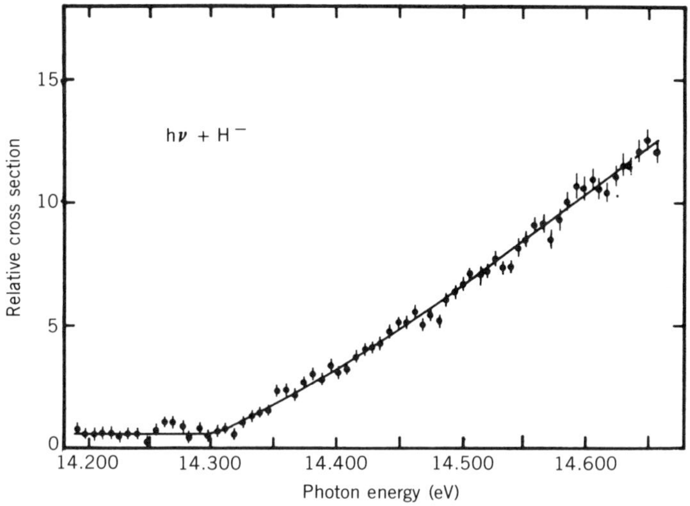

Figure 6-8-2. The relative cross section for the photodetachment reaction $h\nu + H^- \rightarrow H^+ + e + e$ [cf. (6-8-4)].[71] The indicated errors show only the statistical contributions. The curve shows the power law $\sigma \sim (\Delta E + 47)^{1.15}$, where the energy is expressed in meV [see (6-8-8)].

6. Wannier-ridge resonances[73] associated with the high degree of electron-electron correlation in near-threshold ionization and near-threshold excitation. Here the two excited electrons both have large values of the principal quantum number, and they reside on or near the "Wannier ridge" on the potential function surface (i.e., with $\mathbf{r}_1 \approx -\mathbf{r}_2$).

We should mention that numerous other semiclassical and quantal theories have been proposed to describe threshold ionization. (The Coulomb-dipole theory of Temkin[68] will be discussed in Section 6-11.) Some of them give the Wannier law; others give slightly different results. The conclusion that correlation is very important is inescapable. The condition $r_1 = r_2$ implies a radial electron correlation. The correlation is unstable because if one of the electrons lags behind, it senses a stronger attractive ionic field than that experienced by the faster electron, and the slower electron gets still farther behind. If there were no radial correlation, the threshold law would be linear. Stable angular correlation results from the electron-electron repulsion, which makes them move in nearly opposite directions ($\dot{\mathbf{r}}_1 = -\dot{\mathbf{r}}_2$). Because of this effect, the two electrons can have large orbital angular momenta, although the total orbital angular momentum is small. The reviews by Rau,[68,69] Heideman,[68] and Klar[68] contain additional information on correlation.

Finally, we should take note of the triple differential cross section measurements (see Section 6-9) made by Schubert et al.,[74] and Fournier-Lagarde and his colleagues.[75] Their studies of helium provide a close look at the ionization threshold problem and reveal differences with the simple structure associated with the Wannier law. Fournier-Lagarde et al.[75] give experimental evidence that even close to threshold the two outgoing electrons have nonzero angular momentum.

6-9. DIFFERENTIAL CROSS SECTION MEASUREMENTS; ENERGY AND ANGULAR DISTRIBUTIONS OF ELECTRONS[12,76–78]

Consider single, outer-shell ionization of an atom X:

$$e(\boldsymbol{\kappa}_0) + X \rightarrow X^+ + e(\boldsymbol{\kappa}_a) + e(\boldsymbol{\kappa}_b), \tag{6-9-1}$$

where $\boldsymbol{\kappa}_0$, $\boldsymbol{\kappa}_a$, and $\boldsymbol{\kappa}_b$ are the momenta (in units of $\hbar$) of the free electrons, and where we assume that the residual ion X^+ is in its ground state. The simplest way in which to study reaction (6-9-1) is to collect all the residual ions and measure their charge as a function of the energy of the projectile electrons, E_0, where

$$E_0 = E_{\text{th}} + E_a + E_b, \tag{6-9-2}$$

and E_{th} is the threshold energy (i.e., the ionization energy). We thus obtain the *integral ionization cross section* discussed in Part $\mathscr{A}$; it represents an integration over all angles and energies of the scattered and ejected electrons.

More information is provided by the *single differential cross section* $d\sigma/dE$, which is a measure of the probability that, at given E_0 and E_{th}, an electron is detected with energy E between 0 and $E_0 - E_{\text{th}}$, irrespective of the angle at which it was scattered or emitted. Measuring this cross section is equivalent to performing two integrations over the angles of the two emerging electrons in (6-9-1). Additional information comes from the *double differential cross section* $d^2\sigma/dE\,d\Omega$, which corresponds to detection of one of the two emerging electrons with given E and direction of motion Ω regardless of the direction of the second electron. Here only one integration is involved. The *triple differential cross section* $d^3\sigma/dE\,d\Omega_a\,d\Omega_b$ involves no integration, and, consequently, contains the most information about (6-9-1). Here E_a, E_b, Ω_a, and Ω_b must be measured by coincidence techniques in *(e, 2e) experiments*. [We define an (e, 2e) experiment as an electron impact ionization measurement in which the kinematics of all the free electrons are completely determined.[77]] Other sophisticated and difficult experiments are performed by using polarized electron beams and/or state-selected targets to measure singlet and triplet cross sections (Section 6-11). The role of spin in ionizing collisions will not be considered in this section.

Often the integral cross sections for a given process can be calculated with satisfactory accuracy, but the differential cross sections frequently show poor agreement with experiment and are strongly dependent on the theoretical model used for their calculation. Jung et al.[79] suggest that interference effects between different partial waves may be important here. These effects show up immediately in the calculation of differential cross sections, but they cancel out because of the orthogonality of the partial waves when the integral cross section is obtained by summation. Hence the differential cross sections provide much more stringent tests of the models than do integral cross sections.

A. Single Differential Cross Sections

The form of the single differential cross section (SDCS) $d\sigma/dE$ is shown in Fig. 6-9-1.[76] This cross section is a function of the impact energy E_0 and the energy E of the detected electrons, and it represents the energy distribution (integrated over all angles) of the scattered and ejected electrons. The energy distribution is symmetrical about the value $(E_a + E_b)/2$, because the scattered (S) and ejected (E) electrons share the energy $(E_a + E_b)$. Of course, the scattered and ejected electrons in a given event are indistinguishable, but if E_0 is large, the two groups of electrons are widely enough separated to lend validity to our description of them as distinguishable. Moreover, the angular distributions of these two groups are very different, and at high energies the amplitudes for electron and momentum exchange are negligible. The slow electrons have energies ranging from 0 to about 50 eV; the fast electrons have a maximum energy $(E_0 - E_{th})$.

The direct determination of $d\sigma/dE$ is difficult, and this quantity is frequently

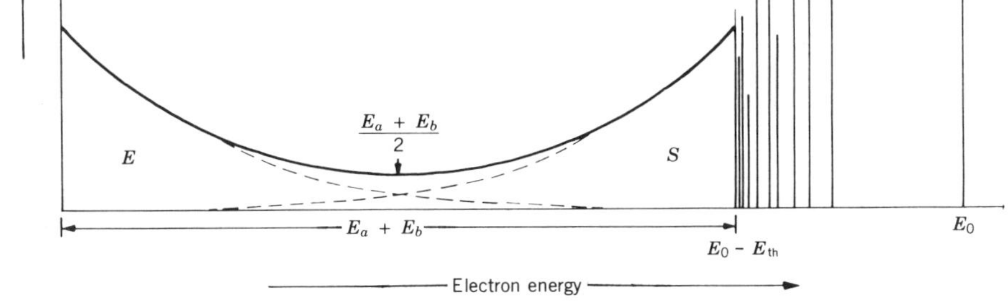

Figure 6-9-1. Schematic representation of the energy-loss spectrum $d\sigma/dE$ integrated over all directions of motion of the postcollision electrons.[76] The impact energy is E_0, the threshold energy is E_{th}, S represents "scattered electrons," and E the "ejected electrons" (see the text for elaboration). The sum of the energies of the postcollision electrons is $(E_a + E_b)$.

obtained by integration of $d^2\sigma/dE\,d\Omega$.[80] The other SDCS, $d\sigma/d\Omega$, appears to have no physical significance and has not been measured.

The single and double differential cross sections contain information on the rate at which projectile energy is deposited in matter and on the energy and angular distributions of secondary electrons that can excite or ionize other atoms and molecules. This fact explains their importance with respect to radiation chemistry, radiation damage, electrical discharges, electron-beam-pumped gas lasers, explosions, and photoelectrons in planetary atmospheres.

B. Double Differential Cross Sections

Double differential cross sections (DDCS) for ionization of helium have been measured by Müller-Fiedler et al.[81] at Kaiserslautern with the electron spectrometer sketched in Fig. 6-9-2; this instrument is similar in many respects to the one shown in Fig. 4-3-8. The energy-selected projectile beam crosses a beam of helium atoms at 90°, and the scattered and ejected electrons are analyzed according to their energies and directions of motion, and detected by devices A and B, respectively. Projectile electrons, which are scattered mainly at small angles, are detected in analyzer A; ejected electrons are detected in B. Measurements were made at impact energies of 100, 200, 300, 400, 500, and 600 eV, and at ejected electron energies of 2, 4, 10, 20, and 40 eV. Detectors A and B are operated independently in the non-coincidence mode, so that $d^2\sigma/dE_a\,d\Omega_a$ and $d^2\sigma/dE_b\,d\Omega_b$ are obtained separately, with A and B, respectively. In order to be able to obtain accurate cross sections from the counting rates of A and B, it is necessary to determine whether an angular correction is required. Scanning the profile of the projectile beam and the field of view of A and B permits calculation of the scattering volume as a function of the angles θ_a and θ_b. No angular correction is necessary because the scattering volume varies with angle by less than 1% over the range 10 to 90°. The metal pin in Fig. 6-9-2 is used in the scanning procedure. The reader should consult Ref. 81 for details of this procedure and for the method used to deduce cross sections from the counting rate data. The cross sections are put on an absolute basis by extrapolating the generalized oscillator strength (Section 5-7-F) for the projectile electrons to the optical limit (zero momentum transfer). The data from B show that the ejected electrons of low energy (a few eV) are almost isotropically distributed, but those with energies of about 15 to 50 eV show a slight maximum around 60° that is explainable by examining triple differential data. The DDCS measurements with A show that the fast electrons are scattered into a narrow cone in the forward direction, the angle of the cone being smaller for larger impact energies. For helium, with $E_0 \geqslant 250$ eV, about 85% of the fast electrons are scattered within about 10% of the total solid angle centered around the forward direction.

Shyn and his colleagues at the University of Michigan have measured double differential cross sections $d^2\sigma/dE_b\,d\Omega_b$ for secondary electron production in several gases at impact energies ranging up to 400 eV. Shyn's data on N_2 for 200-

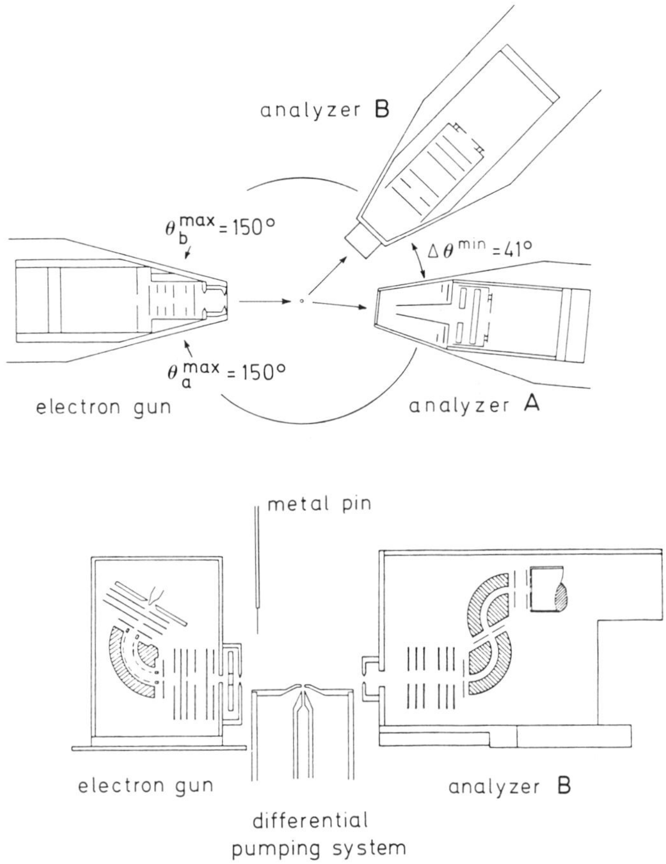

Figure 6-9-2. A crossed electron—slow neutral beam electron spectrometer used in Ehrhardt's laboratory for measurement of differential ionization cross sections.[79,81] The neutral beam passes upward in the lower drawing; in this drawing the analyzer A is not shown.

eV primary electrons are displayed in Fig. 6-9-3 in three-dimensional perspective.[82]

More recently, Bonham's group at Indiana University has also measured DDCS for electrons ejected from N_2 at impact energies from 200 to 2000 eV.[83] The angle of detection was varied from 30 to 150°. Integral ionization cross sections obtained from Platzman plots of their data are in excellent agreement with the values reported by Rapp and Englander-Golden.[26]

Inokuti[84] and Kim[85,86] have analyzed SDCS and DDCS data by scaling them to the Coulomb (Rutherford) cross section or to the high-energy limit of the first Born cross section. In their analyses, they use *Fano plots* [cross section versus $\ln(E_0/R)$, where R is the Rydberg constant] and *Platzman plots* (SDCS versus inverse of E_t, the energy transferred by the primary electron). The high-

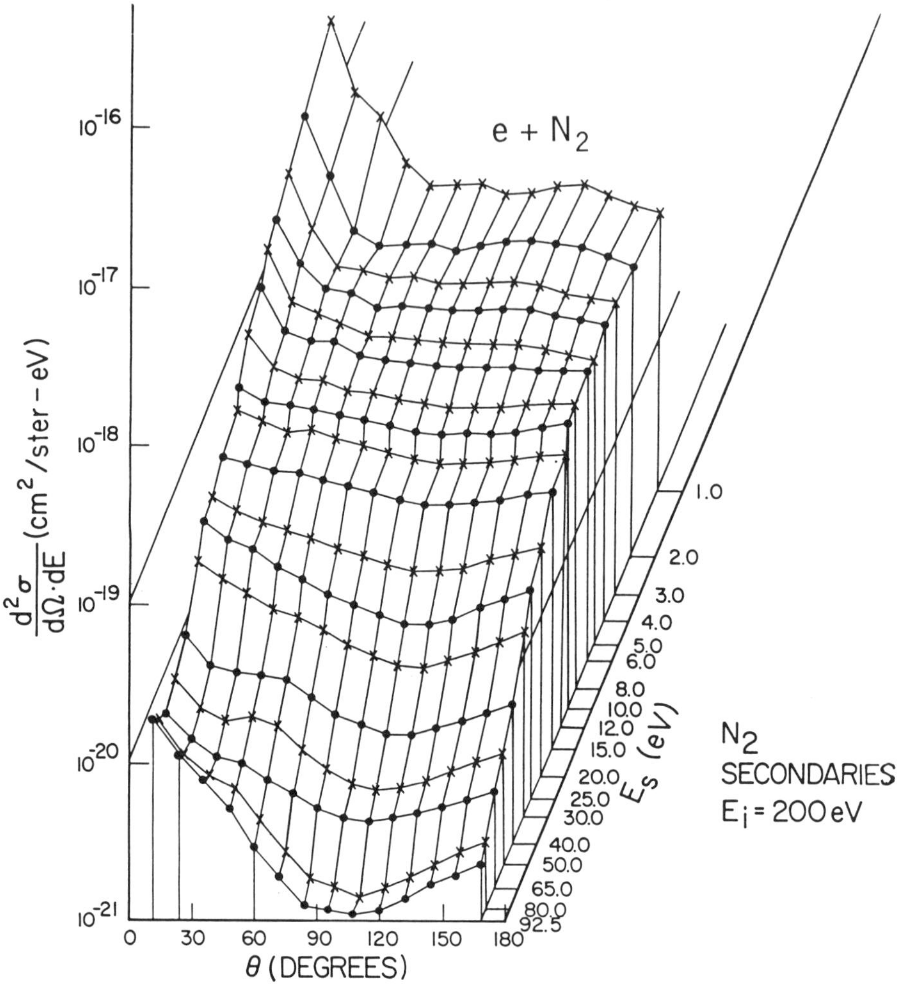

Figure 6-9-3. A three-dimensional representation of the DDCS for secondary electrons ejected from N_2 by 200-eV projectile electrons.[82]

energy portions of these plots provide dipole oscillator strengths and the effective number of outer-shell electrons involved in the ionization. The plots also assist in the critical evaluation of experimental data.[82–86]

C. Triple Differential Cross Sections: (e, 2e) Coincidence Measurements

Triple differential cross sections (TDCS) provide the most complete description of ionization and the most sensitive test of the theory of this process. The TDCS is a measure of the probability that in an (e, 2e) event, a projectile electron with energy E_0 and momentum $\hbar\boldsymbol{\kappa}_0$ will produce two electrons having energies E_A and E_B, and momenta $\hbar\boldsymbol{\kappa}_A$ and $\hbar\boldsymbol{\kappa}_B$ in the solid angles $d\Omega_A(\theta_A, \varphi_A)$ and $d\Omega_B(\theta_B, \varphi_B)$, respectively. The angles are defined with respect to $\boldsymbol{\kappa}_0$. If the observations are restricted to events such that $\boldsymbol{\kappa}_0$, $\boldsymbol{\kappa}_A$, and $\boldsymbol{\kappa}_B$ lie in the same plane ($\varphi_A = 0$, $\varphi_B = 0$ or π), the geometry is said to be *coplanar*; otherwise, it is referred to as *noncoplanar*. The kinematic variables can also be chosen to distinguish between *symmetric geometries* ($E_A = E_B$, $\theta_A = \theta_B = 45°$) and *asymmetric geometries* ($E_A \gg E_B$, $\theta_A \approx 0°$). In the symmetric geometry, very few ionization events are detected; it can be used to determine momentum distributions of the atomic electron before it is ejected.[77,78] In the asymmetric geometry, if E_0 is large and θ_A small ($\leqslant 10°$), most of the events are observable, θ_B is large, and E_B is a few eV. Some of the coincidence techniques that are used in the (e, 2e) measurements are similar to those described in Sections 5-8 and 5-10. The first ionization measurement in which the scattering kinematics of all the free electrons was fully determined was performed by Ehrhardt et al. and published in 1969.[87]

The momentum transfer

$$\mathbf{q} = \hbar(\boldsymbol{\kappa}_0 - \boldsymbol{\kappa}_A) \tag{6-9-3}$$

from the incident electron to the target is a useful quantity in classifying ionization data.[88] As indicated in Figs. 6-9-4 and 6-9-5, if the incident energy is large ($E_0 > 20E_{th}$) and q is small ($q \ll 1$ a.u.), the angular distribution of the ejected electrons resembles that of photoelectrons ejected from atoms by electromagnetic radiation, and a simple first Born approximation provides a

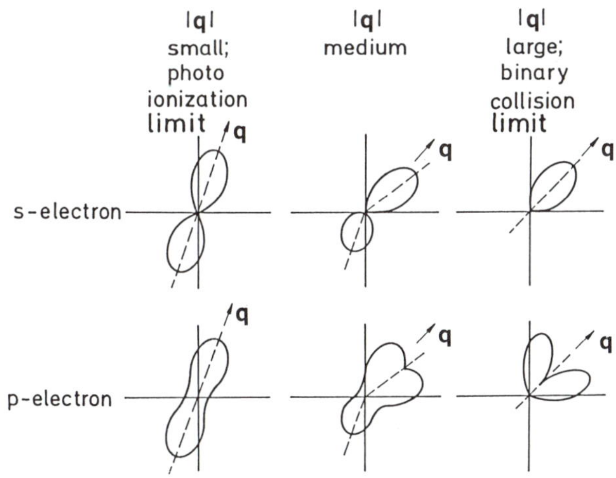

Figure 6-9-4. Angular dependence of the TDCS for ejection of an *s* or a *p* electron for small, medium, and large momentum transfer, *q*, in the range of high (and intermediate) impact energies. (From Ehrhardt et al.[12])

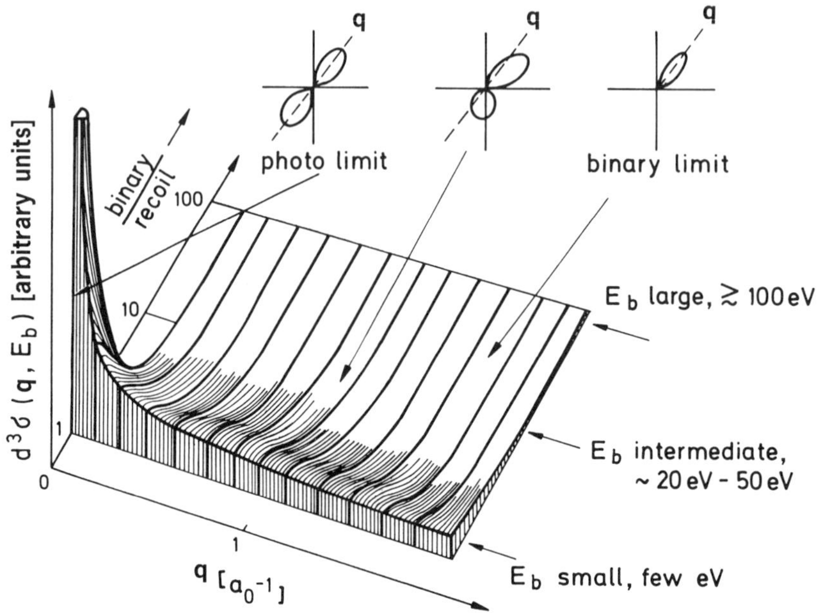

Figure 6-9-5. Schematic plot of the TDCS as a function of the momentum transfer, q, and the ratio of intensities of the binary and recoil peaks. The optical and binary model limits are indicated. (From Ehrhardt et al.[12])

good description. On the other hand, if E_0 is large and q is also large ($q \gg 1$ a.u.), we have essentially a binary electron-electron collision. Here the residual ion serves only as a spectator, and the event resembles an inelastic billiard ball collision that can be described well by classical mechanics. Actually, most of the ionization events, even at high E_0, occur not in these two limiting regimes but in the intermediate range of momentum transfer ($0.3 \leqslant q \leqslant 1$ a.u.). Ionization in this range is difficult to describe theoretically, so reliable experimental data are of special importance here.

Ehrhardt and his colleagues[79,89–91] have made intensive studies of the ionization of helium in coplanar, asymmetric geometry at impact energies ranging from 256 to 600 eV and over a total range of q extending from 0.3 to 1.1 a.u. They used the apparatus of Fig. 6-9-2 with analyzers A and B connected in the coincidence mode. In order to measure $d^3\sigma/dE_B\, d\Omega_A\, d\Omega_B$ for a given E_0, they held E_B and θ_A constant and varied θ_B to get the angular dependence of the slow electrons. (The energy E_A is fixed by the conservation of energy.) Then they made runs at different settings of the angle θ_A and values of the slow electron energy E_B. Figure 6-9-6 shows a sample of their data.[79] There is a strong angular correlation between the two postcollision electrons: the ejected electron leaves the collision site preferentially in a direction that depends on the angle at which the scattered electron is detected. In each panel of Fig. 6-9-6, the angular

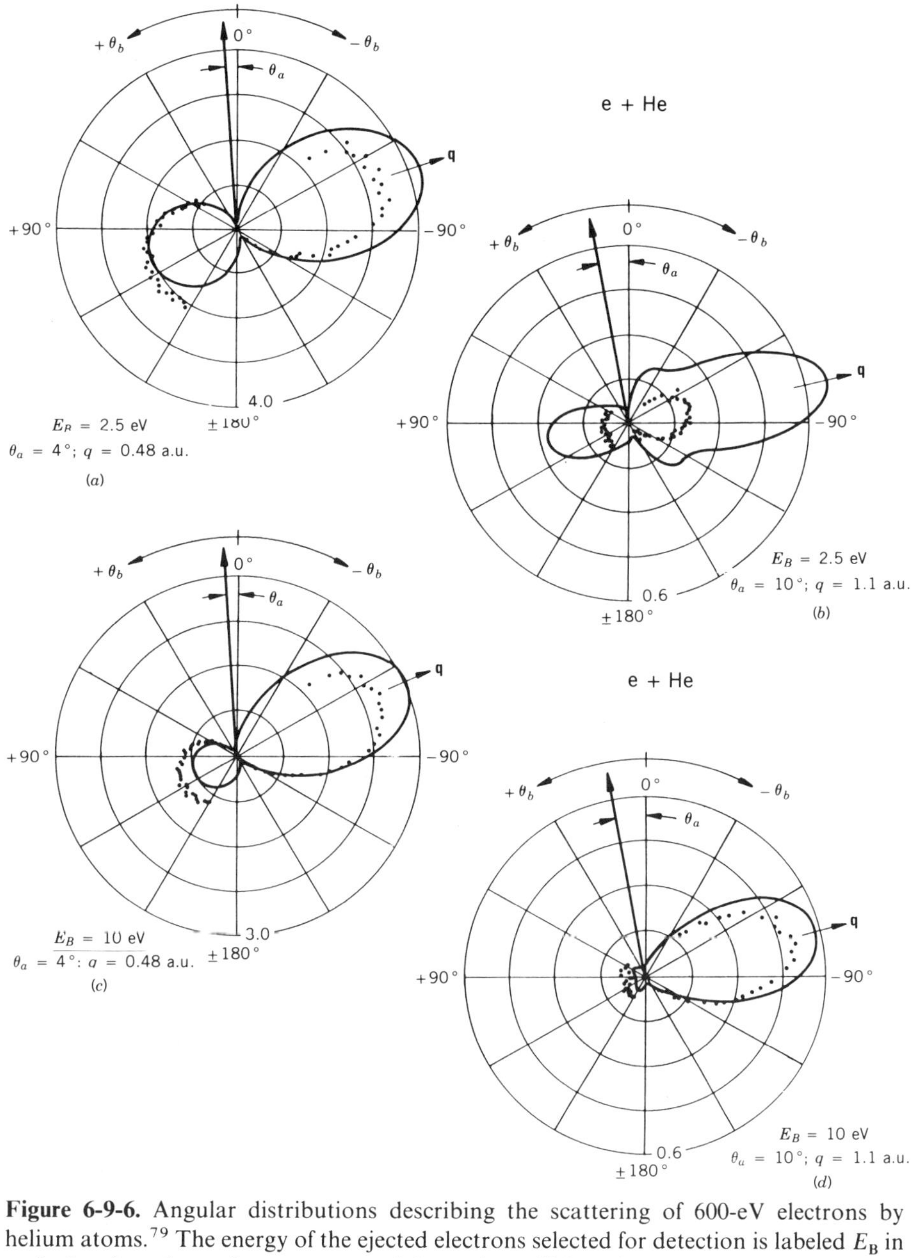

Figure 6-9-6. Angular distributions describing the scattering of 600-eV electrons by helium atoms.[79] The energy of the ejected electrons selected for detection is labeled E_{B} in each drawing; these electrons proceed from the collision site at the angle θ_{B}, and the scattered electrons at the angle θ_{A}. The number inscribed in the outer circle is the absolute value in units of 10^{-22} m^2 sr^{-2} eV^{-1} of the differential cross section. The two vectors show the directions of $\boldsymbol{\kappa}_A$ and $\mathbf{q}$. The dots indicate the experimental data; the full curve gives the results of a first Born approximation calculation performed by Jung et al.[79]

distribution of the ejected electrons has a lobe in the first quadrant, with its maximum close to the direction of **q**. This peak is called the "binary peak" because it is produced mainly by the binary interaction between the projectile and an atomic electron during the collision. There is also a peak in the third quadrant, called the "recoil peak" because it is associated with the interaction of the slow ejected electron with the residual ion and with the momentum transfer to the ion. The recoil peak lies nearly opposite to the binary peak. The first Born approximation (shown as solid line in Fig. 6-9-6) predicts that these two peaks should lie exactly along **q** and −**q**, respectively. In fact, both peaks are observed to be shifted to larger angles, and the binary peak is enhanced in intensity with respect to the recoil peak. Measurements by other groups[92] at much higher values of E_0 give results similar to those observed in Kaiserslautern. Second Born approximation results of Byron et al.[93] are an improvement over the values obtained with the first Born.* Panels (*a*) and (*c*) of Fig. 6-9-6 correspond to the smallest value of q (0.48 a.u.) attainable in the experiments of Jung et al.,[79] and panels (*b*) and (*d*) to the largest value (1.1 a.u.). According to Fig. 6-9-4, experimental data in panel (*a*) represent a situation resembling photoionization, for which the recoil peak would be the mirror image of the binary peak, and the symmetry axis of the peaks would lie along **q**. (Values of q much nearer to zero would be required to reach a close correspondence with photoionization.) Panel (*d*), even though it was obtained with a relatively low value of q (1.1 a.u.), already resembles the angular distribution characterizing the binary collision limit. In this limit, approached at very large values of q, there is a single symmetrical lobe whose axis points along **q**, and whose width is a measure of the momentum distribution of the atomic electron before it is ejected. Panels (*b*) and (*c*) illustrate typical intermediate behavior. The overall absolute accuracy is about $\pm 15\%$.

Measurements on helium are not especially important in the practical sense. However, helium atoms are uniquely suited for the studies described here, where it is essential that the simplest kind of ionization occur, namely direct (nonresonant), single ionization into the ground state of the ion. All other targets (except the H atom) have much larger contributions from autoionization, multiple ionization, and simultaneous ionization and excitation.[12] Experimental difficulties render the hydrogen atom much more difficult to study, but Ehrhardt's group has measured absolute triple differential cross sections for this target.[95] Calculations that have been made for the TDCS of the H atom with various approximations have been reviewed by Curran and Walters,[96] along with their own coupled pseudostate calculation. The first Born approximation gives poor agreement with experiment, but higher-order approximations are in generally good agreement. Coincidence measurements have been made with symmetric geometry by Williams.[97] A number of measurements of this type are described by McCarthy and Weigold[77] and by Teubner.[78]

*However, Ehrhardt et al.[94] have used data on helium, all with $q = 0.5$ a.u. but with E_0 ranging from 150 to 8000 eV, to show convergence for increasing E_0 to a finite limit identical with the first Born prediction. At all impact energies for which deviations from the first Born approximation appear, the most prominent cause is the postcollision interaction (Section 6-6). The role of PCI in (e-2e) experiments has been discussed by G. Stefani, *ICPEAC XV*, Brighton (1987), p. 163.

6-10. ENERGY AND ANGULAR DISTRIBUTIONS OF IONS PRODUCED BY ELECTRON IMPACT

Only a very small amount of momentum can be transferred to an atom or molecule by electron impact, and ions produced in collisions that involve the mere stripping of electrons off the target have essentially the same kinetic energies as the original target structures. If, on the other hand, the target dissociates when ionization occurs, internal energy is transferred to the fragments in the form of kinetic energy, and the target may fly apart rather violently. The breakup of the structure is described in terms of the *Franck-Condon principle*, which states that in an electronic transition the separation and relative velocity of the nuclei of the molecule are altered only to a negligible extent (Section 5-5-A). This principle also governs dissociative electron attachment to molecules.

McConkey et al.[98] in 1972 studied the angular distribution of singly charged ions produced by electron impact on atoms (helium and argon). The observed angular patterns show features similar to those obtained in triple differential cross section measurements on helium performed by Ehrhardt and his colleagues (see Section 6-9). However, the intrinsic problems associated with handling thermal energy ions, and the fact that the desired information about the ions can be obtained from results of (e, 2e) measurements by applying energy and momentum conservation, have curtailed direct observations of the ions produced in nondissociative processes.

This situation is quite different when it comes to dissociative ionization. A considerable amount of work has gone into measurements of the energy distribution of the ionic fragments created in dissociative events. Starting in 1930, Lozier tubes[99,8–10] were used in many of these studies, and most of the remainder were performed with mass spectrometers.[4,5,8] However, in 1962 a paper by Dunn[100] cast serious doubt on the validity of many of the conclusions based on various types of dissociative excitation studies made both on Lozier tubes and on mass spectrometers with conventional ion sources. Both types of apparatus selectively detect particles ejected at right angles to the exciting electron beam, and Dunn shows that the angular distributions of the dissociation products are expected to be anisotropic in a majority of the cases. Dunn argues as follows: Because of the great relative mass of the nuclei, when an electronic transition to an antibonding state occurs in a molecule, the dissociation normally takes place in a time that is short compared with the period of the molecular rotation. Hence, to a good approximation, the dissociation products will be moving in a direction characteristic of the vibratory motion. For this reason, if the excitation probability has a dependence on the relative orientation of the incident electron beam and some axis of the molecule, the dissociation products should have a corresponding distribution.

Anisotropies are clearly revealed in a study made by Dunn and Kieffer[101] of the angular distributions and energy profiles of protons released in dissociative ionization of hydrogen molecules. The apparatus used in this study is illustrated in Fig. 6-10-1. Electrons from a gun rigidly attached to a cylindrical rotatable

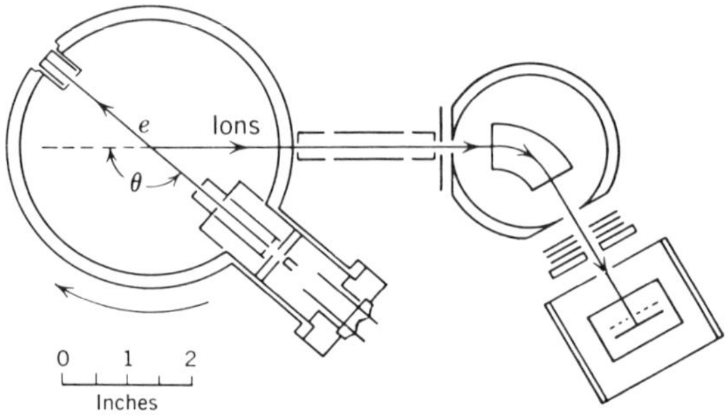

Figure 6-10-1. Apparatus used by Dunn and Kieffer[101] to study anisotropies in the dissociative ionization of hydrogen molecules.

collision chamber produce ionization in the target gas, and ions formed in the collision region drift out through a slot milled along the periphery of the chamber. The slot is in the plane that contains the electron beam and that is perpendicular to the axis of the chamber. Ions leaving the chamber at a selected angle drift through fixed collimating apertures into a sector field magnetic analyzer. They are then accelerated into an electron multiplier and detected. The collision chamber and the electrodes in the drift region are gold-plated to minimize stray electric fields, and shielding is used to reduce the magnetic field in the chamber and drift region to less than $4\,\mu$T. The base vacuum is about 2×10^{-7} torr, and the experiments are conducted at about 2×10^{-5} torr. Two types of measurements are made. First, the magnetic field in the analyzer is set to transmit protons of a selected energy, and the proton current is measured as a function of the angle θ for a given electron energy. This is done for electrons of various discrete energies and for several proton energies. Then the angle and electron energy are fixed, and the proton energy is scanned by varying the magnetic field. These measurements are made for various angles and electron beam energies.

The angular distributions that Dunn and Kieffer obtained for 8.6-eV protons are shown in Fig. 6-10-2. We may note a marked energy dependence, with a transition from forward peaking at lower electron energies to slight peaking at right angles at high energies. (The slight fore-aft asymmetry is due to a fore-aft asymmetry in the scattering chamber.) The energy dependence of the anisotropy is dramatically shown in Fig. 6-10-3, which gives the ratio of the proton current at 20° to that at right angles. The dashed vertical line represents the threshold energy at which this particular ion energy appears. The distribution changes rapidly from being strongly peaked in the forward direction near threshold to isotropic at about 200 eV. At electron energies above about 200 eV the distribution changes very slowly with energy and is slightly peaked at right angles. Figure 6-10-4 shows the proton energy distributions as measured by Lozier[99] using 75-eV electrons, as measured by Dunn and Kieffer,[101] and as predicted by the Franck-Condon principle. (The breakup of the H_2 molecule is

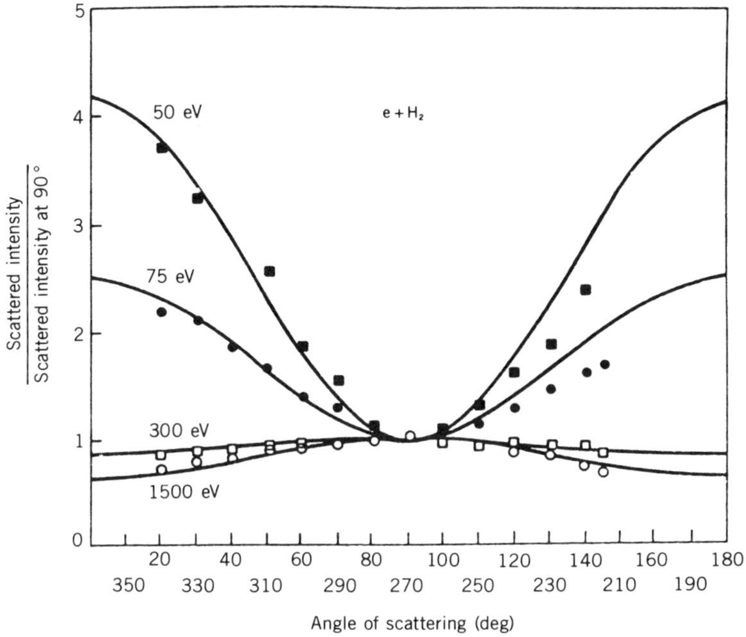

Figure 6-10-2. The angular distribution of 8.6-eV H^+ ions produced in dissociation of H_2 by electrons with energies of 50-, 75-, 300-, and 1500-eV energy.[101] The intensity scale is normalized to unity at 90° for all curves. (Adapted from Massey.[8])

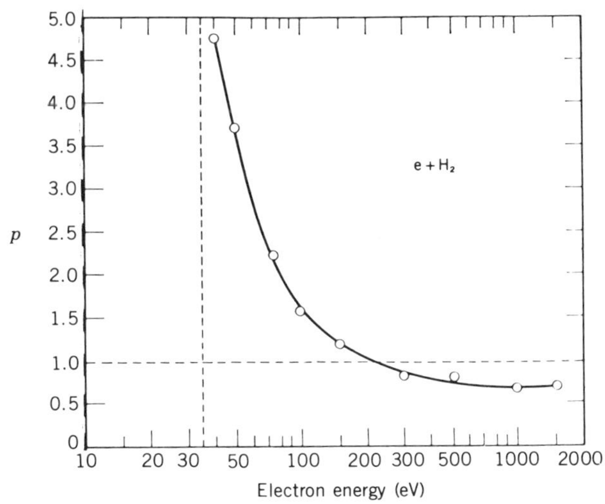

Figure 6-10-3. The ratio p versus electron energy for 8.6-eV H^+ ions produced in dissociative ionization of H_2; p is the ratio of the number of protons received at 20° with respect to the electron beam direction to the number received at 90°.[101]

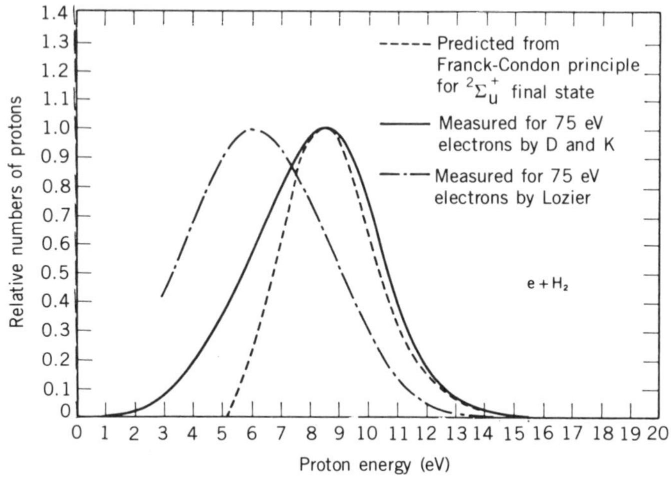

Figure 6-10-4. Proton energy distributions for the reaction $e + H_2 \rightarrow H + H^+ + 2e$.[101]

discussed in terms of the Franck-Condon principle in Section 5-5-A.) Dunn and Kieffer's results on H_2 are seen to be in accord with the Franck-Condon rule. More recent research on the dissociative ionization of H_2 and D_2 is reported in Ref. 102 and discussed on pages 122–130 of the review by Teubner.[78]

Finally, we must return to Ref. 100 for a brief discussion of the *Dunn selection rules*.[8,78] Using group-theoretical arguments, Dunn[100] analyzed dissociative transitions for homonuclear diatomic molecules within sets of states Σ_g^+, Σ_g^-, Σ_u^+, Σ_u^-, Π_g, Π_u, Δ_u, Δ_g. He showed that transitions can take place with the internuclear axis *parallel* to the incident electron momentum only for the following cases:

$$\Sigma_g^+-\Sigma_g^+,\ \Sigma_g^+-\Sigma_u^+,\ \Sigma_g^--\Sigma_g^-,\ \Sigma_g^--\Sigma_u^-,\ \Sigma_u^+-\Sigma_u^+,\ \Sigma_u^--\Sigma_u^-,$$
$$\Pi_g-\Pi_g,\ \Pi_g-\Pi_u,\ \Pi_u-\Pi_u,\ \Delta_g-\Delta_g,\ \Delta_g-\Delta_u,\ \Delta_u-\Delta_u. \qquad (6\text{-}10\text{-}1)$$

If the internuclear axis is *perpendicular* to the incident electron momentum, only these transitions are possible:

$$\Sigma_g^+-\Sigma_g^+,\ \Sigma_g^+-\Pi_u,\ \Sigma_g^+-\Delta_g,\ \Sigma_g^--\Sigma_g^-,\ \Sigma_g^--\Pi_u,\ \Sigma_g^--\Delta_g,$$
$$\Sigma_u^+-\Sigma_u^+,\ \Sigma_u^+-\Pi_g,\ \Sigma_u^+-\Delta_u,\ \Sigma_u^--\Sigma_u^-,\ \Sigma_u^--\Pi_g,\ \Sigma_u^--\Delta_u,$$
$$\Pi_g-\Pi_g,\ \Pi_g-\Delta_u,\ \Pi_u-\Pi_u,\ \Pi_u-\Delta_g,\ \Delta_g-\Delta_g,\ \Delta_u-\Delta_u. \qquad (6\text{-}10\text{-}2)$$

For heteronuclear molecules, the corresponding results for the "*parallel*" case are

$$\Sigma^+-\Sigma^+,\ \Sigma^--\Sigma^-,\ \Pi-\Pi,\ \Delta-\Delta. \qquad (6\text{-}10\text{-}3)$$

The allowed transitions for the "*perpendicular*" case are

$$\Sigma^+{-}\Sigma^+,\ \Sigma^+{-}\Pi,\ \Sigma^+{-}\Delta,\ \Sigma^-{-}\Sigma^-,\ \Sigma^-{-}\Pi,\ \Sigma^-{-}\Delta,\ \Pi{-}\Pi,\ \Pi{-}\Delta,\ \Delta{-}\Delta. \tag{6-10-4}$$

When the target is ionized as well as dissociated, the direction of motion of the ejected electron must be considered, and the symmetries on which Dunn's arguments are based are less definite. Nevertheless, conclusions similar to those presented in (6-10-1) to (6-10-4) apply, although with less certainty. As we have seen, fragment ions resulting from dissociative ionization are ejected along the internuclear axis, so the observed anisotropy can be explained in terms of the Dunn selection rules.

6-11. IONIZATION INVOLVING POLARIZED ELECTRONS AND STATE-SELECTED TARGETS

A. Theoretical Considerations

In Sections 4-6 and 5-10 we discussed elastic scattering and excitation in collisions of polarized electrons with state-selected targets and showed how studies of such collisions can provide much more information about scattering dynamics and mechanisms than conventional studies do. In this section we propose to make a similar demonstration for ionization. We first describe the theory of the measurement of ionization with polarized projectiles and targets.[103–106]

We restrict our attention to collisions of electrons with light one-electron atoms, for which Coulomb direct and exchange interactions, but not spin-orbit effects, are important (Section 4-6). In the notation of Table 4-6-1, the ionization reactions to be considered are

$$\mathrm{e}(\uparrow) + \mathrm{A}(\downarrow) \nearrow \mathrm{A}^+ + \mathrm{e}(\uparrow) + \mathrm{e}(\downarrow) \quad (\text{amplitude } f) \tag{6-11-1}$$

$$\mathrm{e}(\uparrow) + \mathrm{A}(\downarrow) \searrow \mathrm{A}^+ + \mathrm{e}(\downarrow) + \mathrm{e}(\uparrow) \quad (\text{amplitude } g) \tag{6-11-2}$$

$$\mathrm{e}(\uparrow) + \mathrm{A}(\uparrow) \rightarrow \mathrm{A}^+ + \mathrm{e}(\uparrow) + \mathrm{e}(\uparrow) \quad (\text{amplitude } f - g) \tag{6-11-3}$$

These processes are described by the direct ionization amplitude (f) and the exchange ionization amplitude (g). The differential cross sections for the ionization reactions can be written as

$$d\sigma\uparrow\downarrow = |f|^2 + |g|^2 = d\sigma + \mathrm{Re}(f^*g) \tag{6-11-4}$$

$$d\sigma\uparrow\uparrow = |f|^2 + |g|^2 - 2\,\mathrm{Re}(f^*g) = d\sigma - \mathrm{Re}(f^*g) \tag{6-11-5}$$

with

$$d\sigma = d\sigma(E, \boldsymbol{\kappa}_1, \boldsymbol{\kappa}_2) = \tfrac{1}{2}|f|^2 + \tfrac{1}{2}|g|^2 + \tfrac{1}{2}|f - g|^2. \tag{6-11-6}$$

Here $\boldsymbol{\kappa}_1$ and $\boldsymbol{\kappa}_2$ are the wave vectors of the two outgoing electrons. By integrating over all angles, we obtain the (partial) integral ionization cross sections

$$\sigma\uparrow\downarrow = \iint [d\sigma + \mathrm{Re}(f^*g)]\, d\boldsymbol{\kappa}_1\, d\boldsymbol{\kappa}_2 = \sigma + \sigma_{\mathrm{int}}, \qquad (6\text{-}11\text{-}7)$$

and

$$\sigma\uparrow\uparrow = \iint [d\sigma - \mathrm{Re}(f^*g)]\, d\boldsymbol{\kappa}_1\, d\boldsymbol{\kappa}_2 = \sigma - \sigma_{\mathrm{int}}, \qquad (6\text{-}11\text{-}8)$$

expressed in terms of the (*total*) *integral cross section* σ and the *interference cross section* σ_{int}. The difference between (6-11-7) and (6-11-8) gives $2\sigma_{\mathrm{int}} = \sigma\uparrow\downarrow - \sigma\uparrow\uparrow$, or

$$\frac{\sigma_{\mathrm{int}}}{\sigma} = \frac{\sigma\uparrow\downarrow - \sigma\uparrow\uparrow}{\sigma\uparrow\downarrow + \sigma\uparrow\uparrow} = \frac{\sigma\uparrow\downarrow - \sigma\uparrow\uparrow}{2\sigma} = A_I. \qquad (6\text{-}11\text{-}9)$$

A_I is the spin-dependent *ionization asymmetry*. We may alternatively express A_I in terms of the *singlet* (σ_s) and *triplet* (σ_t) ionization cross sections by the relations

$$\sigma_s = \sigma + 2\sigma_{\mathrm{int}}, \qquad (6\text{-}11\text{-}10)$$

$$\sigma_t = \sigma\uparrow\uparrow, \qquad (6\text{-}11\text{-}11)$$

$$\sigma = \tfrac{1}{4}\sigma_s + \tfrac{3}{4}\sigma_t, \qquad (6\text{-}11\text{-}12)$$

$$A_I = \frac{\sigma_s - \sigma_t}{4\sigma} = \frac{1 - \sigma_t/\sigma_s}{1 + 3\sigma_t/\sigma_s}. \qquad (6\text{-}11\text{-}13)$$

For electron-impact ionization, the exchange amplitude (g) is related to the direct amplitude (f) by[107]

$$g(\boldsymbol{\kappa}'_1\boldsymbol{\kappa}'_2) = f(\boldsymbol{\kappa}'_2\boldsymbol{\kappa}'_1) \qquad (6\text{-}11\text{-}14)$$

or

$$|g(\boldsymbol{\kappa}'_1\boldsymbol{\kappa}'_2)| = |f(\boldsymbol{\kappa}'_1\boldsymbol{\kappa}'_2)|. \qquad (6\text{-}11\text{-}15)$$

Our cross sections may be expressed in useful alternative forms:

$$\sigma\uparrow\downarrow = \int \frac{\kappa'_1\kappa'_2}{\kappa}(|f|^2 + |g|^2)\, d\boldsymbol{\kappa}'_1\, d\boldsymbol{\kappa}'_2\, dE'_1, \qquad (6\text{-}11\text{-}16)$$

$$\sigma\uparrow\uparrow = \int \frac{\kappa_1'\kappa_2'}{\kappa} |f - g|^2 \, d\boldsymbol{\kappa}_1' \, d\boldsymbol{\kappa}_2' \, dE_1', \tag{6-11-17}$$

$$\sigma_{\text{int}} = \int \frac{\kappa_1'\kappa_2'}{\kappa} \operatorname{Re} \left[(f^*(\boldsymbol{\kappa}_1'\boldsymbol{\kappa}_2') g(\boldsymbol{\kappa}_1'\boldsymbol{\kappa}_2')\right] d\boldsymbol{\kappa}_1' \, d\boldsymbol{\kappa}_2' \, dE_1' = \tfrac{1}{2}(\sigma\uparrow\downarrow - \sigma\uparrow\uparrow), \tag{6-11-18}$$

$$\sigma_s = \int \frac{\kappa_1'\kappa_2'}{\kappa} |f + g|^2 \, d\boldsymbol{\kappa}_1' \, d\boldsymbol{\kappa}_2' \, dE_1', \tag{6-11-19}$$

$$\sigma_t = \int \frac{\kappa_1'\kappa_2'}{\kappa} |f - g|^2 \, d\boldsymbol{\kappa}_1' \, d\boldsymbol{\kappa}_2' \, \mathrm{d}\mathrm{E}_1', \tag{6-11-20}$$

where κ is the wave number of the incident electron and E_1' the energy of the scattered electron.

The ionization asymmetry A_I is the counterpart of the elastic asymmetry A defined in (4-6-11). As (6-11-9) and (6-11-13) show, measurements of A_I can provide information on the ratio of the interference (σ_{int}) to total (σ) ionization cross sections, and on the ratio of the singlet (σ_s) to triplet (σ_t) cross sections. Another useful feature of the ionization asymmetry is that it permits tests of the Temkin theory of threshold ionization,[108,109] which is distinctly different from the Wannier theory[67] discussed in Section 6-8. The Temkin theory is based on the idea that the two outgoing electrons in the ionization process have very different energies, and that the outer electron sees an electric dipole formed by the inner electron and the residual ion. The dipole moment is regarded as changing adiabatically relative to the speed of the outer electron. The "Coulomb-dipole theory" predicts an undulation of $A_I(E)$ close to threshold, and this prediction should be verifiable with the high-current GaAs sources and high-resolution energy monochromators now available.

B. Experiments

The first experiments involving the ionization of state-selected spin-$\frac{1}{2}$ atoms by spin-polarized electrons were performed at Yale University[110] (on atomic hydrogen) and at the University of Stirling[111,112] (on potassium and sodium). Both groups set out to measure elastic scattering (see Section 4-6), but discovered that the unique residual ion signal in ionization events made the study of ionization much easier. Raith's group at the University of Bielefeld[113,114] has also investigated some of the alkali atoms (Li, Na, and K). Kelley et al.[115] have made a serious search for the oscillations in A_I predicted by the Coulomb-dipole theory in studies of near-threshold ionization of sodium. Their data are consistent with the Wannier theory[67] and showed no oscillations. In all the measurements mentioned here, a polarized electron beam is crossed with a beam of polarized, one-electron atoms, and the residual ions are extracted and counted in a multiplier. The ionization asymmetry A_I is determined from the difference in the ion count rate when the two spins are parallel and antiparallel.

These measurements and related topics are discussed in the review literature.[103–105,116–118]

We conclude this discussion by briefly mentioning the extensive set of experiments being performed at the City University of New York on elastic and inelastic collisions of spin-tagged electrons with spin-tagged H atoms.[119–121] The crossed-beam apparatus being used (see Fig. 6-11-1) is similar in concept to its predecessor described in Section 4-6-A, but it incorporates several important improvements. In particular, a high-intensity GaAs photoemission source is used to produce the polarized electron beam, and an RF source of the Slevin type provides the state-selected H atoms. For the electron beam, $P_e = 0.35$, $I_e = 100$ nA, and $\Delta E < 100$ meV (FWHM). For the atomic hydrogen beam, the low-field polarization is $P_H = 0.5$, the number density is $n_H = 2 \times 10^{10}/\text{cm}^3$, and the dissociation fraction is $f_H = 0.98$. The direction of the H-atom polarization can be reversed periodically by reversing the guide field solenoids along the beam line. Figure 6-11-2 shows preliminary near-threshold results obtained for the measured ionization asymmetry, Δ_I, which is related to the physical

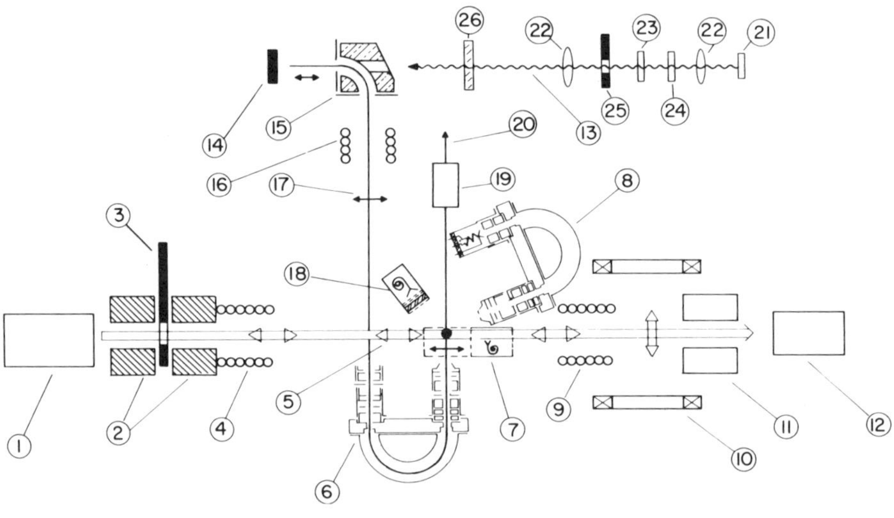

Figure 6-11-1. Sketch of the major components of the polarized e-H apparatus at CUNY[119–121] showing (1) RF hydrogen source; (2) hexapole magnetic high-field state selector; (3) beam chopper; (4) solenoidal adiabatic spin precessor; (5) direction of the electronic polarization vector of hydrogen atoms; (6) molybdenum hemispherical electron monochromator; (7) channel multiplier ion detector; (8) molybdenum hemispherical electron spectrometer; (9) and (10) adiabatic spin precessors; (11) Stern-Gerlach polarimeter; (12) quadrupole mass analyzer; (13) circularly polarized 787-nm laser light for photoemission; (14) GaAs crystal mounted on sapphire block; (15) copper spherical 90° bender; (16) solenoidal spin precessor; (17) direction of polarization vector of electrons; (18) channel multiplier Lyman-α photon detector; (19) movable Faraday cup; (20) electron beam exiting to Mott polarimeter.

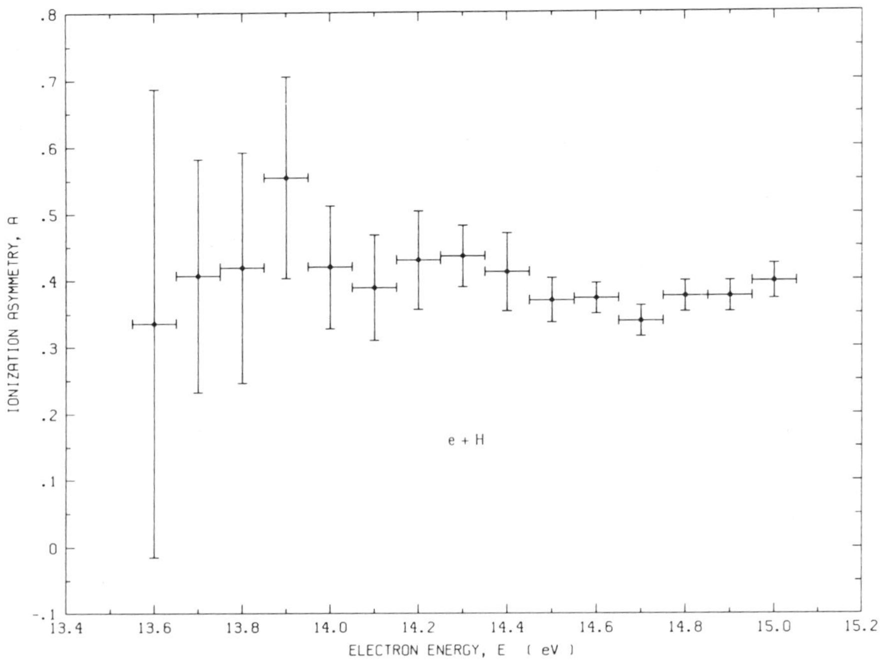

Figure 6-11-2. Measured ionization asymmetry, Δ_I, as a function of electron impact energy, E, from experiments conducted at CUNY on e-H collisions.[119–121]

asymmetry, A_I, by the equation

$$\Delta_I = P_e P_{\mathrm{H}}(1 - F_2) \cos \alpha A_I, \tag{6-11-21}$$

where F_2 is the fraction of events attributable to molecular hydrogen, and α is the angle between the electron and hydrogen polarization vectors. A_I is defined by eq. (4-6-16).

PART $\mathscr{C}$. THE THEORY OF ELECTRON IMPACT IONIZATION

6-12. QUANTUM MECHANICAL THEORY

At present there exists no rigorous quantum theory of the electron-impact ionization of neutrals and ions, and indeed a formal theory has not yet been defined, as the process is inherently a three-body one.[49] In addition to the ubiquitous complications associated with complex coupling effects,[122] target polarization, and resonances,[7] here we have a double-continuum electron final state.[123] The problem of two free electrons in the field of an ion is an especially difficult example of the quantum many-body problem, and is not amenable to computation (recall Section 6-8). Powerful computational techniques have been developed to handle many-electron correlations in bound-state problems, or situations involving many bound electrons plus a single free electron, but these techniques have been much less productive in the double-continuum case. Fadeev[124] reformulated the three-body problem in terms of an infinity of two-body problems, but this device amounts to transforming a *formally* difficult problem into an *operationally* difficult one, so little is gained from the practical standpoint.[49]

Bransden and Joachain[50] offer additional comments on the difficulty of ionization calculations. They point out that near threshold, electron-impact excitation cross sections can be obtained by a close-coupling method, but that this method cannot be extended to the ionization calculation in any straightforward method. The Born approximation can be applied to ionization at higher energies, but it does not acknowledge the Coulomb interaction and correlation between the scattered and ejected electrons, an effect increasingly important at lower energies. Furthermore, the Born approximation does not treat these two electrons on an equal footing, even though they are indistinguishable, so there are logical difficulties in its application here. (It should be pointed out, however, that the Born approximation does give accurate results for the integral cross section at high energies.)

Because of the intrinsic difficulty of the ionization problem, most of the success that has been achieved in calculating ionization cross sections has come via the development and application of approximation methods that provide a compromise between physical realism and computational tractability.[125] In his 1985 review, Younger[49] concentrates on various forms of the partial wave approximation, pointing out that most of the systematic calculations made on many-electron atoms and ions have been based on this approach, although optical potentials and perturbative techniques have also been productive,[126] especially for the H and He atoms. Younger[49] and Bottcher[125] also discuss the level of success obtained in calculations on specific systems.

We now cite a few additional theoretical reviews and books[77,127–130] and move on to classical, semiclassical, and semiempirical approximations (Section 6-13) and to the Bethe-Born approximation (Section 6-14).

6-13. CLASSICAL, SEMICLASSICAL, AND SEMIEMPIRICAL APPROXIMATIONS[7,131–136]

In 1912, J. J. Thomson[137] applied what later became known as the binary encounter method (Section 3-24-B) to the classical calculation of ionization cross sections. He took the atomic electron to be at rest, and used the relation between the impact parameter b and the energy transfer ϵ in a Coulomb collision. Thomson assumed that each collision for which $\epsilon \geqslant \chi$, the ionization energy of the target, would result in ionization. In the general case of a point projectile of mass m_1 and charge $Z_1 e$ with initial kinetic energy E_1, and a point target of mass m_2 and charge $Z_2 e$ with initial kinetic energy $E_2 = 0$, the relation between b and ϵ is

$$b^2 = Z_1^2 Z_2^2 e^4 \frac{m_1}{m_2}\left[\frac{1}{E_1 \epsilon} - \frac{(m_1 + m_2)^2}{4 m_1 m_2 E_1^2}\right]. \tag{6-13-1}$$

This equation leads to the cross section for an energy transfer between ϵ_1 and ϵ_2:

$$\sigma(\epsilon_1, \epsilon_2) = \pi Z_1^2 Z_2^2 \frac{m_1 e^4}{m_2 E_1}\left(\frac{1}{\epsilon_1} - \frac{1}{\epsilon_2}\right). \tag{6-13-2}$$

The cross section $d\sigma(\epsilon)$ for an energy transfer in the range ϵ to $(\epsilon + d\epsilon)$ is

$$d\sigma(\epsilon) = \pi Z_1^2 Z_2^2 \frac{m_1 e^4}{m_2 E_1 \epsilon^2}\, d\epsilon. \tag{6-13-3}$$

If both 1 and 2 are electrons, $Z_1 = Z_2 = -1$, $m_1 = m_2 = m$. Then we may specify that $\epsilon \geqslant \chi$ in (6-13-1) or $\chi \leqslant \epsilon \leqslant E_1$ in (6-13-2) or (6-13-3) and obtain the Thomson ionization cross section:[131]

$$\sigma_{\text{Thomson}} = \frac{N \pi e^4}{E^2}\left(\frac{E}{\chi} - 1\right). \tag{6-13-4}$$

Here N is the effective number of electrons in the atom that contribute to the cross section; it is usually taken to be the number of electrons in the outer shell. In this result we have reverted to the symbol E for the projectile energy.

The Thomson ionization formula immediately leads to the classical scaling law

$$\frac{\sigma_A}{\sigma_B} = \frac{N_A}{N_B}\left(\frac{\chi_B}{\chi_A}\right)^2, \tag{6-13-5}$$

which relates the cross sections of two atoms or ions, A and B, at the same scaled energy E/χ. This scaling law is most useful when $N_A = N_B$, that is, along an isoelectronic sequence such as C, N^+, O^{2+}, and so on.

The Thomson equation (6-13-4) overestimates the cross section at low impact energies. As we have stated, the assumption is made that any collision for which $\epsilon \geqslant \chi$ produces ionization, so that $\sigma_{\text{Thomson}} = N\pi b_c^2$, where the critical impact parameter b_c is given by

$$b_c^2 = \left(\frac{e^2}{E}\right)^2 \left(\frac{E}{\chi} - 1\right). \tag{6-13-6}$$

Actually, the transition probability for ionization seldom approaches unity, even for small b. This is the reason that the Thomson theory overestimates the cross section at low E.[122] At high impact energies, the Thomson theory underestimates the cross section and incorrectly predicts that $\sigma \sim 1/E$. The long-range electron-electron collisions are improperly treated; here the binary interaction cannot be isolated from the interaction with the other electrons and the nucleus. The Bethe-Born approximation (see Sections 6-14-B and 6-14-C) predicts the correct variation of cross section with energy:

$$\sigma \sim \frac{1}{E}(A \ln E + B) \qquad \text{(high } E\text{)}, \tag{6-13-7}$$

where the A term relates to optically allowed transitions and the B term to forbidden. Quantal treatments provide finite contributions to the ionization from impact parameters much larger than those that are significant in the classical treatment at high energies.[122]

The Thomson theory fails badly for the ionization of highly excited states, in that it grossly overestimates the cross section at all energies. However, despite its defects, the simplicity of the Thomson equation has led to its widespread use. Numerous attempts have been made to improve the Thomson theory.[5,134,135] In 1927, L. H. Thomas[138] and E. J. Williams[139] separately attacked the problem. They took the important step of allowing for nonzero target electron energies, considering the electrons to have a spherically symmetric distribution of velocities. (The importance of this impressive work was not appreciated at that time because the impact of the emerging quantum mechanics temporarily made any contributions based on classical mechanics appear dim by comparison.[133]) D. L. Webster et al.[140] in 1933 introduced an external factor in the cross section that allowed for the focusing of the projectiles by the nuclear field before the binary encounter occurs. A considerable upsurge of interest in classical calculations occurred in 1959 when M. Gryziński[141,142] began publishing an account of his analysis of collisions, which proved to be an independent reworking of the Thomas-Williams theory, of which he had been unaware. Gryziński's results appeared to be in amazing agreement with a wide range of experiments. However, in 1963, Ochkur and Petrun'kin[143] showed that Gryziński had made an unjustified mathematical approximation in evaluating an integral that improved his results but rendered them only semiempirical. Further doubt is cast on the significance of his agreement with experiment by the fact that his results are based on an unjustified choice of the velocity distribution function.[131] The velocity distribution of the bound electrons has been more rigorously treated by Ochkur and Petrun'kin[143] in 1963 and by Stabler[144] in 1964. Another improvement was made in 1963 by Burgess,[131,145] who used the Mott cross section for the two-electron interaction[7] instead of the Coulomb cross section, and thereby was able to take the Pauli exclusion principle into account. This approach was further developed by Vriens.[146] An improvement was obtained at low and medium impact energies, but not at high energy. Other refinements and applications of the classical binary encounter theory have been made by Kingston,[147] Gerjuoy,[148] and others.[134]

The binary encounter theories have been surprisingly successful. They are especially well suited for the treatment of electron-impact ionization because relatively large energy transfers are involved, and in fact they yield results that often are superior to the Born approximation cross sections except at high impact energies. Binary encounter theories have also had considerable success in the study of heavy particle collisions of various types.[136,149]

Flannery and his colleagues[150] have applied the binary encounter method to the electron-impact ionization of noble gas metastables and obtained close agreement with experiment. When applied to collisions with *Rydberg atoms*, the binary encounter approach offers the major advantage of automatically including an infinite summation over all angular momenta of the ejected electron. Many angular momenta are required for convergence of the cross section, and contributions from high-order multipoles are important as well as the dipole contribution. Direct application of quantum theory even in the Born approximation is, therefore, time consuming, whereas the binary encounter formulae provide a very efficient method of calculating the ionization cross section for Rydberg atoms.[151]

Percival and his colleagues have taken a different approach to the ionization problem by using Monte Carlo methods to solve exactly the three-body problem (two electrons and an ion). Abrines et al.[152] obtained a result for the ionization of the ground-state hydrogen atom that is superior to the results of any other approximation, although it gives a cross section that is too high at low energies and that peaks at too low an energy. The absence of interference in this classical calculation may account for the overestimation of the cross section at low energies. Quantum effects become less important for excited states, in fact, becoming smaller as the principal quantum number n increases. Independent of these calculations, we know from the correspondence principle that the classical theory should be valid for very large n.

The binary encounter method and the classical three-body calculations yield the incorrect E^{-1} asymptotic energy dependence for the ionization cross section, not the correct $E^{-1}(A \ln E + B)$ dependence in (6-13-7). However, for excitation and ionization from *excited states*, the dipole contribution $E^{-1}A \ln E$ is very small compared to $E^{-1}B$, so that the $E^{-1}A \ln E$ term becomes dominant only at extremely high energies ($\geqslant 1$ keV).[151]

As Massey and Burhop[7] point out, it is difficult to apply classical theory to the excitation of atoms because an ambiguity exists in the relation between a range of classical final energy states and a discrete quantum state.

A number of semiempirical formulas for the ionization cross section have also been proposed, and are displayed by Younger and Märk[134] and by Märk.[5] Perhaps the most widely used is the formula developed by Lotz.[153] He assumed that all electrons in a subshell i contribute with cross sections σ_{ei} to the total ionization cross section σ if the impact energy E exceeds the binding energy χ_i of the bound electrons. He further assumed that all N_i electrons in a subshell i

contribute equally to the subshell cross section σ_i. Thus

$$\sigma_i = N_i \sigma_{ei} \tag{6-13-8}$$

and

$$\sigma = \sum_{i=1}^{n} \sigma_i, \tag{6-13-9}$$

where n is the number of contributing subshells.

Classical theory suggests that below the ionization cross section peak, the cross section is a function of $(E/\chi_i - 1)$. Lotz incorporated this dependence, but took the proper high-energy behavior from quantum mechanics, using the scaled form $\sigma_i \sim \ln(E/\chi_i)/E\chi_i$. Using a trial-and-error approach, and the existing experimental data as a check, Lotz proposed the formula

$$\sigma_i(E) = a_i N_i \frac{\ln(E/\chi_i)}{E\chi_i} [1 - b_i e^{-c_i(E/\chi_i - 1)}], \tag{6-13-10}$$

where for most targets it suffices to include only the three outermost subshells ($i = 1, 2, 3$ and $n = 3$). The best-fit values for the constants a_i, b_i, c_i were determined from measured cross sections and estimated where data were lacking. They are tabulated by Lotz.[153] The values of χ_i, when unavailable, were estimated by Lotz from X-ray data.

The Lotz formula gives reasonable agreement with experimental data on simple atoms, but predicts too slow a rise from threshold for ions. Lotz did not allow for autoionization, and accordingly his formula suffers from this lack when applied to complex systems. Burgess and Chidichimo[154] have developed a formula for the ionization of *ions* in charge states $+2e$ or greater that allows for autoionization. The threshold for inner-shell ionization is taken to be the threshold for excitation to an autoionizing level, and radiative decay is assumed to be negligible.

6-14. EXCITATION AND IONIZATION AT HIGH ENERGIES;* THE FIRST BORN AND BETHE-BORN APPROXIMATIONS[4,7,84,155–157]

In closing this section we shall discuss in some detail the direct, outer-shell ionization of atoms by fast electrons and ions. As a preliminary, we treat the excitation of discrete optical levels in the Born and Bethe-Born approximations.

*For a recent review of electron-impact ionization theory at high energies, see H. R. J. Walters, *ICPEAC XV*, Brighton (1987), p. 147.

A. The Excitation Cross Section in the First Born Approximation

We start with (3-20-27), which gives the differential cross section, in the Born approximation, for direct electron impact excitation of a hydrogen atom from its ground state to state n:

$$I_{0n}(\vartheta)\,d\Omega_{\mathrm{Lab}} = \frac{m_e^2}{4\pi^2\hbar^4}\frac{\kappa_n}{\kappa_0}\left|\int \exp\left[i(\kappa_0\mathbf{n}_0 - \kappa_n\mathbf{n})\cdot\mathbf{r}_b\right]V_{0n}(\mathbf{r}_b)\,d\mathbf{r}_b\right|^2 d\Omega_{\mathrm{Lab}}. \qquad (6\text{-}14\text{-}1)$$

Here κ_0 and κ_n are the wave numbers of the projectile electron before and after the collision, respectively, and $\mathbf{n}_0$ and $\mathbf{n}$ are unit vectors in the directions of incidence and scattering. This equation is expressed in laboratory coordinates and applies to the case in which the target nucleus is considered to remain at rest at the origin of the laboratory frame. We may generalize it to apply to a target atom with N electrons and nuclear charge $+Ne$ by writing the interaction energy matrix element in the form

$$V_{0n}(\mathbf{r}_b) = \int \sum_{i=1}^{N} u_n^*(\mathbf{r}_1,\ldots,\mathbf{r}_N)\left(\frac{e^2}{r_{bi}} - \frac{Ne^2}{r_b}\right)u_0(\mathbf{r}_1,\ldots,\mathbf{r}_N)\,d\mathbf{r}_1\cdots d\mathbf{r}_N \qquad (6\text{-}14\text{-}2)$$

[cf. (3-20-19)]. Here $\mathbf{r}_b$ designates the coordinates of the incident electron, and $\mathbf{r}_1\cdots\mathbf{r}_N$ those of the atomic electrons. The distance between the incident electron and the ith target electron is denoted by r_{bi}. The eigenfunctions of the ground and nth states of the target atom are labeled u_0 and u_n, respectively. We note that the nuclear contribution to this integral vanishes if the eigenfunctions are orthogonal, as we shall assume to be the case.

It is convenient to transform from angular to momentum variables. Let us take the axis of polar coordinates to be in the direction of the change in momentum experienced by the incident electron on being scattered through angle ϑ. Then, if we define the vector

$$\mathbf{K} = \kappa_n\mathbf{n} - \kappa_0\mathbf{n}_0, \qquad (6\text{-}14\text{-}3)$$

the magnitude of the electron's momentum change associated with the scattering angle ϑ is

$$K\hbar = (\kappa_n^2 + \kappa_0^2 - 2\kappa_0\kappa_n\cos\vartheta)^{1/2}\hbar. \qquad (6\text{-}14\text{-}4)$$

Since

$$K\,dK = \kappa_0\kappa_n\sin\vartheta\,d\vartheta, \qquad (6\text{-}14\text{-}5)$$

the cross section for momentum change between $K\hbar$ and $(K + dK)\hbar$ can be written

$$I_{0n}(K)\,dK = \frac{m_e^2}{4\pi^2\hbar^4}\frac{2\pi}{\kappa_0^2}\left|\int e^{iKz}V_{0n}(\mathbf{r}_b)\,d\mathbf{r}_b\right|^2 K\,dK. \tag{6-14-6}$$

This expression may be simplified if we integrate over $d\mathbf{r}_b$. Using Bethe's integration formula,[156]

$$\int \frac{e^{iK\mathbf{n}\cdot\mathbf{r}'}}{|\mathbf{r}-\mathbf{r}'|}\,d\mathbf{r}' = \frac{4\pi}{K^2}\,e^{iK\mathbf{n}\cdot\mathbf{r}}, \tag{6-14-7}$$

we may write

$$\int e^{iKz}V d\mathbf{r}_b = \frac{4\pi e^2}{K^2}\sum_{i=1}^{N} e^{iKz_i} \tag{6-14-8}$$

and thus obtain from (6-14-6) the equation

$$I_{0n}(K)\,dK = \frac{8\pi m_e^2 e^4}{\kappa_0^2\hbar^4}\frac{dK}{K^3}\,|\epsilon_{0n}(K)|^2, \tag{6-14-9}$$

where

$$\epsilon_{0n}(K) = \sum_{i=1}^{N}\int e^{iKz_i}u_n^*(\mathbf{r}_1,\ldots,\mathbf{r}_N)u_0(\mathbf{r}_1,\ldots,\mathbf{r}_N)\,d\mathbf{r}_1\cdots d\mathbf{r}_N. \tag{6-14-10}$$

The integrals in (6-14-10) can be calculated for hydrogenic atoms for all final states n (see Mott and Massey,[155] Chap. XI, Sec. 2).

The total cross section for excitation of the atom from the ground state to its nth state by electron impact is therefore

$$\sigma_{0n}(\kappa_0) = 2\pi\int_0^{\pi} I_{0n}(\vartheta)\sin\vartheta\,d\vartheta = \int_{K_{\min}}^{K_{\max}} I_{0n}(K)\,dK. \tag{6-14-11}$$

Now

$$\kappa_0^2 = \kappa_n^2 + \frac{2m_e}{\hbar^2}(E_n - E_0), \tag{6-14-12}$$

so

$$\kappa_n \approx \kappa_0 - \frac{m_e}{\hbar^2}\frac{(E_n - E_0)}{\kappa_0} \tag{6-14-13}$$

when the impact velocities are high and the energy transfers small, as is the case here. Therefore, the limits of integration for electron impact are

$$K_{\max} = \kappa_0 + \kappa_n \approx 2\kappa_0 \qquad (\vartheta = \pi) \tag{6-14-14}$$

$$K_{\min} = \kappa_0 - \kappa_n \approx \frac{m_e}{\hbar^2} \frac{E_n - E_0}{\kappa_0} \qquad (\vartheta = 0) \tag{6-14-15}$$

The method of generalization of the foregoing results to fast ion impact and center-of-mass motion is obvious. The electron mass m_e is replaced by the reduced mass M_r of the projectile-target combination, and e^2 is replaced by $Z'e^2$, where $+Z'e$ is the charge on the projectile. The lower limit of integration over K now becomes

$$K_{\min} = \frac{M_r(E_n - E_0)}{\hbar^2} \tag{6-14-16}$$

and the upper limit, determined by the requirement of conservation of momentum, becomes

$$K_{\max} = \frac{2\kappa_0 m_e}{M_r + m_e} \approx \frac{2\kappa_0 m_e}{M_r}, \tag{6-14-17}$$

where κ_0 is now the wave number associated with the relative motion of the colliding structures.

Equations (6-14-9), (6-14-10), and (6-14-11) give the Born cross section for excitation of the target atom to its nth bound state. By integrating over final states in the continuum, we could derive instead the cross section for ionization in the first Born approximation. However, the resulting equations have a complicated form and can be solved only by numerical integration. For the present purposes, therefore, it is better to make a further approximation, the Bethe approximation, in order to obtain results that have a simple mathematical form. Cross sections calculated in the Bethe-Born approximation tend to the more precise Born results at very high velocities and thus provide us with insight into the problem we are seeking to solve. As in the preceding discussion, it is convenient first to discuss excitation to bound states by electron impact and then to extend the results to cover ionization by both electrons and ions.

B. Excitation of Discrete Optical Levels in the Bethe-Born Approximation[156]

We shall first obtain the Bethe expression for the total electron impact excitation cross section of a ground state atom to its nth bound state. We shall consider only a single outershell electron (designated by the symbol a) to be involved and drop the summation over the atomic electrons. The results we shall obtain will be valid when only a very small fraction of the impact energy is lost by the projectile in the collision.

The function $I_{0n}(K)$ decreases rapidly as the momentum change $K\hbar$ increases. This conclusion follows from examination of the integral

$$\epsilon_{0n}(K) = \int e^{iKz} u_n^* u_0 \, d\mathbf{r}_a, \qquad (6\text{-}14\text{-}18)$$

which appears in the expression for $I_{0n}(K)$, (6-14-9). The value of this integral is small if many oscillations of e^{iKz} occur within the spatial range in which the product of the wave functions u_n^* and u_0 is appreciable. Now if Z_{eff} is the effective nuclear charge of the target atom in its ground state, the radius of the ground-state orbit will be a_0/Z_{eff}, where a_0 is the radius of the first Bohr orbit of the hydrogen atom. Therefore, we see that $I_{0n}(K)$ will be small when

$$K \gg \frac{Z_{\text{eff}}}{a_0} = K_0. \qquad (6\text{-}14\text{-}19)$$

This condition may also be expressed in the form

$$K^2 \gg \frac{2m_e|E_0|}{\hbar^2} = K_0^2 \qquad (6\text{-}14\text{-}20)$$

since $Z_{\text{eff}}^2 e^2/2a_0$ equals the ionization energy of the normal state, $|E_0|$, and $a_0 = \hbar^2/m_e e^2$. We therefore see that we need to integrate over momentum change only up to the upper limit $K_0\hbar$ to obtain the excitation cross section.

For those values of K that do not satisfy condition (6-14-19) we may make the expansion

$$e^{iKz} \approx 1 + iKz + \frac{(iKz)^2}{2!} + \frac{(iKz)^3}{3!} + \cdots$$

in the integrand of (6-14-18) and obtain the approximate relationship

$$I_{0n}(K)\,dK = \frac{8\pi m_e^2 e^4}{\kappa_0^2 \hbar^4} \frac{dK}{K^3} \left[K^2|z_{0n}|^2 + \tfrac{1}{4}K^4|(z^2)_{0n}|^2 + \cdots\right]. \qquad (6\text{-}14\text{-}21)$$

Then the excitation cross section for electron impact is approximately

$$\sigma_{0n} \approx \frac{8\pi m_e^2 e^4}{\kappa_0^2 \hbar^4} \int_{K_{\min}}^{K_0} \left[K^{-1}|z_{0n}|^2 + \frac{K}{4}|(z^2)_{0n}|^2 + \cdots\right] dK, \qquad (6\text{-}14\text{-}22)$$

where z_{0n}, $(z^2)_{0n}, \ldots$ are the matrix elements of $z, z^2, \ldots$, given by the equation

$$(z^s)_{0n} = \int z^s u_n^* u_0 \, d\mathbf{r}_a. \qquad (6\text{-}14\text{-}23)$$

For an optically allowed ($\Delta l = 1$) transition the first (electric dipole) term in (6-14-22) does not vanish, and the remaining terms in the expansion may be neglected in comparison to it in the high-velocity limit under consideration here. Thus

$$\sigma_{0n}^{\text{dipole}} \approx \frac{4\pi m_e^2 e^4}{\kappa_0^2 \hbar^4} |z_{0n}|^2 \log \frac{2m_e v_0^2}{E_n - E_0}. \tag{6-14-24}$$

For an optically forbidden transition with $\Delta l = 0$ or 2, the dipole moment vanishes and the transition is associated with the quadrupole moment. The cross section is then obtained from the second term in the expansion and has the form

$$\sigma_{0n}^{\text{quad}} \approx \frac{2\pi m_e^3 e^4}{\kappa_0^2 \hbar^6} |(z^2)_{0n}|^2 |E_0|. \tag{6-14-25}$$

We notice that at high impact velocity v_0 the cross section for optically allowed transitions falls off with increasing velocity as $v_0^{-2} \log v_0^2$, whereas the cross section for forbidden transitions decreases slightly faster, as v_0^{-2}. The use of a single term in the series expansion of the exponential in (6-14-18) and the termination of the integration at the upper limit K_0 constitute the *Bethe approximation.* This is a further approximation to those already made in the Born approximation, and gives the form of the Born in the limit of very high impact velocities and small interaction. Equation (6-14-24), which is displayed on page 241 in Mott and Massey,[155] is the *Bethe-Born expression for excitation of optical levels by electron impact.* The cross section corresponding to (6-14-24) for *optical excitation of an atom by a fast ion of charge* $+Z'e$ is

$$\sigma_{0n} \approx \frac{16\pi^4 Z'^2 e^4}{m_e v_0^2} |z_{0n}|^2 \log \frac{2m_e v_0^2}{E_n - E_0} \tag{6-14-26}$$

(cf. Mott and Massey,[155] p. 271, Eq. 3).

C. Outer-Shell Ionization in the Bethe-Born Approximation[156]

The cross section for ejection of an electron from the outer shell of an atom by fast electron impact may be obtained in the same way as the cross section for excitation to a bound state. (Inner-shell ionization is treated in Chapter XI, Section 3.2, of Mott and Massey.[155]) We first need the cross section $\sigma_{nl}^{+}(\kappa)\,d\kappa$ for ionization events in which the wave number of the electron ejected from the outer shell (quantum numbers n and l) lies between κ and $\kappa + d\kappa$, corresponding to a kinetic energy in the range $\kappa^2\hbar^2/2m_e$ to $(\kappa^2 + 2\kappa\,d\kappa)\hbar^2/2m_e$:

$$\sigma_{nl}^{+}(\kappa)\,d\kappa \approx \left(\frac{4\pi m_e^2 e^4 Z_{nl}}{\kappa_0^2 \hbar^4} |z_{nl,\kappa}|^2 \log \frac{2m_e v_0^2}{C_{nl}}\right) d\kappa \tag{6-14-27}$$

[cf. (6-14-24)]. Here Z_{nl} is the number of electrons in the n, l shell, and C_{nl} is an energy of the order of the ionization energy of the n, l shell, $|E_{nl}|$. Then the total cross section for removal of a single electron from the outer shell is

$$\sigma_{nl}^{+} \approx \int_{0}^{\kappa_{\max}} \sigma_{nl}^{+}(\kappa)\, d\kappa. \tag{6-14-28}$$

Using hydrogen-like wave functions with effective nuclear charge Z_{eff}, Bethe[156] showed that (6-14-27) gives for the *electron impact ionization cross section*

$$\sigma_{nl}^{+} \approx \frac{2\pi e^4}{m_e v_0^2} \frac{c_{nl}}{|E_{nl}|} Z_{nl} \log \frac{2m_e v_0^2}{C_{nl}}, \tag{6-14-29}$$

where

$$c_{nl} = \frac{Z_{\text{eff}}^2}{n^2 a_0^2} \int |z_{nl,\kappa}|^2\, d\kappa. \tag{6-14-30}$$

Explicit calculations have been carried out on the hydrogen atom,[156] for which the cross section has the form

$$\sigma_{00}^{+} \approx 0.285 \frac{2\pi e^4}{|E_0| m_e v_0^2} \log \frac{2m_e v_0^2}{0.048|E_0|}. \tag{6-14-31}$$

This result shows that C_{nl} is about one-tenth that of the ionization energy for the ground state of hydrogen.

We see that at high velocity the ionization cross section decreases with increasing impact velocity in the same way that the cross section does for optically allowed excitation. The *cross section for ionization by a fast ion of charge $+Z'e$ is*

$$\sigma_{nl}^{+} \approx \frac{2\pi Z'^2 e^4 c_{nl} Z_{nl}}{m_e v_0^2 |E_{nl}|} \log \frac{2m_e v_0^2}{C_{nl}} \tag{6-14-32}$$

D. Predictions of the Bethe-Born Theory; Comparison with Experiment

The first point to which we wish to call attention is the prediction that at high impact velocities the cross section for ionization of a given target by electron impact should be the same as the cross section for ionization by protons traveling with the same velocity [see (6-14-29) and (6-14-32)]. The experimental verification of this prediction was discussed in Section 6-4 and illustrated for the case of O_2 in Fig. 6-4-8. The electron and proton cross sections for He, Ne, Ar, H_2, N_2, O_2, and CO were shown to be nearly equal at electron impact energies above about 300 eV; the corresponding proton energy is 552 keV.[45]

Equation (6-14-32) also predicts that the cross section for ionization by He^{2+} ions ($Z' = 2$) should be four times as large as the proton cross section at the same projectile velocity, provided that the velocity is sufficiently high. Martin et al.[158] measured the cross sections for free electron production in He, H_2, and N_2 by He^{2+} impact at energies in the range 0.60 to 1.0 MeV and observed this prediction to be obeyed.

Equation (6-14-32) for ionization by ions of charge $Z'e$ applies strictly to point-charge ions, that is, to completely stripped nuclei. Martin et al.[158] compared their data on ion production in several gases by He^+ projectiles with the proton ionization results of McDaniel et al.,[45] to examine the proposition that (6-14-32) is applicable to an ion carrying bound electrons by using an "effective charge" $+Z'_{\text{eff}}e$ lying between the nuclear charge and the actual net charge of the ion. To be a useful concept, the effective charge for a given ionic projectile must be independent of the target gas and of the incident ion energy. It was found that the concept of an effective charge of about $1.2e$ for He^+ does seem to be valid for the ionization of He, Ar, H_2, and N_2.

The experimental proton ionization cross sections of McDaniel et al.[45] for He, Ne, Ar, H_2, N_2, O_2, and CO in the energy range 0.15 to 1.0 MeV have been fitted to the theoretical expression (6-14-32) to obtain empirical values of the parameters appearing in this expression. The data for these gases have the $E^{-1} \ln E$ dependence predicted by the Born approximation at high energies, and these data can be fitted to this energy dependence down to lower incident velocities than the corresponding electron data.

PART $\mathscr{D}$. IONIZATION BY POSITRONS

6-15. IMPACT IONIZATION VERSUS POSITRONIUM FORMATION

As pointed out in Section 4-7, the first reliable data on positron collisions appeared around 1973 in the form of total cross sections (all angles and all open channels). Cross sections for elastic differential scattering and inelastic processes became available about a decade later, and high-quality data on ionization began to appear in the mid-1980s. In the case of positrons, we need to distinguish between two distinct ionization processes. The first is referred to as "impact ionization," with a cross section that we label σ_{ion}^+. (This process is similar to electron-impact ionization, whose cross section we shall designate in this section as σ_{ion}^-.) We write the impact ionization reaction for helium as

$$e^+ + \text{He} \rightarrow \text{He}^+ + e^+ + e^-. \qquad (6\text{-}15\text{-}1)$$

The other ionization process is positronium formation (cross section σ_{Ps}), which is described for a helium target as

$$e^+ + He \rightarrow He^+ + Ps, \qquad (6\text{-}15\text{-}2)$$

followed by decay of the positronium atom, Ps, with 2γ or 3γ emission (recall Section 4-7). Raith and his colleagues at the University of Bielefeld have measured the cross sections of both of these reactions for helium from threshold to 1000 eV.[159] The thresholds of (6-15-1) and (6-15-2) are 24.6 eV and 17.8 eV, respectively. The Bielefeld group utilizes a novel technique in which a given ion is detected in time correlation with the positron that produced the ion. Relative values for the sum $(\sigma_{Ps} + \sigma_{ion}^{+})$ and for σ_{ion}^{+} separately are determined simultaneously. These relative cross sections are then normalized to the accurately known cross section σ_{ion}^{-} (see Ref. 19) at energies sufficiently high that σ_{ion}^{+} and σ_{ion}^{-} merge and no positronium is formed.

6-16. THE BIELEFELD APPARATUS AND CROSS SECTIONS FOR HELIUM

In the Bielefeld instrument,[159] a 70-MBq ^{22}Na source located off axis provides fast positrons that are moderated by an annealed tungsten plate mounted at 45° with respect to the axis (Fig. 6-16-1). The energy of the slow positron beam that enters the collision region in the center of the main drawing is determined by the potential applied to the moderator. Helium target gas is admitted to the differentially pumped scattering tube through holes at its center. This tube is made of glass and has an inside diameter of 11 mm and length of 0.5 m. A tungsten helix on the inner surface of the glass tube carries a current and provides a longitudinal electric field of 20 V/m for extraction of ions produced in the target by incident positrons. Current-carrying coils produce a longitudinal magnetic field that helps guide both the positrons and the ions formed in the target. The flux density is about 35 mT in the scattering tube, but is lower elsewhere. The entrance aperture of the scattering tube has a diameter of 4 mm. The target thickness is set at about 5×10^{14} per cm^2, so double scattering is kept below 1%.

The positrons in the beam that form positronium are annihilated. All of the other beam positrons (except for the very few that are elastically backscattered) pass through the scattering tube, are accelerated to a suitable energy, and enter an $\mathbf{E} \times \mathbf{B}$ mass separator. The positrons experience only a small deflection in the $\mathbf{E} \times \mathbf{B}$ field and proceed essentially straight ahead to the microchannel-plate detector 1, whose counting rate is typically about 1000 per second. Ions formed in the scattering tube are swept out to the right, accelerated, and deflected upward by the $\mathbf{E} \times \mathbf{B}$ separator to the microchannel-plate detector 2. The ion counting rate here is about 10 per second or less. This rate is measured as a function of the positron impact energy. The positron and ion signals are processed by a time-to-amplitude converter (TAC). Ions that are time correlated

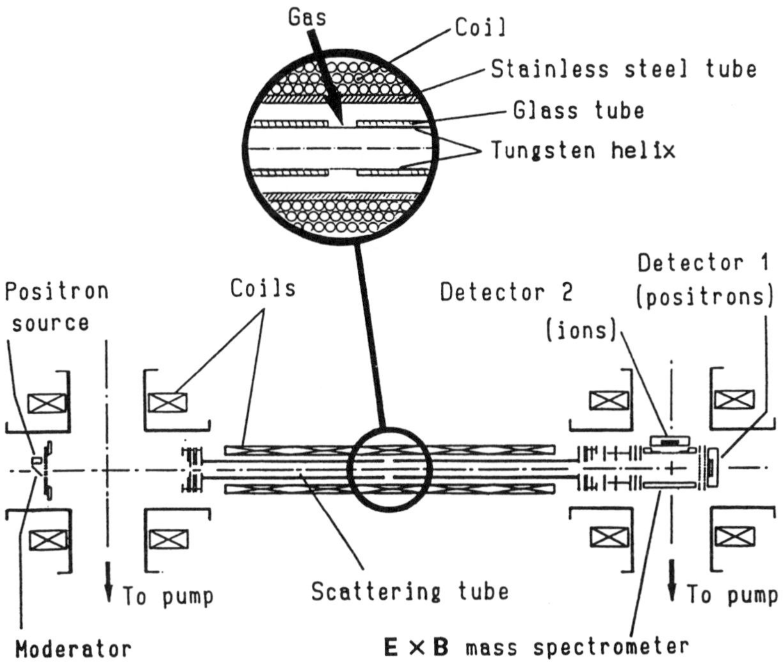

Figure 6-16-1. The Bielefeld instrument for studies of ionization by positrons.[159]

with positrons were produced by impact ionization; uncorrelated ions are associated with positronium formation. Details of background corrections, time correlation measurements, computation of relative cross sections, and normalization to obtain absolute cross sections are provided by Fromme et al.[159]

The Bielefeld data on σ_{ion}^{+} and $(\sigma_{\text{Ps}} + \sigma_{\text{ion}}^{+})$ are shown in Fig. 6-16-2 along with total cross sections for positron (σ_t^{+}) and electron (σ_t^{-}) scattering on helium measured by Kauppila et al.[160] and the cross sections for electron-impact ionization of helium measured by Montague et al.[19] and labeled σ_{ion}^{-}. This figure is taken from the paper by Campeanu et al.[161] It shows that the positron and electron partial cross sections for ionization merge above 600 eV impact energy, whereas the total cross sections merge at a much lower energy, 200 eV. Figure 6-16-2 also indicates that the two ionizing channels in positron scattering provide a large fraction of the total cross section. Finally, σ_{ion}^{+} is larger than σ_{ion}^{-} over a wide energy range, but becomes smaller at low energies, although this feature is not visible in Fig. 6-16-2.* Campeanu and his colleagues have also produced evidence for a cusp in the integral elastic scattering cross section at the threshold for positronium formation. The Campeanu paper is addressed mainly to the

*See also W. N. Lennard, P. J. Schultz, G. R. Massoumi, and L. R. Logan, *Phys. Rev. Lett.* **61**, 2428 (1988).

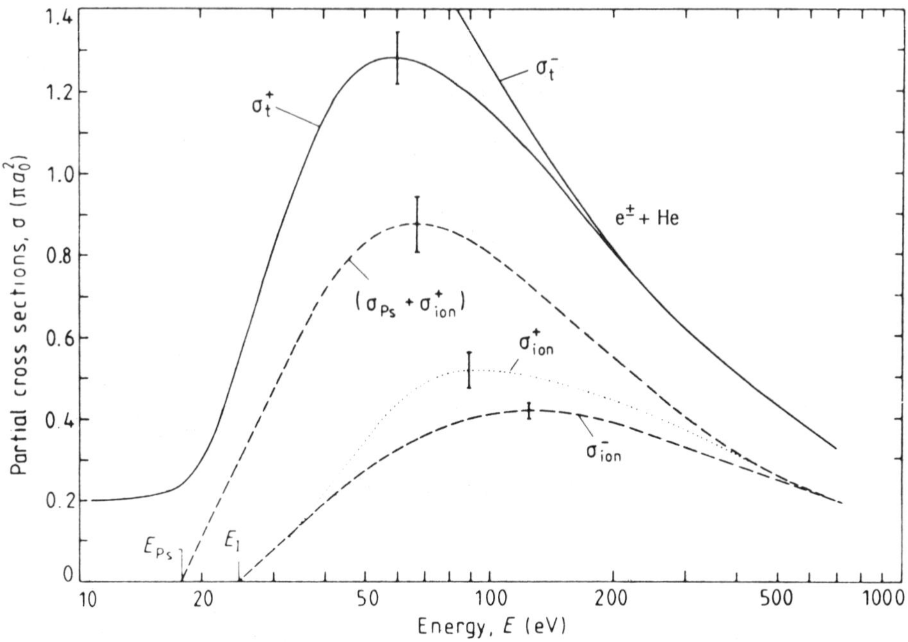

Figure 6-16-2. Cross sections for positrons and electrons on helium atoms.[161]

partitioning of the total cross section for positron-helium scattering and contains a valuable assessment of the recent experimental and theoretical work on elastic scattering, excitation, and ionization.

Other approaches to the measurement of positronium ionization cross sections are discussed in Ref. 162, along with comparisons of experiment and theory. The papers cited in Ref. 163 deal with positron measurements being performed at the Brookhaven National Laboratory with the high-intensity e^+ beam at the High-Flux Beam Reactor.

6-17. NOTES

6-1. Rydberg Atoms.[132,164] Table 6-17-1 gives values for various properties of the hydrogen atom in its ground state ($n = 1$), in a state of arbitrary principal quantum number n, and in the highly excited state $n = 100$. The enormous geometric size of the $n = 100$ atom suggests extremely large collision cross sections, and indeed cross sections as large as $10^{-10}\,\text{cm}^2$ have been recorded for such "Rydberg" states. Atoms other than hydrogen can have properties similar to those in Table 6-17-1 if a single electron is excited to a high value of n and if it has a large angular momentum so that its penetration of the core is negligible. Under these conditions, the remaining electrons shield the Rydberg electron so that it moves in a Coulomb field of unit charge, to a good approximation.

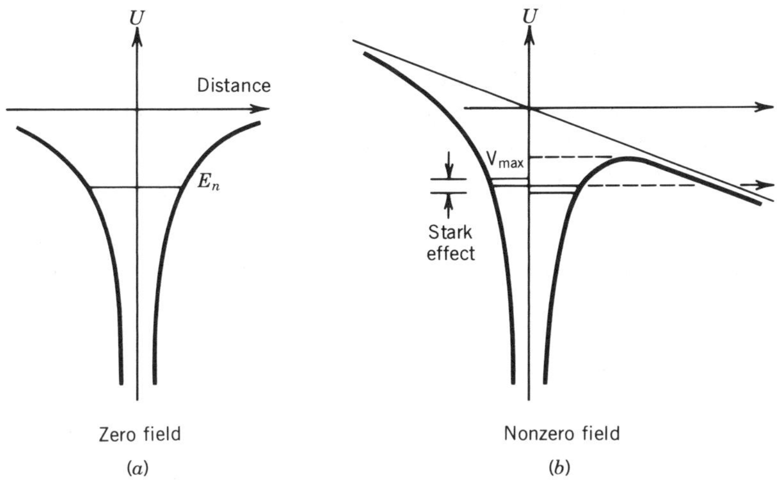

Figure 6-17-1. The potential energy U of the electron in a hydrogen atom, (*a*) with no applied field, and (*b*) with a uniform electric field applied along the axis labeled "distance." The atom in (*a*) is in the energy eigenstate E_n. One should note that the spherical symmetry is destroyed by the electric field.

Table 6-17-1 shows that for the H atom, the separation of adjacent n levels varies as n^{-3}, and $\Delta E \approx 0.2\,\text{meV}$ for $n = 50$. Despite this small separation, single Rydberg levels can be excited if high-resolution optical techniques are utilized. Conventional optical excitation can lead only to excited states with small angular momentum because of the $\Delta l = \pm 1$ selection rule. If the excitation occurs in an electric field, however, this selection rule is not operative, and all Stark components can be excited. Each stark state goes adiabatically into a single angular momentum state at zero electric field, so states with any desired value of l may be populated if the field is slowly reduced to zero.

Table 6-17-1 Properties of the H Atom in Various Quantum States[a]

Property	$n = 1$	Arbitrary n	$n = 100$
Radius of Bohr orbit (cm)	$a_0 = 5.3 \times 10^{-9}$	$n^2 a_0$	5.3×10^{-5}
Geometric cross section (cm^2)	$\pi a_0^2 = 8.8 \times 10^{-17}$	$n^4 \pi a_0^2$	8.8×10^{-9}
Binding energy, $\lvert E_n \rvert$ (eV)	$R_H = 13.6$	R_H/n^2	1.36×10^{-3}
Root-mean-square velocity of electron (cm/sec)	$v_1 = 2.2 \times 10^8$	v_1/n	2.2×10^6
Separation of adjacent n levels, ΔE (eV)		$2R_H/n^3$	2.7×10^{-5}

Source: Stebbings.[164]

[a]n, principal quantum number.

6-2. Field Ionization.[164,165] A property of Rydberg atoms that is of great theoretical and practical interest is their ability to be ionized by an electric field. The basic idea behind this "field ionization" may be understood by reference to Fig. 6-17-1. In part (*a*), the potential energy of the electron in a hydrogen-like atom is plotted for the case of zero applied electric field. In part (*b*) a field is applied, and on the Coulomb potential $-e^2/r$ is superposed the potential energy $-eEz$ associated with an external field E along the Z direction. This field perturbs the atomic energy levels (see the notation "Stark effect" on the drawing), and it also distorts the potential function into the form indicated, so that the electron is now trapped in the atom by a barrier of *finite* strength on the right side. If the atom is in a state of sufficiently high n for a given applied field strength, the electron has a finite probability of tunneling through the barrier and escaping the atom. If the external field is sufficiently strong, the barrier disappears, and nothing impedes the motion of the electron away from the atom. The phenomenon of field ionization is widely used for the detection of Rydberg atoms. Field ionization can also occur when Rydberg atoms pass through a magnetic field, because the magnetic field appears like an electric field to the moving atoms (Section 5-6). It is necessary to emphasize that the simple picture that we used here fails to give an accurate, quantitative description of the process, which is actually quite involved and the subject of active research.[165]

The collisional ionization of Rydberg atoms by neutral projectiles is also of interest here. In a high-n state, the Rydberg electron is widely separated from the nucleus and moving with a time-averaged kinetic energy of perhaps only a few meV. Hence the projectile does not interact simultaneously with both the electron and the remainder of the atom, and the electron may be considered as essentially free. The collision may be viewed as taking place between the excited electron and the projectile, with the remainder of the atom behaving as a passive spectator. Studies of impact ionization have been made using this model.[164]

6-18. PROBLEMS

Solutions for most of these problems appear in the references indicated.

6-1. A particle of mass μ scattered by a nonlocal potential has a Schrödinger equation

$$-\frac{\hbar^2}{2\mu}\nabla^2\psi(\mathbf{r}) + \int V(\mathbf{r}, \mathbf{r}')\psi(\mathbf{r}')\,d^3\mathbf{r}' = E\psi(\mathbf{r}),$$

where

$$V(\mathbf{r}, \mathbf{r}') = \frac{\hbar^2}{2\mu}\lambda u(\mathbf{r})u(\mathbf{r}').$$

(a) Show that only *s* waves are scattered.

(b) Calculate the scattering amplitude.

(c) Calculate the *s*-wave phase shift for the case $u(r) = e^{-r/b}/r$. (ter Haar,[166] p. 59.)

6-2. A particle moving on the surface of a sphere of radius R is hit by a projectile that interacts with this "spherical rotator" by Coulomb forces. Calculate by perturbation theory the differential cross section for inelastic scattering through an angle θ with the simultaneous excitation of the lth energy level of the rotator. (ter Haar,[166] p. 59.)

6-3. Calculate the momentum probability distribution function for a particle in the nth energy level of an infinitely deep, rectangular potential well. (ter Haar,[166] p. 3.)

6-4. Write out the Schrödinger equation in the p representation for a particle moving in a periodic potential of the form $V(x) = V(x + b)$. (ter Haar,[166] p. 8.)

6-5. Calculate the momentum distribution of the electron in a hydrogen atom in the 1*s*, 2*s*, and 2*p* states. (ter Haar,[166] p. 42; Bransden and Joachain,[50] pp. 621–628.)

6-6. An impulse applied to an atomic nucleus gives it a velocity v. The duration τ of the impulse is short compared to the periods of the electronic motions and to D/v, where D is the atomic diameter. Calculate the probability of excitation of the atom. (Landau and Lifshitz,[167] p. 143; ter Haar,[166] p. 50.)

6-7. Calculate the total probability of excitation and ionization of a hydrogen atom that receives an impulse of the type described in Problem 6-6. (Landau and Lifshitz,[167] p. 144; ter Haar,[166] p. 50.)

6-8. A charged harmonic oscillator in the ground state is suddenly subjected to a uniform electric field. Calculate the probabilities of transitions to excited states. (Landau and Lifshitz,[167] p. 143.)

6-9. Calculate the angular distribution of fast electrons inelastically scattered by ground-state hydrogen atoms, for the condition $1 \gg \vartheta \gg v^{-2}$. (Landau and Lifshitz,[167] p. 578.)

6-10. Calculate the differential cross section for electron-impact excitation of hydrogen atoms from the ground state to the state n. (Landau and Lifshitz,[167] p. 578; Rapp,[157] Chap. 28.)

6-11. Calculate the integral cross section for electron-impact excitation of ground-state hydrogen atoms to the state $n = 2$. (Landau and Lifshitz,[167] p. 579; Rapp,[157] Chap. 28.)

6-12. A particle is trapped in the ground state of an infinitely deep rectangular potential well that extends from $x = 0$ to $x = \text{L}$. At time $t = 0$, the right-hand wall is suddenly moved to $x = 2L$. Calculate the probability that, at $t > 0$, (a) the energy of the particle is the same as before $t = 0$, and (b) the energy of the particle is less than before $t = 0$ (ter Haar,[166] p. 3).

6-13. Consider an atom with atomic number Z. Estimate the energy of a K-shell electron both nonrelativistically and relativistically. (ter Haar,[166] p. 16.)

6-14. A tritium nucleus (^3H) undergoes β decay. Calculate the probability that the atom containing the nucleus becomes ionized. Also calculate the total probability for excitation, and the probability of exciting the nth level. (ter Haar,[166] p. 42; Landau and Lifshitz,[167] p. 144.)

6-15. The nucleus of an atom with large Z undergoes α decay. Calculate the probability that a K-shell electron will be ejected in the process. The velocity of the alpha particle is small compared with that of the K-shell electron, but the time of transit of the alpha out of the nucleus is small compared with the orbital period of the electron. (Landau and Lifshitz,[167] p. 145.)

6-16. Calculate the cross section for ionization of a ground-state hydrogen atom with emission of the ejected electron in a given direction (θ, ϕ). Neglect exchange. (Landau and Lifshitz,[167] p. 579.)

6-17. Calculate the total effective retardation of an electron by a hydrogen atom. (Landau and Lifshitz,[167] p. 584.)

6-18. Use the Thomson formula (6-13-4) to calculate the ionization cross section for atomic hydrogen at electron impact energies of 20, 50, 200, and 1000 eV. Compare your results with Fig. 6-4-3.

6-19. Repeat Problem 6-18 using the Lotz semiempirical formula (6-13-10).

6-20. Consider the formation of atomic hydrogen in the reaction $e + H^+ \leftrightarrows H$ as the adsorption of an electron on a proton and an example of *thermal ionization.*

(a) Show that the equilibrium concentrations of the reactants satisfy the relation

$$\frac{[e][H^+]}{[H]} \approx n_Q e^{-I/\tau},$$

where I is the ionization energy of the H atom, $\tau = kT$, and $n_Q = (m\tau/2\pi\hbar^2)^{3/2}$ refers to the electron. Neglect of the spins of the particles does not change the result, which is known as the *Saha equation.* If the concentrations of the electrons and protons are equal, then the electron concentration is given by

$$[e] = [H]^{1/2} n_Q^{1/2} e^{-I/2\tau}.$$

(b) Let $[H(n = 2)]$ be the equilibrium concentration of H atoms in the first excited state. Compare $[H(n = 2)]$ with $[e]$ for the conditions prevailing at the surface of the sun, with $[H] \approx 10^{23}$ per cm^3 and $T \approx 5000$ K. *Ans.:* $[H(n = 2)] \approx 2.07 \times 10^{13}$ per cm^3, and $[e] \approx 1.28 \times 10^{15}$ per cm^3. The concentration of ionized hydrogen is significantly greater than that of excited hydrogen even though the ionization energy exceeds the excitation energy. The reason for this is that the number of states accessible to separate electrons and protons is much greater than the number accessible to a bound electron-proton pair. (Kittel and Kroemer,[168] p. 273; an example dealing with the equilibrium appears on p. 269.)

6-21. Derive (6-13-1) through (6-13-4). (See McDowell and Coleman,[136] Chap. 3.)

6-22. Using (6-14-9) and (6-14-5), show that the differential cross section for electron-impact excitation in the Born approximation may be written

$$I_{0n} = \frac{4m_e^2}{h^4} \frac{\kappa_n}{\kappa_0} \frac{e^4}{K^4} |\epsilon_{0n}(K)|^2, \tag{6-18-1}$$

where $\epsilon_{0n}(K)$ is defined in (6-14-10).

REFERENCES

1. J. M. Meek and J. D. Craggs (Eds.), *Electrical Breakdown of Gases*, Wiley, New York (1978).
2. M. N. Saha, *Philos. Mag.* **40**, 472, 809 (1920). *Proc. R. Soc. London* **A99**, 135 (1921). L. H. Aller, *Astrophysics: The Atmospheres of the Sun and Stars*, 2nd ed., Ronald, New York (1963).
3. H. S. W. Massey, E. W. McDaniel, and B. Bederson (Series Eds.), *Applied Atomic Collision Physics*, Academic, New York (1982–1984), 5 vols. See Vol. 1, Chaps. 3, 4, 10; Vol. 2, Chaps. 3, 5A, 7; Vol. 3, Chaps. 1, 11, 13, 14; Vol. 4, Chaps. 2, 3; Vol. 5, Chaps. 1, 2, 4, 5, 6, 7, 14, 16.
4. T. D. Märk and G. H. Dunn (Eds.), *Electron Impact Ionization*, Springer-Verlag, Berlin (1985). Chapter 9 deals with applications.
5. T. D. Märk, "Ionization of Molecules by Electron Impact," in L. G. Christophorou, (Ed.), *Electron-Molecule Interactions and Their Applications*, Vol. 1, Academic, New York (1984).
6. E. C. Zipf, "Dissociation of Molecules by Electron Impact," in Christophorou, Ref. 5.
7. H. S. W. Massey and E. H. S. Burhop, *Electronic and Ionic Impact Phenomena*, 2nd ed., Vol. 1, Oxford University Press, London (1969).
8. H. S. W. Massey, in Ref. 7, Vol. 2 (1969).

9. J. B. Hasted, *Physics of Atomic Collisions*, 2nd ed., American Elsevier, New York (1972).
10. E. W. McDaniel, *Collision Phenomena in Ionized Gases*, Wiley, New York (1964).
11. I. Shimamura and K. Takayanagi (Eds.), *Electron-Molecule Collisions*, Plenum, New York (1984).
12. H. Ehrhardt, *ICPEAC XIII*, Berlin (1983), p. 21. For an expanded version of this review, see H. Ehrhardt, K. Jung, G. Knoth, and P. Schlemmer, *Z. Phys. D* **1**, 3 (1986).
13. F. J. de Heer and M. Inokuti, "Total Ionization Cross Sections," in Ref. 4.
14. T. D. Märk, "Partial Ionization Cross Sections," in Ref. 4.
15. K. T. Dolder, M. F. A. Harrison, and P. C. Thonemann, *Proc. R. Soc. London A* **264**, 367 (1961).
16. A. J. Dixon, A. von Engel, anld M. F. A. Harrison, *Proc. R. Soc. London A* **343**, 333 (1975).
17. A. J. Dixon, M. F. A. Harrison, and A. C. H. Smith, *J. Phys. B* **9**, 2617 (1976).
18. E. Brook, M. F. A. Harrison, and A. C. H. Smith, *J. Phys. B* **11**, 3115 (1978).
19. R. G. Montague, M. F. A. Harrison, and A. C. H. Smith, *J. Phys. B* **17**, 3295 (1984).
20. R. C. Wetzel, F. A. Baiocchi, T. R. Hayes, and R. S. Freund, *Phys. Rev. A* **35**, 559 (1987). T. R. Hayes, R. C. Wetzel, and R. S. Freund, *Phys. Rev. A* **35**, 578 (1987); *J. Chem. Phys.* **88**, 823 (1988); *J. Chem. Phys.* **89**, 4035, 4042 (1988).
21. M. B. Shah, D. S. Elliott, and H. B. Gilbody, *J. Phys. B* **20**, 3501 (1987). For similar measurements on helium, see M. B. Shah, D. S. Elliott, P. McCallion, and H. B. Gilbody, *J. Phys. B* **21**, 2751 (1988).
22. J. T. Tate and P. T. Smith, *Phys. Rev.* **39**, 270 (1932).
23. P. T. Smith, *Phys. Rev.* **36**, 1293 (1930); **37**, 808 (1931).
24. D. Rapp, P. Englander-Golden, and D. D. Briglia, *J. Chem. Phys.* **42**, 4081 (1965).
25. D. D. Briglia and D. Rapp, *J. Chem. Phys.* **42**, 3201 (1965).
26. D. Rapp and P. Englander-Golden, *J. Chem. Phys.* **43**, 1464 (1965).
27. L. J. Kieffer and G. H. Dunn, *Rev. Mod. Phys.* **38**, 1 (1966).
28. W. Bleakney, *Phys. Rev.* **40**, 496 (1932).
29. W. Bleakney and L. G. Smith, *Phys. Rev.* **49**, 402 (1936).
30. J. T. Tate and P. T. Smith, *Phys. Rev.* **46**, 773 (1934).
31. P. Defrance, W. Claeys, A. Cornet, and G. Poulaert, *J. Phys. B* **14**, 111 (1981).
32. K. L. Bell, H. B. Gilbody, J. G. Hughes, A. E. Kingston, and F. J. Smith, *J. Phys. Chem. Ref. Data* **12**, 891 (1983).
33. M. A. Lennon, K. L. Bell, H. B. Gilbody, J. G. Hughes, A. E. Kingston, M. J. Murray, and F. J. Smith, *J. Phys. Chem. Ref. Data* **17**, 1285 (1988).
34. H. Tawara and T. Kato, ADNDT **36**, 167 (1987).
35. B. L. Schram, F. J. de Heer, M. J. van der Wiel, and J. Kistemaker, *Physica* **31**, 94 (1965).
36. M. V. Kurepa, I. M. Čadež, and V. M. Pejčev, *Fizika* (*Zagreb*) **6**, 185 (1974).
37. J. Fletcher and I. R. Cowling, *J. Phys. B* **6**, L-258 (1973).
38. B. L. Schram, H. R. Moustafa, J. Schutten, and F. J. de Heer, *Physica* **32**, 734 (1966).

39. W. L. Fite and R. T. Brackmann, *Phys. Rev.* **112**, 1141 (1958).
40. E. W. Rothe, L. L. Marino, R. H. Neynaber, and S. M. Trujillo, *Phys. Rev.* **125**, 582 (1962).
41. J. W. McGowan and E. M. Clarke, *Phys. Rev.* **167**, 43 (1968).
42. K. Stephan, H. Helm, and T. D. Märk, *J. Chem. Phys.* **73**, 3763 (1980). See also K. Wiesemann, J. Puerta, and B. A. Huber, *J. Phys. B* **20**, 587 (1987).
43. A. Crowe, J. A. Preston, and J. W. McConkey, *J. Chem. Phys.* **57**, 1620 (1972).
44. T. D. Märk, F. Egger, and M. Cheret, *J. Chem. Phys.* **67**, 3795 (1977).
45. E. W. McDaniel, J. W. Hooper, D. W. Martin, and D. S. Harmer, *Proc. 5th Int. Conf. Ionization Phenom. Gases*, Munich, 1961, North-Holland, Amsterdam, Vol. I, p. 60 (1962). J. W. Hooper, D. S. Harmer, D. W. Martin, and E. W. McDaniel, *Phys. Rev.* **125**, 2000 (1962). See also M. E. Rudd, Y. K. Kim, D. H. Madison, and J. W. Gallagher, *Rev. Mod. Phys.* **57**, 965 (1985).
46. B. Crasemann (Ed.), *X-Ray and Atomic Inner-Shell Physics–1982*, *AIP Conference Proceedings*, No. 94, American Institute of Physics, New York (1982).
47. R. I. Hall and F. H. Read, "Molecular Spectroscopy by Electron Scattering," in Ref. 11.
48. D. J. Fabian, H. Kleinpoppen, and L. M. Watson (Eds.), *Inner-Shell and X-Ray Physics of Atoms and Solids*, Plenum, New York (1981).
49. S. M. Younger, "Quantum Theoretical Methods for Calculating Ionization Cross Sections," in Ref. 4.
50. J. Berkowitz, *ICPEAC XIV*, Stanford (1985), p. 631. T. F. Gallagher, *ICPEAC XIV*, Stanford (1985), p. 609. B. H. Bransden and C. J. Joachain, *Physics of Atoms and Molecules*, Longman Group, Harlow, Essex, England (1983). N. Stolterfoht, *Phys. Rep.* **146**, 315 (1987).
51. J. A. Baxter, P. Mitchell, and J. Comer, *J. Phys. B* **15**, 1105 (1982).
52. J. Comer and F. H. Read, *J. Electron Spectros.* **1**, 3 (1972). P. J. Hicks, J. Comer, and F. H. Read. *J. Phys. B* **6**, L-65 (1973). P. Mitchell, J. A. Baxter, J. Comer, and P. J. Hicks, *J. Phys. B* **13**, 4481 (1980).
53. R. B. Barker and H. W. Berry, *Phys. Rev.* **151**, 14 (1966).
54. P. J. Hicks, S. Cvejanović, J. Comer, F. H. Read, and J. M. Sharp, *Vacuum* **24**, 573 (1974). See also F. H. Read and P. Hammond, *J. Phys. B* **21**, 4225 (1988).
55. F. H. Read, *Radiat. Res.* **64**, 23 (1975).
56. G. C. King, M. Tronc, F. H. Read, and R. C. Bradford, *J. Phys. B* **10**, 2479 (1977). See also I. Harrison, G. C. King, and L. Avaldi, *J. Phys. B* **21**, 4015 (1988).
57. G. C. King, *ICPEAC XIV*, Stanford (1985), p. 215.
58. F. H. Read, *J. Phys. B* **10**, 449 (1977).
59. G. C. King, F. H. Read, and M. Tronc, *Chem. Phys. Lett.* **52**, 50 (1977).
60. R. Hippler, H. Klar, K. Saeed, I. McGregor, A. J. Duncan, and H. Kleinpoppen, *ICPEAC XIII*, Berlin (1983), p. 789.
61. A. Crowe, *ICPEAC XIII*, Berlin (1983), p. 97. See also *AAMP* **24**, 269 (1988).
62. E. G. Berezhko and N. M. Kabachnik, *J. Phys. B* **10**, 2467 (1977). W. Mehlhorn, "Alignment in Ionization by Ion, Electron, and Photon Impact," in Ref. 62, p. 53.
63. C. J. Powell, "Inner-Shell Ionization Cross Sections," in Ref. 4.

64. E. H. S. Burhop, *The Auger Effect and Other Radiationless Transitions*, Cambridge University Press, Cambridge (1953). See also E. H. S. Burhop, *AAMP* **15**, 329 (1979). C. B. O. Mohr, *AAMP* **4**, 221 (1968).
65. F. H. Read, "Threshold Behavior of Ionization Cross Sections," in Ref. 4.
66. E. P. Wigner, *Phys. Rev.* **73**, 1002 (1948).
67. G. H. Wannier, *Phys. Rev.* **90**, 817 (1953).
68. A. R. P. Rau, *ICPEAC XIII*, Berlin (1983), p. 711; see also A. Temkin, p. 755; H. G. M. Heideman, p. 743; H. Klar, p. 767.
69. A. R. P. Rau, *Phys. Rep.* **110**, 369 (1984).
70. S. Cvejanović and F. H. Read, *J. Phys. B* **7**, 1841 (1974). See also F. H. Read and S. Cvejanović, *J. Phys. B* **21**, L-371 (1988).
71. J. B. Donahue, P. A. M. Gram, M. V. Hynes, R. W. Hamm, C. A. Frost, H. C. Bryant, K. B. Butterfield, D. A. Clark, and W. W. Smith, *Phys. Rev. Lett.* **48**, 1538 (1982); **52**, 164 (1984).
72. G. H. Wannier, *Phys. Rev.* **100**, 1180 (1955).
73. F. H. Read, *Aust. J. Phys.* **35**, 475 (1982).
74. E. Schubert, K. Jung, and H. Ehrhardt, *J. Phys. B* **14**, 3267 (1981).
75. P. Fournier-Lagarde, J. Mazeau, and A. Huetz, *J. Phys. B* **17**, L-591 (1984); **18**, 379 (1985). J. Mazeau, A. Huetz, and P. Selles, *ICPEAC XIV*, Stanford (1985), p. 141.
76. H. Ehrhardt, K. H. Hesselbacher, K. Jung, and K. Willmann, *Case Stud. At. Collision Phys.* **2**, 159 (1972).
77. I. E. McCarthy and E. Weigold, *Phys. Rep.* **27**, 275 (1976). E. Weigold and I. E. McCarthy, *AAMP* **14**, 127 (1978).
78. P. J. O. Teubner, "Differential Ionization Cross Sections," in Ref. 4.
79. K. Jung, R. Müller-Fiedler, P. Schlemmer, H. Ehrhardt, and H. Klar, *J. Phys. B* **18**, 2955 (1985).
80. M. A. Bolorizadeh and M. E. Rudd, *Phys. Rev. A* **33**, 882, 888, 893 (1986). The second and third papers cited here provide comparisons of data on ionization of H_2O by H^+ and H impact at 15-150 keV with the data on electron impact ionization in the first paper.
81. R. Müller-Fiedler, K. Jung, and H. Ehrhardt, *J. Phys. B* **19**, 1211 (1986). Also, for measurements of partial doubly differential cross sections for multiple ionization of Ar, Kr, and Xe, see M. A. Chaudhry, A. J. Duncan, R. Hippler, and H. Kleinpoppen, *Phys. Rev. A* **39**, 530 (1989).
82. T. W. Shyn, *Phys. Rev. A* **27**, 2388 (1983).
83. R. R. Goruganthu, W. G. Wilson, and R. A. Bonham, *Phys. Rev. A* **35**, 540 (1987).
84. M. Inokuti, *Rev. Mod. Phys.* **43**, 297 (1971).
85. Y. K. Kim, *Radiat. Res.* **64**, 205 (1975).
86. Y. K. Kim, *Phys. Rev. A* **28**, 656 (1983).
87. H. Ehrhardt, M. Schulz, T. Tekaat, and K. Willmann, *Phys. Rev. Lett.* **22**, 89 (1969).
88. M. J. Van der Wiel, *ICPEAC VIII*, Belgrade (1973), p. 417.
89. H. Ehrhardt, M. Fischer, K. Jung, F. W. Byron, C. J. Joachain, and B. Piraux, *Phys. Rev. Lett.* **48**, 1807 (1982).
90. H. Ehrhardt, M. Fischer, and K. Jung, *Z. Phys. A* **304**, 119 (1982).

91. R. Müller-Fiedler, P. Schlemmer, K. Jung, and H. Ehrhardt, *Z. Phys. A* **320**, 89 (1985).
92. M. Cherid, A. Duguet, and A. Lahmam-Bennani, *J. Phys. B* **20**, L-187 (1987). A. Lahmam-Bennani, M. Cherid, and A. Duguet, *J. Phys. B* **20**, 2531 (1987).
93. F. W. Byron, C. J. Joachain, and B. Piraux, *J. Phys. B* **19**, 1201 (1986).
94. H. Ehrhardt, K. Jung, H. Klar, A. Lahmam-Bennani, and P. Schlemmer, *J. Phys. B* **20**, L-193 (1987).
95. H. Ehrhardt, G. Knoth, P. Schlemmer, and K. Jung, *Phys. Lett.* **110-A**, 92 (1985). H. Klar, A. C. Roy, P. Schlemmer, K. Jung, and H. Ehrhardt, *J. Phys. B* **20**, 821 (1987). P. Schlemmer, T. Rösel, H. Ehrhardt, and K. Jung, *ICPEAC XV*, Brighton (1987), Contributed Papers, p. 234.
96. E. P. Curran and H. R. J. Walters, *J. Phys. B* **20**, 337, 1105 (1987).
97. J. F. Williams, *J. Phys. B* **11**, 2015 (1978).
98. J. W. McConkey, A. Crowe, and M. A. Hender, *Phys. Rev. Lett.* **29**, 1 (1972).
99. W. W. Lozier, *Phys. Rev.* **36**, 1285 (1930).
100. G. H. Dunn, *Phys. Rev. Lett.* **8**, 62 (1962).
101. G. H. Dunn and L. J. Kieffer, *Phys. Rev.* **132**, 2109 (1963).
102. L. J. Kieffer and G. H. Dunn, *Phys. Rev.* **158**, 61 (1967). R. J. Van Brunt and L. J. Kieffer, *Phys. Rev. A* **2**, 1293 (1970). R. J. Van Brunt, *Phys. Rev. A* **16**, 1309 (1977). K. Köllman, *Int. J. Mass Spectrom. Ion Phys.* **17**, 261 (1975); *J. Phys. B* **11**, 339 (1978). C. Backx, G. R. Wight, and M. J. Van der Wiel, *J. Phys. B* **9**, 315 (1976). M. D. Burrows, L. C. McIntyre, S. R. Ryan, and W. E. Lamb, *Phys. Rev. A* **21**, 1841 (1980). M. Landau, R. I. Hall, and F. Pichou, *J. Phys. B* **14**, 1509 (1981). A. Crowe and J. W. McConkey, *Phys. Rev. Lett.* **31**, 192 (1973).
103. K. Blum and H. Kleinpoppen, *AAMP* **19**, 187 (1983).
104. G. F. Hanne, *Phys. Rep.* **95**, 95 (1983).
105. G. F. Hanne, "Spin Polarization in Electron-Atom Collisions," in H. J. Beyer, K. Blum, and R. Hippler, (Eds.), *Coherence in Atomic Collision Physics*, Plenum, New York (1988).
106. J. Kessler, *Polarized Electrons*, 2nd ed., Springer-Verlag, Berlin (1985).
107. R. K. Peterkop, *Proc. Phys. Soc.* (*London*) *A* **77**, 1220 (1961).
108. A. Temkin, *Comments At. Mol. Phys.* **11**, 287 (1982).
109. A. Temkin, *ICPEAC XIII*, Berlin (1983), p. 755.
110. G. D. Fletcher, M. J. Alguard, T. J. Gay, V. W. Hughes, P. F. Wainwright, M. S. Lubell, and W. Raith, *Phys. Rev. A* **31**, 2854 (1985). This paper is the final summary of the Yale experiments, which began in the 1970s.
111. D. Hils and H. Kleinpoppen, *J. Phys. B* **11**, L-283 (1978).
112. D. Hils, W. Jitschin, and H. Kleinpoppen, *J. Phys. B* **15**, 3347 (1982).
113. G. Baum, E. Kisker, W. Raith, W. Schröder, U. Sillmen, and D. Zenses, *J. Phys. B* **14**, 4377 (1981).
114. G. Baum, M. Moede, W. Raith, and W. Schröder, *J. Phys. B* **18**, 531 (1985). See also **21**, L-501 (1988).
115. M. H. Kelley, W. T. Rogers, R. J. Celotta, and S. R. Mielczarek, *Phys. Rev. Lett.* **51**, 2191 (1983).

116. W. Raith, *ICPEAC XIII*, Berlin (1983), p. 107.
117. W. Jitschin, "Inner-Shell Vacancies," in K. Blum et al. (Eds.), Ref. 105.
118. J. Kessler, *ICPEAC XV*, Brighton (1987), p. 21.
119. D. M. Crowe, M. S. Lubell, F. J. Mulligan, J. Slevin, F. C. Tang, and A. Vasilakis, *ICPEAC XV*, Brighton (1987), Contributed Papers, p. 145.
120. M. S. Lubell, in Humberston and Armour, Ref. 162.
121. M. S. Lubell et al., to be published.
122. M. J. Seaton, "The Theory of Excitation and Ionization by Electron Impact," in D. R. Bates (Ed.), *Atomic and Molecular Processes*, Academic, New York (1962), p. 374.
123. U. Fano and A. R. P. Rau, *Atomic Collisions and Spectra*, Academic, New York (1986).
124. L. D. Fadeev, *Sov. Phys.-JETP* **12**, 1014 (1961).
125. C. Bottcher, *AAMP* **20**, 241 (1985).
126. I. E. McCarthy and A. T. Stelbovics, "Theoretical Methods for Ionization," in M. R. C. McDowell and A. M. Ferendeci (Eds.), *Atomic and Molecular Processes in Controlled Thermonuclear Fusion*, Plenum, New York (1980).
127. B. H. Bransden, *Atomic Collision Theory*, 2nd ed., Benjamin-Cummings, Menlo Park, Calif. (1983).
128. I. I. Sobel'man, *Introduction to the Theory of Atomic Spectra*, Pergamon, Oxford (1972).
129. R. K. Peterkop, *Theory of Ionization of Atoms by Electron Impact*, Colorado Associated University Press, Boulder, Colo. (1977).
130. M. R. H. Rudge, *Rev. Mod. Phys.* **40**, 564 (1968).
131. A. Burgess and I. C. Percival, *AAMP* **4**, 109 (1968).
132. I. C. Percival and D. Richards, *AAMP* **11**, 1 (1975).
133. D. R. Bates, *Phys. Rep.* **35**, 305 (1978).
134. S. M. Younger and T. D. Märk, "Semiempirical and Semiclassical Approximations for Electron Ionization," in Ref. 4.
135. L. Vriens, *Case Stud. At. Collision Phys.* **1**, 335 (1969).
136. M. R. C. McDowell and J. P. Coleman, *Introduction to the Theory of Ion-Atom Collisions*, North-Holland, Amsterdam (1970).
137. J. J. Thomson, *Philos. Mag.* **23**, 449 (1912).
138. L. H. Thomas, *Proc. Cambridge Philos. Soc.* **23**, 713, 829 (1927).
139. E. J. Williams, *Nature* **119**, 489 (1927).
140. D. L. Webster, W. W. Hansen, and F. B. Duveneck, *Phys. Rev.* **43**, 839 (1933).
141. M. Gryziński, *Phys. Rev.* **115**, 374 (1959).
142. M. Gryziński, *Phys. Rev. A* **138**, 305, 322, 336 (1965).
143. V. I. Ochkur and A. M. Petrun'kin, *Opt. Spectry. (USSR), English Transl.* **14**, 457 (1963).
144. R. C. Stabler, *Phys. Rev. A* **133**, 1268 (1964).
145. A. Burgess, in M. R. C. McDowell (Ed.), *Atomic Collision Processes*, North-Holland, Amsterdam (1964), p. 237.
146. L. Vriens, *Phys. Rev.* **141**, 88 (1966); *Proc. Phys. Soc (London)* **89**, 13 (1966).

147. A. E. Kingston, *Proc. Phys. Soc.* (*London*) **87**, 193 (1966); *J. Phys. B* **1**, 559 (1968).

148. E. Gerjuoy, *Phys. Rev.* **148**, 54 (1966).

149. D. R. Bates and A. E. Kingston, *AAMP* **6**, 269 (1970).

150. D. Ton-That and M. R. Flannery, *Phys. Rev. A* **15**, 517 (1977). D. Ton-That, S. T. Manson, and M. R. Flannery, *J. Phys. B* **10**, 621 (1977).

151. M. R. Flannery, "Theory of Rydberg Collisions with Electrons, Ions, and Neutrals," in R. F. Stebbings and F. B. Dunning (Eds.), *Rydberg States of Atoms and Molecules*, Cambridge University Press, Cambridge (1983), p. 393.

152. R. Abrines, I. C. Percival, and N. A. Valentine, *Proc. Phys. Soc.* (*London*) **89**, 515 (1966).

153. W. Lotz, *Z. Phys.* **206**, 205 (1967); **216**, 241 (1968); **220**, 446 (1969); **232**, 101 (1970). *J. Opt. Soc. Am.* **58**, 236, 915 (1968).

154. A. Burgess and M. C. Chidichimo, *Mon. Not. R. Astron. Soc.* **203**, 1296 (1983).

155. N. F. Mott and H. S. W. Massey, *The Theory of Atomic Collisions*, 2nd ed., Oxford University Press, Oxford (1952).

156. H. A. Bethe, *Ann. Physik* (*Leipzig*) **5**, 325 (1930).

157. D. Rapp, *Quantum Mechanics*, Holt, Rinehart and Winston, New York (1971), pp. 553–579.

158. D. W. Martin, R. A. Langley, J. W. Hooper, D. S. Harmer, and E. W. McDaniel, *ICPEAC III*, London (1963), p. 679.

159. D. Fromme, G. Kruse, W. Raith, and G. Sinapius, *Phys. Rev. Lett* **57**, 3031 (1986). See also *J. Phys. B* **21**, L-261 (1988) for ionization measurements on H_2.

160. W. E. Kauppila, T. S. Stein, J. H. Smart, M. S. Dababneh, Y. K. Ho, J. P. Downing, and V. Pol, *Phys. Rev. A* **24**, 725 (1981).

161. R. I. Campeanu, D. Fromme, G. Kruse, R. P. McEachran, L. A. Parcell, W. Raith, G. Sinapius, and A. D. Stauffer, *J. Phys. B* **20**, 3557 (1987).

162. T. S. Stein and W. E. Kauppila, *ICPEAC XIV*, Stanford (1985), p. 105, W. E. Kauppila, T. S. Stein, and J. M. Wadehra (Eds.), *Positron* (*Electron*)-*Gas Scattering*, World Scientific, Singapore (1986). G. Sinapius, *ICPEAC XV*, Brighton (1987), p. 73. J. W. Humberston and E. A. G. Armour (Eds.), *Atomic Physics with Positrons*, Plenum, New York (1988), see papers by W. Raith, and by M. Charlton, G. Laricchia, N. Zafar, and F. M. Jacobsen.

163. See papers by M. S. Lubell and G. Sinapius in Humberston and Armour, Ref. 162.

164. R. F. Stebbings, *ICPEAC XI*, Kyoto (1979), p. 459. I. C. Percival, "Collisions between Charged Particles and Highly Excited Atoms," in P. G. Burke, W. B. Eissner, D. G. Hummer, and I. C. Percival (Eds.), *Atoms in Astrophysics*, Plenum, New York (1983). S. A. Edelstein and T. F. Gallagher, *AAMP* **14**, 365 (1978). R. F. Stebbings and F. B. Dunning (Eds.), *Rydberg States of Atoms and Molecules*, Cambridge University Press, Cambridge (1983). D. Kleppner, M. G. Littman, and M. L. Zimmerman, *Sci. Am.* **244**, 130 (May 1981). J. E. Bayfield, *ICPEAC IX*, Seattle (1975), p. 726. F. B. Dunning, *J. Phys. Chem.* **91**, 2244 (1987). T. F. Gallagher, *Rep. Prog. Phys.* **51**, 143 (1988). J. A. C. Gallas, G. Leuchs, H. Walther, and H. Figger, *AAMP* **20**, 413 (1985).

165. G. Casati, B. V. Chirikov, D. L. Shepelyansky, and I. Guarneri, *Phys. Rep.* **154**, 77 (1987). E. J. Galvez, B. E. Sauer, L. Moorman, P. M. Koch, and D. Richards, *Phys.*

Rev. Lett. **61**, 2011 (1988). J. E. Bayfield and D. W. Sokol, *Phys. Rev. Lett.* **61**, 2007 (1988).

166. D. ter Haar (Ed.), *Problems in Quantum Mechanics*, 3rd ed., Pion, London (1975), also published by Academic, New York (1975).
167. L. D. Landau and E. M. Lifshitz, *Quantum Mechanics: Non-relativistic Theory*, 2nd ed., Pergamon, Oxford (1965).
168. C. Kittel and H. Kroemer, *Thermal Physics*, 2nd ed., Freeman, San Francisco (1980).

7

COLLISIONS OF ELECTRONS WITH CHARGED PARTICLES

PART $\mathscr{A}$. ELECTRON-ION COLLISIONS

The subject of the first part of this chapter is collisions of electrons with ions, mainly positive ions, both atomic and molecular. Our discussion concentrates on experimental studies utilizing intersecting beams, but trapped-ion and plasma methods are briefly mentioned, as are certain aspects of theory.

The *electron-positive ion* processes to be discussed here are ionization and excitation of atomic ions, inelastic scattering of electrons by positive ions, collisions of electrons with positive molecular ions (leading to, for example, dissociative recombination), and dielectronic recombination. Electron-impact detachment of electrons from negative ions is the only *electron-negative ion* process that has been studied extensively enough to warrant inclusion here.

Much of the motivation for investigating such collision processes stems from our great interest in astrophysical and laboratory plasmas.[1-4] Such plasmas are seldom dense and hence rarely in complete thermodynamic equilibrium. Details of all the pertinent atomic processes are required for the construction of models of tenuous plasmas and for the determination of the plasma properties from observations. On the other hand, a dense plasma in thermodynamic equilibrium does not require detailed atomic information for its discussion. For example, the measurement of radiated energy, combined with Stefan's law, suffices for the determination of the temperature of the plasma.

We begin with a discussion of electron-impact ionization, the first type of electron-ion collision studied by intersecting-beam techniques.[5] The techniques developed for the study of ionization not only have led to a wealth of accurate information on this phenomenon; they also served as a guide in the development of techniques for the investigation of other types of collisions under consideration here.[6-19] (Here we should note that Dunn[15] has sketched the development of crossed-beam techniques and given many references to the early work that utilized atomic beams. He has also made an interesting comparison between the neutral-neutral and electron-neutral techniques and the electron-ion methods described in this chapter.)

7-1. ELECTRON-IMPACT IONIZATION OF POSITIVE ATOMIC IONS

During the 1950s great interest developed in the ionization of positive ions by electron impact, especially in the context of controlled thermonuclear fusion research, and several laboratories began efforts to measure cross sections for this process. The difficulties proved extremely formidable, and it was not until 1961 that Dolder, Harrison, and Thonemann at the Harwell Atomic Energy Research Establishment in England reported the first successful measurements.* Their experimental technique was summarized in Section 6-2-C; it will be discussed here in more detail because of its seminal importance. Difficulties of a general nature that are inherent in electron-ion collision experiments are described along the way.

A. The Dolder-Harrison-Thonemann Technique[5]

The apparatus used in the first Harwell experiment, on the process

$$e + He^{+} \rightarrow He^{2+} + 2e, \qquad (7\text{-}1\text{-}1)$$

is shown in Fig. 7-1-1. Ions from the source S are accelerated through a potential difference of 5 kV to the electrodes L and then passed through the electromagnet M_1 for e/m analysis. The resulting beam of 5-keV He^+ ions is collimated by the slit h (5 mm high) in the plate P before it enters the interaction space B. Electrons from the gun G cross this beam at right angles and are collected in the Faraday cup C_3. The electron gun, shown in Fig. 7-1-1b, was designed to produce an approximately monoenergetic beam free from secondary electrons ejected from its grids and also to minimize the field penetration from the gun electrodes into the interaction space B. The momentum transferred to the He^{2+} ions which are produced in collisions of electrons with the He^+ ions is insufficient to deflect the He^{2+} ions from the beam. However, after the beam emerges from the interaction space, it is separated into its singly and doubly charged components by the analyzer magnet M_2, which deflects the He^+ ions into C_1 and the He^{2+} ions into C_2.

*During a roundtable discussion at the Boulder 1961 ICPEAC a few months before publication of this work, many participants were pessimistic about ever being able to measure meaningful electron-ion cross sections. Ionization cross sections were reported that ranged from *negative* values up to positive values orders of magnitude larger than those now accepted. It is now known that much of the early trouble stemmed from the use of excessive beam currents and the resulting space-charge effects on the beams.

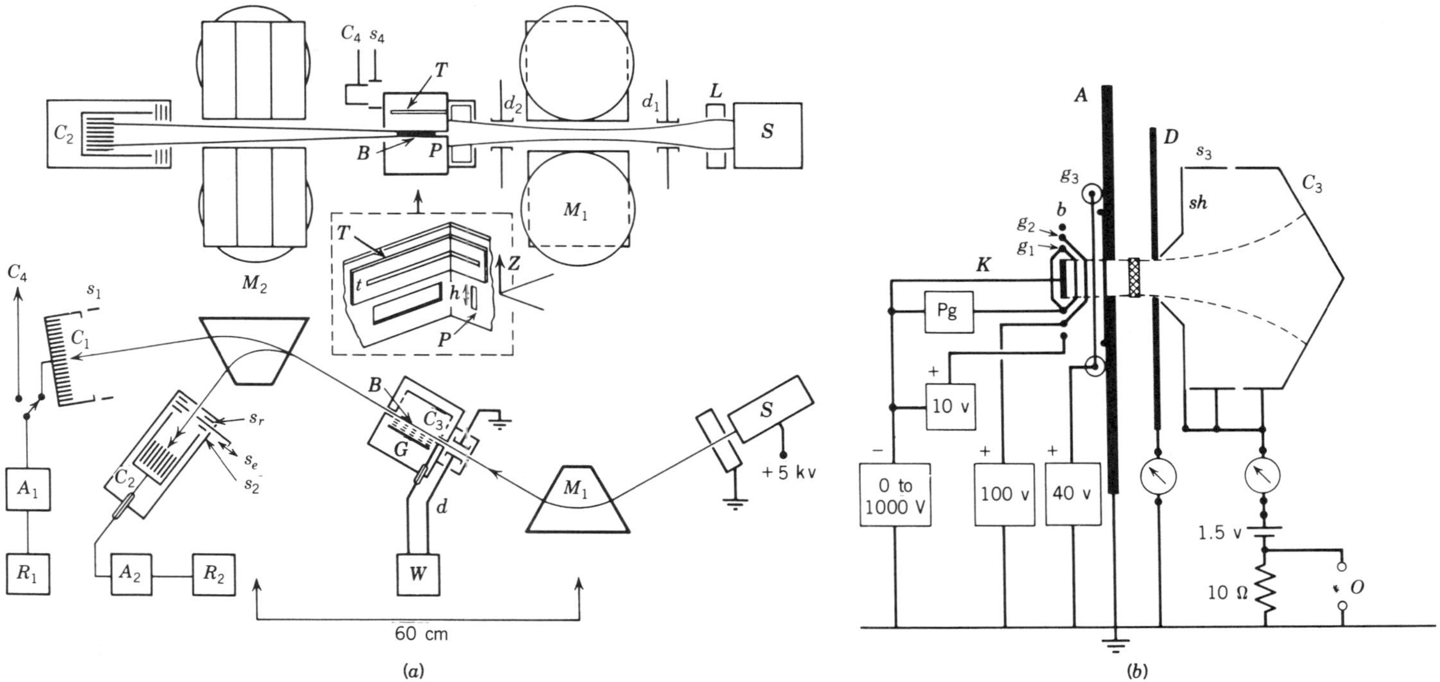

Figure 7-1-1. Apparatus used by Dolder et al.[5] for study of the ionization of He^+ ions by electron impact. (*a*) Schematic views of apparatus: the upper diagram shows the side elevation and the lower the plan view. *S*, ion source; *L*, lens; d_1 and d_2, deflector plates used to align the He^+ ion beam; M_1 and M_2, electromagnets; *B*, interaction space; *G*, electron gun; C_3, electron collector; C_1, C_2, and C_4, ion collectors; A_1, dc amplifier; A_2, vibrating-reed electrometer; R_1 and R_2, pen recorders. (*b*) The electron gun and collector viewed along the axis of the ion beam, showing the cathode (*K*), control grid (g_1), accelerating grid (g_2), and beam-focusing wires (*b*). The grid g_3 draws back secondary electrons from the anode (*A*). Typical electrode potentials are shown. The electron beam has a rectangular cross section approximately 3.7×0.2 cm and is focused vertically.

In terms of the experimentally measured quantities, the ionization cross section for formation of He^{2+} ions from He^+ is given by the equation

$$\sigma^{2+} = \frac{I^{2+}}{I^+} \frac{1}{J} \frac{hevV}{2(v^2 + V^2)^{1/2}} F, \tag{7-1-2}$$

where I^+ and J are the currents of He^+ ions and electrons, respectively, crossing the interaction space, and I^{2+} is the current of He^{2+} ions formed by electron impact. The electron and ion velocities are v and V, respectively, and e is the electron charge. The factor F is given by the expression

$$F = \frac{\int_0^h i(z)\,dz \int_0^h j(z)\,dz}{h \int_0^h i(z) j(z)\,dz}, \tag{7-1-3}$$

where $j(z)\,dz$ and $i(z)\,dz$ are the electron and He^+ currents passing through the areas defined by the height dz and the respective beam widths. The factor F, which takes into account inhomogeneities in the beam number densities, has the value unity when either beam has uniform density; F is evaluated by lowering a slotted shutter T through the beams and simultaneously measuring the electron and He^+ currents passing through the slit t (0.5 mm high). Adjustment of the apparatus always ensures that $0.93 < F < 1.00$. Equations (7-1-2) and (7-1-3) are derived in Section 7-1-C.

The currents measured at the collectors C_1 and C_2 do not correspond exactly to those in the interaction space because of space-charge repulsion losses of ions, mainly in the narrow gap of the magnet M_2. To restrict these losses, the He^+ beam current is kept small, typically about 3×10^{-7} A. Electron beam currents of about 2 mA are used, and the current of He^{2+} ions measured at C_2 does not exceed 3×10^{-15} A. In this experiment it is necessary to pulse both the electron and ion beams. Therefore, the indicated currents refer to mean levels. The electron current is measured by a moving-coil meter, the current to C_1, by dc amplifier, and the He^{2+} current to C_2 by a vibrating-reed electrometer. Secondary electrons produced by ions striking the collectors are suppressed by applying negative potentials to the electrodes s_1, s_2, and s_4. The collector C_2 must be carefully insulated to avoid charge leakage and shielded to prevent the collection of stray particles. To minimize the intrusion of stray particles into C_2, an adjustable entrance slit s_e is set just wide enough to accept the entire He^{2+} beam. Even so, there is a background current to C_2 of about 2×10^{-16} A, which is attributed to fast metastable helium atoms formed when the He^+ beam hits metal surfaces inside the apparatus. Some of these neutral metastables enter C_2 and rebound to release electrons on striking s_2. These electrons are accelerated back into C_2 and give a negative background current.

Dolder, Harrison, and Thonemann found it impracticable to reduce the pressure in their apparatus below 10^{-6} torr, and consequently a current of He^{2+} ions was formed by charge stripping on the background gas which was comparable with the maximum current of He^{2+} produced by electron impact. The background gas pressure depends to some extent on the electron current, but when the electron current is rapidly pulsed the gas pressure does not change appreciably between pulses. Consequently, the rate of charge stripping is steady. On the other hand, the current of He^{2+} arising from electron impacts is produced only during the electron pulses, and it is possible to distinguish between the two sources of the doubly charged ions. Considerations of the volume of the vacuum system and the pumping speed indicate that the gas pressure cannot change appreciably in a time less than about 10^{-2} sec; therefore, the electron beam is pulsed at a frequency of about 5000 Hz. The He^+ beam is also deflected out of the interaction space at this same frequency by voltage pulses applied to the deflector plate *d*. Thus the He^+ beam current crossing the interaction space has a square waveform, and its duty cycle is set to be 50%. The electron current pulses have a shorter duration than the ion pulses, as shown in Fig. 7-1-2. In the "coincidence" mode the electron and ion beams pulses are synchronized, so that both electrons and ions cross the interaction space at the same time. In this mode He^{2+} ions are produced both by charge stripping and electron impact. In the "anticoincidence" mode the He^+ beam is "on" only when the electron beam is "off," so that electron-ion collisions cannot occur and charge stripping is the only mechanism for the formation of He^{2+}. Thus I^{2+}, the ion current produced by electron impacts, is the difference between the mean currents measured at C_2 in the "coincidence" and "anticoincidence" modes. The modulation frequency is sufficiently high that the gas pressure is steady and the rate of charge stripping is the same for both modes.

Dolder, Harrison, and Thonemann's results[5] on the ionization of He^+ from threshold to 1 keV are shown in Fig. 7-1-3 on the curve labeled *M*. The quantity

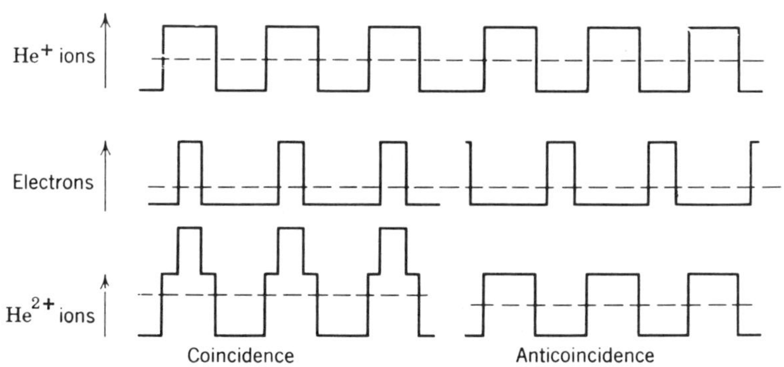

Figure 7-1-2. The waveforms of ion and electron beam currents in the experiment of Dolder et al.[5] The mean levels of the pulsed currents are shown by the dashed lines.

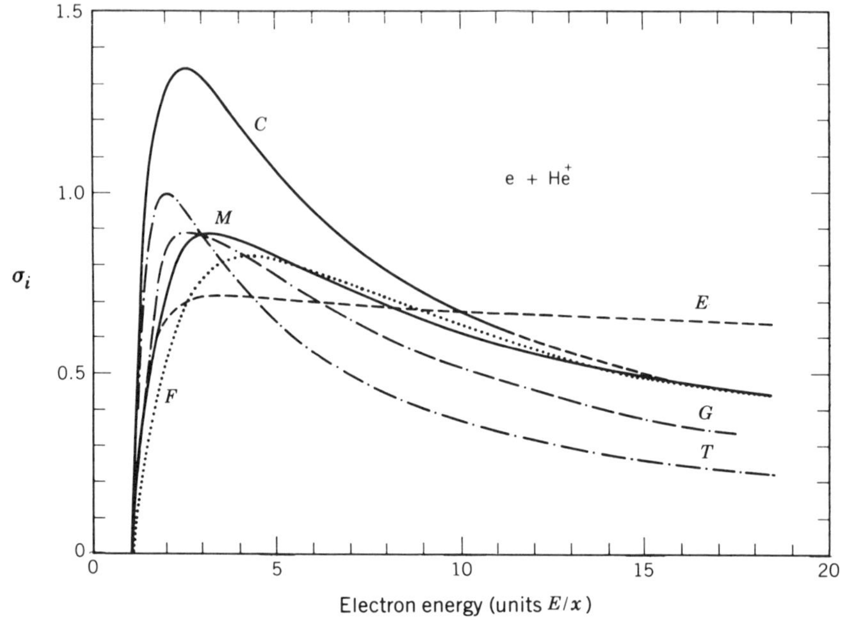

Figure 7-1-3. Cross sections for ionization of He^+ ions by electron impact.[5] Curve M is the experimental cross section multiplied by the scaling factor $(\chi_2/\chi_1)^2$. It is compared with the theoretical curves of Burgess (C), Gryziński (G), Elwert (E), and Thomson (T). Curve F shows Fite and Brackmann's measurements for atomic hydrogen. For much more recent measurements on electron-impact ionization of the H atom, see M. B. Shah, D. S. Elliott, and H. B. Gilbody, *J. Phys. B* **20**, 3501 (1987).

plotted along the ordinate axis is the experimental cross section multiplied by the classical scaling factor $(\chi_2/\chi_1)^2$, where χ_1 is the ionization energy of atomic hydrogen and χ_2 that for He^+ (13.6 and 54.4 eV, respectively). The electron energy is expressed in units of the ionization energy along the abscissa axis. This method of plotting the data facilitates their comparison with the data on atomic hydrogen. [According to the classical theory discussed in Section 6-13, $(\chi_n/\chi_1)^2\sigma_i$ as a function of E/χ_n should be the same for all members of the sequence of ions, of atomic number n, isoelectronic to the H atom.] The curves labeled T, E, and G give estimates for He^+ based on the classical results of Thomson, Elwert, and Gryziński (Section 6-13). The Coulomb-Born calculation made by Burgess is shown by curve C. Since these calculations did not extend past 12 times threshold, the curve C has been extended, as indicated by the dashed line, by use of the plane-wave Born approximation for atomic hydrogen taken from the paper by Fite and Brackmann discussed in Section 6-4. At high energies the two approximations coincide when the cross sections for He^+ ions are scaled by the factor $(\chi_2/\chi_1)^2$. This good agreement between experiment and theory serves to corroborate the experimental results, as the Born approximation is quite accurate at high electron energies. Dolder, Harrison, and Thonemann estimate that their data are accurate to within $\pm 10\%$ at all energies above 150 eV. The effect of the ionic field of the He^+ ion may be assessed by comparing the curve *M* with curve *F*, which shows Fite and Brackmann's experimental data for atomic hydrogen. The curves coincide at high energies, but the He^+ cross section is higher at electron energies less than about five times threshold. This difference may be explained by the fact that at low energies a significant number of electrons are swept in toward the ions by the long-range Coulomb field, and an increase in the probability of ionization results. At high energies the ionic attraction represents only a negligible perturbation on the electronic trajectories.

The 1961 Harwell data on He^+ are compared in Fig. 7-1-4 with more recent experimental data from Peart et al.[20] and from the group at Louvain.[21] The agreement is quite good. Figure 7-1-4, which is taken from the data compilation by Tawara and Kato,[22] also shows the results of several modern theoretical calculations of the ionization cross section. The experimental cross sections of Defrance et al.[21] were obtained by the technique in which the beam interaction

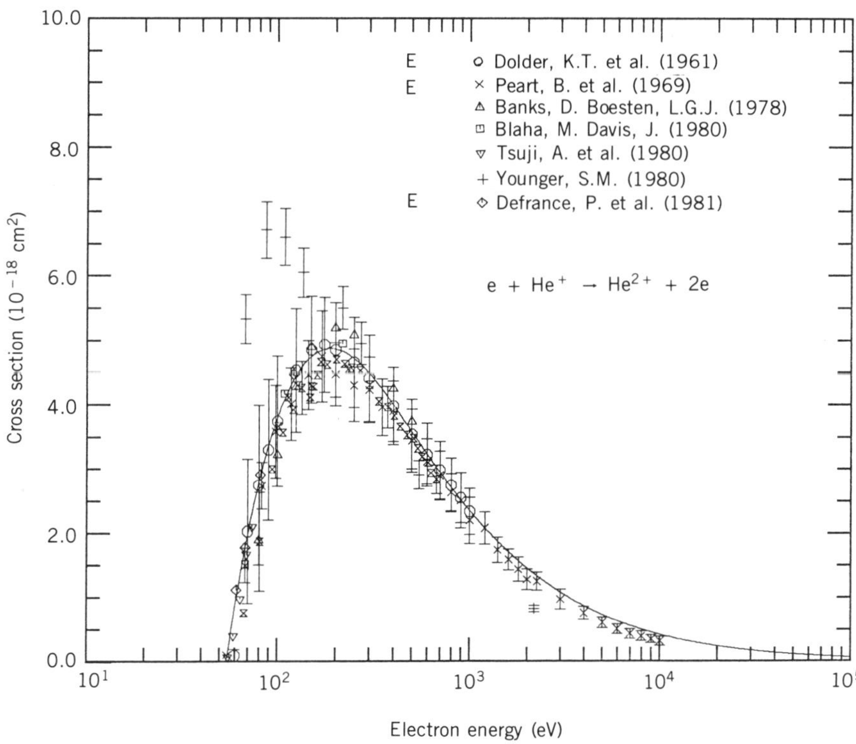

Figure 7-1-4. Integral cross sections for the electron-impact ionization of He^+ ions. (From Tawara and Kato.[22]) The symbols (E) denote experimental values. For measurements on Li^{2+}, see K. Tinschert et al., *J. Phys. B* **22**, 531 (1989).

is modulated by sweeping the electron beam periodically across the neutral beam (see Section 6-3-C). The mathematical basis for this scheme will be presented in Section 7-1-C.

B. Further Developments; Experimental Difficulties and Prospects for Improvement

Following the measurements on He^+, several additional targets were studied at Harwell, and later, the electron-ion intersecting beam research initiated there was continued by Dolder and his colleagues at the University of Newcastle and by Harrison and his coworkers at the Culham Laboratory. Meanwhile, a group at Georgia Tech inaugurated a program of measurements, the first of which involved studies of the single ionization of Li^+, Na^+, and K^+.[23,24] Their ionization apparatus was similar to the Harwell apparatus, but utilized thermionic emitters (in the form of coated filaments) to produce the positive ions to be studied. These emitters yield only singly charged ions, in the ground electronic state, and without the gas input coming from the usual gas-filled ion source.[9,12] (Unfortunately, only alkali ions and Ga^+ can apparently be produced in this fashion.) The avoidance of a gas load from the ion source permitted operation of the apparatus at a pressure of about 10^{-8} torr.* The Georgia Tech cross sections for the ionization of Li^+ are shown together with data from the Newcastle group[25] in Fig. 7-1-5, which is taken from the Tawara-Kato compilation.[22] This figure also shows theoretical results, references to which are provided by Tawara and Kato.[22]

Since the initial Harwell experiments, steady progress has been made in the vacuum techniques used in the electron-ion experiments, and today ultrahigh vacua (about 10^{-10} torr) are usually maintained in the beam intersection region. However, the number densities of the beam particles are limited to the order $10^6/cm^3$ by space charge and are comparable with the residual gas number density in the UHV region traversed by the beams. (By contrast, beam particle number densities in static-gas or atomic-beam experiments usually exceed $10^{12}/cm^3$.) Space charge produces spreading of the beams and also gives rise to electrostatic forces that may alter the beam geometries where they cross.[6,7,9,15] The rate of beam-beam collisions is usually in the range 1 to 1000 per second. In certain kinds of electron-ion experiments, only 1 beam particle in 10^{12} may collide with a particle in the other beam (see Section 7-3), although in ionization measurements, typically 1 target ion in 10^8 becomes ionized.

Background signals can be produced in various ways, the most common being by collisions of the beam particles with background gas. The path length for ion beams is typically about 1 m, so many more collisions occur with the residual gas than with the electron beam. A number of sophisticated beam

*This low pressure allowed ionization cross sections to be measured for Li^+, Na^+, and K^+ ions with continuous, as well as the usual modulated, beams. The agreement obtained between the two sets of results constituted a useful check on the modulated beams technique.

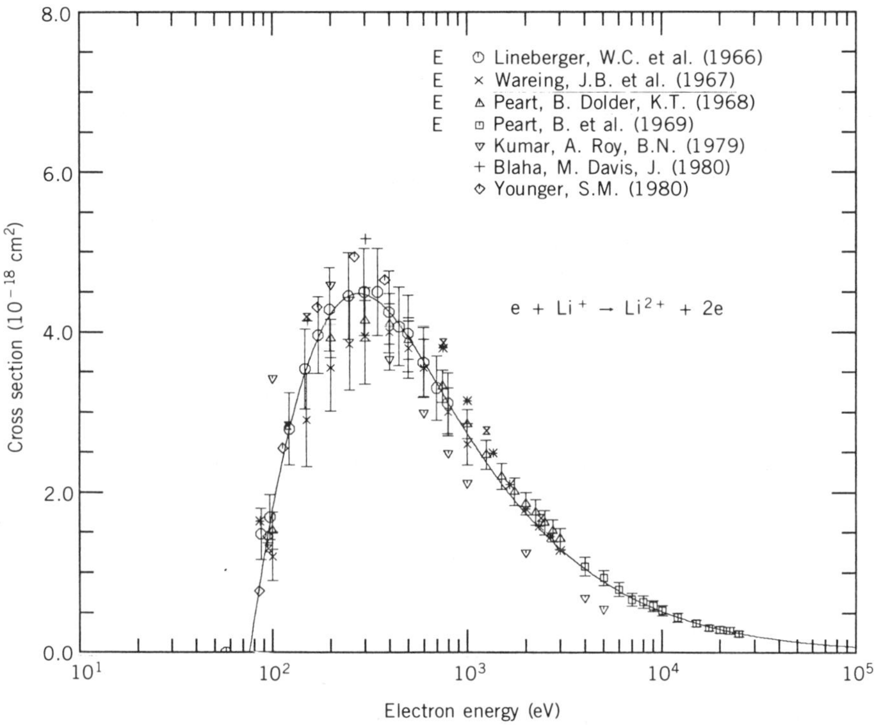

Figure 7-1-5. Integral cross sections for the single ionization of Li^+ ions by electron impact. (From Tawara and Kato.[22]) The symbols (E) denote experimental values.

modulation techniques have been developed to separate the true signal from the spurious contributions,[7,9,10,12,16] but frequently the ratio of the signal to background (SBR) is only 10^{-2} to 10^{-4}.[12] One should attempt to maximize the SBR because, in a given measurement, the required counting time is inversely proportional to the square of the SBR.

Not long after the first electron-ion experiments, electron multiplier and other kinds of pulse counters came into common use to measure the ionization signal, and they provided an enormous increase in detection sensitivity. The sensitivity of the original analog detection scheme was limited by that of the best electrometer available, the vibrating reed, which can measure currents down to about 10^{-16} A, with $\pm 5\%$ accuracy at that level. Recall that the current associated with the detection of one electronic charge per second is approximately 10^{-19} A! The introduction of digital electronics has permitted computer operation of the collision experiments and greatly increased the effective running time that can be conveniently achieved.

Now it is of interest to inquire as to where improvements may be made in electron-ion ionization measurements. Very likely, continuing developments in vacuum technology will bring further reduction in background gas pressures.

However, because the charged particle beams are space-charge limited, the electron and ion number densities in the beams will remain at about the level of $10^6/cm^3$. This limitation is very unfortunate, because most experimental difficulties can be traced to the tenuous character of the colliding beams. Progress should be expected, nonetheless, in the use of energy-resolved electron beams and in the production of useful beams of ions in higher and higher charge states (see Section 7-1-F). Furthermore, in the future, better control may be obtained over the inadvertent production of *excited* ions in the ion source. Excited states, of course, have different cross sections than ground-state ions, and their presence in ion beams has been a serious problem in certain experiments.[9,10,13] To date, one has had to rely mainly on measuring cross sections with various potentials applied within the ion source to detect the production of excited states. The situation is mitigated, however, by the fact that most of the ions formed in ordinary excited states (as opposed to metastables) will decay radiatively before they reach the beam interaction region.

The beam modulation techniques appear quite adequate. The magnetic and electrostatic analyzers[9] that have been used probably cannot be improved significantly. Modern beam particle detectors already provide the ultimate sensitivity: 1 count per incident particle, with negligible noise, if the particle has been accelerated to a sufficiently high energy. However, the electron multiplier is fragile and has a fairly short operating lifetime.

The energy range covered in electron-ion collision measurement is impressively wide. There is no practical upper limit; the lower limit is about 0.1 eV, achieved in merged-beam experiments (Section 7-1-C). This lower limit is imposed by the transverse velocity components of the beam particles produced by thermal energies and other effects.[11]

C. Kinematics of Intersecting Beams[26]

For a wider and more detailed understanding of experiments with intersecting beams of particles, it is necessary to develop some kinematic relationships beyond those already derived in Chapter 1.

1. *Relative Energy and Energy Spread*

For generality, consider two beams intersecting at angle θ. The particles in beam 1 have mass m_1 and Lab velocity $\mathbf{v}_1$; the particles in beam 2 are characterized by m_2 and $\mathbf{v}_2$. The angle between $\mathbf{v}_1$ and $\mathbf{v}_2$ is the angle θ. As in Problem 1-6, the nonrelativistic kinetic energy in the center-of-mass frame is

$$E_R = \tfrac{1}{2} M_r v_R^2, \tag{7-1-4}$$

where M_r is the reduced mass of m_1 and m_2, and

$$v_R = (v_1^2 + v_2^2 - 2v_1 v_2 \cos\theta)^{1/2} \tag{7-1-5}$$

is the relative velocity. Then, if E_1 and E_2 are the Lab kinetic energies of the particles, we have

$$E_R = M_r\left(\frac{E_1}{m_1} + \frac{E_2}{m_2} - 2\sqrt{\frac{E_1E_2}{m_1m_2}}\cos\theta\right). \tag{7-1-6}$$

Note that $E_R = 0$ for nonzero values of E_1 and E_2 only if $\cos\theta = 1$ and $E_1 = (m_1/m_2)E_2$. If there are spreads δE_1 in E_1 and δE_2 in E_2, the corresponding spread in E_R is

$$\delta E_R = M_r\left[\frac{1}{m_1} - \left(\frac{E_2}{m_1m_2E_1}\right)^{1/2}\cos\theta\right]\delta E_1 + M_r\left[\frac{1}{m_2} - \left(\frac{E_1}{m_1m_2E_2}\right)^{1/2}\cos\theta\right]\delta E_2. \tag{7-1-7}$$

δE_R becomes zero under the same conditions as E_R.

2. *Cross Sections*

Now let us suppose that the two particle beams intersect in the X-Y plane, as shown in Fig. 7-1-6, and call beam 1 the "target beam" and beam 2 the "projectile beam." The rate at which collisions occur in the element of height between z and $z + dz$ in each beam is

$$C(z)\,dz = \sigma(E_R)n(z)f(z)L\,dz, \tag{7-1-8}$$

where $\sigma(E_R)$ is the cross section for the process under consideration, $n(z)$ is the number density of the target beam, $f(z)$ is the flux of projectile particles, and L is the effective collision length. The time required for a projectile to cross the target beam is

$$t = \frac{w_1}{v_2\sin\theta}. \tag{7-1-9}$$

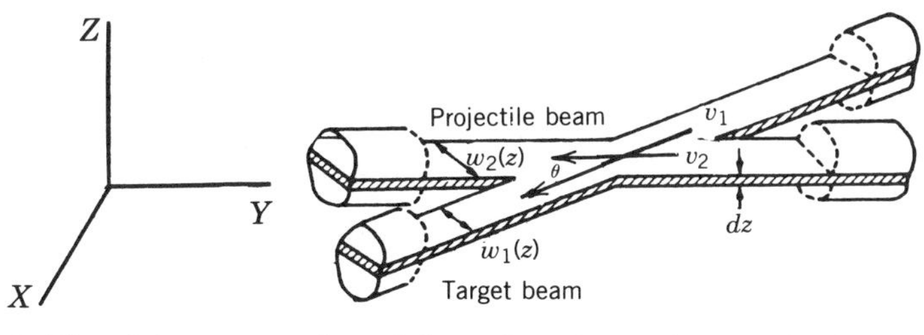

Figure 7-1-6. Pictorial representation of slices of two beams intersecting in the X-Y plane at angle θ with Lab velocities $\mathbf{v}_1$ and $\mathbf{v}_2$. The thicknesses of the two slices at height z are $w_1(z)$ and $w_2(z)$. Each slice has height dz.

Hence the effective collision length is

$$L = tv_R = \frac{w_1(v_1^2 + v_2^2 - 2v_1v_2 \cos \theta)^{1/2}}{v_2 \sin \theta}. \qquad (7\text{-}1\text{-}10)$$

Now

$$n(z) = \frac{i_1(z)}{ew_1v_1} \qquad (7\text{-}1\text{-}11)$$

and

$$f(z) = \frac{i_2(z)}{e}, \qquad (7\text{-}1\text{-}12)$$

where $i_1(z)$ and $i_2(z)$ are the beam currents at height z. The collision rate then becomes

$$C(z)\,dz = \sigma(E_R)\frac{(v_1^2 + v_2^2 - 2v_1v_2 \cos \theta)^{1/2} i_1(z) i_2(z)\,dz}{e^2 v_1 v_2 \sin \theta} \qquad (7\text{-}1\text{-}13)$$

and integration over z gives the cross section in terms of the total collision rate C:

$$\sigma(E_R) = \frac{C}{I_1 I_2} e^2 \frac{v_1 v_2 \sin \theta}{(v_1^2 + v_2^2 - 2v_1v_2 \cos \theta)^{1/2}} F'. \qquad (7\text{-}1\text{-}14)$$

Here F' is a form factor defined as

$$F' = \frac{\int i_1(z)\,dz \int i_2(z)\,dz}{\int i_1(z) i_2(z)\,dz}, \qquad (7\text{-}1\text{-}15)$$

and I_1 and I_2 are the total beam currents. Equations (7-1-2) and (7-1-3) follow immediately from (7-1-14) and (7-1-15) as a special case. Note that the form factor F in (7-1-3) is defined so as to be dimensionless; F' in (7-1-15) lacks the factor h in the denominator and so has the dimensions of length.

The original Dolder-Harrison-Thonemann method, to which the analysis above applies with $\theta = 90°$, continues to provide the principal means of measuring electron-ion ionization cross sections (see Refs. 27–31 for representative recent measurements performed by the Culham, Newcastle, JILA, and Oak Ridge groups).

As first mentioned in Section 6-3-C, DeFrance et al.[21] have introduced an interesting alternative to this method. They sweep one beam through the other while keeping it parallel to its original axis, the sweep speed being held constant at the value u during each pass up and down along the Z axis. If the beam is swept over a large enough distance that at the turning points of its motion the two beams no longer overlap at all, the cross section is easily calculated from the experimental observables. We begin by writing the general expression for the number of collisions per unit time in the intersecting beams, each of which is assumed to consist of singly charged particles:

$$N = \frac{\sigma v_R}{e^2 v_1 v_2 \sin\theta} \iiint j_1(x, y, z) j_2(x, y, z)\, dx\, dy\, dz, \tag{7-1-16}$$

where the current densities for the two beams are given in terms of the number densities and Lab velocities by

$$\mathbf{j}_1 = n_1 e \mathbf{v}_1; \qquad \mathbf{j}_2 = n_2 e \mathbf{v}_2. \tag{7-1-17}$$

Here we take $\mathbf{v}_1 = (0, v_1, 0)$ and $\mathbf{v}_2 = (v_2, 0, 0)$. Now define

$$\mathbf{j}_1(z) = \int \mathbf{j}_1(x, z)\, dx; \qquad \mathbf{j}_2(z) = \int \mathbf{j}_2(y, z)\, dy, \tag{7-1-18}$$

which give the beam profiles at $t = 0$. Then if beam 1 is swept transversely along the Z axis with constant speed $u = dz/dt$, the number of collisions per unit time at t is

$$N(t) = \frac{\sigma v_R}{e^2 v_1 v_2 \sin\theta} \int J_1(z - ut) J_2(z)\, dz. \tag{7-1-19}$$

The number of collisions, K, during one sweep is

$$K = \int_{-\infty}^{\infty} N(t)\, dt = \int_{-\infty}^{\infty} N(t) \frac{dZ}{u} \qquad \text{if } Z = ut.$$

Thus

$$K = \frac{\sigma v_R}{e^2 v_1 v_2 \sin\theta} \int_{-\infty}^{\infty} \int_{-\infty}^{\infty} J_1(z - Z) J_2(z)\, dz \frac{dZ}{u}$$

or

$$K = \frac{\sigma v_R I_1 I_2}{e^2 v_1 v_2 u \sin\theta}. \tag{7-1-20}$$

Solving for σ, we obtain

$$\sigma = \frac{e^2 v_1 v_2 \sin\theta\, uK}{I_1 I_2 v_R}. \tag{7-1-21}$$

As before, I_1 and I_2 are the total beam currents, given here by

$$I_1 = \int_{-\infty}^{\infty} J_1(z - Z)\, dZ; \qquad I_2 = \int_{-\infty}^{\infty} J_2(z)\, dz. \tag{7-1-22}$$

Equation (7-1-21), which resembles (7-1-14) very closely, shows that it is not necessary to determine an overlap integral, or form factor, if the alternative described here is utilized. However, many tests made with beam shutters have demonstrated that the cross sections obtained with the traditional techniques are insensitive to the beam profiles, and since these profiles can now be obtained accurately and quickly with computer-controlled stepping motors, the technique of Defrance et al. is unlikely to see widespread use. The beam sweeping method requires a complicated motion of the electron source, and since this motion is performed *throughout* the period of the measurement, the duty cycle is low. These two factors must be counted as major disadvantages.

3. *Crossed, Merged, Inclined, and Opposed Beams*

As we have seen, the first successful intersecting-beam experiment with two species of charged particles was performed with the beams intersecting at 90°, corresponding to the choice $\theta = 90°$ in Fig. 7-1-6. Such *crossed-beam* experiments[27–31] have remained by far the most common, mainly because of the relative ease with which they can be set up, the clean definition of the collision geometry, and the simplicity of the form factor measurements.

The choice of $\theta = 0°$ in (7-1-6) shows that the *merging beam* geometry permits the study of collisions by beam techniques at extremely low CM energies, the lower limit being about 0.1 eV, as explained in Section 7-1-B. Equation (7-1-7) shows that the spread in the CM collision energy can also be made extremely small. (Numerical examples illustrating the depression of E_R and δE_R are provided in Problems 1-4 and 7-2.) Another desirable feature of the merging-beam geometry is that a long interaction path ($\geqslant 0.1$ m) is possible, so relatively large signals are produced. However, there is some uncertainty in the effective interaction distance, associated mainly with the merging and subsequent splitting apart of the beams. Merging-beam experiments are notoriously difficult to set up and perform properly. One must carefully limit the angular divergence of the beams to avoid ruinous energy shifts and broadening of the energy distribution (see the next paragraph). Inaccuracies in form factor determinations and detector calibration have also led to serious errors in some experiments. For these reasons, the merging-beam method is less well suited to absolute measurements than the crossed-beam method. Most of the successful merged-beam experiments to date have been on heavy particle–heavy particle collisions, rather than on electron–heavy particle collisions. The most important application of merged electron-ion beams has been in the study of dissociative electron-ion recombination.[32] However, as it becomes more refined, the merged electron-ion technique is being used more and more for other kinds of measurements (see, e.g., Section 7-2-B).

Before moving on, we shall estimate the effects of imperfect beam collimation in merging-beam experiments. Expansion of (7-1-6) for small θ gives

$$E_\theta = E_0 + M_r\left(\frac{E_1 E_2}{m_1 m_2}\right)^{1/2} \theta^2 \qquad (\theta \approx 0^\circ), \tag{7-1-23}$$

where E_0 and E_θ are the values of E_R for the interaction angles 0 and θ, respectively. If we now identify θ here with $\delta\theta$, the spread in the angle of intersection of the beams, (7-1-23) provides an approximate expression for the CM energy spread in an experiment with collinear but slightly diverging beams. If we take $E_1 = E_2 = 5000\,\text{eV}$ and $m_1 = m_2$, we find that $\theta \leqslant 1^\circ$ for the second term in (7-1-23) to have a value of 1 eV or less. The angular divergence of the beams not only increases the minimum value of E_R that would otherwise be obtainable; it also increases δE_R because all angles between 0 and the maximum value of θ are involved. Dunn[17] argues that even if space charge and geometrical spreading are ignored, transverse components of the electron velocity characterized by the temperature of the cathode in the electron source will degrade the energy resolution. Unless draconian measures are employed to collimate the electron beam, the best energy resolution attainable with a cathode temperature of 1000 K is about 0.1 eV.

The *inclined beam* geometry ($0^\circ < \theta < 90^\circ$)[33–35] has been less used than the others. It allows conflicting processes to be distinguished in some cases,[33] but at the sacrifice of some energy range and energy resolution (see Problem 1-7). The interaction geometry is sharply defined. The intersection angle θ usually is chosen to have a value between 10 and 20°.

The *opposed-beam* geometry ($90^\circ < \theta \leqslant 180^\circ$)[36] provides access to higher impact energies than would be available otherwise with given beam accelerators. It is seldom used in atomic physics, where "high energy" is not very high, but is standard in elementary particle physics, with $\theta = 180^\circ$. Problem 7-4 deals with opposed beams.

D. Autoionization[13,16,37–42]

Most of the measurements that we have mentioned thus far in this chapter have mainly concerned the *direct, single ionization* of positive ions. The projectile electron transfers an amount of energy to the target electron that exceeds its binding energy, so that it can escape the ion. In this *direct, or "knockout," process*, the bound electron makes a transition directly into the continuum, with no other excited bound levels being involved. Direct ionization usually occurs from the outer (valence) shell, with single electron ejection. Here we have a first-order process; ejection of more than one electron by this mechanism is a higher-order process, which is less likely. Direct ionization from the inner shell leaves the residual ion in a core-excited state. If this excited state decays by photon emission rather than by Auger emission, the process is usually referred to as direct ionization.

We shall now discuss three *indirect* processes that are known to be important in ionization: excitation-autoionization, ionization-autoionization, and resonant-excitation-double-autoionization (also known as resonant-recombination-double-autoionization). The reader may wish to refer to Sections 6-5 and 6-6, where we discussed autoionization with heavy emphasis on neutral atoms and molecules. Figure 7-1-7 may also prove useful.

Figure 7-1-8 shows evidence for *excitation-autoionization* (*EA*) occurring in Ca^+, Sr^+, and Ba^+, ions, for which the ionization cross section rises sharply at an energy exceeding the outer-shell ionization threshold.[28] The EA process is especially important in electron-ion collisions. It is a two-step reaction in which an inner-shell electron is excited, leaving the ion in a core-excited state, which can subsequently decay by autoionization. Depending on the excitation energy of the core state and the branching pattern of its decay, the ejection of more than one electron may be likely. The electron-impact excitation cross section of *neutrals* is characterized by a rise from zero according to $\kappa_1^{(2l_1+1)}$, where κ_1 is the wave number of the scattered electron and l_1 is the angular momentum quantum number. By contrast, the electron-impact excitation of positive *ions* has a finite cross section at threshold, so the cross section for EA of ions can show a steep increase at the threshold for the excitation process, as in Fig. 7-1-8. We see that the EA process can enhance the total ionization cross section by a large factor.

The calculation of EA cross sections is complicated by the presence of numerous target autoionizing states, which may be strongly coupled by interference with the scattered electron. Furthermore, there can be a strong "postcollision interaction" (Section 6-6) between the excited and scattered electrons, especially near the autoionization threshold, where the scattered electron is moving slowly. Finally, interference effects between direct ionization and autoionization may complicate the issue.

As implied above, direct ejection of an inner-shell electron in an electron-ion collision produces a core-excited state from which a second electron may be ejected by autoionization. Multiple, rather than double, ionization may result if the electron expelled in the initial direct step comes from a deep inner shell. Then the resulting state may decay by the Auger process (Sections 4-8, 6-5, and 6-6) with the ejection of several electrons. The *ionization-autoionization* (IA) mechanism described here is responsible for much of the multiple ionization of complex ions, and becomes relatively more important for the higher ionic charge states. As Phaneuf[39] and Gregory[37] point out, the cross sections for inner-shell ionization show little dependence on the charge state of the target, but those for outer-shell ionization decrease rapidly as the ionic charge increases. [As an ion

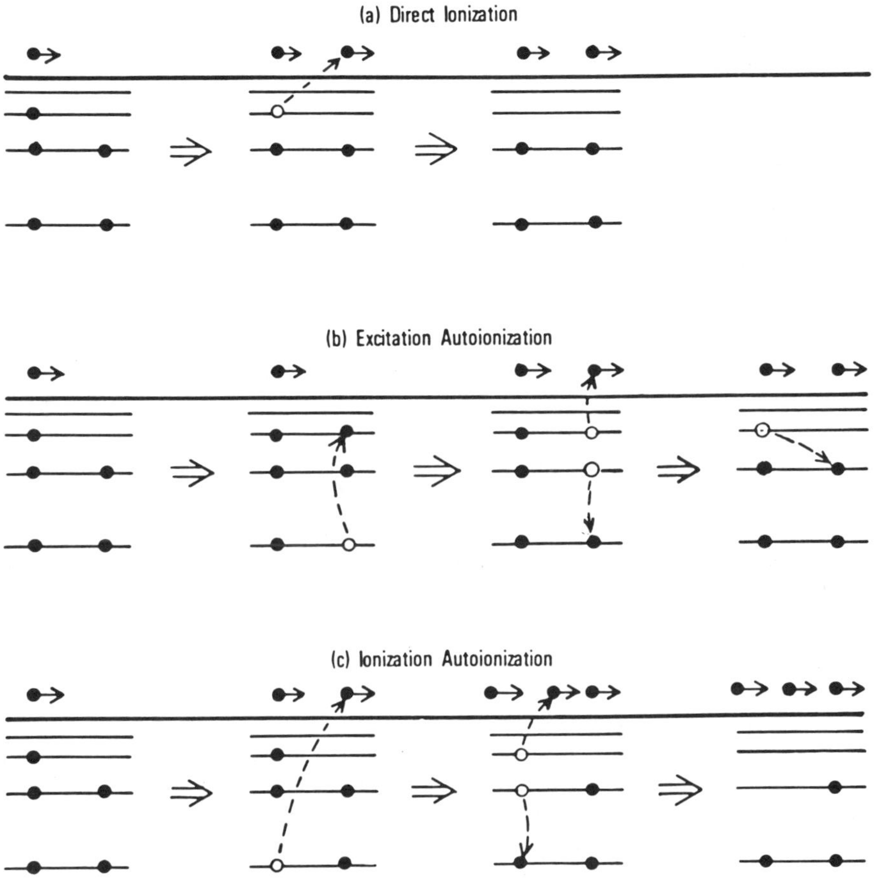

Figure 7-1-7. Three different ionization mechanisms for a five-electron atom or ion.[42] The dots represent electrons, and the circles represent vacancies created as electrons are ejected or promoted (excited) to a state of higher energy. The continuous line at the top of each drawing indicates the ionization limit (i.e., the beginning of the continuum). Here mechanisms (*a*) and (*b*) produce single ionization, whereas (*c*) leads to double ionization. Additional mechanisms are also possible besides the DI, EA, and IA processes shown here.

acquires a higher charge state, it has fewer electrons to screen the nuclear charge, and the electronic binding energy increases (see Fig. 7-1-9). The cross section for direct ionization scales approximately as the inverse square of the binding energy, and thus becomes relatively less important.]

The three ionization mechanisms mentioned up to this point (direct, EA, and IA) are all first-order processes, in that the incoming electron interacts via the two-body Coulomb potential to promote a single target electron. A fourth such process that is important here is *resonant-excitation-double-autoionization* (*REDA*). Let us consider a target ion A^+ with a level $(nl)^*$ whose excitation

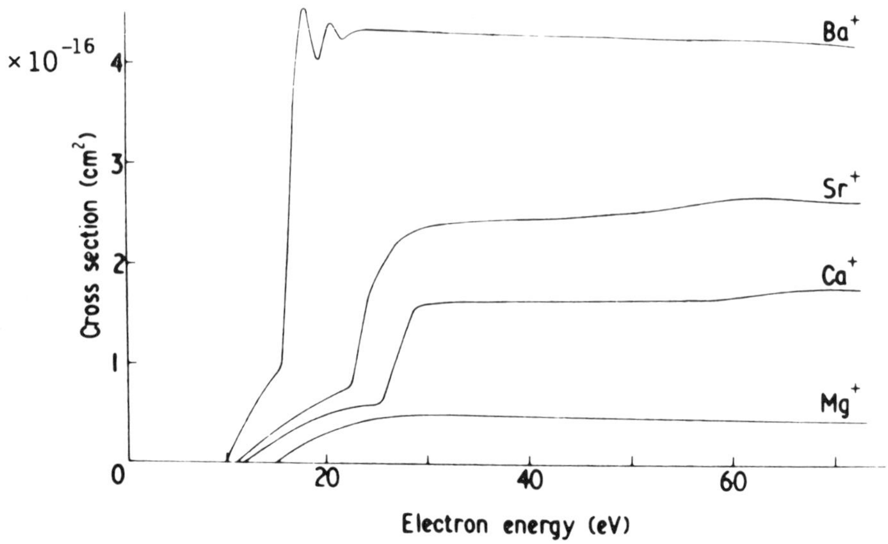

Figure 7-1-8. Experimental integral cross sections for the single ionization of Ba^+, Sr^+, Ca^+, and Mg^+ ions by electron impact.[28] Excitation-autoionization has a strong effect in each case, except for Mg^+.

energy is ΔE, and a projectile electron whose energy at infinity, $\Delta E - \epsilon$, is less than ΔE by the amount ϵ. The projectile will gain kinetic energy as it nears the ion, and at sufficiently close approach, it will be able to excite the level $(nl)^*$. The projectile cannot escape, however; it is bound to the ion with energy ϵ. This process is resonant, because the energy ϵ must match that of a well-defined orbital (nl). The doubly excited ionic state that is produced in this "dielectronic capture" process can decay in several ways. If stabilizing radiation from the state (nl) occurs before autoionization, we say that "dielectronic recombination" (DR) has taken place (see Section 7-4 and especially Fig. 7-4-1). In this section, however, we are interested in the case where the target electron is initially bound in an inner shell, and the ion is left with multiple excited electrons (at least three). Now two electrons are ejected as a third excited electron drops to a vacant inner subshell. There need not be any subsequent emission of radiation if the ion ends in its ground state. This REDA process was proposed by LaGattuta and Hahn in 1981;[43] it gives rise to resonant spikes in the ionization cross section on the low-energy side of the steep rises produced by EA. (It should be noted that if only one electron is ejected in autoionization here, one observes a scattering resonance.)

Additional discussion of autoionization in electron-ion collisions will be deferred until Section 7-1-F. We terminate our present discussion by pointing out that a merged-beams photon-ion technique developed by Lyon et al.[44] provides an extremely powerful method of investigating positive ion autoionizing transitions. This technique is described in Section 8-16-A-9.

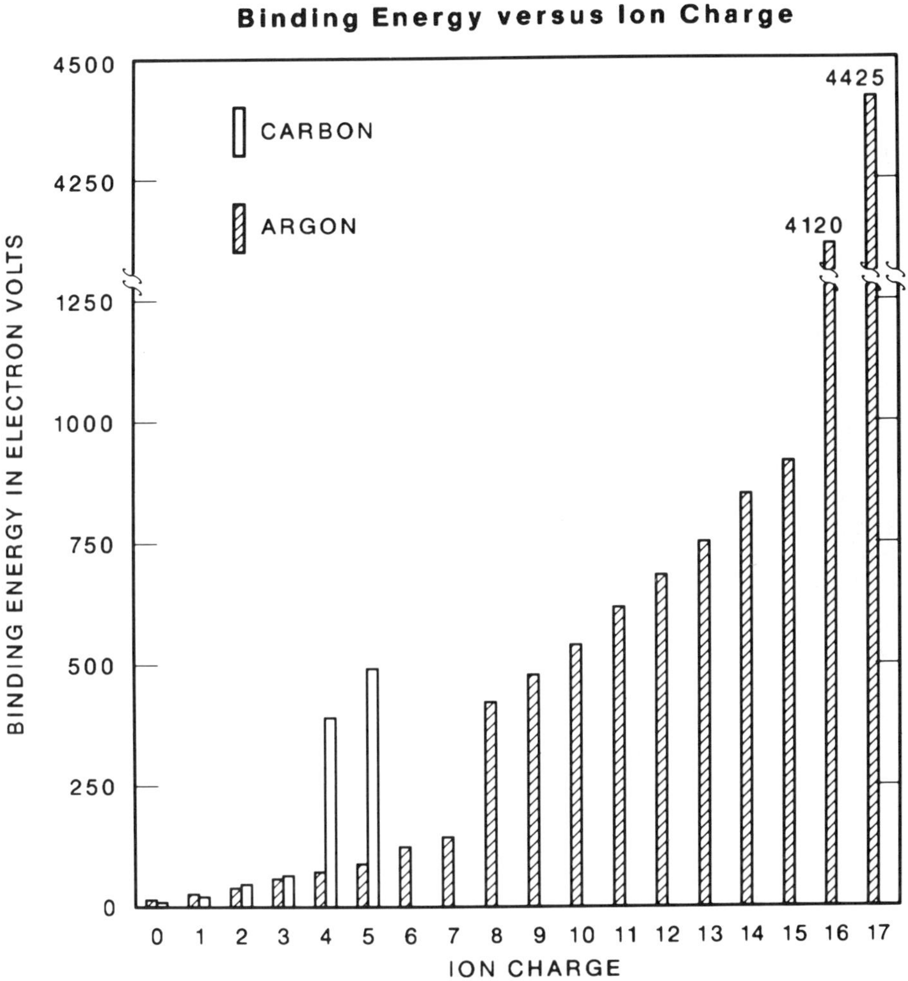

Figure 7-1-9. The electron binding energy versus the charge on a neutral atom or ion, for carbon and argon.[42] The energy required to remove an electron from each structure increases with each successive stage of ionization. The nuclear charge of carbon is +6e, and that of argon is +18e. The minimum total energy required to produce an ion in a given charge state is equal to the sum of the binding energies of the electrons that must be removed. For example, at least 1029 eV is required to produce a bare carbon nucleus from a neutral carbon atom.

E. Multiple Ionization[16,18,39,40,45]

Most of what we need to say about this subject appears in Section 7-1-D, so we confine our discussion here to a few remarks. Cross sections for multiple ionization of ions can be comparable to those for single electron ejection, and consequently the process is quite important in applications. Much of the crossed-beams data on multiple ionization has come from the laboratory at the

University of Giessen.[18,46] In their recent measurements, the Giessen group[47] has utilized a variant of the beam-sweeping technique introduced by Defrance et al.[21]

F. Ionization of Multicharged Ions[13,16,37–41]

In astrophysical and fusion plasmas, multicharged, and even fully stripped ions may be present at high relative concentrations; hence the great interest in collision processes involving such species. Research on electron-impact ionization of multicharged ions is proceeding in several laboratories despite the requirement of large, expensive ion sources. Large PIG sources[48] can provide useful currents ($\geqslant 10^{-8}$ A) of charge-resolved beams of ions with charge states $+5e$, or so. For higher charge states, electron-beam-ionization (EBI)[49] or electron cyclotron resonance (ECR)[50] sources are available. High-power laser sources are also useful.[51]

The apparatus[30] in Fig. 7-1-10 has been used at the Oak Ridge National Laboratory for a number of studies of electron collisions with ions. Of special interest to us is its use for measurement of the cross section for the single ionization of Fe^{15+}.[31] The Fe^{15+} ions were furnished by a large ECR ion source developed by F. W. Meyer.[50] The source is shown in Fig. 7-1-11. The gas that will provide the ions is admitted at the left and is partially ionized by microwave radiation introduced through a wave guide. Atoms and ions (most of which are singly charged) drift slowly out of this "injection stage" and into the "main stage." The latter contains a second microwave input, which heats the electrons

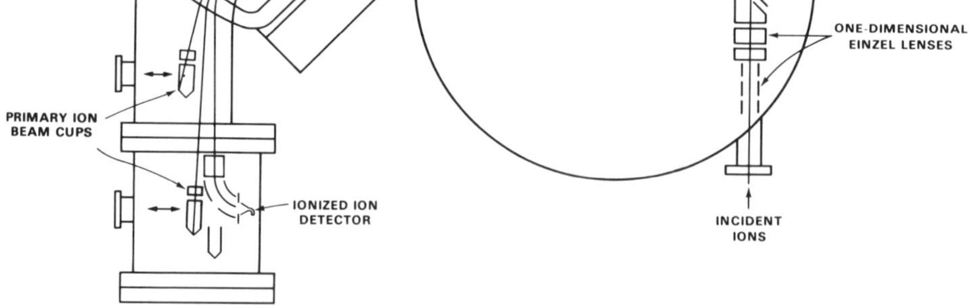

Figure 7-1-10. The Oak Ridge crossed-beam apparatus used for the study of electron-impact ionization of highly charged ions.[30,31] The incident ions are provided by the ion source shown in Fig. 7-1-11.

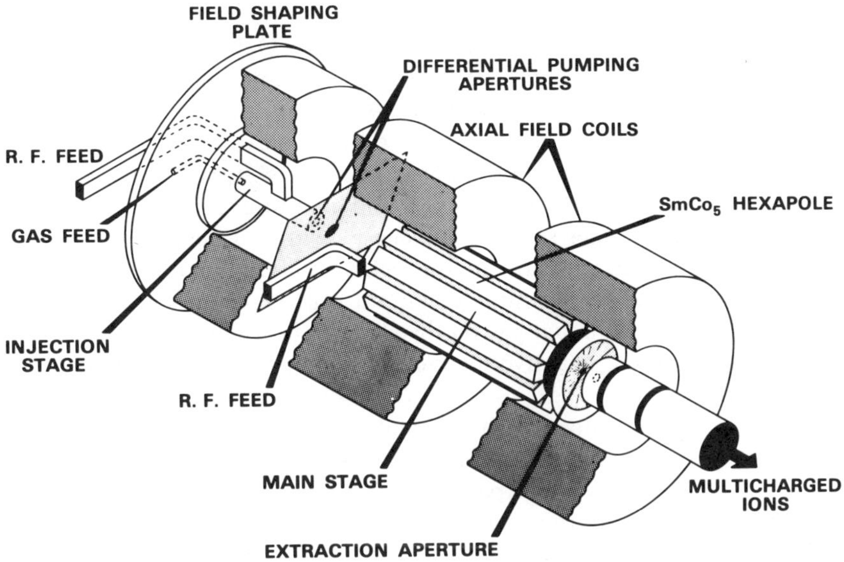

Figure 7-1-11. The Oak Ridge electron cyclotron resonance (ECR) ion source for the production of highly charged positive ions (see Meyer[50] and Phaneuf.[42]) The apparatus is 1.5 m long, and requires about 0.5 kW of microwave power. The magnetic fields reach values as high as about 0.7 T in various regions of the ion source. The total solenoid power is 60 kW. The pressure in the first stage is 10^{-3} to 10^{-4} torr; in the main stage, the pressure is 1 to 6×10^{-6} torr.

that are trapped in this region. A high density of electrons is maintained by an axial magnetic mirror field (which traps the electrons in the axial direction) and a hexapole magnetic field (which inhibits their outward radial motion). The frequency of the microwave radiation in the main stage was set at 10.6 GHz in order to produce resonant heating, the frequency for cyclotron resonance being given by the equation $\omega = eB/m$. About 0.5 kW of microwave power is applied. The atoms and ions require about 1 ms to negotiate the main stage, and during this time they are multiply ionized by successive collisions with the hot, confined electrons. Finally, the multicharged ions are pushed through the extraction aperture by an electric field and focused into a beam for steering and charge-state selection. The gas density in the main stage must be low enough (several 10^{-6} torr) that the ions do not capture electrons from gas atoms before extraction. As indicated above, cross sections have been measured for ions with charge states as high as $+15e$ ($^{26}Fe^{15+}$). Beams of $^{54}Xe^{29+}$ have been accelerated and identified, but their intensity was insufficient for crossed-beams measurements. The difficulty in producing multicharged beams is dramatically illustrated by Fig. 7-1-9.

The crossed-beams apparatus used for the Fe^{15+} measurements and pictured in Fig. 7-1-10 is similar enough to the original Dolder-Harrison-Thonemann apparatus shown in Fig. 7-1-1 that an explanation of its operation is not required. However, we should point out that the Oak Ridge instrument has much improved pumping, and that it uses a pulse-counting electron multiplier for detection of the product ions. The beam-scanning apparatus used to measure the form factor is not shown in Fig. 7-1-10. Typical values of some of the operating parameters may be of interest. The Fe^{15+} current was 0.38 nA, and the electron current 9.8 mA. The background counting rate was 34 per second, and the signal rate 1.4 per second. The energy width of the electron beam was about 1.5 eV, and the pressure in the beam intersection region was $(1.5 \text{ to } 2.0) \times 10^{-9}$ torr. (A lower pressure would not necessarily be beneficial. Positive ions produced in the residual gas by the electron beam provide a space charge that partially offsets the negative space charge of the beam and helps confine the electrons radially.)

The Fe^{15+} ionization measurements[31] showed contributions from excitation-autoionization, but none from resonant-excitation-double-autoionization, contrary to expectation. The role of EA is especially well illustrated in the study of the single ionization of Xe^{6+} made by the Oak Ridge group.[52] Figure 7-1-12 shows their experimental data as dots; the long-dash curve close to the bottom of the figure is a semiempirical estimate of the cross section for direct ionization; the solid curve gives the results of a distorted wave calculation; and the short-dash curve shows the latter results modified to account for the expected CM energy spread. The labeled arrows represent the upper levels of $4d$-nl transitions that autoionize. The experimental data clearly exhibit the characteristic features of excitation-autoionization, and the minor role played here by direct autoionization is in evidence. Similar results were obtained in a recent Oak Ridge study of the single ionization of O^{5+} ions by electron impact.[53]

G. Experimental and Theoretical Data; General Conclusions

Comprehensive compilations of experimental and theoretical cross sections for electron-impact ionization of positive ions have been published by Tawara and Kato[22] and by the Belfast Data Center.[54,55] A useful selection of data appears in the review by Dunn,[16] who discussed calculations that have been made on specific targets and assesses the agreement between theory and experiment. Younger[56] describes theoretical methods (quantal, classical, and semiempirical) and examines the agreement of theory with experiment. General theoretical methods and their application to ionization are treated by Kim.[57] Bottcher[58] discusses numerical methods, and Moores[59] gives a general treatment of the theory of electron-ion collisions. Semiempirical equations for the calculation of ionization cross sections are discussed by Dolder and Peart[13] (see Section 6-13).

Cross sections for the single ionization of singly charged ions have been measured for approximately 40 targets, and a large body of data exists for multiple ionization of ions and for the ionization of multicharged species. Most of these data are of good quality, so it is possible to draw some general

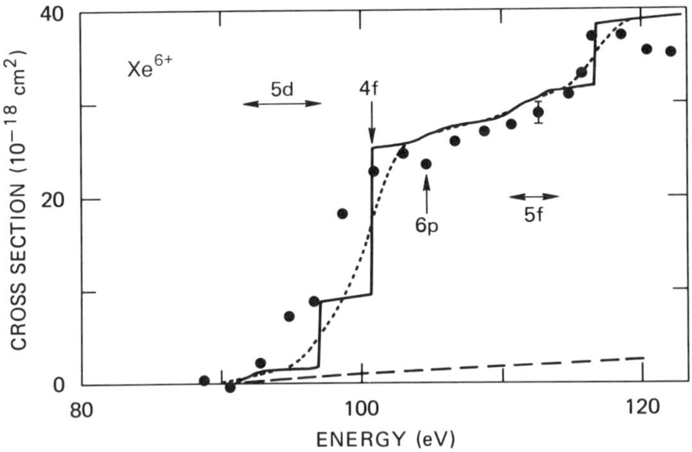

Figure 7-1-12. Integral experimental cross sections for the single ionization of Xe^{6+} ions by electron impact.[31] The large effects of excitation-autoionization that are evident are discussed in the text.

conclusions concerning the electron-impact ionization of ions. According to Phaneuf:[39]

1. Distorted wave theory and semiempirical equations can predict cross sections for direct single ionization with reasonable accuracy.
2. Direct multiple ionization is usually negligible.
3. Excitation-autoionization ranges from negligible to dominant in ionization, depending on the target.
4. Resonant processes may be significant for certain complex ions where the EA process is dominant.
5. Indirect ionization mechanisms usually gain in relative importance as the ionic charge increases along an isoelectronic or isonuclear sequence.
6. Indirect mechanisms generally play a greater role for targets with complex electronic structure than for simple ions.
7. Ionization-autoionization produces most of the ionization in multicharged ions.
8. Multiple ionization of multicharged heavy ions is not always negligible compared to single ionization.

7-2. ELECTRON-IMPACT EXCITATION OF POSITIVE IONS

A. The Radiation Detection Method[13,17,19]

Several years after the first successful electron-ion crossed-beams experiment,[5] on the ionization of He^+, work began on the even more difficult problem of studying electron-impact *excitation* of ions. The first results were published by

Dance et al.[60] in 1966 for the transition He^+ ($1S \to 2S$); cross sections for this process were also published in 1973 by Dolder and Peart.[61] In both experiments, the cross sections were normalized to theory, but they exhibit a much weaker dependence on the electron energy E than theory predicts. This discrepancy apparently has not yet been resolved.[13,17,19]

Following the initial excitation experiments, much effort has gone into developing techniques for measuring absolute excitation cross sections by the radiation detection method. The application of this method to the study of the excitation of neutrals was outlined in Section 5-7-B. All of the difficulties discussed there are present here, and are exacerbated by the weakness of the light signals characteristic of electron-ion crossed-beams measurements. Good absolute excitation cross sections began to be reported in 1973, and we cite the work of Dunn and his colleagues at JILA as especially significant.

An apparatus used by Dunn's group for some very successful measurements is illustrated in Fig. 7-2-1.[62–68] In many respects it closely resembles the ionization apparatus pictured in Figs. 7-1-1 and 7-1-10, so we shall merely point out certain features of the photon handling system. It contains a converging lens to collect the light, an interference filter to disperse it, and a photomultiplier to count the photons. The polarizer is used to measure the degree of polarization of the radiation relative to the electron-beam axis. The scintillator light source permits checks on relative variations in the photomultiplier sensitivity.

As explained in Section 5-7-B, the cross section that is directly determined in excitation measurements of the radiation detection type is the *emission* cross section, σ_{em}. The *excitation* cross section σ_{exc} may be obtained from σ_{em} if various effects such as cascading (Section 5-7-C) can be properly taken into account. In the measurements made by Dunn's group, σ_{em} is calculated from the

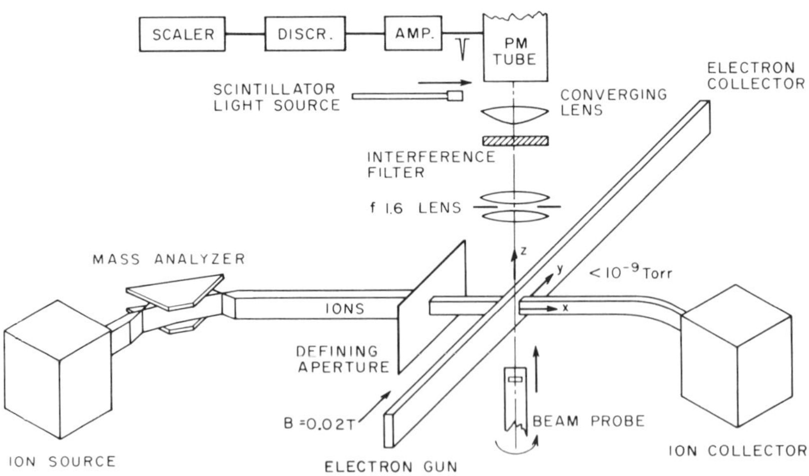

Figure 7-2-1. Apparatus developed by Taylor and Dunn[62] and used by the JILA group to measure cross sections for the excitation of positive ions by electron impact.

experimental parameters by the equation

$$\sigma_{\text{em}} = \frac{S}{Y}\frac{qe^2}{I_e I_i}\frac{v_R F_{\text{exc}}}{D(z_0, \lambda)}, \tag{7-2-1}$$

where S is the photon signal count rate, Y is a radiation anisotropy correction factor, I_e and I_i are the total electron and ion beam currents, qe is the ionic charge, v_R is the relative speed of the beams (assumed to move in the X-Y plane), and F_{exc} is the form factor, which here must include a detection efficiency factor in the overlap integral. Let $\eta(z, \lambda)$ be the relative detection sensitivity profile, which includes corrections for the finite lifetime of the radiating ions; then

$$F_{\text{exc}} = \frac{\int i_e\,dz \int i_i\,dz}{\int i_e i_i \eta(z, \lambda)\,dz}. \tag{7-2-2}$$

In (7-2-1), the factor Y includes allowance for the finite solid angle Ω of the detector system, and is given in terms of the angle θ between the direction of photon emission and the electron-beam axis by the equation

$$Y = \frac{1 - P\langle\cos^2\theta\rangle_\Omega}{1 - \frac{1}{3}P}. \tag{7-2-3}$$

Here P is the polarization of the photons emitted along the observation axis, and $\langle\cos^2\theta\rangle_\Omega$ is the average of $\cos^2\theta$ over the solid angle of detection. (The photons are observed in a cone along the Z axis.) The quantity $D(z_0, \lambda)$ is the absolute average probability that a photon λ emitted in an arbitrary direction from the $z = z_0$ plane inside the collision volume will be recorded. Typically, about 1 count per 10^4 signal photons is the total sensitivity required to perform a satisfactory measurement.

Dunn[17] illustrates the low signal counting rates encountered in excitation measurements by the following example (which he refers to as a "favorable case"!): Take $I_i = 0.5\ \mu\text{A}$, $I_e = 100\ \mu\text{A}$, $F_{\text{exc}} = 0.5\,\text{cm}$, $Y_\Omega = 1$, $D(z_0, \lambda) = 10^{-4}$. Then $S = 1$ count per second for $\sigma_{\text{em}} = 10^{-16}\,\text{cm}^2$, when effects of beam chopping on the duty cycle are taken into account. In a less favorable case, the $2s \rightarrow 2p$ excitation of N^{4+}, the signal counting rate at 52 eV was 0.6 per second against a background of 700 per second, and measurement times of 54 hr were required.[69] In studies of the excitation of the 2^3P level of Li^+, 84 *hr was spent on some points.*[70] Counting times of about 1 hr, however, are required in most measurements.

The apparatus of Fig. 7-2-1 was used by Taylor et al.[66] for studies of the excitation of Be^+ by electron impact. They measured the cross section for emission of the 313-nm resonance radiation doublet that corresponds to the transition Be^+ $(2p) \rightarrow Be^+$ $(2s)$. Their emission cross section shown in Fig. 7-2-2 does not differ greatly from the excitation cross section. If we make the necessary

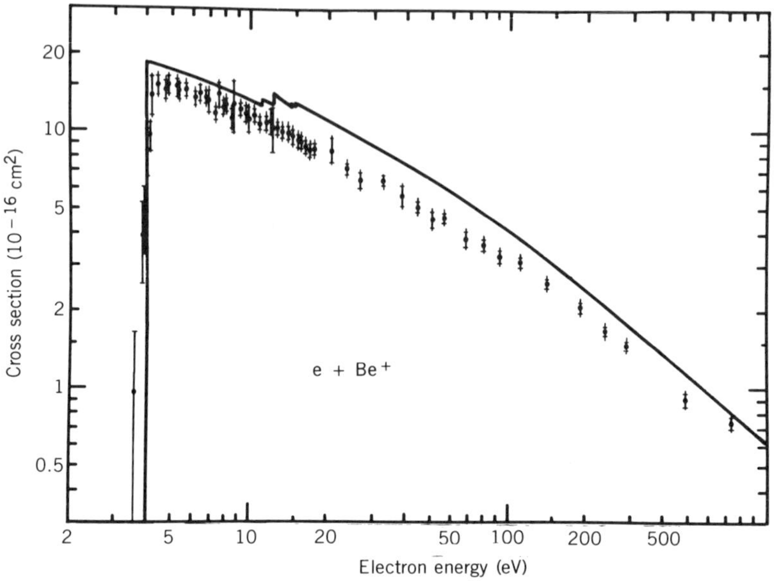

Figure 7-2-2. Cross sections for emission of 313-nm radiation in electron impact on Be^+ ions, as measured by Taylor et al.[66] (symbols). The solid curve shows the results of calculations that include resonance and cascading contributions.[66]

corrections to obtain σ_{exc}, and if we eliminate the effects of the electron energy spread, we obtain an excitation cross section that rises from zero abruptly and has a nonzero value at threshold. This characteristic feature is caused by the Coulomb attraction between the projectile electrons and the target ions, which results in many partial waves affecting the excitation process, even at threshold. The higher-energy behavior of the emission and excitation cross sections here is also characteristic—the cross section usually decreases with increasing electron energy beyond threshold, except for resonances.

A few brief remarks on the theory of electron-impact excitation of positive ions are in order here. At high impact energies, σ_{exc} shows the energy dependence

$$\sigma_{exc} = \frac{A}{E} \ln (BE), \tag{7-2-4}$$

for $\Delta l = 1$, $\Delta s = 0$ transitions. When $\Delta l \neq 1$ and $\Delta s = 0$,

$$\sigma_{exc} = \frac{C}{E}, \tag{7-2-5}$$

but when $\Delta s \neq 0$,

$$\sigma_{exc} = \frac{D}{E^3}. \tag{7-2-6}$$

Here A, B, C, and D are constants for a given excitation process. Notice that (7-2-4) has the form that we encountered in Section 6-14-B in connection with the Bethe-Born approximation. Equation (7-2-4) can be put into another useful form:

$$\sigma_{\text{exc}} = \frac{4\pi a_0^2}{E} \frac{f}{\Delta E} \ln(BE). \tag{7-2-7}$$

Here f is the oscillator strength (Section 5-7-F), ΔE is the energy of the transition, and E is the impact energy, both energies being expressed in rydbergs.

Theoretical results on excitation are usually expressed in terms of the *collision strength* Ω_{if}, which is the particle equivalent of the oscillator strength (f). The collision strength for the transition $i \rightarrow f$ is given in terms of the corresponding excitation cross section σ_{if} by the equation

$$\Omega_{if} = \omega_i E \sigma_{if}, \tag{7-2-8}$$

where E is the impact energy in rydbergs, σ_{if} is the cross section in units of πa_0^2, and ω_i is the statistical weight of the lower (initial) level. The collision strength usually scales as $(Z_{\text{eff}})^{-2}$, where Z_{eff} is an effective charge, frequently taken to be equal to Z, the atomic number of the target.

The "g-bar formula," independently suggested by M. J. Seaton, H. van Regemorter, and A. Burgess, is frequently used to estimate excitation cross sections. If we express E and ΔE again in rydbergs, the g-bar formula is

$$\sigma_{\text{exc}} = \frac{8\pi}{\sqrt{3}} \frac{f}{E} \frac{\bar{g}}{\Delta E} \pi a_0^2, \tag{7-2-9}$$

where $\bar{g}$ is the "effective Gaunt factor." Reliable values of oscillator strengths f are available for many transitions, so the main uncertainty in (7-2-9) is the value of $\bar{g}$ to use for a given transition. Accuracies of a factor of 2 or 3 are generally obtained in the use of the g-bar formula, but in some cases, the accuracy is much poorer, especially if resonances are involved.[13,17,19]

Crossed-beams techniques have been applied to only about 35 electron-ion combinations (mostly atomic ions) in excitation measurements. The results of many of these measurements are discussed and compared with theory by Dunn,[17] Crandall,[19] and Dolder and Peart.[13] Gallagher and Pradhan[71] have compiled all the theoretical data on atomic ions published between about 1955 and the beginning of 1985, and assessed the accuracy of each data set. The amount of theoretical data greatly exceeds that obtained from experiment, and many of the theoretical cross sections (especially those of recent vintage) are of high quality. Comprehensive treatments of the theory have been published by Seaton,[72] Henry,[73] and Itikawa.[74]

B. The Electron Energy-Loss Method

(See Sections 5-3, 5-4, and 5-7-A.) Chutjian and his coworkers at the Jet Propulsion Laboratory[75–78] have developed a technique for studying excitation of ions by electron impact that complements the radiometric method of Section 7-2-A. Using a 90° crossed-beam apparatus, they energy-analyze the electrons emerging from the beam intersection region and obtain energy-loss spectra for electrons inelastically scattered through angles successively set at values between 0 and 17°. They have studied resonance and dipole-forbidden transitions in Cd^+, Zn^+, and Mg^+ at electron energies ranging from 7.5 to 150 eV. The ion beam energy is typically 6 keV, with $I_i = 2$ to $5\,\mu A$ and $n_i = 4$ to $8 \times 10^7/cm^3$. The electron beam current is usually 0.5 to 4.0 μA, with an energy spread of about 0.5 eV. Relative cross sections can be deduced from the peak heights or the areas under the peaks.

Differential cross sections are obtained from a combination of measured and calculated quantities by the equation

$$\frac{d\sigma}{d\Omega} = \frac{R}{I_e I_i} e^2 v_R \frac{F^*}{\Phi\,\Delta\Omega}, \tag{7-2-10}$$

where R is the inelastic count rate and F^* is a form factor. The quantity $\Phi\,\Delta\Omega$ is the product of the transmission-detection efficiency (Φ) of the electron analyzer and the solid angle ($\Delta\Omega$) subtended by the analyzer lens system at the scattering center. The other symbols have the same meaning as in (7-2-1). All the quantities in (7-2-10) are either known or directly measured except for the ratio $F^*/\Phi\,\Delta\Omega$, which is evaluated by calibration with an accurate theoretical result.

The JPL experiments are very difficult because of the small signals—typically, only 1 in 10^{12} of the beam electrons is detected! The crossed-beams inelastic scattering technique is valuable, however, because it provides *differential* cross sections (that provide stringent tests of collision theory) without any cascading effects, and without the need for radiometric measurements. Further, forbidden transitions can be studied, and resonances that autoionize with lifetimes less than about 10^{-12} s can be detected. Disadvantages of the technique include low signal-to-noise ratio, the difficulty of measurements near excitation thresholds (where most interest resides in plasma applications), and the great effort required to obtain integral cross sections from the differential measurements.

Over the angular range covered, the JPL differential excitation cross sections typically drop sharply as the scattering angle is increased (a factor of 100 decrease as θ is varied from a few degrees to 17°). The differential cross sections for resonance transitions usually vary with impact energy in a manner consistent with the predictions of close-coupling theory.

The JPL group has recently developed a new merged-beam apparatus for the measurement of *integral* excitation cross sections.[79] The new instrument, shown in Fig. 7-2-3, provides much greater sensitivity than the crossed-beam apparatus, and permits measurements closer to threshold and on ions with smaller

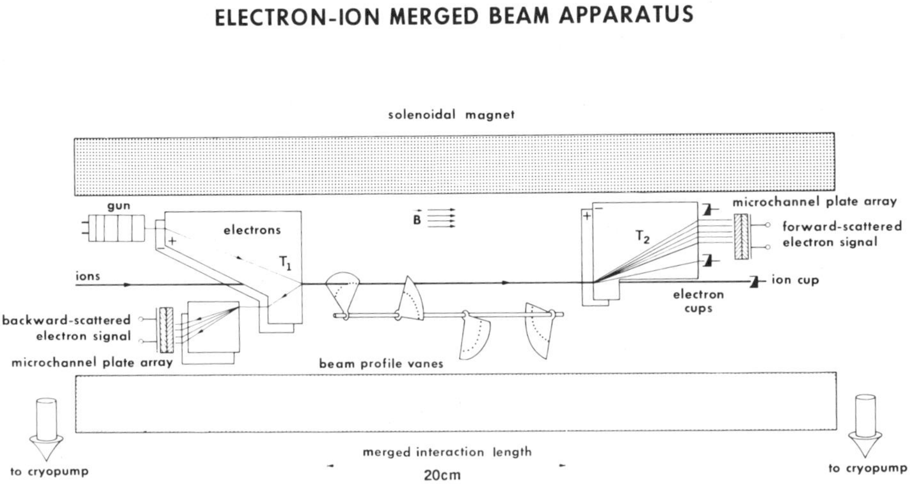

Figure 7-2-3. Merged-beam apparatus used by Chutjian and his associates to measure cross sections for electron-impact excitation of ions.[79]

cross sections. A superconducting solenoid furnishes a uniform and stable magnetic field ($B = 5\,\mathrm{mT}$) for merging the electron and ion beams with the help of trochoidal analyzer T_1. Electrons entering T_1 execute spiraling motion in the combination of the axial magnetic field and a transverse electric field. Under the proper set of conditions, the velocity vector of the emerging electrons is parallel to that of the incoming electrons, but the beam axis is shifted to coincide with the ion beam.[80] After the electrons and ions pass through the merged interaction region, a second trochoidal analyzer T_2 disperses the inelastically scattered electrons with respect to energy, and the resulting energy-loss pattern is detected with a microchannel plate array having a two-dimensional, resistive anode, position-sensitive detector (PSD) (see Section 5-9). The two-dimensional output of the PSD is stored in a large histogramming memory of the computer that controls the experiment.

Additional features of the apparatus in Fig. 7-2-3 are designed to permit absolute cross section measurements to be made. A Faraday cup is set beyond the active area of the PSD to measure superelastic scattering and to determine the number of metastables in the ion beam. Back-scattered electrons are detected by a second PSD, so that true integral cross sections can be obtained. The overlap of the electron and ion beams is measured at four locations along the beam interaction length with a computer-controlled vane system.

Dunn's group at JILA[81] has constructed an apparatus similar to that shown in Fig. 7-2-3. They determine three-dimensional beam profiles with a movable probe consisting of a microchannel plate, a phosphor screen, and a CCD camera connected to the screen via a fiber optics bundle. Cryogenic pumping provides vacuums on the order of 1×10^{-10} torr. The JILA group emphasizes the utility of this apparatus for studies of ionization, as well as excitation.

7-3. DISSOCIATIVE COLLISIONS OF ELECTRONS WITH POSITIVE MOLECULAR IONS

As in Section 5-5, let us consider the various kinds of collisions that may occur, using, for simplicity, the case of electrons colliding with positive diatomic ions (XY^+):

1. $e + XY^+ \rightarrow 2e + X^+ + Y^+$	Dissociative ionization
2. $e + XY^+ \rightarrow X^+ + Y^-$	Ion-pair formation
3. $e + XY^+ \rightarrow e + X + Y^+$	Dissociation into ground-state fragments
4. $e + XY^+ \rightarrow e + X^* + Y^+$	Dissociation into one or more excited fragments (dissociative excitation)
5. $e + XY^+ \rightarrow X + Y$	Dissociative recombination

In all but process 3, the products not indicated as excited (*) may actually be excited. In the case of a polyatomic target ion, dissociation may result in rotationally and vibrationally excited molecular fragments.

Heavy emphasis has been placed on studies of electron collisions with the H_2^+ ion,[82,83] for which measurements have been made on processes 1, 2, 3 + 4, and 5 separately. This ion is of great importance in controlled fusion research, and its simple structure permits accurate calculations to be made of its collision cross sections. Intersecting-beams techniques are especially important for measurements of H_2^+ collisions, because the great chemical activity of this ion renders it unsuitable for afterglow experiments, in which the ions make many collisions in a gas at a pressure of the order of 1 torr. Crossed-, inclined-, and merged-beam geometries have been used in measurements on H_2^+ and its isotopic analogs HD^+, and D_2^+. Beam measurements have the advantage of providing cross sections rather than reaction rates (as in the case of afterglow and other types of swarm experiments). Structure in the energy dependence of the cross section is more easily resolved than for reaction rates, and this fact makes cross sections more valuable for comparisons of experiment with theory. Also, beam methods can cover several orders of magnitude of electron energy, with constant internal target energy, which, of course, swarm methods cannot do.

The matter of the initial state of excitation of the target ion is crucial, as it is known that many cross sections are strongly dependent on the vibrational, as well as the electronic, states of the collision partners. Dunn and his colleagues[83,84] have developed a technique for producing H_2^+ ions with a known distribution of vibrational states. Heppner et al.[85] have used a trapped-ion apparatus in which molecular ions can be confined in a quadrupole ion trap for many hours until they become vibrationally deexcited.

Dolder and Peart[10] have described the early work done on dissociative electron-ion processes up to about the beginning of 1976, and provided a brief update[13] that extends up to the beginning of 1986. Several detailed reviews of dissociative recombination[32,86] have been prepared by Mitchell and McGowan.

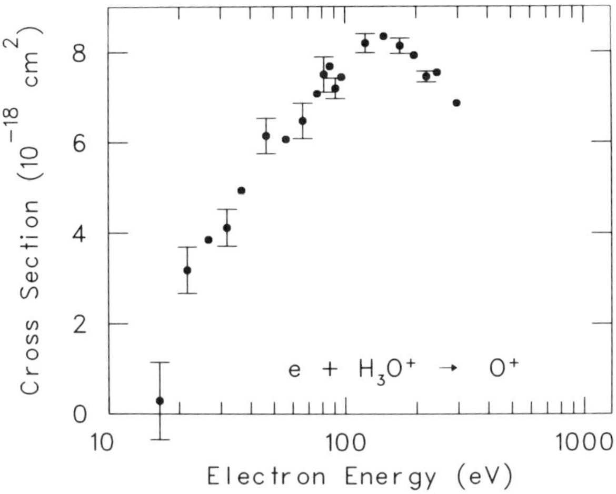

Figure 7-3-1. Cross section for O^+ fragment production in collisions of electrons with H_3O^+ ions, as a function of impact energy.[89]

These reviews give references to experimental measurements, describe experimental techniques, briefly discuss theoretical considerations, and display cross section or reaction rate data for more than 50 ions, including some cluster ions. Their comparison of the many beam and swarm techniques for studying dissociative recombination is especially valuable. The review by Compton and Bardsley, and the discussion of recent theoretical developments by Giusti-Suzor are also recommended.[87]

There are few data on the dissociative ionization process (1)—in his review, Salzborn[18] mentions only CO_2^+ as having been studied in addition to H_2^+. The ion-pair process (2) for H_2^+ has been studied by Peart and Dolder (see the 1975 paper in Ref. 82). The dissociation reactions (3 and 4) have been studied by Dunn and Van Zyl for H_2^+ [83] and for N_2^+ and O_2^+.[88] Several papers in Ref. 82 also deal with dissociation of H_2^+. Schulz et al.[89] have recently studied dissociation of H_3O^+ and D_3O^+ with the crossed-beam apparatus shown in Fig. 7-1-10. They measured cross sections for production of O^+ and OH^+ in e-H_3O^+ collisions, and for D_2O^+ production resulting from impact of electrons on D_3O^+. Their cross section for the formation of O^+ from H_3O^+ is displayed in Fig. 7-3-1.

The dissociative recombination process (5) is of great importance, especially in planetary atmospheres. However, it is fitting to defer further discussion of this topic until the chapter dealing with recombination that appears in the forthcoming sequel to this book.[90] Only at that point will we have at our disposal the necessary understanding of afterglow and swarm techniques, which have contributed so heavily to our understanding of recombination, and only then will we be able to make meaningful comparisons of the data obtained by the various techniques.

In closing this section, we draw attention to the fact that we have not described the details of any experiments on the dissociative processes (1–5). Readers are invited to apply their knowledge of intersecting-beam techniques to the design of such experiments (Problem 7-7).

7-4. DIELECTRONIC RECOMBINATION

A. Introduction

We first encountered dielectronic recombination (DR) near the end of Section 7-1-D in our discussion of the dielectronic capture process. Here a projectile electron with initial kinetic energy $E - \epsilon$ approaches an ion A^+ that has an unoccupied energy level $(nl)^*$ whose excitation energy is ΔE. The projectile is able to excite the ion to level $(nl)^*$ by gaining kinetic energy as it moves in the attractive ionic field. As a result, the electron is trapped in a Rydberg level (nl), so that a doubly excited neutral atom has been produced. Several competing channels are available for the decay of this structure; the one of interest here is dielectronic recombination, which is illustrated schematically in Fig. 7-4-1. The DR process is characterized by (1) the emission of stabilizing photons $h\nu$, (2) the loss of electrons, (3) the loss of ions X^{n+}, (4) the gain of ions $X^{(n-1)+}$, (5) resonant

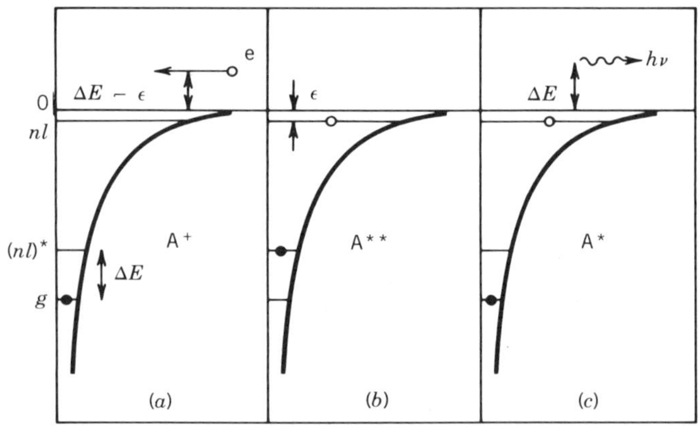

Figure 7-4-1. Sequential steps in the dielectronic recombination of an ion A^+ with a projectile electron having an initial kinetic energy $(\Delta E - \epsilon)$. In (*a*), the ion is in the ground state, *g*. In (*b*), it is in a doubly excited state, the core-excited electron being in the level $(nl)^*$ with the projectile trapped in the Rydberg state (nl). In (*c*), the ion is in a singly excited state, with the projectile electron in level (nl). The ion loses some of its excitation energy by the return of the other electron to the ground level with emission of a photon with energy $h\nu = \Delta E$. Stabilization of the ion by deexcitation of the level (n,l) in (*c*) is less likely. The potential energy curve that determines the energy levels of the structure is shown in each panel of the drawing. Radial distance from the nucleus is measured off to the right from the origin, O. (Adapted from Datz.[97])

structure in the cross section as a function of impact energy, and (6) the ions $X^{(n-1)+}$ being left in specific Rydberg states. In a recent and important review, Dunn[91] cites these characteristics as constituting the "signature" of the DR process. Notice that now we are considering the general case of a multicharged target ion, rather than the singly charged target initially used as an example for simplicity. It is appropriate to deal with dielectronic recombination in this chapter because, thus far, only intersecting-beam measurements have directly yielded cross sections for DR, although beam-gas, beam-plasma, and plasma observation experiments have provided useful information of other types.[91]

The possible importance of the DR process, which is the inverse of autoionization, was first suggested by J. Sayers in 1939, and the first theoretical work on the subject was done by D. R. Bates and H. S. W. Massey following the suggestion of Sayers. These early studies were made in the context of upper atmospheric problems, for which the calculations of Bates and Massey indicated that DR plays an insignificant role. However, about 20 years later, the significance of DR in the solar corona was established by A. Burgess and M. J. Seaton, and now DR is included in essentially all models of hot plasmas. Many theoretical research papers and reviews[92–95] have been written on dielectronic recombination, which is of great importance in astrophysics and controlled fusion because ionization in hot plasmas is balanced mainly by DR.[96] Indeed, in hot plasmas, the ions are atomic, not molecular, so dissociative recombination does not occur. Furthermore, radiative recombination is not important because a fast plasma electron (say, $E \approx 100\,\text{eV}$) passes through the atomic field before the target can emit radiation ($\tau \approx 10^{-8}\,\text{sec}$) and consummate the electron capture. Until 1983, the only experimental information on DR came from spectroscopic observation of plasmas, but subsequently, other methods have provided a new depth of insight into the phenomenon. The recent experimental work has been covered well in several reviews.[13,91,97–100]

B. Theoretical Considerations[91,100]

Measurements of cross sections for DR are extremely difficult, not only because they are generally small, but also, and more important, because the projectile must excite a *discrete* transition in the ion. The high-lying ionic levels are closely spaced, and the DR process is sharply resonant, so just the right impact energy, and a small energy spread, are required to resolve a given excitation. These considerations make DR depend sensitively on the magnitude of plasma microfields and external fields in which the ions are situated. The following discussion will add weight to these qualitative arguments.

The cross section for dielectronic recombination may be written approximately as

$$\sigma_{\text{DR}} = \sigma_c \frac{A_r}{A_a + A_r}, \tag{7-4-1}$$

where σ_c is the dielectronic capture cross section, and the multiplicative factor is a branching ratio between radiative stabilization (rate A_r, in units of 1 per second, and autoionization (rate A_a, in 1 per second). Now note that for exciting the core electron $(nl)^*$ in Fig. 7-4-1, a projectile with energy $(\Delta E + \epsilon)$ will not differ significantly from one with energy $(\Delta E - \epsilon)$ after each has gained kinetic energy in the attractive field of the ion. Hence, in accord with the quantum defect theory,[95] σ_c is simply related to the excitation cross section at threshold, $\sigma_{\text{exc}}^{\text{th}}$, by

$$\sigma_{\text{exc}}^{\text{th}}\, dE = \sigma_c\, \delta E\, dn, \tag{7-4-2}$$

where dE and dn are the differential ranges of energy and principal quantum number, and δE is the resonance width. Use of the Bohr formula for the energy levels of hydrogenic atoms then gives

$$\sigma_c\, \delta E = \frac{2Z^2}{n^3}\, R\sigma_{\text{exc}}^{\text{th}}, \tag{7-4-3}$$

where Z is the ionic charge and R the Rydberg energy.

We can also use detailed balance between capture and autoionization to show that $\sigma_c \sim A_a$, and using (7-4-1), we can demonstrate that

$$\sigma_{\text{DR}}^{nl} = \sigma_0 \cdot 2(2l + 1)\, \frac{A_a(n, l)A_r(n, l)}{A_a(n, l) + A_r(n, l)} \tag{7-4-4}$$

for the cross section for DR into a final Rydberg state (n, l). The quantity σ_0 contains the statistical weights of the initial state and the electron core after the capture of the projectile, as well as the resonance energy width and various constants. The statistical weight, ω_f, of the final Rydberg state (n, l) appears here as $2(2l + 1)$.

As a rule, A_r has little dependence on n, whereas A_a varies as $1/n^3$, as shown by (7-4-3). (The middle panel of Fig. 7-4-1 indicates that electrons captured into high-n states are far from the core-excited electrons, so they should have a small probability of strongly interacting with the core-excited electrons to produce autoionization.) Similarly, A_a will decrease with increasing l, as $1/l^4$ for electric-dipole-allowed excitation. The total cross section for DR involves a sum over many resonances with different n's and l's:

$$\sigma_{\text{DR}}^{t} = \sum_n \sigma_{\text{DR}}^{n} = \sum_n \sum_{l=0}^{n-1} \sigma_{\text{DR}}^{nl}. \tag{7-4-5}$$

For low n's, typically $A_a \approx 10^{15}$ per second and $A_r \approx 10^9$ per second, so that $A_a \gg A_r$ and $\sigma_{\text{DR}}^{n} \sim A_r^n$. Then σ_{DR}^{n} diverges as n^2 because

$$\sum_{l=0}^{n-1} \omega_f = 2n^2. \tag{7-4-6}$$

At higher values of n (of the order of 100), $A_a = A_r$, and for even higher n's, $A_r > A_a$, so that $\sigma_{DR}^n \sim A_a^n \sim 1/n^3$. Similarly, because of the rapid decrease of $A_a(n, l)$ with increasing l, $A_r > A_a$ for high enough values of l. For a given small or medium value of n, and for small values of l, again $A_a \gg A_r$, so that $\sigma_{DR}^{nl} \sim (2l + 1)A_r$. For high l, $\sigma_{DR}^{nl} \sim (2l + 1)\exp(-al^2)$, where a is a constant. Then the cross section becomes negligible for $l > l_c = A_r$. The value of l_c typically is between about 7 and 10.

We see that although the number of resonances that could contribute to DR increases as $2n^2$, only low-angular-momentum states ($l \lesssim l_c$) contribute significantly. However, states with $l > l_c$ form a reservoir of potential contributors that could enhance the DR cross section if their autoionization rates were to be increased. Electric fields in the collision region may, by the Stark effect, mix states of different angular momentum.* States with low A_a in the reservoir mentioned above may be mixed with states of high A_a to produce an effective A_a that is greater than A_r. Thus, to quote Dunn, on whose review[91] this discussion is based, "the field is like a knob by which the participating resonances may be tuned in and out." This effect was predicted for plasma microfields and external fields before it was discovered experimentally. Very large enhancements of σ_{DR} have been observed, as we shall see in the next section. The effects of electric fields on DR are so pronounced as to make dielectronic recombination in fields (DRF) both quantitatively and qualitatively different.

We also note that for $\Delta n \neq 0$, $A_r \sim Z^4$ so that for high enough Z (highly charged ions), $A_r > A_a$ for all n. Even for intermediate values of Z, the important range of n may be restricted to relatively small values. There is no reservoir of unused resonances for these low n, so electric fields will have little or no effect for ions of high enough charge or in situations for which n has low values for other reasons.

Aside from the effects of Stark field mixing on the primary DR process, there are secondary "environmental" effects that can affect DR measurements. They include field and collisional ionization of Rydberg products, and collisional mixing of l states by electrons and ions at high number densities.

C. Experiment

Here we shall briefly review the intersecting-beam methods that have been used to measure dielectronic recombination cross sections and illustrate, with some of the data obtained, the problems posed by electric fields in the collision region. We shall return to dielectronic recombination in the chapter dealing with all mechanisms of recombination in Ref. 90.

*An atom in a given eigenstate in the absence of an electric field will no longer remain in that eigenstate when immersed in a field. Its wave function is then represented as a superposition of eigenfunctions.

1. *Coincidence Experiments*

The first published DR cross section measurement based on the coincidence method was performed by Belić et al.[101] at JILA and reported in 1983. They studied the reaction

$$e + Mg^+(3s) \to Mg(3p, nl) \to Mg(3s, nl) + h\nu \tag{7-4-7}$$

in a 90° crossed-beam geometry with a magnetic field of about 0.02 T to confine the electron beam. This **B** field produced an electric field, $\mathbf{E} = \mathbf{v}_i \times \mathbf{B}$, of about 24 V/cm in the collision region. The Mg^+ ions had a kinetic energy of 2 keV. Product neutral atoms, $Mg(3s, nl)$, were detected in delayed coincidence with the stabilizing photons ($h\nu$) emitted in a direction perpendicular to the plane defined by the electron and ion beams. The cross section was determined by use of the equation

$$\sigma = \frac{(N_c/T)e^2 v_R F}{i_e i_i \xi_\lambda \xi_n}, \tag{7-4-8}$$

where N_c is the number of counts in the coincidence peak recorded during the observation time T, F is the beam overlap factor, and ξ_λ and ξ_n are the detection efficiencies for the photons and neutrals, respectively [cf. (7-1-2), (7-1-14), and (7-2-1)]. The delayed coincidence technique discriminates against the $Mg(3s, nl)$ atoms produced by extraneous processes. Measurement of the flux of the $Mg(3s, nl)$ neutrals was not a simple matter. One problem arose because a weak electric field (36 V/cm) was necessary to separate the neutrals from the parent Mg^+ ions, and Belić et al. calculated that this field would ionize atoms in states with $n \geqslant 64$ (see Notes 6-1 and 6-2). Thus only neutrals with $n < 64$ were believed to be detected. Uncertainties in ξ_n and other factors led to an estimate that the overall accuracy was $\pm 58\%$.

At about the same time, at the University of Western Australia, Williams[102] was measuring cross sections for Ca^+ ions. His apparatus and technique were closely similar to those in the JILA measurement, although he did not use a magnetic field to confine the electron beam. It is believed that only electric fields from space charge were present in Williams's experiment, at the level of about 0.3 V/cm.

The experimental cross sections in both the JILA and the Perth measurements were much greater than those predicted by theory for the no-field case and in much better agreement with theoretical predictions for dielectronic recombination in the presence of electric fields. It was concluded that DRF, not DF, had been observed. To varying extents, this evaluation applies to all beam measurements performed to date.[103]

Another coincidence apparatus has been constructed by Gardner et al.[104] at Harvard and applied to the study of C^{3+}. They used an inclined-beam geometry ($\theta = 45°$) and detected photons in delayed coincidence with product C^{2+} ions.

One possible recombination process here is

$$e + C^{3+}(2s) \rightarrow (C^{2+})^{**}(2p, nl) \rightarrow (C^{2+})^{*}(2s, nl) + h\nu. \tag{7-4-9}$$

The detection efficiency for product ions can be determined much more easily and accurately than can the efficiency of detection of neutrals, unless the neutrals have a high kinetic energy.

2. *Merged-Beam Experiments*

Several laboratories have applied merged-beam techniques to the measurement of dielectronic recombination cross sections. In 1983, Mitchell et al.[105] described measurements at the University of Western Ontario on C^+ ions accelerated to 450 keV energy for merging with an electron beam. The energetic neutral products could be detected with almost 100% efficiency, and the energy resolution was extremely good (about 45 meV). However, the electron beam current was lower, and the signal-to-noise ratio was poorer, than in the contemporaneous coincidence experiments. Some data points required about *100 hours* of integration. The apparatus used here was similar to the merged-beam apparatus shown in Fig. 7-2-3. More recent research at Western Ontario has been described by Mitchell in his 1986 review.[32]

An extensive series of merged-beam recombination measurements on multi-charged ions has been made at the Oak Ridge National Laboratory.[106–109] The first experiment was on B^{2+} and C^{3+} ions and was described by Dittner et al.[106] in 1983.* The ONRL apparatus is shown in Fig. 7-4-2. A unique feature is the extremely high energy of the ion beam; a tandem accelerator is used to produce ions with energies of about 1 MeV/amu (tens of MeV per ion)! A concomitant

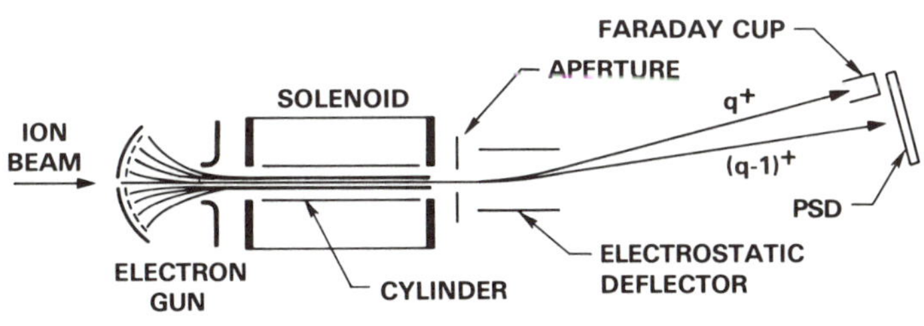

Figure 7-4-2. The ORNL merged-beam apparatus for the study of dielectronic recombination of multicharged ions.[106] Ions enter from the left with tens of MeV kinetic energy and pass through a hole on the axis of the electron gun, which furnishes a current of about 30 mA. The electrons have an energy of approximately 1 keV. The solenoid furnishes an axial magnetic field of about 0.02 T for confinement of the electron beam. The ions are detected in a Faraday cup and with a position-sensitive detector (PSD).

*A preliminary report on this work was given by S. Datz in November 1982 at the IEEE Conference on Application of Accelerators in Research and Industry, North Texas State University.

advantage is that cross sections for charge transfer on background gas by multicharged ions are reduced about four orders of magnitude at MeV/nucleon energies compared to keV/nucleon energies. Also, high ion velocities require high electron velocities, so space-charge limitations on the electron number density in the beam are much less severe. The electron and ion beams interact over a distance of 0.84 m, and the signal-to-noise ratio is high. The only significant problems with the experiment appear to be the low energy resolution (5 eV) and the presence in the collision region of an appreciable electric field whose magnitude is difficult to assess.

All the dielectronic recombination measurements reported to date were on $\Delta n = 0$ transitions, which are characterized by large cross sections. However, recombination events with $\Delta n = 0$ are less important in hot plasmas than events with $\Delta n = 1$, so serious efforts are being directed toward measurement of cross sections for transitions with $\Delta n = 1$.[109]

3. *Current Status and Outlook for the Future*

By this point in our discussion it is clear that our understanding of dielectronic recombination has improved greatly in the last few years, but that it is still in a state of evolution. The effects of fields remain a major obstacle, although they are unimportant for capture into low-n states, which are not at the Rydberg limit. Greater effort to reduce the magnitude of the fields in the collision region and to assess more accurately their strength is called for, along with continued attempts to improve theoretical calculations of dielectronic recombination in fields. We now discuss some of the progress already made along these lines. In Fig. 7-4-3 we display results by Müller et al.[103] that clearly show the dependence

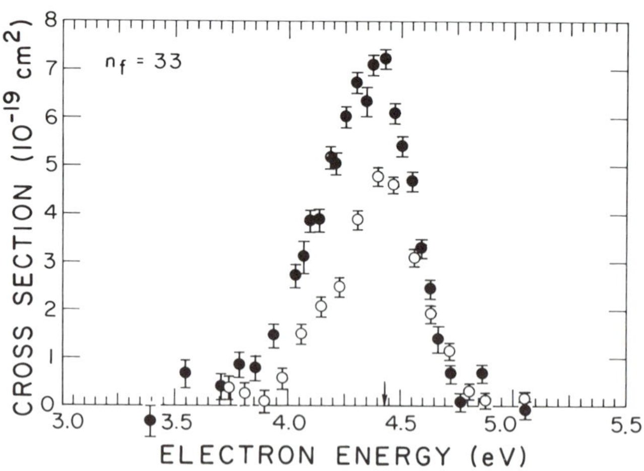

Figure 7-4-3. Measured cross sections versus electron energy for DRF ($n_f = 33$) in Mg^+ for two different electric field strengths in the collision region.[103] The solid symbols are for $E_c = 23.5$ V/cm; the open symbols for $E_r = 7.2$ V/cm.

of the measured (e + Mg^+) DRF cross section on the field strength. Müller et al. also show that the distribution of Rydberg states in the DRF products is strongly dependent on the field. Their experiment utilized a crossed-beam geometry and involved passing the product atoms into a wedge-shaped analyzer/detector that ionizes Rydberg atoms at different positions along their path, depending on their quantum states. The resulting charged particles are then detected with a position-sensitive detector. In discussing their results, Müller et al. use the "quantum number" n_f, defined as

$$n_f = \left(\frac{3.2 \times 10^8}{E_I}\right)^{1/4}. \tag{7-4-10}$$

This number is the principal quantum number of the hydrogenic eigenstate that is barely ionized by the field E_I (here measured in V/cm). Another method of presenting results on the effects of fields is used by Dittner et al.[108] Figure 7-4-4 shows their experimental data on (e + F^{5+}) DRF reaction rate coefficients as a

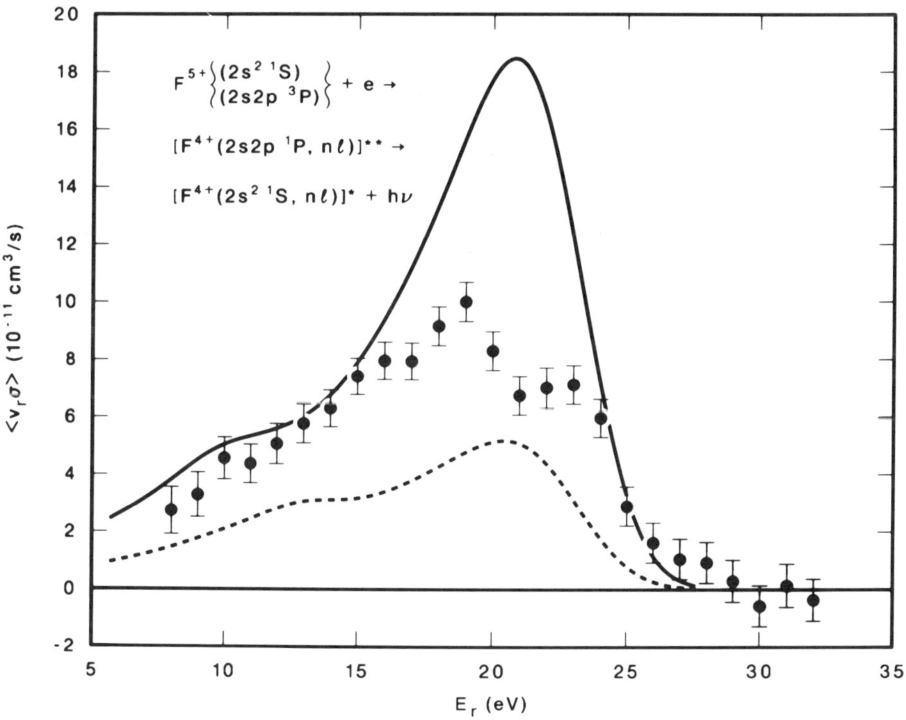

Figure 7-4-4. The DR rate coefficient $\langle v_r\sigma\rangle$ versus the impact energy for F^{5+}.[108] The points show the experimental data. The dashed curve gives the calculated rate coefficient for no field mixing, and the solid curve shows the rate for complete field mixing. The calculations include all resonances up to $n_m = 64$, where n_m is the principal quantum number corresponding to ionization cutoff given semiclassically.

function of impact energy, along with two theoretical curves, the dashed one representing the case of no field mixing of the quantum states and the solid curve the case of complete mixing by an electric field.

In several laboratories, moves are being made to improve the energy resolution of the measurements and to increase the effective ion target density. A significant development is the construction of ion storage rings that will utilize electron-beam cooling of the ions (Note 7-2). The electrons merged with the circulating ion beam are expected to have a transverse energy width of about 0.1 eV and a longitudinal spread of about 1 K.[110] At Oak Ridge, ions will be brought to a very high energy by an electrostatic accelerator with a terminal voltage of 25 MV.

Ion traps will undoubtedly continue to provide uniquely interesting and valuable information, especially on highly charged ions. The Electron Beam Ion Trap (EBIT) at the Lawrence Livermore National Laboratory, for example, has permitted studies of electron-impact excitation as well as dielectronic and radiative recombination of Ba^{46+} ions![111] The experimental observable here is the spectrum of X-rays emitted by the target ions following collisions with energetic electrons.

It is not surprising that observations on plasmas have also been fruitful in the study of dielectronic recombination. Indeed, rate coefficients for DR were first measured in the late 1970s by analyzing the time histories of ions in a θ-pinch plasma and in a tokamak sawtooth discharge. Spectroscopy on the DR satellite lines in highly ionized tokamak plasmas has produced DR rate coefficients for some highly ionized species. Finally, another type of study involves the collision of high-energy ions with gas targets and observations of "resonant transfer and excitation" (RTE). All of these methods are surveyed by Dunn.[91]

In light of the state of flux in the study of dielectronic recombination, it seems appropriate to bring the present discussion to a close. We shall encounter this process again in the chapter on electron-ion and ion-ion recombination in Ref. 90, where we can confidently expect to see substantial advances past the level of understanding described here.

7-5. ELECTRON-IMPACT DETACHMENT OF ELECTRONS FROM NEGATIVE IONS

This process, exemplified by the single- and double-detachment reactions

$$e + H^- \rightarrow H + 2e \tag{7-5-1}$$

and

$$e + H^- \rightarrow H^+ + 3e, \tag{7-5-2}$$

is similar to the ionization of positive ions by electron impact and is studied by similar intersecting-beam techniques. There has been comparatively little research on the electron-impact detachment process, as it apparently has little importance in practical applications, so our treatment will be brief. Electron detachment by photons and heavy particles will receive greater attention, in the chapter on negative ions in Ref. 90.

In H^-, studies have been made both of the single-detachment reaction (7-5-1)[112–115] and the double-detachment reaction (7-5-2).[116,117] Single-detachment cross sections have also been measured for O^-,[118] C^-,[118] and F^-.[119] Interesting structure has been observed in the H^- and O^- cross section dependence on electron energy. Early work, which has been reviewed in detail by Dolder and Peart,[10] showed the sensitivity of theoretical cross sections to the accuracy of the wave functions used in the calculations. Some discrepancies have been uncovered by Defrance et al.[117] in measurements on the reaction (7-5-2), which apparently have not yet been resolved.[13]

Esaulov[120] has provided a scaling law for single detachment that fits the four measured cross sections to within about 20%. He applied the "split-shell model" in which the negative ion is considered to have only one electron that is loosely bound, with binding energy χ (eV). Then classical considerations lead to the scaling law for the cross section σ of the ion:

$$\sigma = \sigma_H \left(\frac{\chi}{0.75} \right)^2, \tag{7-5-3}$$

where σ_H is the cross section for H^- and 0.75 represents the detachment energy of H^- in eV.

7-6. ELECTRON-ION STUDIES WITH TRAPPED IONS AND PLASMAS

Ion traps are used to build up high concentrations of ions and to allow excited ions to become deexcited by providing time for radiation of the excitation energy to occur. Many geometries and many combinations of electric and magnetic fields have been employed in the construction of ion traps having a variety of characteristics. In some cases **E** and **B** fields are used singly; in others, combinations of **E** and **B** fields are employed. Some traps utilize dc fields; others, fields ranging in frequency up into the RF region. The EBIT trap has already been discussed in Section 7-4-C-3. Hasted[121] has described many trap configurations and results obtained by using them in collision experiments. Additional discussion also appears in Refs. 14, 16, 18, 19, 32, and 91. We shall not discuss traps further here. Although they have provided much useful information on collisions of various types, they do not provide absolute cross sections, which are the quantities of main interest to us. Laser cooling and trapping of atoms and ions are discussed in Note 8-4.

Similar considerations apply when we turn to the use of plasma observations for the study of collisions. Sometimes, as in the plasma research described in Section 7-4-C-3, plasma measurements have proved indispensable. However, from our standpoint, direct beam measurements with targets of sharply defined character are to be preferred whenever they can be performed. The reader may consult Refs. 16, 18, 19, 32, and 91 for descriptions of plasma measurements.

PART $\mathscr{B}$. ELECTRON-POINT-CHARGE COLLISIONS

In Part $\mathscr{A}$, we considered electron collisions with targets that possess structure. Here we are interested in collisions of electrons with particles that lack structure (at least as viewed in the energy range relevant to atomic physics), although we must broaden our purview in Section 7-9. In Section 7-7, we sketch the exact derivation of the Coulomb collision cross section as presented by McDowell and Coleman[122], and thereby complete our earlier discussion of Coulomb scattering in Sections 3-8 and 3-17. Symmetry effects that arise when the collision partners are identical are then considered in Section 7-8. Bremsstrahlung are discussed in Section 7-9.

7-7. EXACT QUANTAL DERIVATION OF THE COULOMB SCATTERING CROSS SECTION

Here we study the mutual elastic scattering of two point charges, Ze and ze, with reduced mass M_r. At large separation distance r, their relative velocity is v_0 and their total energy is E. The center-of-mass momentum, k, is related to the wave number of the relative motion, κ, by the equation $k = \hbar\kappa$. Our treatment is nonrelativistic.*

We consider the motion of a single particle of mass M_r in the potential field

$$V(r) = \frac{\beta}{r} \qquad (\beta = Zze^2). \tag{7-7-1}$$

The time-independent wave function ψ_c that describes this motion satisfies the time-independent wave equation

$$\left(\nabla^2 - \frac{2\delta}{r} + k^2\right)\psi_c = 0, \tag{7-7-2}$$

where

$$\delta = M_r\beta = M_r Zze^2; \qquad k^2 = 2M_r E = \hbar^2\kappa^2. \tag{7-7-3}$$

*To avoid confusion about names that appear in the literature, we mention that electromagnetic identical-particle scattering in the lowest order of relativistic quantum field theory is called *Møller scattering*. If one of the identical particles is replaced by its antiparticle, we have *Bhabha scattering*.[123] We shall not be concerned further with these types of collisions.

We make the substitution

$$\psi_c = e^{ikz} f(\mathbf{r}) \tag{7-7-4}$$

and write the resulting equation for $f(\mathbf{r})$ in terms of the parabolic coordinates (ξ, η, ϕ), where

$$\xi = r(1 - \cos\Theta) = r - z, \tag{7-7-5}$$

$$\eta = r(1 + \cos\Theta) = r + z, \tag{7-7-6}$$

$$\phi = \tan^{-1}\frac{y}{x}, \tag{7-7-7}$$

and (r, Θ, ϕ) are spherical polar coordinates. ξ and η range from 0 to ∞, and ϕ from 0 to 2π.

If we take the Z axis to be in the direction of the incoming projectiles, the wave function will be a function of ξ and η only, and

$$\xi \frac{\partial^2 f}{\partial \xi^2} + (1 - ik\xi)\frac{\partial f}{\partial \xi} - \delta f + \eta \frac{\partial^2 f}{\partial \eta^2} + (1 + ik\eta)\frac{\partial f}{\partial \eta} = 0. \tag{7-7-8}$$

Separating variables to require a solution of the form

$$f = f_1(\xi) f_2(\eta) \tag{7-7-9}$$

reduces (7-7-8) to a pair of ordinary differential equations

$$\xi \frac{d^2 f_1}{d\xi^2} + (1 - ik\xi)\frac{df_1}{d\xi} + \gamma_1 f_1 = 0, \tag{7-7-10}$$

$$\eta \frac{d^2 f_2}{d\eta^2} + (1 + ik\eta)\frac{df_2}{d\eta} + \gamma_2 f_2 = 0, \tag{7-7-11}$$

where

$$\gamma_1 + \gamma_2 = -\delta. \tag{7-7-12}$$

Now if we take

$$x = ik\xi, \qquad a = \frac{i\gamma_1}{k}, \qquad b = 1 \tag{7-7-13}$$

in (7-7-10) and

$$x = -ik\eta, \qquad a = -\frac{i\gamma_2}{k}, \qquad b = 1 \tag{7-7-14}$$

in (7-7-11), we can write both members of our pair of equations in the form of the confluent hypergeometric equation[124–127]

$$x\frac{d^2f}{dx^2} + (b - x)\frac{df}{dx} - af = 0. \tag{7-7-15}$$

Seeking a solution of the form

$$f = x^c \sum_{m=0}^{\infty} A_m x^m \tag{7-7-16}$$

leads to the recurrence relation

$$A_{m+1} = \frac{(c + a + m)A_m}{(c + b + m)(c + 1 + m)} \tag{7-7-17}$$

with

$$c = 0 \quad \text{or} \quad 1 - b. \tag{7-7-18}$$

When b is not an integer, two linearly independent solutions of (7-7-15) are the confluent hypergeometric function

$${}_1F_1[a, b, x] \tag{7-7-19}$$

and

$$x^{1-b}{}_1F_1[1 + a - b, 2 - b, x], \tag{7-7-20}$$

where

$${}_1F_1[a, b, x] = \frac{\Gamma(b)}{\Gamma(a)} \sum_{m=0}^{\infty} \frac{\Gamma(a + m)}{\Gamma(b + m)} \frac{x^m}{m!}. \tag{7-7-21}$$

The solution (7-7-19) is a single-valued, analytical function of x and has a finite value at the origin. It is termed the *regular* solution of (7-7-15). For nonintegral values of b, the function (7-7-20) is not analytic at the origin and consequently is an *irregular* solution of (7-7-15).

McDowell and Coleman[122] discuss the more complicated situation where b is an integer. If $b = 1$, (7-7-19) and (7-7-20) are identical, and the regular solution is still given by (7-7-19). Suffice it to say here that in the present context, because the wave function must be finite and single-valued everywhere, only the regular solutions of (7-7-10) and (7-7-11), namely,

$$f_1 = A_1 F_1 \left[\frac{i\gamma_1}{k}, 1, i(kr - \mathbf{k} \cdot \mathbf{r}) \right], \tag{7-7-22}$$

$$f_2 = B_1 F_1 \left[-\frac{i\gamma_2}{k}, 1, -i(kr + \mathbf{k} \cdot \mathbf{r}) \right], \tag{7-7-23}$$

are relevant.

The solution of (7-7-2) that represents asymptotically an incident wave and an *outgoing* scattered wave must contain a term of the form $r^{-1} \exp [ik(r - z)]$, so that f must be a function of ξ but not of η. Hence, for the solution containing an outgoing scattered wave, we have

$$\psi_c^+ = Ae^{i\mathbf{k} \cdot \mathbf{r}} {}_1F_1 \left[-\frac{i\delta}{k}, 1, i(kr - \mathbf{k} \cdot \mathbf{r}) \right], \tag{7-7-24}$$

which is associated with the choice

$$\gamma_1 = -\delta, \qquad \gamma_2 = 0. \tag{7-7-25a}$$

The choice

$$\gamma_1 = 0, \qquad \gamma_2 = -\delta, \tag{7-7-25b}$$

on the other hand, gives the solution that represents asymptotically an incident wave and an *ingoing* scattered wave

$$\psi_c^- = Be^{i\mathbf{k} \cdot \mathbf{r}} {}_1F_1 \left[\frac{i\delta}{k}, 1, -i(kr + \mathbf{k} \cdot \mathbf{r}) \right]. \tag{7-7-26}$$

When the incident wave is normalized to unit amplitude, the two solutions of (7-7-2) become

$$\psi_c^+ = e^{-(1/2)n\pi} \Gamma(1 + in) e^{i\mathbf{k} \cdot \mathbf{r}} {}_1F_1[-in, 1, i(kr - \mathbf{k} \cdot \mathbf{r})], \tag{7-7-27}$$

$$\psi_c^- = e^{-(1/2)n\pi} \Gamma(1 - in) e^{i\mathbf{k} \cdot \mathbf{r}} {}_1F_1[in, 1, -i(kr + \mathbf{k} \cdot \mathbf{r})], \tag{7-7-28}$$

where

$$n = \frac{\delta}{k}. \tag{7-7-29}$$

The confluent hypergeometric function has the asymptotic form

$$
{}_1F_1[a, b, x] \underset{x\to\infty}{\sim} \frac{\Gamma(b)}{\Gamma(b-a)} \exp[-a\ln(-x)]\left[1 - \frac{a(a+1-b)}{x} + \cdots\right]
$$
$$
+ \frac{\Gamma(b)}{\Gamma(a)} \exp[x + (a-b)\ln x]\left[1 + \frac{(1-a)(b-a)}{x} + \cdots\right]. \tag{7-7-30}
$$

When the proper values of a, b, and x are introduced, this solution can be written

$$
\psi_c^+ \underset{r\to\infty}{\sim} \frac{Ae^{(1/2)n\pi}}{\Gamma(1+in)} [I + Sf_c(\Theta)], \tag{7-7-31}
$$

$$
I = \exp[ikz + in\ln k(r-z)]\left[1 + \frac{n^2}{ik(r-z)} + \cdots\right], \tag{7-7-32}
$$

$$
S = \frac{\exp(ikr - in\ln 2kr)}{r}\left[1 + \frac{(1+in)^2}{ik(r-z)} + \cdots\right], \tag{7-7-33}
$$

$$
f_c(\Theta) = \frac{-\delta}{2k^2\, sin^2\,(1/2)\Theta} \exp\left[2i\eta_0^c - in\ln\left(\sin^2\frac{1}{2}\Theta\right)\right], \tag{7-7-34}
$$

with

$$
\eta_0^c = \arg\Gamma(1+in). \tag{7-7-35}
$$

Here I represents the incident wave, S the scattered wave, and $f_c(\Theta)$ is the Coulomb scattering amplitude. The terms in I and S involving $(r-z)^{-1}$ may be neglected for large r except near $\Theta = 0$, $z = r$. Evidently, the excluded angle decreases as the distance from the origin increases. The logarithmic terms in I and S show that the long-range Coulomb potential distorts the incident and scattered waves even at infinity, a property not shared by potentials satisfying the condition

$$
\lim_{r\to\infty} r|V(r)| = 0. \tag{7-7-36}
$$

Finally, we note that the Coulomb scattering intensity obtained quantum mechanically without approximation, namely,

$$
I_c(\Theta) = |f_c(\Theta)|^2 = \frac{\delta^2}{4k^4\sin^4(1/2)\Theta}, \tag{7-7-37}
$$

agrees exactly with the classical differential cross section (3-8-9) and the cross section (3-17-3) obtained with the Born approximation. Recall that δ is defined to be

$$\delta = M_r Zze^2, \tag{7-7-3}$$

where Ze and ze are the charges on the two particles undergoing mutual scattering. The reduced mass is M_r, and the CM momentum is $k = M_r v_0$.

The Coulomb scattering problem can alternatively be solved by making a partial wave expansion

$$\psi_c^+ = \sum_{l=0}^{\infty} (2l+1) i^l \frac{F_l(r)}{kr} P_l(\cos\Theta) \tag{7-7-38}$$

in the wave equation (7-7-2). McDowell and Coleman[122] supply the details of the solution; we merely quote their results:

$$F_l(r) \underset{r\to\infty}{\sim} \exp(i\eta_l^c)\sin(kr - \tfrac{1}{2}l\pi - n\ln 2kr + \eta_l^c), \tag{7-7-39}$$

$$\psi_c^+ = e^{-(1/2)n\pi} \sum_{l=0}^{\infty} i^l (2kr)^l \frac{\Gamma(l+1+in)}{(2l)!} e^{ikr} \times {}_1F_1[l+1+in, 2l+2, -2ikr] P_l(\cos\Theta), \tag{7-7-40}$$

$$\psi_c^+ \underset{r\to\infty}{\sim} \sum_{l=0}^{\infty} (2l+1) i^l \frac{\exp(i\eta_l^c)}{kr} \sin(kr - \tfrac{1}{2}l\pi - n\ln 2kr - \eta_l^c) P_l(\cos\Theta). \tag{7-7-41}$$

Here again, the solution is normalized to unit incident amplitude. There is no profit in the use of partial wave expansions for the case of scattering by a pure Coulomb potential. However, partial wave methods can be useful in the study of "modified Coulomb potentials" (i.e., the Coulomb potential modified by potentials of shorter range).

7-8. COULOMB SCATTERING OF IDENTICAL PARTICLES; THE MOTT FORMULA

As pointed out in Section 3-18 and illustrated by Fig. 7-8-1, a detector in a scattering experiment involving identical particles cannot distinguish between particles that have been scattered and those that have recoiled.[125,128] Hence the differential cross section has the form

$$I(\Theta) = |f(\Theta) \pm f(\pi - \Theta)|^2. \tag{7-8-1}$$

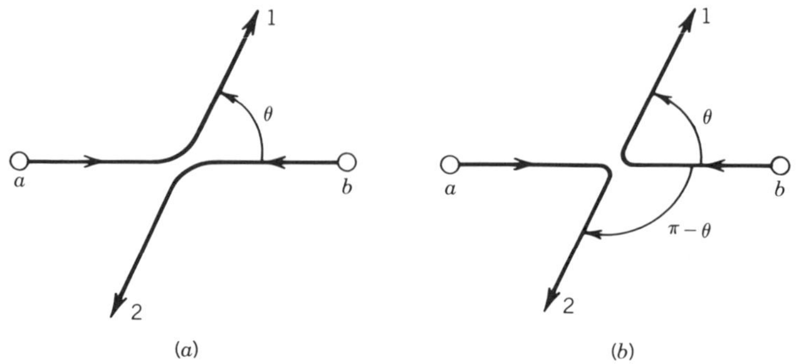

Figure 7-8-1. In the elastic scattering of two identical particles, events (*a*) and (*b*) are indistinguishable.

In the case of Coulomb scattering of two spinless charged particles, the total wave function must be symmetrical, and the minus sign in (7-8-1) must be taken, so that the cross section is given by the *Mott formula*,

$$I_c(\Theta) = \frac{\delta^2}{4k^4 \sin^4 (\Theta/2)} \left[1 + \tan^4 \frac{\Theta}{2} + 2 \tan^2 \frac{\Theta}{2} \cos \left(\frac{\delta}{2k} \log \tan^2 \frac{\Theta}{2} \right) \right], \quad (7\text{-}8\text{-}2)$$

instead of by (7-7-37). The last term in (7-8-2) is produced by interference between the two amplitudes and depends directly on the phase.

7-9. BREMSSTRAHLUNG

A. Classical Treatment

Consider an electron passing a positive point charge and being deflected from its initial trajectory by the Coulomb force that operates. According to classical electromagnetic theory, the accelerated electron emits radiation, usually called *bremsstrahlung* (braking radiation) because it was first observed when fast electrons were stopped in a thick metallic target. The deflected electron has diminished energy, having made a *free-free transition* (i.e., a transition from one unbound state to another). The reverse of the process described here is also a free-free transition, called *inverse bremsstrahlung*. It will be discussed in Section 8-1-C-5.

As described by Fano and Fano,[129] the electron motion during the acceleration constitutes a variable current that is nonperiodic. Hence its Fourier representation contains, in principle, components with all frequencies, and the bremsstrahlung spectrum is continuous. An upper limit $\omega_{\max}$ exists in the frequency spectrum of the emitted radiation because the photon energy $\hbar\omega$ cannot exceed the initial energy of the electron. The emission becomes more intense and the spectrum extends toward higher frequencies as the initial energy of the electron increases. Bremsstrahlung is particularly important in fusion, where the electron energy reaches values of a few keV, and the free-free emission is mainly in the X-ray region. The term "bremsstrahlung" is sometimes used synonymously with *continuous X-rays.* For electron energies above 100 MeV, bremsstrahlung becomes the main mechanism for the dissipation of an electron's energy as it passes through matter.

The classical theory of bremsstrahlung[130–133] is quite involved, and we shall content ourselves here with some qualitative remarks. Jackson[130] launches his discussion with the equation for the energy radiated per unit solid angle $d\Omega$ per unit frequency interval $d\omega$ by a particle of charge ze traveling with velocity $\mathbf{v} = c\boldsymbol{\beta}$ during a collision with a fixed point charge Ze:

$$\frac{d^2 I}{d\omega\, d\Omega} = \frac{z^2 e^2}{4\pi^2 c} \left| \int \frac{d}{dt} \left[\frac{\mathbf{n} \times (\mathbf{n} \times \boldsymbol{\beta})}{1 - \mathbf{n} \cdot \boldsymbol{\beta}} \right] e^{i\omega(t - \mathbf{n} \cdot \mathbf{r}(t)/c)}\, dt \right|^2. \tag{7-9-1}$$

Here $\mathbf{n}$ is a unit vector pointing in the observation direction, and $\mathbf{r}(t)$ is the coordinate of the particle:

$$\mathbf{r}(t) = \mathbf{r}(0) + \int_0^t c\boldsymbol{\beta}(t')\, dt'. \tag{7-9-2}$$

Jackson then proceeds to treat various situations, relativistic and nonrelativistic, and makes contact with the quantal treatment at several points. He emphasizes that the most common process of interest historically has been the collision of a fast electron with a neutral atom, for which the analysis based on a simple two-particle Coulomb interaction does not strictly apply. However, the nucleus is more effective in deflecting the projectile than the atomic electrons, and the point-charge point-charge model, which involves the use of the Coulomb cross section (7-7-37), does have some validity. The main effect of the atomic electrons is in screening the nucleus, and Jackson treats this effect. (The action of the atomic electrons in directly accelerating the projectile can be ignored in the dipole approximation and at low energies.) There is always a frequency of emitted radiation below which screening effects are important. For example, if 100-keV electrons are incident on a gold target, screening is significant for $\omega < 0.04\omega_{\max}$.[130]

Jackson[130,134] also treats the emission of bremsstrahlung in a relativistic collision as the scattering of the virtual quanta in a nuclear Coulomb field by the incident electron in the electron rest frame (see Section 5-6). In this frame of reference, the heavy ion approaches the stationary light electron at high speed, and the nuclear Coulomb field, because of the Lorentz transformation, resembles a pulse of electromagnetic radiation. The transverse **E** and **B** fields of this pulse have a broad frequency spectrum, with virtual quanta present in numbers, per unit frequency interval, inversely proportional to the frequency, ω'.

The virtual photons are scattered by the stationary electron into a wide angular distribution that is proportional to $(1 + \cos^2 \theta')/2$ at low frequencies. For highly relativistic particle motions, the photons, which for the most part have low frequencies (ω') and large emission angles (θ'), are Doppler-shifted to appear in the laboratory at medium to high frequencies $\gamma\omega'$ and small angles (θ) of the order of $1/\gamma$. Figure 7-9-1 illustrates the bremsstrahlung as viewed in the two reference frames used in this discussion.

The exact form of the results provided by the method of virtual quanta depends on the equation used to calculate the cross section for scattering of the virtual quanta in the electron rest frame. If the classical Thomson cross section (see Section 8-1) is employed, the cross section for the energy radiated per unit frequency interval and per unit solid angle *in the projectile rest frame* is

$$\frac{d\chi'}{d\omega'\, d\Omega'} \simeq \frac{1}{\pi} \frac{Z^2e^2}{c} \left(\frac{z^2e^2}{mc^2}\right)^2 (1 + \cos^2 \theta') \ln \frac{\lambda\gamma mc^2}{\hbar\omega'}, \tag{7-9-3}$$

where, for generality, we endow the less massive particle with charge ze and mass m, and the more massive particle with charge Ze. The angle of photon emission is θ', $\gamma = (1 - \beta^2)^{-1/2}$, and λ is a number of order unity. The corresponding *bremsstrahlung radiation cross section in the Lab frame* is

$$\frac{d\chi}{d\omega\, d\Omega} \simeq \frac{16}{3} \frac{Z^2e^2}{c} \left(\frac{z^2e^2}{mc^2}\right)^2 \ln \frac{2\lambda\gamma^2mc^2}{\hbar\omega(1 + \gamma^2\theta^2)} \frac{3\gamma^2(1 + \gamma^4\theta^4)}{2\pi(1 + \gamma^2\theta^2)^4}. \tag{7-9-4}$$

Jackson[130] discusses these two cross sections, along with the results obtained through the use of quantal calculations of the scattering of the virtual quanta.

Jackson[134] also mentions an apparent difficulty in the virtual quanta calculation that was resolved by von Weizsäcker. Among the virtual quanta emitted by the heavy particle are some with energies $\hbar\omega'$ greatly exceeding the rest energy of the light particle. When these quanta are scattered and trans-

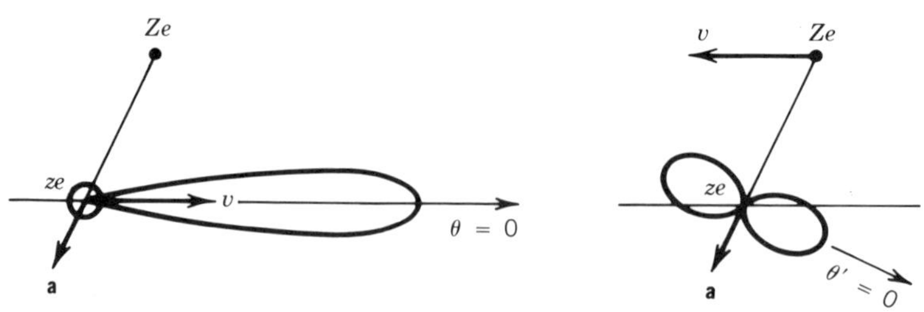

Figure 7-9-1. Bremsstrahlung emission as viewed in (*a*) the Lab frame, that is, the rest frame of the heavy nucleus Ze, and (*b*) the rest frame of the light projectile ze. In (*a*), the radiation is emitted predominantly in the forward direction ($\theta = 0$), whereas in (*b*), the field of the nucleus is represented by a pulse of radiation that is scattered as dipole radiation by the light particle. (Adapted from Jackson.[130,134])

formed to the Lab frame, they would be expected to have energies on the order of $\gamma\hbar\omega'$, much larger than γmc^2. Here we have an apparent violation of energy conservation, because the greatest possible photon energy is $(\gamma - 1)mc^2$. But as von Weizsäcker pointed out, the Compton cross section (see Section 8-1), not the recoilless Thomson cross section, must be used to describe the scattering of quanta with energies $\hbar\omega'$ greater than mc^2 in the light particle's rest frame. When all the details are put into the calculation, energy is indeed conserved in the Lab frame because (1) the Compton cross section decreases with increasing photon energy, (2) recoil reduces the scattered photon energy, and (3) the angular distribution is changed.

Equation (7-9-4) shows that apart from a slowly varying logarithmic factor, the bremsstrahlung intensity is proportional to $z^2(zZ/m)^2 = z^4Z^2/m^2$. Hence the bremsstrahlung energy loss process is most important when the projectile mass, m, is small (i.e., for electrons) and when the medium contains high-Z atoms. The role of bremsstrahlung in hot plasmas is currently receiving a great deal of attention.[135]

B. The Quantal Treatment of Bremsstrahlung

The emission of bremsstrahlung has been treated quantally at various levels up to and including relativistic quantum field theory.[123,136–141] The famous Bethe-Heitler formula, based on the Born approximation, was published by H. A. Bethe and W. Heitler in 1934. This formula and later theoretical results are discussed by Maximon et al.[141] Because the radiative bremsstrahlung process involves the coupling of the electron to the electromagnetic field of the photon that is emitted, the cross sections are of the order of 1/137 times the cross section for elastic scattering. A photon is emitted in only a small fraction of the encounters, whereas in a classical treatment, radiation is expected in every deflection. However, quantum mechanics and classical mechanics give cross sections of the same order of magnitude when averaged over all collisions, although the spectral distributions are quite different. The observed spectral distribution of bremsstrahlung typically has the shape shown in Fig. 7-9-2. Of course, sharp spikes representing *characteristic X-rays* are generally superimposed on the continuous spectra. These X-rays are produced when electrons fill

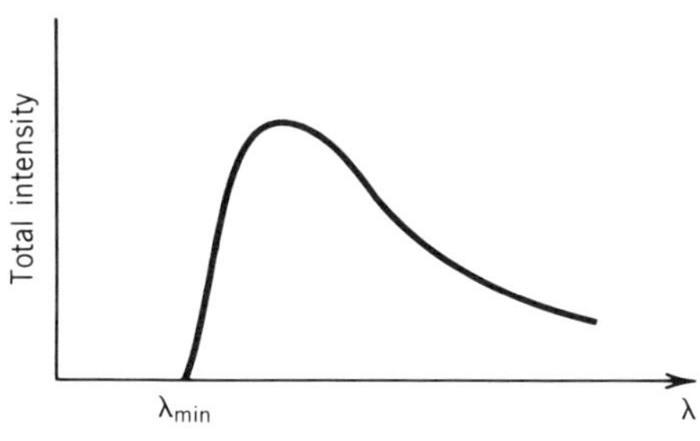

Figure 7-9-2. The general shape of the bremsstrahlung spectrum. [Adapted from F. H. Read, *Electromagnetic Radiation*, Wiley, New York (1980).]

vacancies in the inner atomic shells that are produced by electron impact or some other means. The high-energy photoeffect and bremsstrahlung have been treated recently by Yin et al.[142]

C. Data

Gallagher[143] has published a bibliography on free-free transitions in atoms and molecules that will lead the reader to experimental and theoretical data published before 1977. Seltzer and Berger[144] have tabulated a comprehensive set of theoretical bremsstrahlung cross sections that are differential in the energy of the emitted photons. The set includes data for electrons with energies ranging from 1 keV to 10 GeV incident on neutral atoms with atomic numbers $Z = 1$ to 100. Yamabe et al.[145] have obtained some interesting results with an electron drift tube in which the mean electron energy could be varied from 1.2 to 5.4 eV. Measurements were made of the absolute intensity of the continuum radiation at 500, 650, and 1300 nm. About 1 in 10^8 collisions of the electrons with the argon atoms filling the drift tube results in the emission of a visible photon, and about 1 in 10^4 of these photons is selected by wavelength and reaches the detector. It should be noted that Yamabe et al. did not follow the conventional procedure of expressing the free-free emission in terms of a spectral intensity (i.e., a cross section or excitation coefficient per unit spectral bandwidth). Instead, they define cross sections and rate coefficients per fractional bandwidth. By doing so, they obtain numbers that are independent of the units used to measure bandwidth, and cross sections or excitation coefficients that can be readily compared with data for the production of excited states that emit atomic lines or molecular bands.

Another useful kind of data that appears in the literature on free-free transitions is the *Gaunt factor*. This quantity is the ratio of the full quantal cross section to the cross section provided by the semiclassical formula developed by H. A. Kramers.

7-10. NOTES

7-1. Effects of Space Charge. Dolder[9] has collected some approximate equations that permit numerical estimates of the effects of space charge on beams of charged particles. First consider a *cylindrical beam of ions*, all of which initially are moving in the same direction. The initial beam diameter, d_0, is related to the space-charge-expanded beam diameter, d, a distance $l(m)$ downstream by the approximate expression

$$\frac{d}{d_0} \approx \left(4.1 \times 10^{10} \frac{\sqrt{M/n} I_\rho l^2}{U^{3/2}}\right)^{1/2}, \qquad (7\text{-}10\text{-}1)$$

where M is the ion mass relative to atomic hydrogen, I_ρ the initial current density (in A/cm^2), and U the beam voltage. The charge on each ion is ne. This solution should be used only when $2 \lesssim d/d_0 \lesssim 5$, but Dolder gives references to solutions for more extreme cases. The corresponding solution for the spread of an *ion beam of rectangular cross section* is

$$\frac{h}{h_0} \approx \frac{2.04 \times 10^{10} \sqrt{M/n} I_\rho l^2}{U^{3/2}} + 1, \tag{7-10-2}$$

where h_0 and h are the initial and final beam heights, and it is assumed that the beam width is much greater than h.

For cylindrical electron beams, the corresponding solution is

$$\frac{d_1}{d_0} \approx \left(9.6 \times 10^8 \frac{I_\rho l^2}{U^{3/2}}\right)^{1/2} \tag{7-10-3}$$

when $2 \lesssim d_1/d_0 \lesssim 5$, and for *rectangular electron beams*,

$$\frac{h_1}{h_0} \approx \frac{4.8 \times 10^8 I_\rho l^2}{U^{3/2}} + 1. \tag{7-10-4}$$

Dolder[9] also shows that in crossed-beam experiments, electric fields produced by space charge in the electron beam can deflect the trajectories of ions that pass through the electron beam and prevent them from being collected. In this case the electron beam current must be reduced.

7-2. *The Cooling of Ion Beams.* The high cost of particle physics accelerator research has led to colliding-beam experiments utilizing particles circulating in storage rings, synchrotron-like facilities with magnetic fields fixed in time. The particles are accumulated over time to provide a useful luminosity (event rate per unit cross section), and the quality of the circulating beam is improved by increasing its phase-space density. Three methods of doing so are discussed by Wilson[146] and by Cole and Mills[147]: the emission of synchrotron radiation, stochastic cooling,* and electron cooling. We discuss the latter mechanism here.

Storage rings are constructed with opposing straight sections connected by circular arcs. Electrons from an external gun are injected tangentially into the electron cooling straight section near the point where the circulating ions enter it, and the electrons are deflected back out when they reach the end of the straight section. The electrons have the same speed as the ions, and move parallel to them at a slightly greater distance from the

*An additional review of stochastic cooling is D. Möhl, G. Petrucci, L. Thorndahl, and S. van der Meer, *Phys. Rep.* **58**, 73 (1980).

center of the ring. The electrons increase the phase density of the ion beam and damp its motion through Coulomb interactions, because the electrons take away most of the random ionic energy in collisions with the heavy particles. The electron beam has lower longitudinal and transverse temperatures than the ion beam, and the latter is cooled as the two beams come into temperature equilibrium. As indicated in Section 7-4-C-3, it is possible to obtain, by electron cooling, ion beams with greatly reduced translational and longitudinal energy spreads.

7-11. PROBLEMS

Solutions for most of these problems appear in the references indicated.

7-1. A cylindrical beam of collimated protons travels in a vacuum with velocity v. The beam radius is R, and the number density of the protons is n. Find the forces on a proton at radius r, and investigate the stability of the beam qualitatively. (Cronin et al.,[148] p. 20.)

7-2. A cylinder of radius R and length L is filled with an ionized gas of charge density ρ which moves parallel to the axis with velocity v.

(a) Find the magnetic field at a distance r from the axis.

(b) Consider a parallel beam of fast, but nonrelativistic, protons directed into this cylinder with their initial velocity parallel to its axis. In such an apparatus, the protons can be focused to a point on the axis. Calculate the focal length, f (the distance from the end of the cylinder to the focus). Neglect electrostatic and relativistic effects. Assume that $f \gg L$. (Chen,[149] p. 100.)

7-3. The electrons in a long cylinder of completely ionized hydrogen gas are flowing along the cylinder with a mean velocity 10^5 cm/sec in a circular beam of 50 cm diameter. The total beam current is 10^4 A. What are the magnitude and direction of the force F on an electron at the periphery of the beam? (Chen,[149] p. 115.)

7-4. **(a)** Consider an initially parallel beam of He^+ ions with rectangular cross section 0.1 cm × 1.0 cm. The beam current is 10^{-6} A, and the beam energy is 5 keV. Show that the initial beam width of 0.1 cm expands to about 0.21 cm in a path 1 m long.

(b) Next consider an electron beam whose current is 10^{-3} A at an energy of 100 eV. Its cross section measures 0.2 cm × 2.0 cm. Show that the height of the beam will expand to about 0.224 cm in a path of only 1 cm. (Dolder,[9] p. 258.)

7-5. **(a)** Suppose that H^+ and H^- beams are merged with $\theta = 0$, the energy of one beam (No. 2) being held fixed at 5 keV. Let the energy of the other beam (No. 1) be varied from 3 to 7 keV, and plot the resulting variation in the interaction energy E_R. From your plot, demonstrate how the CM energy spread δE_R in a collision can be much smaller than the spread of the Lab energy of either beam.

(b) Repeat for $\theta = 8.5°$. (See Dolder and Peart,[12] p. 1288.)

7-6. Consider merged beams consisting of ions with $m_1 = m_2$ and $E_1 \approx E_2 \approx 5\,\text{keV}$. Show that an energy resolution of $10^{-3}\,\text{eV}$ implies a collimation angle of 6×10^{-4} rad. (See Dolder and Peart,[12] p. 1289.)

7-7. **(a)** Show that the CM energy of colliding particles of masses m_1 and m_2, energies E_1 and E_2, and momenta p_1 and p_2 is

$$E_R = [2(E_1E_2 + p_1p_2c^2) + (m_1^2 + m_2^2)c^4]^{1/2}. \quad (7\text{-}11\text{-}1)$$

The particles are moving in opposite directions with relativistic speeds.

(b) Now compare E_R for a 400-GeV proton colliding with a proton at rest, with E_R for the same proton colliding with one having 10 GeV of energy and moving in the opposite direction. (Segrè,[150] p. 167.)

7-8. Consider an H_2^+ ion with Lab kinetic energy 10 keV undergoing dissociation in flight, with the dissociation products H^+ and H traveling along the original line of flight. If we take the dissociation energy in the CM frame to be 2 eV, what is the dissociation energy in the Lab frame?

7-9. Consider a molecular ion of mass M that dissociates in flight into fragments of masses m_1 and m_2, with reduced mass M_r. The initial Lab energy of the ion is E_0, and the CM dissociation energy is ΔE_{CM}. The CM velocity of fragment m_1 makes an angle θ with the velocity of the CM. Show that the Lab energies of the fragments are

$$\begin{aligned} \Delta E_1 &= \frac{M_r}{m_1}\,\Delta E_{\text{CM}} + (\cos\theta)\left(\frac{4M_r}{M}\,E_0\,\Delta E_{\text{CM}}\right)^{1/2}, \\ \Delta E_2 &= \frac{M_r}{m_2}\,\Delta E_{\text{CM}} - (\cos\theta)\left(\frac{4M_r}{M}\,E_0\,\Delta E_{\text{CM}}\right)^{1/2}. \end{aligned} \quad (7\text{-}11\text{-}2)$$

(Brouillard and Claeys,[26] pp. 427–428.)

7-10. Analyze the motion of the electrons in the trochoidal analyzer T_1 of Fig. 7-2-3.

7-11. Parallel-plate electrostatic analyzers of the type shown in Fig. 7-1-10 are frequently used in atomic collisions experiments. Suppose that a fast electron enters such a device through a hole in the positive plate at an angle α with respect to this plate. Let the negative plate be at a voltage V with respect to the positive plate, and take d to be the plate spacing. Using relativistic mechanics, calculate the trajectory of the electron inside the analyzer. (Cronin et al.,[148] p. 13.)

7-12. Design experiments to measure cross sections for the dissociative processes 1–5 listed at the beginning of Section 7-3.

7-13. H^+ and He^{2+} ions are passed through a thin gold foil with the same initial kinetic energy. What is the ratio of their nonrelativistic Coulomb scattering? (Chen,[149] p. 278.)

7-14. Show that the height of the Coulomb barrier at the nuclear surface for α-decay of U^{238} is about 30 MeV. (Chen,[149] p. 286.)

7-15. **(a)** Consider a 10-eV electron impinging upon a stationary proton. Calculate $I_c(\Theta)$ from (7-7-37) and plot the result as a function of (Θ).

(b) Repeat for a 1-keV electron.

7-16. **(a)** Consider a 10-eV electron impinging on a stationary electron. Calculate $I_c(\Theta)$ from (7-8-2) and plot the result as a function of (Θ).

(b) Repeat for a 2-keV electron.

7-17. **(a)** Consider two charged particles (ze, m) and (Ze, M), with reduced mass M_r, impact parameter b, and initial relative velocity v_0 interacting through a repulsive Coulomb potential and undergoing mutual scattering. Show that the CM trajectory of each particle is a hyperbola, and sketch the trajectories (see Figs. 1-2-1 and 1-2-2).

(b) Repeat for an attractive potential. (Evans,[151] pp. 836–843.)

7-18. **(a)** In Coulomb scattering, the *collision diameter*, d, is defined as

$$d = \frac{2|Zz|e^2}{M_r v_0^2} \tag{7-11-3}$$

in the notation of Problem 7-17. Show that d is related to the impact parameter, b, and the CM scattering angle, Θ, by

$$b = \frac{d}{2}\cot\frac{\Theta}{2}. \tag{7-11-4}$$

(b) Demonstrate that if $b = d/2$, then $\Theta = 90°$, $\tan\vartheta = M/m$, and $\theta = 45°$ (Figs. 1-2-1 and 1-2-2). (Evans,[151] pp. 843–844.)

7-19. Show that the collision diameter, d, introduced in Problem 7-18 equals the closest distance of approach in a head-on collision of particles interacting through a repulsive Coulomb potential. (Evans,[151] pp. 844–845.)

7-20. In attractive Coulomb scattering, show that the minimum separation distance is

$$r_{\min} = \frac{d}{2}\left(\csc\frac{\Theta}{2} - 1\right). \tag{7-11-5}$$

(Evans,[151] p. 845.)

7-21. Assume that the main effect in a Coulomb interaction occurs during an *effective collision time* τ of the order of $2b/v_0$.

(a) What is the approximate distance through which the target particle M is displaced in Lab coordinates during the time τ?

(b) Show that when $m \gg M$ and $b \gg d$, this displacement of M is of the order of d, and for all other collisions, the displacement of M is less than d. (Evans,[151] p. 847.)

7-22. (a) Show that in an elastic Coulomb collision, the kinetic energy Q_L transferred from a projectile m to a target particle M initially at rest in the Lab frame is

$$Q_L = \frac{2}{M}\left(\frac{Zze^2}{v_0}\right)^2 \frac{1}{b^2 + (d/2)^2}. \tag{7-11-6}$$

(b) Explain physically why $Q_L \sim 1/v_0^2$ when $b \gg d/2$.

(c) What is the kinetic energy transfer, Q_{CM}, from m to M in CM coordinates? (Evans,[151] p. 847.)

7-23. (a) Show that the cross section for Coulomb scattering through CM angle Θ, or greater, is

$$\sigma(\geqslant\Theta) = \pi\frac{d^2}{4}\cot^2\frac{\Theta}{2}. \tag{7-11-7}$$

(b) Show that for collision partners of equal mass, there is no back scattering ($\vartheta > 90°$) in the Lab frame.

(c) Show that for $m \ll M$, the cross section for back scattering becomes $\pi b^2/4$, the area corresponding to the collision diameter, d. (Evans,[151] p. 848.)

7-24. (a) Derive the cross section for Coulomb back scattering in the Lab frame ($\vartheta \geqslant 90°$).

(b) Show that this result reduces to

$$\sigma_{\text{back}} = \pi\left(\frac{Zze^2}{mv_0^2}\right)^2\left[1 - \left(\frac{m}{M}\right)^2\right] \tag{7-11-8}$$

for a projectile (ze, m, v_0) and target (Ze, M, 0). (Evans,[151] p. 20.)

7-25. Show that in Coulomb scattering the probability of an energy transfer Q_L (as in Problem 7-22) with magnitude between Q_L and $Q_L + dQ_L$ is proportional to $1/Q_L^2$, and that the differential cross section is

$$d\sigma = \frac{2\pi Z^2z^2e^4}{Mv_0^2}\frac{dQ_L}{Q_L^2}. \tag{7-11-9}$$

(Evans,[151] p. 849.)

7-26. Demonstrate that in a Coulomb collision, the momentum transfer normal to the initial momentum is

$$\Delta p = \frac{2Zze^2}{bv_0} \frac{1}{1 + (d/2b)^2} \tag{7-11-10}$$

in Lab coordinates.

REFERENCES

1. H. S. W. Massey, E. W. McDaniel, and B. Bederson (Series Eds.), *Applied Atomic Collision Physics*, Academic, New York, Vol. 1 (1982); Vol. 2 (1984).
2. F. Brouillard and J. W. McGowan (Eds.), *Physics of Ion-Ion and Electron-Ion Collisions*, Plenum, New York (1983). For applications, see A. Dalgarno, p. 1 and D. E. Post, p. 37.
3. M. F. A. Harrison, *Phys. Rep.* **37**, 59 (1978).
4. J. M. Shull, in T. D. Märk and G. H. Dunn (Eds.), *Electron Impact Ionization*, Springer-Verlag, Berlin (1985), p. 327.
5. K. T. Dolder, M. F. A. Harrison, and P. C. Thonemann, *Proc. R. Soc. London A* **264**, 367 (1961).
6. M. F. A. Harrison, *Br. J. Appl. Phys.* **17**, 371 (1966).
7. M. F. A. Harrison, *MEP* **7-B**, p. 95 (1968).
8. M. F. A. Harrison, *Inst. Phys. Conf. Ser.* **38**, Chap. 4, 190 (1978).
9. K. T. Dolder, *Case Stud. At. Collision Phys.* **1**, 249 (1969).
10. K. T. Dolder and B. Peart, *Rep. Prog. Phys.* **39**, 693 (1976).
11. K. T. Dolder, in Ref. 2, p. 373.
12. K. T. Dolder and B. Peart, *Rep. Prog. Phys.* **48**, 1283 (1985).
13. K. T. Dolder and B. Peart, *AAMP* **22**, 197 (1986).
14. F. Brouillard (Ed.), *Atomic Processes in Electron-Ion and Ion-Ion Collisions*, Plenum, New York (1986). See P. Defrance, page 157, for a review of electron-impact ionization and excitation of ions.
15. G. H. Dunn, *At. Phys.* **1**, 417 (1969).
16. G. H. Dunn, in T. D. Märk and G. H. Dunn (Eds.), *Electron Impact Ionization*, Springer-Verlag, Berlin (1985), p. 277.
17. G. H. Dunn, in M. Matić (Ed.), *The Physics of Ionized Gases*, *SPIG* (1980), Boris Kidrić Institute of Nuclear Sciences, Belgrade (1981), p. 49.
18. E. Salzborn, in ref. 2, p. 239.
19. D. H. Crandall, in Ref. 2, p. 201.
20. B. Peart, D. S. Walton, and K. T. Dolder, *J. Phys. B* **2**, 1176 (1969).
21. P. Defrance, F. Brouillard, W. Claeys, and G. van Wassenhove, *J. Phys. B* **14**, 103 (1981).
22. H. Tawara and T. Kato, *ADNDT* **36**, 167 (1987).
23. W. C. Lineberger, J. W. Hooper, and E. W. McDaniel, *Phys. Rev.* **141**, 151 (1966).

24. J. W. Hooper, W. C. Lineberger, and F. M. Bacon, *Phys. Rev.* **141**, 165 (1966).

25. J. B. Wareing and K. T. Dolder, *Proc. Phys. Soc. (London)* **91**, 887 (1967). B. Peart and K. T. Dolder, *J. Phys. B* **1**, 872 (1968). B. Peart, D. S. Walton, and K. T. Dolder, *J. Phys. B* **2**, 1347 (1969).

26. F. Brouillard and W. Claeys, in Ref. 2, p. 415.

27. R. G. Montague and M. F. A. Harrison, *J. Phys. B* **16**, 3045 (1983); **17**, 2707 (1984); **18**, 1419 (1985). R. G. Montague, M. J. Diserens, and M. F. A. Harrison, *J. Phys. B* **17**, 2085 (1984). M. J. Diserens, M. F. A. Harrison, and A. C. H. Smith, *J. Phys. B* **17**, L-621 (1984), **21**, 2129 (1988). K. F. Man, A. C. H. Smith, and M. F. A. Harrison, *J. Phys. B* **20**, 1351, 2571, 5865 (1987).

28. B. Peart and K. T. Dolder, *J. Phys. B* **8**, 56 (1975).

29. R. A. Falk, G. H. Dunn, D. C. Gregory, and D. H. Crandall, *Phys. Rev. A* **27**, 762 (1983). D. S. Belić, R. A. Falk, C. Timmer, and G. H. Dunn, *Phys. Rev. A* **36**, 1073 (1987).

30. D. C. Gregory, F. W. Meyer, A. Müller, and P. Defrance, *Phys. Rev. A* **34**, 3657 (1986).

31. D. C. Gregory, L. J. Wang, F. W. Meyer, and K. Rinn, *Phys. Rev. A* **35**, 3256 (1987). See also **39**, 2381, 2397 (1989); *J. Phys. B* **21**, 2117 (1988).

32. J. B. A. Mitchell and J. W. McGowan, in Ref. 2, p. 279. J. B. A. Mitchell, in Ref. 14, p. 185.

33. R. D. Rundel, K. L. Aitken, and M. F. A. Harrison, *J. Phys. B* **2**, 954 (1969).

34. B. Peart, R. Grey, and K. T. Dolder, *J. Phys. B* **9**, L-369 (1976).

35. D. S. Walton, B. Peart, and K. T. Dolder, *J. Phys. B* **4**, 1343 (1971).

36. B. Peart, R. A. Forrest, and K. T. Dolder, *J. Phys. B* **14**, 1655 (1981).

37. D. C. Gregory, *Nucl. Instrum. Methods Phys. Res.* **B-10/11**, 87 (1985); *ICPEAC XIV*, Stanford (1985), p. 205.

38. R. A. Phaneuf, *Nucl. Instrum. Methods Phys. Res.* **A-240**, 461 (1985).

39. R. A. Phaneuf, in Ref. 14, p. 117.

40. R. A. Phaneuf, D. C. Gregory, C. C. Havener, and F. W. Meyer, *At. Phys.* **10**, 265 (1987).

41. R. A. Phaneuf and D. C. Gregory, in H. Tawara and G. H. Dunn (Eds.), *Resonance Effects in Electron-Ion Collisions*, IPPJ-AM-47 (1986), p. 53.

42. R. A. Phaneuf, *Oak Ridge Nat. Lab. Rev.* **3**, 144 (1986).

43. K. J. LaGattuta and Y. Hahn, *Phys. Rev. A* **24**, 2273 (1981). See also Y. Hahn and K. J. LaGattuta, *Phys. Rep.* **166**, 195 (1988).

44. I. C. Lyon, B. Peart, J. B. West, and K. T. Dolder, *J. Phys. B* **19**, 4137 (1986).

45. A. M. Howald, D. C. Gregory, R. A. Phaneuf, D. H. Crandall, and M. S. Pindzola, *Phys. Rev. Lett.* **56**, 1675 (1986).

46. K. Tinschert, A. Müller, R. Becker, and E. Salzborn, *J. Phys. B* **20**, 1823 (1987). A. Müller, K. Tinschert, G. Hofmann, E. Salzborn, and G. H. Dunn, *Phys. Rev. Lett.* **61**, 70 (1988). A. Müller, G. Hofmann, K. Tinschert, and E. Salzborn, *Phys. Rev. Lett.* **61**, 1352 (1988).

47. A. Müller, K. Huber, K. Tinschert, R. Becker, and E. Salzborn, *J. Phys. B* **18**, 2993 (1985).

48. M. L. Mallory and D. H. Crandall, *IEEE Trans. Nucl. Sci.* **NS-23**, 1069 (1976). D. C. Gregory, P. F. Dittner, and D. H. Crandall, *Phys. Rev. A* **27**, 724 (1983). D. C. Gregory and A. M. Howald, *Phys. Rev. A* **34**, 97 (1986). A. Müller, E. Salzborn, R. Frodl, R. Becker, H. Klein, and H. Winter, *J. Phys. B* **13**, 1877 (1980).
49. E. D. Donets, *Phys. Scr.* **T-3**, 11 (1983).
50. F. W. Meyer, *Nucl. Instrum. Methods Phys. Res.* **B-9**, 532 (1985). R. Geller and B. Jacquot, *Phys. Scr.* **T-3**, 19 (1983).
51. R. A. Phaneuf, *IEEE Trans. Nucl. Sci.* **NS-28**, 1182 (1981).
52. D. C. Griffin, C. Bottcher, M. S. Pindzola, S. M. Younger, D. C. Gregory, and D. H. Crandall, *Phys. Rev. A* **29**, 1729 (1984).
53. K. Rinn, D. C. Gregory, L. J. Wang, R. A. Phaneuf, and A. Müller, *Phys. Rev. A* **36**, 595 (1987).
54. K. L. Bell, H. B. Gilbody, J. G. Hughes, A. E. Kingston, and F. J. Smith, *J. Phys. Chem. Ref. Data* **12**, 891 (1983).
55. M. A. Lennon, K. L. Bell, H. B. Gilbody, J. G. Hughes, A. E. Kingston, M. J. Murray, and F. J. Smith, *J. Phys. Chem. Ref. Data* **17**, 1285 (1988).
56. S. M. Younger, in T. D. Märk and G. H. Dunn (Eds.), *Electron Impact Ionization*, Springer-Verlag, Berlin (1985), pp. 1–41.
57. Y. K. Kim, in Ref. 2, p. 101.
58. C. Bottcher, *AAMP* **20**, 241 (1985); **25**, 303 (1988).
59. D. L. Moores, in Ref. 14, p. 1.
60. D. F. Dance, M. F. A. Harrison, and A. C. H. Smith, *Proc. R. Soc. London A* **290**, 74 (1966).
61. K. T. Dolder and B. Peart, *J. Phys. B* **6**, 2415 (1973).
62. P. O. Taylor and G. H. Dunn, *Phys. Rev. A* **8**, 2304 (1973).
63. D. H. Crandall, P. O. Taylor, and G. H. Dunn, *Phys. Rev. A* **10**, 141 (1974).
64. D. H. Crandall, R. A. Phaneuf, and G. H. Dunn, *Phys. Rev. A* **11**, 1223 (1975).
65. R. A. Phaneuf, P. O. Taylor, and G. H. Dunn, *Phys. Rev. A* **14**, 2021 (1976).
66. P. O. Taylor, R. A. Phaneuf, and G. H. Dunn, *Phys. Rev. A* **22**, 435 (1980).
67. W. T. Rogers, G. H. Dunn, J. Ø. Olsen, M. Reading, and G. Stefani, *Phys. Rev. A* **25**, 681 (1982).
68. G. Stefani, R. Camilloni, G. H. Dunn, and W. T. Rogers, *Phys. Rev. A* **25**, 2996 (1982).
69. D. Gregory, G. H. Dunn, R. A. Phaneuf, and D. H. Crandall, *Phys. Rev. A* **20**, 410 (1979).
70. W. T. Rogers, J. Ø. Olsen, and G. H. Dunn, *Phys. Rev. A* **18**, 1353 (1978).
71. J. W. Gallagher and A. K. Pradhan, JILA Information Center Report No. 30 (Oct. 1, 1985).
72. M. J. Seaton, *AAMP* **11**, 92 (1975).
73. R. J. W. Henry, *Phys. Rep.* **68**, 1 (1981).
74. Y. Itikawa, *Phys. Rep.* **143**, 69 (1986).
75. A. Chutjian, *Phys. Rev. A* **29**, 64 (1984); *High Temp. Sci.* **17**, 135 (1984); *ICPEAC XIII*, Berlin (1983), p. 569.
76. A. Chutjian and W. R. Newell, *Phys. Rev. A* **26**, 2271 (1982).

77. A. Chutjian, A. Z. Msezane, and R. J. W. Henry, *Phys. Rev. Lett.* **50**, 1357 (1983).
78. I. D. Williams, A. Chutjian, A. Z. Msezane, and R. J. W. Henry, *Astrophys. J.* **299**, 1063 (1985). I. D. Williams, A. Chutjian, and R. J. Mawhorter, *J. Phys. B* **19**, 2189 (1986).
79. R. J. Mawhorter, I. D. Williams, B. F. Lewis, and A. Chutjian, *ICPEAC XV*, Brighton (1987), Contributed Papers, p. 363.
80. A. Stamatovic and G. J. Schulz, *Rev. Sci. Instrum.* **41**, 423 (1970). D. Roy, *Rev. Sci. Instrum.* **43**, 535 (1972). J. R. Pierce, *Theory and Design of Electron Beams*, 2nd ed., Van Nostrand, New York (1954), pp. 26–28. L. Page and N. I. Adams, *Principles of Electricity*, Van Nostrand, New York (1952), pp. 272–276. L. Evan Bailey, *Rev. Sci. Instrum.* **31**, 1147 (1960). W. L. Barr and W. A. Perkins, *Rev. Sci. Instrum.* **37**, 1354 (1966).
81. C. Timmer, E. Wåhlin, B. DePaola, K. Rinn, D. R. Swenson, D. S. Belić, and G. H. Dunn, *ICPEAC XV*, Brighton (1987), Contributed Papers, p. 805.
82. B. Peart and K. T. Dolder, *J. Phys. B* **4**, 1496 (1971); **5**, 860, 1554 (1972); **6**, 2409, L-359 (1973); **7**, 236 (1974); **8**, 1570 (1975). K. T. Dolder and B. Peart, *Comments At. Mol. Phys.* **5**, 97 (1976).
83. G. H. Dunn and B. Van Zyl, *Phys. Rev.* **154**, 40 (1967).
84. F. von Busch and G. H. Dunn, *Phys. Rev. A* **5**, 1726 (1972).
85. R. A. Heppner, F. L. Walls, W. T. Armstrong, and G. H. Dunn, *Phys. Rev. A* **13**, 1000 (1976).
86. J. W. McGowan and J. B. A. Mitchell, in L. G. Christophorou (Ed.), *Electron-Molecule Interactions and Their Applications*, Vol. 2, Academic, New York (1984), p. 65. J. B. A. Mitchell, *Phys. Rpts.* (1990), to be published.
87. R. N. Compton and J. N. Bardsley, in I. Shimamura and K. Takayanagi (Eds.), *Electron-Molecule Collisions*, Plenum, New York (1984), p. 275. A. Giusti-Suzor, in Ref. 14, p. 223.
88. B. Van Zyl and G. H. Dunn, *Phys. Rev.* **163**, 43 (1967).
89. P. A. Schulz, D. C. Gregory, F. W. Meyer, and R. A. Phaneuf, *J. Chem. Phys.* **85**, 3386 (1986).
90. E. W. McDaniel, *Atomic Collisions: Heavy Particle Projectiles*, Wiley, New York, to be published.
91. G. H. Dunn, in Ref. 14, p. 93.
92. M. J. Seaton and P. J. Storey, in P. G. Burke and B. L. Moiseiwitsch (Eds.), *Atomic Processes and Applications*, North-Holland, Amsterdam (1976), p. 133. This review contains much useful information on the history and general nature of DR, and applications of this process.
93. A. Burgess, *Astrophys. J.* **139**, 776 (1964); **141**, 1588 (1965); *Ann. Astrophys.* **28**, 774 (1965). Y. Hahn, *AAMP* **21**, 123 (1985). Y. Hahn, in Ref. 14, p. 23, M. S. Pindzola, D. C. Griffin, and C. Bottcher, in Ref. 14, p. 75, R. H. Bell and M. J. Seaton, *J. Phys. B* **18**, 1589 (1985). J. Dubau and S. Volonte, *Rep. Prog. Phys.* **43**, 199 (1980). V. L. Jacobs, J. Cooper, and S. L. Haan, *Phys. Rev. A* **36**, 1093 (1987). Y. Hahn and K. J. LaGattuta, *Phys. Reps.* **166**, 195 (1988).
94. M. H. Nayfeh and C. W. Clark (Eds.), *Atomic Excitation and Recombination in External Fields*, Gordon and Breach, New York (1985), see the section on "Dielectronic Recombination: Theory," pp. 339–389.

95. T. Fujimoto, T. Kato, and Y. Nakamura, IPPJ-AM-23 (1982).
96. J. Dubau, in Ref. 14, p. 403. M. F. A. Harrison, in Ref. 14, p. 421.
97. S. Datz, *ICPEAC XIII*, Berlin (1983), p. 795. S. Datz, P. F. Dittner, C. M. Fou, P. D. Miller, and P. L. Pepmiller, *Nucl. Instrum. Methods Phys. Res. B* **23**, 248 (1987).
98. G. H. Dunn, D. S. Belić, T. J. Morgan, D. W. Mueller, and C. Timmer, *ICPEAC XIII*, Berlin (1983), p. 809. G. H. Dunn, *ICPEAC XIV*, Stanford (1985), p. 23.
99. See "Dielectronic Recombination: Experiments," pp. 393–466, and also pp. 497–501, 517–520 in Ref. 94.
100. G. H. Dunn, D. S. Belić, N. Djurić, and D. W. Mueller, *At. Phys.* **9**, 505 (1984).
101. D. S. Belić, G. H. Dunn, T. J. Morgan, D. W. Mueller, and C. Timmer, *Phys. Rev. Lett.* **50**, 339 (1983).
102. J. F. Williams, *Phys. Rev. A* **29**, 2936 (1984).
103. A. Müller, D. S. Belić, B. D. DePaola, N. Djurić, G. H. Dunn, D. W. Mueller, and C. Timmer, *Phys. Rev. Lett.* **56**, 127 (1986); *Nucl. Instrum. Methods Phys. Res. B* **23**, 254 (1987); *Phys. Rev. A* **36**, 599 (1987).
104. L. D. Gardner, J. L. Kohl, G. P. Lafyatis, A. R. Young, and A. Chutjian, *Rev. Sci. Instrum.* **57**, 2254 (1986).
105. J. B. A. Mitchell, C. T. Ng, J. L. Foraud, D. P. Levac, R. E. Mitchell, A. Sen, D. B. Miko, and J. W. McGowan, *Phys. Rev. Lett.* **50**, 335 (1983).
106. P. F. Dittner, S. Datz, P. D. Miller, C. D. Moak, P. H. Stelson, C. Bottcher, W. B. Dress, G. D. Alton, N. Nešković, and C. M. Fou, *Phys. Rev. Lett.* **51**, 31 (1983).
107. P. F. Dittner, S. Datz, P. D. Miller, P. L. Pepmiller, and C. M. Fou, *Phys. Rev. A* **33**, 124 (1986); **35**, 3668 (1987).
108. P. F. Dittner, S. Datz, H. F. Krause, P. D. Miller, P. L. Pepmiller, C. Bottcher, C. M. Fou, D. C. Griffin, and M. S. Pindzola, *Phys. Rev. A* **36**, 33 (1987). P. F. Dittner, *Phys. Scr.* **T-22**, 65 (1988).
109. S. Datz and P. F. Dittner, *Zeit. f. Phys. D* **10**, 187 (1988).
110. S. Datz, private communication (1988).
111. R. E. Marrs, M. A. Levine, D. A. Knapp, and J. R. Henderson, *ICPEAC XV*, Brighton (1987), p. 209; *Phys. Rev. Lett.* **60**, 1715 (1988).
112. D. F. Dance, M. F. A. Harrison, and R. D. Rundel, *Proc. Roy. Soc. London A* **299**, 525 (1967).
113. B. Peart, D. S. Walton, and K. T. Dolder, *J. Phys. B* **3**, 1346 (1970).
114. D. S. Walton, B. Peart, and K. T. Dolder, *J. Phys. B* **4**, 1343 (1971).
115. B. Peart and K. T. Dolder, *J. Phys. B* **6**, L-359, 1497 (1973).
116. B. Peart, D. S. Walton, and K. T. Dolder, *J. Phys. B* **4**, 88 (1971).
117. P. Defrance, W. Claeys, and F. Brouillard, *J. Phys. B* **15**, 3509 (1982).
118. B. Peart, R. A. Forrest, and K. T. Dolder, *J. Phys. B* **12**, 847, 2735 (1979).
119. B. Peart, R. A. Forrest, and K. T. Dolder, *J. Phys. B* **12**, L-115 (1979).
120. V. A. Esaulov, *J. Phys. B* **13**, 1625 (1980).
121. J. B. Hasted, in Ref. 2, p. 461.
122. M. R. C. McDowell and J. P. Coleman, *Introduction to the Theory of Ion-Atom Collisions*, North-Holland, Amsterdam (1970), pp. 239–255.
123. M. D. Scadron, *Advanced Quantum Theory*, Springer-Verlag, New York (1979).
124. B. H. Bransden, *Atomic Collision Theory*, 2nd ed., Benjamin-Cummings, Menlo Park, Calif. (1983), Chap. 3.

125. L. D. Landau and E. M. Lifshitz, *Quantum Mechanics, Non-relativistic Theory*, 2nd ed., Pergamon, Oxford (1965), Chap. XVII, pp. 600–610.

126. N. F. Mott and H. S. W. Massey, *The Theory of Atomic Collisions*, 3rd ed., Clarendon Press, Oxford (1965), Chaps. III, XI.

127. P. M. Morse and H. Feshbach, *Methods of Theoretical Physics*, McGraw-Hill, New York (1953).

128. R. P. Feynman, R. B. Leighton, and M. Sands, *The Feynman Lectures on Physics*, Vol. III, Addison-Wesley, Reading, Mass. (1964), Secs. 3–4, 4–1.

129. U. Fano and L. Fano, *Physics of Atoms and Molecules*, University of Chicago, Chicago (1972), pp. 73, 81.

130. J. D. Jackson, *Classical Electrodynamics*, 2nd ed., Wiley, New York (1975), Chap. 15.

131. W. K. H. Panofsky and M. Phillips, *Classical Electricity and Magnetism*, 2nd ed., Addison-Wesley, Reading, Mass. (1962), pp. 354–376.

132. R. H. Good and T. J. Nelson, *Classical Theory of Electric and Magnetic Fields*, Academic, New York (1971), pp. 522–533.

133. L. D. Landau and E. M. Lifshitz, *The Classical Theory of Fields*, 3rd ed., Pergamon, Elmsford, N.Y. (1971).

134. J. D. Jackson, *Phys. Today* **40**, 34 (May 1987).

135. C. F. Barnett and M. F. A. Harrison (Volume Eds.), *Plasmas*, Vol. 2 of Ref. 1, see R. H. Pratt and I. J. Feng, p. 307; J. C. Weisheit, p. 441.

136. W. Heitler, *The Quantum Theory of Radiation*, 3rd ed., Clarendon, Oxford (1954).

137. H. S. W. Massey, *Electronic and Ionic Impact Phenomena*, 2nd ed., Vol. II, Clarendon, Oxford (1969). M. Ya. Amusia, *Phys. Rep.* **162**, 249 (1988).

138. J. A. Eisele, *Modern Quantum Mechanics*, Wiley, New York (1969).

139. E. M. Henley and W. Thirring, *Elementary Quantum Field Theory*, McGraw-Hill, New York (1962).

140. R. H. Pratt, *Comments At. Mol. Phys.* **10**, 121 (1981).

141. L. C. Maximon, A. de Miniac, T. Aniel, and E. Ganz, *Phys. Rep.* **147**, 189 (1987).

142. R. Y. Yin, O. V. Gabriel, and R. H. Pratt, *Phys. Rev. A* **36**, 1207 (1987).

143. J. W. Gallagher, *Bibliography of Free-Free Transitions in Atoms and Molecules, JILA Information Center Report No.* 16 (Apr. 5, 1977).

144. S. M. Seltzer and M. J. Berger, *ADNDT* **35**, 345 (1986).

145. C. Yamabe, S. J. Buckman, and A. V. Phelps, *Phys. Rev. A* **27**, 1345 (1983).

146. R. R. Wilson, *Sci. Am.* **242**, 42 (Jan. 1980).

147. F. T. Cole and F. E. Mills, *Ann. Rev. Nuc. Part. Sci.* **31**, 295 (1981). See also H. Poth, *Phys. Rep.* (1989), to be published.

148. J. A. Cronin, D. F. Greenberg, and V. L. Telegdi, *University of Chicago Graduate Problems in Physics with Solutions*, Addison-Wesley, Reading, Mass. (1967).

149. M. Chen, *Berkeley Physics Problems with Solutions*, Prentice-Hall, Englewood Cliffs, N.J. (1974).

150. E. Segrè, *Nuclei and Particles*, 2nd ed., Benjamin-Cummings, Menlo Park, Calif. (1977).

151. R. D. Evans, *The Atomic Nucleus*, McGraw-Hill, New York (1955).

8

PHOTON PROCESSES

PART $\mathscr{A}$. INTRODUCTION

We begin this chapter with a catalog of photon processes, some of which are unimportant for our purposes and therefore are mentioned only briefly. Others are discussed at length in Section 8-1 and subsequent sections. Section 8-2 contains a derivation of the principle of detailed balancing and a list of mutually inverse processes that are related by this principle. Particle-particle, as well as photon-particle, collisions are included in this list. Standard International (SI) units are used throughout this chapter.

8-1. SURVEY AND PROSPECTUS

Photon processes may be divided into three categories: emission, scattering, and absorption. Here we make this division and list a number of processes in each.

A. Emission

There are several quite different mechanisms for producing electromagnetic radiation that are relevant to our discussion; each is treated briefly here. Radiation from isolated excited atoms and molecules is discussed further in much greater detail in Parts $\mathscr{B}$ and $\mathscr{C}$. Emission from gas lasers[1-3] and electrical discharges[4] clearly is related to this last-mentioned topic, but is not considered here.

1. *Emission from Excited Atoms and Molecules*[1,2]

Some excited states have extremely long lifetimes, but a quantum mechanical system cannot permanently remain in an excited state; the system will drop, by one or more transitions, to its ground state. The deexciting transitions may involve spontaneous emission of radiation, stimulated (induced) emission, or collisional deexcitation in impact with another particle or system. Figure 8-1-1

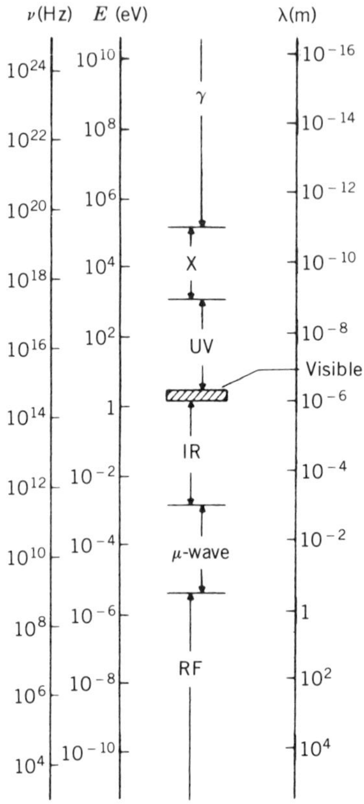

Figure 8-1-1. The spectrum of electromagnetic radiation. (Adapted from Bransden and Joachain.[14])

displays some useful information about the electromagnetic spectrum. Starting at the low-frequency end, we first encounter radio-frequency (RF) radiation, produced by transitions between closely spaced atomic and molecular levels, as well as by laboratory electronics techniques. (The 21-cm line used in radio-astronomy arises from a hyperfine transition in atomic hydrogen.) Similar comments apply to microwave radiation, which is important in molecular physics in connection with rotational transitions in molecules. Infrared radiation is often associated with emission from hot objects (a summary of equations describing blackbody radiation is presented in Section 8-1-A-3), but it is also produced in downward vibrational transitions in molecules. Visible and ultraviolet (UV) light is generally produced in electronic transitions involving outer orbits in both atoms and molecules, while X-rays involve electronic transitions in inner shells. Gamma rays arise from transitions within nuclei. In this volume we are especially interested in the part of the spectrum that extends from microwaves through X-rays.

2. *Emission from Free, Charged Particles*

a. Bremsstrahlung. This type of radiation, treated at length in Section 7-9, involves free-free emission and may be represented by the equation

$$e' \rightarrow e + \hbar\omega \tag{8-1-1}$$

in the most important case, emission by an electron. Both electrons are in continuum states; the energy of [e] is lower than that of [e′] by the photon energy $\hbar\omega$.

b. Synchrotron Radiation.[1,5–10] This type of radiation is emitted by a fast-charged particle undergoing acceleration. We have a special interest in such radiation produced by relativistic electrons moving in a circular path in an accelerator or nearly circular path in a storage ring (Note 7-2) because these arrangements provide a vital laboratory source of UV and X-ray photons. Electrons of energy $\gamma m_0 c^2$ emit a spectrum extending from ω, the frequency of revolution, up to harmonics on the order of $\gamma^3\omega$, where γ is the Lorentz velocity factor $(1 - v^2/c^2)^{-1/2}$, but the spectrum for even a single electron is smeared into a continuum because the particle oscillates in position and energy about its equilibrium orbit. At $\gamma \gg 1$, relativistic effects cause synchrotron radiation to be emitted mainly tangential to the beam direction in a narrow cone of half-angle $\Delta\phi \approx 1/\gamma$, so an observer receives a short burst of radiation once every revolution. The radiation is strongly polarized in the plane of the orbit. Classical electrodynamics provides a rather successful explanation of synchrotron radiation, although quantal effects should be taken into account. If the electrons are moving in a circular accelerator, the wavelength, λ_p, at the peak of the spectrum is given in angstroms by

$$\lambda_p = \frac{2.35R}{E^3}, \tag{8-1-2}$$

where R is the radius of the orbit in meters and the energy E is in GeV.

The electromagnetic power radiated by electrons moving in a circular path is proportional to the fourth power of the electron energy and inversely proportional to the radius of curvature. A 100-mA beam of 3-GeV electrons circulating in a path of 10-m radius radiates more than 70 kW of power. This is an impressive output, but an enhancement of several orders of magnitude is achieved by using a storage ring with "wigglers" or "undulators" inserted into the straight sections to cause the circulating electrons to oscillate about their straight trajectories and radiate. Figure 8-1-2 shows the spectral brilliance expected from the proposed European Synchrotron Radiation Facility (ESRF). The curves marked "bending magnets" relate to the use of only the bending magnets of the accelerator to produce the radiation. The curves labeled "wigglers" and "undulators" show the enormous enhancement of the radiation when devices bearing these names are used. Wigglers and undulators both consist of arrays of magnets of alternating polarity located perpendicular to the electron beam axis. If the maximum beam deflection angle produced by the device is large compared to the natural angular spread of the radiation, the device is called a wiggler; otherwise, it is an undulator. Wigglers provide the highest photon energies, whereas undulators produce the most brilliant photon beams.

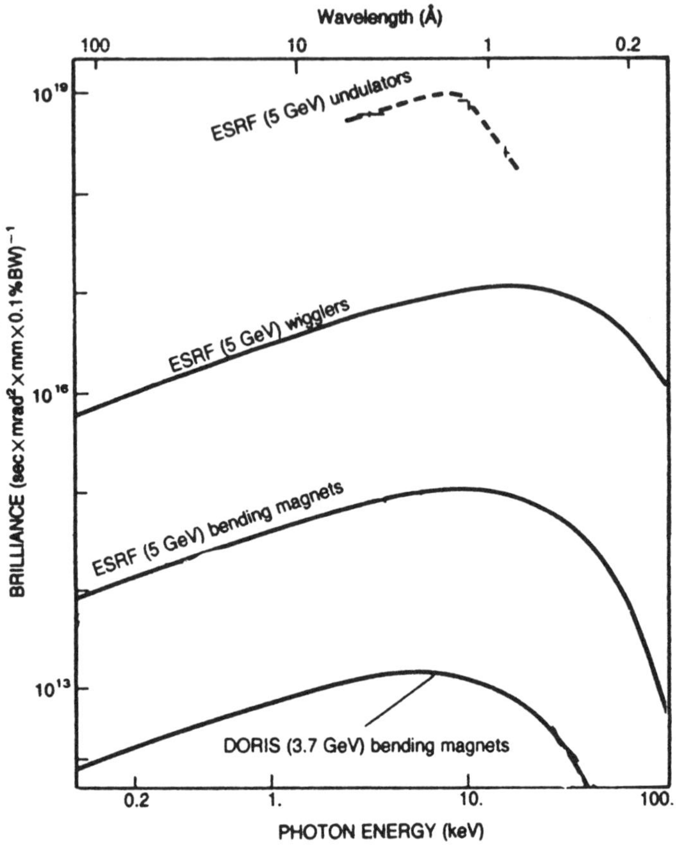

Figure 8-1-2. The spectral brilliance distribution from synchrotron radiation facilities at ESRF and DORIS. The brilliance is defined as the spectral photon intensity per unit area of source. [From B. M. Schwarzschild, *Phys. Today*, Vol. 38, p. 21 (May 1985).]

Synchrotron radiation was first detected experimentally in 1946, and spectroscopic measurements involving its use were first made, by R. B. Madden and K. Codling, in 1961. Since that time there has been an explosive increase in the use of this tool. The highly collimated, intense continuous spectrum provided by synchrotron radiation spans wavelength regions inaccessible at high intensity with other sources. When narrow spectral slices are selected with monochromators, synchrotron radiation furnishes a unique tool for studies that require high resolution in both space and energy simultaneously. In closing, we should point out that *cyclotron radiation* is the term applied to the emission when the electrons are moving with nonrelativistic velocities.

c. Radiation from Free-Electron Lasers and Cyclotron-Resonance Masers.[11,12] Although synchrotron radiation provides an intense, continuous photon source, it has much lower spectral brightness than even moderate-power lasers.

Therefore, considerable effort has been directed toward resonant enhancement, which offers the possibility of unprecedented levels of tunable radiation over a spectrum ranging from millimeter wavelengths to the ultraviolet. Both of the sources discussed here are powered by relativistic electron beams, and both produce coherent radiation. As Sprangle and Coffey[11] point out, they have less in common with modern optical laser techniques than with the earliest sources of coherent radiation. Here we refer to the magnetron, klystron, and traveling-wave tube, devices of World War II vintage that have limited power output and efficiency at short wavelengths.

The *free-electron laser*, whose operation was first demonstrated in 1977 by J. Madey* and his coworkers, employs an electron beam that is generated externally and fired axially into the laser. As it passes through the device, the beam is "wiggled" transversely by a pump field, which is typically a static, periodic magnetic field produced by a set of wigglers spaced along the axis. The transverse accelerations experienced by the electrons cause the emission of radiation down the axis. Also, an "external" wave is directed axially down the entire length occupied by the wigglers, with the **E** vector in the same direction as the wiggle velocity, and the usual laser arrangement of mirrors at the ends of the active path is provided. Although the electrons at the point of injection are randomly phased and radiate incoherently, the ponderomotive wave arising from the beating of the wiggler and radiation fields bunches the electrons and results in stimulated emission. If the axial velocity of the electrons is slightly greater than the velocity of the ponderomotive wave, the average energy of the electrons is reduced, and the radiation field is enhanced. For highly relativistic electrons ($\gamma \gg 1$), the wavelength of the radiation, λ, is approximately $\lambda_w/2\gamma^2$, where λ_w is the "wiggler wavelength," the axial distance between adjacent points at which the wiggler magnetic field repeats itself. The radiation wavelength can be varied by changing the electron beam energy. Free-electron lasers have efficiencies exceeding 20% at millimeter wavelengths and can operate into the UV region.

The *cyclotron-resonance maser* utilizes a nearly monoenergetic beam of electrons that gyrate about an externally produced magnetic field as they pass down the axis of the device. An externally imposed electromagnetic wave of frequency ω approximates the field of the transverse-electric mode of a cavity or waveguide when ω is near one of the cutoff frequencies of the structure. If the radiation frequency slightly exceeds the rotation frequency, and the rotation frequency is energy dependent, phase bunching of the electrons† occurs, and the electrons can lose energy and amplify the field. The mechanism of the cyclotron-resonance maser was first clearly demonstrated by J. Hirschfield and J. Wachtel in 1964. Since that time, the device has been developed to a high degree and now

*The development of the free-electron laser is traced by Madey in W. B. Colson, C. Pellegrini, and A. Renieri (Eds.), *Free Electron Laser Handbook*, North-Holland, Amsterdam (1989).

†For a discussion of phase bunching of electrons in microwave tubes, see O. Gandhi, *Microwave Engineering and Applications*, Pergamon, Oxford (1981).

is among the most efficient instruments for generating high-power coherent radiation at centimeter and millimeter wavelengths. In both the free-electron laser and the cyclotron-resonance maser, transitions occur to a very large number of energy levels, so classical analyses are entirely adequate.

d. Cerenkov Radiation.[1,7–10,13] A charged, structureless particle moving through free space can radiate only while accelerating, but the situation is different if it travels through a dielectric material. In this case, as Cerenkov showed in 1937, the particle can radiate even if it is traveling with essentially constant speed in a straight line, provided that its velocity, v, exceeds the phase velocity of light in the medium, $c' = c/n$. Here n is the refractive index of the medium. A simple Huygens construction involving the spherical wavefronts emitted by the particle shows that a wavefront with half-angle θ is formed and moves with the particle, θ being given by the equation

$$\sin\theta = \frac{c}{nv}. \tag{8-1-3}$$

The radiation travels in a direction normal to this shock wavefront, at an angle $(\pi/2 - \theta)$ with respect to the particle motion (Problem 8-6). In nuclear and particle physics, this angle is measured with Cerenkov detectors in order to determine the velocity of fast charged particles. The number of photons emitted per unit length of travel of the particles in the frequency interval ν to $\nu + d\nu$ is

$$\frac{dN(\nu)}{dx}\,d\nu = \frac{2\pi\alpha Z^2}{c}\left(1 - \frac{c^2}{n^2v^2}\right)d\nu. \tag{8-1-4}$$

Here α is the fine structure constant, which is a dimensionless constant with value 1/137, and Z is the number of charges on the particle.*

3. *Blackbody Radiation*[1,14]

Because this subject is undoubtedly familiar to all readers of this book, we shall merely display a few formulas for future reference. Read[1] and Bransden and Joachain[14] trace the development of the subject and discuss the equations presented here.

Within a blackbody cavity, the radiation energy per unit volume and unit frequency interval at the temperature T is

$$\rho(T, \nu) = \frac{8\pi\nu^2}{c^3}\,\frac{h\nu}{e^{h\nu/kT} - 1}. \tag{8-1-5}$$

*A charged particle can also emit *transition radiation* while moving with constant velocity in an inhomogeneous and/or nonstationary medium [see Jackson[8] and V. L. Ginzburg and V. N. Tsytovich, *Phys. Rep.* **49**, 1 (1979)].

This spectral distribution function is the *Planck radiation law*, which can be expressed alternatively in terms of the energy density in the interval from λ to $\lambda + d\lambda$:

$$\rho(T, \lambda)\, d\lambda = \frac{8\pi ch}{\lambda^5(e^{hc/\lambda kT} - 1)}\, d\lambda. \tag{8-1-6}$$

The *emissive power* of a blackbody surface (i.e., the power emitted per unit area and unit wavelength interval at temperature T) is

$$w(T, \lambda) = \frac{2\pi c^2 h}{\lambda^5(e^{hc/\lambda kT} - 1)}\,. \tag{8-1-7}$$

This function is plotted in Fig. 8-1-3 for several values of the blackbody

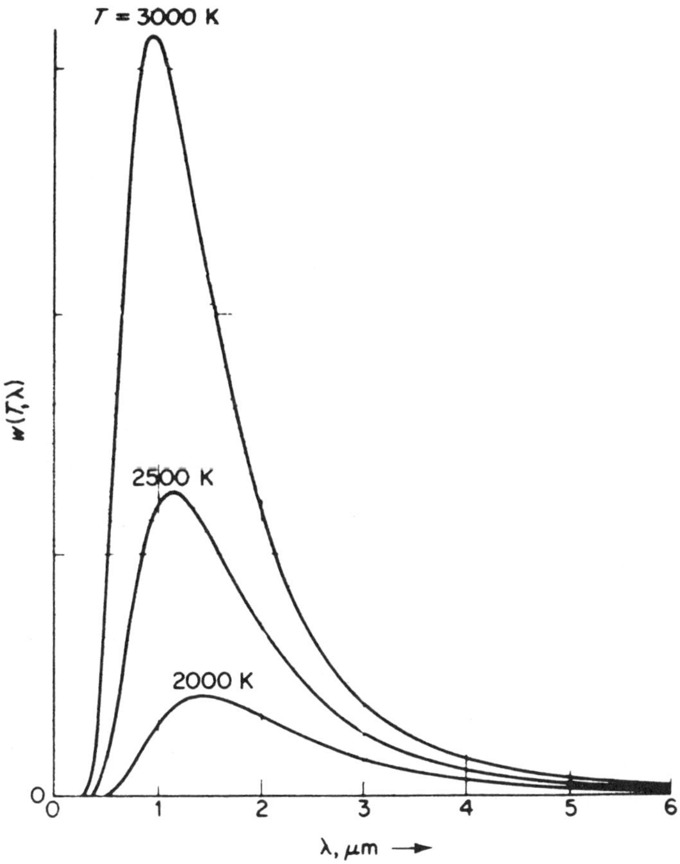

Figure 8-1-3. The emissive power $w(T, \lambda)$ of a blackbody surface as a function of λ for surface temperatures of 2000, 2500, and 3000 K (see Problem 8-7). (Adapted from Read.[1])

temperature. The *Wien displacement law* relates λ_m, the wavelength at which $w(T, \lambda)$ is a maximum, to the temperature:

$$\lambda_m T = \text{constant} = 2.896 \times 10^{-3}\ \text{m K}. \tag{8-1-8}$$

The *Stefan-Boltzmann law* gives the total power radiated by unit area of a blackbody at the temperature T:

$$W = \sigma T^4, \tag{8-1-9}$$

where

$$\sigma = \frac{2\pi^5 k^4}{15c^2 h^3} = 5.67 \times 10^{-8}\ \text{J K}^{-4}\ \text{m}^{-2}\ \text{s}^{-1} \tag{8-1-10}$$

is *Stefan's constant*.

B. Scattering

Both elastic* and inelastic collisions of photons with atoms and molecules are included here. Our interest centers on Thomson, Compton, Rayleigh, and Raman scattering; the other types mentioned here (Tyndall, Brillouin, and Delbrück) are peripheral to our interest and are included mainly for completeness. The scattering phenomena are referred to as second-order processes because in quantum mechanics they are treated by second-order perturbation theory. This method of treatment is a reflection of the view that the overall process consists of two steps: the absorption of a photon with elevation of the system to a virtual excited state, and the subsequent emission of a photon with an energy that may be either equal to or different from the original photon energy, depending on the type of collision being considered. Such processes generally have much smaller cross sections than first-order processes, which are described by first-order perturbation theory.

1. *Thomson Scattering*[1,7–10,15–20]

As indicated in Problem 8-2, a free electron cannot absorb a photon in any sort of permanent sense, but it can be set into oscillatory motion by an electromagnetic wave and radiate as an oscillator. The emitted radiation appears as scattering of the original wave. J. J. Thomson treated this problem of the elastic scattering of electromagnetic radiation by free electrons by classical electromagnetic theory early in this century. His results, which will be reproduced below, are valid only at low frequencies, at which the momentum of the incident

*A survey of theoretical and experimental studies of elastic photon-atom scattering has been prepared by P. P. Kane, L. Kissel, R. H. Pratt, and S. C. Roy, *Phys. Rep.* **140**, 75 (1986).

photon can be ignored. Thomson scattering is the nonrelativistic, nonquantal limit of Compton scattering.

A simple analysis will suffice to obtain the integral Thomson cross section. Consider a free electron subjected to a wave of frequency ω, polarized in the X direction and traveling along the Z axis. In SI units, its equation of motion is

$$m_0 \frac{d^2\mathbf{x}}{dt^2} = -e\mathbf{E} - e\mathbf{v} \times \mathbf{B}. \tag{8-1-11}$$

In this system of units, the field vectors are related by the equation $B = E/c$,[1] so the effect of the magnetic field is negligible when v/c is small, as in the nonrelativistic case treated here. Accordingly, the magnetic component of the Lorentz force will be ignored.

If we take the electric field in the wave to be

$$\mathbf{E} = \mathbf{i}_x E_0 \cos(\omega t - kz), \tag{8-1-12}$$

the solution of (8-1-11) for an electron with z-coordinate z_0 is

$$x = \frac{eE_0}{m_0\omega^2} \cos(\omega t - kz_0). \tag{8-1-13}$$

In response to the incident wave, the electron reradiates as a Hertzian dipole at the incident wave frequency ω, the mean power being given by the equation

$$P_{\text{rad}} = \frac{e^2}{6\pi\epsilon_0 c^3} \overline{\frac{d^2x}{dt^2}} = \frac{e^4 E_0^2}{12\pi\epsilon_0 m_0^2 c^3}. \tag{8-1-14}$$

The integral cross section for Thomson scattering is then given by the ratio of the power of the scattered radiation to the power per square meter in the incident beam:

$$\sigma_T = \frac{P_{\text{rad}}}{P_{\text{beam}}} = \frac{8\pi r_0^2}{3} = 6.652 \times 10^{-29}\ \text{m}^2, \tag{8-1-15}$$

where

$$r_0 = \frac{e^2}{4\pi\epsilon_0 m_0 c^2} = 2.818 \times 10^{-15}\ \text{m} \tag{8-1-16}$$

is the classical electron radius. Note that σ_T has an extremely small numerical value, which is independent of ω.

By considering the angular distribution of the radiation from the Hertzian dipole, Read[1] shows that the differential cross section for Thomson scattering in the case treated here is

$$d\sigma(\theta, \varphi) = r_0^2(1 - \sin^2\theta\cos^2\varphi)\, d\Omega, \tag{8-1-17}$$

where θ is the polar angle between the incident wave and the direction of observation, and φ is the corresponding azimuthal angle. If the incident radiation is unpolarized, the differential cross section is

$$d\sigma(\theta) = \tfrac{1}{2}r_0^2(1 + \cos^2\theta)\, d\Omega. \tag{8-1-18}$$

A second method of calculating the radiated power given in (8-1-14) is to include in the equation of motion a damping term to account for the radiative reaction and to calculate the rate of work done by the incident wave on the electron. Read[1] shows the radiative reaction for an electron oscillator of natural frequency ω_0 to be $-\gamma m_0\dot{x}$, where the damping constant γ (here not the Lorentz velocity factor) is

$$\gamma = \frac{e^2\omega_0^2}{6\pi\epsilon_0 m_0 c^3}. \tag{8-1-19}$$

Rossi[21] considers the motion of such an oscillator, driven by an incident wave and damped by a term of this form. He writes the equation of motion as

$$m_0\frac{d^2x}{dt^2} = -m_0\omega_0^2 x - m_0\gamma\frac{dx}{dt} - eE_0\cos\omega t \tag{8-1-20}$$

and obtains, for the electron displacement, an equation of the form

$$x = \frac{eE_0}{m_0}\frac{\cos(\omega t + \phi)}{[(\omega^2 - \omega_0^2)^2 + \omega^2\gamma^2]^{1/2}}, \tag{8-1-21}$$

where

$$\tan\phi = \frac{\omega\gamma}{\omega^2 - \omega_0^2}. \tag{8-1-22}$$

We can recover eq. (8-1-13) for Thomson scattering by setting the oscillator spring constant k, and hence the frequency $\omega_0 = (k/m_0)^{1/2}$, equal to zero.

2. *Compton Scattering*[1,13–20]

As Read[1] points out, the classical treatment of Thomson scattering is valid only if the length of the oscillating dipole is much smaller than the wavelength of the emitted radiation, and if the mechanical energy of the dipole is much larger than the photon energy. The first condition ensures that the dipole acts as a Hertzian oscillator, the second that the re-radiation of the energy can be treated as a continuous process. These conditions can be combined to give the condition

$$\hbar\omega \ll m_0c^2 = 0.511 \text{ MeV}. \tag{8-1-23}$$

If this condition is not satisfied, important new features arise, which can be treated by considering the scattering in terms of a photon colliding with an electron initially at rest. As A. H. Compton showed in 1923, the electron recoils from the collision site, while the photon is scattered with reduced energy, as indicated in Fig. 8-1-4.

Compton's analysis is based on relativistic mechanics, according to which the mass, m, total energy, E, kinetic energy, T, and momentum, $\mathbf{p}$, of a particle are given by

$$m = \frac{m_0}{(1 - v^2/c^2)^{1/2}}, \tag{8-1-24}$$

$$E = mc^2, \tag{8-1-25}$$

$$T = E - m_0c^2, \tag{8-1-26}$$

and

$$\mathbf{p} = \frac{m_0\mathbf{v}}{(1 - v^2/c^2)^{1/2}}. \tag{8-1-27}$$

The energy and momentum of the particle are related by

$$E^2 = m_0^2c^4 + p^2c^2. \tag{8-1-28}$$

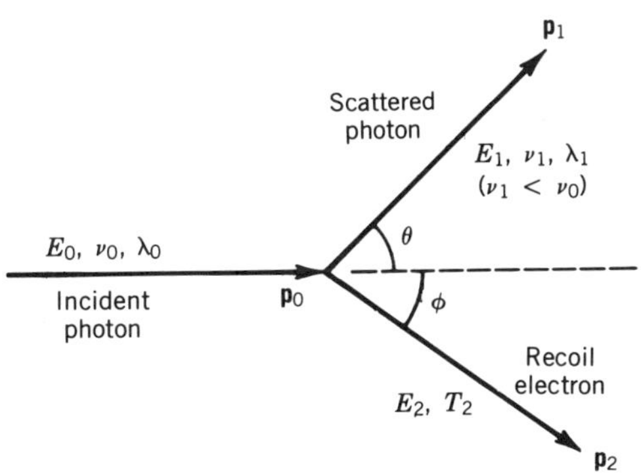

Figure 8-1-4. Schematic representation of a photon undergoing Compton scattering by an electron initially at rest in the Lab frame.

A photon has zero rest mass, and energy $E = h\nu = hc/\lambda$, where ν and λ are the frequency and wavelength of the associated wave. The magnitude of the photon momentum is $p = E/c = h/\lambda$. In the notation of Fig. 8-1-4, with the electron taken to be at rest initially, conservation of momentum is expressed as

$$\mathbf{p}_0 = \mathbf{p}_1 + \mathbf{p}_2, \tag{8-1-29}$$

which gives

$$p_2^2 = p_0^2 + p_1^2 - 2p_0 p_1 \cos\theta. \tag{8-1-30}$$

Conservation of energy requires that

$$E_0 + m_0c^2 = E_1 + (m_0^2c^4 + p_2^2c^2)^{1/2}, \tag{8-1-31}$$

and the kinetic energy of the electron after the collision is

$$T_2 = E_0 - E_1 = c(p_0 - p_1) = (m_0^2c^4 + p_2^2c^2)^{1/2} - m_0c^2. \tag{8-1-32}$$

Hence

$$p_2^2 = (p_0 - p_1)^2 + 2m_0c(p_0 - p_1), \tag{8-1-33}$$

and if we combine this equation with (8-1-30), we obtain

$$m_0c(p_0 - p_1) = p_0p_1(1 - \cos\theta). \tag{8-1-34}$$

Using the fact that $\lambda_0 = h/p_0$ and $\lambda_1 = h/p_1$, we then have the *Compton scattering formula*

$$\Delta\lambda = \lambda_1 - \lambda_0 = \lambda_c(1 - \cos\theta), \tag{8-1-35}$$

where the constant λ_c is the *Compton wavelength*, given by

$$\lambda_c = \frac{h}{m_0c} = 2.426 \times 10^{-12}\ \text{m}. \tag{8-1-36}$$

The kinetic energy of the recoiling electron is related to the scattering angle by

$$T_2 = h\nu\frac{\gamma(1 - \cos\theta)}{1 + \gamma(1 - \cos\theta)}. \tag{8-1-37}$$

Some of the radiation incident on a target is observed to be scattered with no change in wavelength. In this case, the scattering occurs from electrons so tightly bound that the entire atom recoils, and the atomic mass should be used in the analysis, with the result that the Compton shift is negligible. In the case of visible light, no Compton shift is observed because the photon energy is not large compared with the binding energy of even the outermost electrons.

A completely quantum-mechanical treatment is required to calculate the cross section for Compton scattering, and such a calculation was first made in 1928 by Klein and Nishina, who used the Dirac equation with the result

$$d\sigma(\theta) = \frac{r_0^2}{2}\, d\Omega \left(\frac{\nu_1}{\nu_0}\right)^2 \left(\frac{\nu_0}{\nu_1} + \frac{\nu_1}{\nu_0} - \sin^2\theta\right) \tag{8-1-38}$$

for the case of unpolarized incident radiation. This expression is known as the *Klein-Nishina formula.* It can be expressed in terms of variables other than the photon scattering angle θ, and also evaluated for the case of polarized incident radiation.[16,17]

Thus far we have been discussing the scattering of light by isolated particles, a subject that has been extensively studied in many experiments for more than a century. The complementary study of the effect of light on single particles has proved very difficult, but recently Freeman and his colleagues[22] have demonstrated large-angle scattering of free electrons from a focused laser beam in observations of photoelectrons produced in multiphoton ionization of xenon. The laser light had a wavelength of 1064 nm, and the effect was observed when the intensity of the 10-Hz light pulses was raised into the region $(0.5 - 5) \times 10^{13}$ W/cm^2. In quantum electrodynamics, this phenomenon is described in terms of stimulated Compton scattering;[23] classically, it is viewed as the scattering of a charged particle off the electric and magnetic fields of the light.

3. *Rayleigh Scattering*[1,2,15,17–20,24]

As Young[24] points out, this term means different things to workers in different fields. Here we use the simple definition that Rayleigh scattering is the scattering of electromagnetic radiation by atoms and molecules without change of frequency. The main features can be brought out by a simple classical analysis based on (8-1-19) to (8-1-22). We first apply these equations to a model in which an atom or molecule contains a single electron with a natural frequency of oscillation ω_0.[1] The damping constant γ here corresponds to the decay constant of the motion and is reciprocal of the lifetime τ of the bound state in question. The excitation energy of this state is $\hbar\omega_0$. The power radiated by the electron oscillator can be calculated by (8-1-14), with the result

$$P = \frac{e^4 E_0^2 \omega^2}{12\pi\epsilon_0 m_0^2 c^3[(\omega^2 - \omega_0^2)^2 + \omega^2\gamma^2]}. \tag{8-1-39}$$

The cross section for elastic scattering is then calculated as in (8-1-15) and has the form

$$\sigma(\omega) = \frac{8\pi r_0^2}{3} \frac{\omega^4}{(\omega^2 - \omega_0^2)^2 + \omega^2\gamma^2}. \tag{8-1-40}$$

At low frequencies ω, the *Rayleigh law of scattering* applies:

$$\sigma(\omega) = \frac{8\pi r_0^2}{3}\left(\frac{\omega}{\omega_0}\right)^4 = \frac{8\pi r_0^2}{3}\left(\frac{\lambda_0}{\lambda}\right)^4 \qquad (\lambda \gg \lambda_0). \qquad (8\text{-}1\text{-}41)$$

It explains the scattering of visible sunlight by the earth's atmosphere (blue skies and red sunsets) and was first derived by Lord Rayleigh in 1871 on the basis of the elastic solid theory of the luminiferous ether.[24] Generalizing the analysis to target structures with more than one resonant frequency, ω_j, we find that at higher frequencies, such that ω is close to some ω_j associated with a transition from one target state to another, the cross section rises to an enormously high peak. Here we have resonance absorption followed by emission of photons of the original energy. The width of the peak is equal to γ. The scattering here is referred to as *resonance fluorescence.*[1] At still higher energies, for which ω greatly exceeds any ω_j, the target electrons may be regarded as essentially unbound, and we are in the region of Compton scattering.

4. *Tyndall Scattering*[24]

This term refers to the scattering of radiation from particles larger than single atoms or molecules. J. J. Thomson showed that when the size of the scattering particle is large compared with the wavelength of the incident radiation, the scattering is independent of λ. Structure in the cross section appears as the particle size is reduced toward λ. We shall not be further concerned with Tyndall scattering in this book.*

5. *Raman Scattering*[1–3,14,15,25–27]

Rayleigh scattering may be regarded as the absorption of a photon by a target atom or molecule, which is thereby raised to a virtual state that reemits a photon of the same energy in returning to the original state of the target. In *Raman scattering*, a similar mechanism is invoked, but the emitted photon has either a smaller energy than the incident photon (*Stokes scattering*) or a larger energy (*anti-Stokes scattering*). The three cases described here are illustrated in Fig. 8-1-5. If a laser beam is used as the incident radiation source, Raman scattering of extremely high intensity may be produced. The incident beam initially contains only the single frequency associated with the laser, but quickly a coherent set of photons with the frequency of the scattered light appears in the beam and stimulates the Stokes and anti-Stokes scattering processes. In this case we speak of *stimulated Raman scattering (SRS).*[1]

There are several reasons for the great interest in Raman scattering, which include the following:

*The term "Mie scattering" is often applied to Tyndall scattering from large particles.

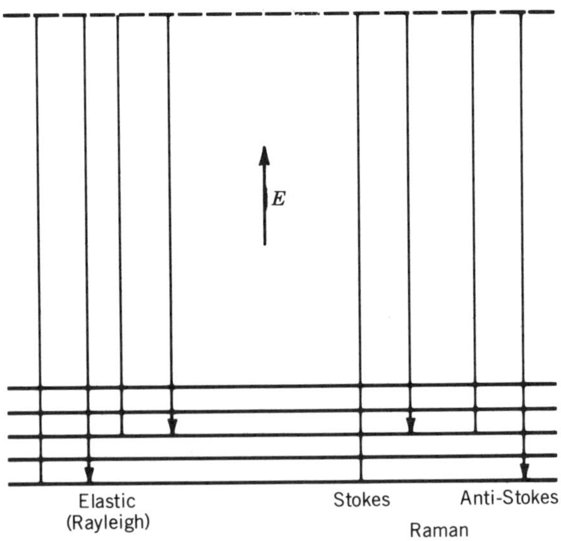

Figure 8-1-5. Illustration of photon energy relationships in Rayleigh and Raman scattering. Transitions occur here between discrete molecular levels (solid lines) and a virtual level (dashed line).

1. Molecules (such as the homonuclear species H_2, N_2, O_2, and F_2) that have no permanent electric dipole moment cannot provide a pure rotation or rotation-vibration spectrum.[2] However, unlike the one-step absorption process, Raman scattering is a two-step mechanism that leaves the molecule in a different rotation-vibration state from the original, so Raman spectroscopy provides the opportunity to study many important molecules that are inaccessible by other techniques.
2. Raman scattering provides an effective probe of molecules in highly excited vibrational levels. It can be used for any kind of molecule and does not require specific resonant transitions. It is also state specific. Intensity problems associated with small Raman scattering cross sections can be overcome by using SRS and coherent anti-Stokes Raman scattering (CARS) techniques.[2]
3. Raman scattering by molecules permits the generation of intense beams of radiation at frequencies not otherwise accessible.

6. *Brillouin Scattering*[1]

This type of scattering occurs in condensed matter when a dielectric nonuniformity is produced by acoustic waves. Brillouin scattering is not of concern to us here.

7. *Delbrück Scattering*[18,19]

This phenomenon is the elastic scattering of a photon by the Coulomb field of a heavy nucleus. It arises from the nonlinear nature of quantum electrodynamics, that is, the possibility of formation of virtual electron-positron pairs that may be scattered by the electromagnetic field before annihilating each other with the emission of a photon. The cross section for Delbrück scattering is extremely small at the energies of interest in atomic and molecular physics, so the process is of no interest here.

C. Absorption

The first four processes listed in this section are of fundamental importance in atomic collision physics, and will receive much attention later, although they are only briefly mentioned here. The remaining processes (inverse bremsstrahlung and pair production) are less important here and are only superficially described.

1. *Photoexcitation*

This well-known process may be described by the equation

$$\hbar\omega + \mathrm{T} \rightarrow \mathrm{T}^*, \tag{8-1-42}$$

where T represents the target (atomic, molecular, or ionic) in some particular quantum state (not necessarily the ground state), and T* the product, which has an energy greater than T by the amount $\hbar\omega$. It is the inverse of the emission process described in Section 8-1-A. Dissociative processes will be considered in the next subsection. Photoexcitation will be treated in great detail in Section 8-14, where the Einstein A and B coefficients, multipole radiation, and selection rules are discussed. Important differences between excitation by electrons and by photons will be brought out.

2. *Photodissociation*

The processes to be considered here are the photon-impact analogs of the electron-impact reactions 1, 2, 3, and 5 listed at the beginning of Section 5-5. The analog of reaction 3, namely, photoionization, is discussed in Section 8-16; the remaining (nonionizing) reactions are treated in Section 8-15.

3. *Photoionization*

We shall restrict this term to denote the photon-impact ejection of one or more electrons from an isolated atom or molecule, the simplest example being the single photoionization of an atom, A:

$$\hbar\omega + \mathrm{A} \rightarrow \mathrm{A}^+ + \mathrm{e}. \tag{8-1-43}$$

Photoejection of bound electrons can also occur in condensed matter, and indeed, this "photoeffect," or "photoelectric effect," is the most important attenuation mechanism for photons passing through matter at energies below about 0.1 MeV.[13,28] (The Compton effect is most important at photon energies between about 0.1 MeV and 1 MeV, as pair production is at higher energies.)

Electron-impact ionization processes were cataloged in Section 6-5, and again it is useful to draw analogies with the corresponding photon-impact reactions. We do this in Section 8-16, and discuss the ways in which they differ.

4. *Photodetachment*[29]

This process consists of the ejection of electrons from negative ions by photon impact; usually, it is the outermost electron that is detached. In the case of the removal of a single electron from an atomic ion, the reaction is described by the equation

$$\hbar\omega + \mathrm{A}^- \rightarrow \mathrm{A} + \mathrm{e}. \qquad (8\text{-}1\text{-}44)$$

This process will be discussed in a later chapter on negative ions. It differs from photoionization (3) primarily in the absence of the Coulomb interaction in the final state. In recent years, photodetachment has provided an extremely valuable tool for the investigation of atomic and molecular structures. Observations on photodetachment have yielded electron affinities accurate to 1 part in 10^7, and even greater accuracy appears to be forthcoming.

5. *Inverse Bremsstrahlung*[30]

In atomic and molecular physics, the important process of this type is the absorption of a photon by a free electron with an increase in its kinetic energy:

$$\hbar\omega + \mathrm{e} \rightarrow \mathrm{e}'. \qquad (8\text{-}1\text{-}45)$$

This reaction is clearly the inverse of the bremsstrahlung process described in Section 7-9 and is another example of a free-free transition.

6. *Pair Production*[1,13]

If a photon has sufficiently high energy, it may create a pair of particles and expire in the process. The total charge of the particles must be zero, the photon energy must exceed the sum of their rest energies, and another particle (or system of particles) must be on hand for conservation of momentum (Problem 8-22). Electron-positron pair production has a threshold of 1.022 MeV or higher, depending on the "particle" P providing the momentum conservation. Thus the photon is properly called a γ ray, and the reaction is written

$$\gamma + \mathrm{P} \rightarrow \mathrm{P} + \mathrm{e}^- + \mathrm{e}^+. \qquad (8\text{-}1\text{-}46)$$

In particular, electron-positron pair production in the field of a free electron,

$$\gamma + e^- \to e^- + e^- + e^+, \tag{8-1-47}$$

has a threshold of 2.044 MeV (Problem 8-23). Additional reactions involving heavier particles belong in the realm of elementary particle physics.

D. Applications

We are concerned in this volume less with applications than with fundamental processes and the mechanisms underlying them. However, the applications of atomic collision phenomena in nature, the laboratory, and in industry are perhaps more numerous and varied than for any other field of physics. Photon processes are no exception here, and the reader is encouraged to consult Ref. 31 for examples and explanations of many applications.

8-2. THE PRINCIPLE OF DETAILED BALANCING

Detailed balancing has its roots in classical statistical mechanics. We shall not trace the history of its development and widespread application, but rather only show how it arises in quantum mechanics from time-dependent perturbation theory. Our discussion is based on the report by Jackson.[32] As he points out, our approach is entirely satisfactory for atomic collision problems and for the emission and absorption of radiation, but it fails for nuclear reactions, in which the differences between the "real" problems and the model problems are so large that the perturbations produce large effects, inconsistent with the assumptions underlying perturbation theory. The results still hold in nuclear physics, but they must be justified in a different fashion. In nuclear physics, the terminology is also different; one usually speaks of "reciprocity"* rather than "detailed balancing."

A. Basis in Perturbation Theory

We first consider a quantum mechanical system that can make a transition from one quantum state i with energy E_i to another state f with energy E_f. If the Hamiltonian that causes the transition is H', and the matrix element of the perturbation is H'_{if}, the *transition probability per unit time* is, to first order,

$$w_{if} = \frac{2\pi}{\hbar} |H'_{if}|^2 \, \delta(E_i - E_f). \tag{8-2-1}$$

*Time-reversal invariance requires that the scattering matrix be symmetrical relative to the exchange of *reaction channels*; there is no symmetry, however, with respect to the exchange of *spin channels* because the spin direction is inverted under time reversal. This distinction is important in nuclear reactions. The property of "almost-symmetry" discussed here is called *reciprocity* [see S. de Benedetti, *Nuclear Interactions*, Wiley, New York (1964), pp. 329–330].

Equation (8-2-1), as well as the closely related equation

$$w_{if} = \frac{2\pi}{\hbar} |H'_{if}|^2 \rho(f), \tag{8-2-2a}$$

is frequently referred to as *Golden Rule No.* 2,[33] so named by Enrico Fermi.* Here $\rho(f)$ is the density of final states.

If, in (8-2-1), we make the usual requirement that H'_{if} be hermitian, so that $H'_{if} = H'^{*}_{fi}$, the resulting symmetry in i and f gives the analog of the microscopic reversibility of classical collisions:

$$w_{if} = w_{fi}. \tag{8-2-3}$$

We must modify (8-2-3) so that it will be useful for particles with internal as well as external degrees of freedom, and allow for the fact that states with the same external, but different internal, quantum numbers may be indistinguishable. The state notations i and f will now refer to a completely specified state, with all internal and external quantum numbers known. We need (8-2-3) averaged over the initial states that are experimentally indistinguishable in the given situation, and summed over the corresponding final states. Let g, called the *statistical weight*, be the number of indistinguishable states. They are usually degenerate in energy, or nearly so. A frequently occurring situation involves n subsystems in a given state, with g being given as the product of the statistical weights of the magnetic substates:

$$g = (2j_1 + 1)(2j_2 + 1)\cdots(2j_n + 1). \tag{8-2-4}$$

Here j_m denotes the total internal angular momentum of the mth subsystem.

We now define the average value of the perturbation matrix element, modulus squared, the average being taken over the indistinguishable substates:

$$\overline{|H'_{if}|^2} \equiv \frac{1}{g_i}\frac{1}{g_f}\sum_i \sum_f |H'_{if}|^2. \tag{8-2-5}$$

The sums are taken over the degenerate substates at the energies E_i and E_f.

B. States in the Energy Continuum

Now suppose that we are interested in a transition from or into a state in the energy continuum, in which case we cannot distinguish between states lying in the same narrow energy range ΔE. Take the final state to be the one in the

*If $H'_{if} = 0$, but there exist one or more states n such that neither H'_{in} nor H'_{nf} equals zero, the transition may occur in second order through the intermediate states. The transition rate is then given by *Golden Rule No.* 1:

$$w_{if} = \frac{2\pi}{\hbar} \sum_n{}' \left|\frac{H'_{in}H'_{nf}}{E_i - E_n}\right|^2 \rho(f). \tag{8-2-2b}$$

continuum, and define $\rho_f(E)\,d\Omega\,dE$ to be the number of states in the energy interval dE at energy E, with momentum direction within the element of solid angle $d\Omega$ (see Note 8-1). To a particle with statistical weight g, there are $g\,\rho_f(E)$ states available per unit energy interval, per unit solid angle.

We define the transition probability $T_{IF}\,d\Omega$ to be the probability per unit time for a transition from any one of the g_i initial states at energy E_i to any one of the $g_f \rho_f(E_f)\,\Delta E\,d\Omega$ states of the final system:

$$T_{IF}\,d\Omega = \frac{1}{g_i} \sum_i \sum_F w_{if}, \tag{8-2-6}$$

where $\sum_F$ indicates a sum over the degenerate states (f) (i.e., $\sum_f$) and an integration over the continuum $\int_{\Delta E} \rho_f(E)\,dE$. Then (8-2-1) in (8-2-6) gives

$$\begin{aligned} T_{IF}\,d\Omega &= \frac{2\pi}{g_i \hbar} \sum_i \sum_f \int_{\Delta E} |H'_{if}|^2 \rho_f(E_f) \delta(E_i - E_f)\,dE_f\,d\Omega \\ &= \frac{2\pi}{\hbar} \frac{1}{g_i} \sum_i \sum_f |H'_{if}|^2 \rho_f(E)\,d\Omega \\ &= \frac{2\pi}{\hbar} \overline{|H'_{if}|^2} g_f \rho_f(E)\,d\Omega. \end{aligned} \tag{8-2-7}$$

In the event that i, as well as f, is a continuum state, the transition probability per unit time for the inverse process is immediately seen to be

$$T_{FI}\,d\Omega = \frac{2\pi}{\hbar} \overline{|H'_{fi}|^2} g_i \rho_i(E)\,d\Omega. \tag{8-2-8}$$

One of the most useful forms of the principle of detailed balancing then follows from (8-2-7) and (8-2-8):

$$\frac{T_{IF}}{g_f \rho_f} = \frac{T_{FI}}{g_i \rho_i}. \tag{8-2-9}$$

This result is of more general validity than its derivation from the approximate perturbation theory might imply.

C. Evaluation of the Density of Continuum States

For application of (8-2-9) to specific problems, a knowledge of the form of the function $\rho(E)$ is required. Consider a single particle enclosed in a large volume V, and de Broglie waves with wave vector $\boldsymbol{\kappa} = \mathbf{p}/\hbar$. Then, as shown in the books cited in Refs. 14 and 34,

$$\rho(E)\,d\Omega\,dE = \frac{V\,d\boldsymbol{\kappa}}{(2\pi)^3}, \tag{8-2-10}$$

or if we use spherical polar coordinates in momentum space,

$$\rho(E) = \frac{V\kappa^2}{(2\pi)^3} \frac{d\kappa}{dE}, \tag{8-2-11}$$

where $d\boldsymbol{\kappa} = \kappa^2\, d\kappa\, d\Omega$. For generality, we use the relativistic relationship between κ and E:

$$E^2 = c^2p^2 + m_0^2c^4 = c^2\hbar^2\kappa^2 + m_0^2c^4. \tag{8-2-12}$$

Then

$$E\, dE = c^2\hbar^2\kappa\, d\kappa \tag{8-2-13}$$

and

$$\frac{d\kappa}{dE} = \frac{E}{c^2\hbar^2\kappa}. \tag{8-2-14}$$

Finally,

$$\rho(E) = \frac{V\kappa E}{(2\pi)^3\hbar^2c^2} = \frac{VpE}{(2\pi\hbar)^3c^2}. \tag{8-2-15}$$

Later we shall need two special forms of (8-2-15): For *nonrelativistic particles* ($E \approx m_0c^2$) we have

$$\rho(E) = \frac{Vm_0p}{(2\pi\hbar)^3}, \tag{8-2-16}$$

while, for *photons* ($E = \hbar\omega = cp$),

$$\rho(E) = \frac{V\omega^2}{(2\pi c)^3\hbar}. \tag{8-2-17}$$

If the final state is composed of more than one particle, ρ_f is the product of the individual densities of states, restricted by the necessary conservation requirements.

D. Detailed Balancing for Collision Cross Sections

Now consider the two "particles" in some initial state i that interact to form two "particles" in some final state f. The product particles are not necessarily the same as the reacting particles. We shall work in the center-of-mass frame and use the single-particle densities of states for the reduced system. If we have one incident particle in volume V, the particle flux is v_i/V, where $v_i = p_ic^2/E_i$ is the velocity of the particle. Then the cross section corresponding to the transition probability per unit time, $T_{IF}\,d\Omega$, is

$$d\sigma_{IF} = \frac{V}{v_i} T_{IF}\,d\Omega. \tag{8-2-18}$$

Using (8-2-18), we may express (8-2-9) in terms of the cross sections:

$$\frac{d\sigma_{IF}}{g_f v_f \rho_f} = \frac{d\sigma_{FI}}{g_i v_i \rho_i}. \tag{8-2-19}$$

But $\rho \sim pE$ and $v = pc^2/E$, so that $v\rho \sim p^2$, and finally,

$$\frac{d\sigma_{IF}}{g_f p_f^2} = \frac{d\sigma_{FI}}{g_i p_i^2}. \tag{8-2-20}$$

Equation (8-2-20) is the *principle of detailed balancing for collision cross sections* and is valid for relativistic, as well as nonrelativistic, velocities. We first displayed this relationship in (5-3-3) and used it several times subsequently.

E. Emission and Absorption of Radiation

The absorption process consists of the absorption of a photon of frequency ν_0 by a quantum system (say, an atom) in an initial state i, with energy E_i and statistical weight g_i, leaving the system in a state f, with energy $E_f = E_i + h\nu_0$ and statistical weight g_f. The reverse emission process consists of the transition from state f to state i with the emission of a photon of energy $h\nu_0$. The description of emission is readily at hand by means of (8-2-8), with ρ_i given by (8-2-17). A complication arises in absorption because (8-2-1) is not strictly valid for the transition probability per unit time w_{if}. The final system (the atom after being excited by absorbing the photon) is not in a stationary state; it has a finite lifetime against the spontaneous emission of radiation and accordingly, a finite energy width through the uncertainty principle. Hence the delta function in (8-2-1) must be replaced by a high, narrow resonance peak that is normalized to unity when integrated over energy. We then need the integrated transition probability per unit time:

$$\int w_{if}(\nu)\,d(h\nu) = \frac{2\pi}{\hbar} \overline{|H'_{if}|^2} g_f. \tag{8-2-21}$$

But the absorption cross section σ_{if} is related to w_{if} by the equation

$$\sigma_{if}(\nu) = \frac{V}{c} w_{if}(\nu), \tag{8-2-22}$$

so

$$\int \sigma_{if}(\nu)\, d(\nu) = \frac{V}{hc} \int w_{if}(\nu)\, d(h\nu) = \frac{V}{hc} \frac{2\pi}{\hbar} \overline{|H'_{if}|^2} g_f. \tag{8-2-23}$$

We may connect this result to the mean radiative lifetime against spontaneous emission (τ_{fi}) by integrating (8-2-8) over all angles:

$$\int T_{FI}\, d\Omega = \tau_{fi}^{-1} = \frac{2\pi}{\hbar} \overline{|H'_{if}|^2} 4\pi g_i \rho_i. \tag{8-2-24}$$

Equations (8-2-23) and (8-2-24) show that the mean radiative lifetime and the absorption cross section are related by

$$\tau_{fi} \int \sigma_{if}(\nu)\, d\nu = \frac{c^2}{8\pi v_0^2} \frac{g_f}{g_i}, \tag{8-2-25}$$

where g_i is the statistical weight for the emitting structure only. The statistical weight for the emitted photon (2, for the two possible states of polarization) is already included in the factor 8π. Only direct transitions from i to f are included here in the lifetime τ_{fi}, which will agree with the observed lifetime of the state f only if radiative decay to other states is negligible. Equation (8-2-25) is closely related to the Einstein A and B coefficients, discussed in Section 8-3.

F. Applications

Among the many pairs of mutually inverse processes to which the principle of detailed balancing has been applied are the following:

1. *The emission and absorption of radiation by an atom or molecule* [see (8-2-25)].
2. *The excitation and deexcitation of an atom or molecule by electron impact.* See Sections 5-3, 5-4, and 5-7 for discussions of rotational, vibrational, and electronic transitions.
3. *Dielectronic recombination and autoionization* (see Sections 7-1-D and 7-4-A, and also Ref. 35, pp. 589–590).
4. *Photoionization and radiative recombination* [see (8-1-43); Ref. 35, pp. 588–589; and Ref. 36, p. 107].
5. *Radiative attachment of an electron (to form a negative ion) and photodetachment* [see (8-1-44) and Ref. 35, pp. 385–386, 414].
6. *Charge transfer in the forward and backward directions* (see Ref. 36, pp. 106–107).
7. *Photodissociation of a deuteron and radiative neutron capture by a proton* (see Ref. 13, pp. 496–498).

8. *The reactions* $p^+ + p^+ \rightleftharpoons \pi^+ + d - 137\ MeV$. As indicated on pages 510, 511 and 756 of Ref. 13, the application of the principle of detailed balancing to these reactions has been used to determine the spin of the pion.
9. *Emission and absorption of radiation by a solid.* Kirchhoff's law states that the emissivity (e) and absorptivity (a) of any object are equal. It can be generalized to any frequency, so that $e(\omega) = a(\omega)$ for any surface in thermal equilibrium. Radiation of a frequency that is emitted by a solid strongly is also absorbed strongly—otherwise, the object's temperature would change because it would not be able to achieve thermal equilibrium with the radiation (see Ref. 37, pp. 96–97, 407–408).
10. *Chemical reactions in the forward and backward directions* (see Ref. 37, pp. 270–271).

PART $\mathscr{B}$. THEORY OF RADIATIVE TRANSITIONS IN ATOMS

Here we use the time-honored semiclassical method to discuss the emission and absorption of electromagnetic radiation by an isolated hydrogen atom, and then indicate briefly how the results can be applied to other atomic species. (No mention is made of molecules. We are not striving for anything approaching completeness in our coverage of the theory, but merely to develop some basic concepts and arrive at some fundamental results.) We are especially interested in transition probabilities and cross sections and in the selection rules for the various radiation multipoles. Our discussion of these topics is based mainly on the book by Woodgate.[38]

For the most part, we follow the semiclassical approach adopted in most texts on quantum mechanics and atomic structure because a full, rigorous treatment based on quantum electrodynamics (QED) would be inconsistent with the level of theory used elsewhere in this book, and could be undertaken only after a great deal of background development. Here we assume that the atom and the radiation field are sufficiently loosely coupled that we can consider the atom to be a separate entity, with stationary states. The wave function of the combined system is considered to be separable into a product of functions, one for the atom and one for the field. We do not quantize the electromagnetic field here, although we do describe this key feature of QED in Note 8-2. Our failure to quantize the electromagnetic field means that we cannot adequately deal with spontaneous emission directly, but relations among the Einstein A and B coefficients provide an indirect means for getting at this mechanism. These relations were developed in Einstein's landmark paper of 1917,[39] and we begin by studying his treatment.

We shall be concerned with transitions between discrete states in Sections 8-3 through 8-8; we shall consider transitions from discrete states into the continuum in Section 8-9.

8-3. EINSTEIN'S *A* AND *B* COEFFICIENTS

For simplicity, first consider a two-level atom, such as described in Fig. 8-3-1. The lower level has energy E_1 and degeneracy $g_1 = 1$, and the upper level E_2 and $g_2 = 3$. (When a magnetic field is applied, the upper level splits into three states.) At any instant t, the populations of the two levels are N_1 and N_2 per unit volume. Einstein considered three radiative processes: (1) *spontaneous emission* of radiation from level 2 to level 1, (2) *induced* (or *stimulated*) *emission* from 2 to 1, and (3) *absorption* of radiation in a transition from 1 to 2. Only emission or absorption transitions that satisfy the relation

$$\hbar\omega = E_2 - E_1 \tag{8-3-1}$$

are permitted, $\hbar\omega$ being the energy of the quantum that is emitted or absorbed (i.e., energy is conserved). Process 1 does not depend on an electromagnetic radiation field being applied. By contrast, the rates of processes 2 and 3 depend on the radiation energy density per unit angular frequency $\rho(\omega)$ to which the atom is coupled. [$\rho(\omega)\,d\omega$ is the electromagnetic energy per unit volume between frequencies ω and $\omega + d\omega$, with ω here satisfying (8-3-1).]

We now define the three Einstein coefficients (which are assumed to be independent of N_1 and N_2):

1. A is the probability of spontaneous decay from level 2 to level 1 per atom in level 2 per unit time.
2. B_{21} is the probability of stimulated emission, and decay from 2 to 1, per atom in level 2 per unit time per unit density of the radiation field at frequency ω.*

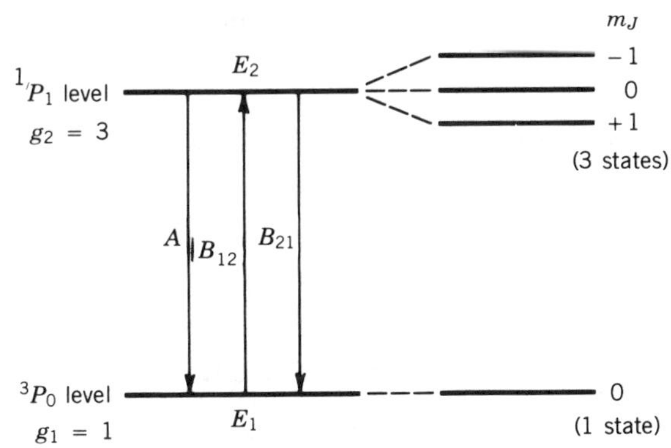

Figure 8-3-1. Illustration of a two-level atom and the Einstein A and B coefficients. The degeneracy is labeled g, and the total magnetic quantum number m_J.

*Here we define the B coefficients in terms of the radiation energy density $\rho(\omega)$. Sometimes they are defined in terms of the radiation intensity $I(\omega)$, and differ from ours by a factor of $4\pi/c$.

3. B_{12} is the probability of absorption, and excitation of the atom from level 1 to level 2, per atom in level 1 per unit time per unit energy density at frequency ω.

Then we can write the rate of change of N_2 as

$$-\dot{N}_2 = \dot{N}_1 = AN_2 - B_{12}\rho(\omega)N_1 + B_{21}\rho(\omega)N_2. \tag{8-3-2}$$

If no field is present ($\rho = 0$),

$$N_2(t) = N_2(0)e^{-At}, \tag{8-3-3}$$

so $A = 1/\tau$, where τ is the lifetime against spontaneous radiative decay of level 2.*

Now it is convenient to define the probabilities of emission and absorption per atom per unit time, P_{em} and P_{ab}, by the equation

$$-\dot{N}_2 = P_{em}N_2 - P_{ab}N_1, \tag{8-3-4}$$

and we see that

$$P_{em} = A + B_{21}\rho(\omega) \tag{8-3-5}$$

and

$$P_{ab} = B_{12}\rho(\omega). \tag{8-3-6}$$

In the steady-state situation $\dot{N}_2 = \dot{N}_1 = 0$, we have

$$\frac{P_{ab}}{P_{em}} = \frac{N_2}{N_1} = \frac{B_{12}\rho(\omega)}{A + B_{21}\rho(\omega)}. \tag{8-3-7}$$

Now if we consider an atom with many energy levels, the principle of detailed balancing requires that in steady state,

$$\frac{P_{ij}}{P_{ji}} = \frac{N_j}{N_i} = \frac{B_{ij}\rho(\omega_{ij})}{A_{ji} + B_{ji}\rho(\omega_{ij})} \tag{8-3-8}$$

for each pair of levels i and j. If the ensemble of atoms is in thermodynamic

*Spontaneous emission is actually not an inherent property of an isolated atom, but of an atom-vacuum system (see Note 8-2). New atomic and optical techniques have made it possible to control and manipulate spontaneous emission. The effect can be largely suppressed or enhanced by placing atoms between mirrors or in cavities. Spontaneous emission can be made to exhibit features of reversibility: an excited atom can exchange energy periodically with a cavity, rather than radiatively decaying to a lower energy state. See S. Haroche and D. Kleppner, *Phys. Today* **42**, 24 (Jan., 1989).

equilibrium with the radiation field at temperature T, the atoms have a Boltzmann distribution

$$\frac{N_j}{N_i} = \frac{g_j}{g_i} e^{-\hbar\omega_{ij}/kT}, \tag{8-3-9}$$

and the radiation field is characterized by the Planck blackbody relation

$$\rho(\omega_{ij}) = \frac{\omega_{ij}^2}{\pi^2 c^3} \hbar\omega_{ij} \frac{1}{e^{\hbar\omega_{ij}/kT} - 1}, \tag{8-3-10}$$

as discussed in Section 8-1-A-3. To obtain consistency among (8-3-8), (8-3-9), and (8-3-10), we require that

$$g_i B_{ij} = g_j B_{ji} \tag{8-3-11}$$

and

$$A_{ji} = \frac{\omega_{ij}^2}{\pi^2 c^3} \hbar\omega_{ij} B_{ji} = \frac{\omega_{ij}^2}{\pi^2 c^3} \hbar\omega_{ij} \frac{g_i}{g_j} B_{ij}. \tag{8-3-12}$$

Note that A_{ji}, B_{ij}, and B_{ji} relate to the structural properties of isolated atoms of whatever type we choose to discuss, and the method used to obtain the relations among these coefficients was based on statistical arguments! We now see that we can get information about A, the rate of spontaneous emission, from either of the B coefficients, which can be calculated without quantizing the electromagnetic field, as we shall show in Section 8-5.

8-4. SEMICLASSICAL INTERACTION OF A HYDROGEN ATOM WITH A RADIATION FIELD

We work with the hydrogen atom because its simplicity allows us to focus upon the interaction with the radiation field. Here we describe the atom quantum mechanically so that it has quantized energy levels and other quantal attributes; on the other hand, we treat the radiation field classically, so our discussion is semiclassical. Our treatment is completely nonrelativistic in that we use the Schrödinger wave equation and ignore spin.

In describing the radiation field, we shall use the electric and magnetic field vectors **E** and **B**, and the scalar and vector potentials ϕ and **A**, which are interrelated by the equations

$$\mathbf{E} = -\nabla\phi - \frac{\partial \mathbf{A}}{\partial t} \tag{8-4-1}$$

and

$$\mathbf{B} = \nabla \times \mathbf{A}. \tag{8-4-2}$$

If we let **p** represent the momentum of the atomic electron, the Hamiltonian of the system is

$$H = \frac{1}{2m}(\mathbf{p} + e\mathbf{A})^2 - e\phi + V(r) \tag{8-4-3}$$

(see Bransden and Joachain,[14] pp. 629–630).* We split H into two parts,

$$H = H_0 + H', \tag{8-4-4}$$

the first of which,

$$H_0 = \frac{p^2}{2m} + V(r), \tag{8-4-5}$$

pertains to the unperturbed atom (no field present), while the second,

$$H' = \frac{e}{2m}(\mathbf{p}\cdot\mathbf{A} + \mathbf{A}\cdot\mathbf{p}) + \frac{e^2}{2m}A^2 - e\phi, \tag{8-4-6}$$

represents the effect of the field. H' is supposed to be a small perturbation, consistent with our assumption of weak coupling between the atom and the field.

We take the radiation field to be an incoherent superposition of plane waves, each of which has the form

$$\phi = 0; \qquad \mathbf{A} = 2\mathbf{A}_0 \cos(\omega t - \mathbf{k}\cdot\mathbf{r}) = \mathbf{A}_0 e^{i(\omega t - \mathbf{k}\cdot\mathbf{r})} + \mathbf{A}_0 e^{-i(\omega t - \mathbf{k}\cdot\mathbf{r})}, \tag{8-4-7}$$

where $\mathbf{A}_0 = \mathbf{A}_0(\omega)$. Each constituent wave may be regarded as a Fourier component of any desired radiation field. Equations (8-4-1) and (8-4-2) show that for each constituent plane wave,

$$\mathbf{E} = 2\omega\mathbf{A}_0 \sin(\omega t - \mathbf{k}\cdot\mathbf{r}) \tag{8-4-8}$$

and

$$\mathbf{B} = 2(\mathbf{k} \times \mathbf{A}_0)\sin(\omega t - \mathbf{k}\cdot\mathbf{r}), \tag{8-4-9}$$

*In this chapter we use the symbol m, rather than M_r, to denote the electron-proton reduced mass.

so the amplitude of the **E** field is $\mathbf{E}_0 = \mathbf{E}_0(\omega) = 2\omega\mathbf{A}_0$ and that of the **B** field is $\mathbf{B}_0 = \mathbf{B}_0(\omega) = 2\mathbf{k} \times \mathbf{A}_0$. Note that the direction of polarization of the **E** wave is the direction of **A**.

We choose the Coulomb, or transverse, gauge,[8,9]*

$$\nabla \cdot \mathbf{A} = 0, \tag{8-4-10}$$

so the wave must be transverse and **p** must commute with **A**:

$$\mathbf{p} \cdot \mathbf{A} - \mathbf{A} \cdot \mathbf{p} = 0. \tag{8-4-11}$$

The perturbation Hamiltonian now reduces to

$$H' = \frac{e}{m} \mathbf{A} \cdot \mathbf{p} + \frac{e^2}{2m} A^2, \tag{8-4-12}$$

but the term in A^2 relates to two-photon processes, which we disallow because of our assumption of weak interaction of the atom with the radiation field. Accordingly, our final expression for the perturbation is

$$H' = \frac{e}{m} \mathbf{A} \cdot \mathbf{p}, \tag{8-4-13}$$

which has the effect of inducing the atom to emit or absorb *one photon*. We note that the Hamiltonian

$$H = \frac{p^2}{2m} + V(r) + \frac{e}{m} \mathbf{A} \cdot \mathbf{p} \tag{8-4-14}$$

is hermitian.

8-5. TRANSITION PROBABILITIES FOR EMISSION AND ABSORPTION BETWEEN DISCRETE STATES

In this section we apply a particular version of time-dependent perturbation theory (Dirac's method of variation of constants) to the problem of the absorption and stimulated emission of radiation by the hydrogen atom. We assume that we have already solved the time-independent Schrödinger wave equation (the energy eigenvalue equation),

$$H_0 \psi_n(\mathbf{r}) = E_n \psi_n(\mathbf{r}), \tag{8-5-1}$$

*The relations preceding (8-4-10) do not fully determine **A**; to pin it down completely, we must select a gauge. We emphasize that the choice of gauge does not affect any physical result.

to obtain the energy eigenfunctions $\psi_n(\mathbf{r})$ and eigenvalues E_n for the unperturbed atom. Then each of the stationary states of the isolated H atom can be described by a time-dependent wave function of the form

$$\psi_n(\mathbf{r}, t) = \psi_n(\mathbf{r})e^{-iE_nt/\hbar}. \tag{8-5-2}$$

For simplicity, we assume that the energy levels are nondegenerate. Because of the completeness of the set of ψ_n, an arbitrary state of the atom can be represented by the expansion

$$\psi(\mathbf{r}, t) = \sum_n c_n(t)\psi_n(\mathbf{r}, t) = \sum_n c_n(t)\psi_n(\mathbf{r})e^{-iE_nt/\hbar}. \tag{8-5-3}$$

The expansion coefficients $c_n(t)$ are time dependent and are normalized, and we set $c_i(0) = 1$ and all the other c_n equal to zero at $t = 0$ (i.e., we assume that the atom is in stationary state i at $t = 0$). We shall calculate the probability $|c_j(t)|^2$ that after a time t has elapsed, the atom is in state j because of the action of the radiation field.

The evolution of the system is described by the Schrödinger time-dependent wave equation

$$i\hbar\frac{\partial\Psi}{\partial t} = H\Psi, \tag{8-5-4}$$

which becomes, upon use of (8-5-3) and (8-5-1),

$$\sum_n i\hbar\dot{c}_n\psi_n \exp\left(\frac{-iE_nt}{\hbar}\right) = \sum_n c_nH'\psi_n \exp\left(\frac{-iE_nt}{\hbar}\right). \tag{8-5-5}$$

Since the perturbation, H', is small, we can approximate the c_n's on the right side of (8-5-5) by their initial values, and if we multiply the resulting equation on the left by ψ_j^* and integrate over spatial coordinates, we obtain

$$i\hbar\dot{c}_j = \langle j|H'|i\rangle \exp\left[\frac{i(E_j - E_i)t}{\hbar}\right] \qquad (j \neq i). \tag{8-5-6}$$

Now introduce the time dependence of H' explicitly from (8-4-13) and (8-4-7), and let $\omega_{ij} = (E_j - E_i)/\hbar$. Then solve the resulting equation subject to the condition $c_{j\neq i}(0) = 0$ to obtain the time-dependent expansion coefficients for transitions from an initial state i to final states j:

$$\begin{aligned} c_j(t) = {} & \left\langle j\left|\frac{e}{m}e^{-i\mathbf{k}\cdot\mathbf{r}}\hat{e}\cdot\mathbf{p}\right|i\right\rangle A_0\frac{1 - e^{i(\omega_{ji}+\omega)t}}{\hbar(\omega_{ji} + \omega)} \\ & + \left\langle j\left|\frac{e}{m}e^{i\mathbf{k}\cdot\mathbf{r}}\hat{e}\cdot\mathbf{p}\right|i\right\rangle A_0\frac{1 - e^{i(\omega_{ji}-\omega)t}}{\hbar(\omega_{ji} - \omega)}, \end{aligned} \tag{8-5-7}$$

in which we have introduced, following Woodgate,[38] the unit vector $\hat{e}$ to describe the direction of polarization. That is, we put $\mathbf{A}_0 = A_0\hat{e}$.

Now examine the *first term* on the right-hand side of (8-5-7): $\omega_{ji} = (E_j - E_i)/\hbar$ is resonant in the denominator $\hbar(\omega_{ji} + \omega)$ when ω_{ji} is negative (i.e., when $E_i > E_j$), corresponding to *stimulated emission.* On the other hand, in the *second term* on the right, $\omega_{ji} = (E_j - E_i)/\hbar$ is resonant in the denominator $\hbar(\omega_{ji} - \omega)$ when ω_{ji} is positive (i.e., when $E_i < E_j$), corresponding to *absorption.* These considerations suggest that we should be able to use the two terms separately, and that each of the two radiative processes might be adequately described by a single term. On page 39 of his book, Woodgate[38] states that making this "rotating wave approximation" (RWA) gives excellent results in the optical region and is adequate for most purposes at radio frequencies. Accordingly, if we examine the case of $E_j > E_i$, and positive ω_{ji}, we can ignore the first (nonresonant) term in (8-5-7) in comparison with the second (resonant) term and obtain, for the *absorption* process,

$$|c_j(t)|^2 = \left|\left\langle j \left| \frac{e}{m} e^{i\mathbf{k}\cdot\mathbf{r}} \hat{e}\cdot\mathbf{p} \right| i \right\rangle\right|^2 A_0^2 \frac{\sin^2 [(\omega_{ji} - \omega)/2]t}{\hbar^2[(\omega_{ji} - \omega)/2]^2}. \tag{8-5-8}$$

This result, however, does not realistically describe the process under investigation here, because the transition probability $|c_j(t)|^2$ is seen to increase not as t, but as t^2, at small t. The difficulty arose because we have been dealing with a single frequency, and no physical wave has this property. (The uncertainty principle would require that a wave train be infinitely long in order that it be purely sinusoidal.) We must integrate over the frequency distribution of the incident wave that exists in reality.

The first step in this integration is to use (8-4-8) to write down the energy per unit volume for each constituent plane wave in the radiation field in terms of the amplitude A_0:

$$\tfrac{1}{2}\epsilon_0 E_0^2 = 2\epsilon_0\omega^2 A_0^2. \tag{8-5-9}$$

Then we replace A_0^2 by $(1/2\epsilon_0)(\rho(\omega)/\omega^2)\,d\omega$ in (8-5-8) and integrate between the limits ω_1 and ω_2 that include the resonant frequency ω_{ji}. Upon dividing the result by t, we obtain the transition rate for absorption

$$\frac{|c_j(t)|^2}{t} = \frac{2}{\epsilon_0\hbar^2}\left|\left\langle j \left| \frac{e}{m} e^{i\mathbf{k}\cdot\mathbf{r}} \hat{e}\cdot\mathbf{p} \right| i \right\rangle\right|^2 \int_{\omega_1}^{\omega_2} \frac{\rho(\omega)}{\omega^2} \frac{\sin^2 [(\omega_{ji} - \omega)/2]t}{(\omega_{ji} - \omega)^2 t}\,d\omega. \tag{8-5-10}$$

Evaluating the integral in the limit of large t, we get the final form for the *transition probability per unit time for absorption of unpolarized incoherent radiation*:

$$\frac{|c_j(t)|^2}{t} = \frac{\pi}{\epsilon_0\hbar^2\omega_{ji}^2}\left|\left\langle j \left| \frac{e}{m} e^{i\mathbf{k}\cdot\mathbf{r}} \hat{e}\cdot\mathbf{p} \right| i \right\rangle\right|^2 \rho(\omega_{ji}) \qquad (\text{H atom}, i \to j). \tag{8-5-11}$$

(Recall that $\mathbf{p}$ is the momentum vector and $\mathbf{r}$ the position vector of the atomic electron.) Note that we did not integrate over the frequency distribution of the plane waves until we had progressed to the point of discussing probabilities, rather than probability amplitudes, because of our assumption that no definite phase exists between the waves of different frequency. If there are no phase correlations between the different Fourier components of the radiation, we may calculate the effect of each component separately, and then sum over the different components. In other words, we assume that there are no interference effects in the interaction of the radiation with the atom. If we add the various frequency components of the radiation before we take the modulus square of $c_j(t)$, we get cross terms between components of different frequency and hence cannot calculate the effect of each component of radiation separately. Differences between coherent and incoherent radiation are discussed by Read,[1] by Bethe and Jackiw on pages 192–201 of their book,[40] and in books on lasers.[3]

The quantity displayed in (8-5-11) is called the *transition rate for absorption*,[14] defined as

$$W_{ji} = \frac{d}{dt}|c_j(t)|^2. \tag{8-5-12}$$

Several comments are in order:

1. The value of the transition rate depends on the modulus square of a matrix element connecting the initial and final states, consistent with Golden Rule No. 2 of perturbation theory, (8-2-2*a*).
2. If, in (8-5-11), we interchange i and j and replace $e^{i\mathbf{k}\cdot\mathbf{r}}$ by $e^{-i\mathbf{k}\cdot\mathbf{r}}$, we obtain the *transition rate for induced emission*:

$$\frac{|c_i(t)|^2}{t} = \frac{\pi}{\epsilon_0 \hbar^2 \omega_{ij}^2} \left| \left\langle i \left| \frac{e}{m} e^{-i\mathbf{k}\cdot\mathbf{r}} \hat{e}\cdot\mathbf{p} \right| j \right\rangle \right|^2 \rho(\omega_{ij}). \tag{8-5-13}$$

3. Noting that $|c_j(t)|^2$ in (8-5-12) is the same as P_{ab} in (8-3-6), we can identify the Einstein B_{ij} absorption coefficient as being

$$B_{ij} = \frac{\pi}{\epsilon_0 \hbar^2 \omega_{ij}^2} \left| \left\langle j \left| \frac{e}{m} e^{i\mathbf{k}\cdot\mathbf{r}} \hat{e}\cdot\mathbf{p} \right| i \right\rangle \right|^2, \tag{8-5-14}$$

   and the other coefficients, A_{ji} and B_{ji}, follow from (8-3-12).
4. The transition rate for absorption or emission is appreciable only if the Bohr frequency condition $\hbar\omega = E_j - E_i$ is obeyed. Thus energy conservation is assured.
5. To treat a multielectron atom, we sum over the electrons s and write $\sum_s e^{i\mathbf{k}\cdot\mathbf{r}_s}\mathbf{p}_s$ in (8-5-11). Here $\mathbf{p}_s$ and $\mathbf{r}_s$ are the momentum and position vectors of the sth electron.

6. Woodgate[38] points out that implicit in our derivation is the assumption of a low radiation power level, and that our results are not valid for an intense laser field.* Furthermore, the bandwidth of the laser radiation in some experiments may be much smaller than the line width of the transition being studied, and different theoretical approaches are required in this case.

8-6. ELECTRIC DIPOLE TRANSITIONS BETWEEN DISCRETE STATES

Consider the absorption of radiation of wavelength $\lambda = 2\pi/k$ by a hydrogenic atom of atomic number Z. If λ is large compared with the size of the atom (which we take as a_0, the first Bohr radius, divided by Z), the value of the electric field intensity at any instant of time is approximately uniform over the dimensions of the atom. Hence all of the electronic charge distribution is driven essentially in phase, so that an electric dipole type of charge oscillation is produced. The exponential in (8-5-11) may be expanded to give

$$e^{i\mathbf{k}\cdot\mathbf{r}} = 1 + i\mathbf{k}\cdot\mathbf{r} + \frac{1}{2!}(i\mathbf{k}\cdot\mathbf{r})^2 + \cdots. \tag{8-6-1}$$

Using only the first term (1) in this multipole expansion corresponds to making the *electric dipole approximation*, and the transitions that are studied in this approximation are known as *electric dipole transitions*. The first term in the expansion is the dominant term in the optical region, in which

$$\frac{\lambda}{\text{size}} \approx \frac{2\pi}{Z\alpha} = 2\pi \times \frac{137}{Z} \tag{8-6-2}$$

according to the Bohr theory, with α being the fine structure constant.[38]

Now suppose that a hydrogen atom is exposed to radiation with the $\mathbf{E}$ vector polarized along the X-axis of the electron coordinate system. For the electric dipole approximation we need $\langle j|(e/m)p_x|i\rangle$, which can be cast into the desired form by using the commutation relation

$$\frac{\hbar}{im} p_x = \frac{1}{2m}[p_x^2, x] = [H_0, x], \tag{8-6-3}$$

where

$$H_0 = \frac{p_x^2}{2m} + \frac{p_y^2}{2m} + \frac{p_z^2}{2m} + V(r). \tag{8-6-4}$$

*In an intense laser field, the atom may be affected so strongly that a perturbation treatment may not even be warranted.

The energies E_j and E_i appear through the use of the eigenvalue equation (8-5-1). The result for the *transition probability per unit time for absorption of X-polarized electric-dipole radiation* by a hydrogen atom assumed to have no degeneracy is

$$P_{\text{ab}} = B_{ij}\rho(\omega_{ji}) = \frac{\pi}{\epsilon_0 \hbar^2} |\langle j|ex|i\rangle|^2 \rho(\omega_{ji}) \qquad (i \to j), \tag{8-6-5}$$

where $\hbar\omega_{ij} = E_j - E_i$.

For *unpolarized electric dipole radiation*, the *absorption transition rate* is

$$P_{\text{ab}} = \frac{1}{3}\frac{\pi}{\epsilon_0 \hbar^2} |\langle j|e\mathbf{r}|i\rangle|^2 \rho(\omega_{ji}) \qquad \text{(unpolarized)}, \tag{8-6-6}$$

where

$$|\langle j|e\mathbf{r}|i\rangle|^2 = |\langle j|ex|i\rangle|^2 + |\langle j|ey|i\rangle|^2 + |\langle j|ez|i\rangle|^2. \tag{8-6-7}$$

It then follows that the *spontaneous emission of unpolarized radiation* in an electric dipole transition between nondegenerate levels of a hydrogen atom is characterized by the Einstein coefficient

$$A_{ji} = \frac{1}{4\pi\epsilon_0}\left(\frac{4}{3}\right)\frac{\omega_{ji}^3}{\hbar c^3} |\langle j|e\mathbf{r}|i\rangle|^2. \tag{8-6-8}$$

Note the presence of the eponymous electric dipole moment ($e\mathbf{r}$) in this equation—electric dipole transitions are indeed aptly named.

If we approximate the matrix element in (8-6-8) by (ea_0), we find that $A \approx 10^8$ per second in the optical region, corresponding to an excited state lifetime of about 10^{-8} s against spontaneous decay by the emission of electric dipole radiation. The strong frequency dependence of A makes spontaneous decay very slow in the microwave and RF regions.

On pages 43–44 of his book, Woodgate[38] shows how our equations must be modified to take degeneracy into account. Also, on pages 44–46, he discusses the classical electron oscillator model of an atom and introduces the oscillator strength, f_{ij}, which we encountered in Section 5-7-F.

8-7. ELECTRIC DIPOLE SELECTION RULES[38]

Let us recall that the transition rates for electric dipole transitions are controlled by the modulus squared of the matrix elements $\langle j|x|i\rangle$, $\langle j|y|i\rangle$, and $\langle j|z|i\rangle$, and that the eigenfunctions of the undisturbed hydrogen atom are

$$\psi_{nlm_l} = R_{nl}(r)P_l^{m_l}(\cos\theta)e^{-im_l\phi}. \tag{8-7-1}$$

The $P_l^{m_l}(\cos\theta)$ functions are the associated Legendre functions, which give the spherical harmonics when multiplied by $e^{-im_l\phi}$. We can now easily indicate how the selection rules for the emission and absorption of electric dipole radiation are obtained. Establish a unique direction by applying an external magnetic field $\mathbf{B}_{\text{ext}}$, and let the Z axis lie in this direction. The resulting Zeeman effect lifts the degeneracy in the orbital magnetic quantum number m_l. We must determine under what conditions the matrix elements containing the separate components of $\mathbf{r}$ do not vanish.

1. For the $z = r\cos\theta$ *component*, the relevant matrix element in (8-5-11) becomes $\langle n'l'm_l'|z|nlm_l\rangle$, which involves three integrals, in r, θ, and ϕ. The ϕ integral vanishes unless $m_l' = m_l$ (i.e., unless $\Delta m_l = 0$).
2. For the $x = r\sin\theta\cos\phi$ and $y = r\sin\theta\sin\phi$ components, the relevant matrix elements are $\langle n'l'm_l'|x|nlm_l\rangle$, and $\langle n'l'm_l'|y|nlm_l\rangle$. In each case, the ϕ integral vanishes unless $\Delta m_l = \pm 1$.

The selection rule $\Delta m_l = 0$ corresponds to the case where the $\mathbf{E}$ vector is parallel to $\mathbf{B}_{\text{ext}}$ ("parallel," or "π," polarization). The selection rule $\Delta m_l = \pm 1$ applies to the situation in which the $\mathbf{E}$ vector is perpendicular to $\mathbf{B}_{\text{ext}}$ ("senkrecht," or "σ," polarization). These relationships arise as follows. The original matrix element in (8-5-7) contained $\mathbf{A}_0 \cdot \mathbf{p}$, but by (8-6-4) and (8-4-8), the matrix element depends essentially on $\mathbf{E}_0 \cdot \mathbf{r}$. Now $\mathbf{B}_{\text{ext}}$ is along the Z axis, so if $\mathbf{E}$ is polarized along Z, then we need the "z" matrix element, and so on. In reality, the electron spin must be taken into account, and we can obtain the selection rules $\Delta M_J = 0$ and $\Delta M_J = \pm 1$, where M_J is the magnetic quantum number corresponding to the total angular momentum J. Figure 8-7-1 shows the connections between each of these selection rules and the directions of $\mathbf{B}_{\text{ext}}$, $\mathbf{E}_0$, and $\mathbf{k}$ for π and σ polarizations, and for left- and right-hand circular polarizations (see Woodgate,[38] pp. 160–166).

To obtain the electric dipole selection rule for the orbital angular momentum quantum number l, we examine the x, y, and z components of the matrix element as before, but now we focus our attention on the θ integral in each case. Using the recursion relations among the $P_l^{m_l}(\cos\theta)$ and the orthogonality of these functions, we find that all three components give the same selection rule: $\Delta l = \pm 1$.

In summary, the *allowed transitions* in the electric dipole approximation are those for which

$$\Delta l = \pm 1 \tag{8-7-2}$$

and

$$\Delta m_l = 0, \pm 1. \tag{8-7-3}$$

All other transitions are said to be *forbidden*, it being understood that such transitions may actually take place by the emission or absorption of higher-order radiation, although generally with much smaller probability.

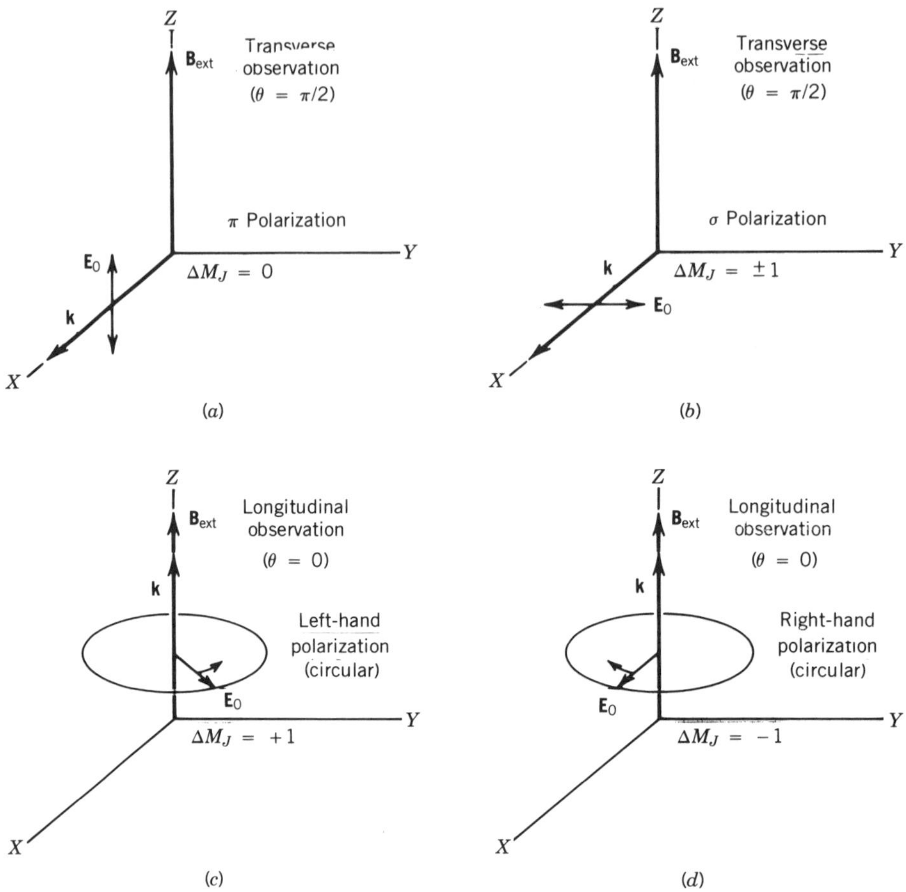

Figure 8-7-1. Connection between selection rules for M_J and the directions of $\mathbf{B}_{ext}$, $\mathbf{E}_0$ and $\mathbf{k}$ for (*a*) π polarization, (*b*) σ polarization, (*c*) left-hand polarization, and (*d*) right-hand polarization. The wave travels in the direction of $\mathbf{k}$, with polarization direction given by $\mathbf{E}_0$. In (*c*) and (*d*), look at the oncoming photon and use the right-hand rule to get the direction of circular polarization. In (*d*), M_J of the atom changes by $-h$, and a right circularly polarized photon carries angular momentum $+h$ along its direction of motion. If $\mathbf{k}$ is along the $-Z$ axis and $\Delta M_J = -1$, a left circularly polarized photon has angular momentum $-h$ along its direction of travel and $+h$ along the $+Z$ axis. A photon emitted in an arbitrary direction is elliptically polarized.

In the central-field approximation of atomic structure (see Woodgate,[38] Chap. 6), each electron in an atom is considered to move in a central field, so that the angular part of its wave function is hydrogenic. Consequently, the selection rules discussed above for the hydrogen atom still apply, because only the angular part of the wave function is used to derive them. For an electron moving in a *noncentral* field, the quantum numbers l and m_l are not well defined, but there are still rigorous selection rules based on parity considerations.

The use of parity when we consider transitions between two states of well-defined l is illustrated by the *Laporte rule*, formulated in 1924 before the advent of quantum mechanics. The parity of each state is odd if l is odd, and even if l is even, while the parity of the radius vector $\mathbf{r}$ is odd. Hence, for states of well-defined l, there is the strict selection rule

$$\Delta l = \text{odd} \tag{8-7-4}$$

for electric dipole transitions.

8-8. HIGHER-ORDER TRANSITIONS BETWEEN DISCRETE STATES

We now proceed to show that inclusion of the second term, $e^{i\mathbf{k}\cdot\mathbf{r}}$, in the multipole expansion (8-6-1) brings in two new higher-order transitions: the magnetic dipole and the electric quadrupole. To demonstrate this fact, we begin by recalling that in the matrix element of (8-5-11), the significant component of $\mathbf{p}$ is that in the direction of $\mathbf{A}_0$. But the wave is transverse, with $\mathbf{k}\cdot\mathbf{A}_0 = 0$, so the components of $\mathbf{p}$ that are of interest are those for which $\mathbf{k}\cdot\mathbf{p} = 0$. Now consider the special case in which $\mathbf{k}$ is along the X axis of the electron coordinate system, and look only at that part of the interaction involving the y component of $\mathbf{p}$. In this case we must deal with the modulus squared matrix element

$$\left|\left\langle j\left|\frac{e}{m}p_y(1 + ik_x x)\right|i\right\rangle\right|^2, \tag{8-8-1}$$

where $k_x = \omega/c$. Looking at the object $(1 + ik_x x)$ in the middle of the matrix element, we see that we do not have to deal with the cross term that arises formally when the matrix element is squared. If the initial and final states have well-defined parity (definite values of l), the electric dipole operator connects only states of the opposite parity, whereas the second-order operator connects only states of the same parity. Hence the first- and second-order terms are mutually exclusive. In any event, we are interested in the second-order term only when the normally larger first-order term vanishes because of the electric dipole selection rules. Consequently, here we are concerned with the quantity

$$\left|\left\langle j\left|i\frac{e}{m}\frac{\omega}{c}p_y x\right|i\right\rangle\right|^2. \tag{8-8-2}$$

Now $[p_y, x] = 0$, so we can write

$$\begin{aligned} xp_y &= \tfrac{1}{2}(xp_y - p_x y) + \tfrac{1}{2}(xp_y + p_x y) \\ &= \tfrac{1}{2}\hbar l_z + \tfrac{1}{2}m(x\dot{y} + \dot{x}y), \end{aligned} \tag{8-8-3}$$

where we have introduced the z component, l_z, of the orbital angular momentum. Associated with l_z is the z component of the orbital magnetic dipole moment of the atom:

$$\mu_{l_z} = \frac{e\hbar}{2m} l_z. \tag{8-8-4}$$

Accordingly, one of the contributions of the second-order multipole term has the form

$$\frac{\omega^2}{c^2} |\langle j|\mu_{l_z}|i\rangle|^2, \tag{8-8-5}$$

which is manifestly of magnetic dipole character, so here we speak of *magnetic dipole transitions.* (If electron spin had been included in our treatment, the z component of the spin magnetic moment $\mu_{s_z} = g_s\mu_B s_z$ would be added to μ_{l_z}. Here $g_s \approx 2$ is the spin g factor,* μ_B is the Bohr magneton, and s_z is the z component of the electron spin angular momentum.)

Now we seek out the electric quadrupole term. The second term of (8-8-3) can be written

$$\begin{aligned}\frac{1}{2}m(x\dot{y} + \dot{x}y) &= \frac{1}{2}m\frac{i}{\hbar}(xH_0y - xyH_0 + H_0xy - xH_0y)\\ &= \frac{1}{2}m\frac{i}{\hbar}(H_0xy - xyH_0),\end{aligned} \tag{8-8-6}$$

so it makes a contribution of the form

$$\frac{1}{4}\frac{\omega^4}{c^2} |\langle j|exy|i\rangle|^2, \tag{8-8-7}$$

where we have put $(E_j - E_i)/\hbar = \omega$. The operator (exy) in the transition matrix element is one of the components of the second-rank electric quadrupole tensor $\mathcal{Q}$, defined in Section 5-2-A. A complete formulation of the problem would

*The g-factor of the free electron was first measured by H. R. Crane and his associates at the University of Michigan in a series of ingenious experiments during the period 1950–1972. They injected pulses of electrons into a long magnetic mirror trap and (in the later experiments) directly measured the small difference between the frequency of spin precession and the frequency of electron cyclotron motion. See H. R. Crane, *Sci. Amer.* **218**, 72 (1968); A. Rich and J. C. Wesley, *Rev. Mod. Phys.* **44**, 250 (1972). More recently, measurements on a single electron in a static Penning trap have been performed at the University of Washington, with greater precision. The electron spin g-factor is now one of the most precisely known physical quantities, and its experimental value provides one of the most sensitive tests of quantum electrodynamics. See P. Ekstrom and D. Wineland, *Sci. Amer.* **243**, 105 (Aug., 1980); G. Gabrielse, H. Dehmelt, and W. Kells, *Phys. Rev. Lett.* **54**, 537 (1985); R. S. Van Dyck, P. B. Schwinberg, and H. G. Dehmelt, *Phys. Rev. Lett.* **59**, 26 (1987).

involve the use of the full tensor. The appearance of the electric quadrupole justifies the use of the term "electric quadrupole transitions."

The presence of *two* spatial coordinates, x and y, in (8-8-7) implies that electric quadrupole transitions are important when the wavelength of the radiation is comparable with the size of the atom,* and one would suppose that the same is true of magnetic dipole transitions since they emerge in the same mathematical treatment. These higher-order transitions are important mainly in the X-ray region. Pure electric quadrupole effects are associated with two parts of the electronic charge distribution oscillating like two dipoles out of phase, the dipole contribution canceling.

Using equations provided by the Bohr model, Read[1] shows that the transition rates for electric dipole, magnetic dipole, and electric quadrupole processes are in the approximate ratio

$$\begin{aligned} ED/MD/EQ &= 1/\alpha^2/(a_0/\lambda)^2 \qquad \left(\alpha = \frac{1}{137}\right) \\ &= 1/10^{-4}/10^{-8} \text{ for visible light.} \end{aligned} \tag{8-8-8}$$

Consequently, for the H atom, the lifetimes of states that can decay only by the emission of MD or EQ radiation are enormously larger than the lifetimes for ED decay.

Good discussions of the selection rules for higher-order multipoles appear in many books.[1,40-43] For *magnetic dipole transitions*, the selection rules are

$$\begin{aligned} \Delta l &= 0, \\ \Delta j &= 0, \pm 1 \qquad (j = 0 \leftrightarrow j' = 0 \text{ forbidden}) \\ \Delta m_j &= 0, \pm 1, \end{aligned} \tag{8-8-9}$$

where the appearance of the total quantum number j indicates that the electron spin must be considered. For *electric quadrupole radiation*, the appearance of products such as (xy) in the matrix elements shows that transitions occur only between states of the same parity. The selection rules are

$$\begin{aligned} \Delta l &= 0, \pm 2 \qquad (l = 0 \leftrightarrow l' = 0 \text{ forbidden}) \\ \Delta m &= 0, \pm 1, \pm 2. \end{aligned} \tag{8-8-10}$$

Transitions of higher order than MD and EQ (magnetic quadrupole, electric hexadecapole, etc.) are of practically no interest in atomic or molecular physics.

*Consider an element of charge with small cross section standing perpendicular to the X-Y plane at the point x, y. Under the influence of the **E** wave, it oscillates back and forth in the Y direction. The phase of its oscillation relative to that of other charge elements depends on its value of x. Effects arising from the arrival of the electromagnetic disturbance at different points within the atom at different times are said to be due to "retardation." There is no retardation in the electric dipole approximation.

8-9. TRANSITIONS INTO THE CONTINUUM; THE PHOTOIONIZATION CROSS SECTION

In this section we shall study photon-induced transitions from a discrete level of an atom into the ionization continuum, and calculate the cross section for photoionization of the hydrogen atom in its ground state. The approach will be the same as in Section 8-5: the atom is treated by nonrelativistic quantum mechanics, with no spin and no degeneracy; the radiation field is considered classically, and is taken to be incoherent; multiphoton processes are disallowed. We shall follow the discussion of Bransden and Joachain[14] and use their notation. The reader may consult pages 189–193 of Ref. 14 for their treatment of photoionization, and pages 155–179 for their discussion of the topics that we covered in Sections 8-3 through 8-8.

We begin with our equation (8-5-11) for the photoabsorption rate in a bound-bound transition, and we replace the electron momentum $\mathbf{p}$ by its operator in configuration space:

$$\mathbf{p} \to -i\hbar\nabla. \tag{8-9-1}$$

The result is, for the transition from state a to state b,

$$W_{ba} = \frac{4\pi^2}{m^2 c}\frac{e^2}{4\pi\epsilon_0}\frac{I(\omega_{ba})}{\omega_{ba}^2}|M_{ba}(\omega_{ba})|^2, \tag{8-9-2}$$

where the matrix element M_{ba} is defined as

$$M_{ba} = \langle\psi_b|e^{i\mathbf{k}\cdot\mathbf{r}}\hat{\epsilon}\cdot\nabla|\psi_a\rangle = \int \psi_b^*(\mathbf{r})e^{i\mathbf{k}\cdot\mathbf{r}}\hat{\epsilon}\cdot\nabla\psi_a(\mathbf{r})\,d\mathbf{r}. \tag{8-9-3}$$

$I(\omega) = c\rho(\omega)$ is the intensity per unit angular frequency of the incident radiation, and $\hbar\omega_{ba} = \hbar kc = E_b - E_a$. Now the *absorption cross section* σ_{ba} may be defined as the transition rate per unit atom, W_{ba}, per unit photon flux, $I(\omega_{ba})/\hbar\omega_{ba}$. Hence

$$\sigma_{ba} = \frac{4\pi^2\alpha\hbar^2}{m^2\omega_{ba}}|M_{ba}(\omega_{ba})|^2 \qquad \text{(discrete states } a \to b\text{)}, \tag{8-9-4}$$

where

$$\alpha = \frac{e^2}{4\pi\epsilon_0\hbar c} \approx \frac{1}{137} \tag{8-9-5}$$

is the fine-structure constant.

The cross section for the bound-free photoionization process can be obtained from (8-9-4) by integrating over the continuum states of the ejected electron. We take the H atom to be in its (1*s*) ground state, with

$$\Psi_a(\mathbf{r}, t) = \psi_{1s}(\mathbf{r})e^{-iE_{1s}t/\hbar} \tag{8-9-6}$$

and assume that the ejected electron can be represented by a plane wave, with

$$\Psi_b(\mathbf{k}_f, \mathbf{r}, t) = \psi_b(\mathbf{k}_f, \mathbf{r})e^{-iE_f t/\hbar} = \frac{1}{V^{1/2}} e^{i\mathbf{k}_f\cdot\mathbf{r}} e^{-iE_f t/\hbar}. \tag{8-9-7}$$

This assumption is reasonable for $\hbar^2 k_f^2/2m \gg |E_{1s}|$. In (8-9-7), V is the volume of the large box in which the system is normalized. We obtain

$$\sigma_{\text{photoion}} = \frac{4\pi^2\alpha\hbar^2}{m^2\omega}\tilde{\rho}_b \int |M_{ba}(\omega)|^2 \, d\Omega, \tag{8-9-8}$$

where

$$\tilde{\rho}_b = \frac{Vmk_f}{(2\pi)^3\hbar^2} \tag{8-9-9}$$

is the density of final states of the ejected electron (see Note 8-1), and $d\Omega\,(\theta, \phi)$ is the element of solid angle for the outgoing electron. Accordingly, the differential cross section for ejection of the bound electron into the solid angle $d\Omega$ in the direction (θ, ϕ) is

$$\frac{d\sigma_{\text{photoion}}}{d\Omega} = \frac{1}{(2\pi)^3}\frac{4\pi^2\hbar\alpha}{m}\frac{k_f}{\omega}|V^{1/2}M_{ba}(\omega)|^2, \tag{8-9-10}$$

where here the matrix element becomes

$$M_{ba} = \frac{1}{V^{1/2}}\int e^{-i\mathbf{k}_f\cdot\mathbf{r}} e^{i\mathbf{k}\cdot\mathbf{r}} \hat{\epsilon}\cdot\nabla\psi_{1s}(r)\, d\mathbf{r}. \tag{8-9-11}$$

We now integrate by parts to obtain

$$V^{1/2}M_{ba} = -i\hat{\epsilon}\cdot(\mathbf{k} - \mathbf{k}_f)\int e^{i(\mathbf{k}-\mathbf{k}_f)\cdot\mathbf{r}}\psi_{1s}(r)\, d\mathbf{r} \tag{8-9-12}$$

and use the fact that $\mathbf{k}\cdot\hat{\epsilon} = 0$ to write

$$\hat{\epsilon}\cdot(\mathbf{k} - \mathbf{k}_f) = -k_f\cos\gamma, \tag{8-9-13}$$

where γ is the angle between the direction of ejection of the electron and the direction of polarization of the incident radiation (see Fig. 8-9-1, in which we

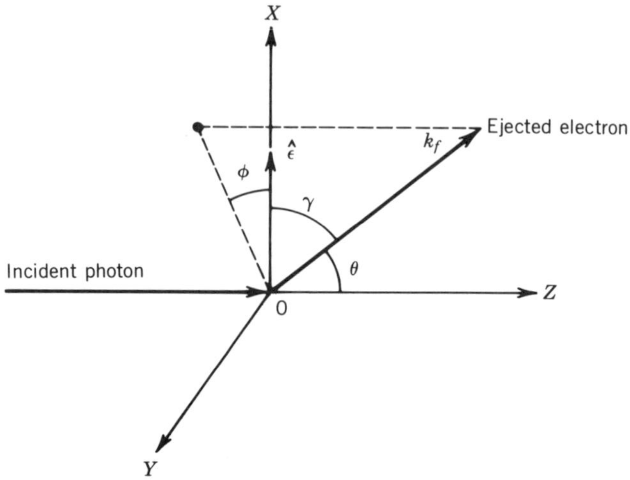

Figure 8-9-1. Schematic representation of the photoionization of a hydrogen atom initially at rest at the origin.

take the Z axis to be along $\mathbf{k}$ and the X axis along $\hat{\epsilon}$). The integral in (8-9-12) is proportional to the Fourier transform of $\psi_{1s}(r)$, so we have

$$\int e^{i\mathbf{K}\cdot\mathbf{r}}\psi_{1s}(r)\,d\mathbf{r} = 8\pi\left(\frac{Z^3}{\pi a_0^3}\right)^{1/2}\frac{Z/a_0}{[(Z/a_0)^2 + K^2]^2}, \tag{8-9-14}$$

where $\mathbf{K}$ is defined as

$$\mathbf{K} = \mathbf{k} - \mathbf{k}_f. \tag{8-9-15}$$

The differential cross section is then

$$\frac{d\sigma_{\text{photoion}}}{d\Omega} = 32\alpha\frac{\hbar}{m}\frac{k_f^3}{\omega}\frac{Z^5 a_0^3 \cos^2\gamma}{(Z^2 + K^2 a_0^2)^4}. \tag{8-9-16}$$

In the geometry of Fig. 8-9-1, we have

$$\cos\gamma = \sin\theta\cos\phi \tag{8-9-17}$$

and

$$K^2 = k^2 + k_f^2 - 2kk_f\cos\theta. \tag{8-9-18}$$

Moreover, for high photon energies, $\hbar\omega \gg |E_{1s}|$,

$$\hbar\omega \approx \frac{\hbar^2 k_f^2}{2m} \tag{8-9-19}$$

and

$$\frac{k}{k_f} \approx \frac{\hbar k_f}{2mc} = \frac{v_f}{2c}, \tag{8-9-20}$$

where v_f is the velocity of the outgoing electron. Hence, in the nonrelativistic regime, we can write

$$K^2 \approx k_f^2\left(1 - \frac{v_f}{c}\cos\theta\right). \tag{8-9-21}$$

Using our assumption of high-energy photons and the fact that

$$E_{1s} = -\frac{Z^2e^2}{8\pi\epsilon_0 a_0}, \tag{8-9-22}$$

we find that

$$k_f^2 a_0^2 \gg \frac{2m}{\hbar^2}|E_{1s}|a_0^2 = Z^2 \tag{8-9-23}$$

and

$$Z^2 + K^2a_0^2 \approx k_f^2 a_0^2\left(1 - \frac{v_f}{c}\cos\theta\right). \tag{8-9-24}$$

Then we have for the differential cross section

$$\frac{d\sigma_{\text{photoion}}}{d\Omega} = 32\alpha\frac{\hbar}{m}\frac{Z^5}{\omega(k_f a_0)^5}\frac{\sin^2\theta\cos^2\phi}{[1-(v_f/c)\cos\theta]^4} \quad (x \text{ polarization}). \tag{8-9-25}$$

The $\cos^2\phi$ factor shows that the outgoing electrons have a cosine-squared distribution about the direction of polarization of the incident radiation.*

If the incident radiation is *unpolarized*, we average over the directions of $\hat{\epsilon}$ to obtain

$$\frac{d\sigma_{\text{photoion}}}{d\Omega} = 16\alpha\frac{\hbar}{m}\frac{Z^5}{\omega(k_f a_0)^5}\frac{\sin^2\theta}{[1-(v_f/c)\cos\theta]^4} \quad (\text{unpolarized radiation}). \tag{8-9-26}$$

The $\sin^2\theta$ factor in both (8-9-25) and (8-9-26) shows that the photoelectrons are ejected preferentially at right angles to the incident photon direction, the factor

*This conclusion is far more general than our approximate treatment indicates [see C. N. Yang, *Phys. Rev.* **74**, 764 (1948)].

containing $\cos\theta$ in the denominator having little importance because $v_f/c \ll 1$. Specifically, to first order in v_f/c, the differential cross section for unpolarized radiation is

$$\frac{d\sigma_{\text{photoion}}}{d\Omega} = 16\alpha \frac{\hbar}{m} \frac{Z^5}{\omega(k_f a_0)^5} \sin^2\theta \left(1 + 4\frac{v_f}{c}\cos\theta\right) \qquad \text{(unpolarized radiation)}$$

(8-9-27)

and the *integral cross section for photoionization of the* (1*s*) *state* is

$$\sigma_{\text{photoion}} = \frac{16\sqrt{2}\,\pi}{3}\alpha^8 Z^5 \left(\frac{mc^2}{\hbar\omega}\right)^{7/2} a_0^2 \qquad \text{(unpolarized radiation).} \qquad (8\text{-}9\text{-}28)$$

Note the strong dependence on photon energy and nuclear charge.

The theory of photoionization is discussed in depth by Bethe and his coauthors[40] and by Starace.[41] Park[44] and Gasiorowicz[45] provide treatments, at about the level of ours, that start with Golden Rule No. 2, our (8-2-1). More advanced discussions consider complex atoms and molecules, exchange effects and correlation, relativistic effects including electron spin, and field quantization. The sources cited in Part $\mathscr{C}$ contain many references to such treatments.

8-10. OSCILLATOR STRENGTHS AND SUM RULES[15,38,40–42,46–48]

The oscillator strength f was introduced in Section 5-7-F, where the emphasis was on inelastic electron collisions, but this quantity is fundamental to any discussion of the emission and absorption of radiation. In quantum mechanics, the oscillator strength, or "f value," for spontaneous emission from level j to level i is

$$f_{ji} = -\left(\frac{2}{3}\right)\frac{m_0}{e^2}\frac{1}{\hbar^2}|\langle j|e\mathbf{r}|i\rangle|^2\hbar\omega_{ji} \qquad \text{(emission } j \to i). \qquad (8\text{-}10\text{-}1)$$

For absorption, the f value is f_{ij}, given in terms of f_{ji} and the statistical weights by

$$f_{ij} = -\frac{g_j}{g_i} f_{ji} \qquad \text{(absorption } i \to j). \qquad (8\text{-}10\text{-}2)$$

We see that the oscillator strength is a dimensionless quantity that is positive for absorption and negative for emission. In the electric dipole approximation, the Einstein coefficients A_{ji} and B_{ij} are expressed in terms of the oscillator strength by the equations

$$A_{ji} = \frac{e^2\omega_{ji}^2}{2\pi\epsilon_0 m_0 c^3}(-f_{ji}) = \frac{e^2\omega_{ji}^2}{2\pi\epsilon_0 m_0 c^3}\frac{g_i}{g_j} f_{ij} \qquad \text{(emission } j \to i\text{)} \quad (8\text{-}10\text{-}3)$$

and

$$B_{ij} = \frac{\pi e^2}{2\epsilon_0 m_0 \hbar\omega_{ij}} f_{ij} \qquad \text{(absorption } i \to j\text{).} \quad (8\text{-}10\text{-}4)$$

The cross section for absorption from a bound state i to another bound state j is

$$\sigma_{ij}(\omega) = \frac{2\pi^2 e^2}{4\pi\epsilon_0 m_0 c} f_{ij}(\omega). \quad (8\text{-}10\text{-}5)$$

SI units are used in all these equations. For transitions from bound states into the ionization or dissociation continuum, (8-10-5) becomes

$$\sigma(\omega) = \frac{2\pi^2 e^2\hbar}{4\pi\epsilon_0 m_0 c}\frac{df}{dE} = 1.097618 \times 10^{-16}\,\text{cm}^2\,\text{eV}\frac{df}{dE}, \quad (8\text{-}10\text{-}6)$$

where df/dE is the differential oscillator strength per electron volt.

Fano and his colleagues[46,47] have stressed the utility of the oscillator strength f and its spectral density df/dE in the study of photoabsorption. Although these quantities are directly related to the cross section by (8-10-5) and (8-10-6), their properties have more general significance and can be interpreted more directly. Furthermore, the use of f and df/dE unifies the treatments of the discrete and continuous spectra and places them on a common basis.

There are many other ways in which the oscillator strength proves useful, one reason being that it contains the modulus squared of the ubiquitous electric dipole matrix element, another being that the multiplicative factor in its definition is exactly right for construction of a number of "sum rules" in simple, compact form. These sum rules express relationships between various properties of the oscillator strength of a structure and theoretical or experimental properties of the structure.[40,41,46,47] The simplest is the *Thomas-Reiche-Kuhn sum rule*

$$\sum_k f_{jk} = Z, \quad (8\text{-}10\text{-}7)$$

where the sum is over all levels, including the continuum.[14] Fano and Cooper[46] discuss eight sum rules, which deal, collectively, with both structural and collisional properties, including even the Lamb shift!

As pointed out by Corney,[42] the starting point in the calculation of oscillator strengths and transition probabilities is the computation of the *line strength*, defined as

$$S_{ki} = S_{ik} = \sum_{m_i m_k} |\langle km_k|e\mathbf{r}|im_i\rangle|^2. \tag{8-10-8}$$

Note that the line strength is symmetrical with respect to the initial and final states. The oscillator strengths for absorption and emission are given in terms of the line strength by the equation

$$g_i f_{ik} = -g_k f_{ki} = \frac{2m\omega_{ki}}{3\hbar}\frac{S_{ki}}{e^2}. \tag{8-10-9}$$

8-11. LINE BROADENING AND LINE SHAPE

These subjects are of prime importance in spectroscopy, since the width and shape of spectral lines must be properly analyzed and understood if spectroscopic data are to be fully utilized, and they are also relevant in atomic collisions involving photons. Hence we must give at least a brief introduction to these topics and refer the reader to the literature for more details.[1-3,14,48,49] We begin by discussing several of the most important mechanisms of line broadening and then examine three types of intensity distributions that they produce.

A. Line-Broadening Mechanisms

1. *Natural Line Broadening*

In Section 8-3 we discussed the spontaneous emission of radiation from excited atoms, which occurs even with no radiation field applied. The existence of spontaneous emission means that excited states have finite natural lifetimes, τ, and implies, through the uncertainty principle, that excited energy levels have nonzero natural widths, ΔE. For a transition from upper level j to ground state i, these quantities are related by

$$\Delta E\,\Delta t = \tau_j\,\Delta E \approx \hbar, \tag{8-11-1}$$

where ΔE is the uncertainty in the energy of the state j and $\Delta t = \tau_j$ is the time available to measure the energy of level j. The *natural width* of the spectral line connecting j and i is then

$$\Gamma = \frac{\hbar}{\tau_j}. \tag{8-11-2}$$

If i were not the ground state (with $\tau = \infty$), the natural line width would be

$$\Gamma = \frac{\hbar}{\tau_j} + \frac{\hbar}{\tau_i}. \tag{8-11-3}$$

The ultimate origin of natural line broadening is the zero-point field fluctuations, which, together with the reaction of the emitted radiation back on the emitter, produce spontaneous emission (see Note 8-2). Natural line broadening is intrinsic to the system and cannot be reduced or eliminated. It is predicted by theory if the electromagnetic field is quantized, even in the nonrelativistic approximation. The natural width of the $2p$ level of the H atom is about 4.1×10^{-7} eV, and if we take this as typical, we see that the natural widths of spectral lines are extremely small* and likely to be obscured by other causes of line broadening (see, however, Note 8-4).

2. *Collision Broadening*

Collision, or pressure, broadening is produced by collisions or near-collisions of excited emitters with molecules of a gas in which they are moving. Additional Fourier components are introduced into the time dependence of the excited-state wave function, and they cause the state to decay more rapidly, with an increase in the line width. Lowering the gas pressure reduces the effect. In certain experiments, it is possible to utilize beam techniques and virtually eliminate the collisions that cause the broadening.

3. *Doppler Broadening*

This type of line broadening results from the thermal motion of molecules and causes the emitted or absorbed photons to have a range of frequencies about the central value. In a gas environment, Doppler broadening can be markedly reduced by cooling the gas. Some elegant techniques involving the use of beams have yielded spectroscopic data that are free of first-order Doppler effects,[50] although second-order Doppler effects arising from time dilation are still in evidence. Other techniques now under development for cooling and trapping neutrals offer exciting new possibilities for spectroscopy at the level of the natural line width (see Notes 8-4 and 8-5).

4. *Transit-Time Broadening*

Excited molecules moving with sufficient speed may escape the region of observation before they can emit photons. In this case the time available to measure the energy of the upper state is not τ_j in (8-11-1), but rather, the shorter transit time of the molecules through the region that can be viewed through the instruments of the observer. Accordingly, the width of the spectral line is increased.

5. *Stark Broadening*

Charged particles near an excited molecule can split and shift its energy levels through the Stark effect and increase the line width. This phenomenon may be

*An exception of considerable interest to us occurs in photodissociation, which will be discussed in Section 8-15.

very important in dense plasmas. Similar effects may be produced by application of strong external fields (see Ref. 51 and Vol. 2 of Ref. 31).

B. Lorentz, Gauss, and Voight Line Shapes

The shapes of spectral lines are determined by the broadening mechanisms that are operative. If natural broadening alone is important, the line shape is Lorentzian and is described by the function

$$F_L(\omega) = \frac{\Gamma/2\pi}{(\omega_0 - \omega)^2 + (\Gamma/2)^2}, \tag{8-11-4}$$

where ω_0 is the central angular frequency of the transition and $\Gamma = A$ is the full width of the peak, A being the Einstein coefficient for the transition. The Lorentzian function $F_L(\omega)$ also characterizes the shape of a collision-broadened peak, with width equal to twice the mean rate at which collisions occur.

On the other hand, if Doppler broadening is dominant, the line has a Gaussian shape and is described by the function

$$F_G(\omega) = (2\pi\delta^2)^{-1/2} e^{-(\omega-\omega_0)^2/2\delta^2}, \tag{8-11-5}$$

where

$$\delta = \left(\frac{\omega_0^2}{\beta M c^2}\right)^{1/2} = \frac{\Delta}{2(2\log_e 2)^{1/2}}, \tag{8-11-6}$$

$$\beta = \frac{1}{kT}, \tag{8-11-7}$$

M is the atomic mass, and Δ is the full width of the Doppler-broadened line at half maximum height, given by

$$\Delta = 2\omega_0 \left(\frac{2\log_e 2}{\beta M c^2}\right)^{1/2}. \tag{8-11-8}$$

The quantity δ is the RMS spread of the Gaussian distribution.

Both F_L and F_G are normalized to unit area. The Gaussian peak has the same width at half-height as the Lorentzian but is more sharply peaked and falls off more rapidly from the center (see Fig. 8-11-1). If both Lorentzian and Gaussian broadening are operative to a significant extent, the spectral line can be fit by a Voight profile obtained from a composite integral of F_L and F_G. The Voight line shape is intermediate between the Lorentzian and Gaussian.

It is now possible to make a distinction between two broad categories of line-broadening processes[48] that are associated with the two basic line shapes, Lorentzian and Gaussian. A *homogeneous* line-broadening mechanism is one in

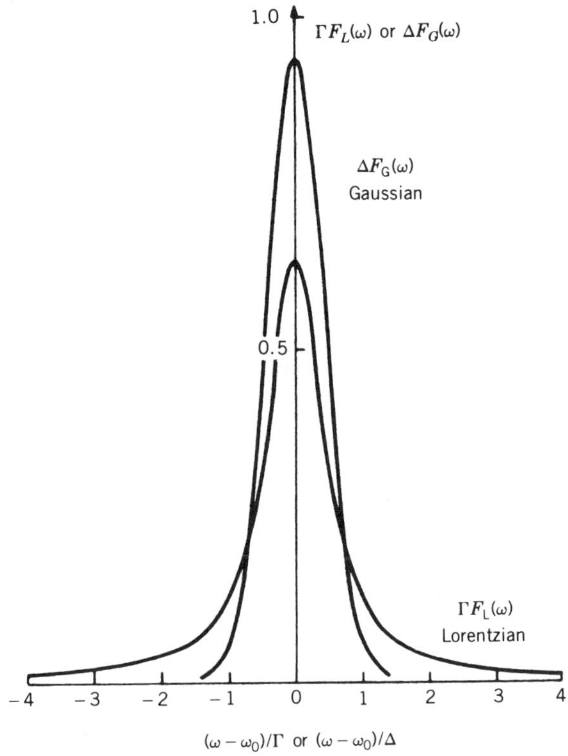

Figure 8-11-1. The Gaussian and Lorentzian line-shape functions, both normalized to unit area under the curve. The abscissa is marked in units of $(\omega - \omega_0)/\Gamma$ for F_L and in units of $(\omega - \omega_0)/\Delta$ for F_G. Both F_L and F_G have the same width at half height (one unit on the abscissa). (Adapted from Loudon.[48])

which all absorbing and emitting atoms play identical roles. Examples are provided by natural and collision broadening: Here there is no experimental method of associating light of a certain frequency with a certain group of atoms. The line shape is Lorentzian. On the other hand, there are situations in which different atoms absorb or emit light at different frequencies because of a statistical distribution in some parameter that determines the transition frequency. Here we have *inhomogeneous* line broadening, and usually a Gaussian line shape. An example is Doppler broadening, in which the atomic velocity is the relevant statistical parameter. In principle, we can distinguish atoms that absorb at a given frequency from the remainder of the atoms by their component of velocity parallel to the velocity of the light beam that is absorbed.[1]

The common feature of natural and collisional line broadening that leads to Lorentzian line shape with homogeneous broadening in both mechanisms is the fact that the width, $\Delta\omega$, of the line arises from the finite mean time, Δt, during which the radiation is undisturbed. Analysis of the two processes shows that in each case, $\Delta\omega\,\Delta t \gtrsim 1$, in accord with Fourier transform theory and the uncertainty principle. Collisional broadening can be interpreted in terms of a model wherein each atom radiates at frequency ω_0, but with abrupt and random changes in the phase of the radiated waves when collisions occur. The spread in detected frequencies is a consequence of the waves being fragmented into finite segments whose Fourier representations include frequencies other than ω_0.[48]

PART $\mathscr{C}$. EXPERIMENT

In this final part of our chapter on photon processes, we describe experimental studies of emission, absorption, excitation, dissociation, and ionization, and briefly mention a few additional theoretical considerations not covered in Part $\mathscr{B}$. We conclude with a discussion of multiphoton processes, restricting our attention to situations in which the target structure is not significantly disturbed by strong external **E** and **B** fields.

8-12. EMISSION

Here we are concerned mainly with experimental methods of determining transition probabilities (or oscillator strengths) for spontaneous emission by atoms, molecules, and ions. We concentrate on a few modern techniques; for more detail and for other techniques, see Corney[42] and Wiese.[52]

A. Line Width Techniques[53]

The discussion in Section 8-11-A-1 shows that if the natural line width of a transition from an excited state to the ground state can be measured, the natural lifetime, τ, of the excited state can be determined. From τ, the transition probability, oscillator strength, and Einstein A coefficient follow. (Branching and cascading would obviously complicate matters.) It is evidently important to reduce the effects of collisional line broadening to a negligible level by operating at a sufficiently low pressure. Doppler broadening effects can be circumvented by the use of either of two spectroscopic methods that developed mainly out of work by W. Hanle (in Germany in the 1920s), by A. Kastler and J. Brossel (in Paris, for two decades following World War II), and by F. Bitter (in the United States during the 1950s). Here we refer to *double-resonance spectroscopy* and *level-crossing spectroscopy*, which have been in widespread use since the 1950s.

Both methods are based on the absorption of electric dipole radiation to achieve a nonstatistical (coherent) population of the excited state of interest. Resonance transitions between sublevels of these states are induced by RF fields, or by level crossings produced by an external **E** or **B** field, and are detected by the change of polarization of the resonance fluorescence. Only information on the polarization, and not the frequency distribution of the light quanta, is utilized, so Doppler broadening is suppressed. The line width is determined by mapping the signal peak, the specifics of which depend on the type of experiment being performed.[53]

B. Lifetime Techniques

1. *The Delayed Coincidence Method*[42]

This technique, widely used since about 1961, utilizes a pulsed electron or photon source to excite the level in question and measures the delay of the photons subsequently emitted in delayed coincidence with the source pulses. Measurements involving initial electron impact excitation were described in detail in Section 5-8. The delayed coincidence method is one of the most accurate and versatile methods available for the measurement of lifetimes.

2. *Beam-Foil Spectroscopy*[54]

This extremely important technique was developed in the mid-1960s by S. Bashkin and others. An ion beam is accelerated to an energy of several MeV and passed through a thin ($\lesssim 10^{-5}$ cm) carbon or metal foil. Many excited neutral and ionic states are produced from the incident beam projectiles by charge transfer and electron stripping, and they continue in the original beam direction with negligible energy loss. Lifetimes are determined, not from the temporal dependence of the decay, but from the spatial variation of the intensity of the radiation along the emerging beam. Typically, measurements of the emitted photons are made over a distance of about 5 cm downstream from the foil. Lifetimes as long as about 10^{-8} s can be measured. Difficulties arising from cascading in the beam-foil excitation can be avoided by replacing the foil with a high-intensity laser beam whose frequency is matched to the excitation desired.[42]

C. Thermal Equilibrium Emission Techniques

Here a Boltzmann distribution of atoms in various excited states is established, and the relative intensities of the different emission lines are measured. Then relative probabilities of transitions from a given level to various lower levels are determined. High-temperature arcs and shock tubes are frequently used to produce the high temperatures that are required.[55]

D. Techniques for Metastables

Metastable states require special treatment. Techniques used for their study are described in Chapter 7 of the book by Corney.[42]

E. Addendum: The Trapping of Resonance Radiation

Under *optically thin* conditions, resonance radiation emitted by atoms in a gas escapes the gas essentially without interacting with it. If these conditions do not obtain, we must consider the process of radiative transfer,[56] wherein absorption and reradiation in resonance transitions may occur many times as the radiation progresses through the gas. Evidently, the trapping, or imprisonment, of resonance radiation that takes place can greatly complicate the analysis of experiments (see Section 5-7-E).

8-13. PHOTOABSORPTION

A. Mechanisms of Photoabsorption

A number of different mechanisms can operate in the absorption of photons by atoms and molecules.*

1. Excitation to upper levels involving resonance transitions without predissociation or preionization (autoionization)
2. Excitation to molecular states adjacent to a dissociation continuum, leading to predissociation (Section 5-5-A)
3. Excitation to upper states of atoms or molecules adjacent to an ionization continuum, leading to preionization (Sections 6-5, 6-6, and 6-7)
4. Continuous absorption due to (a) direct excitation into an atomic ionization continuum; (b) dissociation of molecules into fragments, each of which may be in the ground state or in an excited state; (c) direct excitation into one of the molecular ionization continua, without dissociation or preionization; (d) dissociative ionization into an ion and an atom which may be in an excited state; or (e) dissociative ionization that produces two ions of opposite sign, one of which is in an excited state
5. Multiple ionization (i.e., ejection of two or more electrons by a given photon)
6. Multiphoton absorption [i.e., absorption of two or more photons by a given atom or molecule (see Section 8-17)]

We can define a photoabsorption cross section σ_a as the sum of the cross sections for these processes. The *scattering* processes described in Section 8-1-B do not *absorb* photons in the usual sense, but they do make a contribution to the *attenuation* of a photon beam, because scattering through a large enough angle, or with sufficient energy change, may cause the scattered photon not to be detected in a given experimental arrangement. In the lower photon energy

*Our discussion will deal mainly with absorption by neutrals. We consider positive ions briefly in Sections 8-15 and 8-16; photoabsorption by negative ions is treated in a forthcoming book.[57]

ranges, below a few tens of keV, scattering usually makes an extremely small contribution to the total attenuation of the photon beam, so the attenuation cross section σ_{att} is very nearly equal to the photoabsorption cross section σ_a, and the two terms are regarded as synonymous. Here we shall be concerned very little with scattering, but rather with the combined effects of processes 1–5 above. The excitation, dissociation, and ionization mechanisms will be considered separately, to the extent that they can be decoupled, in Sections 8-14, 8-15, and 8-16.

The multiplicity of absorption mechanisms makes the analysis of absorption data difficult. However, information concerning mechanisms can be inferred from the shape and structure of absorption curves and knowledge of the energy levels of the target molecules. Absorption measurements continue to be useful,[58] although in an increasing fraction of the photon collision experiments, the goal is a direct elucidation of the individual mechanisms that are involved.

There are many applications of data on photon collisions.[31] Photoabsorption is a phenomenon of considerable laboratory interest in connection with studies of photochemistry and the physics of electrical discharges. Its geophysical and astrophysical significance is even more important, as evidenced by the following considerations. Most of the charged particles in the upper atmosphere are produced by solar ultraviolet and X radiation, and many of the reactions occurring there are photochemical in nature. Thus photoabsorption plays a key role in determining the composition and characteristics of the ionosphere. Photoabsorption is also an important factor in the energy balance of the earth's atmosphere and the outer layers of stars. Photoionization cross sections have a bearing on the ionization equilibrium and consequently the chemical abundances in interstellar space. The cross section for highly ionized iron and nickel partly determine the physical conditions of the solar corona. In addition, photoabsorption by H^- makes an important contribution to stellar opacity.

Here we deal mainly with experimental studies of absorption by neutrals in the spectral range extending from the visible region through the vacuum ultraviolet. This range is of great general interest in plasma and astrophysical research, and has been admirably surveyed in books and reviews.[4,15,59–64] A recent critical compilation of data on total molecular absorption and partial-channel photoionization, as well as ionic photofragmentation, will undoubtedly prove indispensable for years to come.[65] This document, prepared at the JILA Data Center by J. W. Gallagher, C. E. Brion, J. A. R. Samson, and P. W. Langhoff, also contains a useful discussion of experimental techniques and theoretical methods and approximations. Only valence-shell processes are considered, inner-shell processes being excluded. A similar compilation for atoms and atomic ions is being prepared.[66]

An extremely valuable source of data on photon processes of many kinds over a very large energy range is the journal *Atomic Data and Nuclear Data Tables*. A sample data compilation for the soft- and hard-X-ray regions (see Figs. 8-1-1 and 8-13-1) appears in Ref. 67. We should also mention compilations by Hubbell and his colleagues of photon cross sections and other quantities for

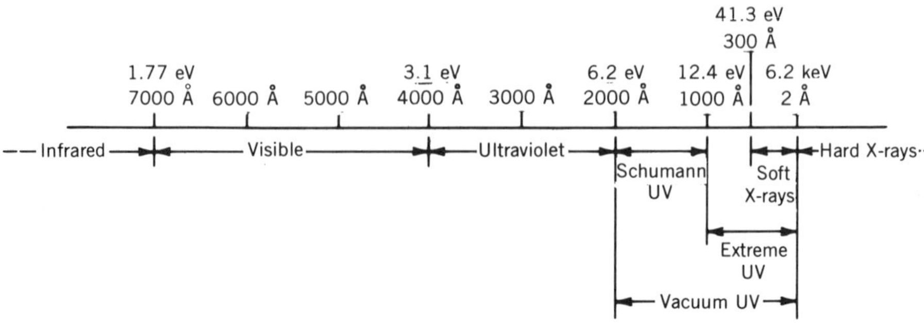

Figure 8-13-1. The spectral range bounded by the infrared and hard X-ray regions. The thresholds for single photoionization of neutrals, with very few exceptions, lie in the vacuum ultraviolet (VUV) region, with He having the highest threshold (24.6 eV).

energies up to 100 GeV.[68] Recent developments in astronomy based on observations of γ rays with energies of 100 GeV and much higher values are reviewed by Weekes.[69]

B. Experimental Methods of Studying Photoabsorption

Many of the experimental studies of photoabsorption have involved measurement of the attenuation of a beam of photons traversing a gas-filled absorption cell. Figure 8-13-2 shows one of the experimental arrangements used for these studies.[70] Light from a source D falls on a reflection grating where it is dispersed. The grating can be displaced along the Rowland focusing circle to direct a beam of variable wavelength toward the absorption cell. With no gas in the cell, a certain intensity I_0 is registered by the photomultiplier P. The measured intensity falls to some lower value I when gas is introduced. If x is the length of the cell, these two intensities are related by the equation

$$I = I_0 e^{-\mu x}, \tag{8-13-1}$$

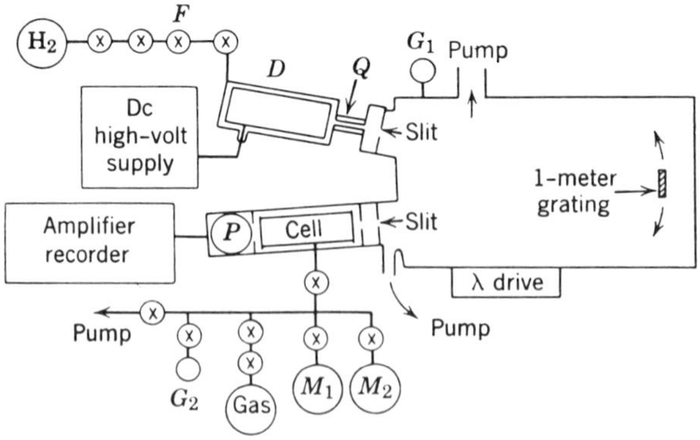

Figure 8-13-2. Apparatus for measurement of photoabsorption cross sections.[70]

where μ is the *absorption coefficient* for the particular wavelength in use; μ, which has units of cm^{-1}, is related to the *photoabsorption cross section*, σ_a, by the equation

$$\mu = \sigma_a N, \tag{8-13-2}$$

where N is the number density of the gas. The published values of μ usually refer to a gas number density corresponding to 0°C and 760 torr. The values of σ_a are often expressed in megabarns (1 Mb $= 10^{-18}$ cm^2). Note the similarity of this method to some of the techniques used to measure electron scattering cross sections (see Section 4-1).

There are two conditions on the validity of (8-13-1). First, the radiation must be so nearly monochromatic that there will be no change in effective absorption as the radiation progresses through the gas. Then the measured value of μ will be independent of the value of x chosen, and *Lambert's law* will be obeyed. Second, μ must be independent of the gas pressure and temperature, so that *Beer's law* is obeyed. The second condition will be satisfied if changes in p and T do not significantly change the composition of the gas, by molecular association or dissociation, for example.

The *double ion chamber method* provides a better means of measuring the absorption cross section at photon energies above the first ionization threshold. It is based on the constancy of the ionization/absorption ratio along the photon beam. The principle is illustrated in Fig. 8-13-3, which shows a parallel plate version of the apparatus. A photon beam of intensity I_0 passes from a monochromator into the first cell, where it is attenuated while it produces an ion current i_1 over a length L in the first section. The attenuated photon beam then

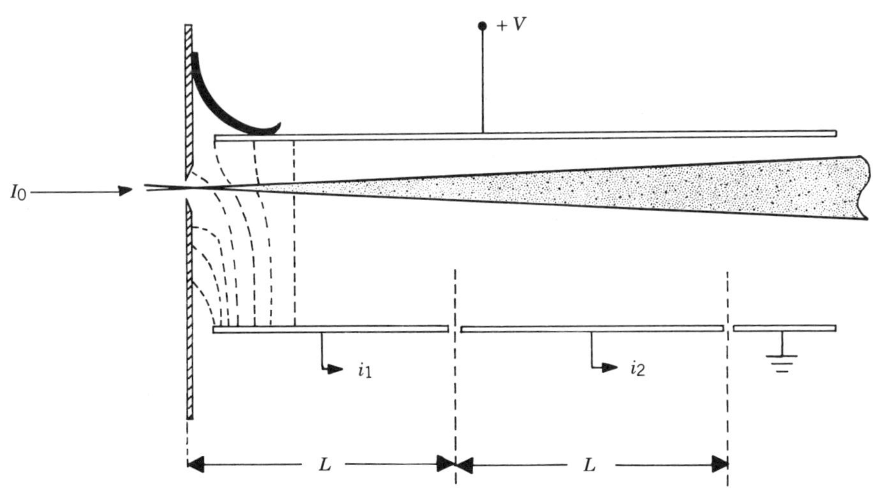

Figure 8-13-3. A double ion chamber used to measure cross sections for photoabsorption and photoionization.[4]

produces ionization over the second section of length L, where a current i_2 is collected. The guard plate at the bottom right, and the geometrical arrangement at the left near the entrance aperture ensure field uniformity at the ends of the active section. It may be shown that

$$\sigma_a = \frac{1}{NL} \ln \frac{i_1}{i_2}, \tag{8-13-3}$$

where N is the gas number density (Problem 8-46). This result is valid even if the photons can produce multiply charged ions and also if the electrons ejected by the photons produce secondary ionization.

The double ion chamber method has several advantages over the conventional absorption method illustrated in Fig. 8-13-2:

1. It can be used to measure accurate ionization yields (coefficients), $\gamma_i \equiv \sigma_i/\sigma_a$, which then provide ionization cross sections, σ_i. [Measurement of σ_i involves comparing the ion signal produced by the target gas to that produced by the rare gases in spectral regions where γ_i for these gases is unity (see Fig. 8-13-4 and Problem 8-47).]
2. Only photons absorbed within the target chamber have an effect, whereas in the conventional absorption method, photons absorbed by gas escaping

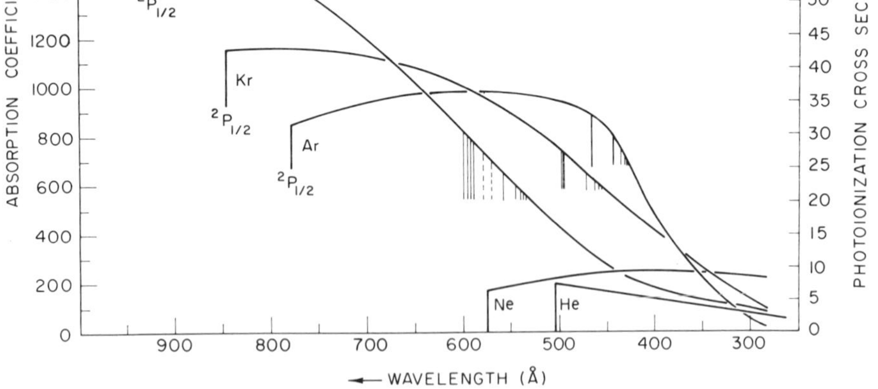

Figure 8-13-4. Experimental cross sections for photoionization of the noble gases, along with the corresponding photoabsorption coefficients.[4] For measurements on atomic oxygen, see G. C. Angel and J. A. R. Samson, *Phys. Rev. A* **38**, 5578 (1988).

through the entrance aperture of the absorption cell produce error in the reading of I at the end of the cell.

3. Fluorescence within the double cell is of no consequence.

Samson[4,59–63] also discusses various other geometries and methods used in measurements of absorption cross sections. His treatment of light sources, monochromators, analyzers, and detectors is highly recommended.

Absorption cross sections for the noble gases at photon energies up to 45 eV are shown in Fig. 8-13-4. In atomic gases, ionization makes essentially the entire contribution to the absorption above the ionization threshold, so here we may take $\sigma_a = \sigma_i$. The vertical lines attached to the curves for Ar, Kr, and Xe at photon energies between 20 and 30 eV indicate positions of discrete structure[59] that cannot be mapped accurately in this figure. It is instructive to compare the curve for Kr here with the curve for Kr in Fig. 6-6-3, which shows autoionizing levels excited by *electron* impact in measurements of ejected electron yield versus electron energy at energies between 24 and 28 eV. On page 164 of his review,[61] Samson displays plots of σ_a for the noble gases that extend up to 140 eV. Xenon shows a broad peak centered at 100 eV, whereas the cross section for krypton has a broad minimum around 80 eV and increases with energy up to the maximum energy reached. No dramatic structure appears for He, Ne, and Ar between 45 and 140 eV.

Haddad and Samson[58] have measured absolute photoabsorption and photoionization cross sections for H_2O at a large number of discrete wavelengths between 100 and 1000 Å by the double ion chamber method. Their data are shown in Figs. 8-13-5 and 8-13-6 as solid dots; the other symbols indicate data obtained by other investigators, most of whom used continuum light

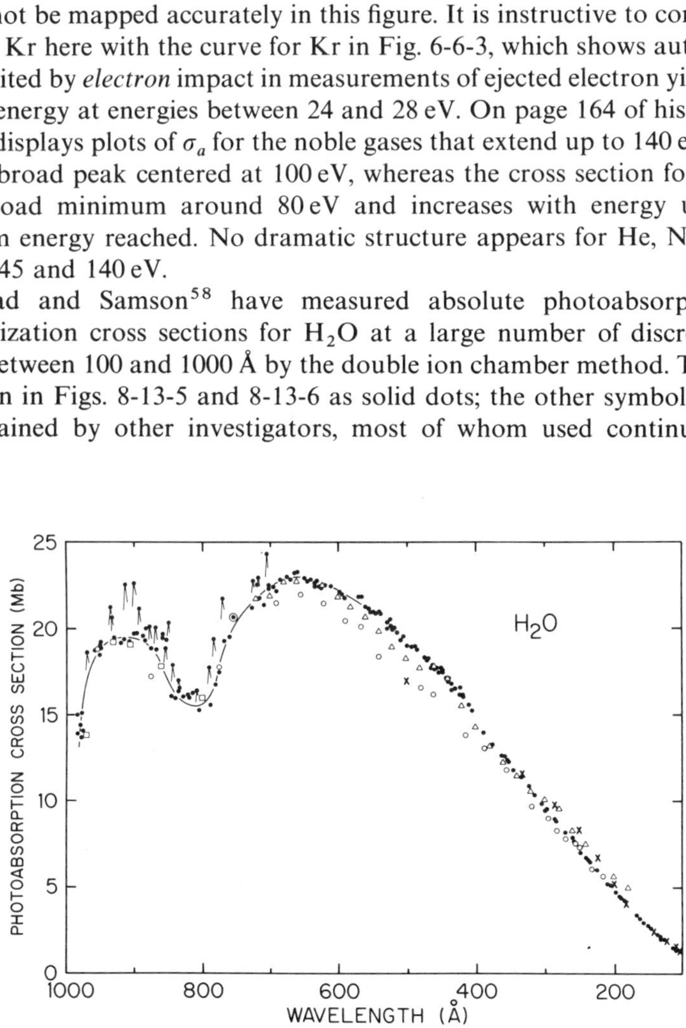

Figure 8-13-5. The photoabsorption cross section for H_2O as a function of wavelength.[58]

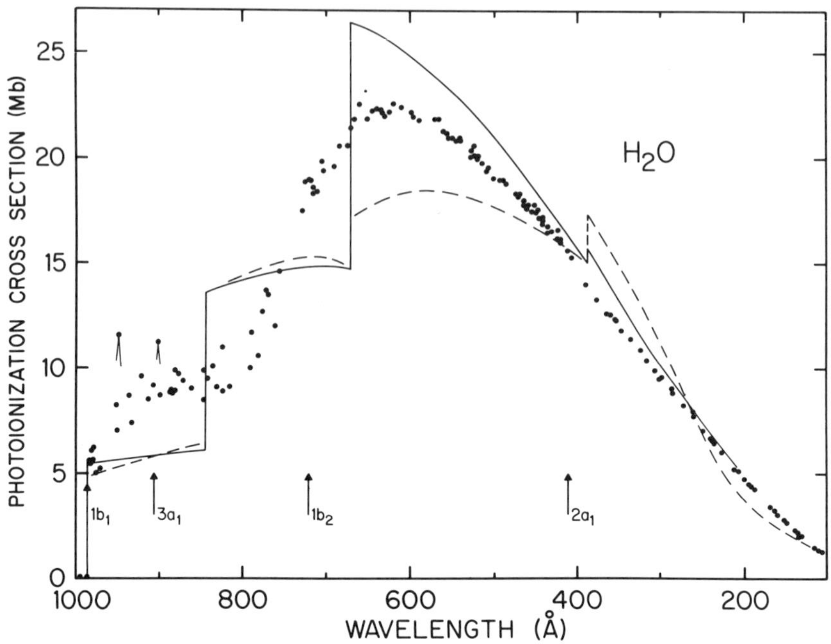

Figure 8-13-6. The photoionization cross section for H_2O as a function of wavelength.[58] The solid and dashed curves represent theoretical results; the symbols, experimental cross sections.

sources. Haddad and Samson[58] estimate the absolute error in their own measurements to be $\pm 3\%$ in regions free from discrete structure. (This level of accuracy represents the state of the art around 1985. Accuracies of 1 to 2% may be forthcoming.[61]) The absorption and ionization cross sections are equal for H_2O at $\lambda < 620$ Å, but for longer wavelengths, the contribution of nonionizing processes to σ_a is large. Near the ionization threshold, the scatter of the data points reflects the presence of numerous autoionizing levels. Two prominent levels are indicated by the inverted V-shaped symbols in Fig. 8-13-6.

High optical resolution is essential in photoabsorption measurements, especially just below the various ionization limits where there are many sharp and diffuse bands and lines. Measured cross sections depend to an extreme degree on the energy of the photons, except in bands that are very diffuse because of predissociation.

8-14. PHOTOEXCITATION

Most of the experimental data on cross sections for photoexcitation were obtained by indirect means or by inference from data of other types. Here we briefly discuss several methods of studying the photoexcitation process and of applying photoexcitation to the study of molecular structure.

A. Laser-Induced Fluorescence

In prelaser times, many studies were made in which radiation from a conventional light source was dispersed, and a beam of well-defined wavelength was directed into a gas cell. The resulting fluorescence radiation was dispersed and detected, and the fluorescence spectra for various incident wavelengths provided information on the excitation processes that were involved.[61] Apparently, this kind of experiment has not been carried to the point of directly providing excitation cross sections, as is done in the experiments on *electron* impact described in Section 5-7-B. An important improvement in the method came with the use of single-frequency dye lasers and laser-induced fluorescence.[71] This type of measurement has resulted in reliable relative photoexcitation cross sections.

Another application of laser-induced fluorescence, one that is both intriguing and important, is in probing product states of gas-phase chemical reactions. This technique was developed by R. N. Zare and his colleagues and has become an indispensable research tool.[72–74]

B. Electron Impact "Photoexcitation"[15,47,61]

In Section 5-6 we demonstrated that a fast electron is similar to a photon in collision processes, and in Section 5-7-F we showed how electrons may be used to generate optical excitation and ionization cross sections. This method has proved to be very useful and is discussed further in Section 8-16.

C. Anomalous Dispersion[75]

In the vicinity of an absorption line, the index of refraction increases, then decreases abruptly, and finally increases again as one sweeps through the line in the direction of increasing frequency. This is the phenomenon of *anomalous dispersion.* Measurement of the variation of the index of refraction with frequency of the light can be used to derive the oscillator strength for the absorption transition associated with the line. An elegant and powerful method called the Rozhdestwensky hook method is described by Ditchburn;[75] references to its application are provided by Hasted.[36]

D. Resonant Raman Excitation Profiles

In Section 8-1-B-5 we discussed Raman scattering in terms of the absorption of a photon that excites the target to a *virtual* state, with subsequent photon emission and deexcitation of the target to a lower bound state. However, if the incident photon energy matches that of a bound excited state of the target, *resonance* Raman scattering can occur, usually with a considerable enhancement of intensity. The theory of the scattering process is worked out by Louisell on pages 296–309 of his book,[76] and by Lee and Heller.[77]

There are several ways in which resonant Raman scattering can be used to obtain information about the geometry of excited states of molecules;[78–80] one involves mapping resonant Raman excitation profiles (RREP). Here light from a tunable dye laser is directed onto a molecular gas, its frequency is swept through the vibrational resonances, and the relative intensities of the emitted spectral lines are measured for each resonance.[81] One can then obtain a set of quantities Δ_{ej}, the shift in the potential energy minimum (the equilibrium position of the normal coordinate Q_j) when the molecule is excited to the eth (resonant) electronic state.

E. Laser Excitation of Dissociating States; Half-Collisions

Kinsey and his collaborators have used absorption and emission spectroscopy of photodissociating molecules to probe details of the dissociation dynamics of extremely short-lived transient species.[82] The study of photodissociation processes via laser excitation offers the advantage of sharply defined initial conditions; also, the experimental results are relatively simple to interpret. The emission spectrum contains information about both the excited-state and ground-state potential surfaces. Properties of the excited-state surface (such as forces and gradients) can be inferred from the intensities of the fundamentals, overtones, and combinations. Characteristic features of the ground-state potential surface can be derived from the observed energies and band contours of the same features.

The process studied might be of the form

$$h\nu + \mathrm{ABC} \rightarrow \mathrm{A}^* + (\mathrm{BC})^*. \qquad (8\text{-}14\text{-}1)$$

The initial photoexcitation into the dissociation continuum creates, essentially instantaneously, a situation in which the products A* and (BC)* are receding from the absorption site, so what is studied is aptly called a *half-collision*.[47]

8-15. PHOTODISSOCIATION

By this point, our discussion has taken us into a very complex field that relates to laser chemistry, laser spectroscopy, molecular dynamics, and chemical kinetics. Not only is the field very broad, but also its external and internal boundaries are diffuse. (Note the overlap between Section 8-14 and what follows in this section.) To keep our discussion within reasonable bounds, we shall concentrate on only two areas: experimental studies of photodissociation dynamics of neutral molecules, and photofragment spectroscopy of positive ions.

A. Photodissociation Dynamics of Neutral Molecules

As implied in Section 8-14-E, photodissociation or photoionization of a molecule produces a distribution of fragments that depends on the nature of the excited molecular states and the interactions between the final state fragments. Useful information can be derived from measurements of the distributions of the various forms of energy (translational, rotational, vibrational, and electronic) among the photofragments. The populations of the different states of the photofragments may be measured by observing fluorescence from electronically excited species, or by probing the ground state by means of laser-induced fluorescence.

The angular distribution of the photofragments is also of interest. If the radiation that produces the dissociation is linearly polarized, the normalized angular distribution of the photofragments from a diatomic molecule is described by the function

$$f(\theta) = \frac{1}{4\pi}[1 + \beta P_2(\cos\theta)], \tag{8-15-1}$$

where θ is the angle between the internuclear axis and the direction of the **E** vector of the incident wave, $P_2(\cos\theta) = \frac{1}{2}(3\cos^2\theta - 1)$ is the Legendre polynomial of order 2, and β is an asymmetry parameter that depends on the dynamics of the dissociation process.[83,84] The photofragments may be probed with a laser that excites some selected transition, and the Doppler profile of the stabilizing radiation that results can be measured. The line shape provides information about the magnitude and direction of the photofragment velocities.[85]

In Section 5-10 we discussed the alignment and orientation parameters[86-88] that prove so useful in describing electron-impact processes. Similar considerations apply to photon[86] and heavy particle[87,88] collisions with heavy particles, to surface impact phenomena,[89] and to drift tubes.[90] For example, the alignment of angular momentum in photodissociation can be determined by observing the variation in fluorescence intensity as the direction of polarization of a probe laser is varied with respect to the direction of polarization of the laser producing the dissociation. If this measurement is conducted for a single rotational level in the photofragment distribution, the second multipole moment of the angular momentum distribution can be determined exactly.[91,92]

Leone[93,94] has pointed out that in the chemical literature the concepts of "alignment" and "orientation" can be characterized in terms of a heteronuclear diatomic molecule as follows. *Alignment* permits the experimenter to place the molecular axis either parallel or perpendicular to a collision partner. *Orientation* means that either of the two different ends of the molecule can be pointed toward the collision partner. The optical preparation of states with lasers can provide alignment, but not orientation. Hexapole fields can produce both alignment and orientation and thus provide complete geometric arrangement of reactants.

The measurement of vector quantities has become very important in the elucidation of photodissociation mechanisms. A noteworthy study has recently been made by Hall et al.,[95] who have developed a technique for determining the angular correlation between a photofragment's angular momentum vector **J**, its recoil velocity vector **v**, and the transition dipole moment of the parent molecule $\boldsymbol{\mu}_p$.

B. Fast Ion Photofragment Spectroscopy[96,97]

Until recently, the inability to achieve suitably high densities of isolated molecular ions impeded the experimental acquisition of spectroscopic information on these species. Only in the 1970s were techniques developed that allowed a marked improvement in the quantity and quality of data on the structure and potential curves of molecular ions. Among the most productive of these new techniques are those based on laser-induced fluorescence,[98] on photoionization with synchrotron radiation and photoion-photoelectron coincidence measurements,[99] on photodetachment,[100] and on photodissociation.[101]

Ion photodissociation has been studied mainly by three experimental methods, all discussed in the reviews by Moseley.[96,97] (For a good review of photofragment dynamics of neutral molecules, see Leone.[102]) These methods are associated with gas-phase techniques, trapped ion approaches, and the use of fast ion beams interacting with photon beams. This section concentrates on the last-mentioned method, which has been the most productive. The fast ion method has been used to study *direct dissociation* in transitions from bound to repulsive states, and it provides information on the vibrational levels of the bound state and on the shape and symmetry of the dissociating state. Direct dissociation occurs rapidly, in a time $t \lesssim 10^{-12}$ s. Ion photofragment spectroscopy is also applicable to *predissociation* of states, with generally much longer lifetimes. Here the technique can yield information of such high resolution as to permit lifetimes of dissociating levels to be obtained, as well as details of the predissociation mechanisms. These photodissociation processes can profitably be viewed in terms of *half-collisions*[103] (see Section 8-14-E), with the system being prepared in a specific state or superposition of states and studied throughout the second half of a "collision."[47]

Fast ion photofragment spectroscopy is usually performed with what amounts to a double-mass spectrometer, inside which a laser beam interacts with a "fast" ion beam between the two spectrometers. An ion source generates ions of various types, which are accelerated to the desired energy (perhaps 10 to 10^5 eV). The species to be studied is selected in the first mass spectrometer and directed into the interaction region. Either intracavity or extracavity photons from fixed frequency or tunable lasers interact with the ion beam, either traveling coaxial with it or perpendicular to it. Charged fragments resulting from photodissociation are selected and energy analyzed in the second spectrometer. High angular collimation on both the primary and photofragment ion beams is required for high resolution in the kinetic energy and angular distribution measurements on the fragments. An apparatus having especially good energy resolution is shown in Fig. 8-15-1.[104–106]

It is of interest to note that if the coaxial laser has a fixed frequency, a limited tuning range can be covered in the coaxial beams mode by varying the primary ion velocity, thereby varying the Doppler shift. At fixed laser frequency, the energy distribution of the photofragments can be obtained by scanning the energy analyzer. In the crossed beams mode, the angular distribution of the

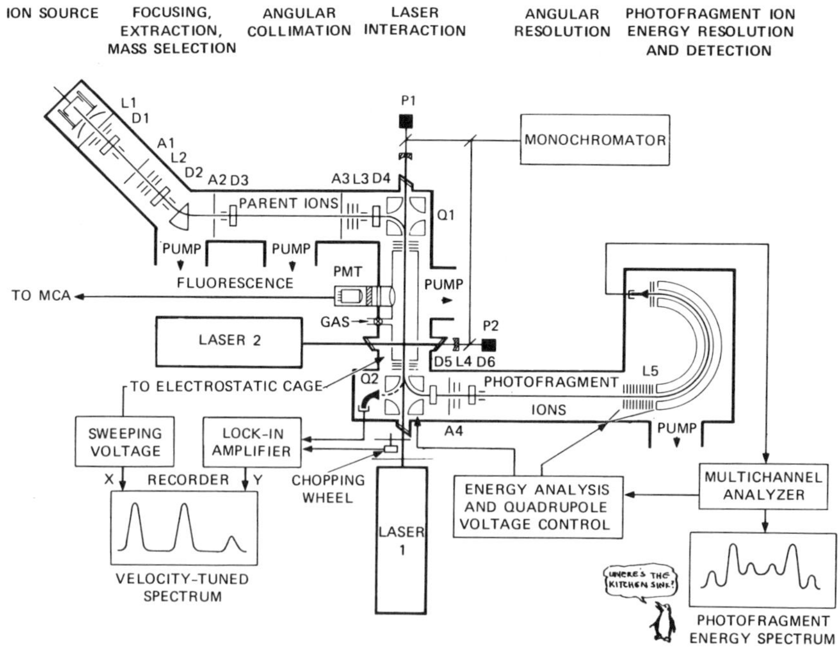

Figure 8-15-1. The fast ion photofragment spectrometer at the Stanford Research Institute.[105]

photofragments can be measured by rotating the laser beam polarization. With coaxial beams and a single-mode laser, it is possible to achieve sub-Doppler resolution because of the narrowing of the velocity distribution in an ion beam when it is brought to a high energy. The apparatus shown in Fig. 8-15-1 has an energy resolution of better than 10 meV for normal photofragment spectroscopy and 0.001 meV for coaxial beams photofragment spectroscopy with a single-mode laser. The kinematics of the photon-molecular ion collisions are discussed in Refs. 97, 105, and 107.

Examination of a special situation will prove informative. Let T_0 be the energy of the parent molecular ion in the Lab frame and W be the total CM kinetic energy release in the dissociation event. Then, if the parent molecule is homonuclear and diatomic, a photofragment that appears at $\theta = 0$ with respect to the primary beam direction will have a Lab kinetic energy, T, given by

$$T = \frac{T_0}{2} \pm (WT_0)^{1/2} + \frac{W}{2}. \tag{8-15-2}$$

Hence, for $T_0 = 3000$ eV, a photofragment corresponding to $W = 0.1$ eV has a Lab energy of $T = 1517$ eV, in contrast to $T = 1500$ eV for $W = 0$. A plot of the relative photofragment yield for O^+ at various values of W and the correspond-

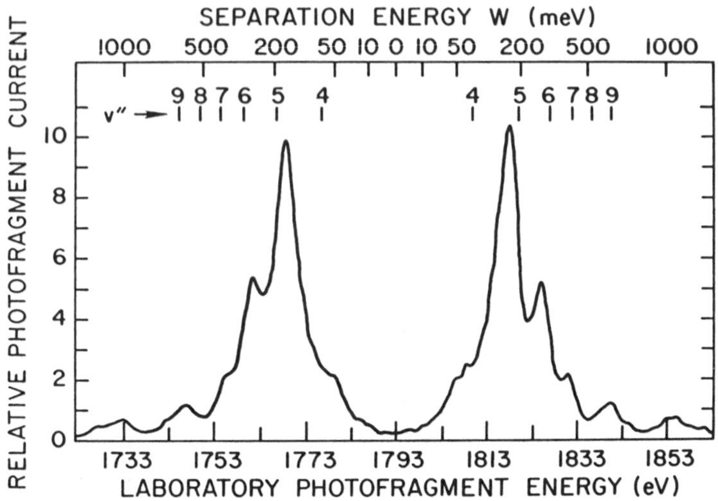

Figure 8-15-2. Kinetic energy spectrum of O^+ photofragments produced from O_2^+ by 5600-Å radiation from a laser with polarization parallel to the primary ion beam.[108]

ing values of T_0 appears in Fig. 8-15-2. This spectrum was taken at 5600 Å for laser polarization parallel to the primary ion beam direction.[108]

Although the emphasis here has been on the investigation of the mechanisms of photodissociation and the interactions between the photofragments, apparatus such as that pictured in Fig. 8-15-1 has also been used to measure photodissociation cross sections (see, e.g., Ref. 104). The cross sections for production of singly charged photofragments from singly charged diatomic ions have peak values, as functions of photon energy, of the order of 6×10^{-18} cm^2.[65] The peak values for dissociation into multiply charged fragments are considerably smaller, whereas those for production of singly charged fragments from neutral diatomics are usually slightly larger. The maximum cross sections for photon-impact processes are typically one or two orders of magnitude smaller than the corresponding electron-impact cross sections.

Additional important experimental work on photodissociation by A. Carrington, J. Durup, H. Helm, L. C. Lee, P. J. Saare, and others is cited in the review by Kirby and van Dishoeck.[109] Theoretical contributions by J. A. Beswick, Y. B. Band, A. Dalgarno, K. F. Freed, P. S. Julienne, D. J. Kouri, F. H. Mies, M. Krauss, and others are also discussed in this review.[109] Photodissociation of cluster ions has been studied by several groups.[110]

8-16. PHOTOIONIZATION

Accurate measurements of absolute cross sections for total photoabsorption and total photoionization of both atoms and molecules in the VUV and soft X-ray regions have been made in abundance since about 1950, with steady improvement in the experimental techniques from that time to the present. However, reliable absolute cross sections for production of individual ionization products (i.e., partial photoionization cross sections) and individual ionic photofragments have been measured only recently. The advent of the electron synchrotron and storage ring as light sources (Section 8-1-A-2) was very important in this development, as were calibrated electron energy analyzers and photomultiplier detectors. Another important development occurred in about 1972 with the introduction of dipole electron inelastic scattering techniques to simulate accurately the conventional photoabsorption and photoionization experiments at continuously tunable energies corresponding to the UV and soft X-ray regions (see Section 8-16-A-4).

In this section we concentrate on experimental aspects of one-photon ionization and the fragmentation that may subsequently occur with molecular targets. We then deal with other topics in less detail. As stated earlier, multiphoton processes are discussed in Section 8-17.

A. Experimental Studies of Single-Photon Ionization

This discussion is based on the review and data compilation by Gallagher et al.[65]

1. *Overview*

To achieve some (but not too much!) generality, we consider the ionization of a diatomic molecule AB according to the equations

$$h\nu + AB \rightarrow [AB^{n+}]^j + ne^- \qquad (8\text{-}16\text{-}1)$$

$$\xrightarrow{n \geq 1} AB^{n+} \qquad (8\text{-}16\text{-}2)$$

$$\xrightarrow{n \geq 1} A^{p+} + B^{q+} \qquad (8\text{-}16\text{-}3)$$

$$\xrightarrow{n = 0} A^+ + B^-. \qquad (8\text{-}16\text{-}4)$$

Here the target molecule AB is assumed to be in the ground vibrational state with a nearly degenerate ensemble of rotational states, and $[AB^{n+}]^j$ is a molecular ion in the electronically excited state j. The electron photoejection process (8-16-1) involves n-fold photoionization with AB^{n+} left in state j, which may be either stable or unstable with respect to dissociation. The process (8-16-1) occurs on the time scale $\tau_e \approx 10^{-17}$ s. The processes (8-16-2, 8-16-3, and 8-16-4) involve relaxation of $[AB^{n+}]^j$ on a time scale $\tau_m \approx 10^{-13}$ s. Equation (8-16-2) refers to production of singly or multiply charged stable parent ions of charge $n \geq 1$; (8-16-3) illustrates the production of fragment ions, with $p + q = n \geq 1$; and (8-16-4) denotes the special case of ion-pair production, with $n = 0$.

Because $\tau_m \gg \tau_e$, the ion $[AB^{n+}]^j$ can be characterized by measuring the kinetic energy of the ejected electrons with the technique of photoelectron spectroscopy (PES), or by dipole (e, 2e) coincidence spectroscopy. The relaxation and fragmentation of the molecular ion $[AB^{n+}]^j$ can be studied by photoionization mass spectrometry (PIMS), or by the equivalent technique of dipole (e, e + ion) coincidence spectroscopy. All of these techniques will be discussed in due course.

In the present discussion, it is convenient to divide the spectrum into several fairly distinct regions:[15]

1. The *discrete absorption region*, which extends from the first electronic excitation energy to the first ionization energy. This region was covered in Section 8-14, where it was emphasized that accurate excitation cross sections are difficult to measure directly.
2. The *structured ionization region*, which extends from the first ionization energy up to about 20 eV. Ionization cross sections are difficult to determine accurately here.
3. The *smooth cross-section region*, from about 20 eV to several hundred eV. Most of the good experimental data refer to this region.
4. The *high-energy region*, from a few hundred eV upward. Inner-shell processes[6,111] are very important in this region but are not covered here.

In our discussion we ignore photon scattering processes (Section 8-1-B) because of their small cross sections in the low energy regime, and write the total photoabsorption cross section as

$$\sigma_{\text{abs}}^T = \sigma_{\text{neut}} + \sigma_{\text{ion}}^T, \tag{8-16-5}$$

where σ_{neut} is the cross section for excitation into stable and/or dissociating neutral states, and σ_{ion}^T is the total photoionization cross section in (8-16-1).

2. *Photoelectron Spectroscopy*

This technique involves passing a "monoenergetic" beam of photons through a gas cell or a molecular beam and mapping the energy spectrum of the ejected electrons. The kinetic energy of an electron ejected from the jth orbit of the target is given in terms of the photon energy ($h\nu$) and the binding energy (I_j) by the equation

$$E_j = h\nu - I_j. \tag{8-16-6}$$

Hence measurements of the kinetic energies of the photoelectrons can provide the binding energies of the electrons in the various orbits.*

*Here it is implicitly assumed that the ion is left in its ground state. If, on the other hand, it is left in an excited state, the electron is ejected with lower energy, and the line in the PES spectrum is referred to as a *satellite line* [see K. Kimura and Y. Achiba, *Phys. Rep.* (1989), to be published].

A beautiful example of a PES spectrum appears in Fig. 8-16-1.[112] It was obtained with a molecular-beam photoelectron spectrometer developed at the Lawrence Berkeley Laboratory.[113] A supersonic SO_2 molecular beam, expanded from 200 torr of pure SO_2 at room temperature through a 100-μm-diameter nozzle and skimmed by a 0.9-mm-diameter conical skimmer, was crossed at 90° by a 584-Å photon beam from a helium discharge lamp. The photoelectron energy analyzer consisted of a double electrostatic deflector that sampled the photoelectrons at an angle of 90° with respect to both the photon beam and the molecular beam. An energy resolution of 13-meV FWHM was achieved. The high energy resolution and the rotational cooling of the molecular beam permitted the vibrational structure to be completely resolved. The Berkeley group has also used synchrotron radiation in many of their PES studies.[114]

Measurements of the areas under the various peaks of PES spectra can provide branching ratios for production of the final ionic states if corrections are made for instrumental effects. Absolute partial cross sections can then be obtained, as we shall now demonstrate. First we need the *total photoionization efficiency*, or *quantum yield*, defined as

$$\gamma = \frac{\text{number of ionizing events per second}}{\text{number of photons absorbed per second}}. \tag{8-16-7}$$

γ assumes different specific forms for different kinds of experiments; the appropriate form for PES is

$$\gamma^{\text{PES}} = \frac{N_1 + N_2/2 + \cdots}{\text{number of photons absorbed per second}}, \tag{8-16-8}$$

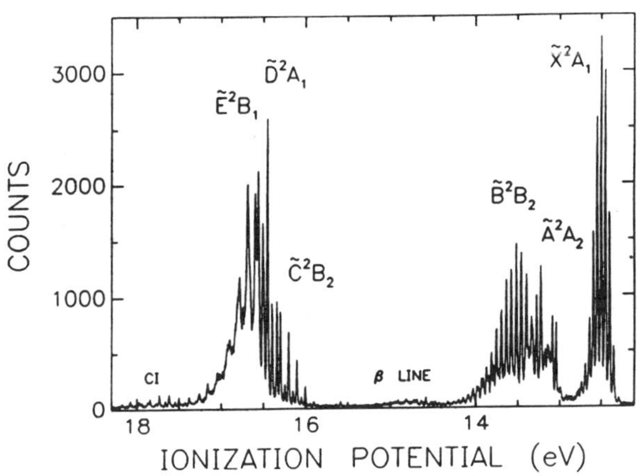

Figure 8-16-1. The complete He I (584 Å) photoelectron spectrum obtained from a supersonic beam of SO_2 with 13-meV resolution.[112]

where $N_1, N_2, \ldots$ are the number of electrons produced per second by all single, double, ... ionization events before photofragmentation occurs (cf. Problem 8-47). Thus we have

$$\sigma_{\text{ion}}^{T} = \gamma^{\text{PES}} \sigma_{\text{abs}}^{T}. \tag{8-16-9}$$

We also need the photoelectron branching ratio, BR^j, for production of a molecular ion $[\text{AB}^{n+}]^j$ in the electronic state j:

$$\text{BR}^j = \frac{N^j/n}{N_1 + N_2/2 + \cdots}. \tag{8-16-10}$$

Here N^j is the number of photoelectrons produced per second during the formation of the ions. Then the partial photoionization cross section for formation of a singly or multiply charged molecular ion in the electronic state j is given by

$$\sigma_{\text{ion}}^{j} = \text{BR}^j \gamma^{\text{PES}} \sigma_{\text{abs}}^{T}. \tag{8-16-11}$$

This equation illustrates the procedure for obtaining partial photoionization cross sections. The total absorption cross section σ_{abs}^{T} is first measured accurately as a function of photon energy, and then the total photoionization efficiency γ^{PES}, or alternatively, the total photoionization cross section σ_{ion}^{T}, is determined. Finally, BR^j is measured, and σ_{ion}^{j} is calculated by (8-16-11). Difficulties arise in conventional PES measurements involving multiple ionization because the ejected electrons share the total energy available, giving rise to a continuous distribution rather than a sharp line.[65]

For electric dipole transitions, the one-electron electronic partial photoionization cross section differential in ejection angle for linearly polarized radiation and unoriented molecular targets is given by[61,115]

$$d\sigma_{\text{ion}}^{j}(\theta) = \frac{\sigma_{\text{ion}}^{j}}{4\pi}[1 + \beta^j P_2(\cos\theta)]\, d\Omega \tag{8-16-12}$$

[cf. (8-15-1)]. Here θ is the angle between the $\mathbf{E}$ vector of the incident radiation and the momentum vector of the outgoing electron, and β^j is the photoelectron asymmetry parameter, or angular anisotropy factor.

In measurements of branching ratios BR^j defined in eq. (8-16-10), it is customary to use the "magic angle" ($\theta = 54.74°$), at which $P_2(\cos\theta) = 0$. Then dN^j (the number of electrons detected in the solid angle element $d\Omega$) is proportional to σ_{ion}^{j}, but independent of β^j, and the total numbers N^j of electrons collected through the total solid angle can be replaced by the numbers received in $d\Omega$. After σ_{ion}^{j} is evaluated from (8-16-11), β^j can be obtained by measuring the angular distribution of the ejected electrons and using (8-16-12). The angular distribution can be determined either by rotating the analyzer in a plane containing the incident photon beam or in a plane perpendicular to it. Another method involves rotating the polarization of the photon beam, with the analyzer held fixed.*

3. *Photoionization Mass Spectrometry*

Here a monochromatized photon beam is passed through a gas target or molecular beam, and the resulting ions are extracted and mass analyzed. The cross section for any fragment produced in a dissociative process is given by

$$\sigma_{\text{ion}}^{\text{frag}} = \text{BR}^{\text{frag}} \gamma^{\text{PIMS}} \sigma_{\text{abs}}^{T}; \tag{8-16-13}$$

σ_{abs}^{T} is assumed to have been measured. The total photoionization efficiency for PIMS is

$$\gamma^{\text{PIMS}} = \frac{\eta_1 + \eta_2 + \cdots + \eta_{pp}}{\text{number of photons absorbed per second}}, \tag{8-16-14}$$

where $\eta_1, \eta_2, \ldots$ are the numbers of singly, doubly, ... charged ions produced in the initial ionization process before ionic photofragmentation takes place, and η_{pp} is the number of ion-pair events per second.

Finally,

$$\text{BR}^{\text{frag}} = \frac{N^{\text{frag}}}{\eta_1 + \eta_2 + \cdots + \eta_{pp}}, \tag{8-16-15}$$

where N^{frag} is the number of ions per second of the single type labeled "frag" that is actually observed. Discrimination effects in ion collection and techniques for circumventing them are discussed by Samson.[61]

An illustration of a technique used by Lee's group at Berkeley for high-resolution measurements of the photoionization spectrum of water molecules is shown in Fig. 8-16-2.[116] A supersonic beam of H_2O or D_2O molecules seeded in neon passes through a pulsed valve and skimmer and into the collision region, where it is intersected by a beam of extreme ultraviolet light (XUV, $\lambda < 1000$ Å) between two ion extraction plates. The XUV light is generated by frequency tripling laser light in a pulsed jet of xenon or argon (see Section 8-17-C). Ions produced in the interaction region are extracted through the quadrupole mass filter and counted. This apparatus provided photoionization spectra for H_2O and D_2O at 1.5-cm^{-1} resolution on molecules with a rotational temperature of about 50 K. The advantage accrued by using a supersonically cooled molecular beam is illustrated by Fig. 8-16-3.[116]

*Mention should be made of new PES techniques that permit angle-resolved and spin-resolved measurements [see, e.g., U. Heinzmann, *ICPEAC XIV*, Stanford (1985), p. 37].

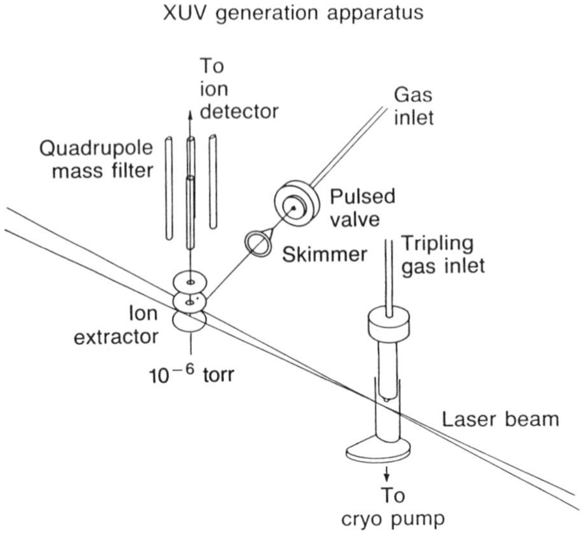

Figure 8-16-2. Apparatus used by Lee's group for photoionization mass spectrometry.[116]

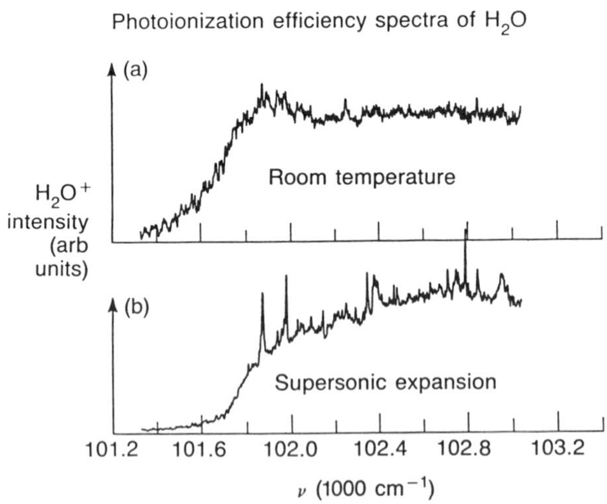

Figure 8-16-3. Photoionization efficiency spectra of H_2O obtained by Lee's group, with and without rotational cooling of the H_2O molecular beam.[116]

4. *Dipole Electron Simulation of Photon Beams*

In Sections 5-6, 5-7-F, and 5-7-G we showed that a beam of fast electrons can simulate quantitatively a beam of photons. In general, electronic projectiles excite both optical and nonoptical transitions in targets, but electrons transfer energy to an atom or molecule in a manner similar to photons if (1) the electron velocity greatly exceeds that of the orbital electrons, and (2) the interaction distance is large. Then the target experiences a spatially uniform electric field that is sharply pulsed in time (a delta function) as the electron passes by. The Fourier transform of the impulse function is a spectrum of constant intensity over a wide range of frequencies of the **E** field, so that the electron simulates a continuum, or white-light spectrum. The width of the region of constant intensity (or flat virtual photon field) increases as the velocity of the projectile electrons increases.

We now show how it is possible to obtain pure optical information from electron collisions with atoms and molecules (i.e., how it is possible to determine dipole oscillator strengths for specific transitions). We start with the relationship between the differential electron scattering cross section $d\sigma(K)/dE_L$ and the generalized oscillator strength density $df(K)/dE_L$,

$$\frac{d\sigma(K)}{dE_L} = \frac{2}{E_L}\frac{k_f}{k_i}\frac{1}{K^2}\frac{df(K)}{dE_L}, \tag{8-16-16}$$

originally established by Bethe in 1930.[117] [This equation is based on the Bethe-Born approximation (see Sections 3-23-B and 6-14).] Here k_i, k_f, and K are the incident electron momentum, the scattered electron momentum, and the momentum transferred from the incident electron to the target, respectively, all in units of $\hbar$. We can make the expansion

$$\frac{df(K)}{dE_L} = \frac{df}{dE_L} + AK^2 + BK^4 + \cdots, \tag{8-16-17}$$

where the first term on the right is the dipole, or optical, oscillator strength density df/dE_L, and we see that in the limit of zero momentum transfer ($K = 0$), the generalized oscillator strength density approaches the optical oscillator strength density. This fact provides the justification for making measurements of inelastic electron scattering to simulate photoabsorption and photoionization.

When $K \to 0$, eq. (8-16-16) becomes

$$\frac{df}{dE_L} = \frac{mc}{\pi e^2 h}\sigma^T_{\text{abs}} = \frac{E_L}{2}\frac{k_i}{k_f}K^2\frac{d\sigma(K)}{dE_L}. \tag{8-16-18}$$

We see that the kinematic factor for conversion of electron scattering cross sections to optical oscillator strengths is

$$\frac{E_L}{2}\frac{k_i}{k_f}K^2. \tag{8-16-19}$$

In practice, this factor must be integrated over the geometry of the experiment.

A key factor in the establishment of the electron simulation is that the electron *energy loss* is equivalent to the photon *energy*. In stimulating any transition, the electron loses a certain amount of energy to the target; in exciting the same transition, a photon imparts the same amount of energy to the target but disappears in the process.

Dipole electron-impact measurements are usually made with electron-beam energies of 1 keV to several tens of keV to simulate photon energies of about 5 to 1000 eV.

The mapping of electron energy-loss spectra for rotational, vibrational, and electronic excitation processes was discussed in Sections 5-3, 5-4, and 5-7-A; high-energy-resolution electron spectrometers such as described in Section 4-3-B are used in such studies. If such a spectrometer is used to measure $d\sigma/dE_L$ for forward-scattered fast electrons, cross sections for total photoabsorption σ_{abs}^T can be obtained through the use of (8-16-18). This electron scattering method is said to be of the dipole (e, e) type. It provides a very useful alternative to the photon beam methods described in Section 8-13-B.

In the simulation of photoelectron spectroscopy, fast forward-scattered electrons undergoing an energy loss E_L are detected in coincidence with electrons ejected at the magic angle $\theta = 54.7°$ in what is termed a (e, 2e) measurement. The coincidence feature ensures that the measurement of the *kinetic energy* of the ejected photoelectron is paired with the measurement of the *energy loss* E_L' experienced by the particular projectile electron producing the ionization event, or alternatively, with the *energy* $h\nu = E_L'$ of the equivalent photon. A typical apparatus for (e, 2e) measurements is shown in Fig. 8-16-4.[65,118]

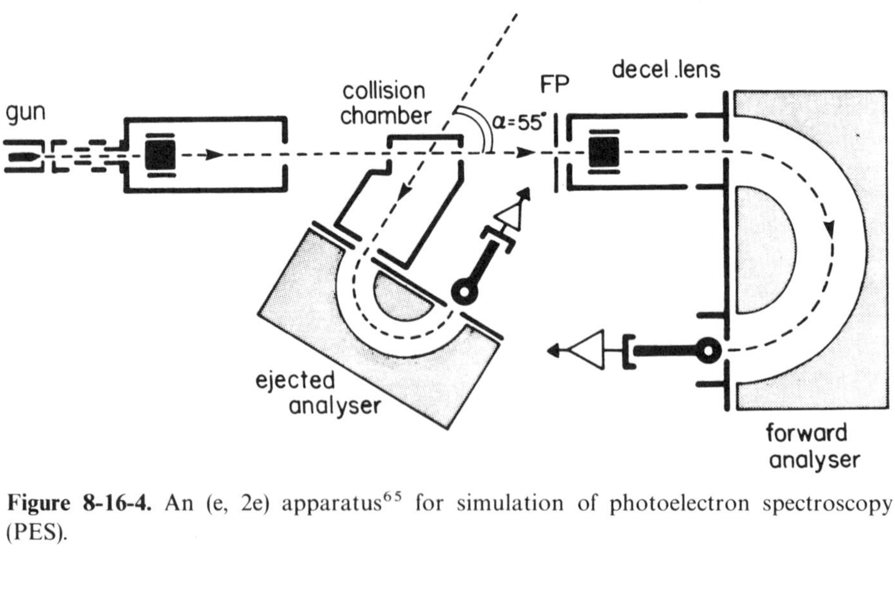

Figure 8-16-4. An (e, 2e) apparatus[65] for simulation of photoelectron spectroscopy (PES).

The (e, e + ion) electron-impact apparatus shown in Fig. 8-16-5 can be used to simulate PIMS measurements.[65,118] The product ions are mass analyzed by time-of-flight (TOF) techniques and detected in delayed coincidence with the forward-scattered, energy-loss-selected electrons. Here the delayed coincidence measurement is required to pair the production of a particular *ionic product* with the particular *energy loss* E'_L by the electron that produced the ionization event, or alternatively, the *energy* of the equivalent photon, $h\nu = E'_L$

The use of dipole electron-beam techniques offers several advantages over the direct photon-beam methods beyond the obvious factor of much lower cost. First, the electron beam is easily tunable in energy over a very wide range, and it provides better energy resolution at energies above about 100 eV. Also, anomalous intensity effects produced by finite line width in photon impact experiments cannot occur in electron-impact simulation measurements because of the nonresonant nature of electron-impact excitation. As Gallagher et al.[65] point out, the entire oscillator strength is encompassed by the measured electron-impact curve, regardless of the line width or the energy resolution. On the other hand, if a photon beam is used, the excitation is resonant, and severe perturbations arise when the natural line widths are narrower than the monochromator bandwidth. Another important advantage associated with the electron-impact simulation method arises from the flat virtual photon field induced in the target by the incident electron. Gallagher et al.[65] discuss the ramifications of this characteristic of the virtual photon field.

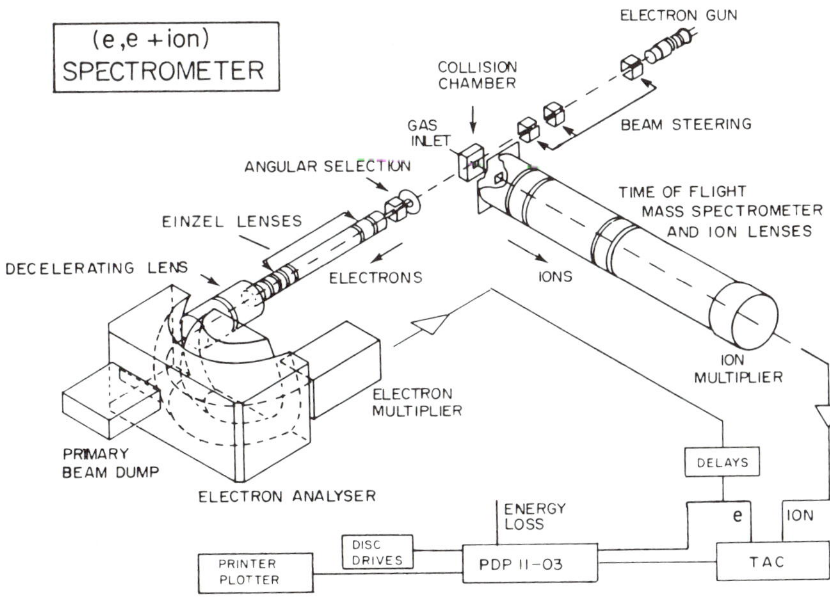

Figure 8-16-5. An (e, e+ion) apparatus[65] for simulation of PIMS measurements.

Data on photoabsorption and photoionization processes obtained by direct photon beams and by dipole electron-beam techniques are collected in numerous reviews,[6,15,61–63,65,119–137] which also contain many comparisons of experiment with theory.

5. *Angular Distribution of Photoelectrons*

We have displayed, in (8-16-12), the angular distribution of photoelectrons ejected in electric dipole, single-electron transitions from unoriented targets by linearly polarized radiation. The angular dependence is characterized by, and completely determined by, the asymmetry parameter β, whose range ($+2$ to -1) is fixed by the requirement that the cross section be nonnegative. If $\beta = +2$, the distribution has a $\cos^2 \theta$ form, peaked along the direction of polarization of the incident radiation (i.e., the direction of the **E** vector). Values of β near $+2$ are common, whereas values of β near $\theta = 0$ (isotropic distribution) are not. Finally, $\beta = -1$ corresponds to a $\sin^2 \theta$ distribution, peaked perpendicular to the polarization direction. If β is close to -1, the outgoing electron is subjected to substantial torques as it leaves the target.[121]

We have indicated in Section 8-16-A-2 how β can be measured; further discussion of the measurement techniques appears in Refs. 61, 121, 138, and 139. The significance of the magic angle $\theta = 54.74°$ warrants further emphasis. Since $P_2(\cos \theta)$ in (8-16-12) is zero at this angle, the measured photoelectron current is proportional to the integrated cross section, σ, irrespective of the polarization of the incident radiation and of the photoelectron angular distribution. Unless measurements are made at the magic angle, or unless all of the photoelectrons are collected, the measured cross section will usually depend on the polarization and the angular distribution.[121]

Theoretical equations appropriate for higher-order multipole transitions and for polarizations other than linear are discussed by Samson,[61] Tseng et al.[115], and Manson and Starace.[138]

Data in the form of β as a function of the photon energy are especially useful, being indicators of the dynamical interaction between the electrons in the different shells of the target. Samson[61] has collected data on β versus photon energy for atoms; Gallagher et al.[65] present similar data for molecules.

6. *Multiple Ionization*

The ratios of multiply charged to singly charged ions in photoionization are similar to those in ionization by electron impact.[61,63] Typically, the maximum value of the ratio σ^{2+}/σ^{+} is several percent, and that for σ^{3+}/σ^{+} is much lower still.

Kossmann et al.[140] have recently studied double photoionization of helium in the threshold region in order to test the Wannier threshold law discussed in Section 6-8. For the process

$$h\nu + \text{He} \rightarrow \text{He}^{2+} + e + e, \tag{8-16-20}$$

the Wannier theory predicts the integral cross section

$$\sigma^{2+} = \sigma_0 \, \Delta E^{1.056}, \qquad (8\text{-}16\text{-}21)$$

where ΔE is the excess energy (the amount by which the photon energy exceeds the threshold for double ionization) and σ_0 is the value of σ^{2+} at $\Delta E = 1$ eV. Kossmann et al. used the BESSY storage ring at Hamburg and worked from threshold (79.013 eV) up to 83 eV photon energy. They obtained a value for the excess energy exponent consistent with the Wannier value in (8-16-21) and found σ_0 to have the value $1.02(4) \times 10^{-21}\,\text{cm}^2$. The Wannier threshold law appeared to be valid up to only about 2 eV above threshold, a limit that is somewhat lower than might be expected.

Another important measurement has also recently been made on double photoionization, by Bae and Peterson[141] on the negative ion K^-. They covered the photon energy range from 70 meV below the threshold to 250 meV above, and found a near-threshold energy dependence consistent with the Wannier theory. Their data also agree nearly as well with the Coulomb-dipole theory of Temkin discussed in Section 6-8.

7. *Photoionization near Threshold*

The studies reported immediately above in Refs. 140 and 141 are representative of the current research on threshold photoionization. As in the case of electron impact threshold ionization, both single and multiple ionization are being studied with refined techniques, and definitive results are being obtained. Recent progress is discussed by King.[142] The theory of threshold ionization outlined in Section 6-8 applies to both electron and photon impact and need not be repeated here. Much of the material on experimental techniques in that section is also applicable to both types of projectiles.

8. *Photoionization of Excited States*

This topic is of considerable importance because effects can arise here that do not appear in photoionization of ground-state targets. Also, the cross sections can be much larger for excited states, and hence they have disproportionately large effects in applications such as plasma physics. Unfortunately, few experimental data obtained by direct methods are available. One important paper to which we wish to direct attention is that by Smith et al.,[143] who used a saturation method to measure absolute photoionization cross sections for the $4D$ and $5S$ excited states of sodium at 1.06 μm. A well-collimated beam of sodium atoms with number density 10^9 atoms/cm^3 was crossed in a vacuum with two collinear continuous-wave dye laser beams that stepwise excited a continuous stable population of the excited state being studied. The number of ions produced by single pulses from a Q-switched Nd:YAG laser was measured as a function of the energy of the ionizing pulse. The ionizing laser beam was directed opposite to the dye laser beams, and Na^+ ions were collected by a weak electric field perpendicular to sodium and laser beams. Observation of the energy at which the ionization process saturated provided the photoionization cross section. Cross sections were derived by matching experimental saturation curves with model curves computed by numerical integration of the rate equations over the duration of the ionizing pulse. The computed curves allowed for the spontaneous decay of the excited state, the production rate of the excited atoms, and the shape of the ionizing pulse. The measured cross sections for photoionization of unaligned sodium atoms were 15.2 ± 1.7 Mb for the $4D$ state and 1.49 ± 0.13 Mb for the $5S$ state at the wavelength used in these measurements ($1.06\ \mu$m). The method described here has the advantage that the only calibration that is required is that of the flux of ionizing photons (F). The number of excited atoms (N) and the number of ions produced (I) do not have to be known. The rate equations describing the experiments are

$$\begin{aligned} \frac{dN}{dt} &= -N\alpha F(t) + A - \Gamma N, \\ \frac{dI}{dt} &= N\sigma F(t), \end{aligned} \tag{8-16-22}$$

where σ is the test photoionization cross section, A is the excited-state production rate, and ΓN is the spontaneous decay rate. Because of dynamic Stark shifts induced by the ionizing laser light, the production of $4D$ and $5S$ excited states is cut off during the ionizing pulse, so A is taken to be zero.

Photoionization cross sections for a number of atoms in excited states have been calculated by S. T. Manson and his colleagues.[144] The paper by Preses et al. cited in Ref. 144 deals with photoionization of the excited $3p$ state of sodium. Experimental measurements, as well as theoretical calculations, were reported, and excellent agreement between them was obtained over the range of measurements (threshold to 2 eV above).

9. *Photoionization of Positive Ions*

Until recently, few cross sections for photoionization of positive ions had been directly measured, and the requirement of photon energies exceeding 10 eV made it impossible to measure absolute cross sections for atomic ions. Furthermore, the particle number density in ion targets is limited by space charge to about 10^6 per cm^3, compared with about 10^{12} per cm^3, or greater, for neutral targets. The availability of synchrotron light sources, and the development of crossed and merged beams techniques described in Section 7-1 have removed these obstacles, as demonstrated by an interesting set of experiments performed by the Newcastle group at the Daresbury synchrotron.[145,146] They studied Ca^+, Sr^+, Ba^+, Zn^+, Ga^+, and K^+ ions in a merged beams apparatus that provided a long interaction path. In measurements on Ba^+, they typically used a 2-keV, 8×10^{-8} A Ba^+ beam and a photon flux of 10^9 per second, with 0.1-Å resolution, and obtained 100 Ba^{2+} ions per second in the region of the largest resonance ($\sigma \approx 2 \times 10^{-15}$ cm^2). There was an additional background count of 40 Ba^{2+} ions per second, produced mainly by charge stripping of the Ba^+ beam by residual gas. At $\lambda = 500$ Å and $\Delta\lambda = 0.1$ Å, the energy resolution is better than 5 meV. The overall absolute error is estimated to be less than $\pm 15\%$. The main motivation of this work was to make a closer study of autoionization* than had been possible in the earlier Newcastle measurements on electron-impact ionization of positive ions (see Section 7-1). In the photoionization studies it was possible to detect resonances smaller than 10^{-17} cm^2 (Fig. 8-16-6), and in the case of Ba^+, approximately 75 peaks were observed in the wavelength range 420 to 775 Å (29.5 to 16.0 eV). The results indicated that inner-shell excitation-autoionization (Section 6-5, process 8) made a very large contribution. It is interesting to note that the largest resonances observed are about three orders of magnitude larger than what would be expected from direct, outer-shell ionization. The reader may refer to Ref. 147 for a number of papers by Manson and his coworkers on theoretical calculations of photoionization of positive ions.

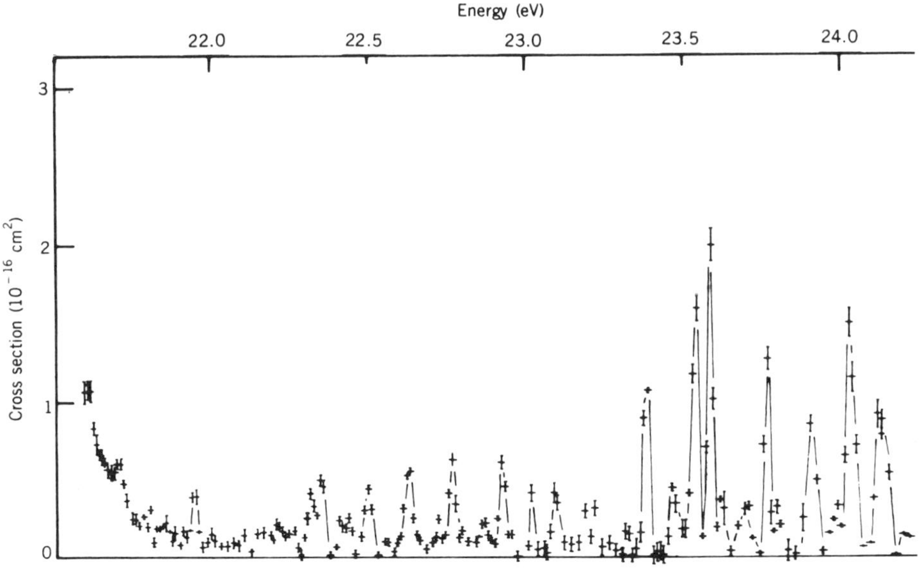

Figure 8-16-6. The photoionization cross section of Ba^+ ions.[145]

*Excellent modern reviews of autoionization in the context of photon experiments are J. Berkowitz, *ICPEAC XIV*, Stanford (1985), p. 631, and *Adv. Chem. Phys.* **72**, 1 (1988).

B. A Few Concluding Remarks on Theory

1. *The Length, Velocity, and Acceleration Forms of the Electric Dipole Transition Matrix Element*[121]

If we refer to (8-6-6) for the rate of absorption of electric dipole radiation by a one-electron atom, we see that the transition matrix element contains **r**, the position vector of the atomic electron. (In a multielectron atom, **r** is replaced by $\sum \mathbf{r}_n$, where n refers to the nth electron.) By using the commutation relations between the electron position and momentum operators, one may replace **r** in the matrix element by the momentum operator, or the gradient operator, and thereby transform from the *length form* of the matrix element to the *velocity form*. (See Problem 8-34; note that a multiplicative factor comes into play in the transformation.) One may further transform to a matrix element in which the acceleration operator appears and obtain the *acceleration form* of the matrix element.

The three forms discussed here give the same results if exact wave functions are used in the matrix element, but different results if only approximate wave functions are available. It is sometimes difficult to know which form of the matrix element will give the best results, but the following considerations are often helpful. The length form emphasizes the wave function in the outer regions of the atom, the velocity form accentuates the intermediate region, and the acceleration form emphasizes the region close to the nucleus. Hence if the wave functions are known to be more accurate in one of these regions, the corresponding form of the element might be profitably chosen.

2. *Cooper Minima*[15,61,120,148]

A Cooper minimum is a minimum in a cross section versus photon energy produced by the dipole matrix element going through a zero because of the positive and negative contributions canceling. This effect was first discovered in D. W. Ditchburn's laboratory in the photoionization of alkali atoms in the early 1940s. The basic explanation for these observations was given by D. R. Bates several years later, and extended by M. J. Seaton to include relativistic effects in 1951. In the early 1960s, J. W. Cooper made the important contribution of showing that the minima occur very generally and are not restricted to the outer shells of the alkalis.

3. *Resonances; Beutler-Fano Profiles*

In Section 4-8-B we introduced Feshbach and shape resonances in the context of elastic electron scattering, and in Chapters 5, 6, and 7 we encountered resonances again in connection with various types of inelastic electron collisions. In a *Feshbach resonance*, a projectile virtually excites a target to a configuration that creates a potential well capable of trapping the projectile for a short time in a temporarily bound "compound" state. Evidently, Feshbach resonances do not occur in photon collisions, because the incident photon cannot be trapped. On the other hand, *shape resonances* are quasi-bound states in which a particle is trapped for a short time by a potential barrier, through which it may tunnel and escape. Such barriers may be of the centrifugal type, which block the motion of otherwise free electrons in certain directions and confine them to regions of space of molecular dimensions. Shape resonances* and also *autoionizing resonances* (Sections 6-5, 6-6, and 7-1) are both conspicuous and important in photon collisions.[114,149–151] Beutler[152] made an important early contribution to the development of this subject by his 1935 study of autoionization effects in photoabsorption of UV radiation by rare gases between the first and second ionization energies. Madden and Codling[153] were the first to use synchrotron radiation to obtain high-resolution far-UV absorption spectra of gases. Their studies, during the early 1960s, of autoionizing levels that give rise to resonance structure in absorption spectra were of great importance, especially as tests of the theory by Fano,[154,155] now to be outlined.

In 1961, Fano[154] published a theory of autoionization in which he considered the effect of an isolated resonance, such as a Rydberg level, on the total photoabsorption or photoionization cross section. A discrete level embedded in one or more continua produces interference in the photon absorption because two indistinguishable paths lead to the final state, direct ionization and autoionization. According to Fano,[154,155] the total cross section σ_T in this case is

$$\sigma_T = \sigma_0\left(p^2\frac{(q+\epsilon)^2}{1+\epsilon^2} + 1 - p^2\right), \tag{8-16-23}$$

with

$$\epsilon = \frac{E - E_0}{\Gamma/2}. \tag{8-16-24}$$

The Fano parameters q and ρ^2 are taken to be constant over the resonance, σ_0 is the nonresonant background cross section, E_0 is the energy of the resonance and Γ is its FWHM width, and ϵ is a reduced energy. Let us denote the ground state by g, the discrete state by ϕ, and the continua by μ. The q parameter is then given by

$$q = \frac{\langle\Phi|\mathbf{r}|g\rangle}{\pi\sum_\mu\langle\phi|V|\mu\rangle\langle\mu|\mathbf{r}|g\rangle}; \tag{8-16-25}$$

*Shape resonances derive their name from the fact that the specifics of the resonance behavior stem from the *shape* of a local potential (i.e., the barrier and the associated inner and outer wells). Shape resonance effects are observed in X-ray and VUV absorption spectra, photoelectron branching ratios and angular distributions, Auger electron angular distributions, elastic scattering of electrons, vibrational excitation by electron impact, and numerous other phenomena.[149]

it determines the shape of σ_T. The correlation coefficient ρ^2 is a measure of the strength of the resonance and is expressed as

$$\rho^2 = \frac{\sum_\mu |\langle\phi|V|\mu\rangle\langle\mu|\mathbf{r}|g\rangle|^2}{\sum_\mu |\langle\phi|V|\mu\rangle|^2 \sum_\mu \langle\mu|\mathbf{r}|g\rangle|^2}. \tag{8-16-26}$$

Here **r** and V are the dipole and Coulomb operators, respectively, and Φ is the discrete state as modified by an admixture of continuum states. The width of the resonance is

$$\Gamma = 2\pi \sum_\mu |\langle\phi|V|\mu\rangle|^2, \tag{8-16-27}$$

and the nonresonance cross section is

$$\sigma_0 = \sum_\mu |\langle\mu|\mathbf{r}|g\rangle|^2. \tag{8-16-28}$$

The *Beutler-Fano profile* in (8-16-23) applies to the case of the total absorption cross section in the vicinity of an isolated resonance. Starace[156] has considered the case of partial cross sections, for which it is necessary to consider several outgoing channels in the vicinity of an autoionizing resonance. Similar work by others is discussed in Refs. 114 and 150.

8-17. MULTIPHOTON PROCESSES*

As pointed out by Eberly and Lambropoulos,[157] multiphoton phenomena began to be studied long before the advent of the laser in 1960. Rayleigh and Raman scattering (discussed in Section 8-1-B) are two-photon processes that were first investigated experimentally more than 50 years ago, and their theoretical descriptions pointed the way toward the description of higher-order processes. In 1929 and 1931, M. Göppert-Mayer published a theory of two-photon spontaneous and stimulated processes based on second-order perturbation theory, after a Green's function calculation of two-photon transitions by Podolsky in 1928. The first observations of the two-photon spontaneous decay of metastable (2S) hydrogen were made in the early 1940s in radiation from stellar nebulae, and higher-order processes were detected before 1960 in the microwave region. It goes without saying, however, that the laser has had a tremendous impact on the field. It has made possible studies of stimulated Raman scattering, harmonic generation, wave mixing, multiphoton ionization, high-resolution multiphoton spectroscopy, and selective excitation of atoms and

*A comprehensive review that deals with the theory of most of the topics covered here is N. L. Manakov, V. D. Ovsiannikov, and L. P. Rappoport, *Phys. Rep.* **141**, 319 (1986). Theory and experiment are covered in S. J. Smith and P. L. Knight (Eds.), *Multiphoton Processes*, Cambridge University Press, New York (1988).

molecules. Underlying this remarkable advance were the oft-quoted advantages provided by laser radiation: high intensity, spectra purity, and coherence.

On the theoretical side, multiphoton processes are distinguished from single-photon phenomena in two main ways. First, higher-order terms appear in the perturbation theory of the interaction of radiation and matter, so more than a single matrix element is required. Second, the field strength associated with the laser radiation may be very large, perhaps comparable with the field that binds the atomic electron, so that the perturbation theory itself may become inapplicable. In this chapter we separate the effects of the radiation field into two types: (1) those related to the distortion of the target by the radiation field (splitting and shift of energy levels; perturbation of wave functions), and (2) the absorption of more than one photon that is permitted by the high density of photons in the radiation field. Of course, in some situations, both kinds of effects are important and must be considered together, but here we suppose that a clean separation for reasons of simplicity. Multiphoton processes will be dealt with in this section; "distortion effects" will be considered in a later volume.[57]

Multiphoton processes have been intensively studied for more than a decade, and entire books have been devoted to the subject. Clearly, we must limit our scope, and accordingly we shall discuss only a few topics involving multiphoton dissociation, ionization, and excitation. Treatments of other multiphoton processes can be found in the reviews and books cited below, along with multiphoton emission processes, which we do not cover. We should point out that some mention of excitation inevitably appears in our discussions of dissociation and ionization, but our treatment of excitation processes in their own right is deferred until Section 8-17-C.

A. Multiphoton Dissociation of Molecules

Here we mention briefly two of the numerous types of studies involving multiphoton absorption that lead to dissociation.

1. *Laser Isotope Separation*

Various schemes for separating isotropes by laser radiation are discussed by Letokhov,[158] Ronn,[159] and Zare.[160] The scheme that we choose for discussion here is based on selective infrared multiphoton absorption. As a preliminary, let us examine Fig. 8-17-1, which shows potential energy curves and vibrational energy levels for a typical diatomic molecule. First we note that absorption of a single UV photon can cause a transition from the lowest vibrational level of the ground electronic state (*G*) into the manifold of vibrational levels of the first excited electronic state (*E*), and if the excited molecule can cross over to the repulsive state (*R*) as indicated at the top of the UV arrow indicated in the figure, the dissociation will occur immediately. We now use some poetic license and apply this figure to the *polyatomic* molecule SF_6, letting the "interatomic distance" in the figure refer to the distance between the sulfur nucleus and one of

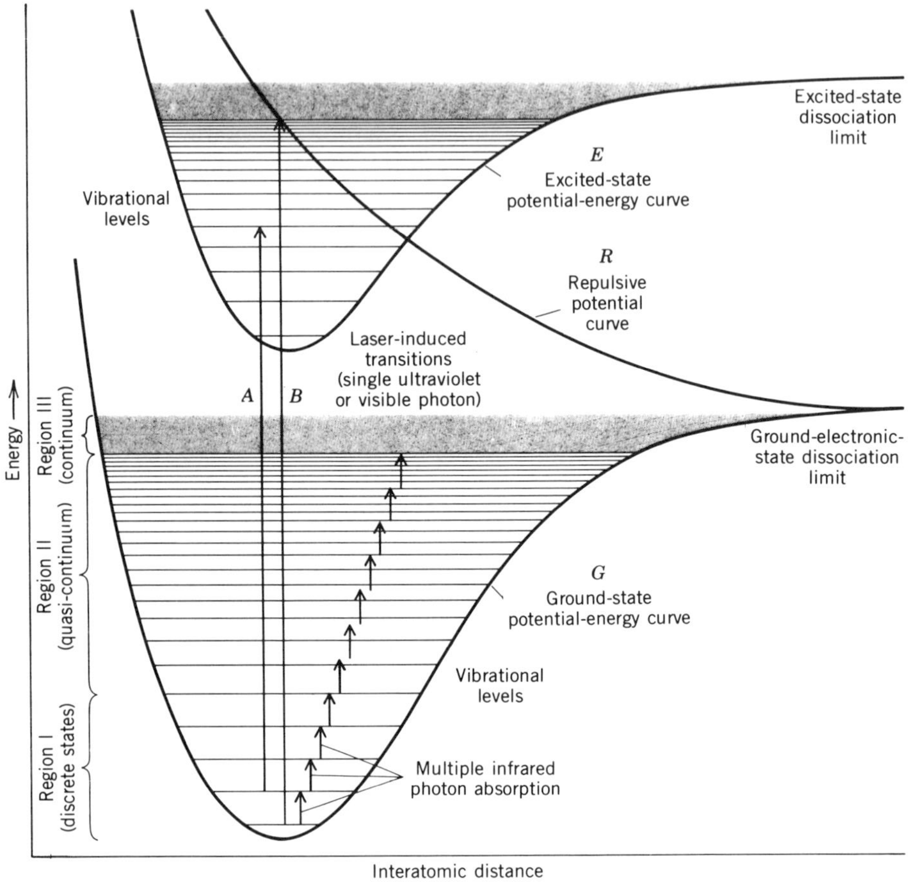

Figure 8-17-1. Potential curves for a typical diatomic molecule.[159] The symbol G denotes the ground state, E a bound excited state, and R an unbound, repulsive state. Vertical arrows indicate transitions induced by photon absorption.

the fluorine nuclei. We argue that a second mode of dissociation is now possible, one involving successive vibrational excitations by single-frequency photons. If infrared photons closely matching the spacing between the first and second vibrational levels of the ground state (G) are directed onto the molecule at a rapid rate, the first photon will be absorbed at resonance, and the next three or four photons close to resonance. The important point is that absorption of the photons will continue past this point even though the vibrational spacing becomes smaller and smaller as we move up the ladder of vibrational states. This continued absorption is possible because by this time the number of energy states of the excited molecule has become enormous ($> 10^9$ states per eV!), and there is no difficulty in matching the photon energy to the state separation energy. If 35 photons are absorbed, a single sulfur-fluorine bond of the SF_6 molecule can be broken.

Now let us suppose that we wish to separate ^{34}S from ^{32}S, which constitutes about 96% of sulfur found in nature. If SF_6 gas is prepared from natural sulfur and irradiated with IR radiation of the frequency required to excite the $^{32}SF_6$ molecules up the vibrational ladder, the molecules containing this isotope will be dissociated, but the $^{34}SF_6$ molecules will not because of their different vibrational spacing. The residual SF_6 gas can be enriched by a factor of several thousand.[159] For the record, the CO_2 laser is ideal for this purpose; it can furnish either 10.61-μm radiation for dissociation of $^{32}SF_6$ or 10.82 μm for dissociation of $^{34}SF_6$. Finally, we emphasize the need for intense laser radiation: At a given height on the excitation ladder, an IR photon must be absorbed before vibrational relaxation of the excited level takes place.

2. *"Thermal" Decomposition of Molecules by Lasers*

Infrared multiphoton dissociation of a complex molecule (RDX) in a molecular beam has been studied in Lee's laboratory at Berkeley.[161] Here a molecular beam was crossed with a pulsed TEA CO_2 laser beam, and the velocity distributions of the various products were measured by time-of-flight techniques as a function of the laboratory angle. It was possible to determine the dissociation channels and their branching ratios, as well as the translational energy distributions of the products. The main objective here was to investigate the mechanism of RDX thermal decomposition. The goal was achieved because the lifetime of the infrared-multiphoton-excited molecule is sufficiently long for vibrational energy randomization to occur before dissociation. Hence the same dissociation mechanism is observed as in the case of thermal decomposition.

B. Multiphoton Ionization

Figure 8-17-2, adapted from the review by Compton and Miller,[162] provides a convenient basis for our discussion. Panel (*a*) schematically shows the direct nonresonance ionization of an atom or molecule by a single photon (NRSPI); no intermediate state of the target is involved. The remaining panels refer to various types of multiphoton ionization, which we shall now discuss.

1. *Nonresonant Multiphoton Ionization*[163-165]

An example of the nonresonant multiphoton ionization (NRMPI) process is shown in panel (*b*) of Fig. 8-17-2, in which ionization is produced by the absorption of five photons of the same frequency, and no real excited state of the target is involved. The photons must arrive closely spaced in time in order to avoid violating the uncertainty principle; for each photon it is required that $\Delta E \lesssim \hbar/\Delta t$, where ΔE is the energy difference between the virtual state reached by the absorption and the nearest real state, and Δt is the time over which this virtual state is considered to be occupied.

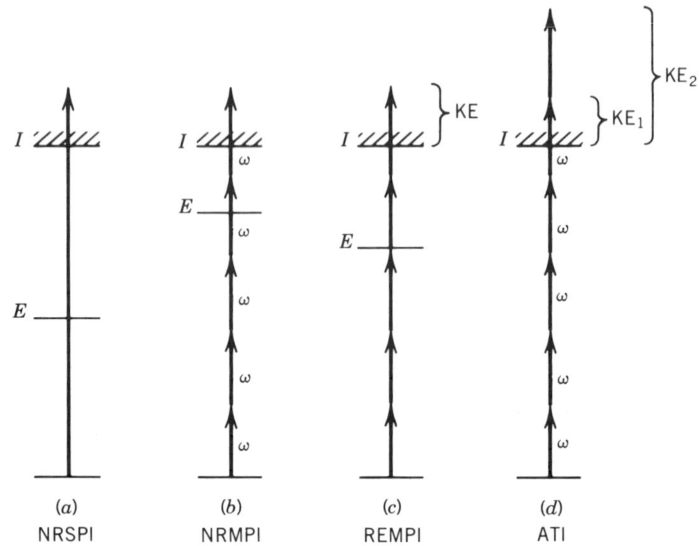

Figure 8-17-2. Illustrations of different photon ionization processes.[162] Resonant energy levels in (*c*) can have an enormous effect, but nonresonant levels in (*b*) are also important because the ionization rate involves a sum over all intermediate states. See A. Dodhy, R. N. Compton, and J. A. D. Stockdale, *Phys. Rev. Lett.* **54**, 422 (1985). (Adapted from Compton and Miller.[162])

2. *Resonantly Enhanced Multiphoton Ionization*[80,162,166–168]

We illustrate the resonantly enhanced multiphoton ionization (REMPI) nonlinear process with the example shown in panel (*c*), a three-photon resonant five-photon ionization event, labeled (3 + 2) REMPI. All five photons here have the same frequency, so we have "one-color REMPI"; if, on the other hand, the first three photons absorbed had one frequency, and the last two photons another, we would have an example of "two-color REMPI." Because of the resonance excitation of the real excited state (*E*), the probability of ionization is greatly enhanced over that for the nonresonant process NREMPI.

One of the most important uses to which REMPI has been put is photoelectron spectroscopy, the first results being reported in 1980.* The conventional single-photon technique (SPI-PES) has provided valuable information on the electronic, vibrational, and rotational states of ions formed from many atoms and molecules (see Section 8-16-A-2), but as Compton and Miller[162] point out, REMPI-PES can provide essentially all the information that comes from SPI-PES, and other kinds of information besides. Instead of using a single photon to ionize a thermal distribution of vibrational and rotational levels of the ground electronic state of a target molecule REMPI-PES excites a single rotational-vibrational level of a resonant intermediate state with one or more photons, and then ionizes this level with one or more additional

*See R. N. Compton, J. C. Miller, A. E. Carter, and P. Kruit, *Chem. Phys. Lett.* **71**, 87 (1980).

photons. With tunable pulsed lasers, it is possible to prepare and then ionize almost any resonant excited intermediate state, at higher-wavelength resolution than provided by SPI-PES. One may probe the products of the decay of this state, or the products of the photodissociation of the resonant intermediate state, by means of photoelectron spectroscopy. In addition, REMPI-PES gives higher-wavelength resolution than does SPI-PES. Finally, REMPI-PES allows the study of the dynamics of photoionization from excited states, the mapping of photoelectron spectra of transient or minority species, the study of intramolecular energy transfer processes, and the observation of resonant intermediate states and final ionic states forbidden in single-photon excitation by symmetry considerations.[167]

General treatments of REMPI, both experimental and theoretical, are provided in the reviews and books cited in Refs. 80, 162, 166, and 167. Compton and Miller[162] describe the various nonlinear effects that can influence photoelectron spectroscopy and angular distribution measurements at the high radiation levels used here. They also describe several REMPI-PES apparatus, including the one shown in Fig. 8-17-3. In this instrument, the plane of polarization of the dye laser can be rotated continuously with a double Fresnel rhomb driven by a stepping motor to obtain continuous photoelectron angular distributions.* Blazewicz et al.[169] made such measurements on xenon with this

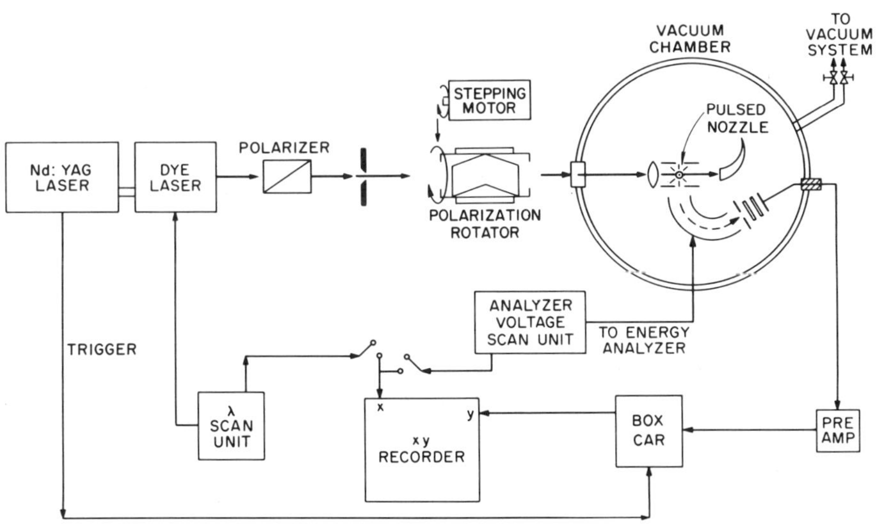

Figure 8-17-3. A REMPI-PES apparatus[162] for photoelectron spectroscopy on resonantly enhanced multiphoton ionization.

*The differential cross section for a common case of single-photon ionization is given by (8-16-12). The corresponding cross section for *n-photon ionization* is a generalization of this equation, namely

$$d\sigma_{\text{ion}}^n(\theta) = \frac{\sigma_{\text{ion}}^n}{4\pi}\left[\sum_{j=0}^{n} \beta_{2j} P_{2j}(\cos\theta)\right] d\Omega. \tag{8-17-1}$$

apparatus at the Oak Ridge National Laboratory. Additional angular distribution measurements are described in the other papers cited in Ref. 169.

Examples of other specialized studies can also be cited: Two-color REMPI-PES has been used to study rotational and vibrational autoionization of Rydberg states of H_2.[170] Multiphoton multiple ionization has been the subject of a number of investigations.[171] Studies of the two-photon ionization of metastable helium have been reported by Haberland et al.[172]. M. Lambropoulos et al.[173] have observed electric quadrupole transitions in three-photon ionization of sodium, and found them to be dominant over the nonresonant electric dipole background for certain frequencies.

3. *Above-Threshold Ionization*[80,162]

Quasi free-free transitions are involved in the above-threshold ionization (ATI) process, as indicated in panel (*d*) of Fig. 8-17-2. The photoelectrons absorb one or more additional photons after entering the continuum, and are ejected with various discrete kinetic energies separated by the photon energy $h\nu$. Described differently, ATI produces electrons that have absorbed more photons than are necessary to overcome the binding energy. Many higher-order nonresonant multiphoton ionization experiments are dominated by ATI, especially at high radiation intensities, say 10^{10} W/cm^2 or greater.

Above-threshold ionization may be observed by focusing an intense pulsed laser beam onto an interaction region in a gas of very low density, and energy analyzing the photoelectrons thus produced. Sharp, equally spaced peaks appear in the spectrum of electron yield versus electron kinetic energy. (Eight ATI peaks were reported by Freeman et al.[22,174] in experiments on Xe atoms.) The angular distributions of the photoelectrons are also of interest and are usually measured. To obtain high-quality data, the laser pulses must be well characterized in time and space, background ionization and space charge must be negligible, and a good electron spectrometer must be used.

In ATI experiments, the laser pulses are said to be "long" if their duration is long compared to the time required for a photoelectron to leave the interaction volume. Typically, a laser pulse longer than 20 ps is in the long-pulse category. Experiments show that in the long-pulse regime, the number and heights of the peaks in the photoelectron spectra, their widths, and the angular distributions all depend on the intensity of the laser radiation, although the electron kinetic energies are nearly independent of the intensity of the radiation. On the other hand, recent measurements by Freeman et al.[175] with short pulses (less than 1 ps long) show that each low-energy ATI peak splits up into a series of narrow peaks. There is insufficient time for the photoionized electron to escape the interaction volume before the termination of the pulse of radiation, so the electron energy spectrum indicates the actual photoelectron energies and angular distributions at the instant that ionization takes place. Eberly and Javanainen[176] have made theoretical calculations that predict "fine structure" splitting of the type observed by Freeman et al.[175]

McIlrath et al.[174] comment on the wide variety of phenomena that are observed in ATI experiments,[177–179] related to both photon absorption and photon scattering. In the former category fall MPI and ATI. In the scattering category lie Stark shifts, ponderomotive forces, light pressure, Thomson scattering, and nonlinear phenomena that relate to the interaction of photons with "dressed atoms."* Considering the wide variety of factors that impinge upon ATI, and the fact that the first experimental paper on the subject was not published until 1979, it is understandable that above-threshold ionization is not well understood. However, the level of understanding will surely improve with the passage of time, and ATI experiments will deepen our insight into multiphoton ionization.

C. Multiphoton Excitation

There have been many interesting experiments involving multiphoton excitation of atoms and molecules besides those described in Sections 8-17-A and 8-17-B, and we cannot do justice to the subject. Here we merely list some representative papers that describe important uses of multiphoton excitation or studies of the phenomenon for its own sake.

Miller and Compton[181] have described their experiments on third-harmonic generation (frequency tripling) in the rare gases, which can be understood in terms of Fig. 8-17-4, plus the selection rule $\Delta l = \pm 1$ and the fact that the photon has spin 1. Hamadani and his colleagues[182] have reported studies of four-wave mixing.

Miller et al.[183] have observed competition between REMPI and third-harmonic generation in Xe, and Jackson and Wynne[184] have shown that destructive interference between two coherent excitation pathways from different optical harmonics can inhibit REMPI.

Two-photon absorption has been used at Bielefeld to produce very highly excited states of Cs.[185] Spontaneous Raman processes have been observed in multiphoton excitation of Na at JILA.[186]

Absolute two-photon absorption cross sections have been measured by Bamford et al.[187] for the $np^3P_{2,1,0} \leftarrow 2p^3P_2$ transitions in atomic oxygen ($n = 4$ and 5). The technique of two-photon-excited fluorescence was used for these measurements.

*We may define "dressed atoms" as atoms under the influence of intense laser light and "dressed" with photons from the laser. This concept is useful in the theoretical treatment of problems involving intense radiation fields, and has been used in a dressed atom approach to resonance fluorescence.[180] At the start one may neglect spontaneous emission and consider only the total isolated system of "atom + laser mode interacting with each other." This system is the dressed atom. Then one can introduce the coupling with the empty modes, which is responsible for the transfer of photons from the laser mode to the empty modes. Thus, for example, resonance fluorescence may be considered as spontaneous emission from the dressed atom. Finally, it may be necessary to introduce the coupling of the laser with the lasing atoms and with the cavity losses, to account for the injection of photons into the laser mode and for the fluctuations of the laser radiation.

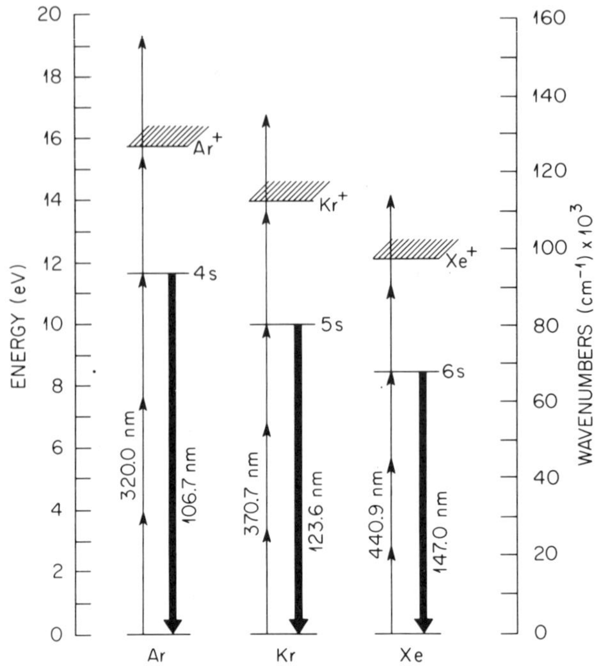

Figure 8-17-4. Illustration of third-harmonic generation in Ar, Kr, and Xe gases.[181]

In Zare's laboratory, a method has been developed for determining the population $A_{0+}^{\{0\}}$, the alignment factors $A_{q\pm}^{\{2\}}$ and $A_{q\pm}^{\{4\}}$, and the orientation factors $A_{q\pm}^{\{1\}}$ and $A_{q\pm}^{\{3\}}$ (cf. Section 5-10) for a ground-state distribution of diatomic (symmetric top) molecules probed by elliptically polarized two-photon nonresonant excitation.[188]

8-18. NOTES

8-1. Density in Phase Space.[14] In Section 8-2 we introduced the density in phase space and defined $\rho(E)\,d\Omega\,dE$ to be the number of states in the energy interval dE at energy E, with momentum direction within the element of solid angle $d\Omega$. Here we shall derive (8-2-11), and in the process of doing so, display another useful expression for the phase-space density.

Let us consider a free particle with a wave function of the form $e^{i\boldsymbol{\kappa}\cdot\mathbf{r}}$, $\boldsymbol{\kappa}$ being the wave vector. We can impose periodic boundary conditions on the wave function by assuming that all space is divided into large cubes, each with side length L, and all containing identical physical systems. The volume V of each cube is required to be extremely large, and it should be noted that the shape of V is immaterial since V is large. The periodic boundary conditions dictate that

$$\kappa_x = \frac{2\pi}{L} n_x, \qquad \kappa_y = \frac{2\pi}{L} n_y, \qquad \kappa_z = \frac{2\pi}{L} n_z, \tag{8-18-1}$$

where n_x, n_y, and n_z are positive or negative integers, or zero. Because L is assumed to be very large, n_x, n_y, and n_z can be treated as continuous variables, and the number of quantum states within the range $d\boldsymbol{\kappa} = d\kappa_x \, d\kappa_y \, d\kappa_z$ is

$$dn_x \, dn_y \, dn_z = \left(\frac{L}{2\pi}\right)^3 d\kappa_x \, d\kappa_y \, d\kappa_z = \left(\frac{L}{2\pi}\right)^3 \kappa^2 \, d\kappa \, d\Omega. \tag{8-18-2}$$

Using the relations $V = L^3$ and $\omega = c\kappa$, we see that

$$\rho(\omega) \, d\omega \, d\Omega = \frac{V}{(2\pi)^3} \frac{\omega^2}{c^3} \, d\omega \, d\Omega, \tag{8-18-3}$$

which is equivalent to

$$\rho(E) = \frac{V\kappa^2}{(2\pi)^3} \frac{d\kappa}{dE}. \tag{8-2-11}$$

In Section 8-2 we showed that (8-2-11) leads to (8-2-16) for nonrelativistic particles. The latter equation reappeared as (8-9-9) in our discussion of photoionization.

8-2. *Quantization of the Electromagnetic Field.* The concept of field quantization dates back to the earliest days of quantum mechanics, but the first productive implementation of the concept was by Dirac in 1927.[189] Our objective here is to gain some insight into the quantization of the electromagnetic field and its consequences in quantum electrodynamics. It is an essential feature of QED, which is widely regarded as the most successful theory in physics. We shall sketch the discussion in Appendix C of the book by Read,[1] leaving the reader to fill in details from this source and from the books by Loudon,[48] Louisell,[76] and Slater.[190] Each of these authors uses a nonrelativistic approach, which suffices for our purposes.

A. *Plan of Attack.* The first step in applying quantum mechanics to the electromagnetic field is to express the field observables **E** and **B**, and the vector potential **A**, in operator form, making use of Maxwell's equations and the properties of the harmonic oscillator enumerated in Section B. We cannot *derive* the expressions for the operators $\hat{E}$, $\hat{B}$, and $\hat{A}$ from classical theory. Rather, we express the classical field energy in a form that resembles the energy of the harmonic oscillator, and proceed by analogy in a way to be indicated in Sections C and D. Then we shall discuss coherent states and fluctuations in Section E, and the absorption and emission of photons by atoms in Section F, where we show how spontaneous emission can be explained in terms of the field quantization. As indicated above, we use a caret to designate an operator, but at times this use can produce a baroque clutter of symbols. Hence we shall frequently omit the caret if confusion is unlikely to result.

B. *The Harmonic Oscillator; Annihilation and Creation Operators* (Park,[191] Loudon,[48] pp. 129–133). The total energy of a one-dimensional harmonic oscillator is

$$E = \frac{p^2}{2m} + \frac{1}{2} Kx^2, \tag{8-18-4}$$

where m is the mass of the oscillating particle, p is its momentum, x is its position coordinate, and K is the spring constant. If p and x are regarded as operators, the right side of (8-18-4) defines the Hamiltonian *operator* for the oscillator. In the coordinate representation, the x operator is just x, and the p operator is $-i\hbar d/dx$, so the Hamiltonian operator is

$$\hat{H} = -\frac{\hbar^2}{2m}\frac{d^2}{dx^2} + \frac{1}{2} Kx^2. \tag{8-18-5}$$

The angular frequency of the oscillator when considered classically is

$$\omega = \left(\frac{K}{m}\right)^{1/2}. \tag{8-18-6}$$

The energy eigenfunctions are the Hermite polynomials, and the eigenenergies are

$$E_n = \hbar\omega(n + \tfrac{1}{2}). \tag{8-18-7}$$

We now introduce two new operators, which are not Hermitian and do not need to be because they do not represent physical observables. The first is the *annihilation* (or *destruction*) *operator*

$$\hat{a} = \left(\frac{m\omega}{2\hbar}\right)^{1/2}\left(\hat{x} + \frac{i}{m\omega}\hat{p}\right) = \frac{1}{\alpha\sqrt{2}}\frac{d}{dx} + \frac{\alpha x}{\sqrt{2}} \tag{8-18-8}$$

and the second is the *creation operator*

$$\hat{a}^\dagger = \left(\frac{m\omega}{2\hbar}\right)^{1/2}\left(\hat{x} - \frac{i}{m\omega}\hat{p}\right) = -\frac{1}{\alpha\sqrt{2}}\frac{d}{dx} + \frac{\alpha x}{\sqrt{2}}, \tag{8-18-9}$$

where

$$\alpha = \left(\frac{m\omega}{\hbar}\right)^{1/2}. \tag{8-18-10}$$

Useful relations involving these operators are

$$a^\dagger a = -\frac{1}{2\alpha^2}\left(\frac{d^2}{dx^2} - \alpha^4 x^2\right) - \frac{1}{2}, \tag{8-18-11}$$

$$aa^\dagger = -\frac{1}{2\alpha^2}\left(\frac{d^2}{dx^2} - \alpha^4 x^2\right) + \frac{1}{2}, \tag{8-18-12}$$

$$aa^\dagger - a^\dagger a = 1, \tag{8-18-13}$$

and

$$H = \hbar\omega(a^\dagger a + \tfrac{1}{2}) = \hbar\omega(aa^\dagger - \tfrac{1}{2}). \tag{8-18-14}$$

If we apply the operator $Ha^\dagger$ to one of the oscillator eigenfunctions ψ_n, we can show that

$$H(a^\dagger \psi_n) = (E_n + \hbar\omega)(a^\dagger \psi_n), \tag{8-18-15}$$

so the function $(a^\dagger \psi_n)$ is also an eigenfunction, with associated energy $(E_n + \hbar\omega)$. The operator $a^\dagger$ has the effect of creating an additional quantum of energy when applied to an eigenfunction of the oscillator, and from this property $a^\dagger$ derives its name. It can also be shown that

$$a^\dagger \psi_n = (n+1)^{1/2}\psi_{n+1}, \tag{8-18-16}$$

and conversely that

$$a\psi_n = n^{1/2}\psi_{n-1}. \tag{8-18-17}$$

The destruction operator $\hat{a}$ reduces the number of energy quanta by one in each application, while also reducing the quantum number of the eigenstate by unity. A special situation arises with

$$\hat{a}\psi_0 = 0. \tag{8-18-18}$$

Here, further application of $\hat{a}$ past the ground state is impossible. Note that

$$E_0\psi_0 = H\psi_0 = \hbar\omega(a^\dagger a + \tfrac{1}{2})\psi_0 = \tfrac{1}{2}\hbar\omega\psi_0, \tag{8-18-19}$$

so $\frac{1}{2}\hbar\omega$ is the zero-point energy of the oscillator.

Another operator that we shall need is the number operator

$$\hat{n} = a^\dagger a. \tag{8-18-20}$$

When applied to an eigenfunction, the number operator gives the same function multiplied by the quantum number of the state, n, which is also the number of quanta in the state:

$$\hat{n}\psi_n = a^\dagger n^{1/2}\psi_{n-1} = n\psi_n. \tag{8-18-21}$$

C. *Standing Waves in a Cavity* (Loudon,[48] pp. 2–8). Read[1] illustrates the connection of the harmonic oscillator to standing electromagnetic waves in a one-dimensional cavity with parallel walls of infinite conductivity. The boundary conditions restrict the normal radiation modes to those with an integral number of half-wavelengths. Each mode is designated by the magnitude of the wave vector **k**. [In a three-dimensional cavity, $\mathbf{k} = (k_x, k_y, k_z)$ specifies the modes, each different set of components of **k** indicating a different mode.] Solving for the **E** and **B** fields, Read calculates the total energy of a single mode of the electromagnetic field contained in a volume V,

$$U = \tfrac{1}{2}(\epsilon_0 E^2 + \mu_0^{-1} B^2)V, \tag{8-18-22}$$

and shows that the energy is equally divided between the electric and magnetic fields. The energy oscillates between the two components in the same way that the energy of a harmonic oscillator oscillates between potential and kinetic energy.

Read then expresses the energy contained within V in terms of the vector potential **A** and its time derivative:

$$U = \frac{1}{2} V\epsilon_0 \left[\left(\frac{dA_x}{dt}\right)^2 + \omega^2 A_x^2\right]. \tag{8-18-23}$$

Comparison of this expression with (8-18-4) establishes the analogies

$$x \sim \left(\frac{V\epsilon_0}{m}\right)^{1/2} A_x \tag{8-18-24}$$

$$p \sim (V\epsilon_0 m)^{1/2} \frac{dA_x}{dt}. \tag{8-18-25}$$

Then Read constructs the appropriate annihilation and creation operators

$$a = \left(\frac{V\epsilon_0}{2\hbar\omega}\right)^{1/2} E_0 e^{i(\omega t + \phi)}, \qquad a^\dagger = a^*, \tag{8-18-26}$$

where E_0 is the amplitude of the single standing wave under consideration and ϕ is its phase. The vector potential operator is given by

$$\hat{A}_x = \left(\frac{\hbar}{2V\epsilon_0\omega}\right)^{1/2} (a + a^\dagger). \tag{8-18-27}$$

We infer that the total electromagnetic energy, in the single mode under discussion, that is contained with the volume V is

$$U = \hbar\omega(n + \tfrac{1}{2}), \tag{8-18-28}$$

irrespective of the volume V. The energy can be changed only by discrete quanta $\hbar\omega$.

D. *Free Traveling Waves* (Loudon,[48] pp. 126–128, 134–137). Let us consider a plane electromagnetic wave having wave vector $\mathbf{k}$ and polarization direction $\mathbf{i}_p$, with $\mathbf{k}\cdot\mathbf{i}_p = 0$. This wave is a single mode of a radiation field and is specified by $\mathbf{k} = (k_x, k_y, k_z)$. Its vector potential is

$$\mathbf{A} = \mathbf{i}_p\{A_k e^{i(\omega t - \mathbf{k}\cdot\mathbf{r})} + A_k^* e^{-i(\omega t - \mathbf{k}\cdot\mathbf{r})}\}. \tag{8-18-29}$$

As Loudon shows on page 135 of his book,[48] the annihilation and creation operators are defined in terms of the free plane waves by the equations

$$a = \left(\frac{2\epsilon_0\omega V}{\hbar}\right)^{1/2} A_k \tag{8-18-30}$$

and

$$a^\dagger = \left(\frac{2\epsilon_0\omega V}{\hbar}\right)^{1/2} A_k^*, \tag{8-18-31}$$

so that

$$\hat{A} = \mathbf{i}_p\left(\frac{\hbar}{2\epsilon_0\omega V}\right)^{1/2}[a e^{i(\omega t - \mathbf{k}\cdot\mathbf{r})} + a^\dagger e^{-i(\omega t - \mathbf{k}\cdot\mathbf{r})}]. \tag{8-18-32}$$

Read calculates the $\hat{E}$ and $\hat{B}$ operators from $\hat{A}$, assuming that $\phi = 0$, and then evaluates the electric field energy contained within a volume V of arbitrary size and shape:

$$\frac{V}{2}\int \psi_n^*(\epsilon_0|\hat{E}|^2)\psi_n\,d\tau = \tfrac{1}{2}\hbar\omega(n + \tfrac{1}{2}). \tag{8-18-33}$$

The corresponding magnetic field energy has the same value, so the total energy in V for the single mode being considered is $\hbar\omega(n + \frac{1}{2})$.

When many modes of a radiation field are allowed, the Hamiltonian operator for the field can be written

$$\hat{H} = \frac{V}{2}(\epsilon_0|\hat{E}|^2 + \mu_0^{-1}|\hat{B}|^2) = \sum_{\mathbf{k},p}(a_{\mathbf{k},p}^\dagger a_{\mathbf{k},p} + \tfrac{1}{2})\hbar\omega_k, \tag{8-18-34}$$

where **k** and p indicate the wave vector and polarization of the various modes. Each of these modes has the total energy

$$E_{n_k} = \hbar\omega_k(n_k + \tfrac{1}{2}). \tag{8-18-35}$$

E. *Coherent States* (Loudon,[48] pp. 141–152). These states are important because they most closely approximate classical electromagnetic waves. A coherent state $\psi(\alpha)$ is an eigenstate of the annihilation operator, as this operator acting on the state function produces the same function multiplied by a constant:

$$a\psi(\alpha) = \alpha\psi(\alpha). \tag{8-18-36}$$

The coherent state $\psi(\alpha)$ is a superposition of pure states ψ_n, each of which has a definite number of quanta (n), and it is defined by

$$\psi(\alpha) = [\exp(-\tfrac{1}{2}|\alpha|^2)] \sum_{n=0}^{\infty} \frac{\alpha^n}{(n!)^{1/2}} \psi_n. \tag{8-18-37}$$

Hence $\psi(\alpha)$ is not a pure state, but it does have a definite *mean number of quanta*

$$\bar{n} = |\alpha|^2. \tag{8-18-38}$$

The *mean square of the number of quanta* in the coherent state $\psi(\alpha)$ is

$$\overline{n^2} = |\alpha|^4 + |\alpha|^2, \tag{8-18-39}$$

so the *root-mean-square (rms) fluctuation in n* is

$$\Delta n = (\overline{n^2} - \bar{n}^2)^{1/2} = |\alpha| = \bar{n}^{1/2}. \tag{8-18-40}$$

Since

$$\frac{\Delta n}{\bar{n}} = (\bar{n})^{-1/2}, \tag{8-18-41}$$

the relative fluctuation $\Delta n/\bar{n}$ becomes smaller as $\bar{n}$ becomes larger.

In the same manner, we can calculate the mean of the electric field strength and its mean square over a volume V to obtain, for the coherent state, the rms fluctuation of the electric field

$$\Delta E = (\overline{E^2} - \bar{E}^2)^{1/2} = \left(\frac{\hbar\omega}{2\epsilon_0 V}\right)^{1/2}. \tag{8-18-42}$$

The relative fluctuation $\Delta E/\bar{E}$ is also proportional to $(\bar{n})^{-1/2}$.

Discussion of the phase of the radiation field proceeds via definition of a *phase operator* $\hat{\varphi}$ that corresponds to the classical phase ϕ in the limit of very large $\bar{n}$. For *any* state of the radiation field with $\bar{n} \gg 1$, the rms fluctuations in n and ϕ are related by the uncertainty relationship

$$\Delta n \, \Delta\phi \geqslant \tfrac{1}{2}. \tag{8-18-43}$$

The product $\Delta n \, \Delta\phi$ has the minimum value $\frac{1}{2}$ only for the coherent state, so $\psi(\alpha)$ is the state of the field that most closely resembles the nonfluctuating, classical electromagnetic field. As Fig. 4-3 on page 151 of Loudon[48] shows, the wave becomes better defined both with respect to amplitude and phase as $\bar{n}$ increases. Read[1] gives an example of a nearly classical wave: a radio transmitter operating at 10 kW of power and a wavelength of 300 m has $\bar{n} \sim 10^{17}$ photons collected in one period by a detector of 10 m^2 effective area at a distance of 10 km. Most laboratory sources of visible light also produce approximately classical fields, while lasers provide beams of very nearly constant and well-defined amplitude and phase.

F. *Interaction of the Radiation Field with an Atom* (Loudon,[48] pp. 162–210). Here the Hamiltonian can be written

$$\hat{H} = \hat{H}_E + \hat{H}_R + \hat{H}_I, \tag{8-18-44}$$

where

$$\hat{H}_E = \sum_i \hbar\omega_i |i\rangle\langle i| \tag{8-18-45}$$

is the Hamiltonian for the isolated atom, expressed in terms of the bra and ket vectors of the atomic states i, with $\sum_i |i\rangle\langle i| = 1$.* The Hamiltonian of the free radiation field is

$$\hat{H}_R = \sum_{\mathbf{k}} \hbar\omega_{\mathbf{k}}(a_k^\dagger a_k + \tfrac{1}{2}) \tag{8-18-46}$$

[cf. (8-18-14)], and H_I is the more complicated part of $\hat{H}$ that describes the interaction of the radiation field with the atom. Loudon works out the general form for $\hat{H}$ and specializes it to the electric dipole case on page 174 of his book.[48] Read sketches some results of the application of the quantized field Hamiltonian to the emission and absorption of radiation by an atom in the electric dipole approximation. He points out that when a photon is absorbed, the state of the radiation field changes from ψ_n to ψ_{n-1} while the atom is excited from $\psi(1)$ to $\psi(2)$, say, and the transition probability for absorption is proportional to

$$\left| \int \psi_{n-1}^* \psi(2)^* \hat{D} \cdot \hat{E} \psi_n \psi(1) \, d\tau \right|^2. \tag{8-18-47}$$

*$\hat{H}_E$, which involves a sum over all bound and free states, is written here in *second quantized form* (Loudon,[48] pg. 173).

Here $\hat{D}$ is an electric dipole moment operator. The part of the transition probability that depends only on the atomic eigenfunctions can be separated out, and it can be shown that the transition probability for absorption is proportional to n, the number of photons in the field:

$$W_{\text{abs}} \sim n. \tag{8-18-48}$$

The zero-point energy, $\hbar\omega(n + \frac{1}{2})$, plays no role in the absorption. On the other hand, the same treatment shows that the transition probability for emission is proportional to $(n + 1)$:

$$W_{em} \sim n + 1. \tag{8-18-49}$$

The first part (n) of the right side of (8-18-49) relates to the process of stimulated *emission*; the second part (1) to *spontaneous emission*, which can occur even if the field contains no photons before the emission takes place from the excited atom. The spontaneous process can be regarded as induced by the zero-point fluctuations of the electromagnetic field. As we have seen, the field can be regarded as equivalent to an infinite set of uncoupled, quantized harmonic oscillators, each of which has a zero-point energy corresponding to there being no photons in that particular mode of the incident field. Although the mean values of the electric and magnetic fields are zero, their mean square values are not, and there are fluctuating fields even when no external fields are applied.

The state of the field in which no photons are excited in any of the field modes is the *vacuum state* of the field (see Boyer[192] for an interesting qualitative discussion of the vacuum). Because the zero-point field fluctuations lead to spontaneous emission of photons in *random* directions, they are the source of "quantum noise" in lasers, parametric amplifiers, and attenuators.[72]* They are also responsible for the natural line width (Section 8-11) and the Lamb shift.[193]

A point of crucial importance for the operation of lasers is clarified when we examine the matrix element for emission. Its form is such that the only photon that can be emitted in the stimulated process is one with the same wave vector and polarization direction as that of the stimulating photons already in the radiation field. That is, the new photon goes into the same mode as the n photons causing the stimulated emission (Loudon,[48] pp. 176–178).

8-3. Superradiance. In the usual situation, the spontaneous decay of atoms is described by a generalization of (8-3-3). Specifically, if N_j is the number of

*Quantum noise in light waves limits the precision of optical measurements. It is sometimes possible to alleviate the problem by "squeezing" the noise in a light beam: the noise is redistributed so that parts of the light wave become less noisy, although other parts are then noisier. For a qualitative discussion of squeezed light, see R. E. Slusher and B. Yurke, *Sci. Amer.* **258**, 50 (May, 1988).

atoms in excited state j at time t,

$$-\frac{dN_j}{dt} = \sum_i A_{ji} N_j, \tag{8-18-50}$$

where the A_{ji} are the Einstein coefficients for spontaneous transitions from j to all lower states i. Accordingly,

$$N_j(t) = N_j(0)e^{-t/\tau_j}, \tag{8-18-51}$$

with

$$\tau_j = \frac{1}{\sum_i A_{ji}} \tag{8-18-52}$$

being the lifetime of state j with respect to spontaneous emission to all lower states i. Under certain conditions, however, an ensemble of atoms can enter a cooperative mode of spontaneous emission, emit an intense burst of radiation, and become deexcited much more rapidly than if the atoms were decaying independently. This process, first proposed by Dicke,[194] is called *superradiance.*

In superradiance, the atoms couple to the radiation field and develop strong correlations, which establish a large macroscopic electric dipole proportional to the number density, n_0, of the atoms in the initial excited state. This dipole radiates with an intensity proportional to n_0^2, and the reaction of this radiation back on the system damps it rapidly. Haroche et al.[195] consider the case of a pencil-shaped active medium of length L and cross-sectional diameter much larger than the wavelength, λ, of the emitted radiation. The evolution of the superradiant burst involves a characteristic time

$$\tau_R = \frac{8\pi}{n_0 L \lambda^2 \gamma_{\text{ord}}}, \tag{8-18-53}$$

as contrasted with the characteristic time for ordinary spontaneous emission, $\tau_{\text{ord}} = 1/\gamma_{\text{ord}}$. The superradiant pulse is emitted after a delay of a few tens of τ_R and lasts for a few τ_R. This effect is observed only if $\tau_R < T_d$, where T_d is the characteristic damping time for the various dephasing mechanisms that tend to disrupt the macroscopic dipole. Accordingly, there exists a threshold for superradiance in terms of the initial excitation number density:

$$n_0 > \frac{8\pi}{\gamma_{\text{ord}} T_d L \lambda^2}. \tag{8-18-54}$$

for n_0 below the threshold value, ordinary isotropic radiation is emitted, with no cooperative effects. If n_0 exceeds the threshold, anisotropic superradiant emission is expected. In the example discussed here, two symmetric lobes of emission from the ends of the pencil-shaped volume would be observed.

Haroche et al.[195] discuss the connection of superradiance with other collective emission effects (e.g., photon echoes and amplified spontaneous emission).

8-4. Laser Cooling and Trapping. In Section 7-6 we discussed briefly the trapping of charged particles in static and time-varying electric and magnetic fields. Many important studies of various types, involving mainly RF and Penning traps (and variants of the latter), have been made on trapped electrons[196] and ions.[197] Also, *neutral* spin-polarized hydrogen[198] atoms, precooled with a dilution refrigerator, have been trapped for more than 20 min in a static magnetic-field-minimum trap. Here, more than 5×10^{12} atoms were cooled to about 40 mK.

Our concern here is with techniques for the cooling and trapping of atoms and ions by *laser radiation*, which had their origin in the scheme proposed in 1975 by Hänsch and Schawlow[199] for the cooling of neutral atoms. To understand the basic mechanism, consider a laser beam directed into a gas of identical atoms at room temperature, and suppose that the laser frequency is just below a resonant frequency corresponding to one of the atomic transitions. Atoms moving antiparallel to the laser beam see photons Doppler-shifted toward their resonant frequency (Note 8-5), and they absorb them at a much faster rate than do atoms with velocities parallel to the beam. The Doppler shift is in the wrong direction for the "parallel" atoms, which are affected by the radiation hardly at all. "Antiparallel" atoms are slowed down by each photon absorption because of the momentum transfer. A typical thermal sodium beam ($T \approx 1000$ K) has a velocity distribution that peaks near $v_0 = 1000$ m/s. The velocity of a sodium atom is degraded by about 3 cm/s in each photon absorption in the usual experimental arrangement (Problem 8-43), so approximately 30,000 absorptions are required by each atom to bring it to rest![200,201] Now, an excited atom can be deexcited by either stimulated or spontaneous emission. In the former case, a photon is always emitted in the beam direction, so the momentum of the atom does not change in the absorption-emission process. In the latter case, the photon emission is isotropic, so the net effect of the laser is to lower the gas energy and cool the atoms (i.e., reduce their velocity spread). In practice, six lasers, antiparallel in pairs, are usually employed to establish an "isotropic" radiation field. Then, the atoms can be considered to move in a viscous, dissipative medium (called "optical molasses") such that their motion is opposed in three dimensions. The cooling mechanism described here operates on what is often called the *light pressure force*, or the *scattering force*, which is an appropriate term if the scattering is viewed by absorption followed by emission.[202,203]

When applied to a *beam* of atoms, this cooling scheme requires only a single laser. A difficulty arises because, as the atoms slow down, the magnitude of the Doppler shift is lowered enough, even at velocity reductions of a few m/s, to degrade the resonance absorption and eventually stop the cooling process. One method of circumventing this effect is to sweep the laser beam to higher frequencies as the atoms slow down and thereby maintain the resonance.[204,205] An alternative technique[200,201,205] involves passing the atoms through a magnetic field whose strength decreases with distance into the field. The variation of the field can be adjusted so that the Zeeman shift keeps the atomic energy levels in resonance with the constant laser frequency. The first technique is referred to as *laser chirping*; the second involves *Zeeman tuning*.

Another type of radiation force, which has been used to trap neutrals, is the *gradient*, or *dipole, force* that arises when the electric field of a spatially inhomogeneous laser beam induces an electric dipole moment in an atom, and the dipole interacts with the field gradient.[200–202,204,206] The atom is attracted to a region of high light intensity if the frequency of the light is below the frequency of the atomic resonance and repelled if the frequency of the light is above the transition frequency.

Research on laser cooling and trapping is being conducted in many laboratories, and rapid progress is being made. Here we describe briefly two recent experiments that give some measure of how far the subject has progressed in a decade. One of these experiments was carried out by Pritchard's group at MIT.[207] They used a beam of sodium atoms from a 550°C ($v_{\text{thermal}} = 800$ m/s) oven, and slowed the atoms to a stop in two stages. In the first stage, the light pressure force from a counter-propagating laser beam reduced the velocity of the atoms, in 20,000 absorptions, to 200 m/s. The Zeeman tuning technique was used to keep the atoms in resonance as they slowed down. In the second stage, the atoms passed continuously into a trap region where they were stopped with a second laser beam and a second tapered magnetic field, which provide a 0.1-K-deep static trap. More than 1×10^9 atoms were trapped at millikelvin energies, with a decay time of 150 s. The trap used here operated on the interaction of the magnetic dipole moment of the sodium atoms with the inhomogeneous magnetic field.[208]

The second experiment was performed in a collaboration between groups at AT&T Bell Labs and MIT.[209] They cooled sodium atoms with a chirped laser beam and loaded them into an optical trap that confined and further cooled the atoms. The trap action was based on near-resonance radiation pressure and involves a *spontaneous light force* (not an induced light force, as had been used in various laboratories). An important feature of the trap is the use of a very weak magnetic field to shift the magnetic sublevels of the atoms differentially across the trap and cause them to scatter photons preferentially toward the field center, the atoms being pushed toward the center of confinement. The trap is about 0.4 K deep and has compressed about 10^8 atoms to densities greater than 10^{11} per cm^3, with trapping times of 2 min and temperatures on the order of a millikelvin. Under these conditions, the width of the absorption line is nearly equal to the natural line width.

Our discussion has centered on the laser cooling and trapping of neutrals, but many interesting studies have also been made on ions. One might note the formation of ion clusters in studies of Mg^+ and Hg^+ ions that were laser cooled and then confined in Paul RF traps.[210] Laser cooling and trapping of atoms and ions will surely have many important applications, such as great increases in spectroscopic resolution,* improvement of atomic clocks,[211] more accurate determinations of fundamental constants, improved studies of atomic collisions,[212] and the manipulation of antimatter. Also these techniques might lead to the observation of Bose condensation.[200] For details of various theoretical aspects of laser cooling and trapping, the reader may consult Ref. 213.

8-5. The Relativistic Doppler Effect for Electromagnetic Radiation.[214] Consider an emitter of sinusoidal radiation embedded in an inertial frame R in which an event is located by the four-vector $\tilde{r} = (\mathbf{r}, ct)$. An observer is fixed in a "primed" inertial frame R' that moves in a vacuum at velocity $\mathbf{u}$ with respect to the "unprimed" frame R, and we align the X axis along the vector $\mathbf{u}$. Let χ be the angle between $\mathbf{u}$ and the direction of propagation of the radiation, with $\chi = \pi$ corresponding to a head-on approach. The frequency of the radiation in the rest frame of the emitter is ν; we wish to find the frequency measured by the observer, ν'.

The phase of the wave,

$$\phi = \mathbf{k} \cdot \mathbf{r} - \omega t, \qquad (8\text{-}18\text{-}55)$$

is invariant under a Lorentz transformation, so ϕ is a four-scalar and can be expressed as the scalar product of the wave four-vector $\tilde{k} = (k_1 = k_x, k_2 = k_y, k_3 = k_z, k_0 = \omega/c)$, with the event four-vector $\tilde{r}$. We now orient the primed axes in R' along the corresponding unprimed axes in R and make a special Lorentz transformation (SLT) from R to R' along the direction of $\mathbf{u}$ (and the X and X' axes). The four-vector $\tilde{k}$ transforms in the same manner as does $\tilde{r}$ [see (5-6-1)], so we obtain

$$\begin{aligned} k_0' &= \gamma(k_0 - \beta k_1), \\ k_1' &= \gamma(k_1 - \beta k_0), \\ k_2' &= k_2, \\ k_3' &= k_3, \end{aligned} \qquad (8\text{-}18\text{-}56)$$

*The first observation of rf-induced transitions of trapped neutral atoms has been reported by A. G. Martin, K. Helmerson, V. S. Bagnato, G. P. Lafyatis, and D. E. Pritchard, *Phys. Rev. Lett.* **61**, 2431 (1988).

where $\beta = u/c$ and $\gamma = (1 - \beta^2)^{-1/2}$. In terms of the polar angle χ and the azimuthal angle ψ, the components of $\tilde{k}$ are

$$\frac{\omega'}{c} = \gamma\left(\frac{\omega}{c} - \beta k \cos \chi\right), \tag{8-18-57a}$$

$$k' \cos \chi' = \gamma\left(k \cos \chi - \beta \frac{\omega}{c}\right), \tag{8-18-57b}$$

$$k' \sin \chi' \cos \psi' = k \sin \chi \cos \psi, \tag{8-18-57c}$$

$$k' \sin \chi' \sin \psi' = k \sin \chi \sin \psi, \tag{8-18-57d}$$

and if we specialize to the case of electromagnetic radiation ($k = \omega/c$), we obtain from (8-18-57*a*) the frequency measured by the observer:

$$\nu' = \frac{\omega'}{2\pi} = \gamma\nu(1 - \beta \cos \chi). \tag{8-18-58}$$

If we put $\chi = \pi/2$, we isolate the "transverse" Doppler effect,

$$\nu' = \gamma\nu, \tag{8-18-59}$$

which is of second order in u/c, and which can be explained in terms of time dilation. There is no first-order Doppler effect at $\chi = \pi/2$.

For antiparallel photon and observer velocities ($\chi = \pi$), we have

$$\begin{aligned} \nu' &= \gamma\nu(1 + \beta) \\ &= \nu \sqrt{\frac{1 + \beta}{1 - \beta}}, \end{aligned} \tag{8-18-60}$$

so that if $u/c \ll 1$,

$$\nu' = \nu(1 + \beta + \tfrac{1}{2}\beta^2 + \cdots) \tag{8-18-61}$$

or

$$\nu' \approx \nu\left(1 + \frac{u}{c}\right). \tag{8-18-62}$$

This is the first-order result that is usually derived by elementary methods, along with the Doppler shift for parallel photon and observer velocities:

$$\nu' \approx \nu\left(1 - \frac{u}{c}\right). \tag{8-18-63}$$

The last three equations in (8-18-57) yield the *aberration equations*, which give the transformed direction of the wave:

$$\psi' = \psi, \tag{8-18-64}$$

$$\tan \chi' = \frac{1}{\gamma} \frac{\sin \chi}{\cos \chi - \beta}. \tag{8-18-65}$$

The azimuthal angle ψ is unchanged in the Lorentz transformation.

8-19. PROBLEMS

Solutions for most of these problems appear in the references indicated.

8-1. Consider the various spectral regions shown in Fig. 8-1-1. Describe the equipment and method that you would use to determine the wavelength of the radiation in each range. (Read,[1] Chap. 9.)

8-2. Prove by a relativistic analysis that a free electron can neither absorb nor emit a photon.

8-3. Show that the energy radiated per revolution by a relativistic electron moving in a circle of radius R is

$$\frac{E}{\text{rev}} = \frac{4\pi}{3} \beta^3 \gamma^4 \frac{r_0}{R} \text{ in units of } m_0 c^2, \tag{8-19-1}$$

where $r_0 = e^2/m_0 c^2$, m_0 is the electron rest mass, and β and γ are the usual relativity parameters. (Cronin et al.,[215] p. 21.)

8-4. Repeat Problem 4-6 for a relativistic electron, with velocity $v = 10^9$ cm/s. (Sard,[214] pp. 182–184.)

8-5. Determine the motion of an electron with initial velocity $v_0 = 10^8$ cm/s that enters a homogeneous, constant electric field of magnitude $E = 10^4$ V/cm at an angle of 45°. The direction of the field is such as to accelerate the electron. (Sard,[214] pp. 187–189.)

8-6. Consider an electron moving through a medium in which the phase velocity of light is $c' = c/n$, where n is the index of refraction of the medium.

(a) First take the velocity of the electron, v, to be less than c', and using the Huygens construction, draw wavefronts of the radiation emitted by the electron.

(b) Now take v to be greater than c', draw wavefronts, and show that (8-1-3) follows.

(c) Consider the acoustical analog of the two electromagnetic cases discussed here, and show how shock waves are generated in one case but not the other. (Read,[1] p. 66.)

8-7. Make the necessary calculations, and provide an absolute scale for the ordinate axis in Fig. 8-1-3. Use SI units. (Read,[1] p. 96.)

8-8. Consider a black, spherical satellite circling the sun at a radius r. Assume that the sun, of radius D, radiates as a blackbody at a known temperature. Given the radius of the satellite (R), show how it is possible to calculate its equilibrium temperature. (Cronin et al.,[215] p. 38.)

8-9. Consider a cubical cavity with perfectly reflecting interior walls. Show that the number of normal modes (i.e., standing waves) of electromagnetic radiation within the cavity, per unit volume and per unit frequency interval, is

$$\rho(\nu) = \frac{8\pi\nu^2}{c^3}. \tag{8-19-2}$$

This quantity, called the *mode density*, is actually independent of the shape of the cavity. (Read,[1] pp. 88–90.)

8-10. Use (8-1-7) to calculate the power per megacycle bandwidth emitted by the sun at 3 cm wavelength. Consider the sun to be a blackbody with a diameter of 10^6 km and a temperature of 6000 K. According to Fig. 8-1-1, in what frequency range does this 3-cm radiation lie? (Cronin et al.,[215] p. 41.)

8-11. Show that for blackbody radiation at temperature T, the number of photons per unit volume is $N = 2.029 \times 10^7\ T^3$ photons/m^3, and that the average energy per photon is $\bar{E} = 2.33 \times 10^{-4}\ T$ eV. (Bransden and Joachain,[14] p. 50.)

8-12. Design an experiment to show whether an object is a blackbody. (Chen,[216] p. 192.)

8-13. Justify the expression $\sigma_T = P_{\text{rad}}/P_{\text{beam}}$ given in (8-1-15). (Evans,[16] App. A.)

8-14. Show that the intensity of radiation from a Hertzian dipole varies as $\sin^2 \chi$, where χ is measured from the axis of the dipole, and then justify (8-1-17) and (8-1-18). (Read,[1] pp. 46–49.)

8-15. Solve the equation of motion (8-1-20) by the method of Laplace transforms to obtain (8-1-21). (Churchill,[217] Chap. 3.) See Rapp[218] for an extensive discussion of the damped oscillator in quantum mechanics. Also look back at Problem 5-14.

8-16. Justify the condition (8-1-23).

8-17. An electron is oscillating in a simple harmonic potential well with an angular frequency of 10^{15} rad/s and an amplitude of 10^{-8} cm.

(a) How much energy is radiated per cycle?

(b) What is the ratio of this energy to the average mechanical energy?

(c) How much time is required for the oscillator to radiate half of its initial energy? (Chen,[216] p. 99.)

8-18. In analyzing Rutherford scattering of alpha particles in a thin metallic foil, one neglects the effect of the atomic electrons, whereas in analyzing Compton scattering, the effect of the nucleus on the X-ray is ignored. Justify the neglect in each case. (Chen,[216] p. 263.)

8-19. Consider visible photons (6900 Å) colliding head-on with 20-GeV electrons, and show that the maximum energy of the scattered photons is essentially equal to that of the incident electrons. (Chen,[216] p. 4.)

8-20. A 1-MeV γ-ray experiences Compton scattering with an electron that is initially at rest. Show that the energy of the scattered electron can range from 0 to 0.796 MeV (Read,[1] p. 260).

8-21. Consider Compton scattering of a photon by an electron moving in the same direction as the photon with energy E and momentum P. Show that the Compton wavelength shift is

$$\Delta\lambda = 2\lambda_0 \frac{(p_0 + P)c}{E - Pc} \sin^2 \frac{\theta}{2} \tag{8-19-3}$$

in the notation of Section 8-1-B-2. Repeat the problem for antiparallel photon and electron velocities. (Bransden and Joachain,[14] p. 50.) For additional problems on the scattering of electromagnetic radiation, see Landau and Lifshitz,[7] pp. 248–250; Jackson,[8] Chaps. 14, 15.

8-22. Show that pair production is impossible unless a particle, or system of particles, is on hand to help conserve energy and momentum.

8-23. Work out the thresholds for the pair production processes (8-1-46) and (8-1-47). (Read,[1] p. 260.) In (8-1-46), take P to be infinitely heavy.

8-24. An electron strikes another electron that is initially at rest and the process

$$e^- + e^- \rightarrow e^- + e^- + e^- + e^+$$

results. What is the threshold kinetic energy for this pair production process? (Chen,[216] p. 284.)

8-25. Show that the quanta corresponding to Coulomb's law have zero mass (cf. Problem 5-29).

8-26. A particle of charge q is trapped in a one-dimensional box of width L. Which radiative transitions are allowed in the dipole approximation, and what are the relative intensities of the emitted lines? (Rapp,[218] p. 504.)

8-27. What are the electric dipole selection rules for light atoms? Which of these selection rules are also applicable for heavy atoms? For nuclei? (Chen,[216] p. 277.)

8-28. Which electric dipole transitions are allowed for a hydrogen atom in the $n = 3$, $l = 2$ state? Show the transitions on an energy-level diagram. (Chen,[216] p. 286.)

8-29. Define a "zero-zero transition" to be one from an initial state of zero angular momentum to a final state of zero angular momentum. Physically, why are such transitions forbidden if only the emission of a single photon is involved? Under what conditions can such transitions occur radiatively? (Cronin et al.,[215] p. 33.)

8-30. Why do radiative transitions that require a large spin change occur slowly? (Cronin et al.,[215] p. 33.)

8-31. **(a)** Suppose that a given state can decay radiatively to another given state only by the emission of electric dipole radiation. Give examples of the angular momentum and parity quantum numbers that might characterize the two states.

(b) Repeat for magnetic dipole radiation only.

(c) Repeat for electric quadrupole radiation only.

(d) Let either magnetic dipole or electric quadrupole radiation be allowed.

(e) Either electric dipole or electric quadrupole radiation allowed.

(f) Either magnetic quadrupole or electric quadrupole.

(g) Electric dipole or electric octupole only. (Read,[1] p. 124.)

8-32. Derive the selection rules for magnetic dipole transitions, (8-8-9), and for electric quadrupole transitions, (8-8-10).

8-33. Consider a particle in a central force field under the influence of a perturbation whose operator is proportional to the gradient operator, ∇. Work out the selection rules for transitions between its stationary states. (ter Haar,[219] p. 36.)

8-34. Work out the following relationship between the dipole velocity and dipole length operators:

$$\left\langle b\left|\frac{\partial}{\partial x}\right|a\right\rangle = \frac{-m}{\hbar^2}(E_b - E_a)\langle b|x|a\rangle. \tag{8-19-4}$$

(Steinfeld,[2] p. 22.)

8-35. The rate of radiative energy loss by an electron undergoing acceleration a is

$$\frac{dE}{dt} = -\frac{a^2e^2}{6\pi\epsilon_0 c^3}, \tag{8-19-5}$$

according to the classical Larmor formula. (Read,[1] p. 54.) Apply this equation to an electron moving in a Bohr orbit of very large radius, and calculate approximately the time required to drop into the first Bohr orbit. (Chen,[216] p. 102.)

8-36. Use (8-19-5) and the correspondence principle to calculate the mean life of a harmonic oscillator initially in a state of large quantum number n. The oscillating particle has mass m and charge e, and the classical angular frequency is ω. Express the result in terms of n, m, e, and ω.

8-37. By means of the principle of detailed balancing, relate the cross section for the photodissociation of the deuteron to that for the radiative capture of a neutron by a proton. (ter Haar,[219] p. 64; S. Gasiorowicz,[220] p. 421.)

8-38. Consider the nonrelativistic collision of two identical, spinless particles, and show that, according to classical electromagnetic theory, neither electric dipole nor magnetic dipole radiation can be emitted in the collision. (Cronin et al.,[215] p. 21.)

8-39. Derive the T-K-R sum rule, eq. (8-10-17). (Bransden and Joachain,[14] pp. 181–182.)

8-40. Work out the commutation relationships in (8-4-11) and (8-6-3). (Park.[221])

8-41. Use the Bohr model of the atom to justify (8-6-2) and (8-8-8). (Read,[1] pp. 119, 120.)

8-42. A hydrogen atom in the $n = 3$ state emits a photon in a transition to the ground state. Find the ratio of the atomic recoil energy to the photon energy, and to thermal energy at 300 K.

8-43. Consider the cooling of sodium atoms by laser radiation.

(a) What frequency must the laser light have in order to degrade the velocity of an atom by 3 cm/s if the momenta of the photon and atom are antiparallel?

(b) If an "antiparallel" atom has an initial velocity of 1000 m/s, by what amount does it see the oncoming light Doppler shifted?

(c) What is the change in the Doppler shift after 10, 100, 1000, and 10,000 photon absorptions?

(d) For the laser light of part (a), and parallel photon and atom momenta, what is the velocity change resulting from the absorption of one photon?

8-44. Prove (8-18-38) and (8-18-39).

8-45. Work out the radio wave example provided at the end of Section E of Note 8-2. (Read,[1] p. 314.)

8-46. Derive (8-13-3), which gives the photoabsorption cross section as determined in the double ion chamber method (see Fig. 8-13-3).

8-47. In the double ion chamber method described in Section 8-13, the total positive charge production efficiency, γ', is defined as

$$\gamma' \equiv \frac{\text{number of positive charges produced per second}}{\text{number of photons absorbed per second}}. \tag{8-19-6}$$

Refer to Fig. 8-13-3 and show that γ' can be expressed as[65]

$$\gamma' = \frac{i_1^2/e}{(i_1 - i_2)I_0}. \tag{8-19-7}$$

If measurements are made on a rare gas in a spectral region in which $\gamma' = \sigma_i/\sigma_a = 1$, then I_0 can be determined from (8-19-7). Then this value of I_0 can be used in (8-19-7), along with measured values of i_1 and i_2 for a molecular gas to determine γ' and σ_i for that gas.

8-48. Derive (8-15-2) for the Lab kinetic energy of a photofragment appearing at $\theta = 0$ with respect to the primary beam direction in fast ion photofragment spectroscopy. The parent ion is assumed to be homonuclear and diatomic.

REFERENCES

1. F. H. Read, *Electromagnetic Radiation*, Wiley, New York (1980).
2. J. I. Steinfeld, *Molecules and Radiation*, 2nd ed., MIT Press, Cambridge, Mass. (1985).
3. A. E. Siegman, *Lasers*, University Science Books, Mill Valley, Calif. (1986). M. Sargent, M. O. Scully, and W. E. Lamb, *Laser Physics*, Addison-Wesley, Reading, Mass. (1974). D. C. O'Shea, W. R. Callen, and W. T. Rhodes, *Introduction to Lasers and Their Applications*, 2nd ed., Addison-Wesley, Reading, Mass. (1990).
4. J. A. R. Samson, *Techniques of Vacuum Ultraviolet Spectroscopy*, Wiley, New York (1967).
5. M. L. Perlman, E. M. Rowe, and R. E. Watson, *Phys. Today* **27**, 30 (July 1974). E. M. Rowe, *Phys. Today* **34**, 28 (May 1981). H. Winick, G. Brown, K. Halbach, and J. Harris, *Phys. Today* **34**, 50 (May 1981). B. Crasemann and F. Wuilleumier, *Phys. Today* **37**, 34 (June 1984).
6. B. Crasemann (Ed.), *Atomic Inner-Shell Physics*, Plenum, New York (1985). H. Winick, and S. Doniach (Eds.), *Synchrotron Radiation Research*, Plenum, New York

(1980). A. A. Sokolov and I. M. Ternov, *Synchrotron Radiation*, Pergamon, New York (1968). E. E. Koch (Ed.), *Handbook on Synchrotron Radiation*, North-Holland, Amsterdam, Vol. I (1983); G. V. Marr (Ed.), Vol. II (1987). C. Kunz (Ed.), *Synchrotron Radiation: Techniques and Applications*, Springer-Verlag, Berlin (1979).

7. L. D. Landau and E. M. Lifshitz, *The Classical Theory of Fields*, 2nd ed., Addison-Wesley, Reading, Mass. (1962).
8. J. D. Jackson, *Classical Electrodynamics*, 2nd ed., Wiley, New York (1975).
9. R. H. Good and T. J. Nelson, *Classical Theory of Electric and Magnetic Fields*, Academic, New York (1971).
10. W. J. H. Panofsky and M. Phillips, *Classical Electricity and Magnetism*, Addison-Wesley, Reading, Mass. (1962).
11. P. Sprangle and T. Coffey, *Phys. Today* **37**, 44 (Mar. 1984). F. Ciocci, G. Dattoli, A. Renieri, and A. Torre, *Phys. Rev.* **141**, 1 (1986). U. Bizzarri et al., *Riv. Nuovo Cimento* **10**, 1 (1987). A. Fruchtman, *Phys. Rev. A* **37**, 4252 (1988). T. C. Marshall, *Free-Electron Lasers*, Macmillan, New York (1985).
12. K. E. Kreischer and R. J. Temkin, *Phys. Rev. Lett.* **59**, 547 (1987).
13. E. Segrè, *Nuclei and Particles*, 2nd ed., Benjamin-Cummings, Menlo Park, Calif. (1977).
14. B. H. Bransden and C. J. Joachain, *Physics of Atoms and Molecules*, Longman Group, Harlow, Essex, England (1983).
15. J. Berkowitz, *Photoabsorption, Photoionization, and Photoelectron Spectroscopy*, Academic, New York (1979).
16. R. D. Evans, *The Atomic Nucleus*, McGraw-Hill, New York (1955).
17. W. Heitler, *The Quantum Theory of Radiation*, 3rd ed., Clarendon, Oxford (1954).
18. J. M. Jauch and F. Rohrlich, *The Theory of Photons and Electrons*, Addison-Wesley, Cambridge, Mass. (1955).
19. F. Mandl, *Introduction to Quantum Field Theory*, Interscience, New York (1959).
20. M. D. Scadron, *Advanced Quantum Theory*, Springer-Verlag, New York (1979).
21. B. Rossi, *Optics*, Addison-Wesley, Reading, Mass. (1957), pp. 360–361.
22. R. R. Freeman, T. J. McIlrath, P. H. Buchsbaum, and M. Bashkansky, *Phys. Rev. Lett.* **57**, 3156 (1986) [Erratum: **58**, 526 (1987)]. See Section 8-17-B-3 for a related discussion.
23. J. H. Eberly, in E. Wolf, (Ed.), *Progress in Optics, Vol. VII*, North-Holland, Amsterdam (1969), p. 361. T. W. B. Kibble, *Phys. Rev.* **150**, 1060 (1966).
24. A. T. Young, *Phys. Today* **35**, 42 (Jan. 1982).
25. G. C. Baldwin, *An Introduction to Nonlinear Optics*, Plenum, New York (1969).
26. Y. R. Shen, *The Principles of Nonlinear Optics*, Wiley, New York (1984).
27. M. Schubert and B. Wilhelmi, *Nonlinear Optics and Quantum Electronics*, Wiley, New York (1986).
28. H. Frauenfelder and E. M. Henley, *Subatomic Physics*, Prentice-Hall, Englewood Cliffs, N.J. (1974).
29. P. S. Drzaic, J. Marks, and J. I. Brauman, "Electron Photodetachment from Gas Phase Molecular Anions," in M. R. Bowers (Ed.), *Gas Phase Ion Chemistry*, Vol. 3, *Ions and Light*, Academic, Orlando, Fla. (1984), p. 167; R. D. Mead, A. E. Stevens, and W. C. Lineberger, "Photodetachment in Negative Ion Beams," p. 213.

30. See, for example, R. Shakeshaft and E. J. Robinson, *Phys. Rev. A* **25**, 1977 (1982).

31. H. S. W. Massey, E. W. McDaniel, and B. Bederson (Series Eds.), *Applied Atomic Collision Physics*, Academic, New York (1982–1984), 5 vols. See Vol. 1, Chaps. 1–12; Vol. 2, Chaps. 1–5, 8; Vol. 3, Chaps. 1–14; Vol. 4, Chaps. 3, 6, 9–11; Vol. 5, Chaps. 2, 6–10, 12–14, 16.

32. J. D. Jackson, National Research Council of Canada, Report N.R.C. No. 2610, Chalk River, Ontario, Canada (Aug. 1951). For a discussion of reciprocity, see S. de Benedetti, *Nuclear Interactions*, Wiley, New York (1964).

33. H. A. Bethe and R. W. Jackiw, *Intermediate Quantum Mechanics*, 2nd ed., Benjamin, New York (1968), p. 7. C. Cohen-Tannoudji, B. Diu, and F. Laloe, *Quantum Mechanics*, Vol. 2, Wiley, New York (1977), Chap. XIII. D. Park, *Introduction to the Quantum Theory*, 2nd ed., McGraw-Hill, New York (1974), p. 299. G. Baym, *Lectures on Quantum Mechanics*, Benjamin, Menlo Park, Calif. (1973), p. 251. D. R. Bates, "Transitions," in D. R. Bates (Ed.), *Quantum Theory*, Vol. I, *Elements*, Academic, New York (1961), Chap. 8. A. Messiah, *Quantum Mechanics*, Vol. II, North-Holland, Amsterdam (1962), p. 736. E. Fermi, *Nuclear Physics*, University of Chicago Press, Chicago (1950), p. 142. L. I. Schiff, *Quantum Mechanics*, 3rd ed., McGraw-Hill, New York (1968), pp. 280–285.

34. D. Park, in Ref. 33, pp. 324–326. E. Fermi, in Ref. 33, pp. 76–142. F. Mandl, *Statistical Physics*, Wiley, New York (1971), App. B.

35. E. W. McDaniel, *Collision Phenomena in Ionized Gases*, Wiley, New York (1964).

36. J. B. Hasted, *Physics of Atomic Collisions*, 2nd ed., American Elsevier, New York (1972).

37. C. Kittel and H. Kroemer, *Thermal Physics*, 2nd ed., Freeman, San Francisco (1980).

38. G. K. Woodgate, *Elementary Atomic Structure*, 2nd ed., Clarendon, Oxford (1980).

39. A. Einstein, *Phys. Z.* **18**, 121 (1917). Translated and reprinted in D. ter Haar, *The Old Quantum Theory*, Pergamon, Oxford (1967).

40. H. A. Bethe and R. W. Jackiw, *Intermediate Quantum Mechanics*, 2nd ed., Benjamin, New York (1968). H. A. Bethe and E. E. Salpeter, *Quantum Mechanics of One- and Two-Electron Atoms*, Academic, New York (1957).

41. A. F. Starace, "Theory of Atomic Photoionization," in *Handbuch der Physik*, Vol. XXXI, Springer-Verlag, Berlin (1982), pp. 1–121.

42. A. Corney, *Atomic and Laser Spectroscopy*, Clarendon, Oxford (1980).

43. D. Park, Ref. 33, pp. 393–398.

44. D. Park, Ref. 33, pp. 398–402.

45. S. Gasiorowicz, *Quantum Physics*, Wiley, New York (1974).

46. U. Fano and J. W. Cooper, *Rev. Mod. Phys.* **40**, 441 (1968).

47. U. Fano and A. R. P. Rau, *Atomic Collisions and Spectra*, Academic, New York (1986).

48. R. Loudon, *The Quantum Theory of Light*, 2nd ed., Clarendon, Oxford (1983).

49. W. R. Hindmarsh and J. M. Farr, *Progress Quantum Electronics*, Vol. 2, pt. 4, Pergamon, Oxford (1973). I. I. Sobelman, *Atomic Spectra and Radiative Transitions*, Vol. 1, *Theories of Spectral Line Shape*, Wiley, New York (1981). F. Rostas (Ed.), *Spectral Line Shapes*, 3 vols., Walter de Gruyter, West Berlin (1985). S. I. Chu and K. Sando, *AAMP* (1988).

50. T. W. Hänsch, A. L. Schawlow, and G. W. Series, *Sci. Am.* **240**, 94 (Mar. 1979). T. W. Hänsch, *At. Phys.* **3**, 579 (1973).
51. H. R. Griem, *AAMP* **11**, 331 (1975); *Plasma Spectroscopy*, McGraw-Hill, New York (1964).
52. W. L. Wiese, *MEP* **7**, Pt. A, p. 117 (1968).
53. G. zu Putlitz, *At. Phys.* **1**, 227 (1969). G. Series, *Rep. Prog. Phys.* **22**, 280 (1959). G. W. Series, "Optical Pumping and Related Topics," in S. M. Kay and A. Maitland (Eds.), *Quantum Optics*, Academic, New York (1970). G. Cohen-Tannoudji and A. Kastler, in E. Wolf (Ed.), *Progress in Optics*, Vol. V, North-Holland, Amsterdam (1966), p. 3. W. Happer, *Rev. Mod. Phys.* **44**, 169 (1972). A. Kastler, in E. U. Condon and O. Sinanoglu (Eds.), *New Directions in Atomic Physics*, Vol. II, Yale University Press, New Haven, Conn. (1972), pp. 1–77. M. Broyer, G. Gouedard, J. C. Lehmann, and J. Vigue, *AAMP* **12**, 165 (1976).
54. S. Bashkin (Ed.), *Beam-Foil Spectroscopy*, Gordon and Breach, New York (1968). S. Bashkin, *At. Phys.* **2**, 43 (1971).
55. W. L. Wiese, *MEP* **7**, Pt. B, 307 (1968). R. A. Hartunian, *MEP* **7**, Pt. B, p. 141 (1968).
56. K. Burnett, *Phys. Rep.* **118**, 339 (1985).
57. E. W. McDaniel, *Atomic Collisions: Heavy Particle Projectiles*, Wiley, New York, to be published.
58. See, for example, G. N. Haddad and J. A. R. Samson, *J. Chem. Phys.* **84**, 6623 (1986). P. M. Dehmer and W. A. Chupka, *J. Chem. Phys.* **62**, 4525 (1975); **65**, 2243 (1976); **66**, 1972 (1977); **79**, 1569 (1983).
59. J. A. R. Samson, *AAMP* **2**, 177 (1966).
60. J. A. R. Samson and G. L. Weissler, *MEP* **7**, Pt. A, pg. 142 (1968).
61. J. A. R. Samson, "Atomic Photoionization," in *Handbuch der Physik*, Vol. XXXI, Springer-Verlag, Berlin (1982), pp. 123–213.
62. J. A. R. Samson, *MEP* **13**, Pt. A, p. 204 (1976).
63. J. A. R. Samson, *Phys. Rep.* **28**, 303 (1976).
64. G. V. Marr, *Photoionization Processes in Gases*, Academic, New York (1967).
65. J. W. Gallagher, C. E. Brion, J. A. R. Samson, and P. W. Langhoff, *J. Phys. Chem. Ref. Data* **17**, 9 (1988).
66. J. W. Gallagher, private communication (1988).
67. E. B. Saloman, J. H. Hubbell, and J. H. Scofield, *ADNDT* **38**, 1 (1988).
68. J. H. Hubbell, H. A. Gimm and I. Overbo, *J. Phys. Chem. Ref. Data* **9**, 1023 (1980).
69. T. C. Weekes, *Phys. Rep.* **160**, 1 (1988).
70. K. Watanabe, E. C. Y. Inn, and M. Zelikoff, *J. Chem. Phys.* **21**, 1026 (1953).
71. H. H. Nelson, L. Pasternack, and J. R. McDonald, *J. Phys. Chem.* **87**, 1286 (1983); *J. Chem. Phys.* **79**, 4279 (1983).
72. A. Schultz, H. W. Cruse, and R. N. Zare, *J. Chem. Phys.* **57**, 1354 (1972).
73. R. N. Zare and P. J. Dagdigian, *Science* **185**, 739 (1974).
74. J. L. Kinsey, *Ann. Rev. Phys. Chem.* **28**, 349 (1977).
75. R. W. Ditchburn, *Light*, 2nd ed., Interscience, New York (1963), pp. 571–573.
76. W. H. Louisell, *Quantum Statistical Properties of Radiation*, Wiley, New York (1973). See also M. Weissbluth, *Photon-Atom Interactions*, Academic, New York (1989).

77. S.-Y. Lee and E. J. Heller, *J. Chem. Phys.* **71**, 4777 (1979).
78. W. Kiefer and D. A. Long (Eds.), *Non-linear Raman Spectroscopy and Its Chemical Applications*, Reidel, Boston (1982).
79. G. L. Eesley, *Coherent Raman Spectroscopy*, Pergamon, Oxford (1981).
80. M. D. Levenson and S. Kano, *Introduction to Nonlinear Laser Spectroscopy*, rev. ed., Academic, New York (1988).
81. See, for example, D. C. Blasej and W. L. Peticolas, *J. Chem. Phys.* **72**, 3134 (1980).
82. D. Imre, J. L. Kinsey, A. Sinha, and J. Krenos, *J. Phys. Chem.* **88**, 3956 (1984).
83. R. N. Zare, *Mol. Photochem.* **4**, 1 (1972).
84. S. Yang and R. Bersohn, *J. Chem. Phys.* **61**, 4400 (1974).
85. R. Vasudev, R. N. Zare, and R. N. Dixon, *J. Chem. Phys.* **80**, 4863 (1984).
86. R. N. Zare, *Angular Momentum*, Wiley, New York (1988).
87. N. Andersen, J. Gallagher, and I. V. Hertel, *Phys. Rep.* **165**, 1 (1988).
88. N. Andersen, E. Campbell, K. Bartschat, J. W. Gallagher, and I. V. Hertel, "Collisional Alignment and Orientation of Atomic Outer Shells II: Quasimolecular Excitation; III: Spin-Resolved Excitation," *Phys. Rep.*, to be published.
89. G. O. Sitz, A. C. Kummel, and R. N. Zare, *J. Chem. Phys.* **87**, 3247 (1987).
90. S. R. Leone, in M. N. R. Ashfold and J. E. Baggott (Eds.), *Gas Phase Bimolecular Reactions*, Royal Society of Chemistry, London (1989).
91. C. H. Greene and R. N. Zare, *J. Chem. Phys.* **78**, 6741 (1983).
92. J. A. Guest, M. A. O'Halloran, and R. N. Zare, *Chem. Phys. Lett.* **103**, 261 (1984).
93. S. R. Leone, *Ann. Rev. Phys. Chem.* **35**, 109 (1984).
94. S. R. Leone, *Science* **227**, 889 (1985).
95. G. E. Hall, N. Sivakumar, D. Chawla, P. L. Houston, and I. Burak, *J. Chem. Phys.* **88**, 3682 (1988).
96. J. T. Moseley, "Determination of Ion Molecular Potential Curves Using Photodissociative Processes," in Vol. 5, of Ref. 31, pp. 269–283.
97. J. T. Moseley, "Ion Photofragment Spectroscopy," in K. P. Lawley (Ed.), *Photodissociation and Photoionization*, Wiley, New York (1985), pp. 245–298.
98. T. A. Miller and V. E. Bondybey, *J. Chim. Phys.* **77**, 695 (1980).
99. T. Baer, P. M. Guyon, I. Nenner, A. Tabche-Fouaille, R. Botter, L. F. A. Ferreira, and T. R. Govers, *J. Chem. Phys.* **70**, 1585 (1979).
100. B. K. Janousek and J. I. Brauman, in M. T. Bowers (Ed.), *Gas Phase Ion Chemistry*, Vol. 2, Academic, New York (1979), pp. 53–86.
101. R. C. Dunbar, in M. T. Bowers (Ed.), *Gas Phase Ion Chemistry*, Vol. 2, Academic, New York (1979), pp. 181–219.
102. S. R. Leone, *Adv. Chem. Phys.* **50**, 255 (1982). For photofragment spectroscopy of O_2, see P. C. Cosby and H. Helm, *J. Chem. Phys.* **90**, 1434 (1989).
103. J. T. Moseley, *J. Phys. Chem.* **86**, 3282 (1982).
104. J. T. Moseley, R. P. Saxon, B. A. Huber, P. C. Cosby, R. Abouaf, and M. Tadjeddine, *J. Chem. Phys.* **67**, 1659 (1977).
105. B. A. Huber, T. M. Miller, P. C. Cosby, H. D. Zeman, R. L. Leon, J. T. Moseley, and J. R. Peterson, *Rev. Sci. Instrum.* **48**, 1306 (1977). For a description of another apparatus of high quality, see L. Andric, H. Bissantz, E. Solarte, and F. Linder, *Z. Phys. D* **8**, 371 (1988).

106. P. C. Cosby, J. B. Ozenne, J. T. Moseley, and D. L. Albritton, *J. Mol. Spectros.* **78**, 203 (1979).

107. M. Tadjeddine, R. Abouaf, P. C. Cosby, B. A. Huber, and J. T. Moseley, *J. Chem. Phys.* **69**, 710 (1978).

108. F. J. Grieman, J. T. Moseley, R. P. Saxon, and P. C. Cosby, *Chem. Phys.* **51**, 169 (1980).

109. K. P. Kirby and E. F. van Dishoeck, *AAMP* **25**, 437 (1988).

110. See, for example, T. D. Märk and A. W. Castleman, *AAMP* **20**, 66 (1985). C. Brechignac, P. Cahuzac, and J. P. Roux, *ICPEAC XV*, Brighton (1987), p. 737. L. I. Yeh, M. Okumura, and Y. T. Lee, *ICPEAC XIV*, Stanford (1985), p. 813. M. L. Alexander, N. E. Levinger, M. A. Johnson, D. Ray, and W. C. Lineberger, *J. Chem. Phys.* **88**, 6200 (1988).

111. D. W. Lindle, P. A. Heimann, T. A. Ferrett, M. N. Piancastelli, and D. A. Shirley, *Phys. Rev. A* **35**, 4605 (1987). P. A. Heimann et al., *J. Phys. B* **20**, 5005 (1987). T. A. Ferrett et al., *Chem. Phys. Lett.* **138**, 607 (1987).

112. L. Wang, Y. T. Lee, and D. A. Shirley, *J. Chem. Phys.* **87**, 2489 (1987). See also *J. Electron Spectros.* **47**, 167 (1988).

113. J. E. Pollard, D. J. Trevor, Y. T. Lee, and D. A. Shirley, *Rev. Sci. Instrum.* **52**, 1837 (1981).

114. See, for example, D. W. Lindle, T. A. Ferrett, P. A. Heimann, and D. A. Shirley, *Phys. Rev. A* **36**, 2112 (1987). Also A. A. Cafolla, J. Comer, and T. Reddish, *J. Phys. B* **21**, 3571 (1988).

115. J. Cooper and R. N. Zare, J. Chem. Phys. **48**, 942 (1968). See also H. K. Tseng, R. H. Pratt, S. Yu, and A. Ron, *Phys. Rev. A* **17**, 1061 (1978).

116. R. H. Page, R. J. Larkin, Y. R. Shen, and Y. T. Lee, *J. Chem. Phys.* **88**, 2249 (1988). For discussions of supersonic beams, beam seeding, and beam cooling, see Part $\mathscr{B}$ of Chapter 9, Ref. 57.

117. H. A. Bethe, *Ann. Phys. (Leipzig)* **5**, 325 (1930). See also M. Inokuti, *Rev. Mod. Phys.* **43**, 297 (1971).

118. C. E. Brion, *ICPEAC XII*, Gatlinburg (1981), p. 579.

119. C. R. Brundle and A. D. Baker (Eds.), *Electron Spectroscopy: Theory, Techniques, and Applications*, Academic, New York, Vol. 1 (1977), Vol. 2 (1978), Vol. 3 (1979), Vol. 4 (1981).

120. S. T. Manson, "Atomic Photoelectron Spectroscopy," Pts. I and II, *AEEP* **41**, 73 (1976); **44**, 1 (1977).

121. S. T. Manson and D. Dill, in Ref. 119, Vol. 2, p. 157.

122. S. B. Hagström, M. O. Krause, and S. T. Manson, "Photoelectron and Auger Spectroscopy," in Ref. 31, Vol. 4, p. 450.

123. J. A. R. Samson, *J. Electron Spectros.* **15**, 257 (1979).

124. T. Rescigno, V. McKoy, and B. Schneider (Eds.), *Electron-Molecule and Photon-Molecule Scattering*, Plenum, New York (1979).

125. D. G. Truhlar (Ed.), *Resonances in Electron-Molecule Scattering, van der Waals Complexes, and Reactive Chemical Dynamics*, American Chemical Society, Washington, D.C. (1984).

126. M. Ya. Amusia, *AAMP* **17**, 1 (1981).

127. D. W. Turner, *Ann. Rev. Phys. Chem.* **21**, 107 (1970).

128. D. A. Shirley (Ed.), *Electron Spectroscopy*, North-Holland, Amsterdam (1972).

129. T. A. Carlson, *Photoelectron and Auger Spectroscopy*, Plenum, New York (1975).

130. T. A. Carlson, *Ann. Rev. Phys. Chem.* **26**, 211 (1975).

131. J. W. Rabalais, *Principles of Ultraviolet Photoelectron Spectroscopy*, Wiley, New York (1976).

132. A. Smith (Ed.), *Very High Resolution Spectroscopy*, Academic, New York (1976).

133. F. J. Wuilleumier (Ed.), *Photoionization and Other Probes of Many-Electron Interactions*, NATO-ASI Series B, Plenum, New York (1976). See also F. J. Wuilleumier, D. L. Ederer, and J. L. Picque, *AAMP* **23**, 197 (1988). C. R. Vidal, *AAMP* **23**, 1 (1987).

134. K. Siegbahn et al., *Electron Spectroscopy for Chemical Analysis*, Almqvist and Wiksell, Stockholm (1967). K. Siegbahn et al., *ESCA Applied to Free Molecules*, North-Holland, Amsterdam (1969).

135. W. C. Price, *AAMP* **10**, 131 (1974).

136. D. W. Turner, C. Baker, A. D. Baker, and C. R. Brundle, *Molecular Photoelectron Spectroscopy*, Wiley, London (1970).

137. H. Siegbahn and L. Karlsson, in *Handbuch der Physik*, Vol. XXXI, pp. 215–467 (1982).

138. S. T. Manson and A. F. Starace, *Rev. Mod. Phys.* **54**, 389 (1982).

139. J. E. Hardis, T. A. Ferrett, S. H. Southworth, A. C. Parr, P. Roy, J. L. Dehmer, P. M. Dehmer, and W. A. Chupka, *Phys. Rev. A*. to be publ. A. C. Parr et al., *Phys. Rev. A* **37**, 437 (1988).

140. H. Kossmann, V. Schmidt, and T. Andersen, *Phys. Rev. Lett.* **60**, 1266 (1988).

141. Y. K. Bae and J. R. Peterson, *Phys. Rev. A* **37**, 3254 (1988).

142. G. C. King, "Threshold Measurements of Electron and Photon Collisions with Atoms and Small Molecules," in P. G. Burke and J. B. West (Eds.), *Electron-Molecule Scattering and Photoionization*, Plenum, New York (1988). See also J. A. R. Samson and G. C. Angel, *Phys. Rev. Lett.* 61, 1584 (1988). M. Zubek, G. C. King, and P. M. Rutter, *J. Phys. B* **21**, 3585 (1988). G. C. King, M. Zubek, P. M. Rutter, F. H. Read, A. A. MacDowell, J. B. West, and D. M. P. Holland, *J. Phys. B* **21**, L-403 (1988).

143. A. V. Smith, J. E. M. Goldsmith, D. E. Nitz, and S. J. Smith, *Phys. Rev. A* **22**, 577 (1980).

144. J. Lahiri and S. T. Manson, *Phys. Rev. Lett.* **48**, 614 (1982); *Phys. Rev. A* **33**, 3151 (1986); *Phys. Rev. A* **37**, 1047 (1988). A. Z. Msezane, J. Lahiri, and S. T. Manson, *Phys. Rev. A* **33**, 4406 (1986). A. Z. Msezane and S. T. Manson, *Phys. Rev. Lett.* **48**, 473 (1982); *Phys. Rev. A* **29**, 1594 (1984). J. M. Preses et al., *Phys. Rev. A* **32**, 1264 (1985).

145. K. T. Dolder, *ICPEAC XV*, Brighton (1987), p. 549.

146. B. Peart, I. C. Lyon, and K. T. Dolder, *J. Phys. B* **20**, 5403 (1987). B. Peart and I. C. Lyon, *J. Phys. B* **20**, L-673 (1987). I. C. Lyon, B. Peart, and K. T. Dolder, *J. Phys. B* **20**, 1925 (1987). I. C. Lyon, B. Peart, K. T. Dolder, and J. B. West, *J. Phys. B* **20**, 1471 (1987). I. C. Lyon, B. Peart, J. B. West, and K. T. Dolder, *J. Phys. B* **19**, 4137 (1986).

147. S. T. Manson and S. K. Bhattacharya, *Aust. J. Phys.* **39**, 799 (1986). G. Nasreen, P.

C. Deshmukh, and S. T. Manson, *Jour. Phys. B* **21**, L-281 (1988). P. C. Deshmukh, G. Nasreen, and S. T. Manson, *Phys. Rev. A* **38**, 504 (1988).

148. S. T. Manson, *Phys. Rev. A* **31**, 3698 (1985). P. C. Deshmukh, B. R. Tambe, and S. T. Manson, *Aust. J. Phys.* **39**, 679 (1986). D. W. Lindle, T. A. Ferrett, P. A. Heimann, and D. A. Shirley, *Phys. Rev. A* **37**, 3808 (1988).
149. J. L. Dehmer, "Shape Resonances in Molecular Fields," in Ref. 125, p. 139. J. L. Dehmer, A. C. Parr, and S. H. Southworth, "Resonances in Molecular Photoionization," in G. V. Marr (Ed.), *Handbook on Synchrotron Radiation*, Vol. 2, North-Holland, Amsterdam (1987), p. 241. J. L. Dehmer, D. Dill, and A. C. Parr, "Shape Resonances in Molecular Fields," in J. S. Briggs, H. Kleinpoppen, and H. Lutz (Eds.), *Fundamental Processes of Atomic Dynamics*, Plenum, New York (1988).
150. T. A. Ferrett, D. W. Lindle, P. A. Heimann, W. D. Brewer, U. Becker, H. G. Kerkhoff, and D. A. Shirley, *Phys. Rev. A* **36**, 3172 (1987).
151. J. M. Bizau, P. Gérard, F. J. Wuilleumier, and G. Wendin, *Phys. Rev. A* **36**, 1220 (1987).
152. H. Beutler, *Z. Phys.* **93**, 177 (1935).
153. R. P. Madden and K. Codling, *Phys. Rev. Lett.* **10**, 516 (1963). For similar work by these and other investigators, see pages 164–182 of Ref. 61.
154. U. Fano, *Phys. Rev.* **124**, 1866 (1961).
155. U. Fano and J. W. Cooper, *Phys. Rev. A* **137**, 1364 (1965).
156. A. F. Starace, *Phys. Rev. A* **16**, 231 (1977).
157. J. H. Eberly and P. Lambropoulos (Eds.), *Multiphoton Processes*, Wiley, New York (1978).
158. V. S. Letokhov, "Principles of Laser Isotope Separation," in R. Balian, S. Haroche, and S. Liberman (Eds.), *Frontiers in Laser Spectroscopy*, North-Holland, Amsterdam (1977), p. 771.
159. A. M. Ronn, *Sci. Am.* **240**, 114 (May 1979).
160. R. N. Zare, *Sci. Am.* **236**, 86 (Feb. 1977).
161. X. Zhao, E. J. Hintsa, and Y. T. Lee, *J. Chem. Phys.* **88**, 801 (1988).
162. R. N. Compton and J. C. Miller, "Multiphoton Ionization Photoelectron Spectroscopy: MPI-PES," in D. Evans (Ed.), *Laser Applications in Physical Chemistry*, Marcel Dekker, New York (1988).
163. J. Morellec, D. Normand, and G. Petite, *AAMP* **18**, 97 (1982).
164. M. D. Perry, O. L. Landen, A. Szöke, and E. M. Campbell, *Phys. Rev. A* **37**, 747 (1988); *Phys. Rev. Lett.* **60**, 1270 (1988).
165. R. Hippler, H. J. Humpert, H. Schwier, S. Jetzke, and H. O. Lutz, *J. Phys. B* **16**, L-713 (1983). S. T. Pratt, P. M. Dehmer, and J. L. Dehmer, *Phys. Rev. A* **35**, 3793 (1987). J. L. Dehmer, S. T. Pratt, and P. M. Dehmer, *Phys. Rev. A* **36**, 4494 (1987).
166. V. S. Letokhov, *Laser Photoionization Spectroscopy*, Academic, New York (1987). S. L. Chin and P. Lambropoulos, *Multiphoton Ionization of Atoms*, Academic, New York (1984). S. H. Lin, Y. Fujimura, H. J. Neusser, and E. W. Schlag, *Multiphoton Spectroscopy of Molecules*, Academic, New York (1984). F. H. M. Faisal, *Theory of Multiphoton Processes*, Plenum, New York (1987). P. Lambropoulos, *AAMP* **12**, 87 (1976). P. Lambropoulos, *ICPEAC XV*, Brighton (1975), p. 533.
167. P. M. Dehmer, J. L. Dehmer, and S. T. Pratt, *Comments At. Mol. Phys.* **19**, 205 (1987).

168. P. M. Dehmer, S. T. Pratt, and J. L. Dehmer, *J. Chem. Phys.* **85**, 13 (1986); *J. Phys. Chem.* **91**, 2593 (1987). S. T. Pratt, P. M. Dehmer, and J. L. Dehmer, *J. Chem. Phys.* **78**, 4315 (1983); **85**, 3379 (1986). M. A. O'Halloran, S. T. Pratt, P. M. Dehmer, and J. L. Dehmer, *J. Chem. Phys.* **87**, 3288 (1987). J. L. Dehmer, P. M. Dehmer, S. T. Pratt, M. A. O'Halloran, and F. S. Tomkins, in J. S. Briggs, H. Kleinpoppen, and H. Lutz (Eds.), *Fundamental Processes of Atomic Dynamics*, Plenum, New York (1988). J. C. Miller and R. N. Compton, *Chem. Phys. Lett.* **93**, 453 (1982); *J. Chem. Phys.* **84**, 675 (1986). L. R. Brewer, F. Buchinger, M. Ligare, and D. E. Kelleher, *Phys. Rev. A* **39**, 3912 (1989).

169. P. R. Blazewicz, X. Tang, R. N. Compton, and J. A. D. Stockdale, *J. Opt. Soc. Am. B* **4**, 770 (1987). A. Dodhy, J. A. D. Stockdale, R. N. Compton, X. Tang, P. Lambropoulos, and A. Lyras, *Phys. Rev. A* **35**, 2878 (1987). S. T. Pratt, J. L. Dehmer, and P. M. Dehmer, *J. Chem. Phys.* **82**, 676 (1985). S. J. Smith and G. Leuchs, *AAMP* **24**, 157 (1988).

170. M. A. O'Halloran, S. T. Pratt, F. S. Tomkins, J. L. Dehmer, and P. M. Dehmer, *Chem. Phys. Lett.*, to be publ.; *J. Chem. Phys.* **89**, 75 (1988).

171. P. Agostini and G. Petite, *ICPEAC XIV*, Stanford (1985), p. 743. U. Johann, T. S. Luk, and C. K. Rhodes, *ICPEAC XIV*, Stanford (1985), p. 751. M. Crance, *ICPEAC XIV*, Stanford (1985), p. 757; *Phys. Rep.* **144**, 117 (1987).

172. H. Haberland, J. Höhne, M. Oschwald, and J. Broad, *ICPEAC XIV*, Stanford (1985), p. 771.

173. M. Lambropoulos, S. E. Moody, S. J. Smith, and W. C. Lineberger, *Phys. Rev. Lett.* **35**, 159 (1975).

174. T. J. McIlrath, P. H. Bucksbaum, R. R. Freeman, and M. Bashkansky, *Phys. Rev. A* **35**, 4611 (1987). See also P. H. Bucksbaum, D. W. Schumacher, and M. Bashkansky, *Phys. Rev. Lett.* **61**, 1182 (1988).

175. R. R. Freeman, P. H. Bucksbaum, H. Milchberg, S. Darack, D. Schumacher, and M. E. Geusic, *Phys. Rev. Lett.* **59**, 1092 (1987).

176. J. H. Eberly and J. Javanainen, *Phys. Rev. Lett.* **60**, 1346, 2458 (1988). **62**, 881 (1989).

177. P. H. Bucksbaum, M. Bashkansky, and D. W. Schumacher, *Phys. Rev. A* **37**, 3615 (1988).

178. M. J. Van der Wiel and H. G. Muller, *ICPEAC XIV*, Stanford (1985), p. 735.

179. W. E. Cooke and T. J. McIlrath (Eds.), *J. Opt. Soc. Am. B* **4**, 701 (1987).

180. C. Cohen-Tannoudji and S. Reynaud, "Dressed Atom Approach to Resonance Fluorescence," in Ref. 157, p. 103, C. Cohen-Tannoudji, "Atoms in Strong Resonant Fields," in Balian, Haroche, and Liberman, Ref. 158, p. 3.

181. J. C. Miller and R. N. Compton, *Phys. Rev. A* **25**, 2056 (1982); *Proc. Int. Conf. Lasers '82*, p. 387.

182. S. M. Hamadani, J. A. D. Stockdale, R. N. Compton, and M. S. Pindzola, *Phys. Rev. A* **34**, 1938 (1986).

183. J. C. Miller, R. N. Compton, M. G. Payne, and W. W. Garrett, *Phys. Rev. Lett.* **45**, 114 (1980).

184. D. J. Jackson and J. J. Wynne, *Phys. Rev. Lett.* **49**, 543 (1982).

185. A. F. J. van Raan, G. Baum, and W. Raith, *J. Phys. B* **9**, L-173 (1976).

186. P. B. Hogan, J. L. Carlsten, F. M. J. Pichanick, S. J. Smith, and W. W. Smith, *Phys. Rev. Lett.* **37**, 990 (1976).

187. D. J. Bamford, R. P. Saxon, L. E. Jusinski, J. D. Buck, and W. K. Bischel, *Phys. Rev. A* **37**, 3259 (1988).

188. A. C. Kummel, G. O. Sitz, and R. N. Zare, *J. Chem. Phys.* **88**, 6707 (1988); also see **88**, 7357 (1988).

189. P. A. M. Dirac, *Proc. R. Soc. London A* **114**, 243 (1927).

190. J. C. Slater, *Quantum Theory of Atomic Structure*, Vol. I, McGraw-Hill, New York (1960).

191. D. Park, *Introduction to the Quantum Theory*, 2nd ed., McGraw-Hill, New York (1974), pp. 123–134.

192. T. H. Boyer, *Sci. Am.* **253**, 70 (Aug. 1985).

193. P. W. Milonni, *Phys. Rep.* **25**, 1 (1976).

194. R. H. Dicke, *Phys. Rev.* **93**, 99 (1954).

195. S. Haroche, C. Fabre, M. Gross, and P. Pillet, *At. Phys.* **5**, 179 (1977). M. Gross and S. Haroche, *Phys. Rep.* **93**, 301 (1982).

196. G. Gabrielse, H. Dehmelt, and W. Kells, *Phys. Rev. Lett.* **54**, 537 (1985). B. M. Schwarzschild, *Phys. Today* **38**, No. 5, 17 (May 1985). R. S. Van Dyck, P. B. Schwinberg, and H. G. Dehmelt, *At. Phys.* **9**, 53 (1984).

197. F. L. Wells and G. H. Dunn, *Phys. Today* **27**, No. 8, 30 (Aug. 1974). G. H. Dunn, *ICPEAC XIV*, Stanford (1985), p. 23. P. E. Toschek and W. Neuhauser, *At. Phys.* **7**, 529 (1981). D. J. Wineland, W. M. Itano, J. C. Bergquist, J. J. Bollinger, and J. D. Prestage, *At. Phys.* **9**, 3 (1984). G. Werth, *At. Phys.* **9**, 28 (1984). D. A. Church, *At. Phys.* **9**, 137 (1984). *Physica Scripta Topical Issue* **22**. B. G. Levi, *Physics Today* **41**, No. 9, 17 (Sept., 1988); J. Hoffnagle, R. G. DeVoe, L. Reyna, and R. G. Brewer, *Phys. Rev. Lett.* **61**, 255 (1988).

198. H. F. Hess, G. P. Kochanski, J. M. Doyle, N. Masuhara, D. Kleppner, and T. J. Greytak, *Phys. Rev. Lett.* **59**, 672 (1987). See also *Phys. Rev. Lett.* **61**, 935 (1988).

199. T. W. Hänsch and A. L. Schawlow, *Opt. Commun.* **13**, 68 (1975). For recent extensions of the ideas presented in this paper, see P. D. Lett, R. N. Watts, C. I. Westbrook, W. D. Phillips, P. L. Gould, and H. J. Metcalf, *Phys. Rev. Lett.* **61**, 169 (1988); A. Aspect, E. Arimondo, R. Kaiser, N. Vansteenkiste, and C. Cohen-Tannoudji, *Phys. Rev. Lett.* **61**, 826 (1988). S. Stenholm, *Eur. J. Phys.* **9**, 242 (1988). W. D. Phillips, *Phys. Today* **42**, S-13 (Jan., 1989). F. Diedrich, J. C. Bergquist, W. M. Itano, and D. J. Wineland, *Phys. Rev. Lett.* **62**, 403 (1989). Y. Shevy, D. S. Weiss, P. J. Ungar, and S. Chu, *Phys. Rev. Lett.* **62**, 1118 (1989). Papers in *Atomic Phys.* **11** (1989). V. I. Balykin and V. S. Letokhov, *Phys. Today* **42**, 23 (Apr., 1989).

200. W. D. Phillips and H. J. Metcalf, *Sci. Am.* **256**, 50 (Mar. 1987).

201. W. D. Phillips, J. V. Prodan, and H. J. Metcalf, *At. Phys.* **9**, 338 (1984).

202. D. J. Wineland and W. M. Itano, *Phys. Today* **40**, No. 6, 34 (June 1987).

203. D. E. Pritchard, E. L. Raab, V. Bagnato, C. E. Wieman, and R. N. Watts, *Phys. Rev. Lett.* **57**, 310 (1986).

204. R. Blatt, W. Ertmer, and J. L. Hall, *Prog. Quantum Electron.* **8**, 237 (1984). W. Ertmer, R. Blatt, J. L. Hall, and M. Zhu, *Phys. Rev. Lett.* **54**, 996 (1985).

205. B. M. Schwarzschild, *Phys. Today* **38**, No. 6, 17 (June 1985).

206. S. Chu, J. E. Bjorkholm, A. Ashkin, and A. Cable, *Phys. Rev. Lett.* **57**, 314 (1986).

207. V. S. Bagnato, G. P. Lafyatis, A. G. Martin, E. L. Raab, R. N. Ahmad-Bitar, and D. E. Pritchard, *Phys. Rev. Lett.* **58**, 2194 (1987).

208. D. E. Pritchard, *ICPEAC XIV*, Stanford (1985), p. 593.

209. E. L. Raab, M. Prentiss, A. Cable, S. Chu, and D. E. Pritchard, *Phys. Rev. Lett.* **59**, 2631 (1987).

210. F. Diedrich, E. Peik, J. M. Chen, W. Quint, and H. Walther, *Phys. Rev. Lett.* **59**, 2931 (1987). D. J. Wineland, J. C. Bergquist, W. M. Itano, J. J. Bollinger, and C. H. Manney, *Phys. Rev. Lett.* **59**, 2935 (1987).

211. N. F. Ramsey, *Am. Sci.* **76**, No. 1, 42 (1988). S. Chu, J. E. Bjorkholm, A. Ashkin, and A. Cable, *At. Phys.* **10**, 377 (1987). D. J. Wineland, W. M. Itano, and R. S. Van Dyck, *AAMP* **19**, 135 (1983).

212. P. L. Gould, P. D. Lett, P. S. Julienne, W. D. Phillips, H. R. Thorsheim, and J. Weiner, *Phys. Rev. Lett.* **60**, 788 (1988). W. D. Phillips, P. L. Gould, and P. D. Lett, *Science* **239**, 877 (1988).

213. R. Blatt, W. Ertmer, P. Zoller, and J. L. Hall, *Phys. Rev. A* **34**, 3022 (1986). S. Stenholm, *Rev. Mod. Phys.* **58**, 699 (1986); *Phys. Rep.* **43**, 151 (1978). A. P. Kazantsev, G. A. Ryabenko, G. I. Surdutovich, and V. P. Yakovlev, *Phys. Rep.* **129**, 75 (1985). G. Nienhuis, *Phys. Rep.* **138**, 151 (1986). V. S. Letokhov and V. G. Minogin, *Phys. Rep.* **73**, 1 (1981).

214. R. D. Sard, *Relativistic Mechanics*, Benjamin, New York (1970).

215. J. A. Cronin, D. F. Greenberg, and V. L. Telegdi, *University of Chicago Graduate Problems in Physics with Solutions*, Addison-Wesley, Reading, Mass. (1967).

216. M. Chen, *Berkeley Physics Problems with Solutions*, Prentice-Hall, Englewood Cliffs, N.J. (1974).

217. R. V. Churchill, *Operational Mathematics*, 2nd ed., McGraw-Hill, New York (1958).

218. D. Rapp, *Quantum Mechanics*, Holt, Reinhart and Winston, New York (1971).

219. D. ter Haar (Ed.), *Problems in Quantum Mechanics*, 3rd ed., Pion, London (1975); also published by Academic, New York (1975).

220. S. Gasiorowicz, *Quantum Physics*, Wiley, New York (1974).

221. D. Park in Ref. 33, Chapter 3.

APPENDIX I

THE DISTINCTION BETWEEN A PLASMA AND AN ORDINARY IONIZED GAS

1. THE DEBYE LENGTH

As indicated in Chapter 1, it is convenient to introduce the concept of the debye length in order to distinguish a true plasma* from an ordinary ensemble of ions and electrons. It was stated that for an ionized gas to be a plasma its physical extent must be much greater than its debye shielding length. Exactly how the debye length serves to define a plasma is shown in the following paragraphs.

The concept of the shielding length arose from studies of strong electrolytes made by Debye and Hückel[1] in 1923. An expression for the shielding length may be readily obtained, as seen in the following derivation.[2] Consider an electrolyte containing several types of positive and negative ions, there being n_{j0} ions per cm^3 of type j with charge Z_je. Let us assume the following:

(a) The ions of each type j are in a Boltzmann distribution at an equilibrium temperature T_j.

(b) The potential energy of the charges due to their separation is small compared with their thermal energy, so that $|Z_jeV| \ll kT$, where V represents the electrostatic potential.

(c) The microscopic variations of potential resulting from the discrete nature of the particles may be neglected.

Since the electrolyte as a whole is electrically neutral, each of the ions must create around itself an ion cloud whose net charge is equal and opposite to that

*With the recent production of positron plasmas in the laboratory, the plasma concept must now be expanded. See, for example, C. M. Surko, M. Levinthal, and A. Passner, *Phys. Rev. Lett.* **62**, 901 (1989).

of the ion itself. This cloud must be spherically symmetric about the ion, and the number density of each species of ion about any particular ion will depend only on the distance r from this ion. We can now determine the potential $V(r)$ effecting this distribution, subject to the boundary conditions that V vanish for large r and approach the Coulomb potential for small r. If ρ_j is the charge density of ions of type j in the cloud, then we may write

$$\rho_j = Z_j e n_{j0} e^{-Z_j eV/kT_j} \tag{1}$$

Assumption (b) allows us to linearize the problem by considering only the first two terms of the series expansion of ρ_j, and we may write

$$\rho_j = Z_j e n_{j0}\left(1 - \frac{eVZ_j}{kT_j}\right). \tag{2}$$

Since the electrolyte as a whole is electrically neutral, $\sum_j Z_j e n_{j0} = 0$. The total charge density ρ is given by

$$\rho = \sum_j \rho_j = -e^2 V \sum_j \frac{n_{j0} Z_j^2}{kT_j}. \tag{3}$$

The charge density and the potential V are related by Poisson's equation

$$\nabla^2 V = -\frac{4\pi\rho}{\epsilon}, \tag{4}$$

where ϵ is the dielectric constant of the electrolyte. Equation (4) may be written in spherically symmetric form as

$$\frac{\partial^2 V}{\partial r^2} + \frac{2}{r}\frac{\partial V}{\partial r} = \left(\frac{4\pi e^2}{\epsilon}\sum_j \frac{n_{j0} Z_j^2}{kT_j}\right) V. \tag{5}$$

Let us define

$$\lambda_D^2 = \left(\frac{4\pi e^2}{\epsilon}\sum_j \frac{n_{j0} Z_j^2}{kT_j}\right)^{-1}, \tag{6}$$

noting that λ_D has the dimensions of length. The general solution to (5) is

$$V(r) = A\frac{e^{-r/\lambda_D}}{r} + B\frac{e^{r/\lambda_D}}{r}. \tag{7}$$

If the origin of coordinates is a particle of charge $Z_a e$, then application of the boundary conditions that $V(r) \to 0$ for large r and $V(r) \to Z_a e/\epsilon r$ for small r

yields as the solution

$$V(r) = \frac{Z_a e}{\epsilon r} e^{-r/\lambda_D}. \tag{8}$$

This screened Coulomb potential becomes quite small at distances much greater than λ_D, which is called the *Debye-Hückel radius* or the *Debye shielding length.* It may be considered to be a measure of the size of the screening ion cloud surrounding the given ion. We see that the shielding length varies directly with the square root of the temperature and inversely with the square root of the number density of the ions. This behavior is in accord with the interpretation that the Debye-Hückel radius is that distance at which the electrostatic forces, which tend to impose charge neutrality, are balanced by the kinetic forces, which tend to produce non-neutrality. Thus an increase in the number density decreases the mean distance between particles and increases the electrostatic forces that maintain electrical neutrality. Evidently, in order to have the requisite electrostatic interaction between particles, the debye length must be considerably less than the minimum dimensions of the ionized medium. If this condition is not satisfied, the medium behaves like a collection of free charges. The foregoing derivation assumes, of course, that the debye length is large compared with the interparticle distances. In a more recent study by Liboff[3] it was shown that a similar argument is also valid for ionized gases.

It should be noted that the assumptions in this derivation result in a single quantity, the Debye-Hückel radius, which characterizes the potential decay about any of the ionized particles in the medium. If we alter our assumptions, it is possible to arrive at what might be called a shielding length for each species of particle. To do this, let us consider an electrically neutral ionized gas consisting of positive ions of one type and electrons. We now replace assumption (a) by the following:

(a′) The electrons are in a Boltzmann distribution at an equilibrium temperature T_e; the ions are uniformly distributed throughout the gas. (Although the assumption of a uniform ion distribution may not be entirely realistic, the result is of sufficient interest to justify the assumption.)

Let $Z_i e$ be the charge of each ion and $-e$ that of each electron; let n_{i0} be the uniform number density of the ions. Choosing one of the ions as our center of coordinates, we see that the electron charge density is given by

$$\rho_e = -e n_{e0} e^{eV/kT_e}. \tag{9}$$

Electrical neutrality requires that

$$Z_i n_{i0} = n_{e0}. \tag{10}$$

Thus the total charge density is given by

$$\rho = \rho_e + \rho_i = en_{e0} - en_{e0}e^{eV/kT_e}. \tag{11}$$

Expansion of the exponential term as before yields

$$\rho = -\frac{e^2 n_{e0}}{kT_e} V. \tag{12}$$

The potential and charge density are again related by Poisson's equation in spherically symmetric form:

$$\frac{\partial^2 V}{\partial r^2} + \frac{2}{r}\frac{\partial V}{\partial r} = \frac{4\pi e^2 n_{e0}}{\epsilon k T_e} V. \tag{13}$$

We now define

$$\lambda_{De}^2 = \left(\frac{4\pi e^2 n_{e0}}{\epsilon k T_e}\right)^{-1}. \tag{14}$$

With this substitution (14) is now in the same form as before, and the solution is the same as (7) with λ_D replaced by λ_{De}. We see that λ_{De} has dimensions of length and that it depends only on the parameters of the electrons. This length is called the shielding length for ions by electrons or, more simply, the electron shielding length. It would, of course, be possible to assume instead that the electrons were uniformly distributed in space and that the ions were in a Boltzmann distribution. Such an assumption leads to an ion shielding length, λ_{Di}, given by

$$\lambda_{Di}^2 = \left(\frac{4\pi e^2 n_{i0}}{\epsilon k T_i}\right)^{-1}. \tag{15}$$

Thus a shielding length can be determined for each of the species of particle in the gas. These lengths are different from the shielding length associated with the gas as a whole, given by (6). In a gas consisting of singly charged positive ions and electrons at the same temperature (T) we see that $n_{e0} = n_{i0} = n$, and the relationship of the various shielding lengths is

$$\lambda_{De} = \lambda_{Di} = \sqrt{2}\,\lambda_D. \tag{16}$$

When in contact with a physical boundary, a plasma forms a protective sheath about itself. The sheath, in effect, separates the main body of the plasma from its environment. Unlike the main body, the sheath is not electrically neutral and strong electric fields may be present in it. The thickness of the sheath is of the order of a debye length. In order to demonstrate this fact, let us consider a thin slablike region, with half-width x_0, perpendicular to the X axis and centered about the origin. We assume that the electron number density greatly exceeds the ion number density. The resulting net charge gives rise to a potential difference between the center of the slab and its boundaries given by the equation

$$\frac{d^2V}{dx^2} = \frac{4\pi n_e e}{\epsilon}, \tag{17}$$

which integrates to

$$V = \frac{2\pi}{\epsilon} n_e e x^2 \tag{18}$$

if we put $V = 0$ at $x = 0$. The potential appears as a "hill" to electrons and as a "trough" to ions. Clearly, the region over which $n_e \gg n_i$ cannot be arbitrarily large or a point would be reached at which the electrical potential energy would exceed the mean thermal energy and ions would then flow in such a way as to restore neutrality. If we seek that value of x that gives rise to a change in potential energy just equal to the mean kinetic energy in one direction, we find

$$\frac{kT_e}{2} = \frac{2\pi n_e e^2 x_0^2}{\epsilon} \tag{19}$$

or

$$x_0 = \left(\frac{\epsilon k T_e}{4\pi n_e e^2}\right)^{1/2}. \tag{20}$$

This expression may be recognized as being equivalent to that derived earlier for the debye length, λ_{De}. We conclude then that the distance over which a plasma can have an appreciable departure from charge equilibrium is of the order of a debye length.

2. MULTIPLE SMALL-ANGLE COULOMB SCATTERING*

The expression for the differential cross section for Coulomb scattering was derived in Section 3-8. Using this cross section, we may obtain an expression for the probability of a charged particle being scattered through an angle greater than 90° in a single Coulomb collision while penetrating to a given depth in a plasma. It is of interest to compare this expression for the probability of a single large angle Coulomb scattering to the probability of many small angle

*For a general discussion of the small-angle scattering of fast particles, see W. T. Scott, *Rev. Mod. Phys.* **35**, 231 (1963).

deflections resulting in a net large angle deflection. An expression for the probability of multiple small-angle scattering resulting in a large-angle deflection may be derived as follows:

Consider a particle of charge ze with mass m and velocity v impinging at an impact parameter b on a scattering center of charge Ze and infinite mass. Since nearly all of the deflection of the projectile occurs in the immediate vicinity of the scattering center, we may assume for computational purposes that the incident particle experiences a lateral deflecting force of magnitude zZe^2/b^2 for a period $2b/v$.* Thus the momentum change of the projectile is

$$\Delta(mv) = \frac{2zZe^2}{bv}, \tag{1}$$

and if we exclude large angle deflections so that $\Delta(mv)$ is small compared with the initial momentum mv, the corresponding angular deflection is

$$\Delta\vartheta = \frac{\Delta(mv)}{mv} = \frac{2zZe^2}{mv^2 b} \equiv \frac{D}{b}. \tag{2}$$

The quantity

$$D = \frac{2zZe^2}{mv^2} \tag{3}$$

defined in (2) is the distance of closest approach of the projectile to the scattering center in the special case of a head-on trajectory.

Now suppose that the projectile, rather than encountering a single scattering center, passes through a plasma containing N scattering centers per cm^3. Since the individual deflections are completely random, the average deflection must be zero. The mean square deflection will not vanish, however, for there exists a random walk in angle away from the direction of initial incidence.[5] The total mean square deflection associated with the projectile penetrating a depth λ into the medium is

$$\overline{\Delta\vartheta^2} = \int_{\Delta\vartheta_{j\min}}^{\Delta\vartheta_{j\max}} \Delta\vartheta_j^2\, n(\Delta\vartheta_j)\, d(\Delta\vartheta_j), \tag{4}$$

where $\Delta\vartheta_j = D/b_j$ (the angular deflection in a collision of impact parameter b_j), $\Delta\vartheta_{j\max} = D/b_{j\min}$, and $\Delta\vartheta_{j\min} = D/b_{j\max}$. The extreme values of the impact parameter b_j used here are defined later. The quantity $n(\Delta\vartheta_j)\, d(\Delta\vartheta_j)$ represents the number of collisions resulting in scattering with changes of direction between $\Delta\vartheta_j$ and $\Delta\vartheta_j + d(\Delta\vartheta_j)$. It is convenient to carry out this integration with

*It is interesting to note that this commonly used approximation gives the same change in momentum obtained when the exact force function is integrated along the entire flight path (see Glasstone and Lovberg[4]).

respect to the variable b_j. Thus, since $N\lambda 2\pi b\, db$ equals the number of scattering centers in a cylindrical shell of length λ, radius b, and thickness db, we have

$$\overline{\Delta\vartheta^2} = \int_{b_{j\max}}^{b_{j\min}} -\left(\frac{D}{b_j}\right)^2 N\lambda 2\pi b_j\, db_j = 2\pi N\lambda D^2 \ln\frac{b_{j\max}}{b_{j\min}}. \tag{5}$$

We must now determine suitable limits for the impact parameter b_j. The upper limit $b_{j\max}$ must reflect the mutual shielding effect of the scattering centers; it is therefore taken to be equal to the debye length characterizing the plasma. The lower limit may be the classical distance of closest approach or the de Broglie wavelength, whichever is smaller. In most cases of interest ln $(b_{j\max}/b_{j\min})$ lies between about 10 and 20.[6]

It is important to note the dependence of $\sqrt{\overline{\Delta\vartheta^2}}$ on $\lambda^{1/2}$. If the rms deflection of a particle is 90° during passage through a distance of $\lambda_{90°}$ in the medium, the particle will undergo about ten 90° deflections in traversing a distance equal to 100 $\lambda_{90°}$.

Now consider N_0 particles incident on a scattering medium of depth λ. The fraction of these particles that have been scattered at a net angle between Φ and $\Phi + d\Phi$ upon exit from the medium is from the theory of statistics[5]

$$\begin{aligned}\frac{N(\Phi)\, d\Phi}{N_0} &= \frac{2\Phi}{\overline{\Phi^2}} e^{-\Phi^2/\overline{\Phi^2}}\, d\Phi \\ &= \frac{\Phi}{\pi N\lambda D^2 \ln(b_{j\max}/b_{j\min})} \exp\left[\frac{-\Phi^2}{2\pi N\lambda D^2 \ln(b_{j\max}/b_{j\min})}\right] d\Phi. \end{aligned} \tag{6}$$

Therefore, the fraction of the incident particles scattered through angles greater than $\sqrt{\overline{\Phi^2}}$ is given by

$$\int_{\sqrt{\overline{\Phi^2}}}^{\infty} \frac{2\Phi}{\overline{\Phi^2}} e^{-\Phi^2/\overline{\Phi^2}}\, d\Phi = \frac{1}{e}. \tag{7}$$

We now wish to compare the probability of multiple Coulomb scattering through large angles to the probability of single-collision large-angle Coulomb scattering. This comparison must of necessity involve the depth of the scattering medium. If the medium is very thin, as in a Rutherford scattering experiment, there can be few multiple deflections through large net angles, and we expect that single large-angle collisions will be dominant. On the other hand, if the medium is very thick, the probability of either type of scattering occurring is unity. In the intermediate range of thickness it turns out that multiple small-angle collisions produce most of the large-angle scattering.

Let us first calculate from (5) the thickness $\lambda_{90°}$ for which $\sqrt{\overline{\Delta\vartheta^2}} = \pi/2$. We find that

$$\lambda_{90^\circ} = \frac{\pi}{8ND^2 \ln (b_{j\max}/b_{j\min})}. \tag{8}$$

According to (7), if a group of particles traverses this distance, $1/e$ of them will be scattered through angles greater than 90° as the result of multiple small-angle scattering.

The probability of a particle undergoing a single Coulomb deflection through an angle greater than 90° while traversing the same distance λ_{90° in the medium is*

$$P_s(\Phi > 90^\circ) = N\lambda_{90^\circ}\frac{\pi D^2}{4} = \frac{\pi^2}{32 \ln (b_{j\max}/b_{j\min})}. \tag{9}$$

Thus the multiple scattering process is $(32/\pi^2 e) \ln (b_{j\max}/b_{j\min}) \approx 12$ to 24 times more probable in this case.

It may be pointed out that some writers[7] use (5) directly as the basis for the "mean free path for a random walk scattering through an angle of $\pi/2$." If we put $\overline{\Delta\vartheta^2}$ equal to unity in (5), we obtain the desired result:

$$\lambda_{90^\circ} = \left[8\pi N \left(\frac{zZe^2}{mv^2}\right)^2 \ln\left(\frac{b_{\max}}{b_{\min}}\right)\right]^{-1}. \tag{10}$$

An "effective cross section" for a 90° deflection by multiple collisions can then be defined by the equation

$$q_{90^\circ} = \frac{1}{N\lambda_{90^\circ}}. \tag{11}$$

This quantity, however, is not a cross section in the usual sense because, according to (5), the relation between the net deflection angle and the number density of scattering centers is not linear.

REFERENCES

1. P. Debye and E. Hückel, *Phys. Z.* **24**, 185, 305 (1923).
2. L. D. Landau and E. M. Lifshitz, *Statistical Physics*, Addison-Wesley, Reading, Mass., 1958, pp. 229–236.
3. R. L. Liboff, *Phys. Fluids* **2**, 40 (1959).

*This expression is derived by computing the area presented by a single target particle for scattering through angles greater than 90° and multiplying this area $\pi D^2/4$ by the total number of scattering centers in a slab of the medium λ_{90° cm thick with a 1-cm^2 cross-sectional area. This number represents the ratio of the total area presented for scattering to the normal surface area 1 cm^2. For fairly thin targets the resulting expression is also the probability of single scattering through angles greater than 90° as given by (9). This expression is not a true probability, however, as it approaches infinity instead of unity as λ approaches infinity.

4. S. Glasstone and R. H. Lovberg, *Controlled Thermonuclear Reactions*, Van Nostrand, Princeton, N.J., 1960, p. 91.
5. R. M. Eisberg, *Fundamentals of Modern Physics*, Wiley, New York, 1961, pp. 94–98.
6. See Ref. 4 and L. Spitzer, *Physics of Fully Ionized Gases*, 2nd ed., Wiley, New York, 1962, Chap. 5.
7. See, for example, A. Simon, *An Introduction to Thermonuclear Research*, Pergamon, New York, 1959, pp. 14–15.

APPENDIX II

GENERAL REFERENCES AND SOURCES OF DATA

1. ABBREVIATIONS AND PUBLICATION DATA FOR SOME OF THE JOURNALS, REPORTS, AND SERIAL PUBLICATIONS CITED HEREIN

AAMP: *Advances in Atomic and Molecular Physics*, bound volumes published yearly by Academic, New York. Vols. 1–9 edited by D. R. Bates and I. Estermann; Vols. 10 onward edited by D. R. Bates and B. Bederson.

AD: *Atomic Data*, journal published by Academic, New York.

ADNDT: *Atomic Data and Nuclear Data Tables*, journal published by Academic, New York. ADNDT came into existence in 1973 as a combination of AD and NDT, which ceased publication at that time.

Adv. Chem. Phys.: *Advances in Chemical Physics*, bound volumes published at irregular intervals by Wiley, New York.

AEEP: *Advances in Electronics and Electron Physics*, bound volumes published at irregular intervals by Academic, New York.

Ann. Rev. Phys. Chem.: *Annual Reviews of Physical Chemistry*, bound volumes published annually by Annual Reviews, Inc., Palo Alto, Calif.

Atomic Physics, bound volumes—see "ICAP" in section on conference series.

ICPEAC, see section on conference series.

IPPJ-AM, reports on atomic and molecular processes published by Institute of Plasma Physics, Nagoya University, Nagoya 464, Japan.

JAERI-M, reports on atomic and molecular processes published by Japan Atomic Energy Research Institute, Tokai-muri, Naka-gun, Ibaraki-ken 319-11, Japan.

MEP: *Methods of Experimental Physics*, bound volumes published by Academic, New York, edited by various individuals.

NDT: *Nuclear Data Tables*, journal published by Academic, New York.

2. MAJOR CONFERENCE SERIES

ICAP: *International Conference on Atomic Physics.* The invited papers presented at each conference are published in volumes entitled *Atomic Physics.* The 11 conferences held to date are listed below together with information on the published volumes.

ICAP	1: New York (1968)	*Atomic Physics* *1* (1969)
	2: Oxford (1970)	*2* (1971)
	3: Boulder (1972)	*3* (1973)
	4: Heidelberg (1974)	*4* (1975)
	5: Berkeley (1976)	*5* (1977)
	6: Riga (1978)	*6* (1979)
	7: Cambridge (1980)	*7* (1981)
	8: Göteborg (1982)	*8* (1983)
	9: Seattle (1984)	*9* (1985)
	10: Kyoto (1986)	*10* (1987)
	11: Paris (1988)	*11* (1989)

Volumes 1–8 published by Plenum, New York; Vols. 9, 10, and 11 by World Scientific, Singapore.

ICPEAC: *International Conference on the Physics of Electronic and Atomic Collisions.* The 16 conferences held to date are listed below.

ICPEAC	I:	New York (1958)	ICPEAC	IX:	Seattle (1975)
	II:	Boulder (1961)		X:	Paris (1977)
	III:	London (1963)		XI:	Kyoto (1979)
	IV:	Quebec (1965)		XII:	Gatlinburg (1981)
	V:	Leningrad (1967)		XIII:	Berlin (1983)
	VI:	Cambridge, Mass. (1969)		XIV:	Stanford (1985)
				XV:	Brighton (1987)
	VII:	Amsterdam (1971)		XVI:	New York (1989)
	VIII:	Belgrade (1973)			

Invited papers and abstracts of contributed papers for the more recent conferences are published by North-Holland, Amsterdam, in separate volumes during the year immediately following each conference. The volumes are entitled *Electronic and Atomic Collisions.* Proceedings of the earlier conferences were published by the host institution and are not generally available. In this book, an entry such as *ICPEAC XV*, Brighton (1987) refers to the bound volume of *invited* papers.

3. GENERAL REFERENCES

ADNDT (*Atomic Data and Nuclear Data Tables*). Contains extensive data compilations, for the most part, with periodic author and subject indices. Essentially all branches of atomic collisions and atomic and molecular structure are covered.

AAMP (*Advances in Atomic and Molecular Physics*). Published annually since 1965. Contains reviews on practically all aspects of atomic physics.

J. Berkowitz, *Photoabsorption, Photoionization, and Photoelectron Spectroscopy*, Academic, New York (1979). General treatise on photon processes.

L. G. Christophorou (Ed.), *Electron-Molecule Interactions and Their Applications*, 2 vols., Academic, New York (1984). Comprehensive treatment of electron-molecule collisions; contains many data and useful data indices.

CFADC (Controlled Fusion Atomic Data Center), Oak Ridge National Laboratory, Oak Ridge, Tennessee. The CFADC data compilations and bibliographies published before 1985 are listed in the bibliography by McDaniel et al. cited below.

JILA Data Center (Joint Institute for Laboratory Astrophysics, University of Colorado, Boulder, Colorado). The JILA data compilations and bibliographies published before 1985 are listed in the bibliography by McDaniel et al. cited below.

J. Phys. Chem. Ref. Data. Exhaustive compilations of data on selected topics in atomic and molecular structure and collisions.

E. W. McDaniel, M. R. Flannery, E. W. Thomas, and S. T. Manson, *ADNDT* **33**, 1–148 (1985). A selected bibliography on atomic collisions: data collections, bibliographies, review articles, books, and papers of particular tutorial value.

A. A. Radzig and B. M. Smirnov, *Reference Data on Atomic and Molecular Physics*, Springer-Verlag, Berlin (1984).

APPENDIX III

TABLES

Table I Physical Constants

Quantity	Symbol	Value[a]	Units
GENERAL CONSTANTS			
Universal constants			
Speed of light in vacuum	c	299 792 458	$m\ s^{-1}$
Permeability of vacuum	μ_0	$4\pi \times 10^{-7}$	$N\ A^{-2}$
		$=12.566\ 370\ 614\ldots$	$10^{-7}\ N\ A^{-2}$
Permittivity of vacuum, $1/\mu_0 c^2$	ϵ_0	$8.854\ 187\ 817\ldots$	$10^{-12}\ F\ m^{-1}$
Newtonian constant of gravitation	G	6.672 59(85)	$10^{-11}\ m^3\ kg^{-1}\ s^{-2}$
Planck constant	h	6.626 075 5(40)	$10^{-34}\ J\ s$
In electron volts, $h/\{e\}$		4.135 669 2(12)	$10^{-15}\ eV\ s$
$h/2\pi$	$\hbar$	1.054 572 66(63)	$10^{-34}\ J\ s$
In electron volts, $\hbar/\{e\}$		6.582 122 0(20)	$10^{-16}\ eV\ s$
Electromagnetic constants			
Elementary charge	e	1.602 177 33(49)	$10^{-19}\ C$
	e/h	2.417 988 36(72)	$10^{14}\ A\ J^{-1}$
Magnetic flux quantum, $h/2e$	Φ_0	2.067 834 61(61)	$10^{-15}\ Wb$
Bohr magneton, $e\hbar/2m_e$	μ_B	9.274 015 4(31)	$10^{-24}\ J\ T^{-1}$
In electron volts, $\mu_B/\{e\}$		5.788 382 63(52)	$10^{-5}\ eV\ T^{-1}$
In hertz, μ_B/h		1.399 624 18(42)	$10^{10}\ Hz\ T^{-1}$
In wavenumbers, μ_B/hc		46.686 437(14)	$m^{-1}\ T^{-1}$
In kelvins, μ_B/k		0.671 709 9(57)	$K\ T^{-1}$
Nuclear magneton, $e\hbar/2m_p$	μ_N	5.050 786 6(17)	$10^{-27}\ J\ T^{-1}$
In electron volts, $\mu_N/\{e\}$		3.152 451 66(28)	$10^{-8}\ eV\ T^{-1}$

Table I (cont.)

Quantity	Symbol	Value[a]	Units
In hertz, μ_N/h		7.622 591 4(23)	MHz T^{-1}
In wavenumbers, μ_N/hc		2.542 622 81(77)	10^{-2} m^{-1} T^{-1}
In kelvins, μ_N/k		3.658 246(31)	10^{-4} K T^{-1}
	ATOMIC CONSTANTS		
Fine-structure constant, $\mu_0 ce^2/2h$	α	7.297 353 08(33)	10^{-3}
Inverse fine-structure constant	α^{-1}	137.035 989 5(61)	
Rydberg constant, $m_e c\alpha^2/2h$	R_∞	10 973 731.534(13)	m^{-1}
In hertz, $R_\infty c$		3.289 841 949 9(39)	10^{15} Hz
In joules, $R_\infty hc$		2.179 874 1(13)	10^{-18} J
In eV, $R_\infty hc/\{e\}$		13.605 698 1(40)	eV
Bohr radius, $\alpha/4\pi R_\infty$	a_0	0.529 177 249(24)	10^{-10} m
Hartree energy, $e^2/4\pi\epsilon_0 a_0 = 2R_\infty hc$	E_h	4.359 748 2(26)	10^{-18} J
In eV, $E_h/\{e\}$		27.211 396 1(81)	eV
Electron			
Electron mass	m_e	9.109 389 7(54)	10^{-31} kg
		5.485 799 03(13)	10^{-4} u
In electron volts, $m_e c^2/\{e\}$		0.510 999 06(15)	MeV
Electron-muon mass ratio	m_e/m_μ	4.836 332 18(71)	10^{-3}
Electron-proton mass ratio	m_e/m_p	5.446 170 13(11)	10^{-4}
Electron-deuteron mass ratio	m_e/m_d	2.724 437 07(6)	10^{-4}
Electron—α-particle mass ratio	m_e/m_α	1.370 933 54(3)	10^{-4}
Electron specific charge	$-e/m_e$	−1.758 819 62(53)	10^{11} C kg^{-1}

Quantity	Symbol	Value	Units
Compton wavelength, $h/m_e c$	λ_C	2.426 310 58(22)	10^{-12} m
$\lambda_C/2\pi = \alpha a_0 = \alpha^2/4\pi R_\infty$	$\bar{\lambda}_C$	3.861 593 23(35)	10^{-13} m
Classical electron radius, $\alpha^2 a_0$	r_e	2.817 940 92(38)	10^{-15} m
Thomson cross section, $(8\pi/3)r_e^2$	σ_e	0.665 246 16(18)	10^{-28} m^2
Electron magnetic moment	μ_e	928.477 01(31)	10^{-26} J T^{-1}
In Bohr magnetons	μ_e/μ_B	1.001 159 652 193(10)	
In nuclear magnetons	μ_e/μ_N	1 838.282 000(37)	
Electron magnetic moment anomaly, $\mu_e/\mu_B - 1$	a_e	1.159 652 193(10)	10^{-3}
Electron g factor, $2(1 + a_e)$	g_e	2.002 319 304 386(20)	
Electron-muon magnetic moment ratio	μ_e/μ_μ	206.766 967(30)	
Electron-proton magnetic moment ratio	μ_e/μ_p	658.210 688 1(66)	
Proton			
Proton mass	m_p	1.672 623 1(10)	10^{-27} kg
		1.007 276 470(12)	u
In electron volts, $m_p c^2/\{e\}$		938.272 31(28)	MeV
Proton-electron mass ratio	m_p/m_e	1 836.152 701(37)	
Proton-muon mass ratio	m_p/m_μ	8.880 244 4(13)	
Proton specific charge	e/m_p	9.578 830 9(29)	10^7 C kg^{-1}
Proton Compton wavelength, $h/m_p c$	$\lambda_{C,p}$	1.321 410 02(12)	10^{-15} m
$\lambda_{C,p}/2\pi$	$\bar{\lambda}_{C,p}$	2.103 089 37(19)	10^{-16} m
Proton magnetic moment	μ_p	1.410 607 61(47)	10^{-26} J T^{-1}
In Bohr magnetons	μ_p/μ_B	1.521 032 202(15)	10^{-3}
In nuclear magnetons	μ_p/μ_N	2.792 847 386(63)	

Table I (cont.)

Quantity	Symbol	Value[a]	Units
	PHYSIOCHEMICAL CONSTANTS		
Avogadro constant	N_A, L	6.022 136 7(36)	10^{23} mol^{-1}
Atomic mass constant, $m_u = \frac{1}{12}m(^{12}C)$	m_u	1.660 540 2(10)	10^{-27} kg
In electron volts, $m_u c^2/\{e\}$		931.494 32(28)	MeV
Faraday constant, $N_A e$	F	96 485.309(29)	C mol^{-1}
Molar Planck constant	$N_A h$	3.990 313 23(36)	10^{-10} J s mol^{-1}
	$N_A hc$	0.119 626 58(11)	J m mol^{-1}
Molar gas constant	R	8.314 510(70)	J mol^{-1} K^{-1}
Boltzmann constant, R/N_A	k	1.380 658(12)	10^{-23} J K^{-1}
In electron volts, $k/\{e\}$		8.617 385(73)	10^{-5} eV K^{-1}
In hertz, k/h		2.083 674(18)	10^{10} Hz K^{-1}
In wavenumbers, k/hc		69.503 87(59)	m^{-1} K^{-1}
Molar volume (ideal gas), RT/p			
$T = 273.15$ K, $p = 101\,325$ Pa	V_m	0.022 414 10(19)	m^3 mol^{-1}
Loschmidt constant, N_A/V_m	n_0	2.686 763(23)	10^{25} m^{-3}
$T = 273.15$ K, $p = 100$ kPa	V_m	0.022 711 08(19)	m^3 mol^{-1}
Stefan-Boltzmann constant, $(\pi^2/60)k^4/\hbar^3c^2$	σ	5.670 51(19)	10^{-8} W m^{-2} K^{-4}
First radiation constant, $2\pi hc^2$	c_1	3.741 774 9(22)	10^{-16} W m^2
Second radiation constant, hc/k	c_2	0.014 387 69(12)	m K
Wien displacement law constant, $b = \lambda_{max}T = c_2/4.965\,114\,23\ldots$	b	2.897 756(24)	10^{-3} m K

Source: E. R. Cohen and B. N. Taylor, *Rev. Mod. Phys.* **59**, 1121 (1987).

[a]The numbers in parentheses are the one-standard-deviation uncertainty in the last digits of the value listed.

Table II Units, Symbols, and Conversion Factors

Quantity	Symbol	Value	Units
Electron volt, $(e/\mathrm{C})\mathrm{J} = \{e\}\mathrm{J}$	eV	1.602 177 33(49)	10^{-19} J
(Unified) atomic mass unit, $1\ \mathrm{u} = m_u = \frac{1}{12}m(^{12}\mathrm{C})$	u	1.660 540 2(10)	10^{-27} kg
Standard atmosphere	atm	101 325	Pa
Standard acceleration of gravity	g_n	9.806 65	$\mathrm{m\ s^{-2}}$

$1\ \mathrm{J} = 10^7$ erg
$1\ \mathrm{C} = 10^{-1}$ e.m.u.
$= 2.99792458 \times 10^{-9}$ e.s.u.
$1\ \mathrm{T} = 10^4$ G
$1\ \mathrm{Ci} = 3.70 \times 10^{10}\ \mathrm{s}^{-1}$
$1\ \text{Å} = 10^{-10}\ \mathrm{m} = 10^{-8}$ cm
$1\ \mathrm{fm} = 10^{-11}\ \mathrm{m} = 10^{-13}$ cm
$1\ \mathrm{b} = 10^{-28}\ \mathrm{m}^2 = 10^{-24}\ \mathrm{cm}^2$
Loschmidt's number (the number density of an ideal gas at 760 torr and 0°C):
$N_L = 2.69 \times 10^{19}$ per cm^3
Number density of an ideal gas at 1 torr and 0°C: $N_{1\,\mathrm{torr}} = 3.54 \times 10^{16}$ per cm^3
1 torr = 1 mm Hg pressure
Thermal energy per K: $k = 8.617 \times 10^{-5}$ eV/K
Thermal energy at 290 K = 17°C: $kT = 1/40$ eV
Temperature associated with 1 eV: $T = 11,604$ K
1 electron volt (eV) = 23.069 kcal/mol
1 Rydberg = 13.606 eV; 1 Hartree = 2 Rydberg
Wavelength of photon with energy, E(eV): $\lambda = 12,398/E$ Å
Wavelength of electron with nonrelativistic kinetic energy, T(eV): $\lambda = 12.26/T^{1/2}$ Å
Energy levels in hydrogen-like atoms: $E_n = -13.60Z^2/n^2\ eV$

Table III Standard International (SI) Units

meter	m	newton	N
kilogram	kg	farad	F
second	s	joule	J
ampere	A	coulomb	C
kelvin	K	tesla	T
mole	mol	hertz	Hz
candela	cd	watt	W

Table IV Atomic Units

Physical Quantity	Unit	Symbol	Numerical Magnitude
Mass	Electronic mass	m_e	9.108×10^{-28} g
Charge	Electronic charge	e	4.8029×10^{-10} esu
Angular momentum	Rationalized Planck's constant	$\hbar$	1.0544×10^{-27} erg-sec
Length	$\hbar^2/m_e e^2$: radius of first Bohr orbit of hydrogen atom with nucleus of infinite mass	a_0	5.2917×10^{-9} cm
Energy	e^2/a_0: twice the ionization energy of hydrogen atom with nucleus of infinite mass	E_0	1 hartree = 2 rydbergs = 27.210 eV
Time	$\hbar/E_0$: time for electron in first Bohr orbit of hydrogen atom to describe one radian	t_0	2.4189×10^{-17} sec
Velocity	a_0/t_0: velocity of electron in first Bohr orbit of hydrogen atom	v_0	2.1877×10^{8} cm/sec
Cross section	Area of first Bohr orbit		$\pi a_0^2 = 0.88 \times 10^{-16}$ cm^2

The velocity of light in a vacuum c has the value $c/v_0 = 137.037$ in atomic units. The reciprocal of this quantity is Sommerfeld's fine-structure constant, $\alpha = v_0/c = e^2/c\hbar$.

Table V Periodic Table of the Elements

IA	IIA	IIIB	IVB	VB	VIB	VIIB	VIIIB			IB	IIB	IIIA	IVA	VA	VIA	VIIA	Noble Gases
1 H 1.008																	2 He 4.003
3 Li 6.941	4 Be 9.012											5 B 10.81	6 C 12.011	7 N 14.007	8 O 15.999	9 F 18.998	10 Ne 20.179
11 Na 22.990	12 Mg 24.305											13 Al 26.982	14 Si 28.086	15 P 30.974	16 S 32.06	17 Cl 35.453	18 Ar 39.948
19 K 39.098	20 Ca 40.08	21 Sc 44.956	22 Ti 47.90	23 V 50.942	24 Cr 51.996	25 Mn 54.938	26 Fe 55.847	27 Co 58.933	28 Ni 58.70	29 Cu 63.546	30 Zn 65.38	31 Ga 69.72	32 Ge 72.59	33 As 74.922	34 Se 78.96	35 Br 79.904	36 Kr 83.80
37 Rb 85.468	38 Sr 87.62	39 Y 88.906	40 Zr 91.22	41 Nb 92.906	42 Mo 95.94	43 Tc (98)	44 Ru 101.07	45 Rh 102.905	46 Pd 106.4	47 Ag 107.868	48 Cd 112.41	49 In 114.82	50 Sn 118.69	51 Sb 121.75	52 Te 127.60	53 I 126.905	54 Xe 131.30
55 Cs 132.905	56 Ba 137.33	57* La 138.905	72 Hf 178.49	73 Ta 180.948	74 W 183.85	75 Re 186.2	76 Os 190.2	77 Ir 192.22	78 Pt 195.09	79 Au 196.966	80 Hg 200.59	81 Tl 204.37	82 Pb 207.19	83 Bi 208.98	84 Po (209)	85 At (210)	86 Rn (~222)
87 Fr (223)	88 Ra 226.02	89† Ac 227.028	104 Unq (261)	105 Unp (262)	106 Unh (263)												

The Nonmetals (B, C, N, O, F; Si, P, S, Cl; As, Se, Br; Te, I; At)

Transition Elements (IIIB–IIB)

The Active Metals (IA, IIA)

Inner Transition Metals

* Lanthanides	58 Ce 140.12	59 Pr 140.907	60 Nd 144.24	61 Pm (145)	62 Sm 150.4	63 Eu 151.96	64 Gd 157.25	65 Tb 158.925	66 Dy 162.50	67 Ho 164.930	68 Er 167.26	69 Tm 168.934	70 Yb 173.04	71 Lu 174.96
† Actinides	90 Th 232.038	91 Pa 231.036	92 U 238.029	93 Np 237.048	94 Pu (244)	95 Am (243)	96 Cm (247)	97 Bk (247)	98 Cf (251)	99 Es (254)	100 Fm (257)	101 Md (258)	102 No (259)	103 Lr (260)

Source: B. G. Segal, *Chemistry, Experiment and Theory*, Wiley, New York (1985), p. 65.

Table VI Ionization Energies (I.E.) of Atoms[a]

Z	Symbol	Ground Configuration[b]	Ionization Energy (eV)
1	H	$1s^1$	13.595
2	He	$1s^2$	24.581
3	Li	[He] $2s^1$	5.390
4	Be	$2s^2$	9.320
5	B	$2s^22p^1$	8.296
6	C	$2s^22p^2$	11.256
7	N	$2s^22p^3$	14.545
8	O	$2s^22p^4$	13.614
9	F	$2s^22p^5$	17.418
10	Ne	$2s^22p^6$	21.559
11	Na	[Ne] $3s^1$	5.138
12	Mg	$3s^2$	7.644
13	Al	$3s^23p^1$	5.984
14	Si	$3s^23p^2$	8.149
15	P	$3s^23p^3$	10.484
16	S	$3s^23p^4$	10.357
17	Cl	$3s^23p^5$	13.01
18	Ar	$3s^23p^6$	15.755
52	Te	$4d^{10}5s^25p^4$	9.01
53	I	$4d^{10}5s^25p^5$	10.454
54	Xe	$4d^{10}5s^25p^6$	12.127
55	Cs	[Xe] $6s^1$	3.893
56	Ba	$6s^2$	5.210
57	La	$5d6s^2$	5.61
58	Ce	$4f5d6s^2$	6.54
59	Pr	$4f^36s^2$	5.48
60	Nd	$4f^46s^2$	5.51
61	Pm	$4f^56s^2$	
62	Fm	$4f^66s^2$	5.6
63	Eu	$4f^76s^2$	5.67
64	Gd	$4f^75d6s^2$	6.16
65	Tb	$4f^96s^2$	6.74
66	Dy	$4f^{10}6s^2$	6.82
67	Ho	$4f^{11}6s^2$	
68	Er	$4f^{12}6s^2$	
69	Tm	$4f^{13}6s^2$	
70	Yb	$4f^{14}6s^2$	6.22
71	Lu	$4f^{14}5d6s^2$	6.15

19	K	[Ar] $4s^1$	4.339	72	Hf	$4f^{14}5d^26s^2$	7.0
20	Ca	$4s^2$	6.111	73	Ta	$4f^{14}5d^36s^2$	7.88
21	Sc	$3d4s^2$	6.54	74	W	$4f^{14}5d^46s^2$	7.98
22	Ti	$3d^24s^2$	6.83	75	Re	$4f^{14}5d^56s^2$	7.87
23	V	$3d^34s^2$	6.74	76	Os	$4f^{14}5d^66s^2$	8.7
24	Cr	$3d^54s$	6.76	77	Ir	$4f^{14}5d^76s^2$	9.2
25	Mn	$3d^54s^2$	7.432	78	Pt	$4f^{14}5d^86s^2$	8.88
26	Fe	$3d^64s^2$	7.87	79	Au	[Xe,$4f^{14}5d^{10}$] $6s^1$	9.22
27	Co	$3d^74s^2$	7.86	80	Hg	$6s^2$	10.434
28	Ni	$3d^84s^2$	7.633	81	Tl	$6s^26p^1$	6.106
29	Cu	$3d^{10}4s^1$	7.724	82	Pb	$6s^26p^2$	7.415
30	Zn	$3d^{10}4s^2$	9.391	83	Bi	$6s^26p^3$	7.287
31	Ga	$3d^{10}4s^24p^1$	6.00	84	Po	$6s^26p^4$	8.43
32	Ge	$3d^{10}4s^24p^2$	7.88	85	At	$6s^26p^5$	
33	As	$3d^{10}4s^24p^3$	9.81	86	Rn	$6s^26p^6$	10.745
34	Se	$3d^{10}4s^24p^4$	9.75				
35	Br	$3d^{10}4s^24p^5$	11.84	87	Fr	[Rn] $7s^1$	
36	Kr	$3d^{10}4s^24p^6$	13.996	88	Ra	$7s^2$	5.277
				89	Ac	$6d7s^2$	6.9

Table VI (cont.)

Z	Symbol	Ground Configuration[b]	Ionization Energy (eV)	Z	Symbol	Ground Configuration[b]	Ionization Energy (eV)
37	Rb	[Kr] $5s^1$	4.176	90	Th	$6d^2 7s^2$	
38	Sr	$5s^2$	5.692	91	Pa	$5f^2 6d 7s^2$	
39	Y	$4d 5s^2$	6.377	92	U	$5f^3 6d 7s^2$	4.0
40	Zr	$4d^2 5s^2$	6.835	93	Np	$5f^4 6d 7s^2$	
41	Nb	$4d^4 5s^1$	6.881	94	Pu	$5f^6 7s^2$	
42	Mo	$4d^5 5s^1$	7.10	95	Am	$5f^7 7s^2$	
43	Tc	$4d^5 5s^2$	7.228	96	Cm	$5f^7 6d 7s^2$	
44	Ru	$4d^7 5s^1$	7.365	97	Bk	$5f^8 6d 7s^2$	
45	Rh	$4d^8 5s^1$	7.461	98	Cf	$5f^{10} 7s^2$	
46	Pd	$4d^{10}$	8.33	99	Es	$5f^{11} 7s^2$	
47	Ag	$4d^{10} 5s^1$	7.574	100	Fm	$5f^{12} 7s^1$	
48	Cd	$4d^{10} 5s^2$	8.991	101	Mv	$5f^{13} 7s^2$	
49	In	$4d^{10} 5s^2 5p^1$	5.785	102	No	$5f^{14} 7s^2$	
50	Sn	$4d^{10} 5s^2 5p^2$	7.342	103	Lw	$5f^{14} 6d 7s^2$	
51	Sb	$4d^{10} 5s^2 5p^3$	8.639	104	Ku	$5f^{14} 6d^2 7s^2$	

Source: R. A. Serway, *Physics for Scientists and Engineers/with Modern Physics*, Saunders, Philadelphia (1983), p. 931.

[a] For term symbols, see B. H. Bransden and C. J. Joachain, *Physics of Atoms and Molecules*, Longman Group, Harlow, Essex, England (1983), pp. 302–303.

[b] The bracket notation is used as a shorthand method to avoid repetition in indicating inner-shell electrons. Thus, [He] represents $1s^2$, [Ne] represents $1s^2 2s^2 2p^6$, [Ar] represents $1s^2 2s^2 2p^6 3s^2 3p^6$, and so on.

Table VII Dissociation Energies (D_0^0) and Ionization Energies (I.E.) of Diatomic Molecules in (eV).

Molecule	D_0^0	I.E.	Molecule	D_0^0	I.E.
Br_2	1.9707	10.52	HF	5.869	16.039
CN	7.7	14.1	HI	3.054	10.38
CO	11.09	14.0139	I_2	1.54238	9.311
CS	7.35	11.33	IBr	1.817	9.85
Cl_2	2.47936	11.50	ICl	2.1531	10.08
F_2	1.602	15.686	N_2	9.759	15.5808
H_2	4.4781	15.4258	NO	6.496	9.26436
HBr	3.758	11.67	O_2	5.115	12.071
HCl	4.433	12.74	SO	5.359	10.29

Source: K. P. Huber and G. Herzberg, *Constants of Diatomic Molecules*, Van Nostrand Reinhold, New York (1979).

Table VIII Dipole (α_d) and Quadrupole (α_q) Polarizabilities of Atoms[a]

Atom	$\alpha_d(\text{Å}^3)$	$\alpha_q(\text{Å}^5)$
H	0.6668	0.6224
He	0.2050	0.1014
Ne	0.3946	0.2665
Ar	1.642	2.084
Kr	2.480	3.965
Xe	4.044	8.82
B	3.03	
C	1.78	
N	1.078	
O	0.734	
F	0.557	
Cl	2.18	
Br	3.05	
I	4.9	
Li	24.3	57.4
Na	23.6	74.7
K	43.4	191
Rb	47.3	248
Cs	59.6	393

Table VIII (cont.)

Atom	$\alpha_d(\text{Å}^3)$	$\alpha_q(\text{Å}^5)$
Be	5.60	12.6
Mg	10.6	34.4
Ca	25.0	113
Sr	27.6	
Ba	39.7	
Hg	5.02	10.1

Source: E. A. Mason and E. W. McDaniel, *Transport Properties of Ions in Gases*, Wiley, New York (1988), p. 531.

[a]A more extensive table of α_d is given by T. M. Miller in *Handbook of Chemistry and Physics*, 67th ed., CRC Press, Orlando, Fla. (1986), pp. E-66 ff.

Table IX Dipole Polarizabilities (α_d) for Molecules[a]

Molecule	$\bar{\alpha}_d(\text{Å}^3)$	$\kappa = (\alpha_d^{\parallel} - \alpha_d^{\perp})/3\bar{\alpha}_d$
H_2	0.807	0.128
N_2	1.74	0.131
O_2	1.57	0.229
CO	1.97	0.090
NO	1.71	0.162
CO_2	2.93	0.266
N_2O	3.08	0.329
NH_3	2.16	0.043
H_2O	1.43	
H_2S	3.95	
SO_2	4.28	0.175
CH_4	2.59	0
CF_4	3.86	0
CCl_4	11.16	0
SF_6	6.55	0

Source: E. A. Mason and E. W. McDaniel, *Transport Properties of Ions in Gases*, Wiley, New York (1988), p. 532.

[a]A more extensive table of α_d is given by T. M. Miller in *Handbook of Chemistry and Physics*, 67th ed., CRC Press, Orlando, Fla. (1986), pp. E-66 ff.

Table X Dipole (μ) and Quadrupole (Θ) Moments for Molecules

Molecule	$\mu\,(10^{-18}$ esu)	$\Theta\,(10^{-26}$ esu)
H_2	0	+0.662
N_2	0	−1.4
O_2	0	−0.39
CO	0.112	−2.5
NO	0.153	−1.8
CO_2	0	−4.3
N_2O	0.167	−3.0
NH_3	1.47	−1
H_2O	1.85	$\sim\pm1$
H_2S	0.97	
SO_2	1.63	±4.4
CH_4	0	0
CF_4	0	0
CCl_4	0	0
SF_6	0	0

Source: E. A. Mason and E. W. McDaniel, *Transport Properties of Ions in Gases*, Wiley, New York (1988), pp. 534–535.

AUTHOR INDEX

SUBJECT INDEX